# Phase Transitions in Liquid Crystals

# NATO ASI Series

**Advanced Science Institutes Series**

*A series presenting the results of activities sponsored by the NATO Science Committee, which aims at the dissemination of advanced scientific and technological knowledge, with a view to strengthening links between scientific communities.*

The series is published by an international board of publishers in conjunction with the NATO Scientific Affairs Division

| | | |
|---|---|---|
| A | **Life Sciences** | Plenum Publishing Corporation |
| B | **Physics** | New York and London |
| C | **Mathematical and Physical Sciences** | Kluwer Academic Publishers |
| D | **Behavioral and Social Sciences** | Dordrecht, Boston, and London |
| E | **Applied Sciences** | |
| F | **Computer and Systems Sciences** | Springer-Verlag |
| G | **Ecological Sciences** | Berlin, Heidelberg, New York, London, |
| H | **Cell Biology** | Paris, Tokyo, Hong Kong, and Barcelona |
| I | **Global Environmental Change** | |

***Recent Volumes in this Series***

*Volume 287*—Coherence Phenomena in Atoms and Molecules in Laser Fields
edited by André D. Bandrauk and Stephen C. Wallace

*Volume 288*—Intersubband Transitions in Quantum Wells
edited by Emmanuel Rosencher, Børge Vinter, and Barry Levine

*Volume 289*—Nuclear Shapes and Nuclear Structure at Low Excitation Energies
edited by Michel Vergnes, Jocelyne Sauvage, Paul-Henri Heenen, and Hong Tuan Duong

*Volume 290*—Phase Transitions in Liquid Crystals
edited by S. Martellucci and A. N. Chester

*Volume 291*—Proton Transfer in Hydrogen-Bonded Systems
edited by T. Bountis

*Volume 292*—Microscopic Simulations of Complex Hydrodynamic Phenomena
edited by Michel Mareschal and Brad L. Holian

*Volume 293*—Methods in Computational Molecular Physics
edited by Stephen Wilson and Geerd H. F. Diercksen

*Volume 294*—Single Charge Tunneling: Coulomb Blockade Phenomena in Nanostructures
edited by Hermann Grabert and Michel H. Devoret

*Series B: Physics*

# Phase Transitions in Liquid Crystals

Edited by

S. Martellucci

The Second University of Rome
Rome, Italy

and

A. N. Chester

Hughes Research Laboratories
Malibu, California

Plenum Press
New York and London
Published in cooperation with NATO Scientific Affairs Division

Proceedings of a NATO Advanced Study Institute/16th Course of the
International School of Quantum Electronics on
Phase Transitions in Liquid Crystals,
held May 2–12, 1991,
in Erice, Sicily, Italy

**NATO-PCO-DATA BASE**

The electronic index to the NATO ASI Series provides full bibliographical references (with key-words and/or abstracts) to more than 30,000 contributions from international scientists published in all sections of the NATO ASI Series. Access to the NATO-PCO-DATA BASE is possible in two ways:

—via online FILE 128 (NATO-PCO-DATA BASE) hosted by ESRIN, Via Galileo Galilei, I-00044 Frascati, Italy.

—via CD-ROM "NATO-PCO-DATA BASE" with user-friendly retrieval software in English, French, and German (© WTV GmbH and DATAWARE Technologies, Inc. 1989)

The CD-ROM can be ordered through any member of the Board of Publishers or through NATO-PCO, Overijse, Belgium.

Library of Congress Cataloging-in-Publication Data

Phase transitions in liquid crystals / edited by S. Martellucci and A.N. Chester.
p. cm. -- (NATO ASI series. Series B, Physics ; v. 290)
"Proceedings of a NATO Advanced Study Institute/16th course of the International School of Quantum Electronics on Phase Transitions in Liquid Crystals, held May 2-12, 1991, in Erice, Sicily, Italy"--T.p. verso.
"Published in cooperation with NATO Scientific Affairs Division."
Includes bibliographical references and index.
ISBN 0-306-44213-2
1. Liquid crystals--Congresses. 2. Phase transformations (Statistical physics)--Congresses. I. Martellucci, S. II. Chester, A. N. III. North Atlantic Treaty Organization. Scientific Affairs Division. IV. NATO Advanced Study Institute/16th Course of the International School of Quantum Electronics on Phase Transitions in Liquid Crystals (1991 : Erice, Italy) V. Series.
QC173.4.L55P53 1992
530.4'29--dc20 92-13797
CIP

ISBN 0-306-44213-2



Printed in the United States of America

PREFACE

The Nato Advanced Study Institute "Phase Transitions in Liquid Crystals" was held May 2-12, 1991, in Erice, Sicily. This was the 16th conference organized by the International School of Quantum Electronics, under the auspices of the "Ettore Majorana" Centre for Scientific Culture.

The subject of "Liquid Crystals" has made amazing progress since the last ISQE Course on this subject in 1985. The present Proceedings give a tutorial introduction to today's most important areas, as well as a review of current results by leading researchers. We have brought together some of the world's acknowledged experts in the field to summarize both the present state of their research and its background. Most of the lecturers attended all the lectures and devoted their spare hours to stimulating discussions. We would like to thank them all for their admirable contributions. The Institute also took advantage of a very active audience; most of the students were active researchers in the field and contributed with discussions and seminars. Some of these student seminars are also included in these Proceedings.

We did not modify the original manuscripts in editing this book, but we did group them according to the following topics:

1) "Theoretical Foundations";

2) "Thermotropic Liquid Crystals";

3) "Ferroelectric Liquid Crystals";

4) "Polymeric Liquid Crystals"; and

5) "Lyotropic Liquid Crystals".

The five sections provide a broad review of the theoretical models which describe phases and phase transitions in liquid crystals as well as the variety of experimental techniques which are employed to investigate phase transitions. The chapters cover the general aspects of the various phases in thermotropic and lytropic liquid crystals and a detailed description of the Ginzburg-Landau theory, which provides information on the values of some important physical quantities which are useful for determinig the universality class of the various phase transitions. There is also a summary of the Renormalization Group Theory, since it provides a relation between the critical exponents of the various physical quantities.

The section on thermotropic liquid crystals includes a description of the ferroelectric and polymeric liquid crystalline compounds, with special emphasis on the different smectic A phases which are found in the polymeric compounds. Frustrated smectics and chiral nematics are also described.

Several chapters are dedicated to the various experimental techniques which are employed to measure the characteristic physical quantities over the phase transitions, including d.c. calorimetric techniques which measure enthalpy changes and specific heat, a.c. calorimetric techniques to measure specific heat and thermal conductivity, and light scattering and high resolution x-ray scattering techniques to provide information on correlation lengths. Other chapters review the most important results to date concerning the smectic C - smectic A, smectic A - nematic, nematic - isotropic transitions, and the phase transitions between the different smectic A phase in the polymeric compounds.

These papers, and the further references therein, should form a useful guide to today's research results, and the basis for future advances. The reader may also wish to consult the 1985 Proceedings ("Progress in Microemulsions", edited by S. Martellucci and A.N. Chester, Plenum Publ. Co., E. Majorana Series N.41). These earlier papers give more emphasis to "Fundamental aspects of surfactant association structures"; thus, the 1985 and 1991 Proceedings are complementary in their content.

Before concluding, we acknowledge the invaluable help of Prof. Anna Maria Scheggi, the Scientific Secretary of the Institute, and of Mrs. Vanna Cammelli for her skillful assistance. We are also grateful to Drs. Leone Martellucci and Elisa Pichini for much of the computer processing work. We also wish to mention with sincere thanks Maria Teresa Petruzzi and Margaret Kyoko Hayashi, secretaries to the ISQE Directors (S.M. and A.N.C.). Finally, we acknowledge the organizations who sponsored the course, especially the generous financial support of the NATO ASI Programme.

The directors of the International School of Quantum Electronics:

Arthur N. Chester
Vice President and Director
Hughes Research Laboratories
Malibu, California (USA)

Sergio Martellucci
Professor of Physics
The Second University of Rome
Rome (Italy)

November 26, 1991

**Note added in proof:** The Editors (A.N.C. and S.M.) wish to note the significance of the 1991 Nobel Prize in Physics, awarded to Professor Pierre-Gilles de Gennes for the discovery that "methods developed for studying order phenomena in simple systems can be generalized to more complex forms of matter, in particular to liquid crystals and polymers".

CONTENTS

## THEORETICAL FOUNDATIONS



## THERMOTROPIC LIQUID CRYSTALS

# THEORETICAL FOUNDATIONS

Chapter 1

# INTRODUCTION TO THERMOTROPIC LIQUID CRYSTAL PHASES

W. H. DE JEU

## 1. Introduction

The basic features that distinguish the various liquid crystalline phases from the other condensed phases will be given, and the type of chemical compounds will be discussed. The various types of thermotropic mesophases formed from rod-like molecules will be described in some detail and the order parameters that characterize them will be defined. Apart from the nematic phase, that is characterized by orientational order and translational symmetry of the molecular positions, various types of smectic phase can be distinguished. Their common feature is the existence of a one-dimensional density wave. Additional symmetry properties distinguish the different types of smectic phase. Thermotropic mesophases with disc-like molecules can also lead to a nematic phase. Reduced positional order now gives so-called columnar phases. The analogies and differences with the phases of rod-like molecules will be discussed. Both symmetry elements can be combined in lath-like molecules, leading to possible transitions between the two types of phase. In the first Chapter of this section a short overview of the various types of liquid crystals that can be observed will be given, including a discussion of the types of molecules from which they can be formed. In addition the order parameters of the most relevant phases will be defined. Emphasis will be on the smectic phases, anticipating what I suppose might be relevant to other lectures in this school. I will restrict myself to *thermotropic* liquid crystals (for which temperature is the important variable for the phase behaviour), and will not consider *lyotropic* liquid crystals (for which the variation of concentration in a solution is a dominant aspect). In spite of their increasing importance liquid crystalline polymers will also not be included. The latter subjects will be well covered in various other lectures at this school. I will rely to some extend on the treatment given in Ref.1, where also further references can be found. The two following Chapters will be referred to as Chapter 2 and 3, respectively.

## 2. Orientational Order: the Nematic Phase

In order to start at the very beginning I recall that a solid crystal is characterized by

W. H. de Jeu - FOM-Institute for Atomic and Molecular Physics, Kruislaan 407, 1098 SJ Amsterdam, The Netherlands and Open University, P.O. Box 2960, 6401 DL Heerlen, The Netherlands

*Phase Transitions in Liquid Crystals*, Edited by S. Martellucci and A.N. Chester, Plenum Press, New York, 1992

long-range positional order. In the case of anisotropic molecules in addition there will be long-range orientational order: the various directions that can be assigned to a molecule will be retained throughout the lattice. Realising this distinction it is evident that not necessarily both types of order disappear at the same melting point. This leads to the following possibilities.

- *Plastic crystalline phases*, in which the positional order is still present, but the orientational order has disappeared or is strongly reduced.
- *Liquid crystalline phases*, characterized by orientational order, while the positional order has disappeared or is reduced.

I will describe first the nematic phase that occurs for substances that have (very roughly) a rod-like or a disc-like shape. It is characterized by a complete loss of positional order at the melting point, so that the centers of mass have translational symmetry as in an isotropic liquid. However, the unique axes of the molecules remain, on average, parallel to each other, leading to a preferred direction in space. This local direction of alignment can be described by a unit vector **n**, the director. The states described by **n** and -**n** appear to be indistinguishable. The actual orientation of the director is imposed by boundaries and/or external fields. In the simplest case this leads macroscopically to an axis of uniaxial symmetry. This orientational order disappears only when on increasing the temperature a second transition point is reached. Fig.1 gives an 'artists view' of the nematic phases of rods and discs, respectively. These objects are far from the exact structure of nematogenic molecules, of which examples of the elongated type are given in Table 1. Note the last two examples that indicate that completely non-polar molecules can form a nematic phase.

To consider the most basic aspects of defining a nematic order parameter let us stick to the simplication of a molecule as a rigid rod. The (lack of) validity of this assumption will be discussed extensively in Chapter 2. To define the local orientation at a point **r**=(x,y,z) we attach to the rod a unit vector **a**(**r**) along the long axis. Now consider the thermal averages of various tensors that are composed of **a**, over a small but macroscopic volume around **r**. As the average $\langle \mathbf{a}\cdot\mathbf{a}\rangle$ is a constant by definition, it is clear that a scalar order parameter is out of the question. The next choice is a vector order parameter $\langle \mathbf{a}\rangle$ analogous to the magnetization in a ferromagnet. However, a non-zero value of this quantity would violate the equivalence of **n** and -**n**. It would describe a ferromagnetic nematic state that has not been observed.

The first relevant choice as order parameter is a second-rank tensor $\underline{\mathbf{S}}$, the elements of which are given by

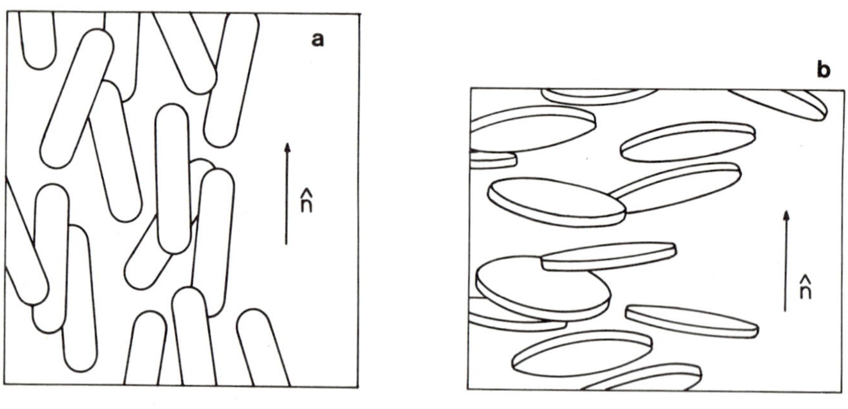


Fig.1. 'Artists view' of: (a) a positive uniaxial nematic phase of rods; and, (b) a negative uniaxial nematic phase of discs.

$$S_{\alpha\beta} = \langle a_\alpha a_\beta \rangle - \frac{1}{3}\,\delta_{\alpha\beta}\ , \qquad \alpha,\beta = x,y,z. \tag{1}$$

Note that the addition of the Kronecker delta ensures that $S_{\alpha\beta}$ is zero in the isotropic phase where $\langle a_\alpha^2 \rangle = 1/3$. The tensor order parameter $\underline{S}$ is a symmetric traceless tensor and thus has in general five different elements. By choosing a suitable coordinate system such a tensor can be brought in a diagonal form. For the uniaxial case considered here then Eq.(1) can be written as

$$S_{\alpha\beta} = S\,(n_\alpha n_\beta - \frac{1}{3}\,\delta_{\alpha\beta}). \tag{2}$$

Table 1. Examples of nematic liquid crystals. Numbers indicate the transition temperature in °C between the adjacent phases.

4,4'-dimethoxyazoxybenzene (*p*-azoxyanisole, PAA)

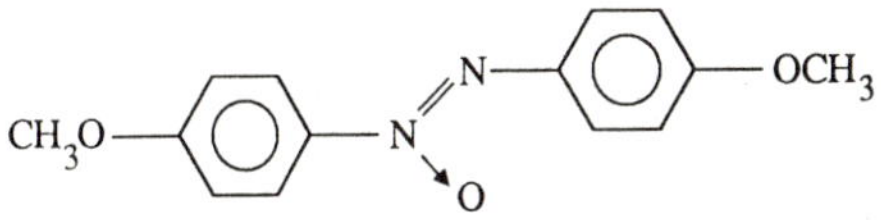


K 118 N 135.5 I

N-(4-methoxybenzylidene)-4'-butylaniline (MBBA)

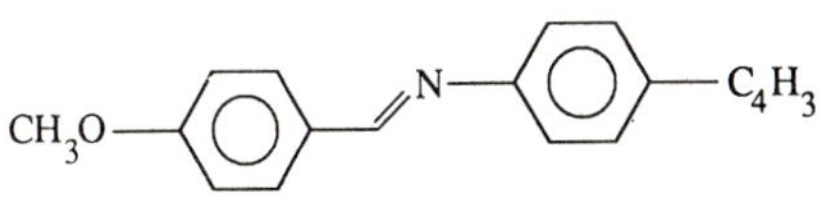


K 22 N 47 I

4-pentyl-4'-cyanobiphenyl (5CB)

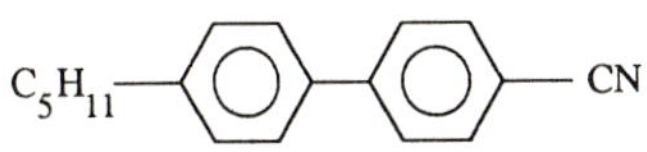


K 22.5 N 35 I

4-pentylphenyl-*trans*-4'-pentylcyclohexylcarboxylate

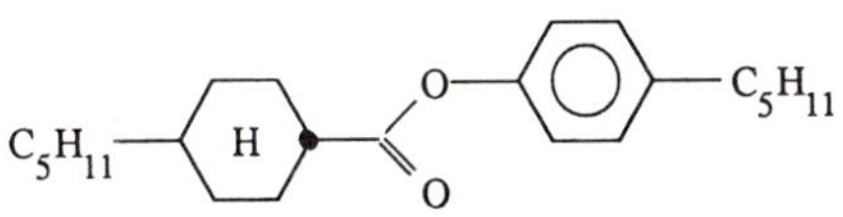


K 37 N 47 I

*p*-quinquephenyl

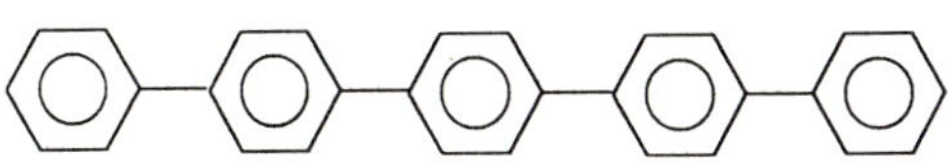

K 380 N 431 I

4,4'-diheptylazobenzene

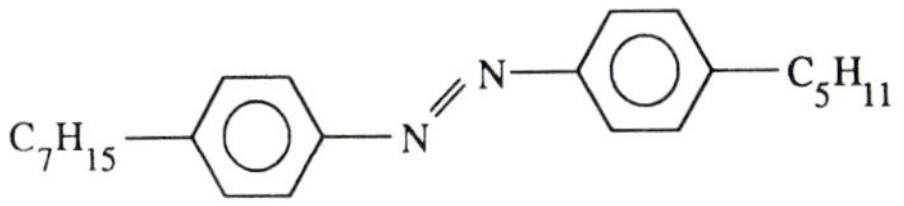


K 40 N 47 I

Table 2. Examples of chiral nematic liquid crystal.

| | |
|---|---|
| N-(4-ethoxybenzylidene)-4'-(β-methylbutyl)aniline | |
| 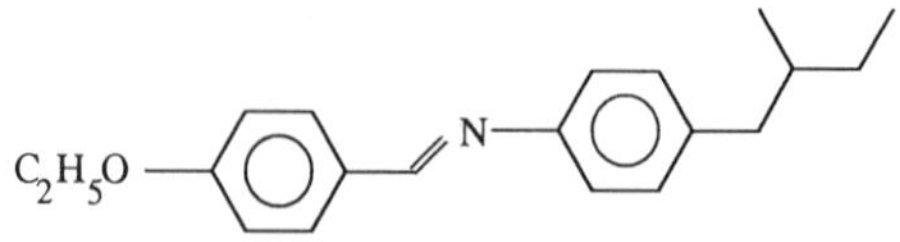 | K 15 N* 60 I |
| cholesteryl myristate | |
| 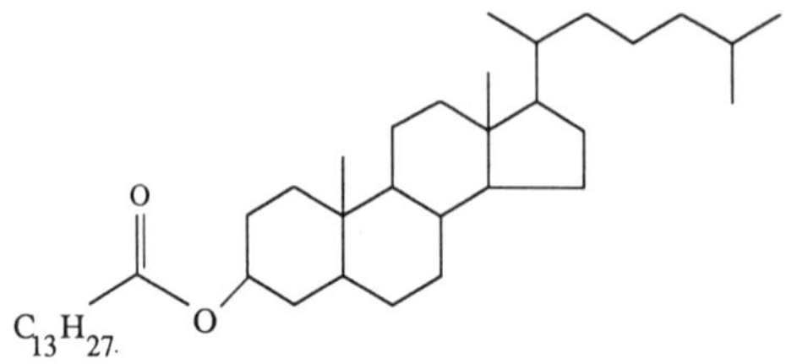 | K 71 $S_A$81 N* 86.5 I |
| 1-butyl-1-d-4-(*p*-cyanobenzylideneamino)cinnamate | |
| 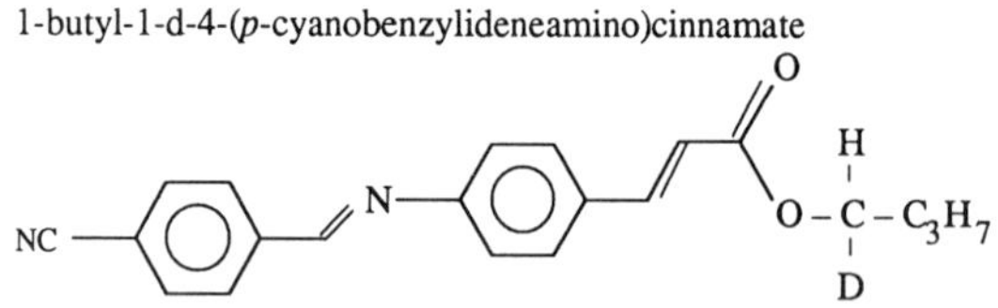 | K 108 N* 133 I |

Choosing **n** along the z-axis the three non-zero diagonal elements are

$$S_{zz} = \frac{2}{3}S, \quad S_{xx} = S_{yy} = -\frac{1}{3}S \tag{3}$$

The scalar S is a measure of the alignment of the molecules and will be considered further in Chapter 2.

So far it has been implicitly assumed that the molecules constituting the liquid crystalline phase have mirror symmetry. In the case of a non-racemic mixture of optically active molecules a *chiral nematic* phase results (abbreviated as N*). In addition to the long-range orientational order there is a spatial variation of the director leading to a helical

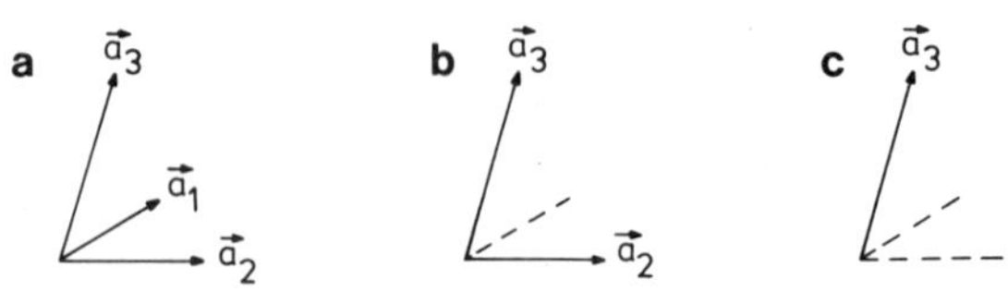


Fig.2. Melting in steps.

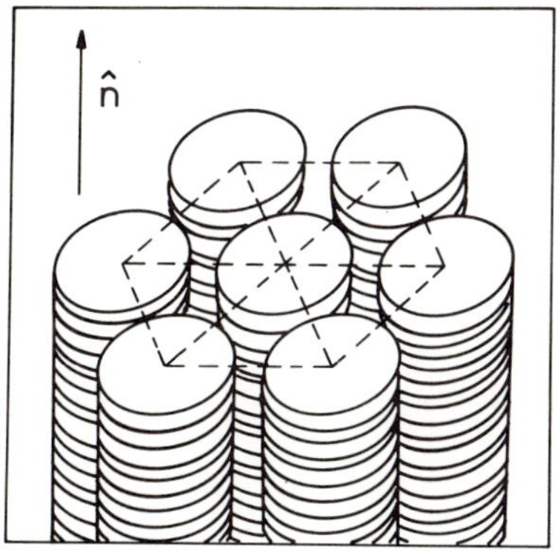


Fig.3. 'Artists view' of a $D_{hd}$ columnar phase (hexagonal, disordered).

structure. Examples of molecules giving an N* phase are shown in Table 2. Note the last example that indicates that isotopic H-D substitution is sufficient to obtain the required effect, resulting in a long-pitch N* phase[2]. This variation on the nematic phase will not be considered here any further.

## 3. Melting in Steps: Smectic and Columnar Phases

To describe the liquid crystalline phases with reduced positional order consider a crystalline unit cell as given in Fig.2a. If the positional melting is complete we arrive at a positive or a negative uniaxial nematic phase depending on whether we deal with a rod-like or a disc-like situation. Intermediate situations are shown in Figs. 2b and 2c. In the first case we assume that only one of the lattice directions becomes disordered. This leads to a columnar phase, which consists of a two-dimensional (2D) lattice of columns in which the disc-like molecules are disordered. This situation is sketched in Fig.3 for a 2D hexagonal lattice ($D_{hd}$ phase, where the indices refer to hexagonal-disordered). A further discussion will be given in paragraph 5. The next step is two disordered directions leading to liquid layers of rod-like molecules (Fig.2c). This so-called smectic A phase is pictured in Fig.4. It should be realized that what is often loosely called a layer structure is usually only a weak density modulation.

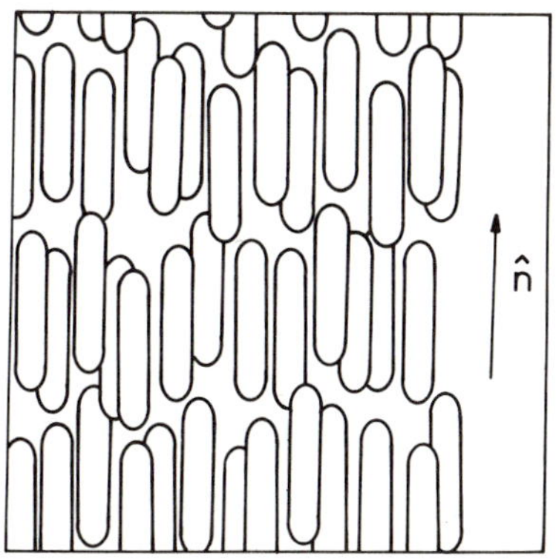


Fig.4. 'Artists view' of a smectic A phase.

Table 3. Examples of smectic liquid crystal.

4,4'-diheptoxyazoxybenzene (HOAB)

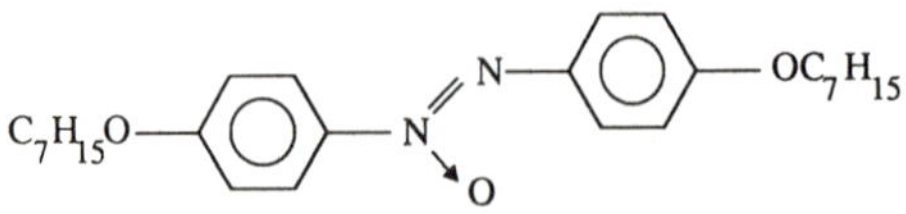


K 74.5 $S_C$ 95.5 N 124 I

N-(4-heptoxybenzylidene)-4'-butylaniline (7O.5)

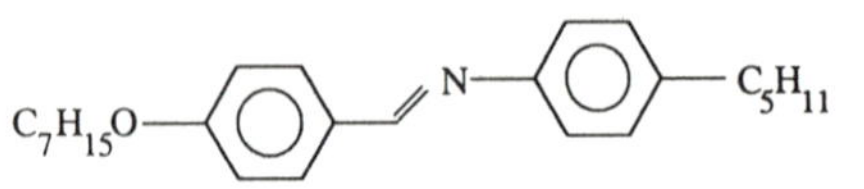


K 23 $S_G$ 58 $S_B$ 64 $S_C$ 68 $S_A$ 80 N 83 I

4-4'-dinonylazobenzene

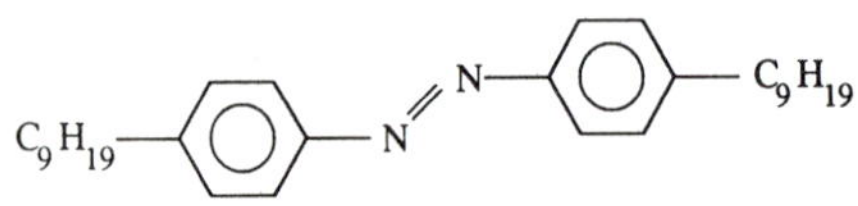


K 37 $S_B$ 40 $S_A$ 53 I

*p*-sexiphenyl

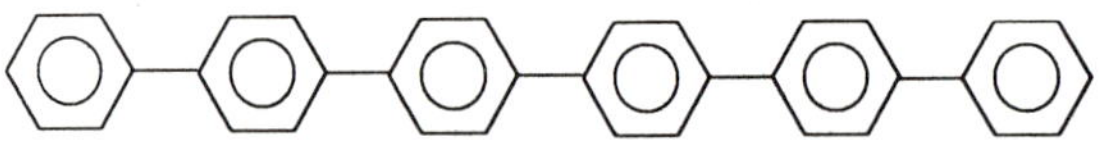

K 435 $S_A$ 470 N >500 I

Examples of molecules giving smectic phases are shown in Table 3. Apart from the smectic A phase described so far there are many more types of smectic phase, indicated by the indices B, C, etc.

In the first place the direction of the density modulation can be parallel to the director (*orthogonal* phases like smectic A) or make an angle with this direction (*tilted* phases). Additional order may arise from the organization of the centers of mass of the molecules within the layers. In a homologous series often a nematic phase is found for short alkyl chain length.

At longer chain length a smectic phase (or phases) may be stable that finally takes over and has a phase transition directly in the isotropic phase. A typical example of this behaviour is given in Fig.5.

A smectic order parameter can be defined as follows. The periodic modulation of the density (taken along the z-axis) can be described as a Fourier-series

$$\rho = \sum_{n=0}^{\infty} a_n \cos(2\pi n z/d), \qquad n \text{ integer},$$
$$= a_0 + a_1 \cos(2\pi z/d) + \ldots \tag{4}$$

The coefficients are determined by the cosine-function averaged over the positional distribution function f(z). Restricting the expansion to the first harmonic only one arrives at *the* smectic order parameter

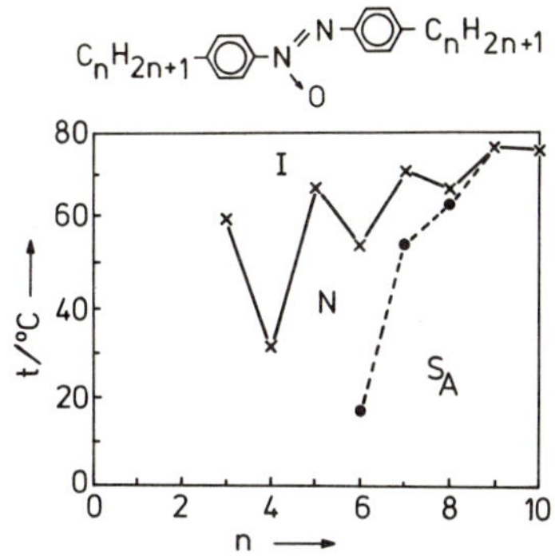


Fig.5. Typical nematic and smectic behaviour in a homologous series.

$$\rho_1 = \tau = \int f(z)\cos(2\pi z/d)\,dz = \langle \cos(2\pi z/d)\rangle \qquad (5)$$

Alternatively the smectic order parameter may be defined as

$$\rho - \rho_0 = \mathrm{Re}[\Psi \exp(iq_0 z)] \qquad \text{with } q_0 = 2\pi/d. \qquad (6)$$

Now the phase factor of the order parameter has been explicitly included. Much of the interest in smectic phases is related to the fact that the 1D ordering of the layers is in fact not truly long-range. While for long-range order the positional correlation between two separate layers approaches a constant in the limit that the layers become infinitely apart, the situation in smectic phases corresponds to an algebraic decay of that correlation proportional to $z^{-\eta}$. Denoting in general the relevant correlation function by G(**r**) the various situations that may occur are summarized in Table 4.

Table 4. Behaviour of the correlation function G(**r**) and the related scattering cross section S(**q**-**q**$_0$).

| | G(**r**) | S(**q**-**q**$_0$) |
|---|---|---|
| Long-range | constant | $\delta(\mathbf{q}-\mathbf{q}_0)$ |
| Quasi long-range | $r^{-\eta}$ | $(\mathbf{q}-\mathbf{q}_0)^{\eta-2}$ |
| Short-range | $\exp(-r/\xi)$ | $[\xi(\mathbf{q}-\mathbf{q}_0)^2 + 1]^{-1}$ |

For liquids G(**r**) decays rapidly to zero, with a typical correlation length of several molecular dimensions. Its Fourier transform, which gives the scattering cross section in x-ray diffraction, is a broad Lorentzian. In the case of long-range order the constant correlation function gives a $\delta$-function type Bragg peak. The interesting point is that in systems characterized by order in less than three dimensions a situation of quasi long-range order can exist, where G(**r**) decays rather slowly as $r^{-\eta}$ ($\eta$ small positive[3]). It leads in x-ray scattering to a singularity that differs in a subtle way from a true Bragg peak. Using special high-resolution techniques the difference has been observed. It follows from the the expression for the free energy of a smectic A phase that such a 1D density wave in a 3D medium is indeed unstable at finite temperatures. The mean-square displacements of the smectic layers from their equilibrium positions diverge logarithmically with the thickness of the sample (Landau-Peierls instability).

## 4. Broken Symmetry and the Organisation of Smectics

The various type of smectic phase that have been observed and their structural characteristics are summarized in Table 5.

Table 5. Classification of the various types of smectic phase.

| Structure within the layer | Orthogonal | Tilted | | |
|---|---|---|---|---|
| | | a>b* | a<b* | |
| Liquid | A | | C | *smectic* |
| Bond-orientational order | hexatix B | hexatic F | hexatic I | *liquid crystals* |
| (Pseudo)-hexagonal packing | B | G | J(G') | *lamellar* |
| Herringbone packing | E | H | K(H') | *plastic crystals* |

* b is the unique axis of the (local) monoclinic cell ($a \neq b \neq c$, $\alpha=\gamma=90^\circ \neq \beta$); see Fig.7.

I will restrict myself here to a discussion of the real-space structure of the various phases as the x-ray aspects will be covered in other lectures. Once the layer structure has been established one should in the first place distinguish between orthogonal and tilted phases. As mentioned already tilted phases are characterized by an angle between the director and the normal to the smectic planes. In general there will be long range order of the tilt directions within a layer. In that cases the tilt directions of adjacent layers will be coupled as well, leading to an overall uniform tilt. Consequently the resulting phase is biaxial. The tilted analogue of a smectic A is called smectic C (first row of Table 5). The tilt angle $\omega$ provides the natural order parameter which can be written as

$$\chi = \omega \exp(i\phi) \tag{7}$$

where the azimuthal angle $\phi$ gives the direction in which the tilt takes place.

In the case of chiral molecules the situation is slightly different: now there is a small angle $\alpha$ between the tilt direction of adjacent layers. In this way a helical structure is formed with a pitch $p=(2\pi/\alpha)d$. Macroscopically the normal to the layers is then again an axis of uniaxial symmetry. Interestingly the helix period is usually not commensurate with the layer periodicity. Much interest in the chiral smectic C* phase is related to the fact that it allows for ferro-electricity[4].

Now I come to the second row of Table 5 which refers to the *hexatic* smectic liquid crystals[5]. These are characterized by: (a) limited positional order within the layers, given by a correlation length $\xi$; and,(b) in addition long range bond-orientational order. To explain the last point consider Fig.6a where two clusters of molecules in a liquid smectic A layer are shown. Because of the assumed short range positional order locally two lattice vectors can be assigned to each cluster. Evidently now the lattice vectors of two different clusters differ in two respects. The distance between the origins is not related to the magnitude of the lattice vectors itself, and additionally they are rotated with respect to each other over an arbitrary angle. Bond-orientational order refers to the situation that the local lattice *directions* are

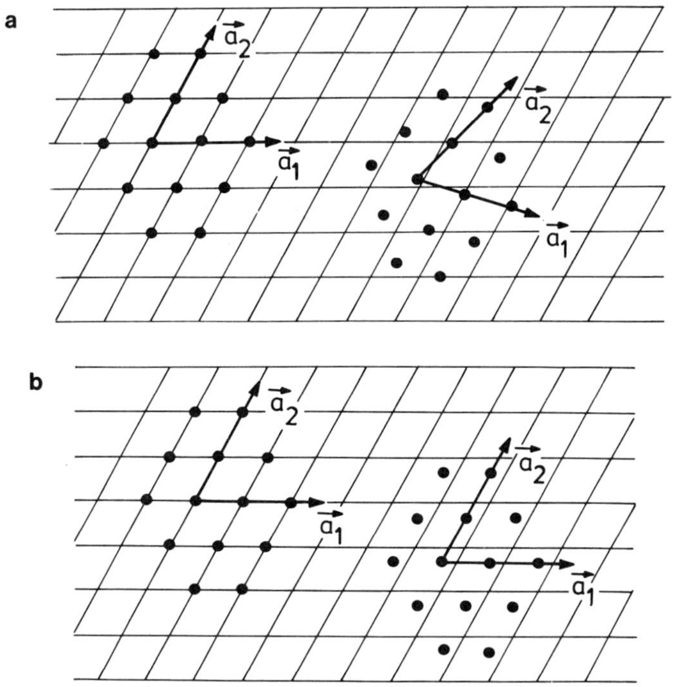


Fig.6. Local 2D lattice for: (a) a liquid; and, (b) a hexatic phase with bond-orientational order.

retained. This is, of course, evident in the case of long range positional order, but leads in combination with limited positional order to new phases, called *hexatic* (see Fig.6b). In practice the bond-orientational order in neighbouring layers is coupled, so that the hexatic smectic phases are 3D from this point of view. This leads to true long range bond-orientational order; in the case of a 2D hexatic phase the bond-orientational order would be expected to decay algebraically. The order parameter characterizing the hexatic order can be written as

$$\Psi(\mathbf{r}) = \exp[6i\theta(\mathbf{r})] \tag{8}$$

where $\theta(\mathbf{r})$ is the angle between a local lattice vector and some reference axis. Finally the subdivision in the column of tilted phases (leading to the division between smectic F and I)

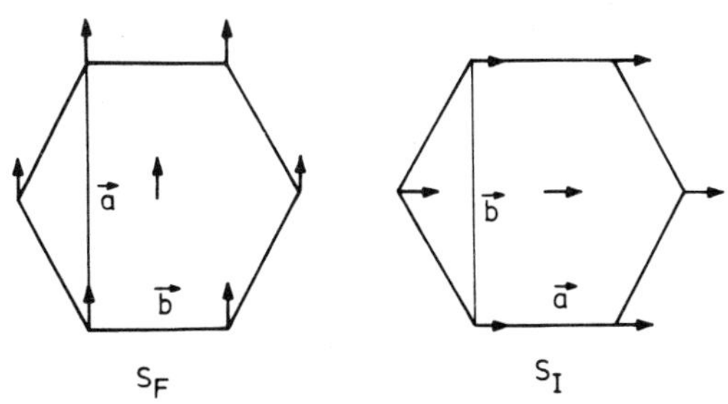


Fig.7. Difference in tilt direction for smectic F and smectic I.

refers to the direction of the tilt with respect to the local lattice (see Fig.7).

At this point the discussion of liquid crystalline phases in the true sense of the word is finished. The remaining rows of Table 5 correspond to what can best be named *lamellar plastic crystals*.

The third row refers to hexagonal order of the molecular centers in the layers, leading to

Table 6. Example of liquid crystals with disc-like molecules.

| | |
|---|---|
| 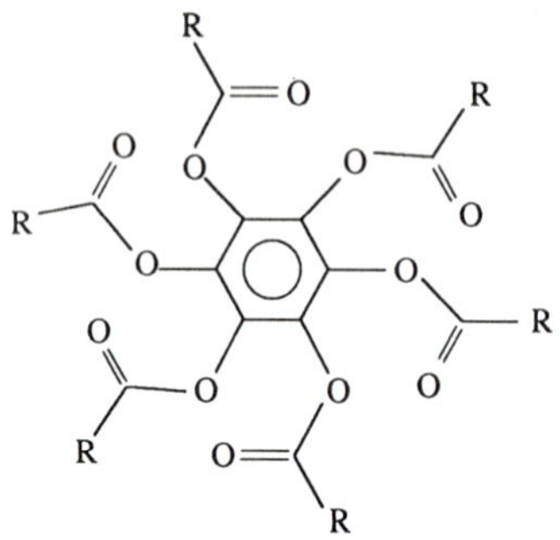 | R is $C_7H_{15}$: K 80 $D_{rd}$ 83 I |
| 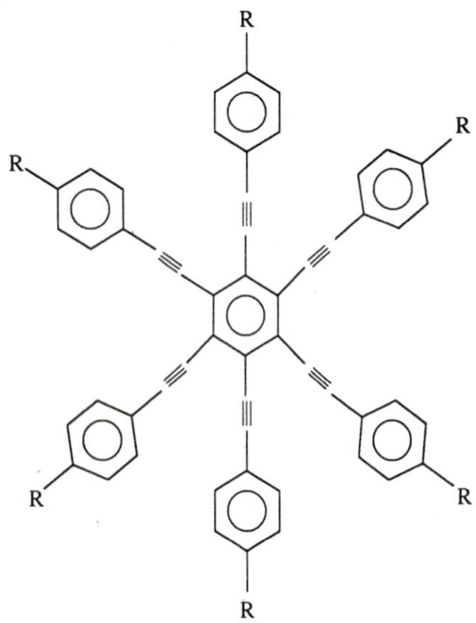 | R is $C_8H_{17}$: K 80 $N_D$ 96 I |
| 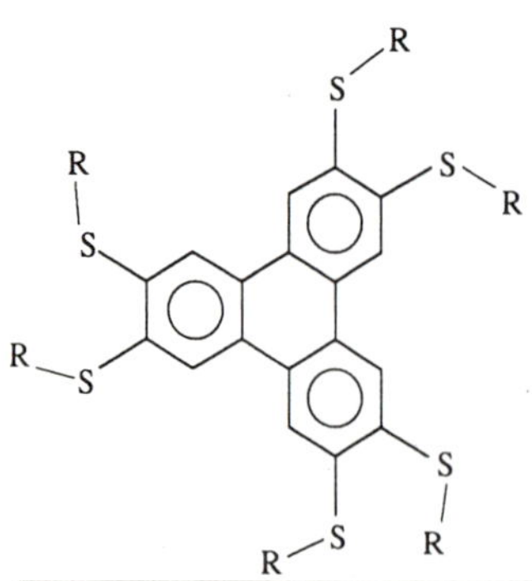 | R is $C_6H_{13}$: K 65 $D_{ho}$ 72 $D_{hd}$ 93 I |

the plastic crystalline B phase. For the tilted phases there is equivalently (pseudo)-hexagonal order in planes perpendicular to the tilt direction. The crucial question is here whether the 1D density wave (leading in x-ray scattering to *00l* reflections) and the 2D lattice in the planes (giving *hk0* reflections) are coupled (mixed reflections). If this is the case the lattice is 3D and the layers cannot translate and/or rotate with respect to one another.

This correlation leading to a 3D crystal – though strongly anisotropic – has been found in many cases.

In combination with the 'softness' in terms of some freedom to reorient (for example evident from dielectric relaxation) this justifies the term *lamellar plastic crystals.* In the last row of Table 5 the hexagonal in-plane structure is replaced by a herringbone type of order. In fact this type of structure is already locally present in the higher temperature phases.

Finally it should be mentioned that various methods should be used in combination to identify the various type of higher ordered phases. Polarizing microscopy can be useful as a first step, but is (apart from some of the simplest smectic phases) not conclusive. Differential scanning calorimetry (DSC) is nowadays a routine tool, but gives at best a very rough indication of whether the structural changes at a phase transition are large or small. Powder x-ray diffraction, preferably at reasonable resolution such as in a Guinier camera, usually provides enough additional information for an unambiguous classification.

## 5. From Discs to Rods and More

In Table 6 some examples are given of liquid crystalline phases of disc-like molecules. In addition to the nematic $N_D$ phase (see Fig.1b) various types of columnar phases are observed. They are indicated by D and further classified as follows:

- An index indicating the type of 2D lattice of the columns, for example r (rectangular) or h (hexagonal).
- A second index indicating order (o) or disorder (d) with regard to the position of the molecules within the columns.

This leads to an evident notation like $D_{hd}$ (see Fig.3) or $D_{ro}$. This scheme does not cover all the cases observed so far: for example, tilted phases also occur.

However, taking the still limited amount of information into account it seems not useful at this stage to attempt a more sophisticated classification. It will be clear from the discussion about positional order in smectic layers that for the phases with ordered columns analogously a coupling between the two sublattices will occur (now between the 2D lattice of the columns and the 1D lattice within each column) leading to a 3D lattice. Hence these phases with ordered columns could be called *columnar plastic crystals.*

The last example of Table 6 is somewhat special in the sense that it shows an order-disorder transition within the columns while the 2D hexagonal lattice of the column remains intact[6]. A review of the structure of columnar phases has been given by Levelut[7]. To define order parameters for the various columnar phases one can write the density as[8]

$$\rho = \rho_{\perp} f(x,y) + \rho_{\parallel} \tag{9}$$

The quantity $\rho_{\parallel}$ can be taken as a quasi-1D density wave that changes to a constant value at a $D_o$-$D_d$ transition. The function f(x,y) governs transitions within the 2D lattice of the columns like $D_r$-$D_h$, while f(x,y) → constant describes the transition D-$N_D$.

Returning to Fig.2 one might ask the question whether systems can be constructed that show transitions from columnar to smectic phases. Evidently this requires molecules that have both elements. Examples are the so-called phasmids[9]:

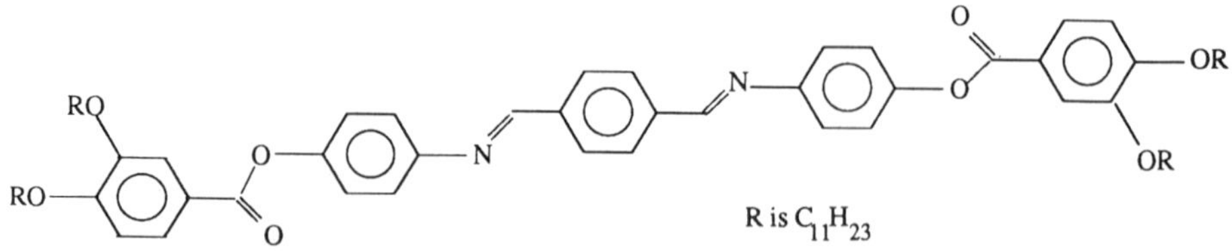


K 130 $S_C$ 155 $\Phi_{ob}$ 169 $\Phi_{hex}$ 186 I

The columnar phases indicated by Φ are build from discs that in turn consists of several molecules. In that respect one could speak of thermotropic micellar phases. Interestingly, in addition to these columnar phasmid phases also a smectic C type phase is observed. X-ray information on the structural aspects can be found in Ref.10. In a similar way a biaxial nematic phase may occur as an intermediate state between a positive and a negative uniaxial nematic phase. Examples can be found in Refs.11-14; the first of these is pictured below.

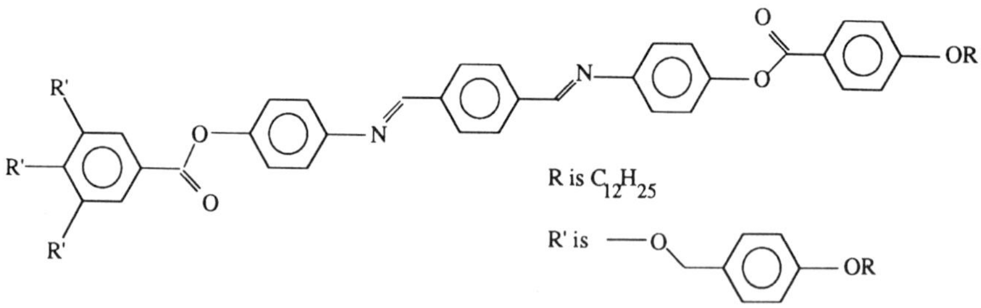


Finally it should be mentioned that various types of optically isotropic liquid crystalline phases can be found. One series of cubic phases is found in short-pitch chiral nematic liquid crystals over a relatively narrow temperature range before the transition to the isotropic phase. These are the so-called blue phases, of which in fact three different types can occur[15]. The optical isotropy can be explained by some form of arrangement of the helices in a cubic superstructure, the lattice being defined in terms of singularities of the director field.

Of course, cubic structures can also be related to positional order, and not surprisingly these are found in the phasmids mentioned above. Evidently a cubic structure can be a solution to a situation where the tendency to form either layer-like or columnar-like order is not dominant. However, cubic liquid crystalline phases are not restricted to such molecules. Classical examples are provided by substituted carboxylic acids:

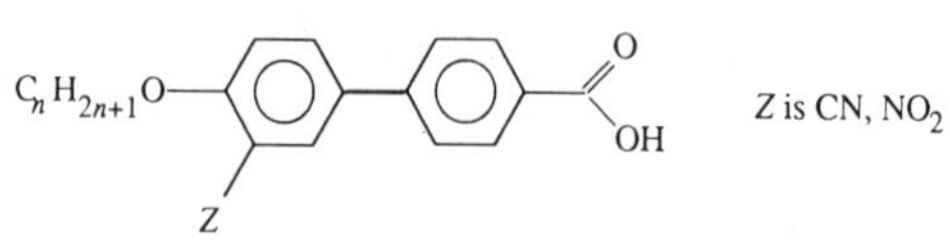


Here a so-called smectic D phase (which is in fact not smectic at all) occurs between an $S_A$ and an $S_C$ phase. An x-ray study for the compound where Z is CN and n=18 lead to a cubic unit cell with a lattice parameter of 86 Å and containing approximately 700 molecules[16]. Other cases can be found in Ref.17.

As a final curiosity I want to mention the liquid crystalline phases recently reported in a compound with tetrahedral symmetry[18]:

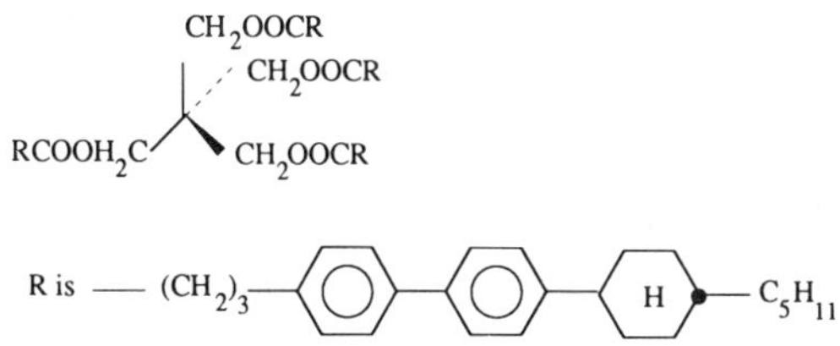

$S_3$ 152 $S_B$ 229 $S_A$ 275 I

Here in spite of the centrosymmetric molecules smectic phases are formed. A model has been given that attributes this to the flexible spacer groups between the central carbon atom and the four mesogenic units, leading to a rod-like structure of the molecule as a whole. This situation is similar to that in liquid crystalline side-chain polymers, the main chain here being reduced to a single carbon atom. It would be interesting to pursue the liquid crystalline behaviour of these types of compound more systematically.

I conclude that the structure and ordering of the various types of smectic phase now is well established. Problems are still present regarding the special types of smectic A phases that occur with molecules with strongly polar end groups (to be discussed in some detail in Chapter 3). Rapid development regarding low-molecular weight thermotropic liquid crystals are still found in the synthesis and the characterization of the phase behaviour of the types of systems discussed in this last paragraph.

ACKNOWLEDGEMENT. This work is part of the research programme of the Stichting voor Fundamenteel Onderzoek der Materie (Foundation for the Fundamental Research of Matter, FOM) and was made possible by financial support from the Nederlandse Organisatie voor Wetenschappelijk Onderzoek (Netherlands Organization for the Advancement of Research, NWO).

## References

1. G. Vertogen and W.H. de Jeu, *Thermotropic Liquid Crystals, Fundamentals,* Springer Series in Chemical Physics vol. 45 (Springer, Heidelberg, 1988).
2. D. Coates and G.W. Gray, Mol.Cryst.Liq.Cryst. **24**, 163 (1973).
3. A very readable treatment is: J. Als-Nielsen, J.D. Litster, R.J. Birgeneau, M. Kaplan and C.R. Safinya, p.57 in *Ordering in Strongly Fluctuating Condensed Matter Systems,* T. Riste, ed., Plenum (New York, 1980).
4. R. Blinc, Chapter 22 of this Volume.
5. R.J. Birgeneau and J.D. Litster, J. Phys Lett. **39**, L-399 (1978).
6. E. Fontes, P.A. Heiney and W.H. de Jeu, Phys.Rev. Lett. **61**, 1202 (1988).
7. A.M. Levelut, J.Chim.Phys. **80**, 149 (1983).
8. E.I. Kats, Sov.Phys. JETP **48**, 916 (1978).
9. Nguyen Huu Tinh, C. Destrade, A.-M. Levelut and J. Malthête, J.Phys. **47**, 553 (1986).
10. A.M. Levelut, J. Malthête, C. Destrade and Nguyen Huu Tinh, Liq.Cryst. **2**, 877 (1987).
11. J. Malthête, L. Liébert, A.-M. Levelut and Y. Galerne, C.R.Acad.Sci.Paris Ser.II, **303**, 1073 (1986).
12. S. Chandrasekhar, V.N. Raja and B.K. Sadashiva, Mol.Cryst.Liq.Cryst.Lett. **7**, 65 (1990).
13. K. Praefcke, B. Kohne, D. Singer, D. Demus, G. Pelzl and S. Diele, Liq.Cryst. **7**, 589 (1990).
14. K. Praefcke, B. Kohne, B. Gündogan, D. Singer, D. Demus, S. Diele, G. Pelzl and U. Bakowsky, Mol.Cryst.Liq.Cryst. **198**, 393 (1991).

15. For a review see, D.C. Wright and N.D. Mermin, Rev.Mod.Phys. **61**, 385 (1989).
16. G. Etherington, A.J. Leadbetter, X.J. Wang, G.W. Gray and T. Tajbakhsh, Liq.Cryst. **1**, 209 (1986).
17. D. Demus, A. Gloza, H. Hartung, A. Hauser, I. Repthel and A. Wiegeleben, Cryst.Res.Techn. **16**, 1445 (1981).
18. R. Eidenschink, F.-H. Kreuzer and W.H. de Jeu, Liq.Cryst. **8**, 879 (1990).

Chapter 2

# MICROSCOPIC ASPECTS OF THE NEMATIC PHASES

W. H. DE JEU

## 1. Introduction

First a critical evaluation will be given of the structure and local order of molecules that typically give nematic behaviour. This allows the approximations made implicitly when real molecules are represented by simple rigid objects to be discussed. Microscopic order parameters will be defined starting from the orientational distribution function of the molecules. Connection will be made with macroscopic anisotropic properties that are often used to determine order parameters experimentally. In terms of intermolecular interactions the role of attractive forces and of excluded volume effects will be considered. The relevance of these molecular models will be discussed.

In the second Chapter of this Section I will describe our present understanding of the various molecular aspects of the nematic phase. I will not go into the details of calculations but rather try to examine critically the ingredients of the various approaches in relation to the final results. This will include some discussion of the often implicit assumptions in analysing experimental results. When required the previous Chapter in this Section will be referred to as Chapter 1.

## 2. Nematogenic Molecules and Local Order

In Chapter 1 models of rods and discs have been used on several occasions to represent molecules that form liquid crystalline phases. Hence it seems useful to investigate in some detail the nature of typical molecules and to address the question how relevant these simple models are. I will take as a classical example MBBA, of which the central core is pictured in Fig. 1. In the first place it should be realized that the molecule is neither rod-like nor lath-like. Within the central core of the molecule three planes can be distinguished associated with the two benzene rings and the CHN-atoms in the bridging group. The various degrees of freedom of the full MBBA molecule are shown in Fig. 2. These comprise:

- Self-diffusion of the center of mass (SD);
- Orientational fluctuations of the long axis of the molecule (OF);

W. H. de Jeu - FOM-Institute for Atomic and Molecular Physics, Kruislaan 407, 1098 SJ Amsterdam, The Netherlands and Open University, P.O. Box 2960, 6401 DL Heerlen, The Netherlands

*Phase Transitions in Liquid Crystals*, Edited by S. Martellucci and A.N. Chester, Plenum Press, New York, 1992

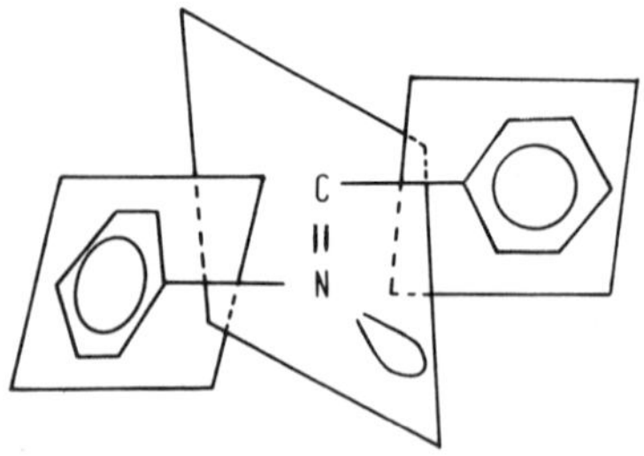


Fig. 1. Stereometric structure of benzylidene-aniline.

- Rotation around the long molecular axis ($R_1$), which rotation may be composed out of various rotations of individual groups within the molecule;
- Rotation around a short axis of the molecule ($R_2$);
- Rotations around various bonds in the end groups ($R_3$, $R_4$, $R_5$), giving rise to different conformations.

It will be clear that many uncertainties arise if we want to come to a simple model representation of a molecule.

Considering first the aromatic core of the molecule it is evident that by itself it will be more lath-like than rod-like. A quantity of major importance in this context is the angle between the planes of the benzene rings. Some typical values are indicated[1] in Table 1. In practice this angle depends on the nature of the end substituents A and B; this is in particular true for the biphenyls. Note that azobenzenes are planar.

A lath-like representation of a molecule is pictured in Fig. 3. Even if we accept such a model as useful, the question remains how the molecular $\xi\eta\zeta$-frame can be related to the directions in the real molecule. Usually the following conventions are used.

- The $\zeta$-axis is taken through the terminal para-positions of the aromatic core, thus making an angle with the para-axis of each of the individual benzene rings. This is not necessarily the major axis of the moment of inertia tensor, which should determine the axis of the rotation $R_1$.
- In order to define the other two axes we need to define a plane that contains the $\zeta$-axis. This plane is usually taken such that it makes equal angles with the planes of the two benzene rings.

With these assumptions the model of Fig. 3 can be related to the geometry of the central core of the real molecule under consideration. It is difficult to test the underlying assumptions explicitly. However, even if this model is accepted two complications arise.

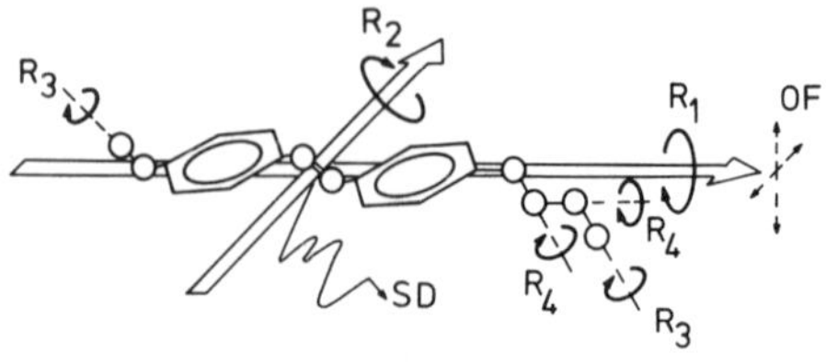


Fig. 2. Various degrees of freedom in MBBA.

Table 1. Angle between the planes of the benzene rings in

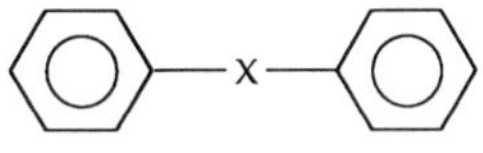


| X | angle |
|---|---|
| – | ~ 30°-38° |
| N=N | 0° (planar) |
| NO=N | ~ 20° |
| CH=N | ~ 40°-45° |

- How should the end groups be incorporated? If we consider a simple alkyl chain in an all-trans conformation, the major axis of the chain does not coincide with the ζ-direction as defined above. Moreover, different conformations will occur, as the trans-gauche energy difference is of the order of kT. In addition the probability of various conformations may be influenced by the presence or absence of orientational order (nematic or isotropic phase). Many of these questions have been studied by magnetic resonance[2]. Detailed analysis of the results shows that a single set of order parameters that can be transformed to characterize the various molecular segments does not in general exist. In other words, the molecular orientation is not independent of the molecular conformation (see Ref. 1 and references therein).
- If one or more benzene rings are replaced by other ring systems and/or side groups are attached to the central core of the molecule, the model as described should be modified and/or extended. Such developments have not yet been attempted.

The crucial question is how important all these complications are in practice. If one realizes that several liquid crystalline phases can be obtained via computer simulations with rigid (sphero)cylinders or ellipsoids[3], one may argue that only the overall elongated shape of the molecule matters. If one wants to describe precisely the orientational order evidently the situation is different, and details of the molecules cannot be avoided. The objective of the summary here has been mainly to become aware of these complications. A consequence is, for example, that an accurate determination of the birefringence of a nematic does not

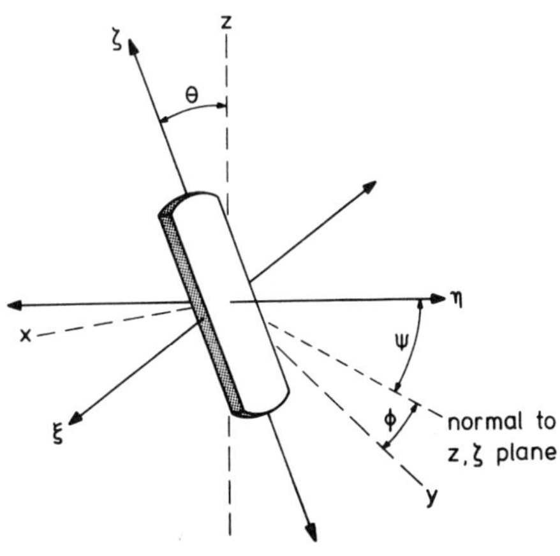


Fig. 3. Lath-like representation of a body; note the existence of two mirror planes.

automatically mean an equally accurate knowledge of the nematic order parameter (see next paragraph).

So far I have discussed individual molecules. Before coming to the orientational distribution and order parameters in the next section, there is one more step to be considered: this concerns the local order. In agreement with many experimental results one usually talks about free rotation around the long molcular axis. However, such rotations are unlikely - if not impossible - for individual molecules. The lateral dimensions of a benzene ring are 6.5 x 3.5 Å, which can be compared with a typical lateral distance of 4.5-5 Å as obtained from x-ray diffraction. Evidently rotations around the long molecular axis must be strongly cooperative. Hence it is important to consider the short-range structure in some more detail. Here two effects can be distinguished.

- *Short-range positional order.* X-ray results lead to an anisotropic positional distribution function. From the linewidth one obtains laterally a coherence length of a few molecules. This indicates ordered regions of typically 10 molecules[4].
- *Short-range orientational order.* Here again two aspects can be distinguished. In the first place one might expect almost complete parallellism of the long molecular axis within the above-mentioned ordered regions. In order words, the perfect local orientational order decays with a correlation length larger than the x-ray coherence length[1,4]. Secondly it should be realized that benzene rings prefer to pack with the rings in a herringbone-like fashion, long range in the solid and locally in the liquid state[5]. In fact there is some evidence for similar behaviour in the nematic phase of a (short-chain) cyanobiphenyl[6]. This effect might lead to a coupling of the lateral positional and orientational order.

## 3. The Orientational Distribution Function and Microscopic Order Parameters

Now I will return to Fig. 3. The order of an ensemble of molecules is specified by the distribution function $P(\mathbf{r},\Omega)$, which gives the probability of finding a molecule at a particular position $\mathbf{r}$ and with a particular orientation $\Omega$, where[1] $\Omega=(\phi,\theta,\psi)$. For the nematic phase evidently $P(\mathbf{r},\Omega)=\rho f(\Omega)$, where $\rho$ is the uniform mass density and $f(\Omega)$ the orientational distribution function, which is normalized , i.e.

$$\int d\Omega\, f(\Omega) = \int_0^{2\pi} d\phi \int_0^{\pi} d\theta \sin\theta \int_0^{2\pi} d\psi\, f(\phi,\theta,\psi) = 1 \tag{1}$$

Thus for the isotropic phase one finds $f(\Omega)=1/(8\pi^2)$=constant. Here I want to restrict the discussion to the uniaxial nematic phase, in which case $f(\Omega)$ is independent of $\phi$:

$$f(\Omega) = f(\theta,\psi)/(2\pi) \tag{2}$$

The resulting distribution function can be expanded in spherical harmonics:

$$f(\theta,\psi) = \sum_{L=0}^{\infty} \sum_{m=-L}^{L} \left(\frac{2L+1}{4\pi}\right)^{1/2} a_{Lm} Y_{Lm}(\theta,\psi) \tag{3}$$

Because of the up-down symmetry of the nematic phase it must hold that

$$f(\theta,\psi) = f(\pi-\theta,\psi) \tag{4}$$

Consequently only even terms in L can occur in Eq.(3). The expansion coefficients $a_{Lm}$ are given by

$$a_{Lm} = \left(\frac{2L+1}{4\pi}\right)^{-1/2} \left\langle Y^*_{Lm}(\theta,\psi)\right\rangle, \tag{5}$$

i.e. for each L there are 2L+1 order parameters. This number can be reduced considerably by taking the two symmetry planes shown in Fig. 3 into account. Mathematically this is expressed by

$$f(\theta,\psi) = f(\theta,\pi-\psi) \ , \ f(\theta,\psi) = f(\theta,\pi+\psi) \tag{6}$$

Consequently for each L the number of order parameters is reduced from 2L+1 to 1+L/2. In lowest order this means two parameters for L=2 and three parameters for L=4.

Let us now consider the most relevant order parameters explicitly. The distribution of the long molecular axis is described by the order parameters with m=0:

$$a_{L0} = \langle P_L(\cos\theta)\rangle = \int_0^1 P_L(\cos\theta)\, f(\theta)\, d(\cos\theta) \tag{7}$$

The first non-trivial order parameters are

$$S = \langle P_2(\cos\theta)\rangle = \frac{1}{2}\langle 3\cos^2\theta - 1\rangle$$

$$\langle P_4(\cos\theta)\rangle = \frac{1}{8}\langle 35\cos^4\theta - 30\cos^2\theta + 3\rangle \tag{8}$$

For L=2 there is in addition to S an additional order parameter that describes the relative distribution of the $\xi$ and $\eta$ axes of the molecule. It can be written as

$$D = \frac{3}{2}\langle \sin^2\theta \cos 2\psi\rangle$$

The order parameter D is related to the degree of 'flatness' of the molecules. A finite D means that there is a difference in tendency of the two transverse molecular axis to project on the director. It does not mean that the phase is biaxial. There is no preference of either the $\xi$ axis or the $\eta$ axis of different molecules to be parallel. D can only be finite if $S\neq 1$, as otherwise such a projection is zero. On the other hand, in the other limit where S=0 also D is necessarily zero. Hence D will show approximately a behaviour like $D\approx S(S-1)$ and will become less important at low temperatures where S is large.

Now I will consider the consequences of these definitions in terms of measurements of physical properties of nematic liquid crystals[1]. Let us consider a macroscopic tensor property such as the magnetic susceptibility $\boldsymbol{\chi}$. Taking the z axis along the director, in agreement with Eq.(3) of Chapter 1 such a tensor can be written as

$$\begin{pmatrix} \chi_\perp & 0 & 0 \\ 0 & \chi_\perp & 0 \\ 0 & 0 & \chi_\parallel \end{pmatrix} \tag{9}$$

In the molecular frame the magnetic polarizability tensor $\underline{\boldsymbol{\kappa}}$ can be written as

$$\begin{pmatrix} \kappa_{\xi\xi} & 0 & 0 \\ 0 & \kappa_{\eta\eta} & 0 \\ 0 & 0 & \kappa_{\zeta\zeta} \end{pmatrix} \tag{10}$$

To obtain the relation between the macroscopic and the molecular properties the following steps are required:

- transform $\underline{\boldsymbol{\kappa}}$ from the molecular frame ($\xi\eta\zeta$) to the macroscopic frame (xyz);
- take the average over $f(\theta,\psi)$.

The result is

$$\begin{aligned} \chi_{//} &= N\,\langle\underline{\boldsymbol{\kappa}}\rangle_{zz} \\ \chi_\perp &= N\,\langle\underline{\boldsymbol{\kappa}}\rangle_{xx} = N\,\langle\underline{\boldsymbol{\kappa}}\rangle_{yy} \end{aligned} \tag{11}$$

where N is the number of molecules. The macroscopic order parameter which can be taken as the anisotropy of the magnetic susceptibility $\Delta\chi = \chi_{//} - \chi_\perp$ can now be related to the microscopic order parameters via

$$\Delta\chi/N = [\kappa_{\zeta\zeta} - \tfrac{1}{2}(\kappa_{\xi\xi} + \kappa_{\eta\eta}]S + \tfrac{1}{2}(\kappa_{\xi\xi} - \kappa_{\eta\eta})D \tag{12}$$

Consequently, even if the molecular tensor properties are known (for example from solid state experiments) the measurement of a single macroscopic anisotropy is not sufficient to determine the microscopic order parameters. Only for axially symmetric molecules (in which case $\kappa_{\xi\xi}=\kappa_{\eta\eta}$) or when D=0 this reduces to

$$\Delta\chi = (\chi_l - \chi_t)S \tag{13}$$

where $\chi_l=N\kappa_{\zeta\zeta}$ and $\chi_t=N\kappa_{\xi\xi}=N\kappa_{\eta\eta}$. Eq.(13) is often used in the literature without any discussion wether these conditions are met. It will be clear that in that case errors will be made that depend on the relative importance of the two terms at the right hand side of Eq.(12), and thus on both D/S and the relative values of the two important molecular anisotropies. Moreover the first of these two factors will vary with temperature.

An accurate determination of D has been made for 5CB using nuclear magnetic resonance[7]. It leads to D/S=0.03. If this result is typical for other mesogenic molecules the conclusion can be drawn that a description of the molecules as cylindrical rods is not unreasonable.

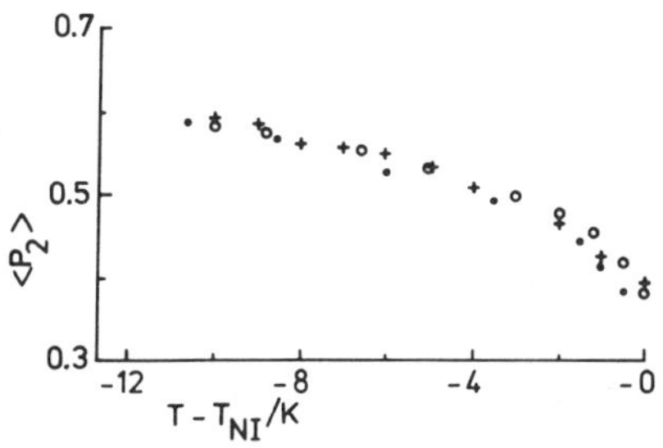


Fig. 4. Order parameters of 5CB from depolarized Raman scattering (+, Ref.9), NMR (●, Ref.10) and $\Delta\chi$ measurements (o). The latter data are scaled to the Raman data at the lower end of the temperature scale.

Using polarized Raman spectroscopy the next higher order parameter $\langle P_4(\cos\theta)\rangle$ can also be determined. These types of experiment are complicated and many erroneous results can be found in the literature. Only for 5CB results are available from independent measurements of two different groups[8,9]; fortunately these agree rather well. The various results for $<P_2(\cos\theta)>$ of 5CB are summarized in Fig. 4. To the best of my knowledge for no other compound such a complete set of data is available, which makes 5CB a useful model compound. The variation of $S=\langle P_2(\cos\theta)\rangle$ with temperature is rather well described by the full curve derived from mean-field theory (see next section). However, close to the phase transition the experimentally observed variation with temperature is somewhat stronger than theoretically predicted, which seems to be quite generally the case. The low value of $\langle P_4(\cos\theta)\rangle$ in 5CB (close to zero) is somewhat special. It indicates deviations from mean-field like behaviour that predicts approximately $\langle P_4(\cos\theta)\rangle \approx \langle P_2(\cos\theta)\rangle/2$. It is also found for 6CB, while for the higher homologues the normal mean-field like behaviour is observed.

A low value of $\langle P_4(\cos\theta)\rangle$ can be interpreted as a broadened orientational distribution function. Such an effect for 5CB and 6CB has been attributed to antiparallel dipole association in which that the aromatic cores 'overlap' to an extend that is appreciable compared to the length of the end chains[9]. This idea is substantiated by preliminary molecular dynamics calculations[11] of 5CB. The full distribution function of a nematic liquid crystal can in principle be obtained from the intensity contours observed in x-ray or neutron scattering[12,13]. For a simple interpretation of the data perfect orientational order within the short-range structure (see the end of paragraph 2) is assumed.Unfortunately a systematic comparison of this method with results for the order parameter from other techniques (for example for 5CB) is not available.

## 4. Molecular-Statistical Theories: the Role of Attractive Forces and Excluded Volume Effects

A priori two effects can be expected to contribute to the orientational order in the nematic phase: anisotropic attractive interactions and anisotropic repulsive forces. Ultimately a successful theory is expected to contain both ingredients. I will first consider the Maier-Saupe theory which is still at the basis of many discussions of orientational order in the nematic phase. In a mean-field approximation each molecule is assumed to experience an average attractive potential $U(\theta)$ given by[1]

$$U(\theta) = -\varepsilon \langle P_2(\cos\theta)\rangle P_2(\cos\theta) \tag{14}$$

where $\varepsilon$ is the strength of the potential.The distribution function is related to the potential by

$$f(\theta) = Z^{-1} \exp[-U(\theta)/k_B T] \quad (15)$$

where Z serves to normalize $f(\theta)$. The predicted behaviour of $\langle P_2(\cos\theta)\rangle$ versus temperature is shown in Fig. 5. Under appropriate conditions the attractive anisotropic energy dominates over counteracting orientational (rotational) entropy. Calculation of both contributions to the free energy leads to a first-order phase transition from the nematic to the isotropic liquid phase where $\langle P_2(\cos\theta)\rangle$ changes discontinuously from 0.43 to 0.

In its original form Maier and Saupe considered anisotropic Van der Waals forces between the molecules. A simple representation of the ingredients of the model is given in Fig. 6. The molecules are supposed to posses an anisotropic polarizability determinedby $\alpha_l$ and $\alpha_t$ or alternatively by $\bar{\alpha}=(\alpha_l + 2\alpha_t)/3$ and $\delta\alpha=\alpha_l - \alpha_t$ localized in the molecular centre. In a perturbation expansion this leads to three types of contribution, which are proportional to the quantities

$$\alpha{\cdot}\alpha\,, \quad \alpha{\cdot}\delta\alpha \quad \text{and} \quad \delta\alpha{\cdot}\delta\alpha \quad (16)$$

In spite of the assumed anisotropy Maier and Saupe take a spherical average. As a consequence only the last term gives an anisotropic contribution to the free energy and thus to $\varepsilon$, stabilizing under appropriate conditions the nematic phase. For such an average any contribution to the anisotropy of the interactions from the first two terms is zero. This is a tricky point, as it turns out that already for a relatively small deviation from a sphere the second term destabilizes the nematic phase. The conclusion is that the functional form of $\langle P_2(\cos\theta)\rangle$ predicted by the Maier-Saupe theory is not bad at all (compare Figs. 4 and 5), but that the mechanism involved cannot be correct.

Alternatively one might consider excluded volume models of the nematic phase, to which the name of Onsager is attached. The principle is illustrated[1] in Fig.7. It will be clear that a favorable packing can be obtained if the anisotropic bodies are parallel. The basic ingredients of the two types of model can be summarized as follows:

Maier-Saupe: stabilizing $U_{rot}$ (rotational energy),
destabilizing $S_{rot}$ (rotational entropy).

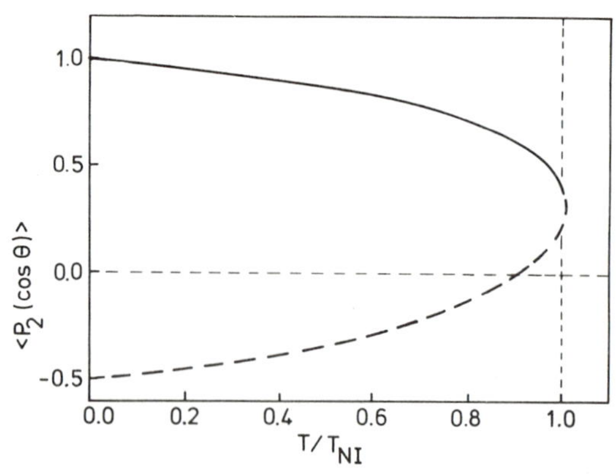


Fig. 5. Order parameter $\langle P_2(\cos\theta)\rangle$ as obtained from Maier-Saupe type mean-field theory.

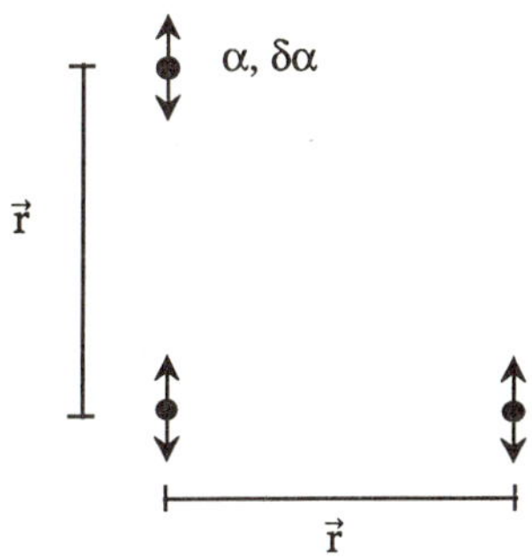


Fig. 6. Schematic drawing of the attractive interactions in the Maier-Saupe model.

Excluded volume: stabilizing $S_{pack}$ (packing entropy),
destabilizing $S_{rot}$ (rotational entropy).

As a consequence excluded volume models are in principle athermal; a phase transition isotropic-nematic can only be accomplished by a variation of the density.

A combination of both types of ingredients has been given in the so-called generalized Van der Waals theory. The interesting point is that this combination leads to an essential new element. The principle is best illustrated by returning to Fig. 7 and assuming that the full polarizability of each molecule is concentrated in its centre. It will immediately be clear that for parallel molecules not only the excluded volume is more favourable, but that also the attractive forces contribute to the stability of the orientaional order, because of the shorter distance between the attractive centers. This demonstrates the principle of a coupling between excluded volume effects and the isotropic part of the Van der Waals interactions. It means that the first of the contributions given in Eq.(16) will be dominant, making the above-mentioned destabilizing effect of the second term over the third one irrelevant. Gelbart[14] has shown that the resulting form of the generalized Van der Waals interactions can still be well approximated by the functional form of Eq.(14). This explains the success of the Maier-Saupe theory, provided that the original meaning of the coefficient $\varepsilon$ is disregarded and instead $\varepsilon$ is taken as an empirical quantity.

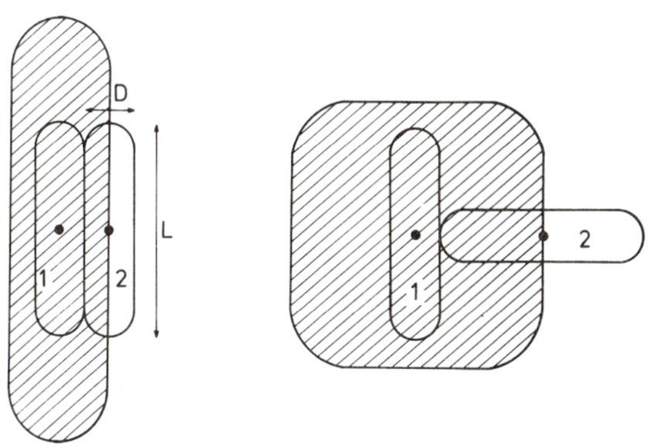


Fig. 7. Illustration of the excuded volume of two spherocylinders. Note the effect of an anistropic excluded volume on the attraction between the centers.

Ultimately one would like to correlate the molecular structure with $T_{NI}$. Especially if systematic variations in the molecules are considered, often regular trends can indeed be observed[15]. In general these trends are well compatible with the ideas expressed above. This means that there is some correlation of $T_{NI}$ with the length-to-width ratio of a molecule, while a relatively rigid polarizable core also is favourable for nematic stability. Nevertheless it is not possible to make definite rules that would provide an unambiguous proof of the theory. From the variation of $T_{NI}$ with pressure one can obtain some idea of the relative importance of attractive and repulsive forces. For PAA a $\Delta p$ of 1000 atm leads to an increase in $T_{NI}$ of about 45 K. This is still relatively small and indicates that the repulsive interaction cannot be dominant. More detailed information can be obtained by considering the quantity

$$\gamma = \frac{\rho(\partial S/\partial \rho)_T}{T(\partial S/\partial T)_\rho} \tag{17}$$

which measures the relative dependence of S on density and temperature. For the case of only hard-core interactions $\gamma = \infty$, in the other limit of only angle-dependent attractions $\gamma=1$. Measurements on PAA indicate $\gamma \approx 4$, in reasonable agreement with estimates from the generalized Van der Waals theory. However, the discontinuities at the nematic-isotropic phase transition (in density, entropy, order parameter) are invariably predicted too large in comparison with the experimental values. The causes of these deviations can be attributed to the following points that are not incorporated into the theory:

- Distribution of the centers of interaction over the molecules.
- Flexibility of the molecules.
- Short range ordering of molecules.

Nevertheless it seems fair to conclude that the generalized Van der Waals model of the nematic phase provides at least a good *qualitative* insight into the factors that contribute to the orientational ordering.

ACKNOWLEDGEMENT. This work is part of the research programme of the Stichting voor Fundamenteel Onderzoek der Materie (Foundation for the Fundamental Research of Matter, FOM) and was made possible by financial support from the Nederlandse Organisatie voor Wetenschappelijk Onderzoek (Netherlands Organization for the Advancement of Research, NWO).

## References

1. G. Vertogen and W.H. de Jeu, *Thermotropic Liquid Crystals, Fundamentals*, Springer Series in Chemical Physics vol. 45 (Springer, Heidelberg, 1988).
2. See, for example, J.W. Emsley (ed.), *Nucelar Magnetic Resonance of Liquid Crystals*, (Reidel, Dordrecht, 1985).
3. D. Frenkel, Chapter 5 of this Volume.
4. A.J. Leadbetter, Chap. 13 in *The Molecular Physics of Liquid Crystals*, G.R. Luckhurst and G.W. Gray, eds. (Academic Press, London, 1979).
5. M. Misawa and T. Fukunaga, J.Chem.Phys. **93**, 3495 (1990), and references therein.
6. I.W. Hamley, R.W. Date, G.R. Luckhurst and I.M. Seddon, to be published.
7. J.W. Emsley, G.R. Luckhurst, G.W. Gray and A. Mosley, Mol.Phys. **35**, 1499 (1978).
8. K. Miyano, J.Chem.Phys. **69**, 4807 (1978).
9. L.G.P. Dalmolen, S.J. Picken, A.F. de Jong and W.H. de Jeu, J. Phys. **46**, 1443 (1985).

10. J.W. Emsley, G.R. Luckhurst and C.P. Stockley, Mol.Phys. **44**, 1708 (1981).
11. S.J. Picken, W.F. van Gunsteren, P.Th. van Duijnen and W.H. de Jeu, Liq.Cryst. **6**, 357 (1989).
12. A.J. Leadbetter and E.K. Norris, Mol.Phys. **38**, 669 (1979).
13. R.M. Richardson, J.M. Allman and G.J. McIntyre, Liq.Cryst. **7**, 701 (1990).
14. W.M. Gelbart, J.Phys.Chem. **86**, 4298 (1982), and references therein.
15. See, for example, D. Demus, Mol.Cryst.Liq.Cryst. **165**, 45 (1988).

Chapter 3

# MICROSCOPIC ASPECTS OF THE SMECTIC PHASES

W. H. DE JEU

## 1. Introduction

First the empirical information on the molecular factors that influence smectic behaviours will be summarized. The various types of molecular model and commonly accepted concept that are supposed to lead to smectic layering will be evaluated. Introduction of a coupling between the smectic order parameter and any other relevant order parameter can drive the nematic-smectic A phase transition from second order to first order. Next the breaking of up-down symmetry will be considered, which can have either a steric or a polar origin. The presence of strong terminal permanent dipoles leads to various types of smectic A phase. The role of dipolar interactions will be discussed in some detail, including its importance for the occurrence of reentrant nematic behaviour. However, the latter effect is not restricted to strongly polar systems. Finally the origin of tilted smectic phases will be discussed.

In the third Chapter of this Section I will explain where we stand in understanding the various molecular aspects of smectic behaviour. Again I will not go into the details of calculations, but rather try to examine critically the ingredients of the various approaches in relation to the final results. Emphasis will be almost completely on the smectic A phase, including the various types of variation that occur in the case of molecules with strongly polar end groups. A short discussion on the possible origin of tilted smectic structures will also be given. The two previous Chapters of this Section will be referred to as Chapter 1 and 2, respectively, where required.

## 2. The McMillan Type of Model for Smectic A

'Classical' smectic behaviour is usually associated with the type of observation shown in Fig.5 of Chapter 1 for a homologous series and already discussed there. This behaviour of smectic phases occuring with increasing alkyl chain length is at the basis of the McMillan model[1,2] for the nematic-smectic A phase transition, though qualitatively the underlying ideas had been around already for a long time[3]. Basic is the assumption that attractive intermolecular interactions can be localized in regions with dimension $r_0$, the central

W.H. de Jeu - FOM-Institute for Atomic and Molecular Physics, Kruislaan 407, 1098 SJ Amsterdam, The Netherlands and Open University, P.O. Box 2960, 6401 DL Heerlen, The Netherlands

*Phase Transitions in Liquid Crystals*, Edited by S. Martellucci and A.N. Chester, Plenum Press, New York, 1992

(aromatic) cores of the molecules. The rest of the molecules is just assumed 'to take up space' so that the smectic periodicity d is approximately equal to the molecular length *l*. Hence two length scales come into play, and in fact the ratio $r_0/d$ determines the phase behaviour. Furthermore, only the first harmonic of the Fourier-expansion of the smectic density wave is considered, leading to a smectic order parameter

$$\rho_1 = \tau = \langle \cos(2\pi z/d) \rangle \tag{1}$$

The range of applicability of the model is restricted by these assumptions, and can be summarized as follows.

- The model can only be applied if any contribution from excluded volume effects (packing entropy) can be considered as a constant background. This is approximately true for the type of homologous series which I have just discussed. However, one should not ask the model to predict smectic behaviour in cases where a variation in packing can be expected to be important. This applies for example to the effect of substitution of side-groups, or to mixtures. In the latter case induced smectic phases may occur (reentrant nematic behaviour as a function of concentration), due to either molecular complexing[4] or to the combination of very specific geometrical aspect of the molecules. Examples of the latter are provided by the so-called filled smectic phases[5]. In terms of the McMillan type of model the smectic phase observed for sexiphenyl mentioned in Table 3 of Chapter 1 can be considered as part of the 'background' of a system of rods that serves as a reference system to which the attractive interactions are added.
- The validity of the assumption of neglecting higher-order Fourier components can be determined from measurements of $I_{002}/I_{001}$, the intensity ratio of the second to the first harmonic of the x-ray reflection corresponding to the smectic density wave. The relation is given by[6]

$$\frac{I_{002}}{I_{001}} = \frac{1}{2} \left| \frac{S(2q_0)}{S(q_0)} \frac{\rho_2}{\rho_1} \right|^2 \tag{2}$$

where $S(q_0)$ is the molecular structure factor at $q_0=2\pi/d$, and the $\rho_i$ have been defined in Chapter 1. Typically $I_{002}/I_{001} \leq 10^{-3}$ leading to $\rho_2/\rho_1 \leq 0.1$. However, for particular compounds this ratio may be much higher, invalidating the present approach.

Details of the mean-field calculation of McMillan can be found in the original papers[2]. If the orientational order is considered as a constant background only the order parameter

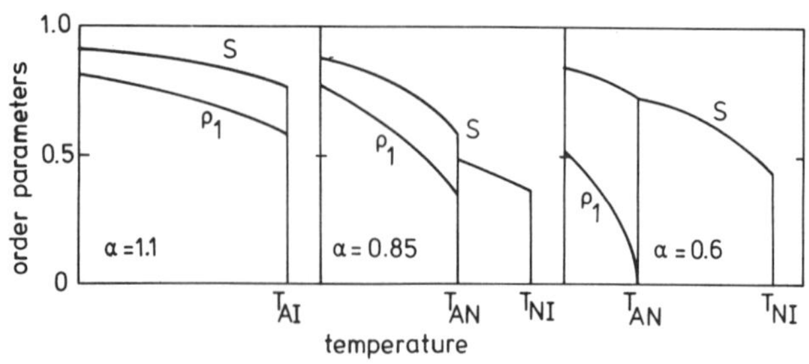


Fig.1. Phase diagram with smectic A, nematic and isotropic phases according to McMillan.

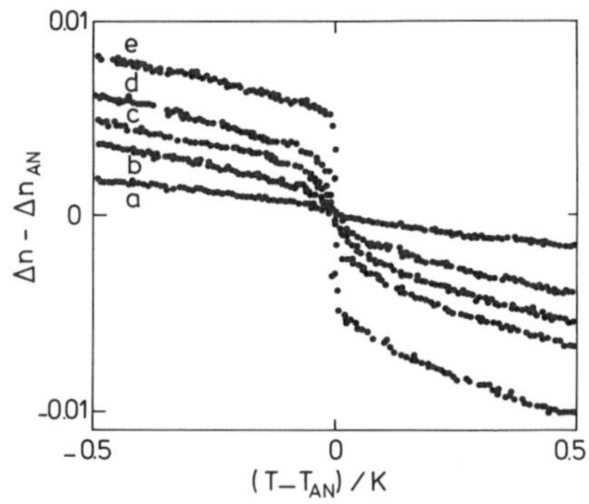


Fig.2. Birefringence of mixtures of 7AB and 8AB (see Fig.5 of Chapter 1) around the $S_AN$ phase transition. Different curves correspond to different mole fractions x of 8AB in 7AB: (a) x=0, (b) x=0.345, (c) x=0.464, (d) x=0.573, (e) x=0.689. The tricritical point is approximately reached for case (c).

given by Eq.1 is required, otherwise also a mixed (orientational-positional) order parameter comes into play. Furthermore the attractive forces are not considered explicitly, but mimicked by Gaussians of range $r_0$. The final steps involve as usual a mean-field calculation of the free energy $F = U - TS_{rot}$, in which (to be emphasized once more) any possible contribution to the entropy by variation in the packing is disregarded. The final phase diagram (Fig.1) depends strongly on the ratio $r_0/d$ via the parameter $\alpha=-u_0\exp[-(\pi r_0/d)^2]$. For small values of $r_0/d$ the attractive interactions are strongly localized and the smectic phase clears directly into the isotropic liquid. For larger values a nematic phase appears that is separated from the smectic phase by a first order phase transition. At the McMillan number $T_{NI}/T_{SN}=0.886$ a tricritical point is predicted where this transition becomes second order.

The predicted first order behaviour is due to a coupling of the smectic order parameter to the orientational order. This leads to a tricritical point at temperatures where the orientational order starts to saturate. The behaviour predicted by the McMillan model is qualitively in agreement with observations for homologous series. The tricritical point is in practice not at a universal ratio of $T_{NI}/T_{SN}$. It seems that the coupling with the orientational order is indeed relevant. In leads to the possibility of determining the location of a tricritical point indirectly via measurements of the orientational order (see Fig.2). The existence of a tricritical point is well established now for a variety of systems. It should be noted, however, that in fact any type of coupling of the smectic order parameter with a relevant other parameter could drive the transition first order. Among the possibilities that have been studied are a coupling with the second harmonic of the density wave[7] and with director fluctuations[8].

There is ample evidence that with asymmetric molecules dipolar interactions are important for the liquid crystalline behaviour. For example in the following series

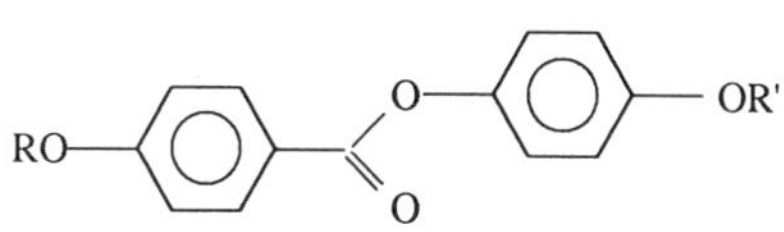


smectic phases are found for R long and R' short. For molecules of the same total length R+R' only a nematic phase is found if the asymmetry is reversed (R short, R' long)[9]. Even more extreme examples are provided by the series

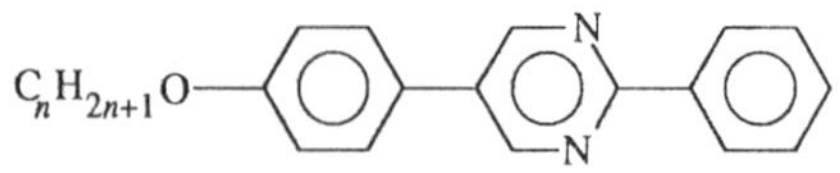


(I)

and

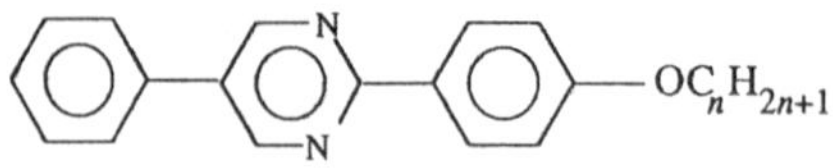


(II)

For series (I) all compounds studied (from n=2 to n=10) show only a smectic A phase. For series (II) all compounds studied (up to n=12) show nematic behaviour. These additional dipolar aspects of the intermolecular interactions cannot be accounted for in the McMillan model. Its influence on the phase behaviour is still largely unexplained.

## 3. The Breaking of Up-Down Symmetry: the $S_{A2}$ Phase and Related Phases

In the case of asymmetric molecules new phases with a broken up-down symmetry, i.e. long-range order of the heads and tails of the molecules, can be expected. In principle such an effect can have two different origins as pictured in Fig.3: (i) dipolar effects, and (ii) steric effects. For rod-like molecules the dipolar case has only been found to occur. This in contrast to disc-like molecules (also pictured in Fig.3) where molecules with a bowl-like shape have been synthesized[10], while it is difficult to introduce strong dipole moments perpendicular to the disc. Of course the division between steric and dipolar effects is somewhat artificial; for example, the introduction of steric differences between molecular heads and tails necessarily implies also a (small) dipole moment. Still it is usefull as a starting point.

Let us restrict ourselves now to the dipolar effects in the smectic A phase of rod-like molecules. When the up-down symmetry is broken within a smectic layer, the layer will be ferro-electric and carry a net polarization. In addition to the dispersion forces considered so far (induced dipole-induced dipole interactions), two additional types of interaction come into play. In the first plase the layer structure will be destabilized because of the repulsions between the permanent dipoles. Secondly, dipole-induced dipole interactions may add somewhat to the stabilizing influence of the dispersion forces that was already present. Most probably the destabilizing dipolar repulsions will be dominant. Note that for a single polarized layer the position of the dipole moment along the molecule is not important.

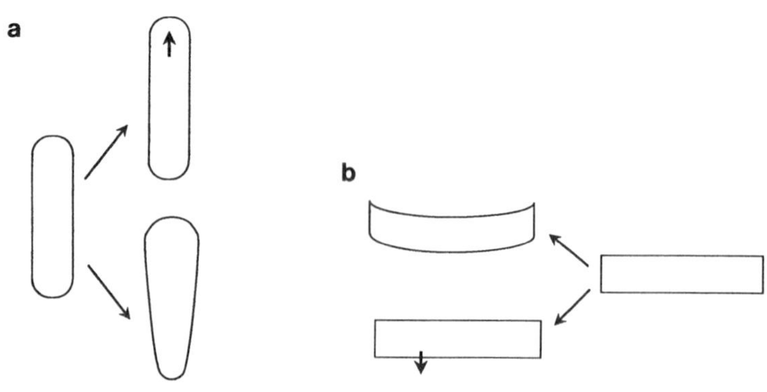


Fig.3. Breaking of the up-down symmetry for rods and discs, respectively, via polar and steric effects (side view).

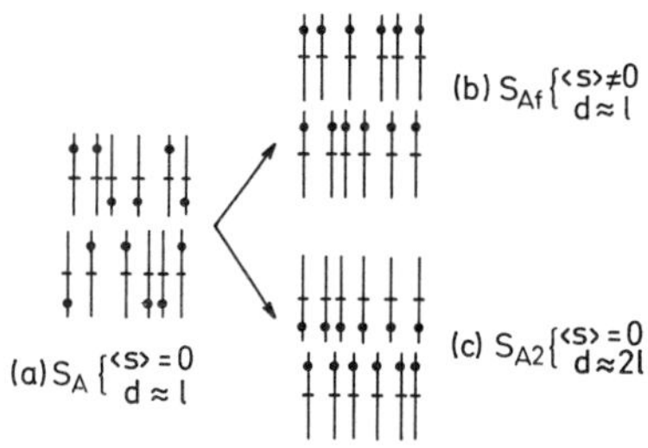


Fig.4. The two ways in which smectic layers with a broken up-down symmetry can be combined; $S_{Af}$ is ferroelectric, $S_{A2}$ is antiferroelectric.

Polarized layers may be combined in two ways (see Fig.4). The $S_{Af}$ phase would carry an overall electric polarization; its possible stability is directly related to that of a single polarized layer. Hence it is not surprising that the $S_{Af}$ phase has not been observed experimentally. The $S_{A2}$ phase has an anti-ferroelectric organisation of neighbouring layers and consequently a periodicity of $d=2l$. In this case the position of the dipoles within a single layer can be expected to be important for the interactions within the bilayer. Experimentally the $S_{A2}$ phase does actually occur, but only for compounds with a strongly polar terminal group like CN or $NO_2$. A typical example is

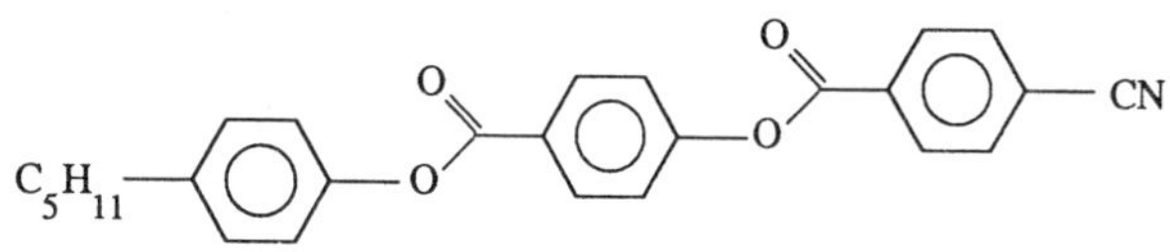


$S_{A2}$ 139 N 256 I

A molecular model for the $S_{A2}$ was been given some time ago[11]. The essential elements are summarized in Fig.5. The role of the (destabilizing) dipole repulsion within the framework of the model is shown in Fig.6. In agreement with the qualitative arguments given above the $S_{Af}$ phase is strongly destabilized, almost independently of the position of the dipole moment. More interestingly, the dipole repulsions become quite small in the bilayer

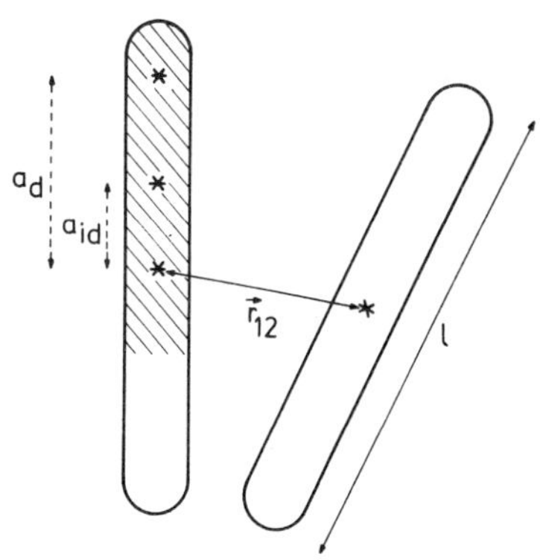


Fig.5. Molecular model for calculating the interactions in asymmetric molecules; a dipole moment and a polarizable center are situated at distances $a_d$ and $a_{id}$ from the molecular center, respectively.

configuration of the $S_{A2}$ phase, provided the dipoles are located at the end of the molecules. This is due to the favourable interactions between the dipoles in neighbouring layers. In combination with the optimal packing (compared with alternatives to be discussed in the next section) this causes the $S_{A2}$ phase to be the lowest temperature smectic A phase, as observed experimentally. Note that in the example given above the dipole moments of the bridging groups and that of the terminal cyano-group point in opposite directions. This causes the terminal dipole to be strongly localized. Reversal of the bridging groups leads to one conjugated dipole moment distributed over a large part of the molecule. This should destabilize the $S_{A2}$ phase, and indeed in this situation the phase behaviour is much more complex[12].

Analogous to Eq.1 the $S_{A2}$ phase can be characterized by an order parameter

$$\xi = \langle s \cos(\pi z/d) \rangle \tag{3}$$

where $s=\pm1$ indicates whether the polar head is up or down, respectively. In agreement with the model the dielectric permittivity is strongly reduced compared with the nematic phase at higher temperatures[13]. This provides direct evidence of the long-range anti-ferroelectricity.

One final point should still be explored regarding the $S_{A2}$ phase. So far I have assumed that the breaking of the up-down symmetry has a purely dipolar origin. Let us now assume that the dipole has a steric effect associated with it: for example that the polar heads are somewhat bulkier than the rest of the molecule. The optimalization of the effects of packing and dipolar interactions in the $S_{A2}$ phase is then disturbed, as the ferroelectric layers must be curved due to the bulky heads (see Fig.7). A possible solution to this problem is alternation of regions of opposite polarity within one layer. Provided the anti-ferroelectric bilayer structure is retained locally, this leads to the introduction of kinks at which the bilayer structure is shifted over one molecular length. When these local bilayer regions have an appreciable size such defects will cost little energy. Now two situations may occur.

- The bilayer regions are randomly distributed. This would lead overall to an $S_A$ phase that has again a periodicity $d=l$, but with strong $S_{A2}$ type fluctuations. I shall strictly use the symbol $S_{A1}$ only for this type of monolayer phase. Macroscopically it cannot be distinguished

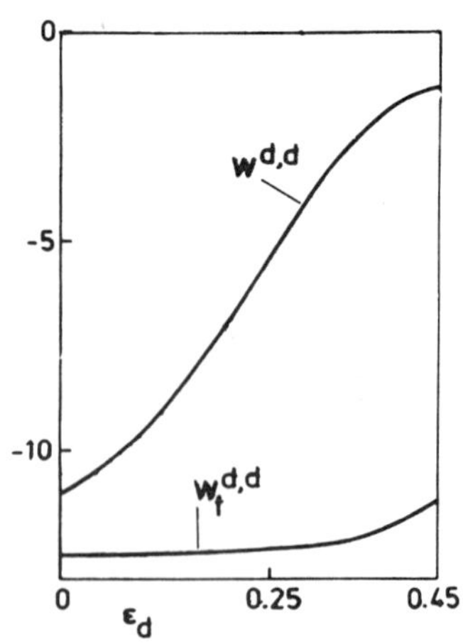


Fig.6. Dipole repulsions for the $S_{Af}$ and $S_{A2}$ phases as a function of the dipole position $\varepsilon_d=a_d/l$ along the axis.

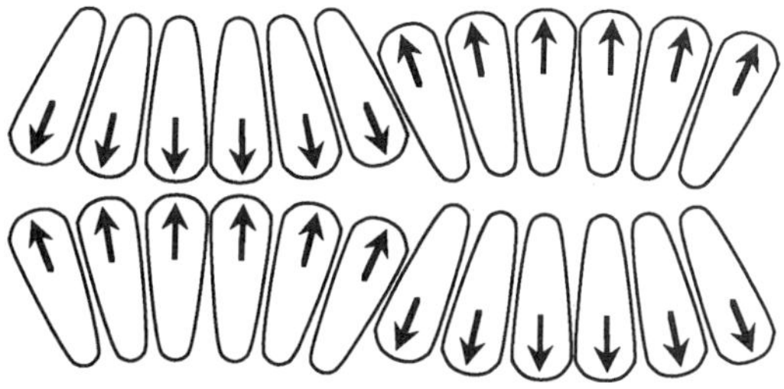

Fig.7. The principle of introducing a modulation to the $S_{A2}$ structure by adding a steric difference between molecular heads and tails.

from a 'classical' $S_A$ phase in which the up-down symmetry is fully retained. Apart from the bilayer fluctuations the difference will be seen in the dielectric permittivity which in the $S_{A1}$ phase should already show the onset of anti-ferroelectricity. An order parameter

$$\xi_1 = \langle s \cos(2\pi z/d) \rangle \ , \quad s=\pm 1, \tag{4}$$

can be associated with the $S_{A1}$ phase[11].

– An alternative possibility is that the kinks, at which the bilayer structure is shifted over one layer, minimize their energy by forming walls that are regularly spaced. This leads to a modulation of the $S_{A2}$ structure perpendicular to the direction of the density wave[11,14]. In principle this modulation could be 2D (grid-like) or 1D. In the latter case the resulting so-called $S_{\tilde{A}}$ phase will be biaxial.

It seems that several of these ideas are relevant for the experimental situation. Both the $S_{A1}$ phase and the $S_{\tilde{A}}$ phase have been found, and always in the region of the phase diagram where the $S_{A2}$ phase also occurs. In fact the order is with decreasing temperature $S_{A1}$-$S_{\tilde{A}}$-$S_{A2}$. In this order the anti-ferroelectric tendencies in the dielectric permittivity increase[13]. The modulation in the $S_{\tilde{A}}$ phase is 1D; the phase is biaxial. $S_{A2}$-type fluctuations have been found in the $S_{A1}$ phase, though often with a periodicity somewhat smaller than $2l$. Especially the polar $NO_2$ end group can be expected to have the assumed bulkiness; for the CN-group this is not the case. The possible consequences of having two types of $S_A$ phase with $d=l$ and the same macroscopic symmetry ('classical' $S_A$ and $S_{A1}$) but a different origin in terms of intermolecular interaction resulting in different microscopic order parameters [Eqs.1 and 4], have not been further explored yet.

## 4. The $S_{Ad}$ Phase and Reentrant Nematic Behaviour

So far the periodicity of the smectic density modulation has always been dictated by the molecular length, even in the $S_{A2}$ phase where a simple doubling occurs. Now I come to the situation that a different period arises. The principle is shown in Fig.8. Here a situation is sketched in which optimal dispersion interactions (in the sense of McMillan) are combined with a favourable situation in terms of dipolar interactions. It should be realized that in 3D the situation is less favourable than pictured: a spin system on a trangular lattice is frustrated (compare Ref.15). Evidently the more favourable dipolar interactions can only occur at the expense of the packing which, on average, must be less favourable than in the $S_{A2}$ phase. Consequently this type of partial double layer phase (called $S_{Ad}$) can in principle be expected at temperatures higher than the $S_{A2}$, $S_{\tilde{A}}$ and $S_{A1}$ phases. This is indeed as observed.

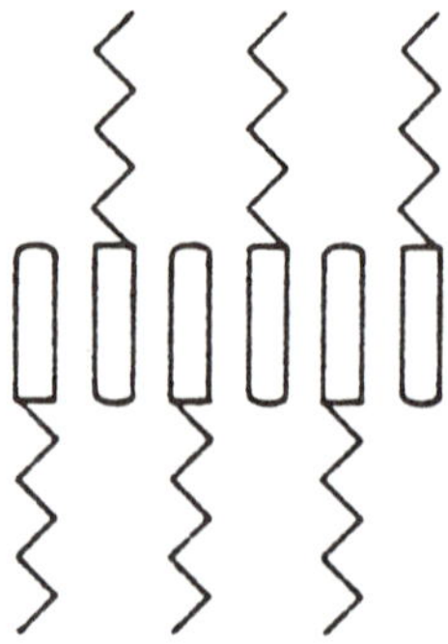

Fig.8. Idealized sketch of the possibility to optimize the dipole-dipole interactions in an $S_{Ad}$ phase.

Experimentally one finds for the layer spacing d of $S_{Ad}$ phases values $l<d<2l$. If the aromatic and aliphatic parts of a molecule have length a and b, respectively, one finds[16] for many cases approximately d=a+2b. However, this should not be generalized too easily to state that always the aromatic cores exactly overlap. The following examples show that replacing a benzene ring by a cyclohexane ring does reduce the overlap, but not to the extend expected from the simple considerations given above:

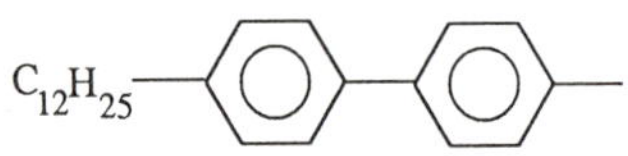

d = 39.1Å

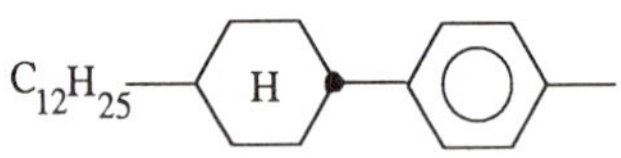

d = 41.2Å

The antiparallel dipole correlation in the $S_{Ad}$ phase is confirmed by data on the dielectric permittivity[13]. Always the permittivity is reduced, indicating an effective dipole moment $\mu_{eff}^2=g\,\mu_{free}^2$, where the correlation factor g is of the order of 1/2 and increases with decreasing temperature. This approximately compensates for the usual $\mu^2/T$ dependence of the permittivity, leading sometimes to an almost constant permittivity over a large temperature range[13]. Two effects are of further importance regarding the information from the dielectric permittivity.

- The antiparallel dipole correlation is not restricted to the $S_{Ad}$ phase but is also observed in the nematic phase (if present) at higher temperatures. In fact one sometimes has to go well into the isotropic liquid phase to recover the usual $\mu^2/T$ dependence.
- Though the dielectric permittivity is strongly reduced, this effect is by no means as dramatic as in the case of the anti-ferroelectric $S_{A2}$ phase. The two cases can be readily distinguished from the dielectric point of view.

When the temperature is further decreased the somewhat unfavourable packing of the $S_{Ad}$ phase will lead to a decreased stability, and a transition to one of the phases discussed in the previous section could be expected. Experimentally indeed $S_{Ad}$ -$S_{A1}$ and $S_{Ad}$ -$S_{A2}$ transitions have been observed[12]. However, other phases might also appear, including a so-called reentrant nematic phase. A classical example is reproduced in Fig.9. In a dimer model for the N-$S_{Ad}$-$N_{re}$ sequence (Ref.18, see Fig.10) the free energy describing the system contains in addition to the rotational entropy an extra packing entropy

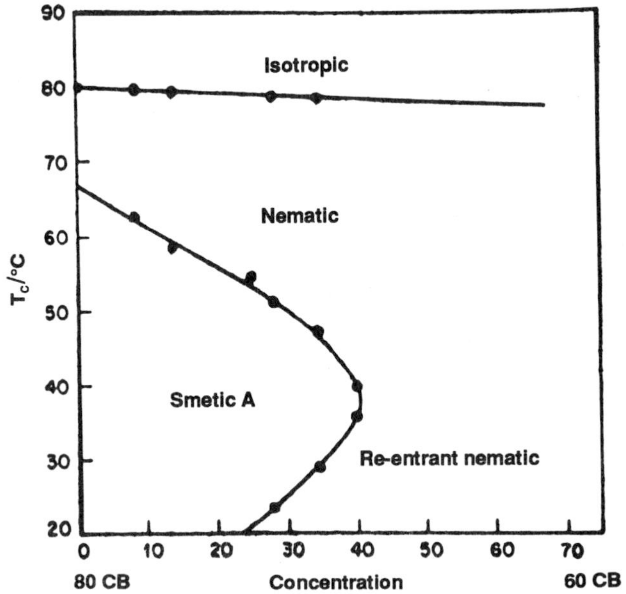


Fig.9. Example of reentrant nematic behaviour in a mixture of two 4-cyano-4'-alkylbiphenyls (Ref.17).

$$F = U - TS_{rot} - TS_{pack} \tag{5}$$

It is the latter factor that causes with decreasing temperature the dimers to move out of the smectic planes again, which destabilizes the $S_{Ad}$ phase and leads to reentrant nematic behaviour. Interesting variations on the model are possible that have not been further explored yet. If the potential minimum for the relative position of the two monomers forming a dimer is relatively broad, a spectrum of relative positions is possible, leading to a variations of the layer spacing in the $S_{Ad}$ phase with temperature. If more than one sharp minimum exists one could expect more than one $S_{Ad}$ phase (with different layer spacings) each separated by a nematic phase (multiple reentrance). This type of behaviour has indeed been observed experimentally[12]. Realizing the importance of $S_{pack}$ - independent of the details of the model - one can easily imagine other mechanisms of reentrance, provided they give rise to a similar contribution to the free energy[15,19]. In agreement with this view not unexpectedly reentrant nematic behaviour has been found in systems where no strongly polar end groups are present (see Fig.11). Full insight into the relevant mechanisms can probably only be

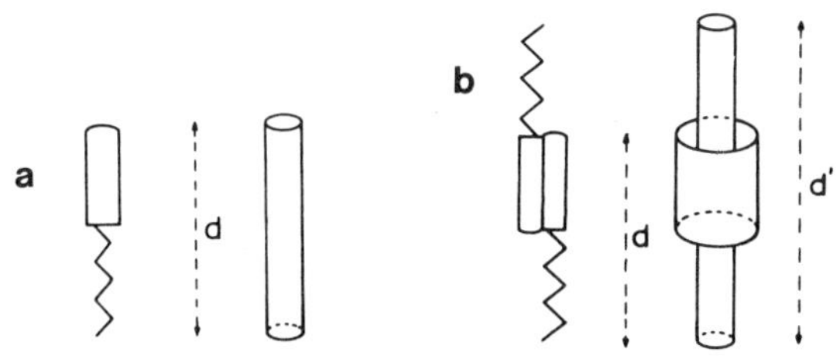


Fig.10. Monomer and dimer and their representation in the dimer-model of reentrant nematic behaviour.

obtained from computer simulations that take the interaction between the molecules more completely into account.

## 5. On the Origin of Tilted Smectic Structures

I will finish with a few remarks concerning the various types of tilted smectic phases, in particular smectic C. The available information concerning non-polar molecules indicates that these only form orthogonal smectic phases[21]. Hence it seems very plausable that dipolar interactions are essential to produce a tilted structure. Hence we can select from the various models that have been put forward only those involving dipolar interactions[22,23,24]. Of these the first two consider outboard dipoles at some angle with the long molecular axis; the last one considers dipoles parallel to this direction. Experimentally the first situation seems to apply to compounds that form smectic C phases. This is nicely demonstrated by the following series of compounds[25]:

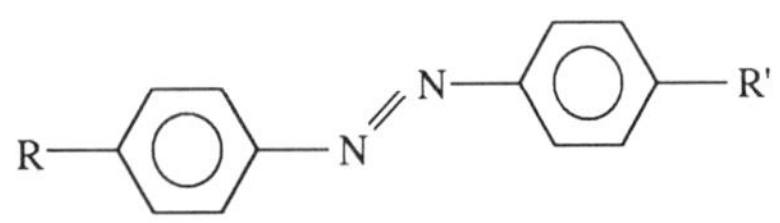


with: $R = R' = OC_{n-1}H_{2n-1}$ , Smectic C,
$R = OC_{n-1}H_{2n-1}$ , $R' = C_nH_{2n+1}$ , Smectic A and C,
$R = R' = C_nH_{2n+1}$ . Smectic A.

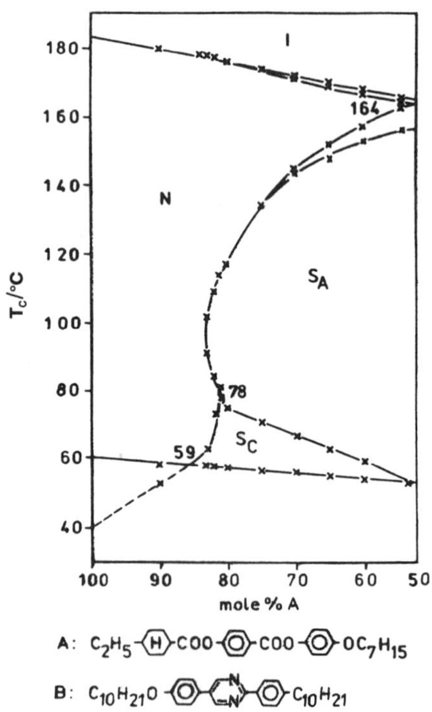


Fig.11. Example of reentrant nematic behaviour in a mixture of two molecules without strongly polar groups (Ref.20).

From the bond distances and bond angles the differences in conformation between these series, and thus in steric interactions, can be expected to be small. It is clear from these results that the introduction of the outboard dipole moments associated with the oxygen atoms promotes a tilted smectic C phase. This behaviour does not change when an oxygen atom is added to the bridging group (azoxybenzenes), thus introducing a centrally positioned dipole moment. As moreover end-dipoles parallel to the long molecular axis do not promote smectic C behaviour, models involving parallel dipole moments will not be considered any further. Then only two models are left. The model due to McMillan[22] involves interactions between permanent dipoles, and hence requires some bias to the the rotation around the long molecular axis in order to give a net result for the dipolar interactions. The model of Vertogen and Van der Meer[23], who considered the interaction dipole-induced dipole, provides a mechanism that allows rotation of the molecules around the molecular axis. Of course, these two proposed mechanisms are not mutually exclusive. Both might contribute and depending on the actual situation one of them may be dominant.

ACKNOWLEDGEMENT. This work is part of the research programme of the Stichting voor Fundamenteel Onderzoek der Materie (Foundation for the Fundamental Research of Matter, FOM) and was made possible by financial support from the Nederlandse Organisatie voor Wetenschappelijk Onderzoek (Netherlands Organization for the Advancement of Research, NWO).

**References**

1. See, for example, G. Vertogen and W.H. de Jeu, *Thermotropic Liquid Crystals, Fundamentals*, Springer Series in Chemical Physics vol. 45 (Springer, Heidelberg, 1988).
2. (a) W.L. McMillan, Phys.Rev. A **4**, 1238 (1971); (b) W.L. McMillan, Phys.Rev. A **6**, 936 (1972).
3. See, for example, H. Arnold, Z.Chem. **4**, 211 (1964).
4. See, for example, W.H. de Jeu, L. Longa and D. Demus, J.Chem.Phys. **84**, 6410 (1986), and references therein.
5. S. Diele, G. Pelzl, W. Weissflog and D. Demus, Liq.Cryst. **3**, 1047 (1988).
6. E.F. Gramsbergen and W.H. de Jeu, Liq.Cryst. **4**, 449 (1989).
7. R.B. Meyer and T.C. Lubensky, Phys.Rev. A **14**, 2307 (1976).
8. B.I. Halperin, T.C. Lubensky and S.-K. Ma, Phys.Rev.Lett. **32**, 292 (1976).
9. J. Malthête, J. Billard, J. Canceill, J. Gabard and J. Jacques, J.Phys.Colloq. **37**, C3-1 (1976).
10. H. Zimmermann, R. Poupko, Z. Luz and J. Billard, Z. Naturforsch. **40a**, 149 (1984).
11. L. Longa and W.H. de Jeu, Phys.Rev. A **28**, 2380 (1983).
12. S. Shashidar, Chapter 15 of this Volume.
13. S. Shashidar, Chapter 18 of this Volume.
14. W. Helfrich, J.Phys. **48**, 291 (1987).
15. R. R. Netz and N. Berker, Chapter 7 of this Volume.
16. G.J. Brownsey and A.J. Leadbetter, Phys.Rev.Lett. **44**, 1608 (1980).
17. D. Guillon, P.E. Cladis and J. Stamatoff, Phys.Rev.Lett. **41**, 1598 (1978).
18. L. Longa and W.H. de Jeu, Phys.Rev. A **26**, 1632 (1982).
19. F. Dowell, Phys.Rev. A **31**, 2464 (1985); *ibid* **36**, 5046 (1987).
20. G. Pelzl, I. Latif, S. Diele, M. Novak, D. Demus and H. Sackmann, Mol.Cryst. Liq.Cryst. **139**, 333 (1986).
21. W.H. de Jeu and R. Eidenschink, J.Chem.Phys. **78**, 4637 (1983).
22. W.L. McMillan, Phys.Rev. A **8**, 1921 (1973).
23. B.W. van der Meer and G. Vertogen, J.Phys.Colloq. **40**, C3-222 (1979).
24. D. Cabib and L. Benguigi, J.Phys. **38**, 419 (1977).
25. W.H. de Jeu, J.Phys. **38**, 1265 (1977).

Chapter 4

# PHASE TRANSITIONS IN LIQUID CRYSTALS: INTRODUCTION TO PHASE TRANSITION THEORIES

P. BAROIS

## 1. Introduction

The understanding of continuous phase transitions and critical phenomena has been one of the important breakthrough in condensed matter physics in the early seventies. The concepts of scaling behavior and universality introduced by Kadanoff and Widom and the calculation of non gaussian exponents by Wilson and Fisher are undeniably brilliant successes of statistical physics in the study of low temperature phase transitions (normal to superconductor, normal to superfluid helium) and liquid gas critical points.

But no other field in condensed matter physics has shown such a rich variety of continuous or weakly first order phase transitions than liquid crystals: order parameters of various symmetries, anisotropic scaling behaviors, coupled order parameters, multicritical points, wide critical domains, defect mediated transitions, spaces of low dimensionality, multiply reentrant topologies are currently found in liquid crystals, at easily accessible temperatures. Beside their famous technical applications in optics and electronic displays, liquid crystals can certainly be regarded as a paradise of the physics of phase transitions.

An exhaustive review of the theories of phase transitions in liquid crystals would certainly require a whole thick book. The aim of these lectures is to give an introduction to a few theories that have been deeply investigated but are still of actual interest (which often means that they are not fully satisfactory yet!) in the restricted world of smectic phases.

## 2. The Nematic to Smectic A Transition

### *2.1. The Mean Field Description (Mc Millan, de Gennes)*

As first suggested by Mc Millan[1], it is now well established that the N-SA transition can be continuous.

The smectic A phase is characterized by a periodic density modulation along the direction $\hat{z}$ perpendicular to the layers[1,2]:

---

P. Barois - Centre de Recherche Paul Pascal, Avenue A. Schweitzer F-33600 PESSAC, France.

*Phase Transitions in Liquid Crystals*, Edited by S. Martellucci and A.N. Chester, Plenum Press, New York, 1992

$$\rho(r) = \rho(z) = \rho_0 + \sum_{i=1}^{\infty} \rho_i \cos(iq_S z - \phi_i) \tag{1}$$

The fundamental term $\rho_1$ of the Fourier expansion is obviously zero in the nematic phase. It is a natural choice for the N-SA order parameter. In the vicinity of the N-SA phase transition, the Landau free energy density may be expanded in even (i.e. translationally invariant) powers of $\rho_1$:

$$\Delta f_1 = \frac{1}{2} r \rho_1^2 + u_0 \rho_1^4 + \ldots \tag{2}$$

With $r = \alpha(T - T_0)$ and $u_0 > 0$, the N-SA transition is second order at the mean field temperature $T_{AN} = T_0$.

This picture is slightly more complicated if the coupling with the nematic order parameter $S_0(T) = \frac{1}{2} < 3\cos^2\theta_i - 1 >$ is included ($\theta_i$ denotes the angle between the long axis of the $i^{th}$ molecule and the director $\boldsymbol{n} = \hat{z}$ and the brackets <...> represent an average over the volume of the sample). If $\delta S$ measures the deviation of the alignment from its equilibrium value $S_0$ in the nematic phase (calculated from the Maier-Saupe theory[3] for instance), the free energy density reads:

$$\Delta F_2 = \frac{1}{2} r \rho_1^2 + u_0 \rho_1^4 + \frac{1}{2\chi} \delta S^2 - C \rho_1^2 \delta S \tag{3}$$

$\chi(T)$ is a response function (susceptibility) and C is a generally positive constant (the appearence of the smectic order usually increases the average attraction between the molecules and hence reinforces the alignment).

Minimization with respect to dS yields $\delta S = \chi C \rho_1^2 (> 0)$ and the new coefficient of the fourth order term is now:

$$u = u_0 - 2C^2 \chi \tag{4}$$

If the nematic susceptibility $\chi$ is low (i.e. far enough from $T_{NI}$, temperature of the isotropic nematic transition) u is positive and the N-SA transition is second order at $T_{NA} = T_0$ again. If $\chi$ is larger (i.e. $T_0 \approx T_{NI}$) u is negative. A sixth order term $v\rho_1^6$ must be added to Eq. 2.3 and the N-SA transition is first order at a temperature $T_{NA} = T_0 + u^2/2\alpha v > T_0$. $\chi = u_0/2c^2$ (i.e. $u = 0$) defines a tricritical point on the N-SA line.

Mc Millan estimated that the tricritical point corresponds to a value 0.87 of the ratio $T_{NT}/T_{NA}$[4]. Experimental data suggest a slightly higher value.

## 2.2. Analogy with Superconductors

Because the smectic order is one dimensional, the fluctuations of the layers described by the local displacement field u($\boldsymbol{r}$) are known to play an important role, even far away from any phase transition: the square amplitude $<u^2(\boldsymbol{r})>$ diverges like the logarithm of the size of the sample (this is the well known Landau-Peierls instability[5,6]) hence killing a true long range order.

Close to the N-SA transition, critical fluctuations of the smectic order parameter $\rho_1(\boldsymbol{r})$ are expected to be important too. The two fields $\rho_1(\boldsymbol{r})$ and u($\boldsymbol{r}$) can be described by a complex order parameter, first introduced by de Gennes[7] and Mc Millan[4]:

$$\Psi(\boldsymbol{r}) = \rho_1(\boldsymbol{r})e^{i\Phi_1(\boldsymbol{r})} \tag{5}$$

The phase $\Phi_1(\boldsymbol{r})$ is simply related to the layers displacement u($\boldsymbol{r}$) through:

$$\Phi_1(\boldsymbol{r}) = -iq_s u(\boldsymbol{r}) \tag{6}$$

Again a free energy density can be expanded in even powers of $|\Psi|$. Including gradient terms and fluctuation of the nematic director $\delta n_\perp = \boldsymbol{n}(\boldsymbol{r}) - \boldsymbol{n}_0$ ($\boldsymbol{n}_0 = \hat{z}$) yields the following Landau-Ginzburg functional:

$$F_s = \frac{1}{2}\int d^3r\{r|\Psi|^2 + \frac{u}{2}|\Psi|^4 + C_\parallel|\nabla_z\Psi|^2 + C_\perp|(\nabla_\perp - iq_s\delta n_\perp)\Psi|^2$$
$$+K_1(div\ \delta n_\perp)^2 + K_2(\hat{z}.curl\ \delta n_\perp)^2 + K_3(\nabla_z\delta n_\perp)^2\} \tag{7}$$

The first two terms are the Landau part Eq. 2. Because of the nematic anisotropy, the gradient terms exhibit anisotropic coefficients ($C_\parallel \neq C_\perp$) along directions parallel and perpendicular to the director $\boldsymbol{n}$. With the notation $\nabla_z = \partial/\partial z$ and $\nabla_\perp = (\partial/\partial x, \partial/\partial y)$ and at lowest relevant order in ($\delta n_\perp$), these gradients have the form of Eq. 7. The last three terms are the usual Frank-Oseen elastic energy of the nematic[8].

The anisotropy of the elastic coefficients $C_\parallel$ and $C_\perp$ can be removed by a simple anisotropic rescaling (set then $C_\parallel = C_\perp$ in Eq. 7).

If one forgets about the fluctuations of the director (i.e. set $\delta n_\perp = 0$ in Eq. 7) the N-SA problem becomes equivalent to the condensation of superfluid helium (XY model, d=3).

With non zero $\delta n_\perp$, Eq.7 bears a striking resemblance to the Landau-Ginzburg functional describing the normal superconductor transition[4,7,9]:

$$F_{sc} = \frac{1}{2}\int d^3r\{r|\Psi|^2 + \frac{u}{2}|\Psi|^4 + \frac{1}{m}|(h\nabla - i\frac{q}{c}A)\Psi|^2 + \frac{1}{8\pi\mu}(\boldsymbol{curl\ A})^2\} \tag{8}$$

Here, $\Psi$ is the superconductor gap order parameter, $\boldsymbol{A}$ the magnetic vector potential, m and q the mass and charge of the particles, c the velocity of light and $\mu$ the magnetic permittivity.

Interesting behaviors of the smectic state can be deduced from the analogy:

* Just as superconductors expell magnetic fields $\boldsymbol{curl\,A}$ (Meissner effect[10]) smectics expell bend and twist. The bend and twist moduli $K_2$ and $K_3$ should therefore diverge upon approaching the smectic state from the nematic.

* Two important lengths characterize the superconductors: the order parameter coherence length $\xi = (|r|m)^{-1/2}$ over which $\Psi$ can vary and the London penetration length $\lambda = (4\pi\mu|r|/um)^{-1/2}$ of a magnetic field. Similarly, two order parameter coherence lengths

$$\xi_{\parallel,\perp} = \left(\frac{C_{\parallel,\perp}}{|r|}\right)^{\frac{1}{2}} \tag{9}$$

and four penetration lengths

$$\lambda_{\parallel,\perp}^{2,3} = \left( \frac{K_{2,3}u}{C_{\parallel,\perp}|r|} \right)^{\frac{1}{2}} \tag{10}$$

can be identified in the smectic case, which is far too many to allow a simple classification.

The superconductor analogy suggests however two distinct behaviors:
- type I ($\lambda\sqrt{2} > \xi$). The superconducting state is observed with perfect Meissner effect below a critical field $H_c$.
- type II ($\lambda\sqrt{2} < \xi$). Two critical values of the field are found. Vortices bearing a quantum flux $\varphi_0 = h/q$ penetrate the system for $H > H_{c1}$ whereas the normal transition occurs at $H_{c2} > H_{c1}$.

The smectic analog of a vortex is a dislocation. Type II smectics would therefore develop dislocations when submitted to a bending or twisting stress whereas type I would resist until a critical stress eventually induces the nematic state.

The observation of edge dislocation arrays in a wedge (i.e. bending stress)[11] and of screw dislocation arrays in chiral compounds[12] (the so-called Twisted Grain Boundary phase anologous to the Abrikosov flux lattices in superconductors[13]) suggests that most smectics are of type II but type I behaviors have also been reported[14].

At last, it can be noticed that the bare smectic coherence length at 0 temperature $\xi_0 = (m\alpha T_0)^{-1/2}$ with $r = \alpha(T - T_0)$ is significantly shorter than its superconductor equivalent (10 to 20 Å instead of 5000 Å) because of the higher value of $T_0$. An interesting consequence is that the critical domain is expected to be much larger (i.e. more easily accessible) in the smectic case.

Important differences exist however about gauge invariance and the absence of true smectic long range order. How severe these differences are is not fully understood yet.

The superconductor Hamiltonian (Eq. 8) is invariant under the following gauge transformation:

$$\begin{aligned} A' &= A - \nabla L \\ \Psi' &= \Psi \exp\left(-\frac{q}{c}L\right) \end{aligned} \tag{11}$$

so that any gauge choice is physically acceptable. The Coulomb gauge div $A = 0$ is generally used.

In the smectic case, a gauge transformation reads:

$$\begin{aligned} \delta n' &= \delta n - \nabla L \\ \Psi' &= \Psi \exp(-iq_s L) \end{aligned} \tag{12}$$

but the Frank-Oseen energy is not gauge invariant because of the splay term $K_1$: there is in fact only one physical gauge, namely $\delta n_z = 0$ (remember that the condition $n^2 = 1$ implies that $\delta\mathbf{n}$ lies in the x,y plane). In the unphysical Coulomb gauge (div $\delta\mathbf{n} = 0$), the splay term disappears and it can be shown that true long range smectic order is not killed by the layers fluctuations[15]. If the physical $\delta\mathbf{n}$ is expressed in the Coulomb gauge via Eq. 12, the splay term reads:

$$\frac{K_1}{2}\int (\Delta L)^2 d^3x \tag{13}$$

and with the condition $\delta n_z = 0$ ($i.e.\ A_z(q) - iq_z L(q) = 0$):

$$\frac{K_1}{2}\int \frac{q^4}{q_z^2}|A_z(q)|^2 \frac{d^3q}{(2\pi)^3} \tag{14}$$

The splay term is non analytic, which implies that $K_1$ cannot be renormalized by the fluctuations of $\Psi$ at any order in perturbation theory. In a renormalization flow, $K_1$ will simply evolve according to simple power counting.

## 2.3. Critical Exponents and Scaling

Experiments suggest the existence of anisotropic scaling laws: $\xi_{\parallel} \propto t^{-\nu_{\parallel}}$ and $\xi_{\perp} \propto t^{-\nu_{\perp}}$ with $\nu_{\parallel} \neq \nu_{\perp}$ for instance for the correlation lengths parallel and perpendicular to the director $\boldsymbol{n}$ ($t = (T-T_{NA})/T_{NA}$ is the reduced temperature).

Anisotropic homogeneous functions can indeed be defined to describe the scaling behavior of the correlation functions of the order parameters:

$$<\Psi(q)\Psi^*(q)> \; = G(q_{\parallel}, q_{\perp}, t) = t^{-(2-\eta_{\parallel})\nu_{\parallel}} G\left(t^{-\nu_{\parallel}} q_{\parallel}, t^{-\nu_{\perp}} q_{\perp}\right) \tag{15}$$

$$<\delta n_{\alpha}(q)\delta n_{\beta}^*(q)> \; = G_{\alpha\beta}(q_{\parallel}, q_{\perp}, t) = t^{-(2-\eta_n)\nu_{\perp}} G_{\alpha\beta}\left(t^{-\nu_{\parallel}} q_{\parallel}, t^{-\nu_{\perp}} q_{\perp}\right) \tag{16}$$

G and $G_{\alpha\beta}$ are the two point correlation functions of the smectic order parameter $\Psi$ and of the director $\delta n_{\perp}$ respectively (indices $\alpha$,$\beta$ can be x or y).

The exponent $\gamma$ of the susceptibility ($G(q=0)t^{-\gamma}$ comes immediately from Eq. 15:

$$\gamma = (2-\eta_{\perp})\nu_{\perp} = (2-\eta_{\parallel})\nu_{\parallel} \tag{17}$$

The free energy density f(t) classically scales like $k_BT$ over the volume of a "block" of condensed phase $\xi^3$. In the smectic case, this yields:

$$f(t) \sim \frac{k_BT}{\xi_{\parallel}\xi_{\perp}^2} t^{2\nu_{\perp}+\nu_{\parallel}} \sim t^{2-\alpha} \tag{18}$$

in which $\alpha$ is the critical exponent of the singular part of the specific heat $\delta C_P$.

The critical behavior of the twist and bend elastic constants on the nematic phase follows from the equipartition theorem

$$<\delta\eta_1(q)\delta\eta_1(-q)> \; = \frac{k_BT}{K_1q_{\perp}^2 + K_3q_{\parallel}^2}$$

$$<\delta\eta_2(q)\delta\eta_2(-q)> \; = \frac{k_BT}{K_2q_{\perp}^2 + K_3q_{\parallel}^2} \tag{19}$$

Axes 1 and 2 in Eq. 19 are the eigendirections of the Frank-Oseen elastic energy in the Fourier plane $(q_x, q_y)$ The corresponding unit vectors $\mathbf{e}_1$ and $\mathbf{e}_2$ are such that $\mathbf{e}_2$ is normal to q and $\mathbf{e}_1$ is normal to $\mathbf{e}_2$[15]. With the structure Eq. 16, one gets:

$$K_2 \propto t^{-\eta_n\nu_{\perp}}, K_3 \propto t^{2(\nu_{\perp}-\nu_{\parallel})-\eta_n\nu_{\perp}} \tag{20}$$

On the smectic side:

$$< \delta n_\perp(q)\delta n_\perp(-q) > = \frac{k_B T q_\perp^2}{B q_z^2 + K_1 q_\perp^4} \tag{21}$$

Eq. 16 and Eq. 21 yield:

$$B \sim t^{(4-\eta_n)\nu_\perp - 2\nu_\parallel} \tag{22}$$

The final scaling laws are then:

$$(a) \quad \xi_\parallel = \xi_0 t^{-\nu_\parallel} \qquad \xi_\perp = \xi_0 t^{-\nu_\perp}$$

$$(b) \quad \delta K_2 \propto \xi_\perp^2/\xi_\parallel \qquad \delta K_3 \propto \xi_\parallel$$

$$(c) \quad B \propto \xi_\parallel/\xi_\perp^2$$

$$(d) \quad 2-\alpha = \nu_\parallel + 2\nu_\perp \qquad \gamma = (2-\eta_\perp)\nu_\perp \tag{23}$$

Eqs. 23 (b) and (d) are valid above and below $T_{NA}$.

### 2.4. Renormalization Group Procedures

According to the classical Wilson-Fisher renormalization process[16], the parameters of the Hamiltonian Eq. 7 will evolve under an anisotropic rescaling of the wavevectors

$$q_\perp \to e^l q_\perp \qquad \text{and,} \qquad q_z \to e^{\nu_\parallel/\nu_\perp l} q_z$$

The recursion relations defining the renormalization flow of the physical parameters are controlled by simple dimensional analysis (power counting) and gaussian integration of the fluctuations of large wavevectors (perturbation theory).

The splay constant $K_1$ plays a particular role since its flow depends on power counting only (see Section 2.2.) which yields:

$$\frac{dK_1}{dl} = -\eta_n K_1(l) = -\frac{(2\nu_\perp - \nu_\parallel)}{\nu_\perp} K_1(l) \tag{24}$$

There are therefore only three possibilities for the fixed point:

-i- $K_1^* = 0$: for $2\nu_\perp > \nu_\parallel$, the $K_1$ eigendirection is stable. The problem reduces exactly to the normal superconductor transition[17] which implies $\nu_\perp = \nu_\parallel$. There is no such stable fixed point however in the smectic case (n=2, d=3)[18].

-ii- $K_1^* = \infty$: the fixed point can be stable if $2\nu_\perp < \nu_\parallel$ only, but none is found[17].

-iii- $K_1^*$ finite and $2\nu_\perp - \nu_\parallel = 0$: this anisotropy is too large to fit any experimental behavior.

Classical theories fail to find a stable fixed point (not even isotropic) and consequently cannot corroborate the experimental evidence that the transition can be second order. Two possible situations can be imagined upon moving away from the mean field tricritical point towards negative values of the Landau coefficient $u$: either the transition is still first order but the discontinuities are to small to be detected (this would agree with the lack of stable fixed point of the renormalization flow) or beyond some original multicritical point the transition becomes truly second order. The theory of a transition induced by dislocation loops provides an example that fits in this last situation.

## *2.5. Dislocation Loops Theory*

Dislocations are the most elementary excitations of a smectic A phase. Since they cannot end in the smectic material, they must form closed loops (at least in the thermodynamic limit of an infinite sample). The elastic energy of a dislocation line is proportional to its length[19]. It is then characterized by a line tension $\gamma_0$. The total free energy per unit length is:

$$\gamma = \gamma_0 - \frac{k_B T}{x} \tag{25}$$

$k_B/x$ measures the entropy contribution that favors spontaneous nucleation of dislocation loops (x has the dimension of a length). Clearly, $\gamma$ becomes negative above some temperature and the density of dislocations will diverge, hence destroying the smectic order. Nelson and Toner[20], Dasgupta and Halperin[21] and Toner[22] have formalized this problem: their most salient result is the existence of a stable fixed point which for the first time gave the theoretical possibility of a true second order N-SA transition.

This fixed point is isotropic and belongs to the so-called inverted XY universality class. Inverted refers to the inversion of the high and low temperature sides of the transition . The universal ratio of the amplitudes $A^+$ and $A^-$ of the heat capacity singularity for instance ($\delta Cp = A^{\pm} t^{-\alpha}$) is the inverse of the superfluid helium value.

## *2.6. The Experimental Situation*

The main observation is that the experimental sets of critical exponents depend on the nature of the compounds and on the ratio $T_{NA}/T_{NI}$ (see Table 1). Anisotropic correlation lengths exponents are often found ($\nu_{\parallel} > \nu_{\perp}$) which may be consistent with gauge transformations[23].

The latest experiments on very large nematic domains[24,25] give critical exponents very close to the helium analogy but with non inverted temperatures.

# 3. The Smectic A to Smectic C Transition

## *3.1. The Superfluid Helium Analogy*

The smectic C state differs from the smectic A by a tilt $\theta$ of the director $\boldsymbol{n}$ with respect to the direction $\hat{z}$ normal to the layers. The director $\boldsymbol{n}$ is totally specified by $\theta$ and the azimuthal angle $\varphi$ (Fig. 1). The smectic C order can thus be described by to real angles $\theta$ and $\varphi$ or equivalently by the complex order parameter:

$$\Psi(\boldsymbol{r}) = \Theta(\boldsymbol{r}) e^{i\phi(\boldsymbol{r})} \tag{26}$$

Table 1. Sets of critical exponents measured for several N-SA system from Refs. 2 (new edition, to appear), 24, 25, 26. Note the absence of true universal behavior but a continuous crossover from nearly XY (but non inverted) to nearly tricritical exponents when the ratio $T_{NA}/T_{NI}$ increases. Anisotropic scaling of the correlation lengths is observed.

| | $T_{NA}/T_{NI}$ | $\alpha^{\pm}$ | $\gamma^{+}$ | $\nu_{\perp}^{+}$ | $\nu_{\parallel}^{+}$ | $\alpha+2\nu_{\perp}+\nu_{\parallel}$ |
|---|---|---|---|---|---|---|
| T8 | 0.66 | 0.07 | 1.22 | 0.65 | 0.70 | 1.93 |
| T7 | 0.71 | 0.05 | 1.22 | 0.61 | 0.69 | 2.04 |
| 60CB-80CB | 0.895 | | 1.75 | 0.73 | 0.95 | |
| 8OPCBOB | 0.897 | -0.008<br>$A^-/A^+ = 0.983$ | 1.39 | 0.59 | 0.75 | 1.92 |
| 60CB-8OCB | 0.898 | | 1.77 | 0.70 | 0.93 | |
| 7S5-8S5 | 0.908 | | 1.21 | 0.67 | 0.64 | |
| $\bar{7}$S5-$\bar{8}$S5 | 0.911 | | 1.52 | 0.68 | 0.82 | |
| 80CB-$\bar{7}$S5 | 0.911 | -0.55 | 1.58 | 0.78 | 0.90 | 1.92 |
| 6OCB-8OCB | 0.920 | | 1.61 | 0.61 | 0.81 | |
| $\bar{7}$S5-$\bar{8}$S5 | 0.924 | | 1.45 | 0.68 | 0.81 | |
| 4O7 | 0.926 | -0.03 | 1.46 | 0.65 | 0.78 | 2.05 |
| $\bar{8}$S5 | 0.936 | 0 | 1.53 | 0.68 | 0.83 | 2.19 |
| CBOOA | 0.940 | 0.15 | 1.30 | 0.62 | 0.70 | 2.09 |
| 4O8 | 0.958 | 0.15 | 1.31 | 0.57 | 0.70 | 1.99 |
| $\bar{6}$O$\overline{10}$-$\bar{6}$O$\bar{8}$ | 0.960 | 0 | | | | |
| 8OCB | 0.963 | 0.2 | 1.32 | 0.58 | 0.71 | 2.09 |
| $\bar{9}$S5 | 0.967 | 0.22 | 1.31 | 0.57 | 0.71 | 2.07 |
| 8CB | 0.977 | 0.31 | 1.26 | 0.51 | 0.67 | 2.00 |
| $\bar{6}$O10 | 0.980 | 0.24 | | | | |
| $\overline{10}$S5 | 0.984 | 0.45 | 1.1 | 0.51 | 0.61 | 2.08 |
| 9CB | 0.994 | 0.53 | 1.1 | 0.37 | 0.57 | 1.84 |
| XY | | -0.007<br>$A^-/A^+ = 0.971$ | 1.32 | 0.67 | 0.67 | 2.00 |
| inverted XY | | $A^-/A^+ = 1.029$ | | | | |
| Tricritical | | 0.5 | 1.0 | 0.5 | 0.5 | 2.00 |

A modulus $\varphi = 0$ corresponds to the smectic A state. A Landau-Ginzburg functional similar to Eq. 7 with $\delta \boldsymbol{n} = 0$ and therefore to the superfluid-normal helium problem can be constructed to describe the SA-SC transition.

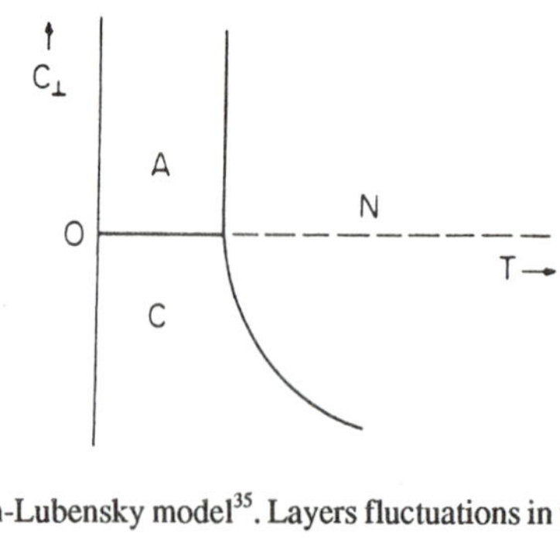


Fig. 1. Phase diagram of the Chen-Lubensky model[35]. Layers fluctuations in the nematic phase switch from SA-like to SC-like across the dashed line.

The straightforward consequence of this analogy is that the SA-SC transition may be continuous at a temperature $T_{CA}$ with XY critical exponents. Below $T_{CA}$ the tilt angle $\theta$ for instance should vary as $\theta = \theta_0 \mid t \mid^{\beta}$ with $\beta = 0.35$. Above $T_{CA}$ an external magnetic field can induce a tilt $\theta$ proportional to the susceptibility $\tilde{\chi} \propto t^{-\gamma}$ with $\gamma = 1.33$.

Experiments (heat capacity measurements in particular[27]) rather show a mean field behavior which may be due to the narrowness of the critical domain or to the influence of a close by tricritical point.

The width of the XY critical domain can be estimated from the usual Ginzburg criterion: equating the mean field heat capacity discontinuity and the contribution of the fluctuations in the gaussian approximation leads to a Ginzburg temperature $T_G$[28]:

$$T_G - T_{CA} \approx T_{CA}/64\pi^2 \qquad (27)$$

so that the critical regime extends over a fraction of a degree. It may easily switch from observable to non observable with a slight change of the roughly estimated numerical constant that lead to Eq. 27. Furthermore, the crossover regime may be different from one observable to another. Recent estimates of the Ginzburg criterion for instance indicate that elastic constants measurements are one hundred times more sensitive to fluctuations than heat capacity ones[28]. Experiments tend to confirm this point: a mean field behavior of the heat capacity has been reported[27] whereas fluctuations seem to be important in some tilt, susceptibility or bulk modulus measurements[29,30].

At last, a first order SA-SC transition is always possible.

### *3.2. The N-A-C Point*

The existence of a NAC multicritical point (i.e. a point where the N-SA, SA-SC and N-SC lines meet) was demonstrated in the late seventies[31,32]. Unlike the N-SA transition, the N-SC line is expected to be first order because of the continuous degeneracy of the smectic C fluctuations in the nematic phase[33,34]: the layers fluctuations are concentrated on two points in reciprocal space in the N-SA case ($\pm q_s$) but on two rings in the N-SC case (Fig.2).

The crossover from SA-like ($\theta = 0$) to SC-like ($\theta \neq 0$) fluctuations of the mass density $\rho(\mathbf{r})$ in the nematic phase can be described by the phenomenological model of Chen and Lubensky[35]:

$$H[\delta\rho] = \int \frac{d^3q}{(2\pi)^3} \left\{ \frac{1}{2}\left(r + C_{\parallel}(q_o^2 - q_{\parallel}^2)^2 + C_{\perp}q_{\perp}^2 + D_{\perp}q_{\perp}^4\right) \mid \delta\rho(q) \mid^2 \right\} + 4^{\text{th}} \text{ order terms} \qquad (28)$$

$\delta\rho(r) = \rho(r) - \bar{\rho}$ describes the smectic modulation of the mass density, $C_{\parallel}$ and $C_{\perp}$ are the elastic coefficients parallel and perpendicular to the director respectively, $q_o$ is the projection of the

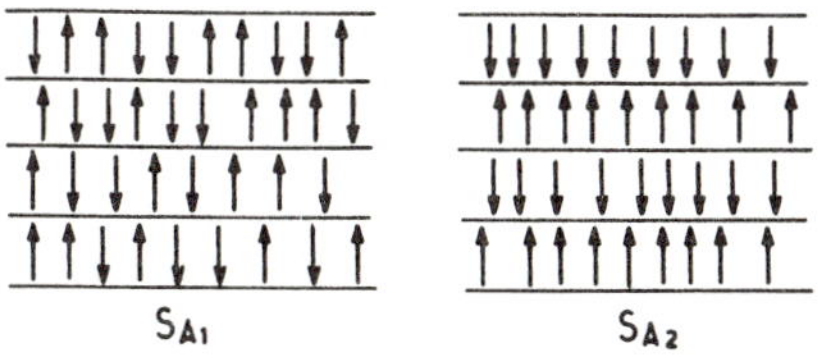


Fig. 2. Schematic representation of the monolayer ($SA_1$) and bilayer ($SA_2$) smectic A phases.

wavevector of the smectic (A or C) order along the director.

The N-SA-SC phase diagram of this model is given on Fig. 3. The nature of the condensed phase depends on the sign of $C_\perp$. In a temperature-pressure or temperature-concentration phase diagram, both r and $C_\perp$ can be controlled and the NAC point (r = 0 and $C_\perp$= 0) can be observed. It is a Lifshitz point in the Chen-Lubensky model.

In the nematic phase, for $C_\perp > 0$ the fluctuations are of SA type (i.e. centered around $q_\perp = 0, q_\parallel = \pm q_o$). For $C_\perp < 0$ they are centered on two rings $q_\perp = (-C_\perp/2D_\perp)^{1/2}, q_\parallel = \pm q_0$).

Complications arise from the vanishing of the N-SC latent heat at the NAC point and from the difficulties connected to the smectic state (Landau-Peierls instability) the N-SA transition (lack of gauge invariance) and the SA-SC transition (proximity of a tricritical point).

The description of the Chen-Lubensky model is reasonably well borne out by experiment[36,37,38] the universal topology of the phase diagram and the existence of a $C_\perp = 0$ line in particular are well established. The exact situation in the very close vicinity of the NAC point where fluctuations are important is not so clear: the nature of the transitions (first or second order) and the existence of a tricritical point are still unknown. An interesting possibility pointed out by Grinstein and Toner[39] with a model of dislocation unbinding is the existence of a biaxial nematic phase: if the N-SA and SA-SC lines are second order with XY critical exponents, the NAC point should be tetracritical[40] and a mixed phase (i.e. a biaxial nematic) should show up. It has however not been observed so far.

## 4. The Smectic A - Hexatic Smectic B Transition

The *crystalline* smectic B differs from the SA phase by the existence of a regular hexagonal packing of the molecules in the plane of the layers. The SB-SA transition corresponds to the melting of a two dimensional crystalline order in a three dimensional material whereas the layered structure along the third direction switches from true to quasi long range order. In two dimensions (i.e. in a hypothetic isolated single layer) a Kosterlitz Thouless[41] mechanism of dislocations unbinding may happen to make the transition continous with no singularities. In three dimensions, a direct transition SB-SA is expected to be first order.

An intermediate layered structure $SB_{hex}$ however exists in which the twodimensional ordering of the molecules within a layer is organized as follows: the correlations of the position of the center of mass of the molecules decay exponentially (like in a liquid) but the direction of the bonds linking two adjacent molecules exhibit a long range order with sixfold symmetry. Halperin, Nelson and Young[42,43] have pointed out that dislocations can create such a two dimensional hexatic order. The existence of this hexatic order in smectics is a beautiful success of the Halperin-Nelson-Young concept of bond ordering.

Accounting for the six-fold symmetry of the SB and $SB_{hex}$ phases, the modulation of the mass density within the (x,y) plane of the layers can be expanded in a Fourier series:

$$\rho(\boldsymbol{r}) - \bar{\rho} = \rho_0(r) + \rho_6(r)\cos 6(\theta - \phi) + \text{higher harmonics} \tag{29}$$

$\boldsymbol{r}$ is a 2-d vector of cartesian coordinates x = r cos θ and y = r sin θ. The correlation function of the radial distribution $\rho_0(r)$ falls off exponentially in the smectic $B_{hex}$ and smectic A phases. The amplitude $\rho_6(r)$ describing the six-fold order vanishes in the smectic A phase only. A possible order parameter of the $SA\text{-}SB_{hex}$ transition is therefore:

$$\Psi_6 = \rho_6 e^{6i\phi} \tag{30}$$

The phase $\phi$ accounts for spatial fluctuations of the orientation of the local six-fold axes.

The order parameter $\Psi_6$ has two components and the SA-SB$_{heX}$ transition is expected to belong to the XY universality class.

Experiments do confirm the second order nature of the transition but high resolution a.c. calorimetry gives[44] values of the specific heat exponent $\alpha$ about 0.6 inconsistent with the XY class ($\alpha$ = 0.06). High sensitivity heat capacity measurements on freely suspended thin films (down to four layer) show a crossover from bulk ($\alpha$ = 0.59) to a two dimensional cusp-like behavior[45] with $\alpha = -0.26$.

## 5. Frustrated Smectics

### *5.1. Polar Smectics*

Liquid crystal molecules with long aromatic cores and strongly polar head groups exhibit a rich smectic A (and smectic C) polymorphism. Since the discovery of the first SA-SA transition by Sigaud, Hardouin and Achard in 1979[46], seven different smectic phases have been identified in pure compounds or in binary mixtures of molecules of generic formula:

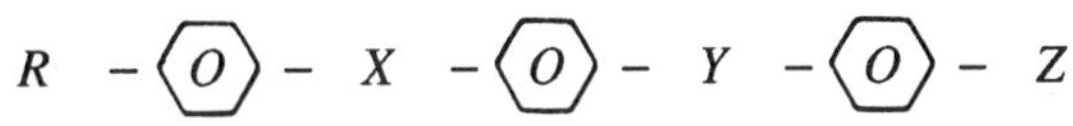

where R is an alkyl or alkoxy group, X,Y are linking groups such as CH=CH, C≡C, CH=N or COO and Z is a strongly dipolar end group CN or $NO_2$.

Extensive experimental studies over the past ten years led to a considerable amount of knowledge about the structures, phase diagrams and physical properties of these thermotropic liquid crystals[47,48].

The variety of structures arises from the asymmetry of the molecules: in addition to the classical N-SA-SC polymorphism, the long range organization of the position of the polar heads generates new phases (Fig. 2). An antiferroelectric stacking of polarized layers for instance generates the bilayer smectic A SA$_2$. If the dipoles are randomly oriented, the asymmetry can be forgotten and a monolayer SA$_1$ phase is obtained. Although it is not different from the classical SA phase, the subscript 1 indicates the potential ability to form SA$_2$ phases. X-ray diffraction patterns clearly show the doubling of the lattice spacing at the SA$_1$-SA$_2$ transition (Fig. 3).

Another experimental characteristic of polar mesogens is the intrinsic incommensurability of their structures. Nematic phases of polar compounds often exhibit diffuse x-ray scattering corresponding to a short range smectic order (Fig. 3). Two sets of diffuse spots centered around incommensurate wavevectors $\pm q_1$ and $\pm q_2$ with $q_1 < q_2 < 2q_1$ are usually found. The wavevector $q_2$ associated with the classical monolayer order is clearly of order $2\pi/l$ where $l$ is the length of a molecule in its most extended configuration. The wavevector $q_1$ associated with the head to tail association of the polar molecules reveals the existence of another natural length $l'$ such as $l < l' < 2l$. $l'$ is commonly identified with the length of a pair of antiparallel partially overlapping molecules, although microscopic approaches suggest that the emergence of $l'$ involves more than just two molecules[49,50]. Condensed smectic phases also exhibit incommensurate behaviors: the socalled partially bilayer smectic phase SAd with a lattice period d×$l$ is commonly observed (Fig. 3). It is indeed amazing to notice that the whole range of variation of d from 1 to 2 has been observed in different compounds.

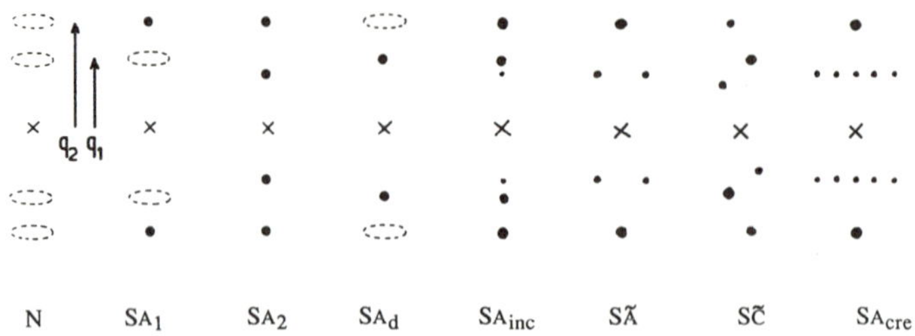


Fig. 3. Sketches of the x-ray scattering patterns for polar smectic A phases. Axes $q_x$ and $q_z$ are horizontal and vertical respectively. Note the appearance of the natural incommensurate wavevectors as diffuse spots in the nematic phase.

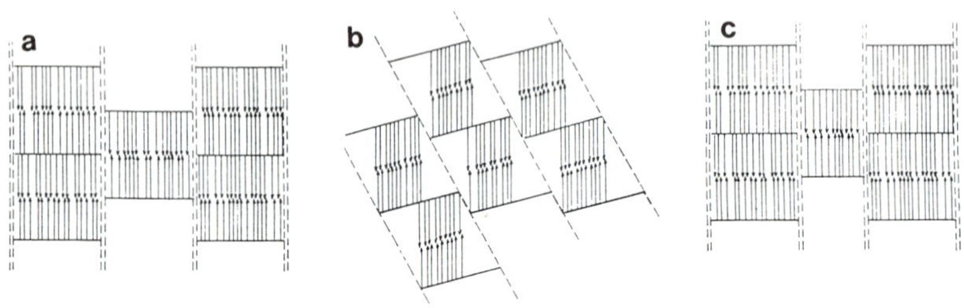


Fig. 4. Schematic representation of the biaxial polar smectics: (a) antiphase SÃ (b) tilted anti-phase SC̃ and (c) crenelated smectic A $SA_{Cre}$.

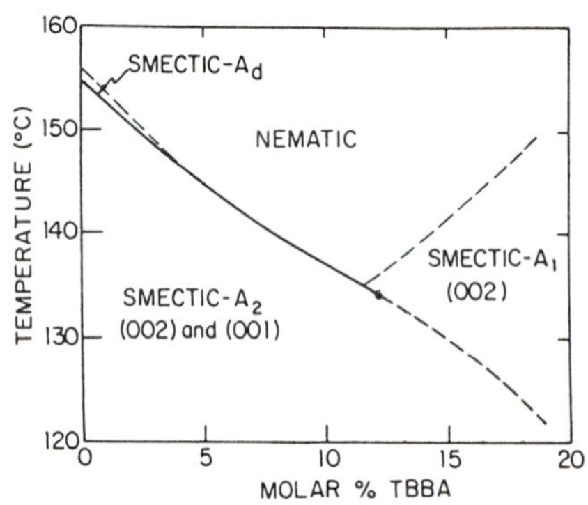


Fig. 5. Experimental phase diagram of the binary mixture DB6-TBBA[57]. Dashed (solid) lines represent second (first) order transitions. Note the tricritical point on the $SA_1$-$SA_2$ line.

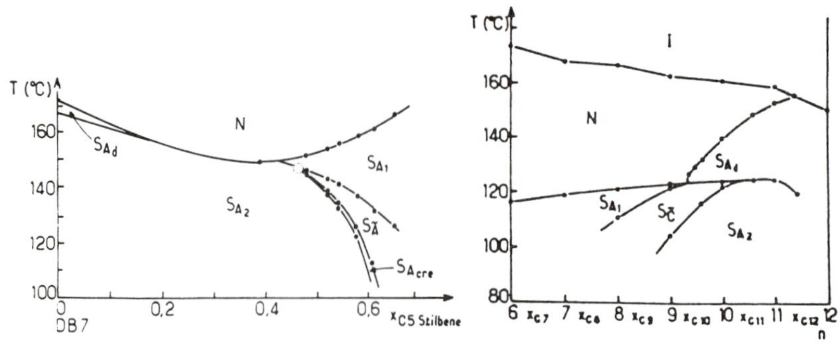


Fig. 6. Typical experimental phase diagrams exhibiting biaxial smectic phases: binary mixture DB7-C5 stilbene[56]. The crenelated phase appears over a very narrow range of concentration.

More rarely encountered but definitely revealing of the incommensurate nature of polar smectics are the incommensurate smectic A phases SAinc in which the phase of the bilayer modulation shifts periodically (with a period Z) with respect to the monolayer order[51] Note that the structure is truly incommensurate if the ratio Z over the period of the underlying smectic lattice ($Z/l$) is irrational. If it is rational ($Z/l = m/n$), the structure is rather modulated and has a period nZ. The coexistence of two incommensurate smectic waves with no apparent coupling has also been reported[48,52].

The four structures described above are uniaxial since all their modulations have collinear wavevectors but polar mesogens can form biaxial structures too.

The smectic antiphase SÃ, first discovered by Sigaud, Hardouin, Achard and Levelut[53] exhibits a periodic modulation of the antiferroelectric order along a direction x parallel to the plane of the layers (Fig. 4a). The natural incommensurability is again clear in reciprocal space: tilted incommensurate wavectors can lock in easily in two dimensions (Fig. 3).

The tilted antiphase (or ribbon phase) SC̃[54] arises from an asymmetric 2-d lockin of the wavevectors. The denomination C emphasizes the fact that both the layers and the antiferroelectric modulations are tilted with respect to the director ***n*** (Fig.s 3 and 4b).

At last, careful studies of the $SA_2$-SÃ and $SA_2$-SC̃ transitions have revealed the existence of the socalled crenelated $SA_{cre}$ phase over a very narrow range of temperature[55,56]. $SA_{cre}$ exhibits the basic transverse modulation of SÃ but with non equal up and down domains in the plane of the layers (Fig. 4c).

We should mention at last that the smectic C state, if it exists in polar compounds, may reproduce some features of the smectic A polymorphism. Bilayer smectic C phases ($SC_2$) are indeed observed.

The experimental phase diagrams of polar compounds are usually represented in axes temperature-pressure or temperature-concentration in binary mixtures. Although the whole set of structures described above is not found in one single system, most phase diagrams fit in a common topology: N, $SA_1$, $SA_2$ and $SA_d$ form the generic phase diagram of polar systems[46,57] (Fig. 5). If biaxial phases (SÃ and SC̃) are present, their domain opens up between $SA_1$ and $SA_d$[56,58] (Fig. 6). Incommensurate structures are observed in a triangular region surrounded by $SA_1$, $SA_2$ and $SA_d$[48] (Fig. 7). Tricritical point are observed on the N-$SA_1$ and N-$SA_2$ lines[57] whereas reentrant behavior is often associated with the triple (or multicritical) point N-$SA_1$-$SA_2$. The $SA_2$-$SA_d$ line may end up on a critical point beyond which no transition is detected.

## *5.2. The Model of Frustrated Smectics (Prost)*

As shown by Prost[51,59], the properties and structures of frustrated smectics can be described by two order parameters. The first $\rho(\boldsymbol{r})$ measures mass density modulations familiar in traditional smectic A phases[2]. The second $P_z(\boldsymbol{r})$, often referred to as a polarization wave, describes long range head-to-tail correlations of asymmetric molecules along the z axis (normal to the smectic layers). Although emphasized by the notation $P_z$, the antiferroelectric nature of this modulation is not a necessary input of the model.

In the absence of coupling between $\rho$ and $P_z$, $\rho$ would develop spatial modulation along the z axis at wavevector $q_2 = 2\pi/l$ where $l$ is of the order of a molecular length whereas $P_z$ would develop modulations at wavevector $q_1 = 2\pi/l'$ where $l'$ is identified with the length of a pair of antiparallel partially overlapping molecules as discussed in the previous section.

To describe the appearance of modulated order, two complex fields $\psi_1$ and $\psi_2$ are introduced:

$$P_z(\mathbf{r}) = Re(\psi_1(\mathbf{r})) \quad \text{and} \quad \psi_1(\mathbf{r}) = |\psi_1| e^{i(Q_1.r+\phi_1)}$$

$$\rho(\mathbf{r}) = Re(\psi_2(\mathbf{r})) \quad \text{and} \quad \psi_2(\mathbf{r}) = |\psi_2| e^{i(Q_2.r+\phi_2)} \tag{31}$$

In terms of these fields, the Landau free energy of the model reads:

$$\Delta F = \frac{1}{V}\int_v dr\{\frac{A_1}{2}|\psi_1|^2 + \frac{D_1}{2}|(\Delta+q_1^2)\psi_1|^2 + \frac{C_1}{2}|\nabla_\perp\psi_1|^2 + \frac{U_1}{2}|\psi_1|4$$

$$+\frac{A_2}{2}|\psi_2|^2 + \frac{D_2}{2}|(\Delta+q_2^2)\psi_2|^2 + \frac{C_2}{2}|\nabla_\perp\psi_2|^2 + \frac{U_2}{2}|\psi_2|^4$$

$$+\frac{U_{12}}{2}|\psi_1|^2|\psi_2|^2 - wRe(\psi_1^2\psi_3^*)\} \tag{32}$$

where V is the volume, $A_1 = a_1(T-TC_1)$ and $A_2 = a_2(T-TC_2)$ measure the temperatures from the non interacting mean field transition temperatures $TC_1$ and $TC_2$ of the fields $\psi_1$ and $\psi_2$. $\nabla_\perp$ is a derivative in the plane perpendicular to the director $\boldsymbol{n}$. The terms in $D_1$ and $D_2$ favour $Q_1^2 = q_1^2$ and $Q_2^2 = q_2^2$ respectively whereas the coupling term w $Re(\psi_1\psi_2^*)$ favours lockin conditions $2\mathbf{Q}_1 = \mathbf{Q}_2$ in the case of a weak overlap ($l'$ close to $2l$). A linear coupling term w′ $Re(\psi_1\psi_2^*)$ would be more appropriate[59] in the strong overlapping limit $l' \approx l$.

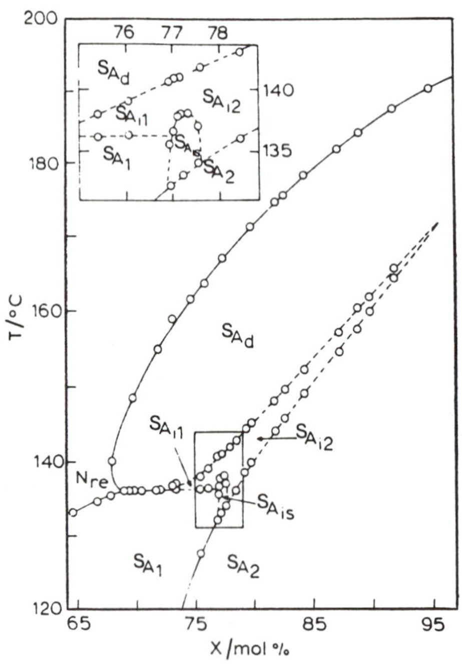


Fig. 7. Experimental phase diagram for 80BCAB-DB80CN mixtures exhibiting $SA_{inc}$ phases[48]. Three different behaviors are identified within the incommensurate region. A soliton like modulation with diverging periodicity is found close to the $SA_1$-$SA_2$-$SA_{inc}$ point in agreement with theoretical expectation (see text section 5.3.).

Frustration arises from the impossibility to satisfy simultaneously all these tendencies.

Note that the fluctuations of the director and the coupling with the nematic order are not included.

### 5.3. The Mean Field Model

To study the different structures and the phase diagrams in mean field, the free energy (Eq.32) has to be minimized with respect to the smectic amplitudes $|\psi_1|$ and $|\psi_2|$ and the wavevectors $\mathbf{Q}_1$ and $\mathbf{Q}_2$. The following phases are expected:

1) The nematic phase (N) with $|\psi_1|=|\psi_2|=0$. The director $n$ defines the z axis.

2) Uniaxial layered structures with $\mathbf{Q}_1$ and $\mathbf{Q}_2$ along z. $|\psi_1|=0$, $|\psi_2|\neq 0$ and $Q_2 = q_2$ defines the monolayer smectic phase $SA_1$. $|\psi_1|\neq 0$, $|\psi_2|\neq 0$ and $Q_2 = 2Q_1$ defines the bilayer antiferroelectric smectic phase $SA_2$.

3) Biaxial layered structures: $|\psi_1|\neq 0$, $|\psi_2|\neq 0$ and at least $\mathbf{Q}_1$ non collinear with $\mathbf{Q}_2$.

4) Uniaxial modulated structures: $|\psi_1|\neq 0$, $|\psi_2|\neq 0$ and modulated phases $\varphi_1(z)$ and $\varphi_2(z)$.

It may be worth noting that the number of external parameters in 5.2 can be reduced by an appropriate straighforward rescaling of variables:

$$\left[\frac{2w^2}{u_{12}D_1}\right]^{3/4}\frac{u_{12}^3}{64w^4}\frac{D_1}{D_2}\frac{\Delta F}{V}=\Delta F[\theta_1,\theta_2]$$

$$=\frac{1}{V}\int_v d^3x\{y_1|\theta_1|^2+|(\Delta+k_1^2)\theta_1|^2+\gamma_1|\nabla_\perp\theta_1|^2+\delta u_1|\theta_1|^4+y_2|\theta_2|^2+$$

$$+|(\Delta+k_2^2)\theta_2|^2+\gamma_2|\nabla_\perp\theta_2|^2+\delta u_2|\theta_2|^4+(|\theta_1|^2+|\theta_2|^2)^2-(\theta_1^2\theta_2^*+\theta_1^{2*}\theta_2)\} \quad (33)$$

with complex order parameters:

$$\theta_1 = x_1 exp[i(k_0 z + k_x x + \alpha_1(z))],\ \theta_2 = x_2 exp[i(2k_0 z + k'_x x + \alpha_2(z))] \quad (34)$$

and

$$x_1 = x\cos\theta,\ x_2 = x\sin\theta \quad (35)$$

The phase diagrams are classically calculated in the plane $y_1, y_2$.

Only the difference $\tilde{Z} = k_1^2 - k_2^2/4$ of the reduced wavevectors $k_1$ and $k_2$ turn out to be relevant. The physical significance of $\tilde{Z}$ is clear with the original parameters $\tilde{Z}\sim(q_1^2 - q_2^2/4)w$: incommensurability over coupling strength. The frustration is thus basically controlled by $\tilde{Z}$, usually referred to as the incommensurability parameter. Incommensurate structures for instance are expected at high values of $\tilde{Z}$.

The elastic coefficients of tilt $\gamma_1$ and $\gamma_2$ control the appearance of the biaxial phases $S\tilde{A}$ and $S\tilde{C}$.

The last two coefficients $\delta u_1$ and $\delta u_2$ account for anisotropic fourth order terms. Although

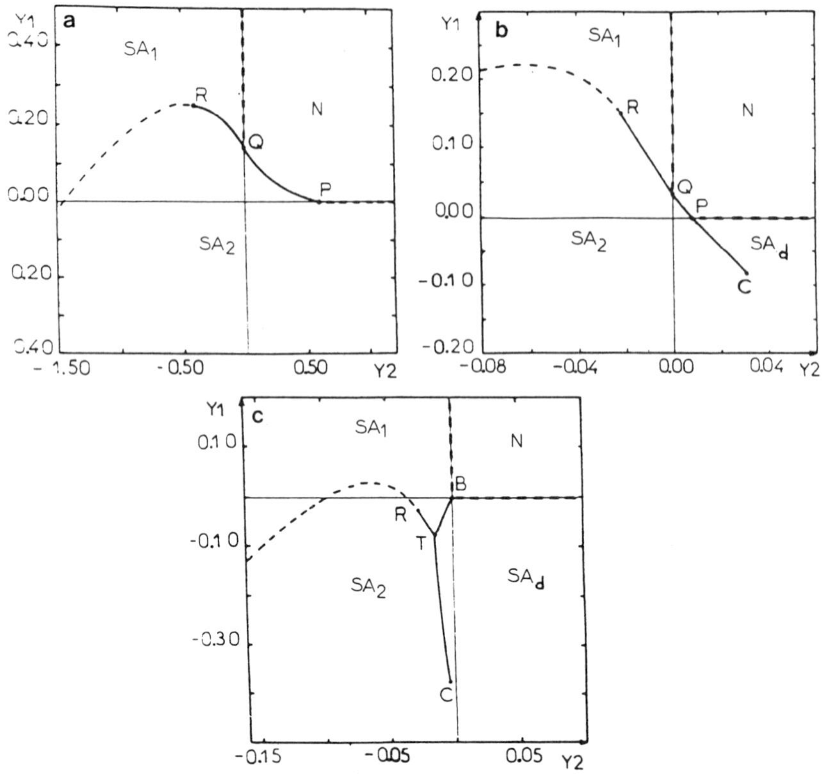


Fig. 8. Theoretical phase diagrams in mean field in the case of weak (a) medium (b) and strong (c) incommensurability. Case (a) is very similar to the experimental diagram of Ref. 46 whereas case (b) is to be compared to Fig. 5. Dashed (solid) lines are second (first) order.

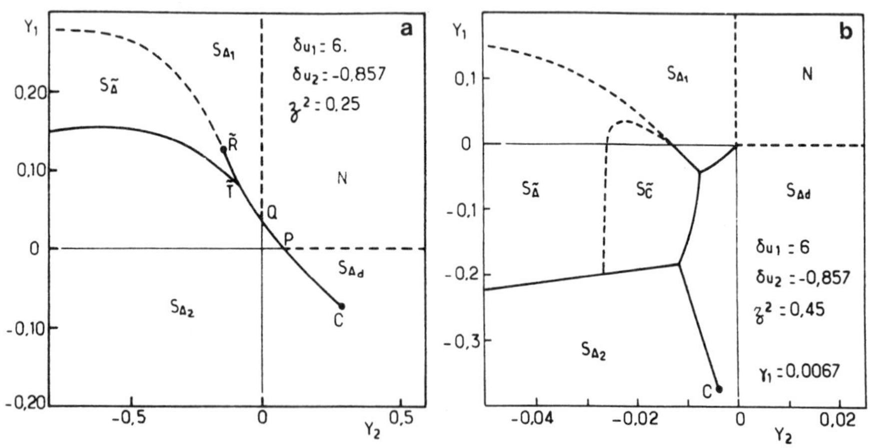


Fig. 9. Theoretical phase diagrams in mean field showing biaxial phases in the case of a weak (a) and strong (b) incommensurability. Compare to the experimental topologies of Fig. 6a and 6b. Dashed (solid) lines are mean field second (first) order lines.

they can have a significant effect on the shape of the phases diagrams in the plane $y_1$,$y_2$ they do not change the qualitative features of their topology.

On a technical point of view, minimization with respect to the wavevectors of Eq. 34 and the amplitude x of Eq. 35 is straightforward. The last minimization with respect to the parameter $\theta$ can be done numerically. The most significant phase diagrams calculated from this model are given in Fig.s 8, 9 and 10.

- 1) Uniaxial structures:

For a small incommensurability parameter $\tilde{Z}$ and symmetric fourth order coefficients ($\delta u_1 = \delta u_2$) the phase diagram shown on Fig. 8a is similar to the very first diagram calculated by Prost in which incommensurability was not considered[59]. A second order N-$SA_1$ line terminates at a mean field critical end point Q where the N, $SA_1$ and $SA_2$ phases meet. A second order N-$SA_2$ line terminates at a tricritical point P. The N-$SA_2$ line QP is first order and continues into the smectic region as a first order $SA_1$-$SA_2$ line. Beyond a tricritical point R, the $SA_1$-$SA_2$ line is second order. Because of the coupling term w, the phase of the bilayer order parameter $\Psi_1$ is locked in the monolayer smectic $SA_1$ so that the amplitude $x_1$ only is critical at the $SA_1$-$SA_2$ transition . It is therefore expected to be in the Ising universality class[59,60].

For higher incommensurability parameter $\tilde{Z}$ (and/or asymmetric fourth order terms $\delta u_1 > \delta u_2$)[61] a new phase boundary separating two $SA_2$ phases appears (Fig. 8b). The two $SA_2$ phases are distinguished by different values of the amplitudes of the order parameters $x_1$ and $x_2$ and therefore of the wavector $k_0$.

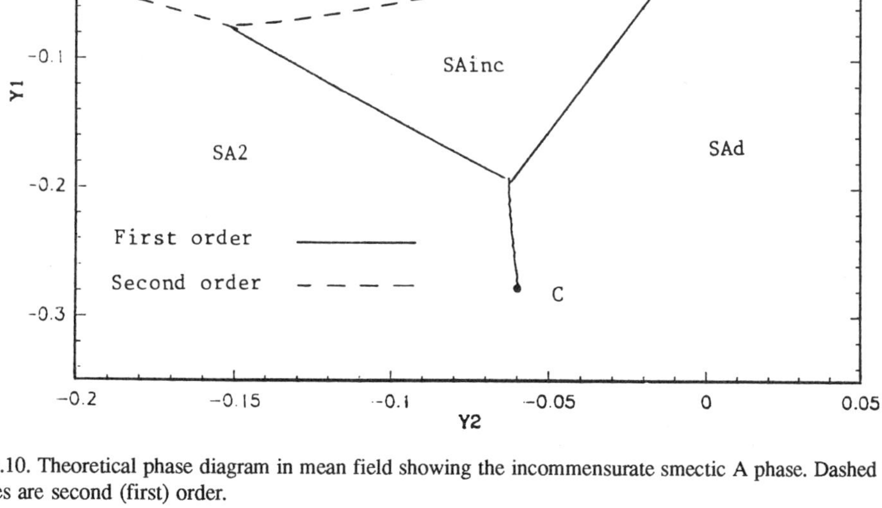


Fig.10. Theoretical phase diagram in mean field showing the incommensurate smectic A phase. Dashed (solid) lines are second (first) order.

For $x_1 >> x_2$, $k_0$ is of order $k_1$ (i.e. the smectic period is close to $l' = d \times l < 2l$) and the modulus $x_1$ is much larger than $x_2$. The $SA_2$ phase is identified with the partially bilayer $SA_d$ phase.

The new phase boundary is tangent to the first order N-$SA_2$ line in P and terminates at a critical point C where the jump in wavevector goes to zero. The wavevector of the fundamental smectic wave is plotted against $y_2$ in the vicinity of C on Fig. 6. Although the fluctuating parameter is a scalar (layer thickness) it will be shown in Section 5.4. that the new critical point C is not expected to belong to the Ising universality class[62].

When the incommensurability parameter $\tilde{Z}$ is further increased, a new $SA_1$-$SA_d$ line appears (Fig. 8c) terminating at a mean field bicritical point B where the N, $SA_1$ and $SA_d$ phases meet, and a triple point T where the $SA_1$, $SA_2$, and $SA_d$ phases coexist.

- 2) Biaxial structures:

Two different structures will be investigated:

$$\theta_1 = x_1 exp[i(k_0 z + k_x x)] + x_1 exp[i(k_0 z - k_x x)]$$

$$\theta_2 = x_2 exp[2ik_0 z] \tag{36}$$

describes the smectic antiphase $S\tilde{A}$[63] and

$$\theta_1 = x_1 exp[i((k_0 + \delta k_0)z + k_x x)] + x_1' exp[i((k_0 - \delta k_0)z - (k_x' - k_x)x)]$$

$$\theta_2 = x_2 exp[i(2k_0 z - k_x' x)] \tag{37}$$

the tilted antiphase $S\tilde{C}$[64].

Note that the name of "smectic" is improperly but commonly used to denote these two dimensionally ordered phases.

Fig. 9a shows the phase diagram in the case of weak/medium incommensurability parameter. A biaxial domain opens up for $\gamma_1 < 2\tilde{Z}$. The $SA_2$-$S\tilde{A}$ transition is always first order since the free energies cannot link up continuously when $k_x$ goes to zero. The $SA_1$-$S\tilde{A}$ line is found second or first order in mean field with a tricritical point $\tilde{R}$. It meets the first order $SA_1$-$SA_2$ line at a triple point $\tilde{T}$. If variations of the elastic coefficient $\gamma_1$ with temperature are allowed, the second order $SA_1$-$S\tilde{A}$ line may meet the second order $SA_1$-$SA_2$ line. As pointed out by Benguigui[65], such a $SA_1$-$SA_2$-$S\tilde{A}$ would be a Lifshitz point. Because of the continuous degeneracy of the fluctuations of the $S\tilde{A}$ order parameter on a cone in reciprocal space, the $SA_1$-$S\tilde{A}$ transition, like N-SC, is however expected to be always first order[34].

At higher incommensurability parameter, the biaxial domain reaches the $SA_d$ region (Fig. 9b). The tilted antiphase $S\tilde{C}$ is stable in between $SA_d$ and $S\tilde{A}$ so that two new lines are expected: a first order $S\tilde{C}$-$SA_d$ and a second order $S\tilde{A}$-$S\tilde{C}$.

- 3) Uniaxial modulated structures:

Incommensurate smectics with modulated phases $\alpha_1(z)$ and $\alpha_2(z)$ were among the first to be predicted by Prost[51]. An exact functional minimization of Eq. 33 with respect to the phases $\alpha_1(z)$ and $\alpha_2(z)$ is possible in terms of elliptic integrals[51,66,67]. The problem is in fact isomorphous to the cholesteric-nematic transition induced by a magnetic field[68]. In the modulated smectic, the phase difference $\alpha_1(z)$ - $\alpha_2(z)$ undergoes $\pi$ jumps (discommensurations or solitons) with a period Z. The free energy of such modulated structure can be calculated from Eq.33. The corresponding phase diagram is shown on Fig. 10[67,69]. The $SA_2$-$SA_{inc}$ and $SA_d$-$SA_{inc}$ lines are first order, whereas the $SA_1$-$SA_{inc}$ line is second order.

The period Z of the discommensurations does not diverge except at the $SA_1$-$SA_2$-$SA_{inc}$ mean field bicritical point. Further away from this point, Z is not much larger than the layer thickness so that the phase difference $\alpha_1(z) - \alpha_2(z)$ is close to a linear function of z (weak coupling limit)[67]. An unfortunate consequence of this remark is that x-ray diffraction experiments may not distinguish $SA_{inc}$ from a simple coexistence of $SA_2$ and $SA_d$.

Experimental phase diagrams[56,57,70] turn out to compare remarkably well to the calculated ones. Actually most of the experimental topologies exhibiting uniaxial phases only fall in one of the theoretical schemes of Fig.s 8, 9 and 10.

Binary mixtures of TBBA (terephtal bis butyl aniline) and homologous compounds of the DBn (alkylphenyl cyanobenzoyloxy benzoate) series for instance illustrate well cases 8a and 8b (compare to Fig.s 5a and 5b). The DB6-TBBA system in particular (Fig. 5b)[57] exhibits the $SA_d$ phase, the tricritical $SA_1$-$SA_2$ point and the order of all the phase transitions agrees well with Fig. 8b.

The DBn-$C_5$ stilbene binary mixtures[56] series (Fig. 6) reproduce well the theoretical diagram of Fig. 8a .

The critical $SA_2$-$SA_d$ point was first unambiguously identified by Shashidar, Ratna, Krishna, Somasekhar and Heppke[71].

The existence of coexisting incommensurate smectic modulations was reported by Brownsey and Leadbetter[72], Ratna, Shashidar and Raja[52] and Shashidar and Ratna[48]. The experimental phase diagram of Fig. 7 compares well to Fig. 10.

The critical properties of the various transitions were also investigated by high resolution X-ray diffraction or a.c. calorimetry.

3-d XY critical exponents with non inverted temperatures are measured for the N-$SA_1$ transition[25].

The Ising nature of the $SA_1$-$SA_2$ transition is confirmed[57,73,74] with Fisher-renormalized exponents[75].

The $SA_1$-$S\tilde{A}$ transition is found to be weakly first order with unusually large pretransitional fluctuations[76].

The stability of the crenelated phase in the general theoretical diagram (Fig.s 9a, 9b) has not been discussed yet. $SA_{cre}$ can be roughly described by the following order parameter:

$$\theta_1 = (x_0 + x_1 exp[ik_x x] + x_1 exp[-ik_x x])exp[ik_0 z]$$

$$\theta_2 = x_2 exp[2ik_0 z] \tag{38}$$

If this structure (Eq. 38) is inserted in the free energy density (Eq. 33), $SA_{cre}$ is found to be always unstable with respect to both $SA_2$ and $S\tilde{A}$. If a more realistic profile of the polarization wave along the direction x is considered (i.e. include higher harmonics or compute the profile of lowest energy) $SA_{cre}$ is found to be only marginably stable at the $SA_2$-$S\tilde{A}$ transition: it has exactly the same free energy as $SA_2$ and $S\tilde{A}$ right on the $SA_2$-$S\tilde{A}$ line[77]. Whether the experimental stability is due to higher order terms in the free energy density[78] or to fluctuations is still an open question. The mean field result suggests anyway that the $SA_{cre}$ domain should be narrow, as indeed observed[55,56].

A conclusion to this Section could be that the mean field analysis of the Prost's model of frustrated smectics allows to understand most of the reported experimental observations on polar smectics. Situations where fluctuations are important are however not correctly described by the mean field approximation: the understanding of multiply reentrant behavior, the appearance of nematic bubbles and the scaling properties of the $SA_2$-$SA_d$ critical point for instance require more elaborate analysis. Some of these points will be discussed in the following Sections.

### 5.4. Critical Properties of the Isolated SA-SA Critical Point

A plot of the smectic wavevector $q_0$ in the vicinity of the $SA_2$-$SA_d$ critical point C (Fig. 11) clearly suggests a close similarity with a liquid gas critical point.

In order to describe the fluctuations around C, it is convenient to express the order parameter m(**r**) in the following way:

$$m(\mathbf{r}) = [S_{01} + a_1 S(\mathbf{r})] e^{i\theta} e^{iq_0[z+u(\mathbf{r})]} + [S_{02} - a_2 S(\mathbf{r})] e^{2i\theta} e^{2iq_0[z+u(\mathbf{r})]} \tag{39}$$

where $S_{01}$ and $S_{02}$ are the amplitudes of $\Psi_1$ and $\Psi_2$ at the critical point. The equilibrium wavevector is $q_0 = q_{0c}(1 + (\nabla_z u))$. $S(r)$ measures the deviation of the amplitudes from the critical values.

The part of the Landau-Ginzburg Hamiltonian depending on the order parameter $S(r)$ has the liquid-gas form:

$$H_s = \int d^d r \left\{ h_s S + \frac{1}{2}[rS^2 + c(\nabla S)^2] + \frac{w_s}{3!} S^3 + \frac{u_s}{4!} S^4 \right\} \tag{40}$$

where $r = a_s\,(T\text{-}Ts)$. In absence of coupling, $H_s$ is critical when r vanishes.

The energy of elastic deformations in smectic A phases $H_{sm}[u(\mathbf{r})]$ can be developed likewise with the constraint that it must be invariant with respect to both rigid translations and rigid rotations. The former invariance is assured if $H_{sm}$ is a function only of the gradients of u(**r**), the latter by requiring that the compressional term[79] appears only in the rotationally invariant combination $E[u(\mathbf{r})] = \nabla_z u + 1/2(\nabla u)^2$. The elastic Hamiltonian is thus:

$$H_{sm}[u] = \int d^d r \{ h_0 E(u) + B_0 E^2(u) + \frac{w_0}{3!} E^3(u) + \frac{v_0}{4!} E^4(u)$$

$$+ \frac{1}{2}\left[K_1^0 (\nabla_\perp^2 u)^2 + K_2^0 (\nabla_z^2 u)^2 + K_3^0 (\nabla_z \nabla_\perp u)^2\right]\} \tag{41}$$

The classical physics of smectic A phases is recovered by expanding Eq. 41 in powers of the gradients of u(**r**). The quadratic terms in $(\nabla_z u)^2$ and $(\nabla_\perp^2 u)^2$ are responsible for the Landau-Peierls instability[5,6,79]. The non-harmonic terms in $(\nabla_z u)(\nabla_\perp u)^2$ and $(\nabla_\perp u)^4$ lead to a breakdown in the conventional elasticity: B(q) and $K_1(q)$ respectively vanishes and diverges as powers of Lnq at small wavevectors q[79]. Other anharmonic terms are irrelevant in the renormalization group sense in the smectic phase.

The Hamiltonian coupling S to $u$ is:

$$H_{int} = \int d^d r \{ \lambda_{11} SE + \lambda_{12} S^2 E + \lambda_{21} SE^2 + \lambda_{22} S^2 E^2 \} \tag{42}$$

The final Hamiltonian is $H[S,u] = H_s + H_{sm} + H_{int}$. The physics of the critical point can be understood from the Gaussian truncation of H in reciprocal space:

$$H_G[S,u] = \frac{1}{2} \int \frac{d^d q}{(2\pi)^d} \{ (B_1 q_z^2 + K_1 q_\perp^4 + K_2 q_z^4 + K_{12} q_z^2 q_\perp^2) \,|\, u(q) \,|^2$$

$$+ (r + Cq^2) \,|\, S(q) \,|^2 - \lambda_{11} (i q_z u(q) S(-q) + conj.complex) \} \tag{43}$$

with $B_1 = B_0 + h_0$.

The linear coupling of S and u in Eq. 43 can be removed via the transformation:

$$S(q)=\sigma(q)+\frac{i\lambda_{11}q_z u(q)}{r+Cq^2} \tag{44}$$

The Gaussian correlation functions of the new variables are:

$$G^{(2)}_{uu}(q,r)=\frac{1}{B_1(1-r_c/r)q_z^2+K_1 q_\perp^2+K'_2 q_z^4+2K'_{12}q_\perp^4}$$

$$G^{(2)}_{\sigma\sigma}(q,r)=\frac{1}{r+Cq^2} \tag{45}$$

with $r_c = \lambda^2_{11} > 0$. The layer compression is controlled by an effective elastic constant $B_{eff}(q)$

$$B_{eff}(q=0)=B_1\frac{(r-r_c)}{r} \tag{46}$$

which goes to zero for a positive value of $r$. $G^{(2)}_{\sigma\sigma}$ is therefore not critical at C and a Gaussian integration of the fluctuations of $\sigma$ is acceptable. u($r$) is the critical variable and the critical point is reached when $B_{eff}$ vanishes. This behavior is to be compared to the divergence of the compressibility in the liquid-vapor problem: 1/B is analogous to a smectic compressibility along the direction z.

The critical Landau-Ginzburg Hamiltonian is then:

$$H=\int d^d r\{h\nabla_z u+\frac{B_1}{2}(\nabla_z u)^2+\frac{B_2}{2}(\nabla_\perp u)^2+\frac{1}{2}[K_1(\nabla_z u)^2+K_2(\nabla_\perp u)^2+2K_{12}(\nabla_z\nabla_\perp u)^2]$$

$$+\frac{w_1}{3!}(\nabla_z u)^3+\frac{w_2}{3!}(\nabla_z u)(\nabla_\perp u)^2+\frac{v_1}{4!}(\nabla_z u)^4+\frac{v_1}{4!}(\nabla_\perp u)4+\frac{v_{12}}{12}(\nabla_z u)^2(\nabla_\perp u)^2\} \tag{47}$$

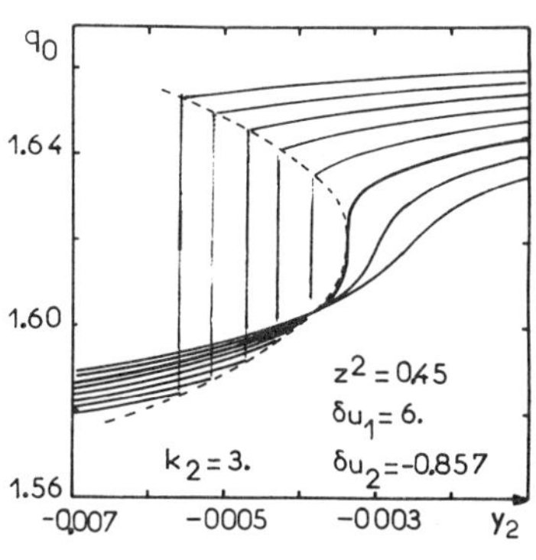


Fig. 11. Plots of the smectic wavevector qo versus the Landau temperature $y_2$ for different values of the field $y_1$ in the vicinity of the $SA_2$-$SA_d$ critical point C of Fig. 8c (mean field calculation). Note the close similarity with the isotherms in the pressure volume plane near the liquid gas critical point.

In the absence of external field, h = $B_2$ = 0. In mean field theory, $< \nabla_\perp u >$ is zero, the order parameter is $< \nabla_z u >$ and the critical point is found for $Bl = wl = 0$.

When fluctuations are included, $< \nabla_\perp u >$ cannot be ignored. In the liquid-gas Hamiltonian, there is only one third order invariant $\varphi^3$ which can be removed by shifting $\varphi$. This implies that the liquid-gas transition belongs to the same universality as the Ising model with additional higher order irrelevant potentials. In the SA-SA Hamiltonian (Eq. 47), there are two distinct third order potentials $(\nabla_z u)^2$ and $\nabla_z u(\nabla_\perp u)^2$. They cannot be removed by shifting the order parameter $< \nabla_z u >$. A renormalized theory should therefore take them into account explicitely. The general $\varphi^3$-field theory suggests that the critical dimension is $d_c = 6$.

The complete perturbation theory of the Hamiltonian (Eq. 47) has been developped by Park, Lubensky, Barois and Prost[62]. The inverse susceptibility $\tilde{B}$ is found to be related to the coefficients of Eq. 47 as follows:

$$B_1 - B_{1c} = \tilde{B}(1 - w_{2c}^2 B^{d/2-3} + v\tilde{B}^{d/2-2} + \ldots) \tag{48}$$

since $w_{2c} \neq 0$, the expansion (Eq. 48) breaks down for spatial dimensions d lower than 6.

For $4 < d < 6$, fourth order terms are irrelevant in Eq. 48. An anisotropic $\varphi^3$-renormalized theory was constructed[62]. The Gaussian and Ising fixed point are unstable, as expected. A stable nontrivial fixed point was found and anisotropic critical exponents were calculated in 6 - $\varepsilon$ dimensions.

Although a number of theoretical questions remain (unknown equation of state, non analytic corrections to the third order vertex for $6 < d < 8$) the main result is that the SA-SA critical point is expected to belong to a new universality class. Experiments confirm this point: a high resolution calorimetric study on a binary mixture[71] gives a critical exponent $\gamma/\Delta$ of $0.45 \pm 0.08$[80] whereas elastic constant measurements show a vanishing of B with a value $0.4 \pm 0.2$ of the same exponent $\gamma/\Delta$[81]. These numbers are only marginally consistent with one another but definitely different from the Ising value 0.79. On the other hand, a mean-field behavior ($\gamma/\Delta = 0.67$) would not explain the asymmetry of the calorimetric and X-ray data.

It should be mentioned at last that the mean field theory of frustrated smectics in the limit of a strong overlap ($l' \approx l$) predicts several critical points. All of them (including the present $SA_2$-$SA_d$ point) are in fact of $SA_d$-$SA_d$ type (with $1 < d < 2$) and belong to the same universality class. It is worth emphasizing the unusual possibility to go from $SA_1$ to $SA_2$ via a second order transition, a first order jump or without any phase transition at all by a continuous growth of the smectic wavevector since all smectic A phases have the same symmetry. Second order transitions may only occur with exact doubling (or tripling) of the period by continuous condensation of a subharmonic modulation.

## *5.5. The Reentrant Phenomenon*

Although not specific of liquid crystals, the reentrant phenomenon (i.e. reappearence of the phase of higher symmetry upon cooling) is often observed in frustrated smectics[82,83]. Theoretical analyses suggest that there is no universal explanation for reentrance in the N-SA problem.

Single reentrance may occur in mean field: the coupling of two order parameters may generate a curved N-SA phase boundary and thus produce a reentrant phenomenon if the physical temperature axis is a suitable combination of the two Landau control parameters.

Experiments suggest that double reentrance is associated with the bi or tetracritical N-$SA_1$-$SA_d$ point (Fig. 5c). Prost and Lubensky[84,85] have argued that the correlation function of the smectic order parameter in both $SA_1$ and $SA_d$ obeys a scaling relation such as:

$$G(t,p) = \tilde{t}^{-\gamma} f\left(\frac{\tilde{p}}{\tilde{t}^{x}}\right) \tag{49}$$

where t, p are the reduced temperature and pressure in the vicinity of the critical point (defined by t = p = 0) $\tilde{t}$ and $\tilde{p}$ are linear combinations of t and p. Eq. 49 implies that the phase boundaries obey a relation

$$\tilde{p} = w^{\pm} \tilde{t}^{x} \tag{50}$$

which gives a doubly reentrant behavior with suitable rotation of the physical axes p, t with respect to $\tilde{p}$ and $\tilde{t}$.

Beside these phenomenological approaches that say nothing about the microscopic origin of reentrance, molecular theories have been proposed. The frustrated spin glass model of Indekeu and Berker[49.50] has been particularly successful.

The general theory of reentrant phases will be developed in the Chapter by Netz and Berker in this Volume.

The appearance of closed nematic domains (or nematic "bubbles") deep in the smectic region seems to be closely related to the existence of a $SA_2$-$SA_d$ or $SA_1$-$SA_d$ critical point[86,87] and may be discussed here.

As stressed in the last Section, the compressional elastic constant B vanishes at the SA-SA critical point so that the system is very close to a nematic. Prost and Toner[86] have shown that depending on bare parameters of a particular system, either the nematic bubble or the critical point could be observed. In their model of dislocation loop melting, the free energy density of the smectic A - nematic system can be expanded in powers of the disorder parameter $\psi$($|\psi| \neq 0$ in the nematic phases) as:

$$F = \frac{1}{V}\int_{v} dr \left\{ \frac{1}{2} r^{eff} |\psi|^2 + \text{higher orders} \right\} \tag{51}$$

in which $r^{eff}$ is the bare Landau coefficient r corrected by a Gaussian integration of the dislocation field fluctuations. The result is:

$$r^{eff} \approx r + c_{\perp} K_1 q_{1c}^3 + c_{\perp} B q_{2c} \tag{52}$$

$q_{1c}$ and $q_{2c}$ are cut off wave vectors of order $d^{-1}$ and $c_{\perp}$ a positive Landau coefficient. The vanishing of B at the critical point leads to a downward dip in $r^{eff}$. In the vicinity of the critical point, B is controlled by two scaling fields r and $\delta\tilde{r}$ ($\tilde{r}$ = O defines the critical isochore of constant layer spacing d = $d_c$) and behaves like:

$$B \sim |\tilde{r}|^{\gamma} g_{\pm}^{B}\left(\frac{\delta\tilde{r}}{|r|^{\phi}}\right) \tag{53}$$

The critical exponents $\gamma$ and $\varphi$ have been calculated in 6 - $\varepsilon$ dimensions (Section 5.4.) but are still unknown for d = 3. Two cases are to be considered:

- if $\gamma < 1$, $r^{eff}$ has a downward cusp and the nematic island forms if $r_{cp}^{eff}(\tilde{r} = \delta\tilde{r} = 0)$ is negative.

The critical point is observed otherwise.

- if $\gamma > 1$, the nematic bubble is stable again if the minimum of $r^{eff}$ is negative. This may happen away from the critical point so that the nematic island can detach itself from the SA-SA first order line.

Prost and Toner have argued that combining these results with the mean field topologies of the model of frustrated smectics in the limit of a strong overlap ($l' \approx l$) generates more than 100 possible phase diagrams[86].

Beside this fluctuation corrected mean field theory, an exact model has been developped by Pommier and Prost[87] in the limit of an infinite number of components n of the order parameter (recall n = 2 for the smectic case). With the same physics of frustration, exact phase diagrams have been calculated that reproduce fairly well the continuous inclusion of the reentrant nematic phase that leads to a closed island (Fig 12).

## 6. Conclusion

The study of phase transitions in thermotropic liquid crystals has made considerable progress during the past ten years. Strong theoretical efforts coupled with high resolution and high sensitivity experimental techniques have provided a good understanding of the physics of the phase transformations that occur in mesophases. Clear experimental universal behaviors have been identified and most exceptions have received a reasonable theoretical explanation (Ginzburg criterion for the SA-SC transition, crossover towards tricritical behavior).

A number of open questions remain: the nematic to smectic A transition for instance is almost understood but not quite which suggests more efforts have to be done.

Many other aspects have not been developped here. The isotropic-nematic transition for instance has been extensively studied in the past.

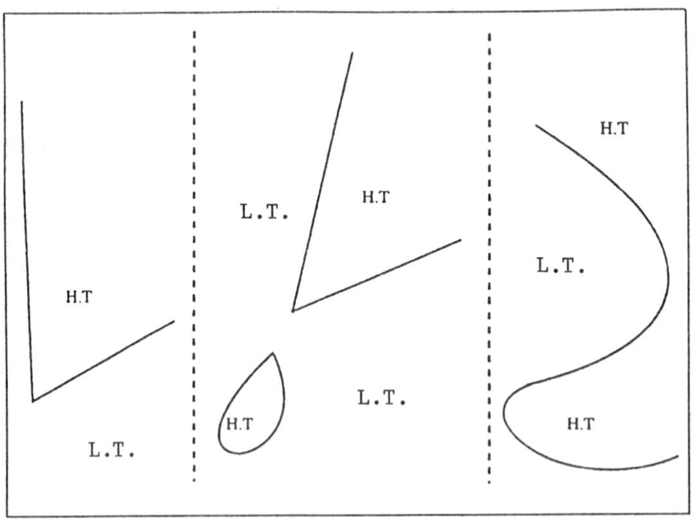


Fig. 12. Sketches of the theoretical phase diagrams calculated by Pommier and Prost[87] with a model of two order parameter coupled with frustration in the limit of an infinite number of components n. H.T. and L.T. would be the nematic and the smectic A phase respectively in the case of liquid crystals (n = 2).

Transitions involving disk-like molecules or mesogenic polymers have certainly not received such a great deal of attention and should constitute an active field of research in the future.

ACKNOWLEDGMENTS. The author is indebted to J. Prost for having communicated the manuscript of the revised edition of Ref. 2 prior to publication. Sections 2 to 4 of the present Chapter owe a lot to this manuscript. The work presented in Section 5 has benefited from helpful and enriching discussions with C. Coulon, C.W. Garland, F. Hardouin, A.-M. Levelut, T.C. Lubensky, J.P. Marcerou, H.T. Nguyen, Y. Park, J. Pommier, J.Prost, R. Shashidar and G. Sigaud.

## References

1. W.L. Mc Millan, Phys. Rev. **A4**: 1238 (1971)
2. P.G. de Gennes, "The Physics of Liquid Crystals", Clarendon Press, Oxford (1974)
3. W. Maier and A. Saupe, Z. Naturforsh. **14A**: 882 (1959) and **15A**: 287 (1960)
4. W.L. Mc Millan, Phys. Rev. **A6**: 936 (1972), Phys. Rev. **A7**: 1673 (1973)
5. L.D. Landau, Phys. Z. Sowj. Un. **2**: 26 (1937)
6. R.E. Peierls, Annls Inst. H. Poincarre **5**: 177 (1935)
7. P.G. de Gennes, Solid State Commun. **10**: 753 (1972)
8. C.W. Oseen, Trans. Faraday Soc. **29**: 883 (1933) F.C. Frank, Disc. Faraday Soc. **25**: 19 (1958)
9. V.L. Ginzburg and L.D. Landau, J. Exptl. Theoret. Phys. (USSR) **20**: 1064 (1950)
10. W. Meisner and R. Ochsenfeld, Naturwiss. **21**: 787 (1933)
11. S.T. Lagerwall, R.B. Meyer and B. Stebler, Ann. Phys. (France) **3**: 249 (1978)
12. J. Goodby, M. Waugh, S. Stein, R. Pindak and J. Patel, Nature **337**: 449 (1988)
13. S.R. Renn and T.C. Lubensky, Phys. Rev. **A38**: 2131 (1988)
14. P.E. Cladis and S. Torza, J. Applied Phys. (USA) **46**: 584 (1975)
15. F. Jahnig and F. Brochard, J. Phys. (France) **35**: 301 (1974)
16. K.G. Wilson and J. Kogut, Phys. Rep. **12**: 77 (1974)
17. T.C. Lubensky, Chen Jing-Huei, Phys. Rev. **A17**: 366 (1978)
18. B. Halperin, T.C. Lubensky and Ma Shang-Keng, Phys. Rev. Lett. **32**: 292 (1974)
19. C.E. Williams and M. Kleman, J. Phys. Colloq. **36**: C1315 (1975), M. Kleman, "Points, Lignes, Parois", Les Editions de Physique, Orsay, France (1977)
20. D.R. Nelson and J. Toner, Phys. Rev. **B24**: 363 (1981)
21. C. Dasgupta and B.I. Halperin, Phys. Rev. Lett. **47**: 1556 (1981)
22. J. Toner, Phys. Rev. **B26**: 462 (1982)
23. T.C. Lubensky, J. Chim. Phys. (France) **80**: 31 (1983)
24. K.W. Evans-Lutterodt, J.W. Chung, B.M. Ocko, R.J. Birgenau, C. Chiang, C.W. Garland, E. Chin, J. Goodby and H.T. Nguyen, Phys. Rev. **A36**: 1387 (1987)
25. C.W. Garland, G. Nounesis, K.J. Stine and G. Heppke, J. Phys. (France) **50**: 2291 (1989), C.W. Garland, G. Nounesis and K.J. Stine, Phys. Rev. **A39**: 4919 (1989)
26. M.E. Huster, K.J. Stine and C.W. Garland, Phys. Rev. **A36**: 2364 (1987), J.P. Hill, B. Keimer, K.W. Evans-Lutterodt, R.J. Birgeneau and C.W. Garland, Phys. Rev. **A40**: 4625 (1989)
27. M.A. Anisimov, V.P. Voronov, A.O. Kulkov and F. Kholmurodov, J. Phys. (France) **46**: 2157 (1985)
28. L. Benguigui and P. Martinoty, Phys. Rev. Lett. **63**: 774 (1989)
29. Y. Galerne, J. Phys. (France) **46**: 733 (1985)
30. D. Collin, J.L. Gallani and P. Martinoty, Phys. Rev. Lett. **61**: 102 (1988)
31. G. Sigaud, F. Hardouin and M.F. Achard, Solid State Commun. **23**: 35 (1977)
32. D. Johnson, D. Allender, D. Dehoff, C. Maze, E. Oppenheim and R. Reynolds, Phys. Rev. **B16**: 470 (1977)
33. P.G. de Gennes, Mol. Cryst. Liq. Cryst. **21**: 85 (1975)
34. S.A. Brazowskii, Soviet Phys. JETP **41**: 85 (1975)
35. Chen JingHuei and T.C. Lubensky, Phys. Rev. **A14**: 1202 (1976)
36. C.R. Safinya, L.J. Martinez-Miranda, M. Kaplan, J.D. Litster and R.J. Birgeneau, Phys. Rev. Lett. **50**: 56 (1983)
37. L.J. Martinez-Miranda, A.R. Kortan and R.J. Birgeneau, Phys. Rev. Lett. **56**: 2264 (1986)

38. L.J. Martinez-Miranda, A.R. Kortan and R.J. Birgeneau, Phys. Rev. **A36**: 2372 (1987)
39. G. Grinstein and J. Toner, Phys. Rev. Lett. **51**: 2386 (1983)
40. M.E. Fisher and D.R. Nelson, Phys. Rev. Lett. **32**: 1350 (1974)
41. J.M. Kosterlitz and D.J. Thouless, J. Phys. (France) **C6**: 1181 (1973)
42. B.I. Halperin and D.R. Nelson, Phys. Rev. Lett. **41**: 121 (1978), D.R. Nelson and B.I. Halperin, Phys. Rev. **B19**: 2457 (1979)
43. A.P. Young, Phys. Rev. **B19**: 1855 (1979)
44. C.C. Huang, J.M. Viner, R. Pindak and J.W. Goodby, Phys. Rev. Lett. **46**: 1289 (1981)
45. R. Geer, C.C. Huang, R. Pindak and J.W. Goodby, Phys. Rev. Lett. **63**: 540 (1981)
46. G. Sigaud, F. Hardouin and M.F. Achard, Phys. Lett. **72A**: 24 (1979)
47. F. Hardouin, A.M. Levelut, M.F. Achard and G. Sigaud, J. Chim. Phys. **80**: 53 (1983)
48. R. Shashidar and B.R. Ratna, Liq. Cryst. **5**: 421 (1989)
49. J.O. Indekeu and A.N. Berker, J. Phys. (France) **49**: 353 (1988)
50. A.N. Berker and J.S. Walker, Phys. Rev. Lett. **47**: 1469 (1981)
51. J. Prost, "Proceedings of the Conf. on Liq. Cryst. of One and Two Dimensional Order, Garmisch Partenkirchen", Springer Verlag, Berlin (1980)
52. B.R. Ratna, R. Shashidar and V.N. Raja, Phys. Rev. Lett. **55**: 1476 (1985)
53. G. Sigaud, F. Hardouin, M.F. Achard and A.M. Levelut, J. Phys. (France) **42**: 107 (1981)
54. F. Hardouin, H.T. Nguyen, M.F. Achard and A.M. Levelut, J. Phys. Lett. (France) **43**: L327 (1982)
55. A.M. Levelut, J. Phys. Lett. (France) **45**: L603 (1984)
56. G. Sigaud, M.F. Achard and F. Hardouin, J. Phys. Lett. (France) **46**: L825 (1985)
57. K.K. Chan, P.S. Pershan, L.B. Sorensen and F. Hardouin, Phys. Rev. Lett. **54**: 1694 (1985) and Phys. Rev. **A34**:1420 (1986)
58. H.T. Nguyen and C. Destrade, Mol. Cryst. Liq. Cryst. Lett. **92**: 257 (1984)
59. J. Prost, J. Phys. (France) **40**: 581 (1979)
60. Wang Jiang and T.C. Lubensky, Phys. Rev. **A29**: 2210 (1984)
61. P. Barois, J. Prost and T.C. Lubensky, J. Phys. (France) **46**: 391 (1985)
62. Y. Park, T.C. Lubensky, P. Barois, J. Prost, Phys. Rev. **A37**: 2197 (1988)
63. P. Barois, C. Coulon and J. Prost, J. Phys. Lett. (France) **42**: L107 (1981)
64. J. Prost and P. Barois, J. Chim. Phys. (France) **80**: 66 (1983)
65. L.G. Benguigui, J. Phys. (France) **44**: 273 (1983)
66. P. Barois and J. Prost, Ferroelectrics **58**: 193 (1984)
67. P. Barois, J. Pommier and J. Prost, Chapter 6: Frustrated smectics in "Solitons in Liquid Crystals", Lui Lam and J. Prost eds. Springer Verlag, Berlin (1991).
68. P.G. de Gennes, Solid State Commun. **6**: 163 (1968)
69. P. Barois, Phys. Rev. **A33**: 3632 (1986)
70. F. Hardouin, M.F. Achard, C. Destrade and H.T. Nguyen, J. Phys. (France) **45**: 765 (1984)
71. R. Shashidar, B.R. Ratna, S. Krishna, S. Somasekhar and G. Heppke, Phys. Rev. Lett. **59**: 1209 (1987)
72. G.J. Brownsey and A.J. Leadbetter, Phys. Rev. Lett. **44**: 1608 (1980)
73. C. Chiang and C.W. Garland, Mol. Cryst. Liq. Cryst. **122**: 25 (1985)
74. C.W. Garland, C. Chiang and F. Hardouin, Liq. Cryst. **1**: 81 (1986)
75. D.A. Huse, Phys. Rev. Lett. **55**: 2228 (1985), M.A. Anisimov, A.V. Voronel and Gorodetskii, Sov. Phys. JETP **33**: 605 (1971)
76. K. Ema, C.W. Garland, G. Sigaud and H.T. Nguyen, Phys. Rev. **A39**: 1369 (1989)
77. J. Pommier, unpublished
78. L.G. Benguigui, Phys. Rev. **A33**: 1429 (1986)
79. G. Grinstein and R.A. Pelcovits, Phys. Rev. Lett. **47**: 857, Phys. Rev. A26:915 (1982)
80. Y.H. Jeong, G. Nounesis, C.W. Garland and R. Shashidar, Phys. Rev. **A40**: 4022 (1989)
81. J. Prost, J. Pommier, J.C. Rouillon, J.P. Marcerou, P. Barois, M. Benzekri, A. Babeau and H.T. Nguyen, Phys. Rev. **B42**: 2521 (1990)
82. P.E. Cladis, Phys. Rev. Lett. **35**: 48 (1975)
83. H.T. Nguyen, F. Hardouin and C. Destrade, J. Phys. (France) **43**: 1127 (1982)
84. J. Prost, Advances in Physics **33**: 1 (1984)
85. T.C. Lubensky, unpublished
86. J. Prost and J. Toner, Phys. Rev. **A36**: 5008 (1987)
87. J. Pommier, Thèse de l'Universite de Bordeaux 1, no 274 (1989)

Chapter 5

# COMPUTER SIMULATIONS OF PHASE TRANSITIONS IN LIQUID CRYSTALS

D. FRENKEL

## 1. Introduction

The aim of these lectures is twofold. First, I wish to give a simple introduction to the computer simulations of classical many-body systems, with special emphasis on those technical aspects that are of particular relevance for simulation of liquid crystals.

To this end, I first give a brief elementary introduction to Molecular Dynamics and Monte Carlo simulations of classical many-body systems. Thereupon (section 4) I discuss the choice of technique and simulation ensemble. Section 5 deals with 'measurements' in a computer simulation. In Section 6, techniques to locate first order phase transitions are discussed. In this context, a brief discussion of free energy calculations is included. This 'technical' introduction is quite sketchy. The material presented in this Chapter is based on articles that have been published elsewhere. For a more detailed discussion, the reader is referred to one of the many excellent textbooks on this topic, e.g. the book of Allen and Tildesley[1].

Next, we apply the tools that we have thus introduced to computer simulation of simple models that exhibit liquid-crystalline behavior. In particular, I shall discuss computer-simulation studies of simple models that exhibit nematic, smectic and columnar phases. The isotropic-nematic transition in two-dimensions is rather different from its 3D counterpart and is therefore dicussed separately. I conclude with a discussion of recent developments and their possible implications for the numerical study of liquid-crystalline phases and phase transitions.

## 2. Monte Carlo Simulation

The prime purpose of the kind of Monte Carlo or Molecular Dynamics simulations that we shall be discussing, is to compute equilibrium properties of classical many-body systems. Let us first look at Monte Carlo (MC) simulations. Such simulations are used to compute ensemble averaged static properties of classical many-body systems. In the canonical (i.e. constant-NVT) ensemble, the ensemble average of a function $A(q^N)$ of the particle coordinates $\{q^N\}$ is defined as:

---

D. Frenkel - FOM Institute for Atomic and Molecular Physics, Kruislaan 407, 1098 SJ Amsterdam, The Netherlands.

*Phase Transitions in Liquid Crystals*, Edited by S. Martellucci and A.N. Chester, Plenum Press, New York, 1992

$$< A >_{NVT} = \frac{\int dq^N A(q^N) \exp(-\beta U(q^N))}{\int dq^N \exp(-\beta U(q^N))} \quad (1)$$

In Eq. 1, $\beta = 1/k_B T$ (where T is the absolute temperature and $k_B$ the Boltzmann constant), and $U(q^N)$ is the potential energy function. In the Metropolis Monte Carlo method, $< A >$ is estimated as the *unweighted* average of the values $A(q^N)$ sampled during a random walk through configuration space. The trick of the Metropolis MC method is to construct this random walk in such a way that the probability to visit a particular point $q^N$ is proportional to the Boltzmann factor $\exp(-\beta U(q^N))$. There are many ways to construct such a random walk. In the approach proposed by Metropolis et al.[2] the construction of a step in the random walk consists of two steps. First a random trial move is made from the current position in configuration-space $q^N$ to a trial position $q'^N$. Usually, such a trial move corresponds to the displacement of a single particle. But other moves are acceptable, as long as the probability of attempting a move from $q^N$ to $q'^N$ is equal to the probability of a trial move from $q'^N$ to $q^N$. Whether or not a trial move is in fact accepted depends on the change in potential energy, $\Delta U$, associated with the trial move. If $\Delta U < 0$, the move is always accepted. If $\Delta U > 0$, the move is accepted with a probability $\exp(-\beta\Delta U)$, and rejected otherwise. The quantity A is computed at the position that results after accepting or rejecting the trial move. The ayerage of all these 'measurements' of A during the random walk yields the desired ensemble average $< A >_{NTV}$.

## *2.1. Periodic boundary conditions*

For a system with short-range forces, the precise nature of the boundary conditions should be unimportant in the thermodynamic limit. The problem is that the number of particles in a normal simulation is of the order of $10^2$ to $10^3$. If *real* boundaries are used in a simulation (e.g. hard walls), then the particles at the boundaries experience interactions very different from those in the bulk. This results in a correction to all equilibrium properties. Sure enough, such a correction would vanish in the thermodynamic limit, but only as $N^{-(1/D)}$, where D is the dimensionality of the system. In contrast, periodic boundary conditions mimic the situation where the system is embedded in an infinite, homogeneous sample of the same phase. There are still system-size effects, but these are much weaker (e.g. of order $N^{-1}$ or $(\ln N)/N$). In a 3D simulation, one should also consider the shape of the simulation box. For simulations on liquids, a cubic shape is simplest, and in most cases perfectly adequate. For the perfectionist, other shapes are sometimes preferable (see Ref. 1).

In contrast, for crystalline solids and smectic or columnar liquid crystals, the choice of the shape of the unit box is not simply a matter of taste. It is clearly essential that the simulation box be commensurate with the periodicity of the ordered phase. For highly symmetric solid phases this is easy to achieve. However, for phases with lower than cubic symmetry, the shape of the unit-cell of the crystal depends on temperature and pressure, and this dependence is not known *a priori*. The most extreme example of such a dependence is a phase transition from one solid phase to another. At such a transition, the crystal unit cell may take on a completely different shape. If the simulation box cannot adapt to the new structure, such a phase transition would be artificially suppressed. A Molecular Dynamics technique that overcomes this problem was introduced in 1980 by Parrinello and Rahman[3,4]. The Parrinello-Rahman technique can easily be incorporated in a Monte Carlo program[5].

So much for the good news. Now some of the problems. In small systems, periodic boundary conditions may not be quite as harmless as they seem. They may induce artificial cubic order in a system that would otherwise be isotropic. Usually this induced order has only a minor effect on 'scalar' properties of atomic fluids (such as the energy or the pressure). However, it may have a pronounced effect on the tensor properties of atomic fluids, in particular those that depend on correlations between three or more atoms[6,7]. In molecular fluids the periodic boundaries may influence the orientational distribution function of the molecules, and thereby all properties that depend on this quantity[8]. Obviously, this effect of periodic boundaries is of considerable relevance for the numerical study of liquid crystals.

### 2.2. Intermolecular Interactions

Selecting the most convenient form for the intermolecular interactions to be used in a simulation is an art in itself. On this topic, I will limit myself to a few general remarks. The first may be superfluous, but I believe that it cannot be said too often. There is a fundamental distinction between simulations that aim to model *real* substances, and those that focus on testing theoretical concepts. For the latter kind of simulation it is essential to use the simplest possible interaction potential that reproduces the essential physics of the problem. Any more complicated potential just makes comparison with theory harder and the simulation slower. In contrast, if the aim is to understand the behaviour of real systems, then it is advisable to use the best potential energy function that you can get. Again, the reason is simple. If you are not using the best potential and you find that simulation and experiment are at odds then *you have learned nothing*. But if you use the best available potential, then a discrepancy between simulation and experiment is meaningful. Often, the nature of the discrepancy suggests how to improve the potential function.

### 2.3. Monte Carlo Moves and Myths

Monte Carlo trial moves should be generated in such a way that, in the absence of the Boltzmann factors, the probability to go from point $\alpha$ to point $\beta$ in configuration space should be equal to the probability to return from $\beta$ to $\alpha$. For translational moves (i.e. displacements of the center-of-mass of a molecule) this condition is so easy to satisfy that it almost requires an effort to get it wrong. However, for orientational moves, the choice of the random displacement requires a little more care: it is only too easy to generate orientational moves that lead to a distortion of the orientational distribution function of the molecules. This point is discussed in detail in Ref. 1. It is clearly of crucial importance for the numerical study of liquid crystals that the sampling scheme does not bias the orientational distribution function. In the case of large flexible molecules with constraints on some bondlengths and bondangles it is still possible to perform Monte Carlo sampling of the internal degrees-of-freedom[9]. However, for such model systems the conventional Monte Carlo sampling becomes very cumbersome and other schemes are needed to perform efficient numerical simulations. I will return to this point later.

How large should a Monte Carlo trial move be? If it is very large, it is likely that the resulting configuration will have a high energy and the trial move will probably be rejected. If it is very small, the change in potential energy is probably small and most moves will be accepted. In the literature one often finds the mysterious state that an acceptance of approximately 50 % should be optimal. This statement is not always true. For a rational discussion of the acceptance criterion it is necessary to state what exactly is meant by 'optimal'.

My preferred definition of 'optimal' is as follows: *That Monte Carlo sampling scheme is optimal, which yields the lowest statistical error in the quantity to be computed for a given expenditure of 'computing budget'*. Usually, 'computing budget' is equivalent to CPU time. From this definition it is clear that, in principle, a sampling scheme may be optimal for one quantity but not for another. Actually, the above definition is all but useless in practice (as are most deflnitions). It is just not worth the effort to measure the error estimate in, for instance, the pressure, as a function of the Monte Carlo step-size for a series of runs of fixed length. However, it is reasonable to assume that the meansquare error in the 'observables' is inversely proportional to the number of 'uncorrelated' configurations visited in a given amount of CPU time. And the number of independent configurations visited is a measure of the distance covered in configuration space. This suggests a more manageable criterion to estimate the efficiency of a Monte Carlo sampling scheme, namely, the sum of the squares of all displacements in configuration space divided by computing time. This quantity should be distinguished from the meansquare displacement per unit of computing time, because the latter quantity goes to zero in the absence of diffusion, whereas the former does not.

If we try to translate the present criterion into a rule for the optimal acceptance ratio, it is easy to see that different Monte Carlo codes will have different optima. The reason is that it makes a crucial difference if the amount of computing required to test whether a trial move is accepted depends on the magnitude of the move. More generally, for continuous potentials where *all* interactions have to be computed before a move can be accepted or rejected, the amount of computation does not depend on the size of a trial move. However, the situation is very different for simulations of molecules with a hard repulsive cores. Here a move can be rejected as soon as overlap with *any* neighbor is detected. In this case, a rejected move is cheaper than an accepted one, and hence the average computing time per trial move goes down as the step-size is increased. As a result, the optimal acceptance ratio for hard-core systems is appreciably lower than for systems with continuous interactions. Exactly how much depends on the nature of the program, in particular on whether it is a scalar or a vector code (in the latter case hard-core systems are treated much like continuous systems), how the information about neighbor-lists is stored, and even on the computational 'cost' of random numbers and exponentiation. The consensus seems to be that for hardcore systems the optimum acceptance ratio is closer to 20 % than to 50 %, but this is just another rule-of-thumb. As computers and models change, old rules-of-thumb may have to change too.

## 3. Molecular Dynamics

The structure of a Molecular Dynamics program differs only little from that of a Monte Carlo program, yet the two approaches are very different. In a Monte Carlo simulation equilibrium averages are estimated by sampling an integral over configuration space, such as in Eq. 1. The result of such a simulation is an estimate of the ensemble-average of the quantity of interest. The order in which points in configuration space are sampled has no physical meaning, and hence a Monte Carlo simulation yields no dynamical information. In contrast, a Molecular Dynamics simulation follows the natural time-evolution of a classical many-body system along its path in phase space. Molecular Dynamics simulations yield time-averages rather than ensemble-averages. An important consequence of this fact is that the method can be used to measure time-dependent quantities (e.g. time-correlation functions). This last feature of Molecular Dynamics, combined with the fact that an MD program is as simple as an MC program, is responsible for the fact that the majority of all simulations on classical many-body systems employ MD rather than MC. Nevertheless, there are situations where MC is more convenient. The choice of technique will be discussed in Section 4. In the present section, I present a very elementary description of the Molecular Dynamics method.

### *3.1. Algorithm*

In a Molecular Dynamics program, the time-evolution of a classical many-body system is simulated by numerical integration of Newton's equations of motion. The central part of any MD program is therefore the algorithm to carry out this integration. However, it should be stressed that the actual time spent on integrating the equations of motion is negligible compared with the time it takes to compute all intermolecular forces.

A large number of algorithms exist that can be used to integrate the equations of motion of systems of particles with continuous intermolecular forces. These algorithms integrate the equations of motion using a finite difference method. Clearly, it is important to choose the best available algorithm. In order to do so, it is necessary to specify what we mean by a 'good' algorithm. There are several criteria that a good algorithm should satisfy. For the sake of comparison, we assume that all algorithms can be made to do an equally good job in reproducing thc trajectory of the system through phase-space , and that they keep all 'conserved' quantities (e.g. energy, total momentum) constant to the same degree of accuracy. The comparison which then remains is simply: which algorithm is cheapest. As before, what is 'expensive' on one computer (e.g. memory) may be 'cheap' on another.

The idea behind many of the more sophisticated algorithms is that it is possible to use a larger integration step (and thereby to gain in speed), by utilizing stored information about the higher derivatives of the particle-coordinates. Although this is certainly true in principle, it turns out that, unless very high accuracy is needed, the very simplest MD algorithm, named after Verlet[10], is as good as most higher-order schemes[11]. The Verlet algorithm can be derived in the following way. First, we express the positions of a particle at times t + $\Delta$t and t-$\Delta$t in terms of its position, velocity and acceleration at time t:

$$x(t+\Delta t) = x(t) + \dot{x}\Delta t + \frac{1}{2}\ddot{x}\Delta t^2 + \ldots$$

$$x(t-\Delta t) = x(t) - \dot{x}\Delta t + \frac{1}{2}\ddot{x}\Delta t^2 + \ldots \qquad (2)$$

Adding these two equations, and subtracting x(t-$\Delta$t) from both sides, yields:

$$x(t+\Delta t) = 2x(t) - x(t-\Delta t) + \ddot{x}\Delta t^2 \qquad (3)$$

where we have dropped all terms of order $\Delta t^4$ and higher. Note that Eq. 3 does not explicitly contain the particle-velocities. For more details on the integration of the equations of motion of a classical many-body system, the reader is referred to[1]. It should be noted that finite difference schemes such as Eq. 3 cannot be used to solve the equations of motion of 'hard-core' systems. For a review of the numerical techniques that can be used in the latter case, the reader is referred to Ref. 12.

## 4. Choice of Technique

Having introduced both the Monte Carlo and the Molecular Dynamics methods, we must now address the question which technique should be used when. Actually, this question is not as clear-cut as it may seem, because there exist, in fact, a great number of modified MD and MC techniques to compute averages in various ensembles. All these techniques aim to estimate the equilibrium-average of a particular observable in one ensemble or another. However, in an actual simulation the observable is sampled over a finite time-interval along a trajectory in phase-space (MD), or along a random walk in configuration-space (MC). If a simulation technique is to be useful, the average accumulated in such a sampling procedure should approach the correct ensemble-average, in the limit that the length of the run tends to infinity.

Usually, the equivalence of the MD and MC averages is simply assumed. In the case of Molecular Dynamics, this assumption is equivalent to the 'ergodic hypothesis' of statistical mechanics. For convenience, we shall use the same terminology in the case of Monte Carlo simulations. Non-ergodic behaviour should be distinguished from differences between time and ensemble-averages that may be observed when the computer simulations are too short to sample the accessible phase-space adequately. True non-ergodic behaviour is observed when some parts of the 'permissible' phase-space simply cannot be reached at all, even in an infinitely long simulation. In practice, it is often difficult to distinguish between inadequate sampling due to the limited duration of a simulation and true non-ergodicity, if only because simulations are not infinitely long.

Both Monte Carlo and Molecular Dynamics simulations may suffer from incomplete sampling of the accessible phase-space. The most common example is the simulation of a meta-stable phase, such as for instance an undercooled liquid or a meta-stable crystalline phase. We shall come back to this point in Section 6. There are, however, a number of cases where phase-space is sampled much less efficiently by Molecular Dynamics simulation than by Monte Carlo. This happens, for instance, if the system has certain modes of vibration that are weakly coupled to the remaining degrees of freedom. This is quite a common phenomenon in a low-temperature solid, where long-wavelength phonons may have very long life-times. Another example is a high-frequency internal vibration of a molecule. Energy exchange between such a mode and the other degrees of freedom may be extremely slow in a MD simulation. In contrast, Monte Carlo does not suffer from this particular equilibration problem.

Another situation where the 'unphysical' nature of Monte Carlo moves can be exploited is in (binary) mixtures, in particular when the interdiffusion of the two species is slow. An example is a solid solution. In such solutions the local composition around an atom of species 1 will differ from the overall composition. It is almost impossible to study such changes in local composition using Molecular Dynamics, because particle diffusion in a solid can be a very slow process. In contrast, in a Monte Carlo simulation one may define a trial move which swaps a randomly selected pair of particles of species 1 and 2. If the particles are not too dissimilar, such moves will have a reasonable chance of acceptance, and local compositions can equilibrate rapidly.

However, Monte Carlo is not always the most efficient technique to sample phase-space. There are many cases where the route from one pocket in phase-space to another requires a collective rearrangement of the coordinates of many particles: examples are conformational changes in large molecules and structural phase-transitions in solids. In such cases, Molecular Dynamics often finds a 'natural' reaction-path from one state to the other, where random and uncorrelated Monte Carlo trial moves are much less successful.

### *4.1. Other Ensembles*

In a conventional MD simulation, the total energy $E$ and the total linear momentum P are constants of motion. Hence, MD simulations measure (time-) averages in an ensemble that is very similar to the microcanonical (see Ref. 13), namely the constant-NVE-p ensemble. In contrast, a conventional Monte Carlo simulation probes the canonical (i.e. constant-NVT) ensemble. The fact that these ensembles are different leads to observable differences in the statistical averages that are computed in MD and MC simulations. Most of these differences disappear in the thermodynamic limit and are already relatively small for systems of a few hundred particles. However, the choice of ensemble does make a difference when computing the mean-square value of *fluctuations* in thermodynamic quantities. Fortunately, techniques exist to relate fluctuations in different ensembles[14]. Moreover, it is nowadays common practice to carry out Molecular Dynamics simulations in ensembles other than the microcanonical. In particular, it is possible to do Molecular Dynamics at constant pressure[15], at constant stress[3] and at constant temperature[16]. The choice of ensembles for Monte Carlo simulations is even wider: isobaric-isothermal[17,18], constant-stress-isothermal[5], Grand-canonical (i.e. constant-$\mu VT$)[19,20] and even micro-canonical[21]. A recent addition to this list is a Monte Carlo method which employs the 'Gibbs'-ensemble[22]. The latter technique was developed to study phase-coexistence in moderately dense (multi-component) fluids. The 'Gibbs'-method maintains the coexisting phases at equal temperature, pressure and chemical potential.

Clearly, the sheer number of different MC and MD techniques makes it impossible to discuss them in any detail in the present introductory Chapter. However, it is important that the reader be aware that this wide choice of techniques exists. Technical details about most of these simulation methods can be found in Ref.s 1 and 23. I wish to add one cautionary remark: in MD simulations at constant pressure, stress or temperature, additional dynamical variables are introduced that act as manostat/thermostat. The time-evolution of the particle-coordinates in such simulations is governed by equations-of-motion that contain these artificial variables. Although the effect of these extra variables on the particle-dynamics may be small, it is nevertheless advisable to stick to conventional micro-canonical MD if one is primarily interested in the study of dynamical properties. For static equilibrium properties, all of the above methods (MC and MD) should be fine.

### *4.2. Molecular Dynamics or Monte Carlo ?*

From the discussion in Section 4.1 it is probably clear that the distinction between Molecular Dynamics and Monte Carlo simulations is not all that sharp. Most ensembles of practical importance can be simulated using both techniques. Thus, the choice of one or the other must be dictated by other considerations. All other things being equal, the Molecular Dynamics method is certainly preferable, because it yields information about the dynamical properties of the system under consideration. As was discussed above, MD simulations may sometimes run into 'ergodicity' problems, where MC simulations do not. In such a case it may be preferable to use the Monte Carlo technique, or a hybrid method.

Sometimes the reason to prefer Monte Carlo simulations is much more mundane. It may simply be that, as the expressions for the potential energy function become more complex, the explicit evaluation of the forces and torques for a Molecular Dynamics program becomes quite cumbersome. This implies that there is a distinct risk of introducing errors in the MD code, unless computer-algebra is used to derive the correct expressions. In such cases it is safest to start with a Monte Carlo simulation.

## 5. "Measurements"

In the previous sections we have described the core of a program to simulate classical many-body systems. In the present section we discuss 'measurements' during such a simulation.

Deciding what quantity to measure during a simulation is a problem, not because such measurements are difficult (in fact, most are quite trivial), but because there is such a wide choice of quantities to measure. Any quantity that can be expressed as an average over phase-space of some function of the particle coordinates and momenta can be measured during a simulation. For instance, in a Monte Carlo simulation, we can measure, apart from the primary thermodynamic variables *(E,V,T,P,N)*, the heat-capacity $C_v$, the isothermal compressibility $X_T$ and the radial distribution function *g(r)* (or, equivalently, the structure factor *S(k)*). Molecular Dynamics simulations offer, in addition, the possibility of measuring transport-properties such as the self-diffusion constant D, the shear-viscosity $\eta$ (and the bulk-viscosity $\zeta$), the thermal conductivity $\lambda_T$ and the dynamical structure-factor $S(k,\omega)$. This list is far from exhaustive. Moreover, we are not limited to the computation of quantities that can be measured in real experiments. We can gain insight into the microscopic structure and dynamics of a system by making 'snapshots' of the molecular configurations, or by computing any function of the particle-coordinates and momenta that we consider illuminating. A discussion of computer-'measurements' is therefore open-ended, and an enumeration is not particularly useful. Rather, I wish to make a few general remarks about the accuracy of measurements in a simulation. The actual examples of computer 'measurements' will follow in the sections that deal with computer simulations of liquid crystals.

## *5.1. Error Estimates*

Before discussing error estimates in MC and MD simulations, it is useful to recall what quantities cannot be measured at all in a standard MC or MD simulation. In a simulation one measures averages over phase-space of functions of the particle coordinates and momenta. However, it is not possible to measure directly the *volume* of the accessible phase-space. This volume determines the entropy S of a system, and thereby the Helmholtz free-energy F and the chemical potential $\mu$. We refer to such quantities as 'thermal', to distinguish them from the 'mechanical' quantities that can be expressed as a function of the phase-space coordinates. Methods to compute thermal quantities are discussed in Section 6. Another situation where MC and MD simulations on a small system cannot be used is in the study of critical phenomena or, for that matter, any situation where correlations over distances much larger than (or incommensurate with) the periodic box play an important role. And finally, *classical* MC and MD will clearly fail if quantum effects become important.

Let us now consider the measurement of a dynamical quantity A in a Molecular Dynamics simulation (the present discussion applies, with minor modifications, to Monte Carlo simulations). During a simulation of length (that is, duration) T, we obtain the following estimate for the equilibrium-average of A

$$A_T = (1/T)\int_0^T A(t)dt \tag{4}$$

where the subscript on $A_T$ refers to averaging over a finite time T. If the ergodic hypothesis is justified then $A_T \to \langle A \rangle$, as $T \to \infty$, where $\langle A \rangle$ denotes the ensemble-average of $A$.

Let us now estimate the variance in $A_T$ $\langle (\Delta A_T)^2 \rangle$

$$\begin{aligned} \langle (\Delta A_T)^2 \rangle &= \langle A_T^2 \rangle - \langle A_T \rangle^2 \\ &= (1/T^2)\int_0^T\int_0^T \langle (A(t) - \langle A \rangle)(A(t') - \langle A \rangle) \rangle \, dt\,dt' \end{aligned} \tag{5}$$

Note that $<(A(t) - <A>)(A(t')-<A>)>$ in Eq. 5 is simply the time-correlation function of fluctuations in the variable A. Let us denote this correlation function by $C_A(t'-t)$. If the duration of the sampling T is much larger than the characteristic decay time $\tau_A$ of $C_A$ then we may rewrite Eq. 5 as:

$$<(\Delta A_T)^2> \approx (2/T)\int_0^{\infty} C_A(t'')dt'' \approx (2\tau_A/T)C_A(0) \tag{6}$$

In the last equation we have used the definition of $\tau_A$ as the integral from 0 to $\infty$ of the normalized correlation function $C_A(t)/C_A(0)$. The relative variance in $A_T$ is therefore given by:

$$\frac{<(\Delta A_T)^2>}{<A>^2} \approx (2\tau_A/T)\frac{<A^2>-<A>^2}{<A>^2} \tag{7}$$

Eq. 7 clearly shows that the root-mean-square error in $A_T$ is inversely proportional to $(\tau_A/T)^{1/2}$. This result is hardly surprising. It simply states the wellknown fact the variance in a measured quantity is inversely proportional to the number of uncorrelated measurements. In the present case, this number is clearly proportional to $T/\tau_A$. This result may be trivial, but it is nevertheless very important, because it shows directly how the lifetime and amplitude of fluctuations in an observable $A$ affect the statistical accuracy. This is of particular importance in the study of fluctuations associated with hydrodynamical modes or pretransitional fluctuations near a symmetry-breaking phase transition. Such modes usually have a characteristic life-time that is proportional to the square of their wavelength. This is another reason why it is so difficult to study continuous phase transitions by MD. In order to minimize the effects of the finite system-size on such phase transitions, it is preferable to study systems with a box-size L that is large compared with all relevant correlation-lengths in the system. However, due to the slow decay of long-wavelength fluctuations, the length of the simulation needed to keep the relative error fixed should be proportional to $L^2$. As the CPU time for a run of fixed length is proportional to the number of particles (at best), the CPU-time needed to maintain constant accuracy increases quite rapidly with the linear dimensions of the system (e.g. as $L^5$ in three dimensions).

There is another aspect of Eq. 7 that is not immediately obvious, namely that it makes a difference whether or not the observable $A$ can be written as a sum of uncorrelated single-particle properties. If this is the case, then it is easy to see that $\langle A^2>-<A>^2)/<A>^2$ is inversely proportional to the number of particles, N. To see this, consider the expressions for $<A>$ and $<A^2>-<A>^2$ in this case:

$$<A> = \sum_{i=1}^{N} <a_i> = N<a> \tag{8}$$

and

$$<A^2>-<A>^2 = \sum_{i=1}^{N}\sum_{j=1}^{N} <a_i-<a>> <a_j-<a>> \tag{9}$$

If the fluctuations in $a_i$ and $a_j$ are uncorrelated, then we find:

$$\frac{<A^2>-<A>^2}{<A>^2} = (1/N)\frac{<a^2>-<a>^2}{<a>^2} \tag{10}$$

From Eq. 10 it is clear that the statistical error in a single-particle property is inversely proportional to $\sqrt{N}$. Hence, for single-particle properties the accuracy improves as one goes to larger systems (for fixed length of the simulation). In contrast, no such advantage is to be gained when computing truly collective properties. This is one more reason why bigger is not always better. A more detailed discussion of statistical errors in collective and single-particle time-correlation functions can be found in Ref.s 24 and 25. Systematic techniques to measure statistical errors in a simulation are discussed in Ref.s 1 and 26.

## 6. Phase Transitions

As was mentioned in Section 5.1, 'thermal' quantities such as the free energy F, the entropy S and the chemical potential μ cannot be measured directly in a computer simulation because they depend explicitly on the accessible volume in phase-space. However, knowledge of such thermal quantities (in particular μ) is usually necessary to locate the coexistence line for a first-order phase-transition. At first sight this knowledge may appear superfluous. After all, a computer simulation mimics the behaviour of a real solid or liquid. If the simulation is ergodic it should spontaneously transform to whatever phase is thermodynamically most stable, and then we would know all there is to know. Unfortunately, this approach does not work, at least not for strong first-order phase transitions such as freezing. At a solid-solid or solid-liquid phase transition very strong hysteresis effects are usually observed in a simulation. In fact, it is very difficult to nucleate a crystal from a liquid during a simulation. Hence, to locate the point where two phases coexist, we must compute the chemical potential of the homogeneous phases at the same temperature and pressure and find the point where the two μ's are equal. General methods to compute μ and related thermal quantities by computer simulation are discussed in Ref. 27. Here I wish to focus exclusively on those techniques that should work for dense systems. In these techniques it is usually the free energy F rather than the chemical potential μ that is computed.

When discussing techniques to measure free energies, it is useful to recall how such quantities are measured in real experiments. In the real world free energies cannot be obtained from a single measurement either. What can be measured, however, is the derivative of the free energy with respect to volume V and temperature T:

$$\left(\frac{\partial F}{\partial V}\right)_{NT} = -P \tag{11}$$

and

$$\left(\frac{\partial F/T}{\partial 1/T}\right)_{NV} = E \tag{12}$$

Here $P$ is the pressure and $E$ the energy of the system under consideration. The trick is know to find a reversible path that links the state under consideration to a state of known free energy. The change in F along that path can then simply be evaluated by integration of Eq.s 11 and 12.

In the real world the free energy of a substance can only evaluated directly for a very limited number of thermodynamic states. One such state is the ideal gas phase, and another is the perfectly ordered ground state at $T = 0\ K$. In computer simulations, the situation is quite similar. In order to compute the free energy of a dense liquid, one may construct a reversible path to the very dilute gas phase. It is not really necessary to go all the way to the ideal gas. However, at least one should reach a state that is sufficiently dilute that the free energy can be computed accurately, from knowledge of the first few terms in the virial expansion of the compressibility factor $PV/Nk_BT$, or that the chemical potential can be computed by other means (see Ref. 27).

For solids, the ideal gas reference state is less useful (although techniques have been developed to construct a reversible path from a dense solid to a dilute (lattice-) gas[28]). The obvious reference state for solids is the harmonic lattice. Computing the absolute free energy of a harmonic solid is relatively straightforward, at least for atomic and simple molecular solids. However, not all solid phases can be reached by a reversible route from a harmonic reference state. For instance, in molecular systems it is quite common to find a strongly anharmonic plastic phase just below the melting line. This plastic phase is not (meta-) stable at low temperatures.

Fortunately, in computer simulations we do not have to rely on the presence of a 'natural' reversible path between the phase under study and a reference state of known free energy. If such a path does not exist, we can construct an artificial path. This is in fact a standard trick in statistical mechanics (see e.g. Ref. 29). It works as follows: Consider a case where we need to know the free energy $F(V,T)$ of a system with a potential energy function $U_1$, where $U_1$ is such that no 'natural' reversible path exists to a state of known free energy. Suppose now that we can find another model system with a potential energy function $U_0$ for which the free energy *can* be computed exactly. Now let us define a generalized potential energy function $U(\lambda)$, such that $U(\lambda{=}0) = U_0$ and $U(\lambda = 1) = U_1$. The free energy of a system with this generalized potential is denoted by $F(\lambda)$. Although $F(\lambda)$ itself cannot be measured directly in a simulation, we can measure its derivative with respect to $\lambda$:

$$\left(\frac{\partial F}{\partial \lambda}\right)_{NVT\lambda} = \left\langle \frac{\partial U(\lambda)}{\partial \lambda} \right\rangle_{NVT\lambda} \tag{13}$$

If the path from $\lambda = 0$ to $\lambda = 1$ is reversible, we can use Eq. 13 to compute the desired $F(V,T)$. We simply measure $\langle \partial U / \partial \lambda \rangle$ for a number of values of $\lambda$ between 0 and 1. Typically, 10 quadrature points will be sufficient to get the absolute free energy per particle accurate to within 0.01 $k_BT$. It is, however, important to select a reasonable reference system. One of the safest approaches is to choose as a reference system an Einstein crystal with the same structure as the phase under study. This choice of reference system makes it extremely improbable that the path connecting $\lambda = 0$ and $\lambda = 1$ will cross an (irreversible) first order phase transition from the initial structure to another, only to go back to its original structure for still larger values of $\lambda$. Nevertheless, it is important that the parametrization of $U(\lambda)$ be chosen carefully. Usually, a linear parametrization (i.e. $U(\lambda) = \lambda U_1 + (1-\lambda)U_0$ ) is quite satisfactory. But occasionally such a parametrization may lead to weak (and relatively harmless) singularities in Eq. 13 for $\lambda \to 0$. More details about such free energy computations can be found in Ref.s 27, 30.

Before proceeding to the actual simulations of liquid crystals, I wish to point out that it is often not trivial to construct a reversible path that will link a liquid crystalline phase to a state of known free energy. Usually, the liquid-crystalline phase of interest will be separated by first order phase transitions from both the dilute gas and the low temperature (harmonic) solid. In the case of the nematic phase this problem has been resolved by switching on a strong ordering field. In the presence of such a field, the first-order isotropic-nematic transition is suppressed and a reversible expansion to the dilute gas becomes possible[31]. For the calculation of free energies of smectic and columnar phases, other techniques have to be used. Although I do not wish to go into any detail here, I wish to point out two fairly general approaches: the first is based on the fact that a smectic phase can be consider as a 1 dimensional solid stacking of 2D fluid layers, while a columnar phase resembles a 2D crystal of 1D fluid columns. In Ref. 30, a technique is described that makes it possible to reversibly decompose such smectic (columnar) phases into isolated fluid layers (columns). However, this approach has, to my knowledge, not yet been applied to such liquid crystals.

It should be stressed that such absolute free energy calculations need not be repeated for *every* model that we may care to study. For instance, if we have computed the absolute free energy of one state point in the smectic phase of rod-like molecules with an aspect ratio of 5 (say), then we can compute the free energy of the smectic phase of similar molecules with another aspect ratio simply by computing the reversible work needed to change the shape of our model particles from the initial aspect ratio to the desired value. An example of such a technique, in a slightly different context, can be found in Ref. 32.

### *6.1. Isotropic-nematic Transition*

With computer simulation it is, in principle, possible to compute the properties of *any* model for a classical many-body system. The choice of the model is determined by the nature of the question that we wish to answer. In the present case we are interested in crystalline and liquid-crystalline ordering in simple models for molecular liquids.

What is the most appropriate model? There is no unique answer to this question. Several mechanisms have been invoked to explain the onset of liquid crystalline ordering (see e.g. Ref. 33). Onsager[34] showed that nons-pherical excluded-volume effects are all that is needed to induce an isotropic to nematic transition in a system of thin, rigid rods. However, the Maier-Saupe mean-field theory shows that the same transition may also be caused by anisotropic attraction forces[35]. In addition, there are other more factors that can play a role. For instance, for thermotropics, the tendency of molecules to form liquid crystals depends strongly on the nature of the flexible side-chains attached to the rigid core of the molecule[36], while the angle-dependence of the effective attractive forces depends on the shape of the non-spherical hard core of the molecule[37]. For lyotropics, molecular flexibility and polydispersity both have a pronounced effect on the tendency to form a liquid crystal. All these factors are real. However, it would be very unwise to try to take everything into account at once. It is inevitable that a choice must be made between primary and secondary factors.

In the case of freezing of molecular liquids the situation is similar: attraction, repulsion and flexibility all play a role. For instance, we know that the presence of flexible side-chains in a molecule makes the formation of a molecular crystal entropically less favourable. Only for *atomic* liquids is the picture clear. Since the work of Alder and Wainwright[38] it has been known that the freezing of atomic liquids is primarily an excluded volume effect. That does not imply that attractive forces do not affect the location of the freezing point (they do). But the structure of the coexisting phases is largely determined by the harsh repulsive interactions between the atoms. The effect of attraction can be considered a perturbation[39].

If freezing can be understood in terms of excluded volume effects alone, it is natural to ask how far we can push this idea in the case of liquid crystalline ordering. Our aim then is to study the simplest possible hard-core models that may form both crystalline and liquid-crystalline phases.

It should be noted that an alternative approach has been followed by Luckhurst and collaborators[40]. Following the pioneering work of Lebwohl and Lasher[41], these authors focused on the effect of anisotropic intermolecular interactions of the form $v_{ij} = -J_{ij}P_2(\cos\theta_{ij})$. The choice of the Lebwohl-Lasher model was inspired by the original justification of the Maier-Saupe theory in terms of anisotropic dispersion forces. It should be noted that in simulations of models such as the one introduced by Lebowhl and Lasher it is assumed that the molecules reside on a periodic lattice. In that case the 'I-N' transition may also be interpreted as a rotational order-disorder transition in a crystalline solid.

Computer simulations of hard-core models for two-dimensional liquid crystals were pioneered by Vieillard-Baron in the early seventies[42]. Vieillard-Baron also made much progress towards the study of three-dimensional model systems[43], but did not observe spontaneous nematic ordering in 3D. The first systematic simulation study of a three-dimensional hardcore nematogen was performed by Frenkel and Mulder[31] who studied a system of hard ellipsoids-of-revolution for a range of length-to-width ratios.

The shape of hard ellipsoids of revolution is characterized by a single parameter, $x$, the ratio of the length of the major axis *(2a)* to that of the minor axis *(2b)*: $x = a/b$. Prior to the simulations reported in Ref. 31, the phase behaviour of hard spheroids[29] was only known for a few special values of $x$, viz:

(1) $x = 1$ : hard spheres,which freeze at 2/3 of close packing[44];
(2) $x \to \infty$ : thin hard needles, because this limit is equivalent to the Onsager model; this latter system has a transition to the nematic phase at vanishing volume fraction;
(3) $x \to 0$ : thin hard platelets, which also form a low-density nematic[45].

The simulations of Ref. 31 were performed for values of $x$ between 3 and 1/3. In order to locate all phase transitions, the absolute free energy of every phase was computed. Fig. 1 shows how the stability of the different phases of hard ellipsoids depends on their length-to-width ratio.

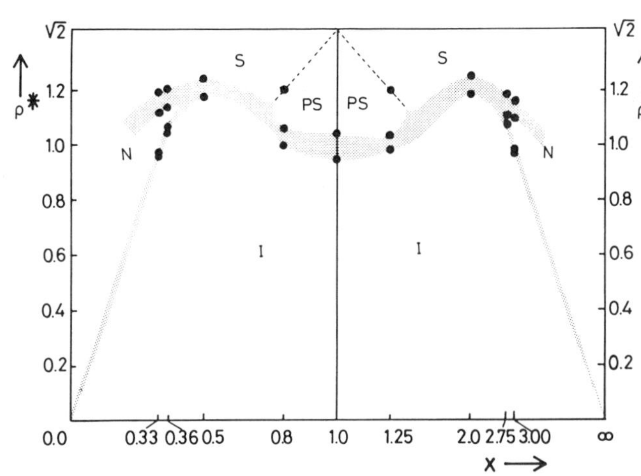


Fig.1. 'Phase diagram' of a system of hard ellipsoids of revolution[31]. The ratio of the length of the semi-major to the semi-minor axis is denoted by $x$. The reduced density $\rho^*$ is defined such that the density of regular close packing is equal to $\sqrt{2}$ for all $x$. The shaded areas indicate two-phase regions associated with a first-order phase transition. The following phases can be distinguished: I: Isotropic fluid. S: Orientationally ordered crystalline solid. PS: Orientationally disordered ('plastic') crystal, and N: Nematic liquid-crystalline phase. The densities of coexisting phases at a first order phase transition (black dots) were computed in a free-energy calculation. Note that no stable nematic is possible for $0.4 < x < 2.5$.

Four distinct phases can be identified, namely the low-density isotropic fluid, an intermediate-density nematic liquid crystalline phase, which is only stable if the length-to-width ratio of the ellipsoids is larger than 2.5 or less than 0.4, and a high-density orientationally ordered solid phase. In the case of weakly anisometric ellipsoids, an orientationally disordered solid phase was also observed. One thing to note about the phase transitions in the hard-ellipsoid system is that for particles with $3 \geq x \geq (1/3)$ the relative density-jump at the I-N transition is much smaller than for the Onsager model. Typically, the density changes only by some 2% at the I-N transition, hence the very large density discontinuity at the I-N transition in the Onsager model (more than 20%) is peculiar to long rods and not to hard-core models in general.

Perhaps the most striking feature of the phase diagram in Fig. 1 is the near symmetry between the behavior of oblate and prolate ellipsoids with inverse length-to-width ratios. Prolate-oblate ($x \to 1/x$) symmetry of ellipsoids is to be expected at *low*) densities because the second virial coefficient $B_2(x)$ equals $B_2(1/x)$. However, no such relation holds between the third and higher virial coefficients. To give a specific example: in the limit $x \to \infty$ (the Onsager limit), $B_3/B_2^2$, whereas for $x \to 0$ (hard platelets[45]) $B_3/B_2^2 \to 0.4447(3)$. Hence there is no reason to expect any exact symmetry in the phase diagram of hard ellipsoids of revolution. For larger anisometries than studied in the simulations of Ref. 31 one should expect to see asymmetric behaviour in the location of the isotropic-nematic transition. In fact, Allen has performed[46] simulations of ellipsoids with aspect ratios 5,10,0.2 and 0.1. These simulations show that, as the molecular anisometry increases, the Isotropic-Nematic transition continues to shift to lower densities. This is to be expected in view of the known limiting behavior of infinitely thin hard platelets and infinitely thin hard rods (see above). However, in Ref. 46 the exact location of the isotropic-nematic transition is not computed.

Even though we expect to see appreciable prolateoblate asymmetry in the location of the isotropic-nematic transition for highly anisometric spheroids, it is doubtful if the near symmetry

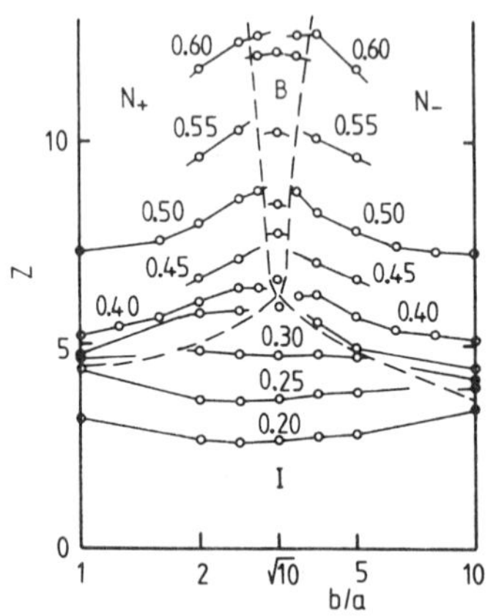


Fig.2. $Z = PV/Nk_BT - 1$ of a system of biaxial ellipsoids with axial ratio $c/a = 10$ while $b/a$ varies between 1 (prolate limit) and 10 (oblate limit)[48]. In this figure $b/a$ is represented on a logarithmic scale. The dashed curves separate state points helonging to the different phases: $I$ denotes the isotropic phase, $N+$ the 'rod-like' nematic phase, $N-$ the 'plate-like' nematic phase and $B$ the biaxial phase. The drawn curves connect the measured state-point data within one phase at a given reduced density $\rho/\rho_{close\text{-}packed}$. Note that the biaxial phase ends in a tri-critical point at an aspect ratio $a/b \approx \sqrt{10}$. This behavior is in agreement with the theoretical predictions of Mulder[49]. This figure was reproduced with permission of Dr. Allen.

of the melting line will be much affected. Strongly aligned rods and platelets follow the same equation of state ($P = 3\rho$) and a simple estimate of the melting point of very anisometric ellipsoids[47] suggests that in the limit $x \rightarrow \infty$ the symmetry between oblate and prolate ellipsoids is still present.

More recently, Allen has studied the effect of molecular biaxiality on the mesogenic properties of hard ellipsoids[48]. In particular, Allen studied the nature of the liquid-crystalline phase as a prolate spheroid was made increasingly biaxial and was finally transformed into an oblate spheroid. In this case it was found that the rod-like and plate-like nematic phases are separated by a biaxial phase. Fig. 2 shows how the stability of the different liquid-crystalline phases depends on the molecular biaxiality.

## *6.2. Theoretical Description*

The numerical simulation of the phase diagram of hard ellipsoids of revolution provided theoreticians with an opportunity to compare analytical theories for the isotropic-nematic transition with 'exact' numerical data. Several rather different theoretical approaches have been tested in this way. Actually, this work was predated by the scaled-particle for the isotropic-nematic transition constructed by Cotter and Martire[50,51]. However, this scaled-particle theory was worked out for hard sphero-cylinders and could only be compared with simulation data for the isotropic phase.

The first statistical mechanical theory for the I-N transition in a fluid of hard ellipsoids was developed by Mulder[53], who followed the so-called 'y-expansion' approach of Barboy and Gelbart[52]. Mulder found that the y-expansion led to a slight over-estimate of the pressure in the isotropic phase and that the isotropic-nematic transition was predicted to occur at too low a density. Moreover, the density-jump at this transition was predicted to be larger than observed in the simulations. However, the very symmetric appearance of the phase-diagram was well reproduced by the theory of Ref. 53.

Subsequently, several authors have applied density-functional theory to the study of the isotropic-nematic transition of hard ellipsoids. The first such theory was formulated by Singh and Singh[54]. These authors also discussed the freezing transition of hard ellipsoids and used density-functional theory to estimate the stability of the plastic crystalline state. Subsequently, alternative density-functional theories for hard ellipsoids of revolution were presented by Baus et al.[55] and by Marko[56]. A good discussion of the relative merits of these theories can be found in the article by Colot, Wu, Xu and Baus[57].

A rather different approach was followed by Perera et al.[58]. These authors studied the generalization to convex hardbody fluids of the Percus-Yevick (PY) and Hypernetted Chain (HNC) integral equations that are known to be quite successful for simple fluids[29]. Perera et al. found that for hard ellipsoids of revolution the PY approach failed to predict the existence of a stable nematic phase. In contrast, the HNC theory yielded a fair estimate of the location of the isotropic-nematic transition (or, to be more precise, of the density where the isotropic phase becomes mechanically unstable).

## *6.3. Two-dimensional Nematics*

Two-dimensional liquid crystals are very different from their 3D counterparts. This can be clearly demonstrated by considering the isotropic-nematic transition in two dimensions. On basis of the Landau theory of phase transitions, we expect that the isotropic nematic transition in two dimensions should be of second order. Density-functional theory[63] predicts the same.

However, the actual situation is much more subtle than that. The point is that two-dimensional nematics are very similar to the two-dimensional Heisenberg system ('2D-*xy* model') and hence there is a possibility that topological defects have a pronounced effect on the nature of the phase transitions[59].

In order to see this, let us consider the expression of the Frank free-energy density of a 2D-nematic. We choose the average director along the *y*-axis. We denote the angle between the average director and the instantaneous local director by $\theta$. By analogy to the three-dimensional case[33], the expression for the deformation free energy is of the form:

$$F_D = \frac{1}{2}K_1(\partial_x n_x)^2 + \frac{1}{2}K_3(\partial_y n_x)^2 = \frac{1}{2}K_1(\partial_x \theta)^2 + \frac{1}{2}K_3(\partial_y \theta)^2 \tag{14}$$

where we have assumed that $\theta$ is small. We shall simply postulate that the deformation free energy of a 2D nematic is given by Eq. 14. Moreover, we shall assume for the sake of convenience that $K_1 = K_3$. In that case we obtain a very simple expression for the deformation free-energy density:

$$F_D = \frac{1}{2}K(\nabla\theta)^2 \tag{15}$$

Using this expression, it is easy to compute the elastic contribution to the free energy of a single $\pi$-disclination in a 2D-nematic. Consider a circular path (circumference $2\pi r$) around the dislocation core. Along this path, the director rotates over an angle $\pi$. Hence $(\nabla\theta)^2 = (1/2r)^2$. If we insert this expression in Eq. 15 and integrate from the dislocation core (radius $a_0$, say) to $L$ (the linear dimension of the system) then we find that the elastic energy of an isolated disclination is:

$$F_{el} = \frac{1}{2}K\int_{a_0}^{L} \frac{2\pi r}{4r^2}dr = \frac{\pi K}{4}\log(L/a_0) \tag{16}$$

Clearly, $F_{el} \to \infty$ if $L \to \infty$.

This would seem to suggest that no free disclinations are possible in a 2D nematic. However, we should also consider the 'configurational entropy' of a single disclination, i.e. the entropy $k\log\Omega$ associated with the number of distinct ways in which we can place a dislocation in a two-dimensional area $L^2$. If we use $a_0$ as our unit of length, then the configurational entropy is given by $k\log(L/a_0)^2$ (where we have neglected an additional constant, independent of system size). Combining this expression for the configurational entropy with our expression for the elastic free energy (Eq. 16) we obtain the following expression for the total free energy of a single disclination in a 2D nematic:

$$F_{total} = \left(\frac{\pi K}{4} - 2kT\right)\log(L/a_0) \tag{17}$$

Eq. 17 suggests that if $kT < (\pi K/8)$ no free disclinations are possible, whereas for $kT > (\pi K/8)$ spontaneous generation of free disclinations may take place. However, if a nematic contains a finite concentration of free disclinations, orientational correlations are destroyed over distances longer than the characteristic separation of the free defects and the resulting phase is an isotropic fluid. This simple version of the Kosterlitz-Thouless scenario for defect mediated phase transitions predicts that the nematic phase cannot be stable above a critical temperature $kT^* = (\pi K/8)$. At that temperature there is a continuous phase transition (of 'infinite' order) from the nematic to the isotropic phase. However, there is an alternative possibility: namely that the I-N transition is simply first order. But if the I-N transition is of first order then this transition must occur *before* the nematic phase has reached the point where it becomes unstable with respect to the formation of free disclinations: i.e. at a first-order I-N transition the following inequality must hold:

$$K > \frac{8kT}{\pi}$$

This condition also follows from the more rigorous version of the KT-theory.

Note that our discussion of the disclination-mediated I-N transition was based on the assumed from of the Frank free energy (Eq. 14). It should be stressed that this form of the deformation free energy has quite drastic consequences for the nature of orientational order in 2D nematics. In particular, it implies that there exist no true long-ranged orientational order in such systems. We define the $l$-th orientational correlation function as:

$$g_l(r) \equiv \langle \cos 2l(\theta(0)-\theta(r)) \rangle = Re \langle \exp[-2il(\theta(0)-\theta(r))] \rangle \qquad (18)$$

Using the fact that the free energy (Eq. 15) is quadratic in $\theta(k)$, it is easy to show that $g_l(r)$ has the following form:

$$g_l(r) = \left(\frac{r}{a_0}\right)^{-\frac{2l^2kT}{\pi K}} \equiv \left(\frac{r}{a_0}\right)^{-\eta_l} \qquad (19)$$

where the last term on the right-hand side of Eq. 19 defines the exponent $\eta_l$. Note that this equation implies that, provided that Eq. 14 is valid, there is no true long-range orientational order in a 2D nematic, but algebraic or 'quasi long-range' order. Similarly, it can be shown that the order parameter $\langle \cos 2\theta \rangle$ also vanishes algebraically with increasing system size:

$$\langle \cos 2\theta \rangle \propto \left(\frac{L}{a_0}\right)^{-\frac{kT}{\pi K}} \qquad (20)$$

Now recall that a 2D nematic is only expected to be stable against the spontaneous generation of free disclinations, when $K$ is larger than $\pi/(8kT)$. Hence, at the $K$-$T$ transition the orientational correlation functions and the nematic order parameter must satisfy the following relations:

$$g_l(r) = \left(\frac{r}{a_0}\right)^{-\frac{l^2}{4}} \qquad (21)$$

$$\langle \cos 2\theta \rangle \propto \left(\frac{L}{a_0}\right)^{-\frac{1}{8}} \qquad (22)$$

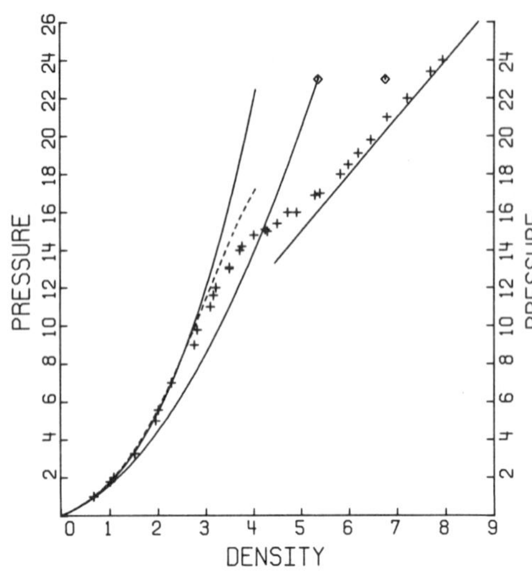


Fig.3. Equation of state of two-dimensional fluid of infinitely thin hard needles of length 1, as obtained by Monte Carlo simulation[61]. Note that in this figure the reduced pressure is the independent variable. The reduced density $\rho L^2$ is indicated by crosses, the chemical potential $\mu$ by triangles. The curves at low pressure were computed using a 5-term virial series. For more details, see Ref. 61.

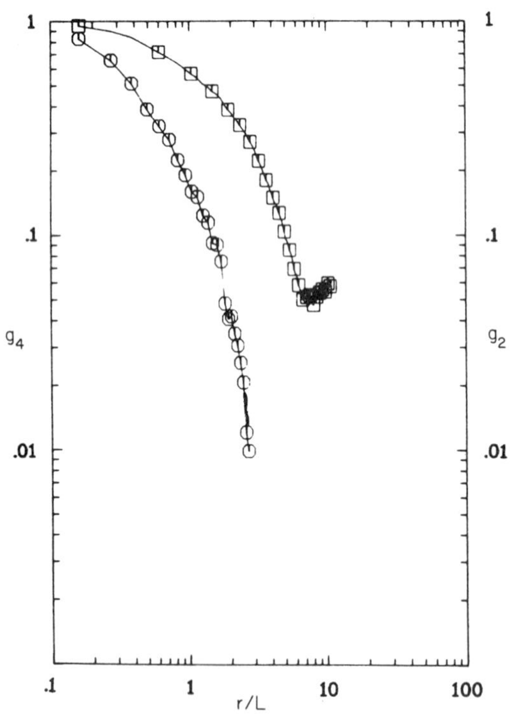


Fig.4. Orientational correlation functions $g_2 \equiv \langle\cos 2(\phi(0)-\phi(r))\rangle$ and $g_4 \equiv \langle\cos 4(\phi(0)-\phi(r))\rangle$ for a two dimensional system of hard needles of length L = 1. This figure shows that at a reduced density $\rho L^2 = 6.75$ the orientational order decays exponentially. In other words, the phase is isotropic.

Two points should be stressed:

[1] If the I-N transition is first order then, at the transition the exponents of $g_l$ and $< \cos 2\theta >$ must be less than the critical values given by Eq. 21; and,

[2] The above arguments rest on the assumption that the deformation free energy is of the form given by Eq. 15.

If this expression is valid, 2D nematics can only have algebraic orientational order. However, it has thus far only been possible to prove the absence of true long-range orientational order for a certain class of short ranged potentials called *separable*[60]. A pair-potential is called separable if the interaction energy of two molecules at fixed center-of-mass separation $\mathbf{r}_{ij}$ depends only on the relative orientation of the two molecular axes $\mathbf{u}_i \cdot \mathbf{u}_j$, but not on $\mathbf{r}_{ij} \cdot \mathbf{u}_i$ or $\mathbf{r}_{ij} \cdot \mathbf{u}_j$. Realistic pair-potentials are hardly ever separable.

There are therefore two obvious questions that one could ask about 2D nematics:

[1] If the pair potential is non-separable, do we find algebraic or true long-range order? and,

[2] If we find algebraic order, do we observe a first order I-N transition or a continuous one of the Kosterlitz-Thouless type?

To start with the first question: a good starting point would be to choose a pair-potential that is as non-separable as possible. An obvious candidate is a two-dimensional model of infinitely thin hard needles[61]. This pair potential is very non-separable in the sense that, at fixed $| r_{ij} |$ and fixed $\mathbf{u}_i \cdot \mathbf{u}_j$ the pair potential is *not* constant, but may vary between 0 and $\infty$. The equation of state of this system is shown in Fig. 3. According to the bifurcation analysis of the corresponding Onsager limit[62] a second-order isotropic-nematic transition is expected at a density $\rho L^2 = (3\pi/2) = 4.712...$ and a pressure $PL^2 = 11.78...$ At first sight this seems to be quite a reasonable estimate of the I-N transition, because very close to this point the equation of state appears to exhibit a change of slope. However, analysis of the long-range behaviour of the orientational correlation functions and of the system-size dependence of the order-parameter $< \cos 2\theta >$ indicate that the higher density phase is not a stable nematic. The orientational correlation functions decay either exponentially (see Fig. 4) or with an apparent algebraic exponent that is larger than the critical value given in Eq. 21. Only at a density that is almost twice the Onsager transition point does the observed behaviour conform to what is expected for a stable nematic with algebraic order (see Fig. 5). However, at this density the equation of state is completely featureless. Such behaviour is to be expected if the I-N transition is in fact of the *K-T* type.

Subsequently, Cuesta and Frenkel[64] have studied the isotropic to nematic transition in a system of 2D ellipses with aspect ratios 2, 4 and 6. It is found that in all cases where a stable nematic phase is found (aspect ratios 4 and 6), this phase exhibits algebraic orientational order. However, whereas the I-N transition appears to be of the *K-T* type for aspect ratio 6 (and larger), the transition is found to be first order for aspect ratio 4. This implies that in the latter case the 2D nematic undergoes a first-order transition *before* it has reached the point where it becomes absolutely unstable with respect to diclination unbinding.

A very puzzling feature is the nature of the high-density phase of 2D ellipses. A snapshot of such a phase is shown in Fig. 6. This phase does not appear to have true crystalline order, nor for that matter, smectic order, yet it does clearly have a large amount of local order. The precise nature of this high density phase is currently under investigation.

I am not aware of any 2D hard-core systems that show anything but quasi-long range nematic order, and I would be very surprised to see true long range orientational order in a 2D liquid crystal in the absence of some kind of positional order. For a discussion of nematic ordering in systems with continuous intermolecular interactions, see Ref. 65.

## 7. Beyond Nematics

The existence of a nematic phase in a system of hard-core molecules is not surprising. In fact, the earliest analysis of any statistical-mechanical model for a liquid-crystalline phase, i.e. Onsager's study of a system of of thin hard rods, shows that this simple hard-core system *must* form a nematic phase at sufficiently high density. It would of course be nice if there existed something like the Onsager model for smectics: an exactly solvable model system that exhibits a transition to the smectic A phase. Unfortunately, no such model is known.

Hence, the only way to test approximate 'molecular' theories of the smectic phase is to compare with computer simulations. In the spirit of the Section 6.1 we look for the simplest possible model that will form a smectic phase. In the case of nematics, convex hard-core models are the natural candidates because these constitute the natural generalization of the Onsager model. However, for smectics it is not obvious that hardcore models will work. In fact, in existing textbooks on liquid-crystal physics the possibility of a hard-core smectic is not even discussed. To my knowledge the only pre-simulation article discussing the possibility of hard-core smectics is a paper by Hosino et al.[66] The 'traditional' approach was to ascribe the smectic ordering to

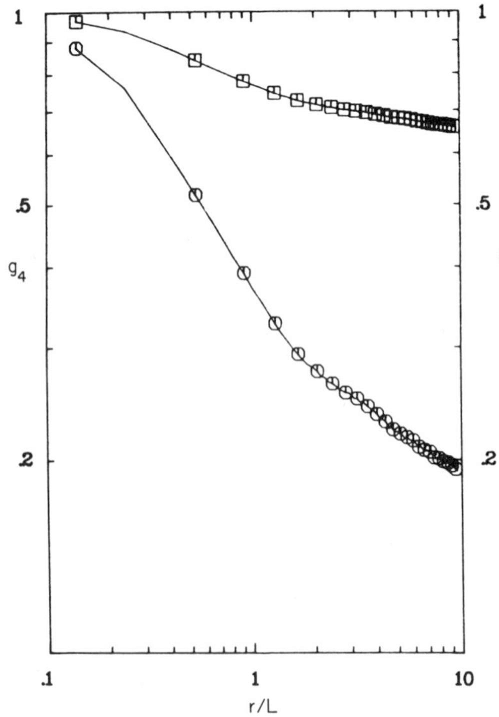


Fig.5. Orientational correlation functions $g_2 \equiv <\cos 2(\phi(0)-\phi(r))>$ and $g_4 \equiv <\cos 4(\phi(0)-\phi(r))>$ for a two-dimensional system of hard needles of length L = 1. This figure shows that at a reduced density $\rho L^2 = 8.75$ the orientational order decays algebraically. From the values of the algebraic exponents $\eta_2$ and $\eta_4$ the effective Frank elastic constant can be computed. At $\rho = 8.75$ this Frank constant is large enough to make the 2D nematic stable with respect to disclination unbinding. The isotropic-nematic transition is estimated to occur somewhere around $\rho = 7.5$. Note that if we had only used Fig. 3 to locate the isotropic-nematic transitionl we might have concluded that this transition occurs at a density of approximately 5.0 in reduced units.

attractive interactions between the molecular cores or, alternatively, to the change in packing entropy of the flexible tails of the mesogenic units[36].

## 7.1. *Parallel Molecules*

Whereas essentially *any* fluid of sufficiently non-spherical convex hard bodies will form a nematic phase, non-sphericity alone is not enough to from a smectic phase. This is best demonstrated by the following simple example. We know from experiment that in many smectic phases the orientational order parameter $S \approx 1$. Let us therefore first consider the possibility of forming a smectic phase in a fluid of *perfectly aligned* molecules ($S = 1$). We know that sufficiently nonspherical hard ellipsoids can form a nematic phase (see Section 6.1). It is natural to ask whether a perfectly aligned nematic of hard ellipsoids can transform into a smectic phase. The answer to this question is: *no*. The reason is quite simple. Consider a fluid of ellipsoids with lengthtowidth ratio $a/b$ all aligned along the $z$-axis (say). Now we perform an affine transformation that transforms all $z$ coordinates into coordinates $z'$, such that $z' = (b/a)z$. At the same time we transform to new momenta in the $z$-direction: $p'_z = (a/b)p_z$. Clearly, this transformation does not change the partition function of the system, and hence all thermodynamic properties of the system are unchanged. However, the effect of this affine transformation is to

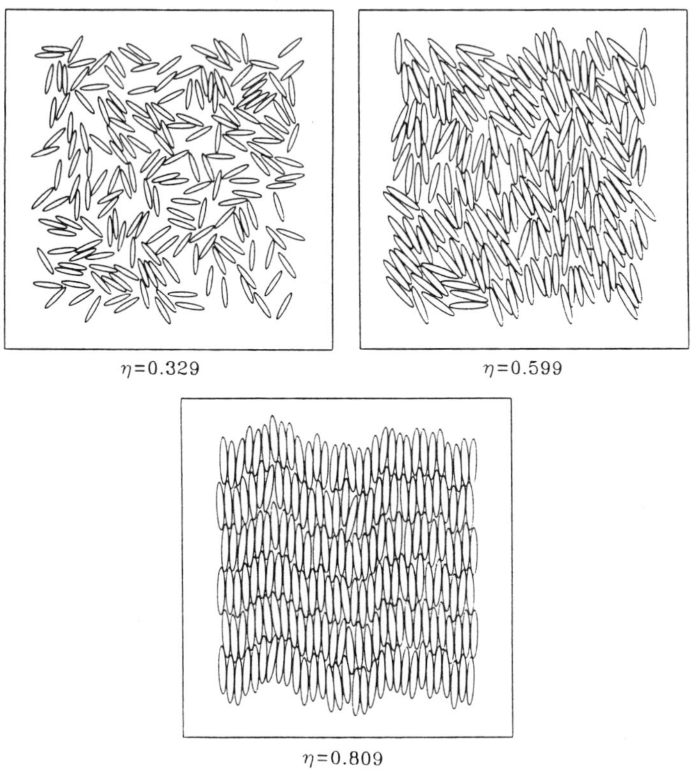


Fig.6. Snaphots of typical configurations of a system of hard ellipses in the isotropic phase (η = 0.329), near the estimated isotropic-nematic transition (η = 0.599) and in the high-density phase (η = 0.809). Although the latter phase exhibits local solid-like ordering, it is not a true (two-dimensional) solid.

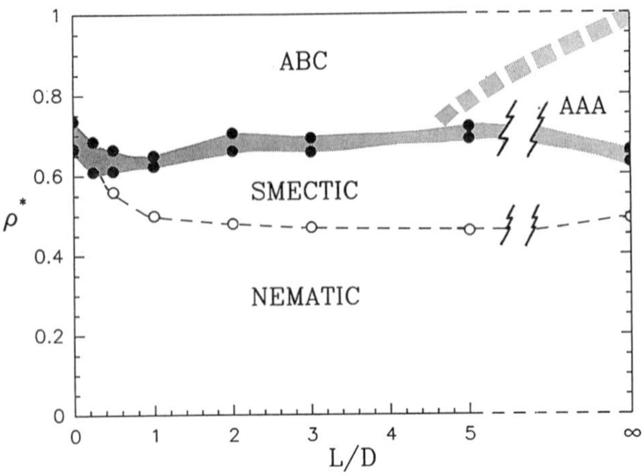


Fig.7. Schematic 'phase diagram' of a system of parallel hard spherocylinders as obtained by computer simulation[71]. The abcissa indicates the length-to-width ratio *L/D*. The ordinate measures the density referred to the density at regular close packing.The dashed area indicates the two-phase region at the first-order freezing transition. For *L/D* < 5,the solid consists of 'ABC'-stacked triangular planes. For larger values of *L/D*, we find evidence for a hexagonal ('AAA') stacking of the molecules (i.e. triangular lattices stacked on top of one another). At very high densities and large *L/D* values we find a pocket where the system appears to form a colulnnar phase. However, the range of stability of this phase is strongly dependent on the size of the system studied. Although we still observed this columnar phase for a system of 1080 particles, it is conceivable that this phase will disappear altogether in the thermodynamic limit. The dashed curve indicates the nematic-smectic transition.

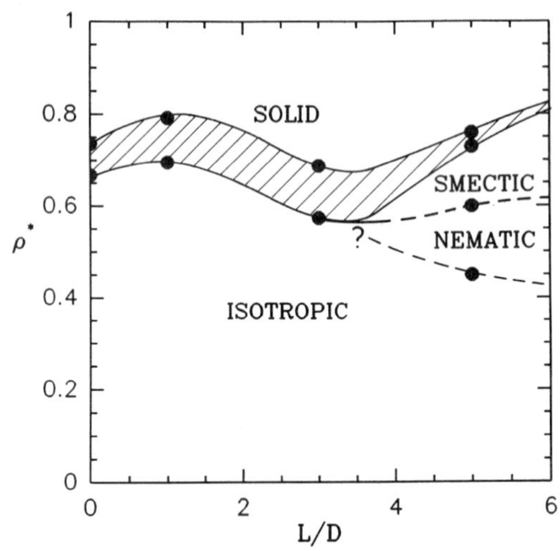


Fig.8. Phase diagram of a system of freely rotating spherocylinders as a function of the ratio between the length of the cylindrical part (*L*) and the diameter (*D*). The ordinate $\rho^*$ measures the density divided by the density at regular close packing. The grey (dashed) area is the two-phase region separating the densities of the coexisting solid and fluid phases. The black dots indicate computed phase-coexistence points. The nematic-smectic transition is indicated by a dashed curve, as is the isotropic-nematic transition. Although the latter transition is expected to be of first order, the resolution of the present simulation was insufficient to determine the density discontinuity at this transition. The location of the isotropic-nematic-smectic triple point can only be estimated approximately and is indicated in the figure by a question mark.

change a fluid of parallel ellipsoids into a system of hard spheres. But, as far as we know, hard spheres can only exist in two phases: fluid and crystal. Hence parallel ellipsoids can only occur in the (nematic) fluid phase and in the crystalline solid phase. In particular: no smectic phase is possible. This makes it extremely improbable that a fluid of non-parallel ellipsoids will form a stable smectic. Such a phase is only expected in the unlikely case that the orientational fluctuations would *stabilize* smectic order[67].

This example demonstrates that we should be careful in selecting possible models for a hard-core smectics. Surprisingly (and luckily) it has turned out that another very simple hard-core model system, namely a system of parallel hard sphero-cylinders, does form a smectic phase[69,70]. A stable smectic phase is possible for length-to-width ratios $L/D \geq 0.5$. In addition, we find that another phase appears at high densities and larger $L/D$ values. In small systems, the phase appears to be columnar[70], but in larger systems the range of stability of the columnar phase shrinks (see Fig. 7) and is largely replaced by a hexagonal solid phase. In order to tell whether the latter phase is indeed truly solid or, for example, smectic-B, would require simulations on systems that contain many more particles than the 1000-2000 that we have thus far been able to study systematically.

At first sight it seems surprising that pure excluded volume effects can give rise to smectic ordering. However, stimulated by the computer-simulation results a number of authors have reexamined[72,73,74,75,76,77,78,80,81] the theoretical description of hard-core liquid crystals and have come up with quite simple models that do in fact predict smectic and, in some cases, colummnar phases.

## 7.2. The Effect of Rotation

Of course, a model system consisting of *parallel* spherocylinders is rather unphysical. It is therefore of considerable interest to know whether if a system of freely rotating hard-core molecules can form a smectic phase. This question is of some practical interest, in view of the recent experimental evidence that smectic[82] and columnar[83] ordering may take place in concentrated solutions of rod-like DNA molecules.

Simulations[84,85] of a system of freely rotating spherocylinders with length-to-width ratio $L/D = 5$ revealed the presence of a stable smectic phase, in addition to the expected isotropic, nematic and solid phases. This work was recently extended to other aspect ratios by Veerman and Frenkel[86]. These authors show that the smectic phase disappears at $L/D = 3$. At this aspect ratio, the nematic phase is no longer (meta-)stable. Fig. 8 shows a tentative phase diagram of freely rotating hard spherocylinders.

Density-functional theories for hard spherocylinder systems have been proposed by Holyst and Poniewierski[76,77] and by Somoza and Tarazona[87,88]. Both theories predict the presence of a stable smectic phase if the length-to-width ratio $L/D$ exceeds a minimum value of ~ 3. However, the two theories differ in the their estimate of the point where the nematic-smectic transition has its tricritical point.

## 7.3. Columnar Phases

If hard-core models exhibit smectic phases, one may wonder if excluded volume effects can also induce the formation of the even more ordered columnar phase. In this case, it is natural to look for a convex, plate-like molecule. Oblate ellipsoids are not expected to form columnar phases. The argument why this should be so is essentially the same as the one that 'explains' why prolate ellipsoids should not form smectic phases. Rather, we should look for the oblate equivalent of the spherocylinder.

For ellipsoids the transition from prolate to oblate shapes is controlled by a single parameter (the axial ratio $a/b$). In contrast, spherocylinders cannot be changed into oblate particles simply by changing $L/D$ (unless we allow for the possibility of negative $L/D$). It turns out that a particularly convenient 'oblate spherocylinder' model is the so-called *cut sphere*[89] (see Fig. 9).

The cut sphere is a hard convex body. Using the standard techniques applicable to such objects (see e.g. Ref. 90), the second virial coefficient of cut spheres can be evaluated for arbitrary $L/D$ ratios:

$$B_2 = \frac{\pi D^3}{6}\left\{\cos\theta_M\left(1+\frac{\sin^2\theta_M}{2}\right)+3\left(\cos\theta_M+\frac{\theta_M\sin\theta_M}{2}\right)\left(\cos\theta_M+\frac{\sin^2\theta_M}{2}\right)\right\} \tag{23}$$

where $\theta_M \equiv arcos(L/D)$.

At high densities, cut spheres can be stacked in a regular close-packed lattice. The volume fraction at regular close-packing is: $\eta_{cp} = (\pi/6)\sqrt{3-(L/D)^2}$. Note that for $L/D = 1$ (hard spheres), this reduces to the well-known hard-sphere result $\eta_{cp} = \pi/\sqrt{18}$. For $L/D \rightarrow 0$ (flat, cylindrical platelets), we obtain the *2D* hard-disk value $\eta_{cp} = \pi/\sqrt{12}$. Monte Carlo simulations were carried out on a system of cut spheres with $L/D = 0.1$ and $L/D = 0.2$, over a range of densities between dilute gas and crystalline solid[89,91]. Surprisingly, it turned out that the systems with $L/D = 0.1$ and $L/D = 0.2$ behaved completely differently.

For the system with $L/D = 0.1$ it was observed that the system spontaneously ordered to form a nematic phase at a reduced density of 0.335 (i.e. at 33.5% of regular close packing). At a density corresponding to 49% of regular close packing, this nematic phase undergoes a strong first-order transition to a columnar phase (at a reduced density $\rho^* = 0.534$). The columnar-crystalline transition occurs at much higher density ($\rho^* > 0.80$).

Next, we turn to the system of platelets with $L/D = 0.2$. At first sight, the behavior of this system looks quite similar to that observed for the thinner platelets. In particular, the equation-of-state for cut spheres with $L/D = 0.2$ looks similar to the one corresponding to $L/D = 0.1$. At low densities the compressibility factor is well described by a 5-term virial series and at higher densities the actual pressure is lower. In other hard-core systems this is usually an indication of a precursor to some kind of ordering transition (e.g. the I-N transition in the $L/D = 0.1$ case). And, indeed, at $\rho \approx 0.475$ a change of slope in the equation of state, as is to be expected in a simulation near a weakly first-order (or, possibly, second-order) phase transition. The most natural assumption is that an isotropic-nematic transition takes place. After all, this is what happens in the $L/D = 0.1$ case, and we also know that oblate hard ellipsoids with a axial

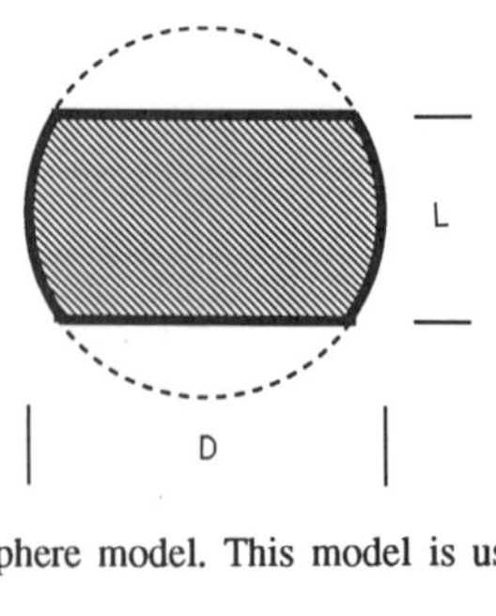


Fig.9. Schematic drawing of the cut-sphere model. This model is used as an oblate counterpart of the hard spherocylinder model.

ratio less than 0.4 have a stable nematic phase. However, if we measure the orientational correlation function $g_2(r) \equiv < P_2(\mathbf{u}(0)\cdot\mathbf{u}(r)) >$ we find that it decays to zero within one molecular diameter, even at the highest densities of the 'fluid' branch (see Fig. 10). In a nematic phase, $g_2(r)$ should tend to a finite limit: $g_2(r) \rightarrow S^2$ as $r \rightarrow \infty$, where $S$ is the nematic order parameter.

It should be stressed that the absence of nematic order in the $L/D$ = 0.2 system is not a consequence of the way in which the system was prepared. Even if we started with a configuration at a reduced density $\rho/\rho_{cp} = 0.50$ with all the molecules initially aligned, the nematic order would rapidly dissapear. In other words, at that density the nematic phase is *mechanically* unstable.

The great surprise comes when we consider the higher-order orientational correlation function $g_4(r) \equiv < P_4(\mathbf{u}(0)\cdot\mathbf{u}(r)) >$. Usually, when $g_2(r)$ is short ranged, the same holds *a fortiori* for $g_4(r)$. However, Fig. 11 shows that for densities $\rho/\rho_{cp} > 0.55$, $g_2(r)$ is much longer ranged than $g_2(r)$. This suggests that the system has a strong tendency towards orientational order with *cubic* symmetry (we reserve the term 'cubatic' for a phase with cubic orientational order but no translational order. In contrast, 'cubic' is used to denote a phase with cubic translational order). In computer simulations one should always be very suspicious of any spontaneous ordering with cubic symmetry, because such ordering could be induced by the (cubic) periodic boundary conditions.

In order to test whether the boundary conditions were responsible for the cubatic order we did a number of long simulations with systems of up to 2048 particles. These simulations strongly suggest that the onset of cubatic orientational order is not an artifact of the boundary conditions. Another indication that the boundary conditions are not the cause of the observed ordering is that still higher order correlations ($g_6$ and $g_8$) that could also be induced by the periodic boundaries, are in fact rapidly decaying functions of $r$. If we make a log-log plot of $g_4(r)$ in the large system for several densities between $\rho = 0.51$ and $\rho = 0.63$ (see Fig. 12), it appears that the cubatic order is not truly long-ranged but quasi-long-ranged, i.e. $g_4(r) \sim r^{-\eta}$, where $\eta$ depends on the density $\rho$. This observation should, however, be taken with a large grain of salt, as the range over which

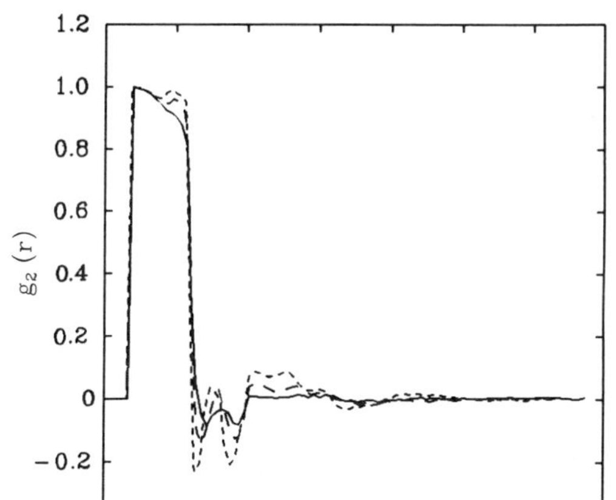


Fig.10. Density dependence of the orientational correlation function $g_2 \equiv < P_2(\mathbf{u}(0)\cdot\mathbf{u}(r)) >$ in a system of hard cut spheres with a lengthtowidth ratio $L/D$ = 0.2. Drawn curve: $\rho/\rho_{cp} = 0.51$, long dashes: $\rho/\rho_{cp} = 0.57$, short dashes: $\rho/\rho_{cp} = 0.63$. Note that even at the highest densities studied $g_2(r)$ is short ranged.

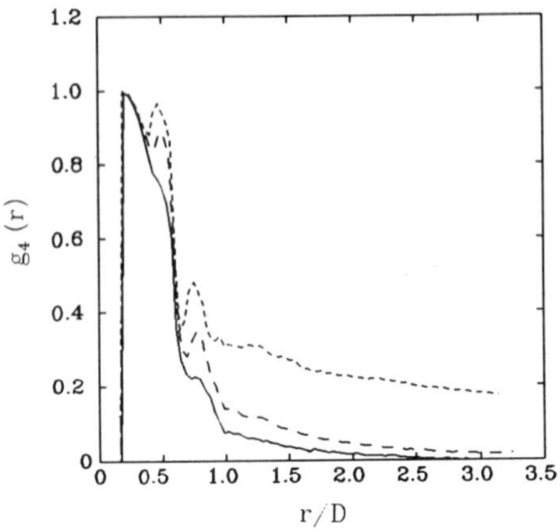


Fig.11. Density dependence of the orientational correlation function $g_4 \equiv \langle P_2(\mathbf{u}(0)\cdot\mathbf{u}(r))\rangle$ in a system of hard cut spheres with a length-to-width ratio $L/D$=0.2. Symbols as in Fig. 10. Note that, unlike $g_2(r)$, $g_4(r)$ appears to be long-ranged at high densities.

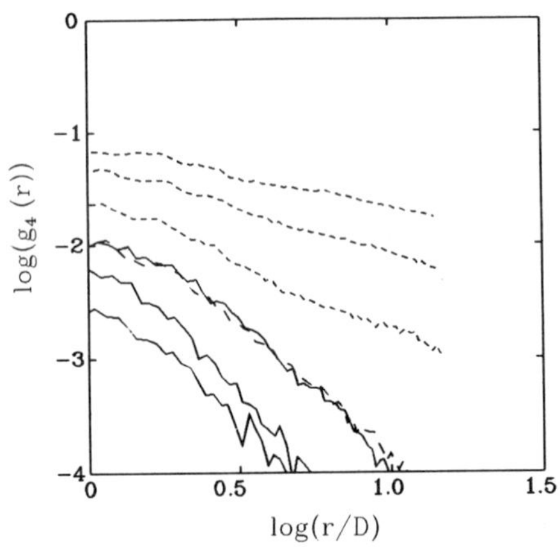


Fig.12. Log-log plot of the orientational correlation function $g_4 \equiv \langle P_2(\mathbf{u}(0)\cdot\mathbf{u}(r))\rangle$ in a system of hard cut spheres with a length to width ratio $L/D$=0.2 as a function of density. With increasing density, the amplitude of this correlation function increases. The lowest curve corresponds to $\rho^* = 0.51$, followed by curves for $\rho^* = 0.54$ and $\rho^* = 0.56$. The long-dashed curve corresponds to $\rho^* = 0.57$. At higher densities $\rho^* = 0.58, 0.60$ and 0.63 (long-dashed curves), $g_4(r)$ appears to decay algebraically over the narrow range of distances ($1 < r/D < 3.2$) where we could observe monotonic decay of $g_4(r)$.

linear behavior in the loglog plot is observed corresponds to less than 1 decade in *r*.

Simulations by Veerman[91] indicate that, at least for the model with $L/D = 0.2$, the cubatic phase may be largely meta-stable. This observation does not diminish the interest of this phase. After all, many liquid-crystalline phases that occur in nature are only metastable, but still very real. I am, however, not aware of any experimental observations of *cubatic* liquid crystals, although *cubic* liquid-crystalline phase of disklike molecules have been observed experimentally[92].

## 8. Conclusion

In this review, I have almost exlusively discussed phase transitions in hard-core models. However, I do not wish to suggest that excluded volume effects are the whole story. On the contrary, such models provide only a moderately realistic description of some lyotropic liquid crystals, and they are totally unrealistic for thermotropic liquid crystals. It is therefore of crucial importance to investigate the effect of longer-ranged attractive forces and of molecular flexibility on the stability of liquid crystalline phases. The recent work of Rull et.al.[68] shows clearly that attraction can have a drastic effect on the stability of liquid-crystalline phases in simple model systems. Most importantly, the results of Ref. 68 suggest that thermodynamic perturbation theory may be much less successful for liquid-crystalline phases than for simple liquids. If so, this would imply that we may have to revise the concept of a hard-core reference system for a liquid crystal.

Thus far, there have been no direct numerical studies of the location of phase-transitions involving meso-phases of flexible molecules. As was explained in the beginning of this paper, the reason is simply that simulations of flexible molecules are quite time-consuming. However, this situation is likely to change due to recent advances in the simulation of flexible molecules[93,94,95]. Once we can perform efficient simulations flexible mesogens, it becomes possible to connect the numerical study of simple mesogens with that of liquid-crystalline polymers and of ordering in self-assembling structures.

ACKNOWLEDGMENTS. The work of the FOM Institute is part of the research program of FOM and is made possible with financial help from the Nederlandse Organisatie voor Wetenschappelijk Onderzoek (NWO). Part of the work reported in this paper was carried out at the FOM Institute by Jan Veerman and Jose Cuesta. I thank both for their assistance in preparing some of the illustrations in this paper. I thank Mike Allen for permission to reproduce Fig. 2.

### References

1. M.P. Allen and D.J. Tildesley, "Computer Simulation of Liquids", Clarendon, Oxford, 1987.
2. N. Metropolis, A.W. Rosenbluth, M.N. Rosenbluth, A.H. Teller and E. Teller, J. Chem. Phys. **21**: 1087 (1953).
3. M. Parrinello and A. Rahman, Phys. Rev. Lett. **45**: 1196 (1980).
4. M. Parrinello and A. Rahman, J. Appl. Phys. **52**: 7182 (1981).
5. R. Najafabadi and S. Yip, Scripta Metall. **17**: 1199 (1983).
6. M.J. Mandell, J. Stat. Phys. **15**: 299 (1976).
7. P.J. Steinhardt, D.R. Nelson and M. Ronchetti, Phys. Rev. B **28**: 784 (1983).
8. R.W. Impey, P.A. Madden and D.J. Tildesley, Mol. Phys. **44**: 1319 (1981).
9. M. Fixman, Proc. Nat. Acad. Sci. **71**: 3050 (1974).
10. L. Verlet, Phys. Rev. **159**: 98 (1967).
11. H.J.C. Berendsen and W.F. van Gunsteren, in: "Molecular Dynamics Simulations of Statistical Mechanical Systems", Proceedings of the 97th International School of Physics 'Enrico Fermi', G. Ciccotti and W. G. Hoover, editors. NorthHolland, Amsterdam, 1985, p.43.

12. M.P. Allen, D. Frenkel and J. Talbot, Computer Physics Reports **9**: 301 (1989).
13. J.J. Erpenbeck and W.W. Wood in: "Statistical Mecahnics", Part B, ed. B.J. Berne, Plenum Press, New York, 1977.
14. J.L. Lebowitz, J.K. Percus and L. Verlet, Phys. Rev. **153**: 250 (1967).
15. H.C. Andersen, J. Chem. Phys. **72**: 2384 (1980).
16. S. Nose, J. Chem. Phys. **81**: 511 (1984).
17. W.W. Wood, J. Chem. Phys. **48**: 415 (1968).
18. I.R. McDonald, Mol. Phys. **23**: 41 (1972).
19. G.E. Norman and V.S. Filinov, High Temp. Res. USSR **7**: 216 (1969).
20. D.J. Adams, Mol. Phys. **28**: 1241 (1974).
21. M. Creutz, Phys. Rev. Lett. **50**: 1411 (1983).
22. A. Panagiotopoulos, Mol. Phys. **61**: 813 (1987).
23. G. Ciccotti, D. Frenkel and I.R. McDonald, "Simulation of Liquids and Solids", NorthHolland, Amsterdam, 1987. This reprint collection contains, among others, Ref.s 2, 4, 14, 15, 16, 18, 10, 24, 28.
24. R. Zwanzig and N.K. Ailawadi, Phys. Rev. **182**: 280 (1969).
25. D. Frenkel in: "Intermolecular Spectroscopy and Dynamical Properties of Dense Systems", Proceedings of the 75th International School of Physics 'Enrico Fermi', J. van Kranendonk, ed., Soc. Italiana di Fisica, Bologna,1980, p.156.
26. G. Jacucci and A. Rahman, Nuovo Cimento D **4**: 341 (1984).
27. D. Frenkel in: "Molecular Dynamics Simulations of Statistical Mechanical Systems", Proceedings of the 97th International School of Physics 'Enrico Fermi', G. Ciccotti and W. G. Hoover, editors. NorthHolland, Amsterdam, 1985, p.151.
28. W.G. Hoover and F.H. Ree, J. Chem. Phys. **47**: 4873 (1967).
29. J.P. Hansen and I.R. McDonald, "Theory of Simple Liquids", 2nd edition, Academic Press, London, 1986.
30. E.J. Meijer, D. Frenkel, R.A. LeSar and A.J.C. Ladd, J. Chem. Phys. **92**: 7570 (1990).
31. D. Frenkel and B.M. Mulder, Mol. Phys. **55**: 1171 (1985).
32. W.G.T. Kranendonk and D. Frenkel, Mol. Phys. **72**: 699 (1991).
33. P.G. de Gennes, "Physics of Liquid Crystals", Oxford University Press, Oxford, England (1974).
34. L. Onsager, Proc. NY. Acad. Sci. **51**: 627 (1949).
35. W. Maier and A. Saupe, Z. Naturforsch. A **13**: 564 (1958).
36. F. Dowell and D.E. Martire, J. Chem. Phys. **68**: 1088(1978), ibid.: **68**: 1094 (1978), ibid.: **69**: 2322 (1978). F. Dowell, Phys. Rev. A **28**: 3526 (1983).
37. W.M. Gelbart and A. Gelbart, Mol. Phys. **33**: 1387 (1977).
38. B.J. Alder and T.E. Wainwright, J. Chem. Phys. **27**: 1208 (1957).
39. J.D. Weeks, D. Chandler and H.C. Andersen, J. Chem. Phys. **54**: 5237 (1971).
40. See, for instance: C. Zannoni in "The molecular physics of liquid crystals", G.R. Luckhurst and G.W. Gray editors, Academic Press, London, 1979, p.191.
41. P.A. Lebwohl and G. Lasher, Phys. Rev. A **6**: 426 (1972).
42. J. Vieillard-Baron, J. Chem. Phys. **56**: 4729 (1972).
43. J. Vieillard-Baron, Mol. Phys. **28**: 809 (1974).
44. W.G. Hoover and F.H. Ree, J. Chem. Phys. **49**: 3609 (1968).
45. R. Eppenga and D. Frenkel, Mol. Phys. **52**: 1303 (1984).
46. M.P. Allen and M.R. Wilson, J. Computer-Aided Molec. Design **3**: 335 (1989).
47. D. Frenkel, Mol. Phys. **54**: 145 (1985).
48. M.P. Allen, Liquid Crystals, **8**: 499 (1990).
49. B.M. Mulder, Phys. Rev. A **39**: 360 (1989), Liquid Crystals **8**: 527 (1990).
50. M.A. Cotter and D.E. Martire, J. Chem. Phys. **52**: 1902 (1970), J. Chem. Phys. **53**: 4500 (1970).
51. M.A. Cotter, Phys. Rev. A **10**: 625 (1974), J. Chem. Phys. **66**: 1098 (1977).
52. B. Barboy and W.M. Gelbart, J. Chem. Phys. **71**: 3053 (1979).
B. Barboy and W.M. Gelbart, J. Stat. Phys. **22**: 685 (1980).
53. B.M. Mulder and D. Frenkel, Mol.Phys. **55**: 1193 (1985).
54. U.P. Singh and Y. Singh, Phys. Rev. A **33**: 2725 (1986).
55. M. Baus, J.L. Colot, X.G. Wu and H. Xu, Phys. Rev. Lett. **59**: 2148 (1987).
56. J.F. Marko, Phys. Rev. Lett. **60**: 325 (1989). J.F.Marko, Phys. Rev. A **39**: 2050 (1989).
57. J.L. Colot, X.G. Wu, H. Xu and M. Baus, Phys. Rev. A **38**: 2022 (1988).
58. A. Perera, P.G. Kusalik and G.N. Patey, J. Chem. Phys. **87**: 1295 (1987), **89** :5969(1988). A. Perera, G.N. Patey and J.J. Weis, J. Chem. Phys. **89**: 6941 (1989). A. Perera and G. N. Patey, J. Chem. Phys. **89**: 5861 (1989).

59. J.M. Kosterlitz and D. Thouless, J. Phys. **C6**: 1181 (1973).
60. M. Romeiro, J. Math. Phys. **19**: 802 (1978).
61. D. Frenkel and R. Eppenga, Phys. Rev. A **31**: 1776 (1985).
62. R.F. Kayser and H.J. Raveche, Phys. Rev. A **17**: 2067 (1978).
63. J.A. Cuesta, C.F. Tejero and M. Baus, Phys. Rev. A **39**: 6498 (1989).
64. J.A. Cuesta and D. Frenkel, Phys. Rev. A **42**: 2126 (1990).
65. M.J.P. Gingras, P.C.W. Holdsworth and B. Bergersen, in "Proceedings of the 13th International Liquid Crystal Conference" (in press), and M.J.P. Gingras, P.C.W. Holdsworth and B. Bergersen, Phys. Rev. A **41** :6786 (1990).
66. M. Hosino, H. Nakano and H. Kimura, J.Phys. Soc. Japan **46**: 1709 (1979).
67. Recently, it has been shown that this argument does not extend to models that include attractive forces. In particular, it has been shown that the Gay-Berne model, i.e. the 'ellipsoidal' generalization of the Lennard-Jones model for atomic fluids, does indeed exhibit one or more stable smectic phases[68].
68. E. de Miguel, L.F. Rull, M.K. Chalam, K.E. Gubbins and F. van Swol, Mol. Phys. **72**: 593 (1991).
69. A. Stroobants, H.N.W. Lekkerkerker and D. Frenkel, Phys. Rev. Lett. **57**: 1452 (1986).
70. A. Stroobants, H.N.W. Lekkerkerker and D. Frenkel, Phys. Rev. A **36**: 2929 (1987).
71. J.A.C. Veerman, thesis, University of Utrecht, 1991. J.A.C. Veerman and D.Frenkel, Phys. Rev. A (in press).
72. B.M. Mulder, Phys. Rev. A **35**: 3095 (1987).
73. X. Wen and R.B. Meyer, Phys. Rev. Lett. **59**: 1325 (1987).
74. A.M. Somoza and P. Tarazona, Phys. Rev. Lett. **61**: 2566 (1988).
75. R. Holyst and A. Poniewierski, Phys. Rev. **39**: 2742 (1988).
76. A. Poniewierski and R. Holyst, Phys. Rev. Lett. **61**: 2461 (1988).
77. R. Holyst and A. Poniewierski, preprint.
78. M.P. Taylor, R. Hentschke and J. Herzfeld, Phys. Rev. Lett. **62**: 800 (1989), ibid: **62**: 1577 (1989).
79. M.P. Taylor, R. Hentschke and J. Herzfeld, Phys. Rev. A **40**: 1678 (1989).
80. J.M. Caillol and J.J. Weis, J. Chem. Phys. **90**: 7403 (1989).
81. T.J. Sluckin, Liquid Crystals **6**: 111 (1989).
82. T.E. Strzelecka, M.W. Davidson and R.L. Rill, Nature, **331**: 457 (1988).
83. F. Livolant, A.M. Levelut, J. Doucet and J.P. Benoit, Nature **339**: 724 (1989).
84. D. Frenkel, J. Phys. Chem. **92**: 3280 (1988).
85. D. Frenkel, H.N.W. Lekkerkerker and A. Stroobants, Nature **332**: 822 (1988).
86. J.A.C. Veerman and D. Frenkel, Phys. Rev. A **41**: 3237 (1990).
87. A.M. Somoza and P. Tarazona, Phys. Rev. A **41**: 965 (1990).
88. A.M. Somoza and P. Tarazona, J. Chem. Phys. **91**: 517 (1989).
89. D. Frenkel, Liquid Crystals, **5**: 929 (1989).
90. T. Boublik and I. Nezbeda, Coll. Czechoslovak. Chem. Commun. **51**: 2301 (1986).
91. J.A.C. Veerman, thesis, University of Utrecht, 1991. J.A.C.Veerman and D. Frenkel, to be published.
92. J. Billart, C.R. Acad. Sci. Paris, **305**: 843 (1987).
93. J.I. Siepmann, Mol. Phys. **70**: 1145 (1990).
94. D. Frenkel, J. Phys. Condensed Matter **2(SA)**: 265 (1990).
95. D. Frenkel, Physica A, submitted for publication.

Chapter 6

# RENORMALIZATION GROUP STUDIES OF STATICS AND DYNAMICS OF LIQUID CRYSTAL PHASE TRANSITIONS

C. DASGUPTA

## 1. Introduction

Liquid crystals exhibit a wide variety of continuous phase transitions between different mesophases. Due to divergent critical fluctuations, simple theoretical descriptions such as mean field theory and perturbation theory break down in the vicinity of these transitions. Static and dynamic behavior in the critical region may be analyzed by using Renormalization Group (RG) methods. The starting point of a RG calculation of the static critical behavior is the Ginzburg-Landau (GL) free energy expressed as a functional of the order parameter field. In some cases, due to symmetry or other considerations, terms coupling the order parameter to other non-ordering fields may have to be included in the GL functional. Standard methods[1] are used to obtain the RG recursion relations for the coupling constants appearing in the GL functional. Information about the critical properties is obtained from an analysis of these recursion relations. Such calculations have been carried out for several liquid crystal transitions, including the nematic to smectic A transition[2-5], the nematic-smectic A-smectic C multicritical point[6], various hexatic transitions in two and three dimensions[7-9] and transitions between different smectic A phases[10,11]. Results obtained from these studies are reviewed in Section 2.

The relaxation time associated with time-dependent fluctuations of the order parameter diverges at a continuous phase transition. This so-called "critical slowing down" is reflected in singular behavior of certain transport coefficients at the transition. The relaxational dynamics of the order parameter in the critical region may be studied by a RG analysis of appropriate time-dependent Ginzburg-Landau (TDGL) equations[1] derived from the GL free energy functional. Such calculations, however, are more complicated than studies of the static critical behavior because couplings of the order parameter field to all other slow (hydrodynamic) variables have to be taken into consideration. For this reason, the dynamic behavior at liquid crystal transitions has not been studied as extensively as the static critical behavior. Most of the existing studies of the critical dynamics[12-15] use approximate methods such as mode-coupling

C. Dasgupta - Department of Physics, Indian Institute of Science, Bangalore 560 012, India; and Jawaharlal Nehru Centre for Advanced Scientific Research, Bangalore 560 012, India.

*Phase Transitions in Liquid Crystals*, Edited by S. Martellucci and A.N. Chester, Plenum Press, New York, 1992

theory[16] and dynamic scaling theory [17]. Only a few RG studies of the dynamic properties near liquid crystal transitions have been carried out[18-20]. These calculations are discussed in Section 3.

## 2. Statics

In this Section, the general procedure for carrying out RG calculations of the static critical behavior is briefly described. Then, the results obtained from RG studies of various phase transitions in liquid crystal systems are summarized.

### *2.1. General Formalism*

An RG transformation maps a thermodynamic system into another one with the same structure, but with a smaller number of degrees of freedom and an increased length scale. The link between the two systems is obtained by the requirement that the partition function should remain unchanged under the RG transformation. Each application of the RG transformation constitutes a partial solution of the problem because it corresponds to summing over a fraction of the degrees of freedom. A full treatment is obtained by applying this transformation iteratively. In nearly all RG studies of phase transitions in liquid crystals, one starts by writing the partition function as a functional integral over continuous momentum-space variable $\{\psi_i(\vec{q})\}$:

$$Z = \int \prod_i \prod_{q<\Lambda} D\psi_i(\vec{q}) \exp[-f(\{\psi_i(\vec{q})\})] \tag{1}$$

Here, $\{\psi_i(\vec{q})\}$ are the Fourier transforms of the order parameter field, $i$ is a component index, $f$ is the reduced GL free energy functional, and $\Lambda$ is an appropriately chosen high momentum cut-off. An RG transformation is constructed by performing the following three steps:

1) All variables $\{\psi_i(\vec{q})\}$ with $\Lambda/b < q < \Lambda, b > 1$, are integrated over. The scale factor is chosen to be different in different direction in space [e.g. $b_\perp = b, b_\parallel = b^{1+\mu}$] for systems in which anisotropic scaling is expected. Denoting

$$\exp[-f''(\{\psi_i(\vec{q})\})] = \int \prod_i \prod_{\Lambda/b<q<\Lambda} D\psi_i(\vec{q}) \exp[-f(\{\psi_i(\vec{q})\})] \tag{2}$$

one then obtains

$$Z = \int \prod_i \prod_{q<\Lambda/b} D\psi_i(\vec{q}) \exp[-f''(\{\psi_i(\vec{q})\})] \tag{3}$$

where $f''$ is a functional of the variables $\{\psi_i(\vec{q})\}$ with $q < \Lambda/b$.

2) All momenta are rescaled by the factor $b$:

$$\vec{q} \rightarrow \vec{q}' = b\vec{q} \tag{4}$$

3) All the variables $\{\psi_i(\vec{q})\}$ are rescaled by a factor $b^{\zeta}$:

$$\psi_i(\vec{q}) \rightarrow \psi_i'(b\vec{q}) = \psi_i'(\vec{q}') = b^{-\zeta}\psi_i(\vec{q}) \tag{5}$$

The factor $\zeta$ is determined through the requirement that the transformation should preserve the magnitude of the basic order-parameter fluctuation. Introducing these two rescalings in $f''$ yields the renormalized free energy functional $f'$:

$$f'(\{\psi_i'(\vec{q}')\}) = f''(\{\psi_i(\vec{q})\}) \tag{6}$$

The coupling constants $\{K'_\alpha\}$ appearing in $f'$ are determined in terms of the ones of the original free energy functional through the RG recursion relations. These recursion relations may be written in a differential form by choosing $b = \exp(\delta l), \delta l \rightarrow 0$:

$$\frac{dK_\alpha}{dl} = g_\alpha(\{K_\beta\}) \tag{7}$$

The scale invariance at critical points implies that every critical point should correspond to a fixed point $\{K_\alpha^*\}$ of these recursion relations, determined from a solution of the equations

$$g_\alpha(\{K_\beta^*\}) = 0 \forall \alpha \tag{8}$$

The eigenvectors and eigenvalues of the derivative matrix $\partial g_\alpha / \partial K_\beta$ evaluated at the fixed point determine the scaling fields and their scaling dimensions. Scaling fields with eigenvalues greater than, less than, and equal to zero are respectively called relevant, irrelevant, and marginal. The values of the critical exponents are determined from the scaling dimensions of the relevant scaling fields.

A second order critical point corresponds to a fixed point with one relevant scaling field (the temperature) in the symmetric subspace. The absence of any accessible fixed point of this type is usually interpreted as the indication of a first order transition.

The results of a RG calculation of the type described above are usually obtained through an expansion in $\varepsilon = d_u - d$, where d is the dimension of space and $d_u$ is the so-called upper critical dimension. For $d > d_u$, the critical behavior is governed by a Gaussian fixed point, leading to mean field values of the critical exponents. The critical behavior for $d < d_u$ is controlled by a different fixed point whose location and values of critical exponents can be calculated as expansions in $\varepsilon$. Since $d_u \geq 4$ in most cases, $\varepsilon \geq 1$ for the physical dimension, d = 3. For this reason, most RG calculations using $\varepsilon$-expansion cannot provide quantitative information about the critical properties observed in experiments. Also, there exist a few cases (e.g. the nematic - smectic A transition discussed below) for which the critical behavior in three dimensions is qualitatively different from the prediction of RG studies carried out near the upper critical dimension.

Nevertheless, such calculations do, in most cases, provide valuable information about the general phase structure and about the qualitative effects of various factors on the nature of the transition.

## 2.2. *Survey of Results*

RG calculations of the static critical behavior have been carried out for a large number of transitions found in liquid crystal systems. The results obtained from some of these studies are summarized below.

### 2.2.1. *The Nematic to Smectic A(NA) Transition*

This is possibly the most extensively studied liquid crystal phase transition. Most theoretical studies of this transition have been based on the de Gennes free energy functional[21] which has the form

$$F = \int d\vec{r}[a\,|\psi|^2 + \frac{1}{2}u\,|\psi|^4 + C_{\parallel}\,|\nabla_z\psi|^2 + C_{\perp}\sum_{\mu=x,y}|(\nabla_\mu - iq_0\delta n_\mu)\psi|^2 +$$

$$+\frac{1}{2}K_1^0(\vec{\nabla}\cdot\delta\vec{n})^2 + \frac{1}{2}K_2^0(\nabla_x\delta n_y - \nabla_y\delta n_x)^2 + \frac{1}{2}K_3^0\{(\nabla_z\delta n_x)^2 + (\nabla_z\delta n_y)^2\}] \qquad (9)$$

where the complex scalar field $\psi(\vec{r})$ is the smectic A order parameter (the complex amplitude of a one-dimensional density wave with wave number $q_0$ in the z-direction), $\delta\vec{n}$ is a two-component real vector field describing small fluctuations of the nematic director about its equilibrium orientation in the z-direction, $K_1^0, K_2^0$ and $K_3^0$ are the "bare" Frank elastic constants, $C_{\parallel}$ and $C_{\perp}$ are "bare" stiffness constants, $u > 0$ is a phenomenological parameter and $a = (T - T_0)/T_0$ where $T_0$ is the mean field transition temperature. The coupling between $\psi$ and $\delta\vec{n}$ is introduced in order to make the free energy invariant under a small global rotation. Since the symmetry of the smectic A order parameter is the same as that describing the superfluid transition, the NA transition would belong to the universality class of the $\lambda$-transition in liquid He if the coupling to the director fluctuations were not present. Thus, the interesting question is how this coupling affects the critical behavior at the NA transition. The form of the coupling of $\psi$ to $\delta\vec{n}$ is very similar to the coupling of the vector potential to the superconducting order parameter in the GL theory of superconductivity. In fact, the de Gennes free energy for $K_1^0 = 0$ can be transformed by a change of gauge[3] to an anisotropic version of the GL free energy for superconductors. This analogy with the superconductig transition has been used extensively in theoretical studies of the NA transition. It is, however, necessary to understand the effects of the $K_1$ term on the critical behavior before one can draw any conclusion about the nature of the NA transition from the superconductor analogy.

Many years ago, Halperin, Lubensky and Ma[2] argued that the coupling of the order parameter to the vector potential should make the superconducting transition weakly first order. Their conclusion was based on fluctuation-corrected mean field theory and a RG calculation carried out near four dimensions on a generalized GL model in which the order parameter was assumed to have n/2 complex components. The RG calculation showed that for $n < 365.9$, there is no fixed point for a non-zero value of the coupling of the order parameter to the vector potential. This result, together with an analysis of the flow in the parameter space under iterations of the RG transformation, indicated a weakly first order transition. A similar RG analysis[3] of the de Gennes model predicted the same behavior for the NA transition. Experimentally, however, the NA transition often appears to be continuous.

A possible explanation of this discrepancy between theory and experiments was provided by Dasgupta and Halperin[22] who presented compelling analytic and numerical evidence indicating that type-II superconductors in three dimensions undergo a continuous phase transition belonging to the so-called "inverted XY" universality class. This study also showed that the nature of this phase transition may be understood in terms of the statistical mechanics of vortex loops which are the topological defects of three dimensional superconductors. The topological defects of the smectic A phase are dislocation loops. A continuum version of the dislocation loop model[23] of the NA transition was analyzed by Toner[4] using RG methods. His analysis showed that $K_1$ is irrelevant in the RG sense at the inverted XY fixed point, indicating that the NA transition should exhibit inverted XY critical behavior, at least for small values of $K_1^0$. Lubensky and co-workers investigated the possibility of anisotropic scaling and the gauge-dependence of the critical behavior in a series of papers[5,24,25]. These calculations showed that the experimentally observed fact that $K_l$ remains finite at the transition allows only two possibilities for the nature of the critical point: a) an isotropic critical point with inverted XY exponents and b) a critical point described by anisotropic scaling, with the coherence length exponents for fluctuations parallel and perpendicular to the direction of smectic ordering having the ratio 2:1. They also showed that the observed anisotropy of the coherence length exponents measured in X-ray scattering experiments does not necessarily imply anisotropic scaling for the underlying critical point. More recently, a detailed Monte Carlo study[26,27] of a lattice version of the de Gennes model has presented strong evidence indicating that the phase transition in this model belongs in the inverted XY universality class.

To summarize, the present theoretical view is that the de Gennes model in three dimensions exhibits a continuous phase transition in the inverted XY universality class. Unfortunately, all the experimental results are not in agreement with this prediction. A number of recent experiments[28,29] suggest that this discrepancy between theory and experiments may be caused by crossover effects associated with the presence of a nearby tricritical point in materials with narrow nematic ranges.

### 2.2.2. Smectic A - Smectic C (AC) Transition

In the smectic C phase, the molecules are tilted away from the direction of the layer normal. Since the azimuthal angle of the tilt direction is arbitrary, the order parameter describing the smectic C phase is a complex scalar. The AC transition, therefore, is expected[30] to be in the 3-d XY universality class. A detailed analysis[31] with a full treatment of both elastic and director fluctuations at this transition confirms this expectation. Experiments, however, show a behaviour which is well-described by mean field theory with an unusually large sixth order term[32]. A theoretical understanding of this behaviour is lacking at the present time.

### 2.2.3. The Nematic-Smectic A - Smectic C (NAC) Multicritical Point

In some liquid crystal mixtures exhibiting nematic, smectic A and smectic C phases, the NA, AC and NC transition lines meet at the NAC multicritical point. Several years ago, Chen and Lubensky[33] proposed a model for the critical behavior in the neighborhood of this special point. This model uses the fact that the X-ray scattering intensity in the nematic phase near the NA transition shows strong peaks at wave numbers $\vec{q}_a = \pm q_0 \vec{n}$, whereas near an NC transition, the peaks spread into rings at $\vec{q}_c = (\pm q_{\parallel}, q_{\perp}\mathrm{Cos}\,\phi, q_{\perp}\mathrm{Sin}\,\phi)$, where the subscript symbols refer to directions parallel and perpendicular to the nematic director $\vec{n}$, and $0 \le \phi \le 2\pi$. The model free energy, expressed in terms of the mass density field and the director field, is constructed in a way that ensures that the free energy is minimized if the mass density is periodic with wave number $\vec{q}_a$ for the NA transition, and with wave number $\vec{q}_c$ for the NC transition. This free energy reduces to the de Gennes form in the vicinity of the NA transition, the AC transition in this model is in the 3-d XY universality class and the NC transition in this model is first order. The NAC multicritical point in this description corresponds to the special point at which both the "mass" term and the coefficient of the $q_\perp^2$ term appearing in the free energy functional vanish. In the terminology of critical phenomena, this point corresponds to a d = 3, m = 2 Lifshitz point. The critical behavior at a Lifshitz point of this kind has been analyzed[34] by RG methods and the critical exponents have been calculated as an expansion in $\varepsilon = 5 - d$.

An alternative description on the NAC point was proposed by Grinstein and Toner[6] who constructed a model in which the NA transition is described by the dislocation loop model of Toner[4] and the smectic C order is described by the usual complex tilt order parameter. RG arguments suggest that the term that couples the two sets of order parameter fields is irrelevant in three dimensions, indicating that the NA and AC transition lines cross at the NAC point without affecting each other. Another interesting aspect of this theory is the prediction of a new phase that intervenes between the nematic and the smectic C phases on one side of the NAC point (the smectic A phase lies between the nematic and smectic C phases on the other side of the NAC point). In this so-called biaxial nematic ($N'$) phase, the tilt order parameter is non-zero, but the smectic layering is absent. The $NA$ and $N'C$ transitions in this model are in the inverted XY universality class, and the AC and $NN'$ transitions exhibit regular XY critical behavior. The NAC point is predicted to be a decoupled tetracritical point, with no special critical behavior.

Most of the results obtained from experiments[35] in the vicinity of the NAC point agree with the main features of the Chen-Lubensky model, although there are some discrepancies which are not explained. The $N'$ phase predicted in the theory of Grinstein and Toner has not yet been observed in experiments.

### *2.2.4. Hexatic Transitions in Two Dimensions*

The theory of two dimensional melting developed by Halperin and Nelson[36] and by Young[37] predicts the existence of a new phase that intervenes between the solid and isotropic liquid phases. In this so-called hexatic phase, the positional order is short ranged, but the bond-angle correlation function decays algebraically with distance. The transition from the hexatic to the isotropic liquid phase in two dimensions has been analyzed by using a RG procedure which is somewhat different from the general one discussed above. This phase transition is driven by the unbinding of pairs of oppositely "charged" point disclinations which are the topological defects of the bond-angle field. The RG treatment, which is similar to the Kosterlitz-Thouless theory[38] of the phase transition in the two-dimensional XY model, involves deriving RG recursion relations for a generalized "stiffness" parameter for the bond-angle field and a fugacity parameter for the point defects. In the hexatic phase, the fugacity parameter renormalizes to zero at long length scales, indicating that all the defects are paired. Above the transition temperature, the fugacity increases under RG iterations, signalling a proliferation of unbound disclinations and a consequent loss of bond orientational order. This theory predicts a broad heat capacity peak and an universal jump in the value of the stiffness constant at the transition.

In smectic films where the molecules are tilted away from the layer normal (e.g. films of materials which exhibit a smectic C phase in the bulk), the coupling between the two-component vector field that describes the projection of the tilt axis on the plane of the layer and the bond-angle field leads to a rich variety of tilted hexatic phases. The phase diagram for this system has been analyzed by Nelson and Halperin[7] by using RG methods. The starting point of their analysis is an effective Hamiltonian functional of the bond-angle field $\theta(\vec{r})$ and the tilt-azimuthal-angle field $\phi(\vec{r})$:

$$H/k_BT = \frac{1}{2}\int d^2r[K_6\,|\,\vec{\nabla}\theta\,|^2 + K_1\,|\,\vec{\nabla}\phi\,|^2 + 2g\vec{\nabla}\theta\cdot\vec{\nabla}\phi] - h\int d^2r\,\mathrm{Cos}[6(\theta-\phi)] \qquad (10)$$

Kosterlitz-Thouless type recursion relations are then constructed for the stiffness constants $K_6$ and $K_1$, the gradient cross coupling g, the direct coupling h (the coupling term has this particular form because both $\theta(\vec{r})$ and $\phi(\vec{r})$ feel a sixfold symmetric potential if one field is rotated with the other one held fixed), and the fugacity parameters for the disclinations and the point defects (vortices) of the $\phi$ field. Analysis of these recursion relations leads to a phase diagram with an isotropic liquid phase, an untilted hexatic phase in which the $\theta$ field has quasi-long-range order and the $\phi$ field has only short-range order, and two tilted hexatic phases in which both $\theta$ and $\phi$ fields have quasi-long-range order. In one of these two tilted hexatic phases, the long-wavelength fluctuations of the $\theta$ and $\phi$ fields are locked together; in the other phase, these fluctuations are independent. The phase transition between any two of these phases is predicted to be of the Kosterlitz-Thouless type.

The theory of hexatic-to-hexatic transitions in two dimensions has been refined by Selinger and Nelson[8] who studied the effects of including a term of the form $h'\,\mathrm{Cos}[12(\theta(\vec{r})-\phi(\vec{r}))]$ in the Hamiltonian of Eq. 10. They find three different locked tilted hexatic phases: 1) the hexatic-I phase in which the the local tilt-azimuthal angle is along one of the local bonds, 2) the hexatic-F phase in which the local tilt-azimuthal angle is 30° away from a local bond, and 3) the hexatic-L phase in which the tilt angle varies continuously between 0 and 30° away from the local bond. From a RG analysis of the "spin wave" part of the Hamiltonian, they obtained a phase diagram which is in qualitative agreement with experiments.

### *2.2.5. Hexatic transitions in three dimensions*

All the hexatic phases described above, except the unlocked tilted one, have their three dimensional counterparts. The presence of interlayer coupling in the three dimensional systems changes the quasi-long-range order found in two dimensions to true long-range order. The three dimensional versions of the isotropic liquid, untilted hexatic and locked tilted hexatic phases found in two dimensions are, respectively, the smectic A, hexatic B and smectic C, I and F phases (the last three phases have the same macroscopic symmetry because the presence of long-range tilt-angle order induces a long-range bond-orientational order). Since the bond orientational order parameter $\psi(\vec{r})$ (= $\exp[6i\theta(\vec{r})]$ where $\theta(\vec{r})$ is the angle a nearest-neighbor bond makes with a reference direction) is a complex scalar, the smectic A to hexatic B transition is expected to belong to the three-dimensional XY universality class. Heat capacity measurements[39] at this transition, however, consistently show a specific heat exponent $\alpha\sim 0.6$, in contradiction with the 3-d XY value, $\alpha\sim -0.02$. Also, X-ray scattering studies[40] show the existence of local herring-bone packing of the molecules near this transition. Motivated by these facts, Bruinsma and Aeppli[9] proposed a model in which the bond-orientational order parameter is coupled to a second complex scalar field $\phi(\vec{r})$ that describes local herring-bone order ($\phi(\vec{r}) = \exp[2i\theta'(\vec{r})]$, where $\theta'(\vec{r})$ is the angle between the local herring-bone axis and the reference direction). The model free energy they propose has the form

$$F/k_BT = \int d\vec{r}[\frac{1}{2}|\vec{\nabla}\psi|^2 + \frac{1}{2}r_6|\psi|^2 + u_6|\psi|^4 + \frac{1}{2}|\vec{\nabla}\phi|^2 +$$

$$+\frac{1}{2}r_2|\phi|^2 + u_2|\phi|^4 + w|\phi|^2|\psi|^2 + hRe(\psi^*\phi^3)] \tag{11}$$

A RG analysis carried out near four dimensions on the behavior of this model suggests that the fluctuations in the $\phi$ field may drive the smectic A-hexatic B transition first order, thus leading to the existence of a tricritical point on the transition line. This calculation, thus, suggests that the observed large value of the exponent $\alpha$ may be a consequence of the proximity to a tricritical point for which $\alpha = 0.5$. It is interesting to note in this context that recent experiments[41] on the smectic A-hexatic B transition in three and four layer free-standing films show a divergence of the heat capacity with an exponent $\alpha \sim 0.3$, in contradiction with the prediction of the disclination unbinding theory.

Another interesting application[42] of RG methods to the study of hexatic transitions in three dimensions involves a calculation of the 6n-fold order parameters $C_{6n} = Re < \psi^n >$. From an expansion in $\varepsilon = 4 - d$, it is found that

$$C_{6n} = [C_6]^{\sigma_n}, \qquad \sigma_n = n + \lambda n(n-1) \tag{12}$$

where $\lambda$ depends weakly on $n$. This prediction has been beautifully confirmed in X-ray scattering experiments[43].

### *2.2.6. Transitions between different modulated smectic A phases*

There exist a number of liquid crystals in which the bare molecular length $l$ is not the sole determining factor of the period d of the one-dimensional density wave in the smectic A phase. In these materials, different smectic A phases with $d = l$ (smectic $A_1$), $d = 2l$ (smectic $A_2$) and $l < d < 2l$ (smectic $A_d$) have been found and transitions between these phases have been studied experimentally. All these phases have the same macroscopic symmetry. However, first order transitions in which the layer spacing changes discontinuously are always possible. Since the layer spacing in the $A_2$ phase is an integral multiple of the layer spacing in the $A_l$ phase, a continuous $A_l$ - $A_2$ transition is also allowed. Some of these transitions have been studied by RG methods [10,11], using a free energy functional expressed in terms of two complex scalar fields, representing the complex amplitudes of one-dimensional smectic density waves with the two periods, and the director field. The predictions of these calculations include a continuous smectic $A_l$-smectic $A_2$ transition in the Ising universality class and a critical end-point terminating a first order transition line between two diffferent smectic A phases. Some of the qualitative predictions of these calculations are in agreement with experimentally observed results[44].

## 3. Dynamics

In this Section, the general procedure of performing dynamic RG calculations is outlined and the difficulties involved in carrying out such calculations for liquid crystal phase transitions are discussed. The few existing RG studies of the critical dynamics at liquid crystal transitions are then reviewed.

### 3.1. General Formalism

The relaxation time of order parameter fluctuations diverges at a continuous phase transition. This divergence is characterized by the dynamic critical exponent z:

$$\tau \sim (\xi)^z \sim |t|^{-\nu z} \tag{13}$$

where $\xi$ is the coherence length, $\nu$ is the coherence length exponent and t is the reduced temperature. The dynamic exponent z also appears in the dynamic scaling hypothesis[17] which states that the frequency and wave-number dependence of a general correlation function is given by the following scaling form in the critical region:

$$g(\vec{k},\omega) = \xi^{\mu} g'(k\xi, \xi^z \omega) \tag{14}$$

where the exponent $\mu$ is related to the scaling dimensions of the fields whose correlation function is being considered. Using this hypothesis in conjunction with Kubo formulae which relate transport coefficients to appropriate correlation functions, one may obtain useful information about the critical behavior of certain transport coefficients. For a calculation of the exponent z, it is necessary to carry out a dynamic RG calculation.

The starting point of a RG calculation of the critical dynamics is a set of kinetic equations which describe the time evolution of the slow (hydrodynamic) variables of the system. In the simplest situation, one considers only the order parameter field whose relaxational dynamics is described by a TDGL equation of the form

$$\frac{\partial \psi_i(\vec{q},t)}{\partial t} = -\Gamma(q)\frac{\delta F(\{\psi_i(\vec{q},t)\})}{\delta \psi_i(-\vec{q},t)} + \xi_i(\vec{q},t) \tag{15}$$

where $\Gamma(q) = \Gamma_0$ (constant) if the order parameter is not conserved and $\Gamma(q) = \Gamma_0 q^2$ for a system with a conserved order parameter, F is the appropriate GL free energy and the $\xi_i$'s are random Gaussian noise variables with the statistics

$$< \xi_i(\vec{q},t)\xi_j(\vec{q}',t') >= 2k_B T\Gamma(q)\delta_{ij}\delta(\vec{q}+\vec{q}')\delta(t-t') \tag{16}$$

The parameter space on which the dynamic RG transformation operates includes the parameter $\Gamma_0$ in addition to all those appearing in F. A RG transformation with a scale factor b consists of the following two steps:

1) All variables with $\Lambda/b < q < \Lambda$ are eliminated. This involves solving the kinetic equations for the variables $\psi_i(\vec{q})$ with $q > \Lambda/b$, substituting the solutions in the remaining equations for $\psi_i(\vec{q})$ with $q < \Lambda/b$, and then integrating over the noise variables with $q > \Lambda/b$.

2) In the remaining equations for $\psi_i(\vec{q}), q < \Lambda/b$, the momentum, time, field and noise variables are rescaled as follows:

$$\vec{q} \to \vec{q}' = b\vec{q}$$

$$t \to t' = b^{-z}t$$

$$\psi_i(\vec{q},t) \to \psi'_i(\vec{q}',t') = b^{-\zeta}\psi_i(\vec{q},t)$$

$$\xi_i(\vec{q},t) \to \xi_i'(\vec{q}',t') = b^{-\zeta+z}\xi_i(\vec{q},t) \tag{17}$$

The resulting kinetic equations define the renormalized parameters. The dynamic exponent is determined from the requirement that the RG transformation in the expanded parameter space must have non-trivial fixed points.

The procedure described above is a straightforward generalization of static RG. However, dynamic RG calculations for physical systems are usually more complicated than their static counterparts. The difficulties arise from the fact that the simple TDGL equation of Eq. 15 is almost never sufficient for a complete description of the dynamics. Several factors not included in Eq. 15, which are usually not important in a description of the static critical behavior, may affect the critical dynamics in significant ways. These factors include:

1) Reactive terms: In many cases, reactive (reversible) terms, such as the one that describes the precession of the local magnetization in an isotropic ferromagnet, have to be included in the kinetic equation in addition to the dissipative TDGL term.

2) Couplings to other slow variables: For a complete description of the dynamics at long time scales, one should, in principle, take into account the couplings of the order parameter to all other hydrodynamic variables which, depending on the system, may include the mass density, the momentum density and the energy density. The presence of these couplings lead to singular behavior of transport coefficients such as viscosities and thermal conductivity at a critical point. The form of non-dissipative mode-coupling terms may be derived from the Poisson bracket formalism[17]. Often, the specific forms and strengths of the couplings appearing in the free energy cannot be determined from first principles. The standard procedure is to include all couplings allowed by symmetry and to treat their coefficients as phenomenological parameter.

Thus, for a complete description of the critical dynamics, it is, in principle, necessary to consider a set of coupled kinetic equations which describe the time evolutions of the order parameter field and all other hydrodynamic variables of the system. All reversible and dissipative terms, including all cross-coupling terms allowed by symmetry, should be incorporated in these equations. A RG treatment of these equations may be formulated by generalizing the procedure described above. In practice, however, the calculations become rather complicated because the parameter space on which the RG transformation acts acquires a large dimension.

This problem is particularly severe in liquid crystal systems because the anisotropic nature of most liquid crystalline phases leads to a further increase in the number of parameters one has to keep track of.

For example, the equation of motion of the momentum density in the nematic phase involves five independent viscosity coefficients instead of the two needed for an isotropic fluid. For this reason, most studies of the critical dynamics at liquid crystal transitions [12-15,45,46] have been based on approximation methods such as mode-coupling theory (which is basically a perturbation theory) and dynamic scaling. The few existing RG studies of the dynamic critical behavior are discussed below.

### 3.2. Survey of Results

A few dynamic RG calculations of the critical dynamics at the NA transition have been reported in the literature. The first calculation was performed by Shiwa[18] who considered a simplified version of the de Gennes model with $K_1^0 = 0, C_\perp = C_\parallel$ and $K_2^0 = K_3^0$. This model can be mapped exactly onto the GL theory of superconductivity by a gauge transformation. Shiwa considered TDGL equations for the order parameter and the director field in this so-called "superconducting" gauge, neglecting all couplings of these fields to other hydrodynamic variables. From a dynamic RG calculation near four dimensions on a generalization of this model in which the order parameter is assumed to have n/2 complex components, he calculated the dynamic exponent z as an expansion in $\varepsilon = 4 - d$ at the "superconducting" fixed point which exists for $n > 365.9$, and at a second fixed point which is obtained if fluctuations in the phase of the order parameter field are neglected. Since none of these two fixed points is believed to describe the NA transition in three dimensions, this calculation is not relevant to the experimental situation.

A more detailed RG calculation on the dynamics at the NA transition was performed by Hossain and Swift[19] who considered the full de Gennes model and took into account the couplings of the $\psi$ and $\delta\vec{n}$ fields with the velocity field. They argued that calculations of the dynamics should be performed in the physical liquid crystal gauge. Divergent fluctuations of the phase of the order parameter prevented them from analyzing the critical behavior near fixed points with $K_1^* = 0$. They calculated the exponent z to first order in $\varepsilon = 4 - d$ at a fixed point with $K_1^* = \infty$. This fixed point, which exists for $n > 238.2$, exhibits anisotropic scaling with different exponents for the coherence lengths parallel and perpendicular to the direction of smectic order. Since the current theoretical understanding of the NA transition in three dimensions suggests that it is described by a fixed point with $K_1^* = 0$, the results obtained in this study are probably not applicable to experimental observations. Also, this calculation did not take into account the density-order parameter coupling which was later shown[13,14] to be important in understanding the experimentally observed critical behavior of the sound velocity and attenuation at the NA transition.

Another application of dynamic RG theory to liquid crystal transitions involves the uniaxial to biaxial transition[20] in nematics. In this calculations, RG techniques are used to argue that the critical dynamics at this transition should correspond to that of a system with a non-conserved order parameter which is not coupled to any other slow variable. Therefore, the transport coefficients are not expected to show any singularity at this transition.

It is clear from this survey that applications of dynamic RG to phase transitions exhibited by liquid crystals have been rather limited. More work in this direction would be most welcome.

## References

1. S.K. Ma, "Modern Theory of Critical Phenomena", W.A. Benjamin Inc., Reading, Massachusetts, (1976).
2. B.I. Halperin, T.C. Lubensky and S.K. Ma, Phys. Rev. Lett. **32**: 292 (1974).
3. B.I. Halperin and T.C. Lubensky, Solid St. Comm. **14**: 997 (1974).
4. T.C. Lubensky and J.H. Chen, Phys. Rev. B **17**: 366 (1978).
5. J. Toner, Phys. Rev. B **26**: 462 (1982).
6. G. Grinstein and J. Toner, Phys. Rev. Lett. **51**: 2386 (1984).
7. D.R. Nelson and B.I. Halperin, Phys. Rev. B **21**: 5312 (1980).
8. J.V. Selinger and D.R. Nelson, Phys. Rev. Lett. **61**: 416 (1988).

9. R. Bruinsma and G. Aeppli, Phys. Rev. Lett. **48**: 1625 (1982).
10. J. Wang and T.C. Lubensky, Phys. Rev. A **29**: 2210 (1984).
11. Y. Park, T.C. Lubensky, P. Barois and J. Prost, Phys. Rev. A **37**: 2197 (1988).
12. K.A. Hossain, J. Swift, J.H. Chen and T.C. Lubensky, Phys. Rev. B **19**:432 (1979).
13. J. Swift and B. Mulvaney, J. Phys. (Paris) Lett. **40**: L-287 (1979).
14. J. Swift and B. Mulvaney, Phys Rev. B **22**: 4523 (1980).
15. B.S. Andereck and J. Swift, Phys. Rev. A **25**: 1084 (1982).
16. K. Kawasaki, Ann. Phys. (N.Y.) **61**: 1 (1970).
17. P.C. Hohenberg and B.I. Halperin, Rev. Mod. Phys. **49**: 435 (1977).
18. Y. Shiwa, Prog. Theor. Phys. **55**: 629 (1976).
19. K.A. Hossain and J. Swift, (Paris), **40**: 1055 (1979).
20. C.A. Cajas, J.B. Swift and H.R. Brand, Phys. Rev. A **30**: 1579 (1984).
21. P.G. de Gennes, Solid St. Comm. **10**: 753 (1972).
22. C. Dasgupta and B.I. Halperin, Phys. Rev. Lett. **47**: 1556 (1981).
23. C. Dasgupta, Phys. Rev. A **27**: 1262 (1983).
24. T.C. Lubensky, S.G. Dunn and J. Issacson, Phys. Rev. Lett. **47**: 1609 (1981).
25. S.G. Dunn and T.C. Lubensky, J. Phys. (Paris) **42**: 1201 (1981).
26. C. Dasgupta, Phys. Rev. Lett. **55**: 1771 (1985).
27. C. Dasgupta, J. Phys. (Paris) **48**: 957 (1987).
28. J. Thoen, H. Marynissen and W. van Dael, Phys. Rev. Lett. **52**: 204 (1984).
29. B.M. Ocko, R.J. Birgeneau, J.D. Litster and M.E. Neubert, Phys. Rev. Lett. **52**: 208 (1984).
30. P.G. de Gennes, Mol. Cryst. Liq. Cryst. **15**: 49 (1973).
31. G. Grinstein and R.A. Pelcovits, Phys. Rev. A **26**: 2196 (1982).
32. C.C. Huang and J.M. Viner, Phys. Rev. A **25**: 3385 (1982).
33. J.H. Chen and T.C. Lubensky, Phys. Rev. A **14**: 1202 (1976).
34. R.M. Hornreich, M. Luban and S. Shtrikman, Phys. Rev. Lett. **35**: 1678 (1975).
35. L.J. Martinez-Miranda, A.R. Kortan and R.J. Birgeneau, Rev. Lett. **56**: 2264 (1986).
36. B.I. Halperin and D.R. Nelson, Phys. Rev. Lett. **41**: 121, **41**: 519(E) (1978).
37. A.P. Young, Phys. Rev. B **19**: 1855 (1979).
38. J.M. Kosterlitz and D.J. Thouless, J. Phys. C **6**: 1181 (1973).
39. C.C. Huang, S.M. Viner, R. Pindak and J.M. Goodby, Phys. Rev. Lett. **46**: 1289 (1981).
40. R. Pindak, D.E. Moncton, S.C. Davey and J.W. Goodby, Phys. Rev. Lett. **46**: 1135 (1981).
41. R. Geer, T. Stoebe, C.C. Huang, R. Pindak, G. Srajer, J.W. Goodby, M. Cheng, J. T. Ho and S.W. Hui, Phys. Rev. Lett. **66**: 1322 (1991).
42. A. Aharony, R.J. Birgeneau, J.D. Brock and J.D. Litster, Phys. Rev. Lett. **57**: 1012 (1986).
43. J.D. Brock, A. Aharony, R.J. Birgeneau, K.W. Evans-Luttrodt, J.D. Litster, P. M. Horn, G.B. Stephenson and A.R. Tajbaksh, Phys. Rev. Lett. **57**: 98 (1986).
44. K.K. Chan, P.S. Pershan, L.B. Sorensen and F. Hardouin, Phys. Rev. A **34**: 1420 (1986).
45. F. Brochard, J. Phys. Colloq. **37**: C3-85 (1976).
46. W.L. McMillan, Phys. Rev. A **9**: 1720 (1974).

Chapter 7

# MICROSCOPIC LIQUID CRYSTAL THEORY OF NEMATIC REENTRANCE, SMECTIC C ORDERING, AND IN-PLANE DOMAIN FORMATION

R.R. NETZ and A.N. BERKER

## 1. Introduction

The frustrated spin-gas model with dipolar, van der Waals, and benzene-ring steric interactions is studied by Monte Carlo. Reentrant phase diagrams are obtained, with the sequence nematic-smectic $A_d$-nematic-smectic $A_1$-smectic C. The reentrance mechanism is in terms of interpenetrating order and disorder. Layer tilting is found to be due to permeation-rotation lock-in. The smectic $A_1$ and C phases occur in two versions, one pointing to in-plane domain formation. The phase diagrams are obtained by adapting the Lindemann melting criterion. Layer thicknesses, tilts, specific heats, and dimer concentrations are evaluated. The model explains three types of smectic C phases distinguished by tilt saturation. Results qualitatively agree with experiments.

## 2. Reentrance Phenomena

The occurrence of a less ordered phase at temperatures below those of a more ordered phase is, at first glance, surprising. Elementary thermodynamics has imprinted in us that the free energy, F=U-TS, is minimized at low temperatures T by minimizing the internal energy U, which is usually accomplished by the system forming an ordered array, and at high temperatures by maximizing the entropy S, which is accomplished by the system sampling all of its possible states, abandoning the subspace of order. We should hasten to reassure that elementary thermodynamics still works: the paradox mentioned in the first sentence above is only an apparent one, and is conceptually dissolved by identifying long-range order and short-range order, which replace each other in these systems.

Liquid crystals are the physical systems in which this phenomenon of reentrance (the disordered phase of high temperatures "reenters" at temperatures below the intermediate temperatures of the ordered phase) has received much attention, after the experimental discovery by Cladis.[1] Reentrance also occurs in very diverse other systems such as liquid mixtures,

---

R.R. Netz and A.N. Berker - Massachusetts Institute of Technology, Cambridge, Massachusetts 02139, USA

spin-glasses, surface adsorbates, and magnets with crystal fields. It has turned out that the theoretical challenge posed by reentrant phase transitions is the converse of the challenge of critical exponent universality of two decades ago: Whereas universality of critical exponents in very diverse systems was eventually understood by the single mechanism of irrelevant thermodynamic fields in renormalization-group flows toward a fixed point,[2] reentrant phase transitions have required a detailed microscopic understanding and model solving for each diverse system:

(1) In liquid mixtures, the orientational correlations of the lower temperatures change the effective net interspecies interactions from repulsive to attractive, via hydrogen-bonding in mutually oriented interspecies pairs, thereby causing the phase separation of intermediate temperatures to reverse itself.[3]

(2) In spin-glasses, the saturation of short-range ferromagnetic order destroys long-range antiferromagnetic correlations and, simultaneously and conversely, the saturation of short-range antiferromagnetic order destroys long-range ferromagnetic correlations.[4]

(3) In krypton adsorbed on a graphite substrate, cooling is necessary for enough adsorption from the vapor to establish connectivity and (commensurate) order on the surface; but further cooling and, thereby, further adsorption introduces heavy and superheavy domain walls that disrupt and eventually destroy this order; even further cooling and adsorption creates a network of superheavy domain walls, which is incommensurate order.[5]

(4) In magnets with a quenching crystal field, ferromagnetic order can disappear at low temperatures due to the removal of local moments by the crystal field.[6]

In liquid crystals, the closepacking inherent to the liquid inescapably leads to frustration with molecules that have dipole moments:[7] in many local positional configurations of the molecules, different local orientational configurations equally minimize the energy. Another microscopic characteristic is molecular corrugation, amounting to a mutual substrate effect in nearest-neighbor molecules. The frustrated spin-gas model[7] is a microscopic model that begins with incorporating these two characteristics, dipolar frustration and molecular corrugation. An approximate solution of the model has yielded reentrant phase diagrams,[7] in qualitative agreement with experiments, up to the quadruply reentrant case of nematic-smectic $A_d$-nematic-smectic $A_d$-nematic-smectic $A_1$ as temperature is lowered.[8] Calculated relative specific heats[9] and layer thicknesses,[8] and molecular dimer concentrations[10] have also been in qualitative agreement with experiments.

We have considerably further developed[11] the frustrated spin-gas model:

(1) The solution of the model has been made less ad hoc, at the price of increased computation, by the use of Monte Carlo sampling. The phase diagrams are calculated by adapting to the nematic-smectic transition the Lindemann criterion of melting. Reentrant phase diagrams are obtained, and the microscopic mechanism for the intermediate-temperature existence of the smectic $A_d$ phase has become apparent, in terms of the interpenetrating coexistence of order and disorder. Specific heats, layer thicknesses, and molecular dimer concentrations are evaluated.

(2) The model itself has been further developed by incorporating the aspect ratio of the benzene rings in the molecular cores and the resultant steric effects, in addition to the dipolar frustration. Thus, the smectic C phase, in the sequence nematic-smectic $A_d$-nematic-(smectic $A_1$-)smectic C as temperature is lowered, has been obtained and the microscopic mechanism for smectic C layer tilting has been found. It is the lock-in of permeational and rotational degrees of freedom, which are respectively influenced by dipolar interactions and benzene-ring steric hindrances. Layer tilt angles are evaluated. The microscopic model explains the long-standing catalogued experimental data[12] of three types of smectic C phases distinguished by tilt saturation behavior.

(3) Moreover, the van der Waals intermolecular interactions have been more realistically included, in addition to the dipolar interactions. The consequence is two separate versions of the smectic $A_1$ and C phases, based on the competition between the dipolar and van der Waals interactions, one version of each phase pointing to in-plane domain formation ($\tilde{A}$ and $\tilde{C}$), in agreement with the experimental trend. All these results are in good qualitative agreement with experiments.

## 3. A Microscopic Model for Dipolar Liquid Crystals: The Frustrated Spin-Gas Model

The model has been constructed to investigate the minimal microscopic ingredients needed for each qualitatively distinct experimental feature. Thus, it was possible to address separately the microscopic mechanisms for distinct phenomena (Table I). It was also possible to test the model, satisfactorily, by correlating the occurrence of these phenomena with molecular dipolarity and molecular structure. For an alternate and complementary approach[13,14], using Landau and renormalization-group theories, Chapters 4 and 6 of this volume should be consulted.

Table I. The Frustrated Spin-Gas Model

| | Minimal microscopic requirements | Results | Mechanisms | References |
|---|---|---|---|---|
| (1): | Dipolar interactions and molecular corrugation | $N$-$A_d$-$N$ reentrance[a] | Relief of dipolar frustration by interpenetrating coexistence of order and disorder[b] | [a]Ref. 7 [b]Ref. 11 |
| (2): | (1) with van der Waals modification | $N$-$A_d$-$N$-$A_1$ reentrance | Relief of dipolar frustration by (1) and by local tilt of molecules | Ref. 7 |
| (3): | (2) with librational permeation | $N$-$A_d$-$N$-$A_d$-$N$-$A_1$ reentrance | Relief of dipolar frustration by (2) and by librations | Ref. 8 |
| (4): | (1) with van der Waals interactions | In-plane domains | Two versions of $A_1$, C from dipolar/van der Waals competition | Ref. 11 |
| (5): | (4) with core benzene aspect ratio | Smectic C order | Lock-in of permeations and rotations | Ref. 11 |

### *3.1. Basic Ingredients: Dipolar Frustration and Corrugation*

Consider, in a system that is in the nematic or smectic phases, a cross-section that is normal to the average molecular axis. Since the system is a liquid (a con*dense*d phase), it is closepacked, locally approximating a triangular structure. Accordingly, for molecules with a dipole, the system is necessarily frustrated: In each nearest-neighbor triplet of molecules, for many positions of the molecules along the average axis direction, different orientational configurations equally minimize the dipolar energy.

To start with this basic property, consider molecules that occupy n consecutive sites along the z direction in a three-dimensional stacked triangular lattice. In calculations to-date, each molecule has been assigned a fixed dipole pointing along the axis and located at an end segment. This choice can readily be modified in future calculations, as called by actual molecular structures. The molecules interact via the dipole-dipole interaction

$$V_d(\vec{r},\hat{s}_1,\hat{s}_2) \quad = \quad B[\hat{s}_1 \cdot \hat{s}_2 - 3(\hat{s}_1 \cdot \hat{r})(\hat{s}_2 \cdot \hat{r})] \,/\, r^3 \tag{1}$$

where $\vec{r}$ is the separation between dipoles 1 and 2, $\hat{r} = \vec{r}/|\vec{r}|$, and $\hat{s}_1$ and $\hat{s}_2$ are unit vectors giving the orientations of the respective dipoles. The other two types of interactions are the van der Waals interactions (Secs. 3.2, 4.1.2) and the benzene-ring steric hindrances (Sec. 4.1.3).

The segmentation of the molecules into n units does have a physical basis: Due to their chemical structure, two neighboring molecules have preferred positions of mutual displacement along the molecular axes. This corrugation can be due to either the carbon atoms in the tail or the benzene rings in the core, depending on the molecular species.

The above two ingredients are sufficient to yield singly reentrant phase diagrams (nematic-smectic $A_d$-nematic), as an approximate calculation using a mapping onto the distorted Ising model indicates.[7]

### *3.2. Additional Ingredients: van der Waals Interactions and Librational Permeation*

In addition to the dipole-dipole interactions, the molecules interact via van der Waals interactions. An approximate account of this was first taken by modifying[7] the potential of Eq.1 as

$$V_{dt}(\vec{r},\hat{s}_1,\hat{s}_2) \quad = \quad [A\hat{s}_1 \cdot \hat{s}_2 - 3B(\hat{s}_1 \cdot \hat{r})(\hat{s}_2 \cdot \hat{r})] \,/\, r^3 \tag{2}$$

which reflects the addition of the simple tail-tail interaction

$$V_t(\vec{r},\hat{s}_1,\hat{s}_2) \quad = \quad (A-B)\,\hat{s}_1 \cdot \hat{s}_2 / r^3 \tag{3}$$

so that A>B or B<A correspond to net tail-tail repulsion or attraction, respectively. The spatial dependence of Eq. 3 was chosen for calculational convenience. More complicated functions can be used. This development is sufficient to obtain,[7] by the approximate mapping procedure, doubly reentrant phase diagrams (nematic-smectic $A_d$-nematic-smectic $A_1$). Good qualitative agreement is obtained with experiment, as seen in Figs. 1 of Ref. 10. The original approximation predicted a zero-temperature smectic $A_d$-smectic $A_1$ transition, which moves to finite temperature with improved mapping approximations.[15]

Further attention has been given to the displacements of the molecules, away from the stacked-triangular lattice positions. Firstly, deviations within the xy plane from the triangular positions were studied. It was found[7] that these in-plane positional fluctuations (1) by themselves do not produce reentrance, (2) do not remove the reentrance caused by permeation fluctuations. Thus, they have been ignored, justifiably by the current purposes of the theory. On the other hand, out-of-plane positional fluctuations, called *permeations*, are already accounted in the basic theory (Sec. 3.1). In these, the discrete lattice positions along the z direction are justified by the corrugation of the molecules and the resultant preferred permeation positions. These energy minima of displacements along the z direction are separated by atomic length scales. Thus, this is called *atomic permeation*. However, for each atomic permeation configuration, permeation fluctuations on smaller length scales, corresponding to thermal fluctuations within a single corrugation minimum, can occur and affect dipolar frustration. Such fluctuations are called *librational permeation* and were included in the model.[7]

The latter development is sufficient to obtain,[8] by the same approximate mapping procedure, quadruply reentrant (nematic-smectic $A_d$-nematic-smectic $A_d$-nematic-smectic $A_1$) phase diagrams, in qualitative agreement with experiments[16] as shown in Figs. 2 of Ref. 10. It is noteworthy that, both among all the compounds that exhibit nematic/smectic reentrance and in the broad range of these model systems (spanned by A/B, n values), quadruple reentrance occurs in a very narrow range. In fact, very good agreement is obtained, in the occurrence of quadruple reentrance, between the number of segments n in the model and the number of tail carbon atoms, if it is presumed that the latter cause the molecular corrugation for these systems.[8]

### 3.3. Other Results

Using the model developed thus far and the approximate mapping, in addition to the phase diagrams already mentioned, reentrances below smectic $A_1$ (namely, nematic-smectic $A_1$-nematic-smectic $A_d$-nematic-smectic $A_1$, as temperature is lowered) were obtained,[17] also in qualitative agreement with experiment. Relative specific heats[9] and layer thicknesses,[8] and molecular dimer concentrations[10] were calculated in good agreement with experiments.

### 3.4. Missing Ingredients

As mentioned in Sec. 3.2, lateral positional fluctuations away from the triangular structure of the xy plane have been omitted, after (and due to) exploratory calculations.[7] These calculations indicated that lateral fluctuations, in contrast to the included permeation fluctuations, are unimportant to the reentrant liquid crystal phase transitions. This theoretical prediction has been confirmed by x-ray scattering experiments.[18]

The permeation fluctuations themselves are simplified in the model, by not defining the corrugation minima strictly by the relative positions of the neighboring molecules, but by referring the entire system to a lattice where the vertical lattice constant is determined by the atomic permeation scale. Note that this distinction exists only when librational permeations are also included (and, thus, not in Secs. 4 and 5 below). In a similar vein, an uneven distribution of energy minima as a function of mutual permeation could more realistically reflect molecular structure. However, these developments would encumber the calculations without addressing theoretical issues of currently pressing interest.

More in relevance to the current theoretical issues, fluctuations of the molecular axes away from the z direction are missing.

Other microscopic characteristics (realistic van der Waals interactions, core benzene-ring aspect ratios) have recently been included (Sec. 4) with important physical consequences (in-plane domain formation, smectic C ordering - Sec. 5).[11] Still more microscopic characteristics (e.g., different locations, orientations, and numbers of dipoles on molecules) are readily accessible to inclusion to the current theory, but have not yet been included.

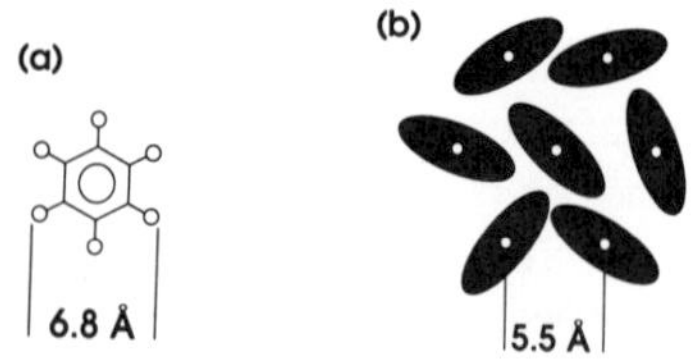


Fig. 1. Benzene-ring width and typical nearest-neighbor molecular separation dictate rotational steric hindrance, incorporated into our model. A view along the molecular axes is seen in (b).

Fig. 2. Nearest-neighbor configurations, viewed along the molecular axes, exhibiting different amounts of steric interference.

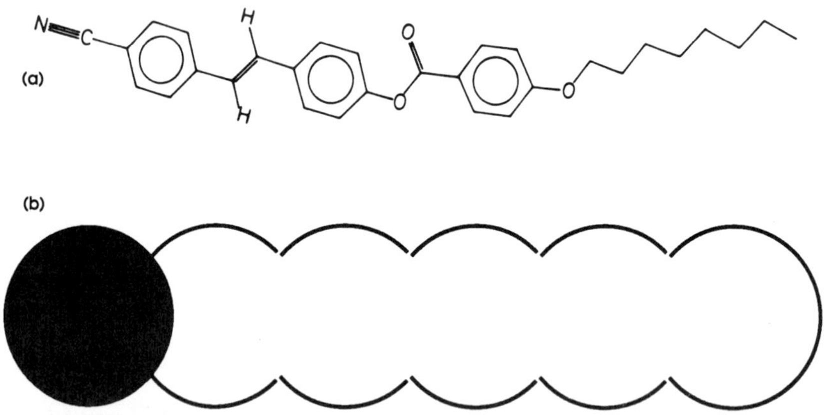


Fig. 3. Molecular corrugation in (a) the physical system of compound "$T_8$" and (b) model molecules. $T_8$ has a doubly reentrant phase diagram Ref. 22.

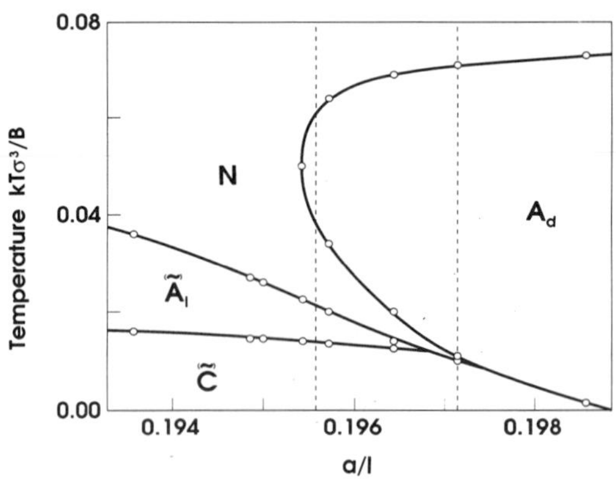


Fig. 4. Calculated phase diagram for a strongly dipolar system ($B\sigma^3/\varepsilon = 14.29$) with $n=7$, $n_o=10$, $\gamma_1=1.091$, $\gamma_2=1.048$, and $\gamma_3=1.050$ . The horizontal axis is the ratio of in-plane nearest-neighbor separation a to effective molecular length l. The in-plane separation a should be monotonic decreasing with pressure (or mixture concentration).

# 4. New Improved Frustrated Spin-Gas Model

## 4.1. Intermolecular Interactions

### 4.1.1. Dipolar Interactions

As before, these are pure dipolar interactions between nearest-neighbor molecules, as given in Eq. 1. Each molecule has one dipole fixed to its end segment and pointing along the molecular axis. (These choices can readily be modified.) The frustration of the dipolar interactions underlies the intermediate-temperature existence of the smectic $A_d$ phase (Sec. 5.1).

### 4.1.2. van der Waals Interactions

Each molecular segment interacts with any segment, that occupies a nearest-neighbor position, of another molecule, via the Lennard-Jones potential

$$V_{LJ}(r) \quad = \quad 4\varepsilon\,[(\sigma/r)^{12} - (\sigma/r)^{6}] \tag{4}$$

where $r$ is the separation between the segments. This improvement (over Eqs. 2 and 3) has led to the identification of smectic $A_1$ and C phases in which in-plane domain formation is indicated (Sec. 5.3).

### 4.1.3. Benzene-Ring Steric Hindrance

As seen in Fig. 1a, the width of a benzene ring is 6.8 Å, whereas the typical lateral nearest-neighbor intermolecular separations are between 4.8 Å and 5.5 Å. Thus, the rotations about the molecular axes must be correlated, as depicted in Fig. 1b. Each molecule, being in a local near-sixfold-symmetric rotational environment, must have three rotational minima. This is introduced to our model by assigning three rotational states to the molecule, with the longer transverse axis (6.8 Å) pointing between pairs of nearest-neighbors. [An alternate definition of rotational minima yields the herringbone packing.] The resulting rotational configurations of two nearest-neighbor molecules, shown in Figs. 2, have different amounts of steric interference. This is accounted by augmenting the intermolecular interaction of the pair in Fig. 2d by a factor of $\gamma_1$~1.1 over that of the pair in Fig. 2c, which itself is augmented by a factor of $\gamma_2$~1.05 over that of Figs. 2ab. This is a simplified overall accounting, justified by the fact that in regimes of our model where rotational correlations are important, the intermolecular interactions of dominant configurations are monotonic decreasing, in absolute value, with increasing lateral separation, and, of course, sterically less hindered pairs pack closer. This is a way of incorporating into the model the effect of systematic local deformations from the triangular structure.

Two more specifications complete the description of our intermolecular interactions. Firstly, for the benzene-ring steric hindrance to be effective, the molecular pair must be in a congested environment. Thus, the steric factors mentioned in the previous paragraph are used only for pairs in which a given molecule overlaps by at least $n_o$ segments with the nearest-neighbors in opposite directions. Secondly, when this overlap condition is satisfied, the intermolecular interaction of the pair in Figs. 2ab is augmented by a factor of $\gamma_3$~1.05. We found that the latter specification is necessary only to obtain the experimentally observed relative specific heats of the various phase transitions.[9] This is an empirical indication of the non-linear character of the induced dipoles underlying the van der Waals interaction.

### 4.2. Molecular Corrugation

Molecular corrugation can be due to either tail carbon atoms, or core benzene rings (Fig. 3), depending on molecular structure. The molecules in the model occupy n consecutive sites, one for each molecular segment, along the z direction of a three-dimensional stacked triangular lattice. The lattice constant along the z direction is thus determined by the molecular structure, and was accordingly chosen to be $\sigma$ (from Eq. 4).

### 4.3. Solution of the Model

Rather than using the approximate mapping of the previous studies, the partition function of the model is calculated by Monte Carlo sampling. A single xy cross-section of molecules, imbedded and evolving in the three-dimensional lattice, is considered.

We develop[11] a new Lindemann criterion that determines the nematic-smectic transition, in terms of the layer thickness fluctuation due to the molecular positions along the z direction:

$$L \quad = \quad \sqrt{\langle(\delta - <\delta>)^2\rangle}/l \tag{5}$$

where <> denotes averaging over Monte Carlo samplings, $l$ is the effective molecular length, and

$$\delta \quad = \quad \sqrt{\langle\langle (z_i - z_j)^2 \rangle\rangle} \tag{6}$$

where << >> denotes averaging over the system, and $z_i$ and $z_j$ are coordinates of nearest-neighbor molecules. A singularity in $L$ establishes (e.g., see Fig. 7) the criterion for the nematic-smectic transition for a given compound (i.e. $B\sigma^3/\varepsilon$, n, $n_o$, $\gamma_1$, $\gamma_2$, and $\gamma_3$ values). This singularity is found to occur typically for $L$ between 0.15 and 0.25 . The smectic C phase boundaries, on the other hand, are determined by the onset of layer tilt.

The individual molecules have discrete positional (along the z direction, but maintaining a nearest-neighbor overlap of at least one molecular segment to be part of the studied set), orientational (dipole up or down), and rotational (threefold) degrees of freedom. In the sampling, N=24x24=576 molecules were used. For each data point, typically between 10,000 and 20,000 Monte Carlo trials per molecule were taken, after discarding typically 1,000 Monte Carlo trials per molecule. At each trial, the molecule was allowed to sample its entire permeation range and, in a certain fraction f of the trials, its orientational (f=0.5) and rotational (f chosen between 0.1 and 0.5) degrees of freedom.

## 5. Microscopic Mechanism for Nematic Reentrance, Smectic C Layer Tilting, and In-Plane Domain Formation

### 5.1. Reentrance around the Smectic $A_d$ Phase

Our calculations show that the reentrant phase diagrams, exhibiting the nematic-smectic $A_d$-nematic-smectic $A_1$ sequence as temperature is lowered, are maintained in the new improved frustrated spin-gas model. [Since librational permeation has not yet been included, quadruple reentrance (Sec. 3.2) is not seen.] As a new result, smectic C phases are now seen below the smectic $A_1$ phase. The mechanism for this will be given in Sec. 5.2. Calculated phase diagrams for strongly and less strongly dipolar systems are respectively given in Figs. 4 and 5, and typical calculated molecular configurations for each phase appearing in these phase diagrams are shown in Fig. 6.

The important differentiation between the two cases will be discussed in Sec. 5.3. Figs. 7, 8, and 9 show examples of calculated temperature scans of the specific heat, molecular dimer concentration, layer thickness[8]

$$d = (l + <\delta>)\cos\phi \tag{7}$$

tilt angle $\phi$ (determining the smectic C boundaries), and layer thickness fluctuation $L$ (determining the nematic phase boundaries, in agreement with the specific heat singularities).

The mechanism for reentrance around the smectic $A_d$ phase now becomes quite apparent. Fig. 10 shows a top view of the system in the smectic $A_d$ phase. It is seen that only a subset of the molecules correlate positionally and orientationally. The remaining molecules freely permeate and reverse their dipoles (which is equivalent to an uncorrelated molecule leaving the cross-section under consideration and another molecule, with opposite dipole, entering the cross-section). The disordered molecules provide the local filling necessary to relieve the frustration of the other molecules: the latter are no longer closepacked and thereby can correlate. This ordered subset is correlated with respect to permeation positions and dipole directions, for which the correlation directions randomly change across the subset. This subset is, however, positionally disordered in the xy directions, as seen in Fig. 10. This figure is, in fact, a calculational realization of the pictorial scenario originally proposed[7] for the smectic $A_d$ phase.

The smectic $A_d$ layer obtained here has approximately 1/3 of its molecules disordered (Fig. 11c). This contributes an entropy of about ln(2n)/3 per molecule, where generally the number of segments $n \geq 6$, as dictated by molecular structure. Simultaneously, the smectic $A_d$ phase has substantial energy lowering, from the interactions of the correlated subset of molecules. Thus, due to the interpenetrating coexistence of order and disorder, the smectic $A_d$ phase has substantial amounts of both gained entropy and lowered energy. Accordingly, it can be expected to be free energetically favorable to occur at intermediate temperatures, optimizing both entropy and energy effects.

Another way to relieve dipolar frustration is shown in Fig. 11a. In this triplet, all molecules are correlated (in contrast to Fig. 11c). They are permeated with respect to each other, optimizing the combined energies of the three pairs. This configuration has two equivalent distances between the pairs, and one "inequivalent" distance. For the entire system, the one "inequivalent" distance has to be chosen in every nearest-neighbor triplet. This gives the entropy of packing of rhombi into a plane (equivalently, the zero-temperature entropy of the Ising antiferromagnet on the triangular lattice), which is calculated[19] to be 0.323 per molecule in this system.

Another triplet permeation that optimizes the combined energies of the three pairs is shown in Fig. 11b. For the entire system, the three mutually inequivalent pairings have to be chosen in every nearest-neighbor triplet. The entropy is the zero-temperature entropy of the 3-state Potts antiferromagnet on the Kagomé lattice, which is bound below by and probably close to the triangular Ising entropy mentioned previously. The triplet configurations in Figs. 11a and b dominate the smectic $A_1$ phases respectively for the strongly and less strongly dipolar systems (Figs. 4 and 5, with corresponding global molecular configurations shown in Fig. 6).

Thus, in the smectic $A_1$ phases, in which all pair interactions are optimized together, a built-in entropy also occurs, but is less than that of the smectic $A_d$ phase. Accordingly, the smectic $A_1$ phases are expected and calculationally found at lower temperatures than the smectic

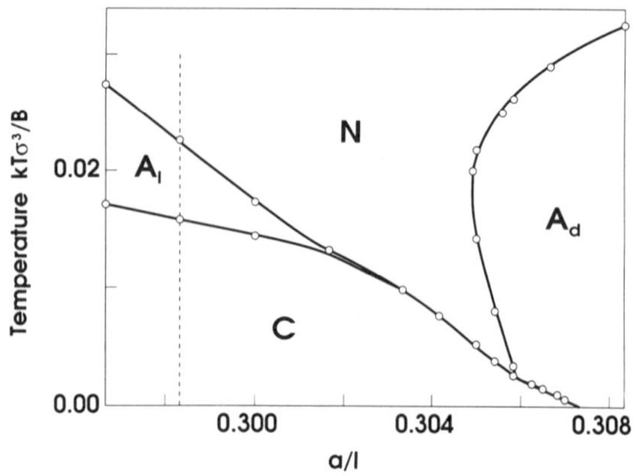


Fig. 5. Calculated phase diagram for a less strongly dipolar system ($B\sigma^3/\varepsilon = 4$) with n=6, $n_o$=9, $\gamma_1$=1.067, $\gamma_2$=1.014, and $\gamma_3$=1.035. The smectic $A_l$ and C phases are distinct from the ones in Fig. 4, as seen from the global molecular configurations in Fig. 6.

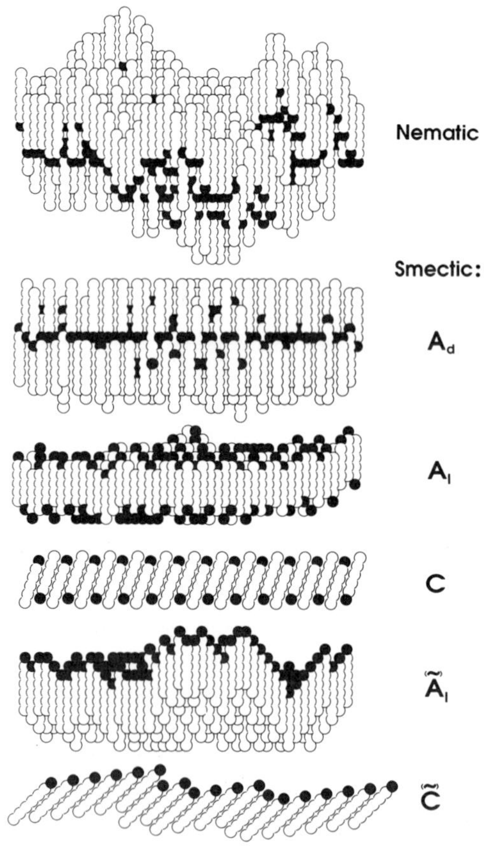


Fig. 6. Global molecular configurations calculated in each of the phases featured in the phase diagrams of Figs. 4 and 5.

$A_d$ phase. The in-between nematic phase is due to the competition and cancellation of the two types of smectic phases;[10] its extent, as well as the actual phase diagrams, are of course obtained from calculations. [This entropy (high degeneracy) of the smectic $A_1$ phases is removed or reduced, respectively for the two versions just discussed, by rotational interactions. The lock-in of the permeational and rotational degrees of freedom causes the smectic C phase (Sec. 5.2).]

In the actual liquid crystal systems, the precise constraints of lattice positions do not exist. However, the entropy arguments above will not be quantitatively modified much, when given in fuzzy versions, by the relaxation of the lattice constraint.

### *5.2. Smectic C Layer Tilting*

Consider the smectic $A_1$ phase, in which the frustration of the dipolar interactions is relieved by permeations in the manner shown in Fig. 11a. The triplet is tilted. However, local tilting does *not* automatically imply global tilting, since a priori local tilts would occur randomly upwards or downwards, averaging out to give the untilted smectic $A_1$ phase and yielding the entropy discussed in Sec. 5.1. The intermolecular energies determined by Eqs. 1 and 4 are negative (favorable) for all pairs in the triplet in Fig. 11a, being more favorable for the (13) pair due to the form of the dipolar interaction. Therefore, steric energy is minimized by the rotational state shown in the top view triangle (123) in Fig. 12a. This dictates the rotational states for the

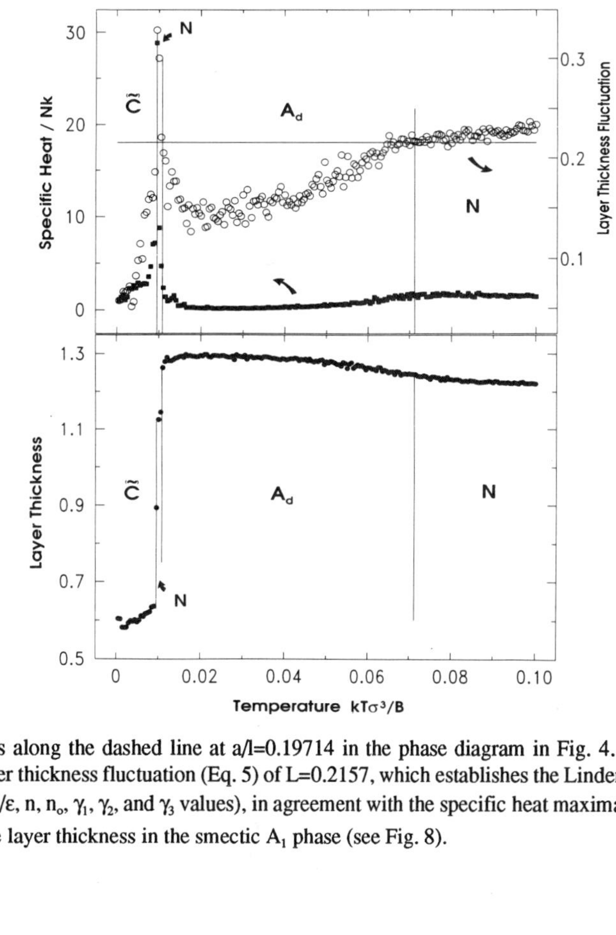


Fig. 7. Calculations along the dashed line at a/l=0.19714 in the phase diagram in Fig. 4. Note the singularity occurring at the layer thickness fluctuation (Eq. 5) of L=0.2157, which establishes the Lindemann criterion for this compound (i.e. $B\sigma^3/\varepsilon$, n, $n_o$, $\gamma_1$, $\gamma_2$, and $\gamma_3$ values), in agreement with the specific heat maxima. The layer thickness is normalized to the layer thickness in the smectic $A_1$ phase (see Fig. 8).

consecutively adjoining triangles, shown in Fig. 12a. This in turn dictates the permeational states shown by the numbering (which represents the z coordinates), repeating in the most favorable way the permeations of Fig. 11a. Clearly this procedure continues to all adjoining triplets. We thus find that the tilts of individual triplets, due to dipolar interactions, are coordinated, by

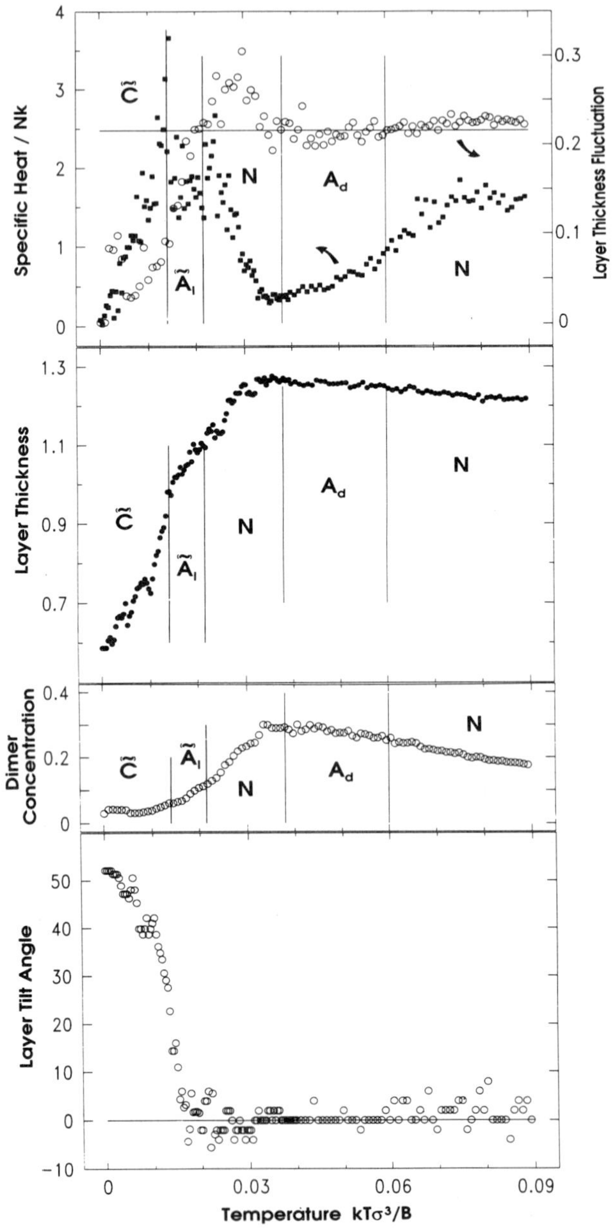


Fig. 8. Calculations along the dashed line at a/l=0.19557 in the phase diagram in Fig. 4. The molecular dimer concentration is the fraction of nearest-neighbor dipole pairs that point oppositely and have the same z coordinate.

rotational steric hindrances, throughout the layer. The lock-in of the dipolar and rotational degrees of freedom is the mechanism for smectic C ordering.

A similar argument with the smectic $A_1$ phase composed of triplets shown in Fig. 11b leads to layer tilting shown in Fig. 12b. Note that in this case the degeneracy is not totally lifted: only every other pair of adjacent triplets can be energetically optimized, leading to similarly oriented molecules forming lines that freely meander along only two in-plane directions and thereby creating an average layer tilt (Fig. 12b inset).

This behavior sets in inside the smectic $A_1$ phase, at a temperature determined by the incremental energies involved in the steric interferences. In fact, by increasing the factors $\gamma_1$ and $\gamma_2$, the smectic C region can be increased at the expense of the smectic $A_1$ region.

Figs. 8 and 9 show our calculations of layer tilt angles as function of temperature, for the strongly and less strongly dipolar systems. The tilt angles saturate, in accord with the mechanism above, at $\tan^{-1}(2\sigma/a)$ and $\tan^{-1}(\sigma/2a)$, which is about 50° and 20°, respectively corresponding to Figs. 12a and b. In fact, the experimentally catalogued saturation angles of different compounds do divide into two such groups![12] Furthermore, according to our microscopic picture, molecules that are not effectively corrugated will not show tilt saturation, yielding the third group of experimentally catalogued smectic C compounds.[12] Finally, the above angles are calculated for corrugation due to the core benzenes (Fig. 3); if the corrugation effective in the smectic C phases were due to tail carbons, the saturation angle values would be approximately halved, pointing to yet another group with about 10°.

## 5.3. *In-Plane Domain Formation*

We have conducted our calculations for the entire range of dipolarity of compounds, as given by the ratio $B\sigma^3/\varepsilon$ of dipolar and van der Waals interaction strengths, from Eqs. 1 and 4. In this global phase diagram, the low-temperature smectic phases, namely $A_1$ and C, appear in two separate and distinct versions, as seen in the molecular configurations in Fig. 6.

In the physical systems, the molecules are structurally up/down asymmetric. Thus, the oriented versions of the smectic $A_1$ and C phases will develop local curvature.[20] Global curvature would imply the loss of layer sliding entropy, so that the layer will tend to form in-plane domains,

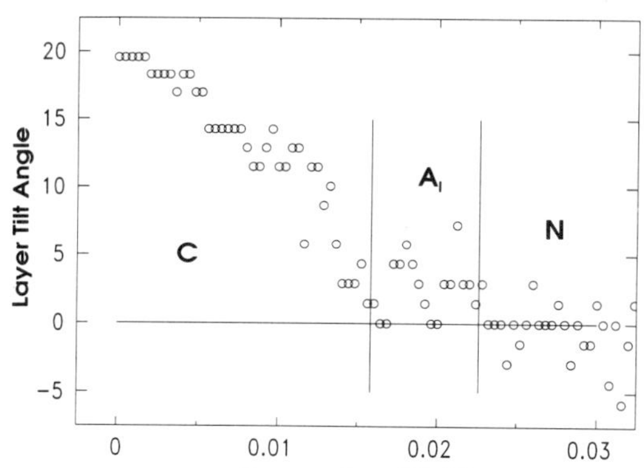


Fig. 9. Calculated layer tilt angle along the dashed line at a/l=0.29833 in the phase diagram in Fig. 5. Note the different saturation angle than in Fig. 8.

reversing its local curvature and thereby avoiding global curvature (Fig. 13). Thus, in-plane domain formation is indicated for one version of the smectic $A_1$ and C phases. This effect would be most likely for strong molecular up/down asymmetry, but away from low temperatures where entropy is not as important. (At the low temperatures, local curvature could be neutralized by a minimal interlayer interdigitation, with loss of layer sliding entropy at every other interface but lowered dipolar energy, thus forming the $A_2$ phase.) Since the smectic C phase is biaxial (Sec. 5.2), the topology of the smectic $A_1$ and C domains, as well as their threshold conditions just mentioned, will be different.

These oriented (therefore indicative of in-plane domain formation) versions of the smectic $A_1$ and C phases occur, in our calculations, for strongly dipolar molecules (Fig. 4), whereas the non-oriented versions occur for less strongly dipolar molecules (Fig. 5), in agreement with experiments.[16,21] Other calculated properties also correlate with molecular dipolarity. For example, layer thickness decreases from the smectic $A_d$ phase to the smectic $A_1$ phase by a factor of about 1.3 for the strongly dipolar molecules (Fig. 7), whereas it stays roughly constant for the weakly dipolar molecules (not shown here).

The above indication of in-plane domain formation due to local curvature is valid for thermotropic systems. In lyotropic systems, where maintaining a sliding entropy of packed layers is not an issue, local curvature controlled by this dipolar/van der Waals competition should influence the formation of micelles and lamellar phases.

ACKNOWLEDGMENTS. A.N.B. would like to thank J.O. Indekeu for many years of fruitful collaboration on liquid crystal problems.We would like to thank P. Cladis and C.W. Garland for many useful conversations. Support by an Evangelisches Studienwerk Fellowship is gratefully acknowledged by R.R.N. This research was supported by NSF Grant No. DMR-90-22933 and by JSEP Contract No. DAAL 03-89-C0001.

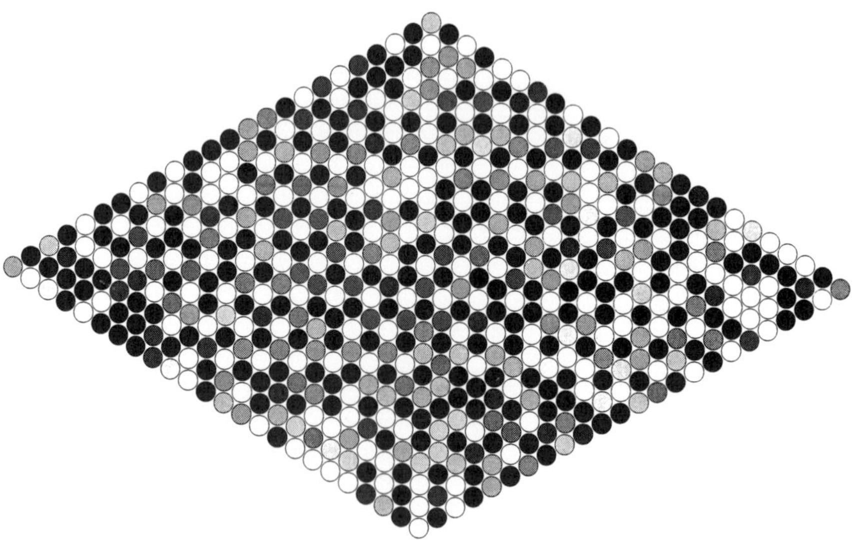

Fig. 10. Top view of the layer in the smectic $A_d$ phase, calculated for $kT\sigma^3/B=0.04$, $a/l=0.195$, $n=7$, $n_o=10$, $\gamma_1=1.048$, $\gamma_2=1.050$, and $\gamma_3=1$ . The black and white molecules remained positionally and orientationally (respectively in opposite directions) fixed over 50 Monte Carlo samplings per molecule. The gray molecules permeated and flipped their orientations, as indicated by the tone of grayness.

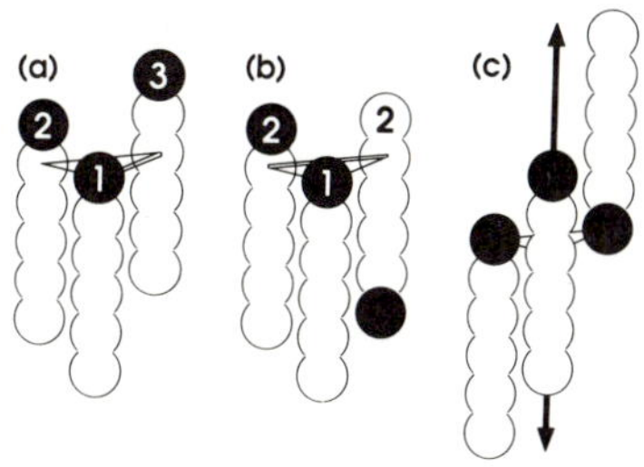


Fig. 11. Nearest-neighbor triplet configurations that dominate, respectively, in the smectic $\tilde{A}_1$, $A_1$, and $A_d$ phases. In the latter, one molecule is free to permeate. The numerals indicate the z coordinates. The double line indicates the most favorable pair (dipolar and van der Waals) interaction in the triplet.

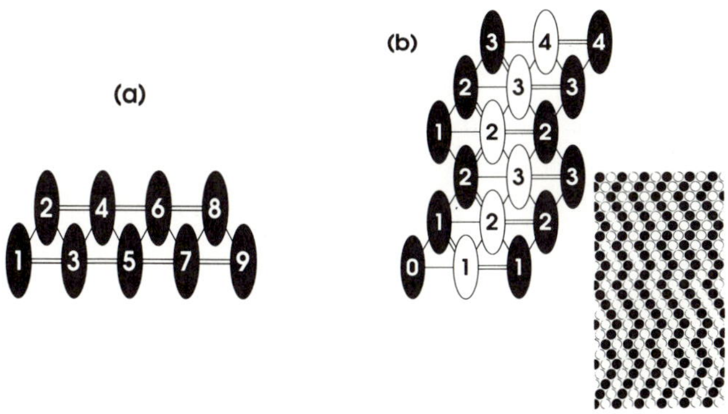


Fig. 12. Top views of layers dominated by permeation configurations correspondingly shown in Figs. 11(a) and (b): Optimizing the steric packing of the most favorable interaction, indicated by the double line, introduces a tilt propagation, as seen by the numerals indicating the z coordinates. In (b), the most favorable interaction can be sterically optimized only in every other pair of adjacent triplets, leading to similarly oriented molecules forming lines that freely meander along only two in-plane directions, as seen in the inset. The net tilt, thus, is from right to left.

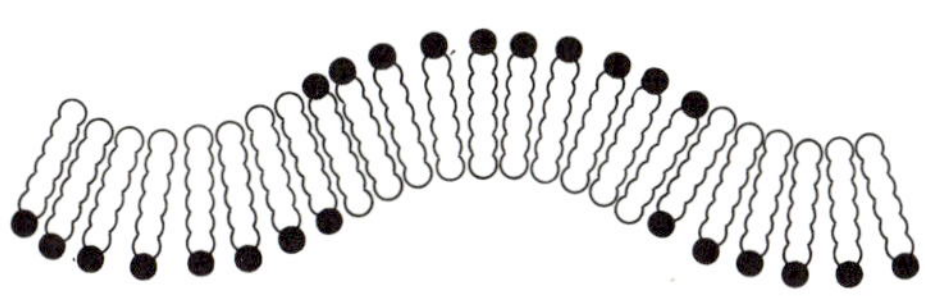

Fig. 13. Oriented layers, like those occurring in the smectic $A_1$ and C phases of Fig. 4, will tend to form in-plane domains, in order not to have global curvature, in spite of local curvature due to molecular up/down asymmetry.

## References


1. P. Cladis, Phys. Rev. Lett. **35**: 48 (1975).
2. K.G. Wilson, Phys. Rev. B **4**: 3174, 3184 (1971).
3. G.R. Andersen and J.C. Wheeler, J. Chem. Phys. **69**: 2082 (1978).
4. S.R. McKay, A.N. Berker and S. Kirkpatrick, Phys. Rev. Lett. **48**: 767 (1982).
5. R.G. Caflisch, A.N. Berker and M. Kardar, Phys. Rev. B **31**: 4527 (1985).
6. W. Hoston and A.N. Berker, Phys. Rev. Lett. **67**: 1027 (1991).
7. A.N. Berker and J.S. Walker, Phys. Rev. Lett. **47**: 1469 (1981).
8. J.O. Indekeu and A.N. Berker, Phys. Rev. A **33**: 1158 (1986).
9. J.O. Indekeu, A.N. Berker, C. Chiang, and C.W. Garland, Phys. Rev. A **35**: 1371 (1987).
10. J.O. Indekeu and A.N. Berker, J. Phys. (Paris) **49**: 353 (1988).
11. R.R. Netz and A.N. Berker, Phys. Rev. Lett. **68**: 333 (1992).
12. A. de Vries, in: "Liquid Crystals, the Fourth State of Matter," by F. D. Seva, ed., Marcel Dekker, New York, (1979), p. 1.
13. J. Prost, Adv. Phys. **33**: 1 (1984).
14. P. Barois, Chapter 4 of this Volume (1991).
15. A.N. Berker, K. Hui, and J. Marko, to be published.
16. F. Hardouin, A.M. Levelut, M.F. Achard, and G. Sigaud, J. Chim. Phys. **80**: 53 (1983).
17. J.O. Indekeu, Phys. Rev. A **37**: 288 (1988).
18. A.R. Kortan, H. von Känel, R.J. Birgeneau, and J.D. Litster, J. Phys. (Paris) **45**: 537 (1984).
19. G.H. Wannier, Phys. Rev. **79**: 357 (1950).
20. C.K. Bagdassarian, D. Roux, A. Ben-Shaul, and W.M. Gelbart, J. Chem. Phys. **94**: 3030 (1991).
21. N.H. Tinh, Mol. Cryst. Liq. Cryst. **127**: 143 (1985).
22. F. Hardouin and A.M. Levelut, J. Phys. (Paris) **41**: 41 (1980).

Chapter 8

# EFFECTS OF ANISOTROPY IN THE ORDERING MECHANISMS OF TWO DIMENSIONAL ROTORS

H. LEITÃO and M.M. TELO DA GAMA

## 1. Introduction

The study of two dimensional systems has been an active field of research for many years[1]. These systems exhibit a rich, but subtle, phase behaviour, which renders them quite distinct from their three dimensional counterparts, and a complete understanding of its physical behaviour has been a challenge to theoreticians and experimentalists alike.

Recently there has been an increased interest in the behaviour of liquid crystals in two dimensions (see: this Volume as well as Ref. 2 for a survey of the experimental situation). From a theoretical point of view, two recent studies have addressed the question of explaining the stability of mesophases in two dimensions and have motivated for the work presented here. Using a density functional approach, Cuesta et al[3] showed that a two dimensional system of hard ellipses exhibits an orientational ordered state, disordering through a continuous transition to an isotropic state, in contrast with the three dimensional fluid, which exhibits a first order transition. Subsequently, a Monte Carlo simulation of a system of hard ellipses[4] suggested that the order of the nematic-isotropic transition depends sensitively on the anisotropy (aspect ratio) of the molecules.

This last result indicates that the role played by the anisotropy in these systems is not yet clearly understood. In particular, one would like to know if the order of the nematic-isotropic transition in two dimensions depends, in general, on the anisotropy of the molecules. From a theoretical point of view the simplest way to address this problem is (following Maier and Saupe[5]) to consider a system of anisotropic molecules on a lattice.

We have chosen to study a simple lattice model consisting of plane rotors located at the sites of a square lattice and interacting via a nearest neighbour anisotropic pair potential:

$$U_{ij}(\phi,\phi') = J[a \cos 2\phi \cos 2\phi' + b \sin 2\phi \sin 2\phi'] \tag{1}$$

H. Leitão and M.M. Telo da Gama - Departamento de Fisica, Universidade de Lisboa, Centro de Fisica da Materia Condensada, P-1699 Lisboa Codex, Portugal.

*Phase Transitions in Liquid Crystals*, Edited by S. Martellucci and A.N. Chester, Plenum Press, New York, 1992

In this expression, $\phi$ and $\phi'$ are measured with respect to an arbitrary direction, $J$ is a positive coupling constant, and $a$ and $b$ are parameters describing the potential anisotropy. The potential was chosen for essentially two reasons:

*(i)* Exact results[6] indicate that, when $0 \leq |b| \leq a = 1$, the system possesses an orientationally ordered phase, which disorders at a temperature $T_c = O(1-b)$. When $a = b$, on the other hand, the system disorders at all finite temperatures, but exhibits a Kosterlitz-Thouless transition[7] at finite temperature.

*(ii)* A recent Monte Carlo simulation of the model[8], with $b = 0$, confirmed the existence of a low temperature nematic ordered state, which disorders through a continuous transition. In addition, the simulation results provide a complete characterization of the thermodynamic and structural properties of the model, which can be used to test the accuracy of our approximations.

In the following we present preliminary results of our study on the ordering behaviour of this lattice model, using two simple techniques.

## 2. Cluster Variational Method

The Cluster Variational Method (CVM)[9] is a hierarchy of closed form approximations for the configurational entropy of classical lattice systems. In this method, the entropy is approximated by a function of the probabilities for the different configurations of a cluster of $n$ lattice sites. The resulting free energy is then minimized, yielding the equilibrium cluster probabilities. Although originally developed to study lattice models with discrete degrees of freedom, CVM was recently generalized to lattices with continuous degrees of freedom and it has been applied to models of adsorbed molecules on crystal surfaces[10] and to liquid crystalline ordering[11].

In order to improve on simple Mean Field results we used the CVM in the pair approximation. The CVM expression for the free energy in the pair approximation reads[10]:

$$\beta F = \sum_{(ij)} \int\int d\phi d\phi' \rho_{ij}(\phi,\phi') \ln[\rho_{ij}(\phi,\phi')] - (\nu - 1) \sum_{i} \int d\phi \rho_i(\phi) \ln[\rho_i(\phi)] +$$

$$+ \sum_{(ij)} \int\int d\phi d\phi' \rho_{ij}(\phi,\phi') \beta U_{ij}(\phi,\phi') \tag{2}$$

where $\rho_{ij}(\phi,\phi')$ are the pair distributions functions, $\rho_i(\phi)$ the single-spin distribution function, and $\nu$ is the lattice coordination number. The first and second terms are the entropy in the pair approximation and the third term is the internal energy. This functional must be minimised with respect to the pair distribution function, $\rho_{ij}(\phi,\phi')$, taking into account the symmetries and normalizations for the system under consideration. For a translationally invariant square lattice the result obtained is[12]:

$$\rho^{(2)}(\phi,\phi') = C \exp(-\beta U(\phi,\phi'))\, (\rho^{(1)}(\phi))^{3/4}\, (\rho^{(1)}(\phi'))^{3/4} \tag{3}$$

The nematic order parameter, defined as the average of $\cos 2\phi$,

$$Q = \int d\phi \rho^{(1)}(\phi) \cos 2\phi \tag{4}$$

is plotted, for $b = 0$, as a function of temperature in Fig. 1. The solid line is the pair approximation result, while the dots are computer simulation results from Ref. 8. For $b \neq 0$ we have found, in agreement with Monte Carlo predictions, the existence of a continuous transition at $T_c = 1.56$. This value is much closer to the simulation $T_c = 1.31$ than the simple Mean Field estimate[8], $T_c = 2$. It should also be noticed the excellent agreement with the simulation results at moderately low temperatures. Results for the specific heat, with $b = 0$, were also obtained and are shown in Fig. 2.

We have found that there is always nematic ordering, with the spins aligning preferentially parallel or anti-parallel to each other at low enough temperatures. The order-disorder transition was found to be continuous for all anisotropies but, due to the presence of a disordering term, the transition temperature decreases as one approaches the isotropic limit, $a = b$. The dependence of the transition temperature on the potential anisotropy, $\eta = b/a$, is plotted in Fig. 3 (dashed line).

Some comments are in order. First of all, one sees the expected drop in transition temperature as the anisotropy of the potential is decreased. In the limit of isotropic systems, $\eta = 1$, the transition temperature obtained within the pair approximation is $T_c = 1.41$. However, for this isotropic potential it is known that low energy, long-wavelength fluctuations, destroy any orientational ordering at all finite temperatures, and a simple spin-wave analysis can show that $T_c \rightarrow 0$ as $\eta \rightarrow 0$.

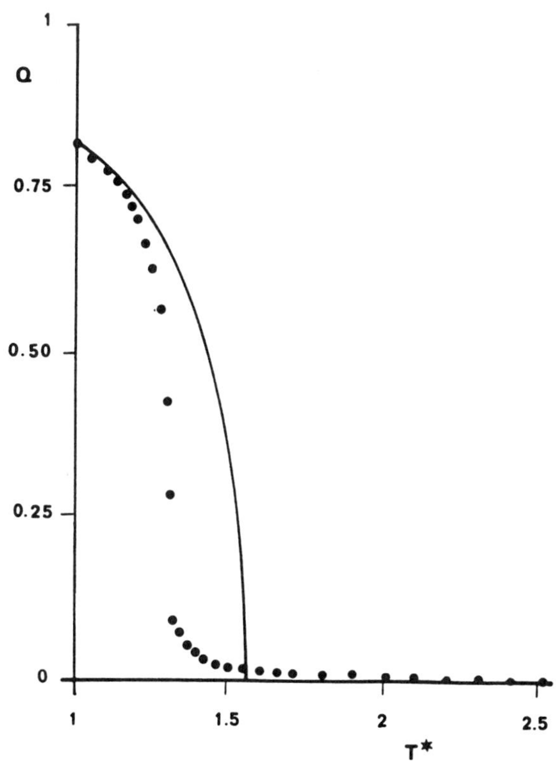


Fig. 1. Nematic order parameter for $\eta = 0$. Dots are computer simulation results from Ref. 8.

## 3. Mean Field Renormalization Group

The simplest way to obtain non-classical estimates for the critical exponents is by a Mean Field Renormalization Group (MFRG) strategy, as developed originally by Indekeu and co-workers[13]. MFRG is basically a simple and versatile technique for constructing a regular renormalization group mapping using closed form classical equations for the order parameter. We have extended MFRG, previously used only in systems with discrete degrees of freedom, in order to apply it to the continuum spin model under consideration.

For a one spin cluster, the partition function is[14]:

$$Z_1 = \int_0^{2\pi} d\phi e^{\beta H} = \int_0^{2\pi} d\phi e^{h\cos\phi} = 2\pi I_0(h) \tag{5}$$

where $h = \nu K b$, is the effective field and $I_0$ is a modified Bessel function. From this expression, the order parameter is trivially obtained, yielding, to linear order in $h$:

$$\langle q \rangle_1 = h/2 \tag{6}$$

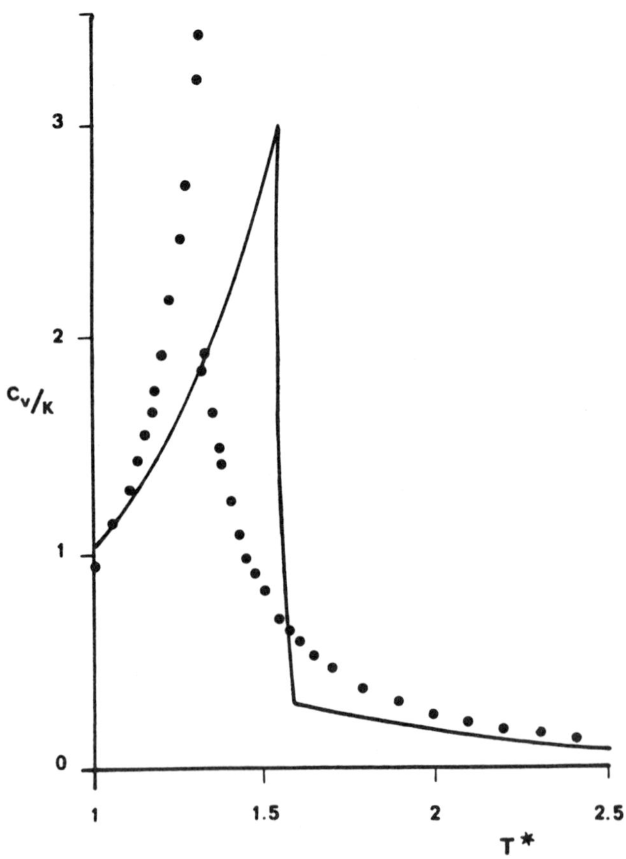


Fig. 2. Specific heat per particle, $C_V = \partial E/\partial T$ for $\eta = 0$. Dots are computer simulation results from Ref. 8.

For a cluster of two spins, the Hamiltonian is:

$$-\beta H_2 = K\cos\phi\cos\phi' + \eta K\sin\phi\sin\phi' + h(\cos\phi + \cos\phi') \tag{7}$$

and the partition function can be shown to be[14], to $O(h^2)$:

$$Z_2 = 2\pi\int_0^{2\pi} d\phi I_0(\sqrt{K^2\cos^2\phi + \eta^2 K^2\sin^2\phi}) \tag{8}$$

It is then straightforward to show[12] that the order parameter for the two-spin cluster is:

$$\langle q\rangle_2 = \frac{\int_0^{2\pi} d\phi\cos^2\phi I_0[f(\phi,K,\eta)] + \int_0^{2\pi} d\phi \frac{\cos^2\phi}{\sqrt{\cos^2\phi+\eta^2\sin^2\phi}} I_1[f(\phi,K,\eta)]}{\int_0^{2\pi} d\phi I_0[f(\phi,K,\eta)]} h \tag{9}$$

The fixed point of the MFRG transformation is obtained by imposing that the order parameters of the two clusters scale as:

$$\langle q\rangle_1 = l^{d-y_H}\langle q\rangle_2 \tag{10}$$

where $l = (N/N')^{1/d}$ is the length re-scaling factor, and $y_H$ is the magnetic exponent.

Results for the critical temperature are shown in Fig. 3 (solid line). The MFRG results for $T_c$ are similar, but slightly higher than those obtained with the pair approximation. This can be understood as a consequence of the approximate way in which the renormalization flow equations were obtained and, in particular, by the fact of having used a single equation to describe a flow with two relevant parameters, $K$ and $\eta$.

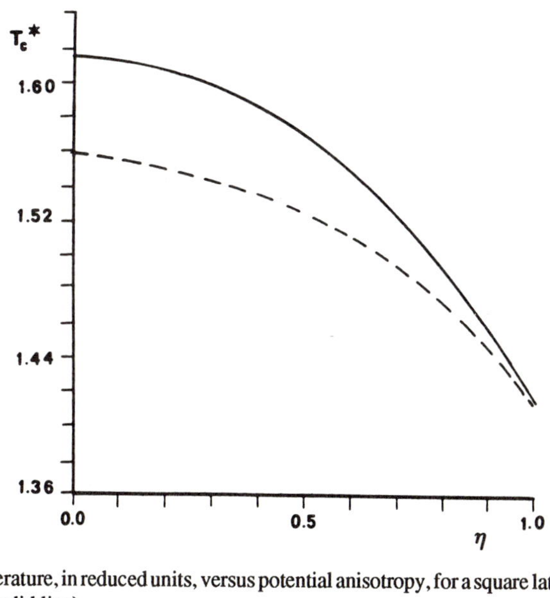


Fig. 3. Transition temperature, in reduced units, versus potential anisotropy, for a square lattice: CVM results (dashed line); MFRG results (solid line).

Estimates for the thermal exponent $y_T$, defined as:

$$\left(\frac{\partial K'}{\partial K}\right)_{K_c} = l^{y_T} \tag{11}$$

were also obtained and are plotted in Fig. 4.

The two dimensional system considered is a notoriously subtle system, particularly in the isotropic case. In an attempt to assess the validity of the techniques used, we have also determined the critical properties of a simpler situation, namely the system of anisotropic planar rotors with the interaction potential (Eq.1), in three dimensional lattices. The results obtained, using CVM and MFRG, for the transition temperature in a cubic lattice are shown in Fig. 5. Identical comments as the ones discussed previously can be applied to these results, but it should be stressed that now the estimate for the transition temperature in the isotropic case, $T_c = 2.45$, is in reasonably good agreement with the estimates obtained with more sophisticated studies[15], $T_c = 2.20$.

## 4. Conclusions

We have studied a very simple model of anisotropic plane rotors using two very simple, but complementary, techniques: CVM and MFRG. The former is a simple and versatile closed form approximation for thermodynamic and structural properties of classical continuous spin systems, while the later is a simple version of a Renormalization Group transformation which has the merit of accounting for non-classical critical exponents. The main objective of our study was motivated by the simulations of Cuesta and Frenkel of the hard ellipse system, which uncovered a non-trivial dependence of the ordering mechanism in two-dimensional nematics on the anisotropy of the molecules, which seems to be absent in the simple model considered in this work.

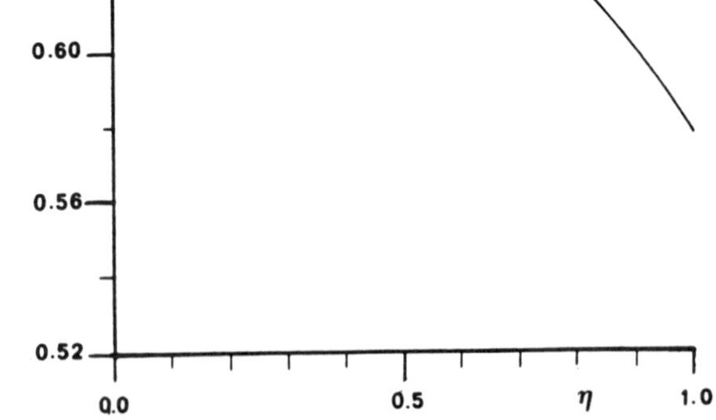


Fig. 4. Thermal exponent, $y_T$, versus potential anisotropy.

The types of approximations used in our work are too crude to describe (even qualitatively) the rather subtle physics of isotropic two-dimensional systems. They can, however, be applied with some confidence in other physical situations. For example, as we have shown, MFRG and CVM give a reasonably good qualitative as well as quantitative description of three dimensional systems, and CVM (in the pair approximation) can be easily applied to two-dimensional systems in an external potential, mimicking the effects of the substrate-adsorbate interactions.

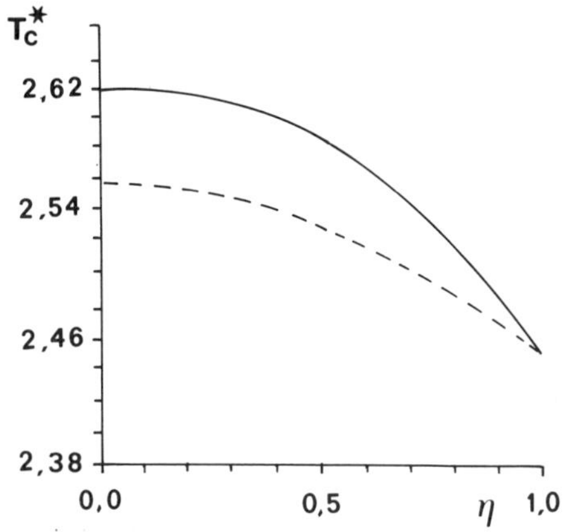


Fig. 5. Transition temperature, in reduced units, versus potential anisotropy, for a simple cubic lattice: CVM results (dashed line); MFRG results (solid line).

## References


1. S. Sinha, ed. "Ordering in Two Dimensions", North Holland, New York (1980).
2. B. Jérôme, Surface Effects and Anchoring in Liquid Crystals, Rep. Prog. Phys. **54**: 391 (1991).
3. J.A. Cuesta, C. Tejero and M. Baus, Phys. Rev. A **39**: 6498 (1989).
4. J.A. Cuesta and D. Frenkel, Phys. Rev. A **42**: 2126 (1990).
5. W. Maier and A. Saupe, Z. Naturf. A **14a**: 882 (1959).
6. J. Fröhlich and E.H. Lieb, Commun. Math. Phys. **60**: 233 (1978).
7. J.M. Kosterlitz and D.J. Thouless, J. Phys. C **6**: 1181 (1973).
8. S. Romano, Liq. Crystals **6**: 457 (1989).
9. R. Kikuchi, Phys. Rev. **81**: 988 (1951).
10. E. Chacòn and P. Tarazona, Phys. Rev. B **39**: 7111 (1989).
11. M.M. Telo da Gama and P. Tarazona, Phys. Rev. A **41**: 1149 (1990).
12. H. Leitao and M.M. Telo da Gama, (submitted Physica A).
13. J.O. Indekeu, A. Maritain and A.L. Stella, J. Phys. A **15**: L291 (1982).
14. I.S. Gradshteyn and I.M. Ryzhik, "Table of Integrals, Series and Products", Academic Press, New York (1980).
15. Y.H. Li and S. Teitel, Phys. Rev. B **40**: 9122 (1989).

# THERMOTROPIC LIQUID CRYSTALS

Chapter 9

# LIGHT SCATTERING STUDIES OF THERMOTROPIC LIQUID CRYSTALS

S. SPRUNT and J.D. LITSTER

## 1. Introduction

Scattering of narrow band coherent light from a laser source is a powerful tool to elucidate the behavior of thermotropic liquid crystals. That is because one may use light scattering to investigate both the static and dynamic behavior of director modes in liquid crystal phases. The director modes are often modified by the establishment of other types of order in liquid crystals, and their study enables one to investigate the establishment of both positional and bond orientational order in smectic phases.

We review the theory of light scattering, and apply it to understand the information one obtains from light scattering investigations of various thermotropic liquid crystal phases. We discuss the isotropic phase, the nematic phase, the effect of several kinds of smectic order, and conclude by presenting recent results on light scattering from "hexatic" phases with bond orientational order.

## 2. Theory of Light Scattering

A classical model for the interaction of light with matter will be quite adequate to understand how thermotropic liquid crystals scatter light. A good fundamental reference is the Born and Wolf text[1] on optics. We shall begin with the simpler case of scattering in an isotropic medium, and later modify the result to treat liquid crystal phases with uniaxial anisotropy. The light is almost always scattered by changing electric dipoles in the matter. The radiation field at point **r** due to a classical electric dipole **p**($t$) located at the origin is:

$$\mathbf{E}_r(\mathbf{r},t) = \frac{1}{c^2 r}\left[\mathbf{1}_k \times \left(\mathbf{1}_k \times \frac{\partial^2 \mathbf{p}(t')}{\partial t'^2}\right)\right]_{t' = t - r/c} \tag{1}$$

S.Sprunt and J.D.Litster - Department of Physics and Francis Bitter National Magnet Laboratory, Massachusetts Institute of Technology, Cambridge, Massachusetts 02139-4307, U.S.A.

*Phase Transitions in Liquid Crystals*, Edited by S. Martellucci and A.N. Chester, Plenum Press, New York, 1992

where $1_k$ is a unit vector along the direction of propagation of the radiation, which is given by the wave vector **k**. A light wave traveling through matter polarizes the atoms or molecules and induces an electric dipole moment. In c.g.s. units,

$$\mathbf{D} = \mathbf{E} + 4\pi\mathbf{P} = \varepsilon\mathbf{E}$$

where **P** is the induced electric dipole moment per unit volume. Thus, a field $\mathbf{E}(\mathbf{r},t)$ will induce a dipole moment in a volume $dr$ at r that is given by

$$d\mathbf{P}(\mathbf{r},t) = (1/4\pi)\,[\varepsilon(\mathbf{r},t) - 1]\,\mathbf{E}(\mathbf{r},t)d\mathbf{r} \tag{2}$$

One may calculate the scattered field by integrating Eq. 2 over the illuminated volume. If $\varepsilon$ does not depend on **r**, all the infinitesimal dipole fields will cancel except in the "forward" direction, that is unless **k** is along the direction of propagation of the incident light. If we write $\varepsilon(\mathbf{r},t) = \bar{\varepsilon} + \delta\varepsilon(\mathbf{r},t)$ then light will be scattered only by the fluctuations $\delta\varepsilon(\mathbf{r},t)$.

If the incident field at $(\mathbf{r}',t)$ is

$$\mathbf{E}_e^{i\mathbf{k}_i\cdot\mathbf{r}'-i\omega t}$$

to calculate the scattered light at $(\mathbf{r},t)$ we need to integrate the following expression over the illuminated volume:

$$d\mathbf{E}_s(\mathbf{r},t) = \frac{-\omega^2 d\mathbf{r}'}{4\pi c^2\,|\,\mathbf{r}-\mathbf{r}'\,|}\left[\mathbf{1}_{k_s}\times\mathbf{1}_{k_s}\times\delta\varepsilon(\mathbf{r}',t')\mathbf{E}_0 e^{i\mathbf{k}_i\cdot\mathbf{r}'-i\omega t'}\right]_{t'=t-n|\mathbf{r}-\mathbf{r}'|/c} \tag{3}$$

where $n = \bar{\varepsilon}^{1/2}$ is the refractive index of the medium. The generalization of Eq. 3 to situations where $\delta\varepsilon$ is a tensor is straightforward, and is usually required for liquid crystals. We shall do it presently, but continue now for the case when $\delta\varepsilon$ is a scalar.

Frequently one wants to know the correlation function

$$\langle \mathbf{E}_s^*(\mathbf{r},t_1)\cdot\mathbf{E}_s(\mathbf{r},t_2)\rangle \tag{4}$$

whose Fourier transform gives the spectrum of the scattered light. Calculating Eq. 4 requires a double integral over the illuminated volume. We may take the illuminated volume to be centered about the origin, and generally the illuminated volume dimensions are small compared to the distance to the field observation point. Then, the denominator $4\pi c^2\,|\,\mathbf{r}-\mathbf{r}'\,|$ becomes $4\pi c^2\mathbf{r}$. The optics of the experiment usually collect collimated scattered light, so all scattered waves have the same wave vector $\mathbf{k}_s$. The processes that cause the fluctuations $\delta\varepsilon(\mathbf{r},t)$ are *stationary*, which means that the correlation function Eq. 4 depends only on the difference $\tau = |\,t_1 - t_2\,|$. Under these conditions, the double integral becomes:

$$\langle \mathbf{E}_s^*(\mathbf{r},t_1)\cdot\mathbf{E}_s(\mathbf{r},t_2)\rangle$$

$$= \frac{E_0^2\omega^4\sin^2\Phi}{(4\pi)^2c^4r^2}\int d\mathbf{r}'\int d\mathbf{r}''\langle\delta\varepsilon^*(\mathbf{r}',t')\delta\varepsilon(\mathbf{r}'',t'')\rangle e^{i\mathbf{k}_i\cdot(\mathbf{r}'-\mathbf{r}'')-i\omega(t'-t'')} \tag{5}$$

where

$$t' = t_1 - n\,|\,\mathbf{r} - r'\,|\,/c, \qquad t'' = t_2 - n\,|\,\mathbf{r} - \mathbf{r}''\,|\,/c,$$

and $\Phi$ is the angle between $\mathbf{E}_0$ and $\mathbf{k}_s$. The fluctuations $\delta\varepsilon(\mathbf{r},t)$ are usually *homogeneous*, which means that the integrals in Eq. 5 depend only on $\mathbf{r}'-\mathbf{r}''$; one of the volume integrals can then be done immediately. Furthermore, since all scattered light rays have the same wave vector $\mathbf{k}_s$ with magnitude $k_s = n\omega/c$,

$$n\omega\,|\,\mathbf{r}-\mathbf{r}'\,|\,/c = k_s r - \mathbf{k}_s\cdot\mathbf{r}'$$

This means:

$$e^{-i\omega(t'-t'')} = e^{-i\omega\tau - i\mathbf{k}_s\cdot(\mathbf{r}'-\mathbf{r}'')}$$

Everywhere else in Eq. 5 it is an excellent approximation to replace $t'-t''$ with $\tau$. Then we find:

$$\langle \mathbf{E}_s^*(\mathbf{r},0)\cdot\mathbf{E}_s(\mathbf{r},\tau)\rangle$$

$$= \frac{E_0^2\omega^4\sin^2\Phi}{(4\pi)^2c^4r^2}\int d\mathbf{r}'\int d\mathbf{r}''\langle\delta\varepsilon^*(\mathbf{r}',0)\delta\varepsilon(\mathbf{r}'',\tau)\rangle e^{i(\mathbf{k}_i-\mathbf{k}_s)\cdot(\mathbf{r}'-\mathbf{r}'')-i\omega\tau} \tag{6}$$

The momentum transfer in the scattering is $\mathbf{q} = \mathbf{k}_i - \mathbf{k}_s$. Doing one integral over the illuminated volume $V$, the final formula for the scattered field correlation function is:

$$\langle \mathbf{E}_s^*(\mathbf{q},0)\cdot\mathbf{E}_s(\mathbf{q},\tau)\rangle = \frac{VE_0^2\omega^4\sin^2\Phi}{(4\pi)^2c^4r^2}\int d\mathbf{r}'\langle\delta\varepsilon^*(0,0)\delta\varepsilon(\mathbf{r}',\tau)\rangle e^{i\mathbf{q}\cdot\mathbf{r}'-i\omega\tau} \tag{7}$$

The intensity of the scattered light may be obtained by setting $\tau = 0$ in Eq. 7, and is proportional to the Fourier transform of the equal time correlation function $\langle\delta\varepsilon(0)\delta\varepsilon(\mathbf{r})\rangle$.

$$\langle E_s^2\rangle = \frac{VE_0^2\omega^4\sin^2\Phi}{(4\pi)^2c^4r^2}\int d\mathbf{r}'\langle\delta\varepsilon^*(0)\delta\varepsilon(\mathbf{r}')\rangle e^{i\mathbf{q}\cdot\mathbf{r}'} \tag{8}$$

If the intensity of the light is $cE^2/4\pi$, the power scattered into a solid angle $d\Omega$ is

$$dP_{sc} = (cE_s^2/4\pi)r^2 d\Omega$$

If the illuminated volume $V$ has a cross section $A$ and length $l$ along $\mathbf{k}_i$, the incident power is

$$P_0 = A(cE_0^2/4\pi)$$

From Eq. 8 we may calculate the Rayleigh ratio, which is the fraction of the incident power scattered into all directions per unit length of the incident beam.

$$R \equiv \frac{P_{sc}}{P_0 l} = \int d\Omega\frac{\omega^4\sin^2\Phi}{(4\pi)^2c^4}\int d\mathbf{r}'\langle\delta\varepsilon^*(0)\delta\varepsilon(\mathbf{r}')\rangle e^{i\mathbf{q}\cdot\mathbf{r}'} \tag{9}$$

We shall see how to calculate this in terms of the material parameters when we discuss nematic liquid crystals; it is only necessary to calculate:

$$\langle\delta\varepsilon^2(\mathbf{q})\rangle = \int d\mathbf{r}'\langle\delta\varepsilon^*(0)\delta\varepsilon(\mathbf{r}')\rangle e^{i\mathbf{q}\cdot\mathbf{r}'} \tag{10}$$

## 3. Light Scattering from the Isotropic Phase

Light scattering is simplest to discuss for the isotropic phase of nematogens. We shall work through it in some detail, as the principles involved are the same for light scattering from all phases, and it is necessary to be able to carry out such calculations during the analysis of experimental data.

Two typical thermotropic liquid crystal molecules are shown in Fig. 1. The upper molecule, octyloxy-cyanobiphenyl or 8OCB, has nematic and smectic A phases. The lower molecule is 4-(2-methylbutyl) phenyl 4′-(octyloxy)-(1,1′)-biphenyl-4-carboxylate (8OSI) which has the following sequence of phases (temperatures are given in °C):

$$\textit{Isotropic} \xrightarrow{174.5^\circ} N \xrightarrow{170.0^\circ} S_A \xrightarrow{133.4^\circ} S_C \xrightarrow{79.9^\circ} S_I \xrightarrow{75.1^\circ} S_J \xrightarrow{61.9^\circ} S_K$$

8OSI, as can be seen from the branched carbon chain, is chiral and, if the mixture of enantiomers is not racemic, will have a ferroelectric smectic C phase.

The simplest ordering in liquid crystals is that of nematics, in which there is a tendency for all of the molecules to orient along one direction. The ordered phase will be uniaxially birefringent and the ordering direction is the optic axis. There are, as yet, no known ferroelectric nematic phases. Thus one may describe the nematic order by considering the molecules to be rod-like objects which are equally likely to orient parallel or anti-parallel to the optic axis, and we take the molecules to to have a single axis of symmetry, although it is clear from the structure that is an approximation. The figure below illustrates a Cartesian coordinate system with the optic axis along the $z$ axis. It is most convenient to express the molecular order parameter and the dielectric constant tensor in such a coordinate system, illustrated in Fig. 2.

When a unit vector $\zeta$ is chosen to represent the molecular axis, the nematic order parameter, a tensor whose trace is zero[2], may be expressed in terms of the Cartesian components $\zeta_\alpha$, $\zeta_\beta$.

$$Q_{\alpha\beta} \equiv \frac{1}{2}\langle 3\zeta_\alpha\zeta_\beta - \delta_{\alpha\beta}\rangle \tag{11}$$

where the average is over a small macroscopic volume. With biaxial nematic order $Q_{\alpha\beta}$ would have five independent components, but it is simpler in a uniaxial nematic and may be written as

$$Q_{\alpha\beta} \equiv \frac{1}{2}S\langle 3n_\alpha n_\beta - \delta_{\alpha\beta}\rangle \tag{12}$$

where $n_\alpha$, $n_\beta$ are the Cartesian components of a unit vector $\boldsymbol{n}$ along the optic axis; this vector is commonly called the nematic *director*. In this uniaxial case, there is a scalar order parameter $S$ given by

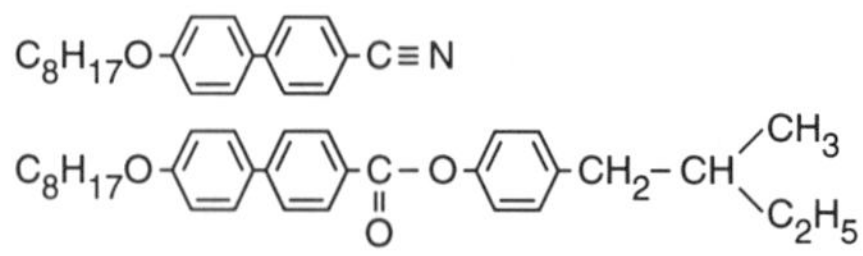


Fig. 1. Two typical thermotropic liquid crystal molecules. The upper one is commonly known as 8OCB, and the lower as 8OSI. Their phases are discussed in the text.

$$S \equiv \frac{1}{2}\langle 3\cos^2\theta - 1\rangle \tag{13}$$

when $\theta$ is the angle between $\boldsymbol{n}$ and $\zeta$. Rotational invariance in the isotropic phase requires that the anisotropy in all other quantities should be a scalar multiple of $Q_{\alpha\beta}$. The dielectric constant tensor may therefore be written as

$$\varepsilon_{\alpha\beta} = \bar{\varepsilon}\delta_{\alpha\beta} + \frac{2\Delta\varepsilon}{3}Q_{\alpha\beta} \tag{14}$$

where $\Delta\varepsilon$ would be the anisotropy in $\varepsilon_{\alpha\beta}$ for a completely ordered nematic. Eq. 14 provides a phenomenological macroscopic definition of the nematic order; however, local field corrections make it difficult to relate the $Q_{\alpha\beta}$ of Eq. 14 to the molecular definition in Eq. 11.

A formula for light scattered by the liquid crystal may be obtained[3] by combining Eq. 14 with Eq.s 9 and 10.

$$\begin{aligned}\frac{1}{P_0 l}\frac{dP_{sc}}{d\Omega} &= \left(\frac{\omega}{c}\right)^4 \frac{1}{(4\pi)^2}\int d\mathbf{r}'\langle\delta\varepsilon^*_{if}(0)\delta\varepsilon_{if}(\mathbf{r}')\rangle e^{i\mathbf{q}\cdot\mathbf{r}'} \\ &= \left(\frac{\omega}{c}\right)^4 \frac{V}{(4\pi)^2}\langle\delta^2_{if}(\mathbf{q})\rangle\end{aligned} \tag{15}$$

In Eq. 15 $\delta_{if}$ is the fluctuation in the component of $\varepsilon_{\alpha\beta}$ which couples incident light polarized along $\boldsymbol{i}$ with scattered light polarized along $\boldsymbol{f}$.

To calculate $\langle\delta^2_{if}(\mathbf{q})\rangle$ requires knowledge of the statistical mechanics of the liquid crystal. In the isotropic phase, we may use the mean field model proposed by de Gennes[2], in which the free energy per unit volume in the vicinity of the nematic-isotropic transition is expanded in a power series:

$$\begin{aligned}\phi = \phi_0(\rho,T) &+ \frac{1}{2}AQ_{\alpha\beta}Q_{\beta\alpha} - \frac{1}{3}BQ_{\alpha\beta}Q_{\beta\gamma}Q_{\gamma\alpha} + \frac{1}{4}C(Q_{\alpha\beta}Q_{\beta\alpha})^2 \\ &+ \frac{1}{2}L_1\partial_\alpha Q_{\beta\gamma}\partial_\alpha Q_{\beta\gamma} + \frac{1}{2}L_2\partial_\alpha Q_{\alpha\gamma}\partial_\beta Q_{\beta\gamma} + \ldots\end{aligned} \tag{16}$$

In Eq. 16 summation over repeated indices is implied, and a single fourth order invariant (a second rank tensor generally has two) suffices because the trace $Q_{\alpha\alpha} = 0$. The notation $\partial\alpha$ means

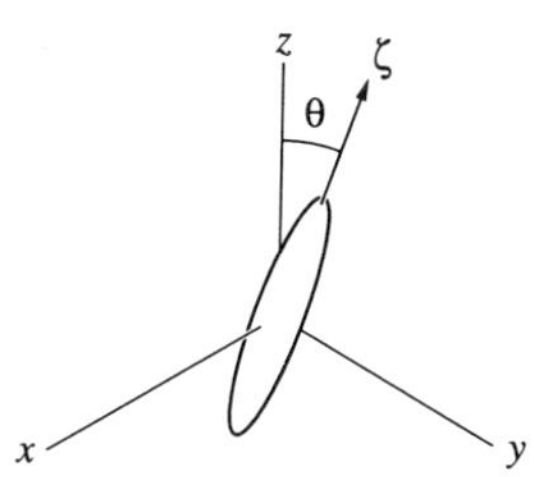


Fig. 2. Coordinates used to describe the orientation of a molecule in a nematic liquid crystal.

$\partial/\partial x_\alpha$. In the mean field approximation, which provides a good description of the nematic to isotropic transition[3], the coefficient $A$ has the form a(T - T$_c^*$). In the isotropic phase the equilibrium values of $Q_{\alpha\beta}$ are all zero, so the free energy cost of fluctuations in a volume $V$ is, to lowest order:

$$\delta\phi_V = \frac{1}{2}\int_V \{A Q_{\alpha\beta}(\mathbf{r})Q_{\beta\alpha}(\mathbf{r}) + L_1\partial_\alpha Q_{\beta\gamma}(\mathbf{r})\partial_\alpha Q_{\beta\gamma}(\mathbf{r}) + L_2\partial_\alpha Q_{\alpha\gamma}(\mathbf{r})\partial_\beta Q_{\beta\gamma}(\mathbf{r})\} \tag{17}$$

We use the following conventions in carrying out our Fourier transforms:

$$\int_V d\mathbf{r}e^{i(\mathbf{q}-\mathbf{q}')\cdot\mathbf{r}} = V\delta(\mathbf{q}-\mathbf{q}') \tag{18$a$}$$

$$Q_{\alpha\beta}(\mathbf{q}) = \int_V d\mathbf{r}Q_{\alpha\beta}(\mathbf{r})e^{i\mathbf{q}\cdot\mathbf{r}} \tag{18$b$}$$

$$Q_{\alpha\beta}(\mathbf{r}) = \sum_{\mathbf{q}} Q_{\alpha\beta}(\mathbf{q})e^{-i\mathbf{q}\cdot\mathbf{r}} \tag{18$c$}$$

The choice of the $z$ axis is arbitrary, so we may choose it to lie along $\mathbf{q}$; then taking the Fourier transform of Eq. 17 we find:

$$\delta\phi_V = \frac{V}{2}\sum_{\mathbf{q}}[(3A/2)\left(1+\xi_1^2q^2+\frac{2}{3}\xi_2^2q^2\right)Q_{zz}^2(\mathbf{q}) + (A/2)(1+\xi_1^2q^2)\{Q_{zz}(\mathbf{q})-Q_{yy}(\mathbf{q})\}^2$$
$$+2A(1+\xi_1^2q^2)Q_{xy}^2(\mathbf{q}) + 2A\left(1+\xi_1^2q^2+\frac{1}{2}\xi_2^2q^2\right)\{Q_{xz}^2(\mathbf{q})+Q_{yz}^2(\mathbf{q})\}] \tag{19}$$

In Eq. 19, two correlation lengths are defined:

$\xi_1 = \sqrt{L_1/A}$

and

$\xi_2 = \sqrt{L_2/A}$

We may apply the equipartition theorem to the different terms in Eq. 19, and thereby calculate the fluctuations in the components of the order parameter tensor. As an example, for the diagonal component along $\mathbf{q}$, we find

$$\langle Q_{zz}^2(\mathbf{q})\rangle = \frac{2k_BT}{3AV\left(1+\xi_1^2q^2+\frac{2}{3}\xi_2^2q^2\right)} \tag{20}$$

This may, in turn, be used to evaluate $\langle\delta\varepsilon_{if}^2(\mathbf{q})\rangle$ by means of Eq. 14. Not surprisingly, the general result is complicated. The simplest way to express it is in terms of the components of the polarization vectors in, and perpendicular to, the scattering plane determined by the wave vectors $\mathbf{k}_i$ and $\mathbf{k}_s$. Remember that $\boldsymbol{i}$ is orthogonal to $\mathbf{k}_i$ and $\boldsymbol{f}$ is orthogonal to $\mathbf{k}_s$. We let the components of $\boldsymbol{i}$ and $\boldsymbol{f}$ normal to the scattering plane be $i_V$ and $f_V$, while those in the scattering plane are $i_H$ and $f_H$. If the scattering angle of the light is $\theta$, we obtain

$$\langle\delta\varepsilon_{if}^2(\mathbf{q})\rangle = [(\boldsymbol{i}\cdot\boldsymbol{f})^2 - 3i_H f_H \cos^2(\theta/2)]^2 \frac{4\Delta\varepsilon^2 k_B T}{54VA\left(1+\xi_1^2 q^2 + \frac{2}{3}\xi_2^2 q^2\right)}$$

$$+ [(i_V f_V)^2 + (i_V^2 f_V^2 + i_H^2 f_H^2)\sin^2(\theta/2) + (i_H f_H)^2 \sin^4(\theta/2)] \frac{4\Delta\varepsilon^2 k_B T}{18VA(1+\xi_1^2 q^2)}$$

$$+ [i_V f_H + i_H f_V]^2 \cos^2(\theta/2) \frac{4\Delta\varepsilon^2 k_B T}{18VA\left(1+\xi_1^2 q^2 + \frac{1}{2}\xi_2^2 q^2\right)} \tag{21}$$

This expression is for $\boldsymbol{i}$, $\boldsymbol{f}$, and $\theta$ all measured in the *liquid crystal medium.* Standard methods for refraction at the interface of anisotropic media[4] should be used to find their values in the laboratory coordinate system. In practice, the correlation lengths $\xi_1$ and $\xi_2$ are so small[5] that they affect the intensity by only about a percent and can be neglected here. If we take an experimental geometry where $i_V = 1$ and $i_H = 0$, $\theta = 90°$ and both $f_V$ and $f_H$ are collected, then we calculate from Eq. 21

$$\frac{1}{P_0 l}\frac{dP_{sc}}{d\Omega} = \left(\frac{\omega}{c}\right)^4 \frac{V}{(4\pi)^2}\left[\frac{16\Delta\varepsilon^2 k_B T}{54Va(T-T_c^*)} + \frac{12\Delta\varepsilon^2 k_B T}{54Va(T-T_c^*)}\right] \tag{22}$$

The first term in Eq. 22 comes from the polarized ($i_V f_V$) scattering, and the second term is depolarized ($i_V f_H$). In Fig. 3 we show scattering obtained in the isotropic phase of p-methoxybenzylidene-p′-butylaniline (MBBA) under the experimental conditions described by Eq. 22. The incident light was from a helium neon laser with $\omega/c = 9.93 \times 10^4$ cm$^{-1}$. For most thermotropic liquid crystals the value of $\Delta\varepsilon$ is very close to 1. The solid line in Fig. 3 has the equation $0.029\,(T - 45.708)^{-1}$; from this we calculate that

$$\frac{A}{T-T_c^*} = 4.8 \times 10^5 \mathrm{erg K^{-1} cm^{-3}}$$

and that a first order nematic-isotropic transition occurs 1 K above $T_c^*$. The transition is weakly

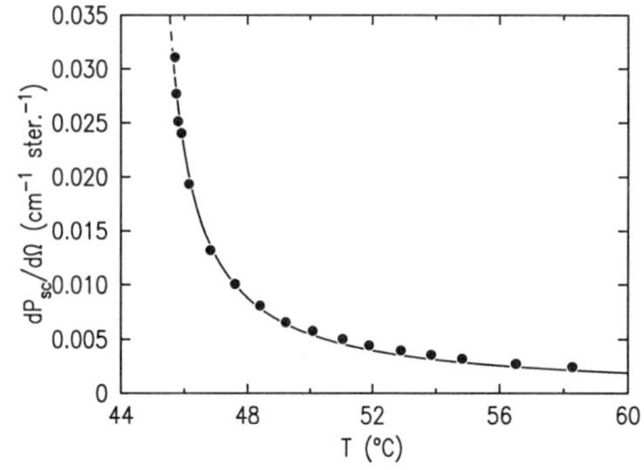


Fig. 3. Light scattered by short-range nematic order fluctuations in the isotropic phase of MBBA. The dashed and solid lines are a fit to Eq. 22.

first order because of the cubic invariant in Eq. 16. If Eq. 22 were integrated over all $4\pi$ steradians, the extinction of the incident laser beam due to scattering by short-range nematic order just one degree above the nematic-isotropic transition would be seen to be about 20% per cm. This scattering is strong compared to most normal transparent liquids, where the extinction due to density fluctuations is about 0.01% per cm. It is weak compared to the scattering from director modes, which we shall calculate in the next section. From the spectrum of the scattered light one may obtain information about the dynamical behavior of the nematic short-range order fluctuations. The experiments are difficult, as the linewidth of the scattered light is a few MHz - a range which still poses some experimental problems. For the interested reader, a discussion of the dynamical behavior of short-range nematic order may be found in the literature[3].

## 4. Scattering from Nematic Director Modes

In the nematic phase long-range molecular orientational order exists and the optic axis is established; hence the director $\boldsymbol{n}$ exists and the dielectric constant tensor may be obtained by substituting Eq. 12 into Eq. 14. The main source of light scattering comes from reorientation fluctuations of the director, and this can be seen from Eq.s 12 and 14 to produce only off-diagonal fluctuations in $\varepsilon_{\alpha\beta}$. It is customary to choose the unperturbed director to lie along the $z$ axis. In such a coordinate system light is scattered by

$$\langle\delta\varepsilon_{xz}^2(\mathbf{q})\rangle = \varepsilon_a^2\langle\delta n_x^2(\mathbf{q})\rangle \quad \text{and} \quad \langle\delta\varepsilon_{yz}^2(\mathbf{q})\rangle = \varepsilon_a^2\langle\delta n_y^2(\mathbf{q})\rangle \tag{23}$$

where $\varepsilon_a = S\,\Delta\varepsilon$, according to the model of the previous section.

The fluctuations $\delta n_x$ and $\delta n_y$ are the two normal modes (the Goldstone modes) associated with the establishment of nematic order. They are discussed in the original literature by the Orsay Liquid Crystal group[6], and the book by de Gennes[7] is also an excellent reference. We shall summarize the salient points here. Since the director is a unit vector field, it can have only three types of distortion which represent small departures from uniformity. These are splay ($\nabla\cdot\boldsymbol{n}$), twist ($\boldsymbol{n}\cdot\nabla\times\boldsymbol{n}$), and bend ($\boldsymbol{n}\times\nabla\times\boldsymbol{n}$). The elastic free energy cost per unit volume of the distortions may be written as

$$F_{nd} = \frac{1}{2}(K_1[\nabla\cdot\boldsymbol{n}]^2 + K_2[\boldsymbol{n}\cdot\nabla\times\boldsymbol{n}]^2 + K_3[\boldsymbol{n}\times\nabla\times\boldsymbol{n}]^2) \tag{24}$$

The phenomenological elastic constants ($K_1$, $K_2$, and $K_3$ for splay, twist, and bend, respectively) were introduced by Frank[8], and have units of energy per unit length (e.g. dynes). As the nematic phase has rotational symmetry about the $z$ axis or director, we are still free to choose the direction of the $x$ axis. It is common to choose it so that the wave vector $\mathbf{q}$ of the fluctuations lies in the $x$-$z$ plane. With this choice of coordinates, the Fourier analysis of Eq. 18 applied to Eq. 24 yields

$$\int_{\delta V}\delta F_{nd}\,d\mathbf{r} = \frac{V}{2}\sum_{\mathbf{q}}[(K_1q_x^2 + K_3q_z^2)\delta n_x^2(\mathbf{q}) + (K_2q_x^2 + K_3q_z^2)\delta n_y^2(\mathbf{q})] \tag{25}$$

By applying the equipartition theorem to 25, we can obtain the fluctuations

$$\langle\delta n_x^2(\mathbf{q})\rangle = \frac{k_BT}{V(K_1q_x^2 + K_3q_z^2)} \quad \text{and} \quad \langle\delta n_y^2(\mathbf{q})\rangle = \frac{k_BT}{V(K_2q_x^2 + K_3q_z^2)} \tag{26}$$

needed for Eq. 23. The two modes are degenerate (pure bend) when $q_z = 0$; generally mode 1 is a mixture of bend and splay, while mode 2 is a mixture of bend and twist.

The three Frank constants are all roughly the same size, about $1 \times 10^{-6}$ dynes. In order to simplify an estimate of the intensity of light scattering by director modes, we shall set them all equal to a single constant $K$. The scattered light will be depolarized; if we assume the incident light is polarized along the director, the scattering is given by

$$\frac{1}{P_0 l}\frac{dP_{sc}}{d\Omega} = \left(\frac{\omega}{c}\right)^4 \left(\frac{\varepsilon_a}{4\pi}\right)^2 \frac{k_B T (f_x^2 + f_y^2)}{K q^2} \qquad (27)$$

Integrating Eq. 27 over all solid angles is complicated because the value of $q^2$ is a complicated function of the scattering angle $\theta$. In an isotropic medium, one would have $q = 2(n\omega/c)\sin(\theta/2)$, which approaches 0 as $\theta$ approaches 0. However, in the liquid crystal the minimum value of $q$ is $\Delta n\,\omega/c$ as a result of the birefringence $\Delta n$. As an estimate, we may evaluate Eq. 27 at $\theta = 90°$ and multiply the result by $4\pi$. The result, using

$$K \sim 10^{-6} \text{ dyne}$$
$$\varepsilon_a \sim 0.5 \text{ and}$$
$$n \sim 1.5$$

is ~ 50 cm$^{-1}$. That is about 250 times greater than for isotropic phase scattering and nearly $10^{-6}$ larger than for normal liquids. The mean free path of a photon in a nematic phase is only a fraction of a mm. This is an important consideration in designing optical devices that use nematic phase liquids.

The dynamical behavior of the nematic director modes is also interesting and of practical interest, since it determines the response time of optical devices. We do not have the time to go into it here.

The original literature of the Orsay paper[6] is an excellent reference.

We should point out that only five viscosity coefficients are needed to describe nematics; it has since been shown[7] that the six coefficients of the original paper can be reduced by an Onsager relation

$$\alpha_6 - \alpha_5 = \alpha_2 + \alpha_3\,.$$

## 5. The Effects of Smectic A Order

In the smectic A (SmA) phase one-dimensional positional order is established in the form of a density wave whose wave vector $\mathbf{q}_0$ is parallel to the director $\boldsymbol{n}$[9]. The value of $q_0$ can be determined by X-ray scattering; typical values are 0.2 Å$^{-1}$. The disappearance of the SmA state on heating may be regarded as the simplest form of melting to occur in three dimensions, yet a satisfactory theoretical understanding still eludes us. It is possible to describe the SmA ordering by the following expression for the density

$$\rho = \rho_0\left[1 + \mathrm{Re}\left(\Psi_A e^{iq_0 z}\right)\right] \qquad (28)$$

where the order parameter $\Psi_A$ is a complex quantity with both a magnitude and phase.

The similarities between the SmA state and superconductors has been discussed by de Gennes[9] and McMillan[10]. One of them can be discussed while referring to Fig. 4, a representation of the molecular ordering in a SmA phase. We should point out that in most thermotropic SmA phases, the layers are not nearly so well defined as would be inferred from Fig. 4. In fact, the density modulation is nearly purely sinusoidal, as implied by Eq. 28. In one SmA sample, very careful measurements[11] showed the X-ray scattering from the second harmonic of the density wave to be at least $10^6$ weaker than the fundamental. There are several kind of SmA order possible. Fig. 4 illustrates the smectic $A_1$ phase, in which the wavelength of the density wave is equal to the molecular length. There is also a $A_2$ with a density wavelength twice that of the molecules. The molecule 8OCB of Fig. 1 has a $A_d$ phase with the wavelength intermediate between one and two molecular lengths; that is thought to arise through short-range antiferroelectric ordering of the molecules. All of these SmA phases are currently believed to have the same symmetry and to belong to the same universality class in statistical mechanics.

If we compute the line integral

$$\frac{1}{d}\int_A^B \boldsymbol{n}\cdot d\mathbf{r}$$

from any point A to any other point B in the liquid crystal, it measures the number of SmA layers crossed. In a sample free of defects, the result must be independent of the path chosen. That can only be so if $\nabla \times \boldsymbol{n} \equiv 0$. This is the analogue of the Meissner effect in the liquid crystal. Referring back to Eq. 24, we see that bend and twist distortions of the director field must be excluded from the SmA phase. Establishing the SmA order will have profound effects on the light scattering from director modes.

The most interesting phenomenon to study is the suppression of bend and twist by short-range SmA order in the nematic phase; this is the analogue of fluctuation diamagnetism in the disordered phase of superconductors, and provides a means to study the evolution of the SmA order as the second order transition is approached. The effects appear as an increase in the elastic constants $K_2$ and $K_3$ above the SmA transition. To discuss this quantitatively, one needs a model for the phase transition. A mean field model has been proposed[9,10] in which the free energy takes the form

$$\Phi = \Phi_N + \frac{1}{2}A\left[\Psi_A^2 + \xi_\parallel^2 \,|\nabla_\parallel \Psi_A|^2 + \xi_\perp^2 \,|(\nabla_\perp + iq_0\boldsymbol{n}_\perp)\Psi_A|^2\right] \tag{29}$$

where $\Phi_N$ is the free energy density of the nematic phase (when $\Psi_A = 0$), $A = a\,(T - T_c)^{-1}$, and $\xi_\parallel$ and $\xi_\perp$ are the correlation lengths along and transverse to $z$. The form of the transverse gradient term ensures that $\Phi$ is minimized when the SmA density wave is parallel to the director. Through a rather complicated calculation[12], Eq. 29 can be used to obtain the effect of the SmA short-range order on the nematic elastic constants. The bend and twist constants diverge proportional to the correlation length for SmA short-range order. However, the full divergence only occurs at $q$=0. Consequently, the result depends on the wavelength of the director mode; for $K_3$ it is

$$K_3 = K_3^0 + \frac{k_B T q_0^2}{8q_z}\left[\left(1 + \frac{1}{X^2}\right)\tan^{-1}X - \frac{1}{X}\right] \tag{30}$$

where

$$X = 1/2q_z\xi_{\parallel}$$

and $K_3^0$ is the normal nematic phase value. This value of $K_3$ may be used in Eq. 24. The SmA order also causes an increase in $K_2$, but does not alter the value of the splay constant $K_1$[13]. Eq. 30 may be used in conjunction with light scattering intensity measurements to determine the correlation length. Absolute measurements are not required; $X$ can be determined from the *shape* of the intensity as a function of temperature, and as $q_z$ is known, $\xi_{\parallel}$ is readily calculated. In the SmA phase, director mode 2 is non-hydrodynamic because there is a restoring force even at $q$=0 if the director is not perpendicular to the smectic layers. This means that Eq. 26 will be replaced by[13]

$$\langle \delta n_y^2(\mathbf{q}) \rangle = \frac{k_B T}{V(D + K_2 q_x^2 + K_3 q_z^2)} \tag{31}$$

where the new elastic constant $D$ is expected to vanish at the transition as $D = D_0\,(T_c/T - 1)^{\varphi}$. In Fig. 5 we show intensity data for mode 2 with $\mathbf{q} = q_z$ near the SmA to nematic transition in p-butoxybenzylidene p-octylaniline (4O.8). The solid line is a fit to the combined Eqs. 30 and 31. It yields

$$T_c = 337.100\ \mathrm{K} \qquad \text{and} \qquad X = X_0\,|\,T/T_c - 1\,|^{-\nu_{\parallel}}$$

with $X_0 = 0.0012$, and $\nu_{\parallel} = 0.72$. For $T < T_c$, one replaces $X_0$ by $1/2\ X_0$ and includes the elastic constant $D$. The exponent $\varphi$ is not well determined, but the fit gives $\varphi = 0.66 \pm 0.1$.

This method of analyzing the intensity of the scattered light leads values of $\xi_{\parallel}$ which are in excellent agreement with direct measurements by high resolution X-ray scattering. A comparison of results by the two methods is shown in Fig. 6. The solid line fit to the data shows

$$\xi_{\parallel} = \xi_{\parallel}^0 (T/T_c - 1)^{-\nu_{\parallel}}$$

with $\xi_{\parallel}^0 = 1.64\mathring{A}$, and $\nu_{\parallel} = 0.78 \pm 0.05$ in 4O.7. The SmA short-range order decays with a different correlation length, $\xi_{\perp}$, in directions perpendicular to the director. This has been measured by X-ray scattering and is typically 10 times smaller than $\xi_{\parallel}$. For this reason, the effects on the twist constant $K_2$ are smaller, and it is more difficult to determine $\xi_{\perp}$ as reliably by light scattering. However it has been confirmed for a number of SmA materials[14] that the values of $\xi_{\parallel}$, and $\xi_{\perp}$ determined independently by X-ray and light scattering are in good agreement.

The agreement of the two methods, as represented in Fig. 6, is a remarkable confirmation of some of the predictions of the theoretical models[9,10] for the SmA to nematic phase change. The models are not, however, in good agreement with many other features observed in a series of careful experiments[15]. The main difficulty lies in the fact that the two exponents $\nu_{\parallel}$ and $\nu_{\perp}$ describing the divergence of $\xi_{\parallel}$ and $\xi_{\perp}$ are not equal, which the model[16] requires. They also do not have the same value in different materials. The latter is now understood to be the result of crossover effects in some materials[17] where reentrant nematic phases are found. Phase transition behavior can also be modified, and crossover effects observed, when two order parameters are coupled. When the SmA and nematic order parameters are coupled, one can have tricritical behavior[10] and it has been predicted[18] that the coupling to the nematic order parameter can alter the observed correlation lengths in X-ray scattering measurements - although one still expects $\nu_{\parallel} = \nu_{\perp}$ in light scattering measurements. This has remained an unsolved problem for a decade.

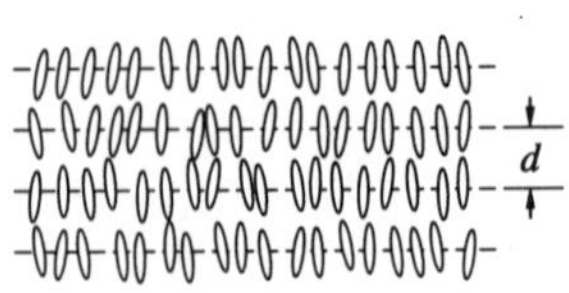


Fig. 4. A representation of the local molecular arrangement in a smectic A liquid crystal. The dashed lines indicate the mean position of the smectic "layers" or the maxima in the density wave, where the layer spacing $q_0$ is $2\pi/d$.

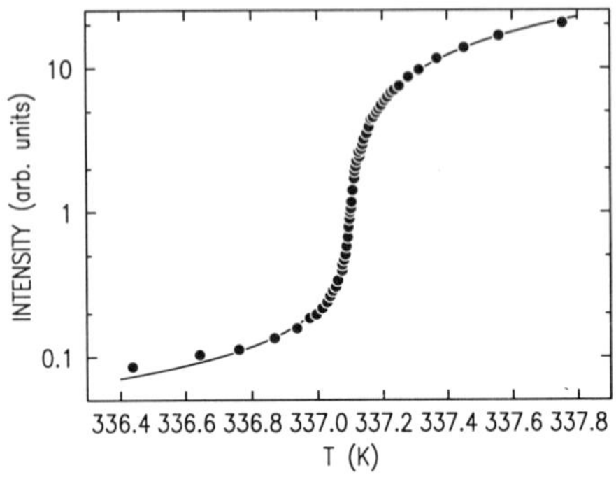


Fig. 5. Light scattered by director mode 2 with $q_x = 0$ in the nematic phase of 4O.8. The solid line is a fit to Eq. 30 combined with Eq. 31.

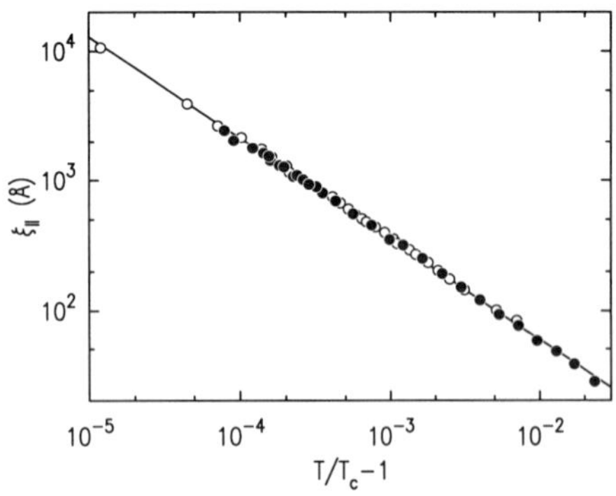


Fig. 6. The open circles are absolute values of the correlation length $\xi_{\parallel}$ for determined from analysis of light scattered by the liquid crystal 4O.7. The filled circles are determined by linewidths in high resolution X-ray scattering[11].

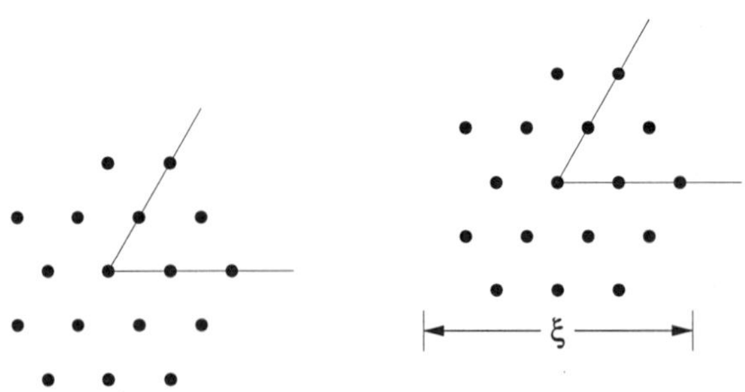


Fig. 7. Local molecular arrangements in a liquid with bond orientational long-range order.

Recent heat capacity studies[19] of materials with the SmA phase preceded by a nematic phase that exists over a very wide temperature range offer some hope. The wide nematic range indicates that the nematic order should be fixed in magnitude and cannot therefore couple significantly to the SmA order, and the heat capacity measurements indicate an exponent $\alpha$ for the specific heat divergence which agrees with the X-Y model prediction[16]. It is not yet known if other properties of these materials agree with the model predictions; if they do, one might expect that a refined theory of coupling between the nematic and SmA order parameters could be successful.

## 6. Light Scattering from Phases with Hexatic Order

Experiments during the past decade have verified that bond orientational, or "hexatic" long range order was the missing idea needed to classify the structure of the currently know thermotropic phases of liquid crystals[20]. The idea was an old one[21], and reemerged from theoretical investigations of melting in two dimensions[22]. An isotropic liquid has complete rotational and translational symmetry, while both of these symmetries have been broken in a crystalline solid. It is theoretically possible to have a phase in which only the rotational symmetry has been broken. This is illustrated in Fig. 7. The figure shows two widely separated clusters of molecular local order. Locally, a hexagonal positional order persists over a correlation length $\xi$. However, the imaginary lines joining neighboring molecules have the same orientation everywhere in the sample, as sketched in the figure. It is a strange liquid in which one does not know precisely how far to go to reach neighboring molecules, but one does know in what directions they lie. The phases which have this kind of order are the smectic I, smectic F, and a subset, called $B_H$, of smectic B phases. The SmI and SmF phases have a director which is tilted from normal to the smectic layers; thus they have a component of the director in the plane of the layers, which can both couple to the bond orientational order and scatter light strongly. We now proceed to discuss light scattered from these phases.

The tilted hexatic phases not only have the "layering" of the SmA phase, they also have molecular tilt (specified by a smectic C tilt order parameter, $\Phi$) and bond orientational (BO) order specified by

$$\Psi_H(\mathbf{r}) = \langle e^{i6\psi(\mathbf{r})} \rangle$$

where $\psi$ is the angle between the "bond" orientation at $\mathbf{r}$ and some reference axis. Both $\Phi$ and $\Psi$ have two degrees of freedom, a magnitude and a phase, and can be represented by a complex number, as was the SmA order in Eq. 28. The phase $\varphi$ of $\Phi$ indicates the tilt direction of the molecules, and the phase $\psi$ of $\Psi_H$ indicates the directions of the "bonds". In the SmI phase the molecules are tilted toward the nearest neighbors, while they are tilted towards next-nearest neighbors in the SmF phase.

A complete theory[23] of the hydrodynamics of smectic phases with BO is sufficiently complicated that we shall simplify matters by ignoring the compressibility of the fluid and consider only situations where $\mathbf{q}$ lies along directions of high symmetry. We begin with a phenomenological free energy expression valid for situations where both the SmC and the SmI order are well established, i.e.

$$<\Phi> \equiv \Phi_0 \approx 1 \quad \text{and} \quad <\Psi> \equiv \Psi_0 \approx 1$$

In a coordinate system where the smectic "layers" are in the $x$-$y$ plane and the tilt direction is along $x$, we might expect an elastic free energy density[24]

$$F_{CH}(\psi,\phi)=\frac{1}{2}(K_1^{\parallel}|\partial_x\phi|^2+K_1^{\perp}|\partial_y\phi|^2+K_6^{\parallel}|\partial_x\psi|^2+K_6^{\perp}|\partial_y\psi|^2$$

$$-2H_6\cos 6(\psi-\phi)-2H_{12}\cos 12(\psi-\phi)+\ldots) \qquad (32)$$

where $K_1^{\parallel}$ and $K_1^{\perp}$ are the two-dimensional bend and splay constants, respectively, for the component of the director in the $x$-$y$ plane. The stiffness constants for the BO Goldstone mode, $\psi$, are $K_6^{\parallel}$, and $K_6^{\perp}$, while $H_6$ and $H_{12}$ provide coupling between the $\psi$ and $\varphi$ modes. This coupling induces some SmI order in the SmC phase and means that, strictly speaking, the SmC to SmI transition has been eliminated. Practically, if the coupling is small, the effect will not be observable. The scattered light is calculated by substituting into Eq. 15:

$$\delta\varepsilon_{xy}(\mathbf{r})=\varepsilon_a\sin^2\Phi_0\phi(\mathbf{r}) \text{ and } \delta\varepsilon_{zy}(\mathbf{r})=-\varepsilon_a\sin\Phi_0\cos\Phi_0\phi(\mathbf{r})$$

As usual for coupled modes, Eq. 32 is diagonalized by two normal modes: an "optic" mode $\theta_- = \psi - \varphi$, and an acoustic mode $\theta_+ = f\psi + (1-f)\varphi$, where

$$f=[1-(K_1^{\parallel}q_x^2+K_1^{\perp}q_y^2)/(K_6^{\parallel}q_x^2+K_6^{\perp}q_y^2)]^{-1}$$

In practice

$$f\approx 1+K_1/K_6$$

which is very close to 1. If the incident light propagates perpendicular to the film (along $z$) and the scattering angle is $\theta_s$ one obtains in pure splay geometry ($\mathbf{q} = q_y$)

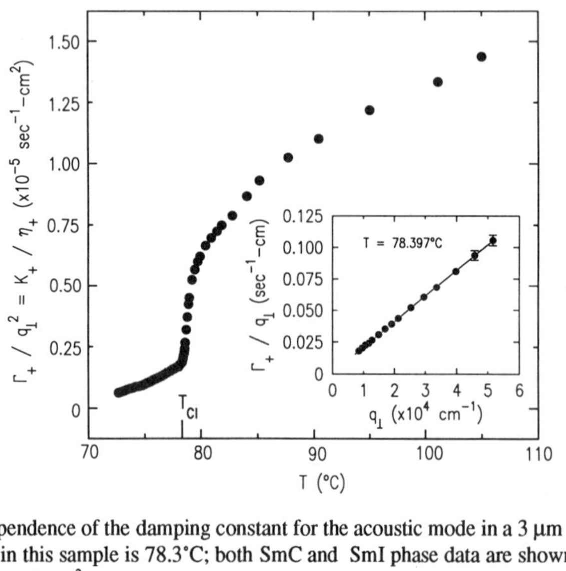


Fig. 8. Temperature dependence of the damping constant for the acoustic mode in a 3 μm thick film of 8OSI. The SmC to SmI transition in this sample is 78.3°C; both SmC and SmI phase data are shown. The insert shows that the damping is proportional to $q^2$.

$$< \delta\varepsilon_{if}^*(q_y,0)\delta\varepsilon_{if}(q_y,\tau) > = \frac{\varepsilon_a^2 \sin^4\Phi_0 \cos^2\theta_s k_B T}{V}\left(\frac{e^{-\Gamma_+^{\perp}\tau}}{(K_1^{\perp}+K_6^{\perp})q_y^2}+\frac{e^{-\Gamma_-^{\perp}\tau}}{H+K_1^{\perp}q_y^2}\right) \tag{33}$$

where $H \equiv 36(H_6+4H_{12})$. In pure bend geometry ($\mathbf{q}=q_x$) one calculates

$$< \delta\varepsilon_{if}^*(q_x,0)\delta\varepsilon_{if}(q_x,\tau) > = \frac{\varepsilon_a^2 \sin^2\Phi_0 \sin^2(\Phi_0+\theta_s) k_B T}{V}\left(\frac{e^{-\Gamma_+^{\parallel}\tau}}{(K_1^{\parallel}+K_6^{\parallel})q_x^2}+\frac{e^{-\Gamma_-^{\parallel}\tau}}{H+K_1^{\parallel}q_x^2}\right) \tag{34}$$

The light scattering is studied by suspending films freely across an opening of ≈3 mm diameter in a glass or metal substrate. The $x$-$y$ plane is then that of the film, and the film thickness is readily controlled between two molecules and several microns. In Fig. 8 the light scattered by a thick film of 8OSI is shown as a function of temperature. Eq.s 33 and 34 may be used to describe scattering in the SmC phase by setting $H$ and $K_6$ to zero. In the SmI phase, the optic and acoustic modes are most easily separated dynamically; both modes are overdamped, but the optic mode relaxes much more rapidly at a rate independent of $q$ for small $q$. The relaxation rate for the acoustic mode has the form

$$\Gamma_+ = (K_1+K_6)q^2/\eta_+$$

as can be seen in Fig. 8. Data for the optic mode, which has

$$\Gamma_- = (H+K_1q^2)/\eta_-$$

are shown in Fig. 9. The former shows a marked slowing-down as $T \rightarrow T_{IC}^+$ in qualitative agreement with a microscopic theory[25] of hexatic hydrodynamics in *two dimensions*, where the damping is predicted to decrease like the square of the bond-orientational correlation length $\xi_6$ in the experimental regime where $q^{-1} > \xi_6$. The theory has not been extended to a three-dimensional

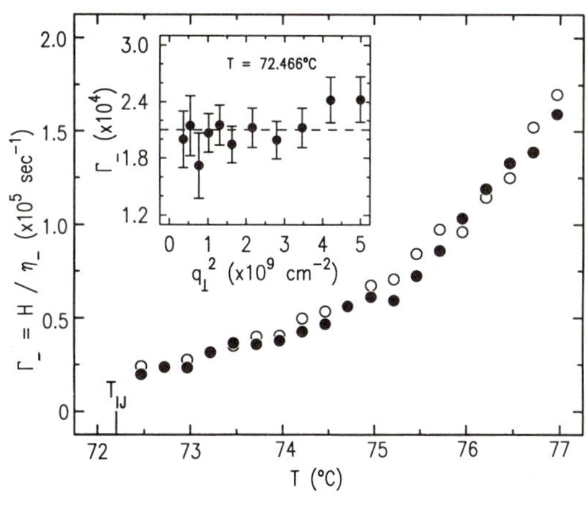


Fig. 9. Temperature dependence of the damping constant for the optic mode in the SmI phase of a 3 μm thick film of 8OSI. The SmC to SmI transition in this sample is 78.3°C. The insert shows the damping time is independent of $q$. The open circles are for $q^2 = 4.20\times10^9$ cm$^{-2}$ and the open circles are for $q^2 = 2.78\times10^9$ cm$^{-2}$.

stack of hexatic layers. Fig. 9 shows that $H >> K_1q^2$ for the range of $q$ studied and that $\eta_-$ decreases significantly as $T \to T_{IC}^-$; at present there is no theoretical calculation for the critical behavior of the optic mode dynamics.

In Fig. 10 we show the temperature dependence of the elastic constant $K_6^{\parallel}$ obtained from light scattering measurements of a thick film of 8OSI. The data have been scaled to compare with X-ray scattering measurements[26] of the positional correlation length, $\xi$. In two dimensions, this elastic constant is predicted[27] to scale as $\xi^2$ which is proportional to the core energy of a dislocation. The result is not anticipated in three dimensions (thick films) and awaits a theoretical explanation.

The light scattered is sufficiently intense that one may, with some care, study films as thin as two molecules[28,29,30]. The films are stable, and persist for days, or even weeks. This makes it possible to study two-dimensional systems without an interfering substrate, as well as the crossover from two to three dimensions.

We show in Fig. 11 intensity data obtained from films of 8OSI which are two molecules thick. The reciprocal of the intensity multiplied by $2q_\perp$ is plotted for splay and bend modes as a function of $q_\perp$ for two different temperatures. The intensity is dominated by the acoustic mode, and the solid lines are fits to Eq. 33 and Eq. 34. The leading terms would give straight lines, such as the dotted line in the upper panel. The curvature in the bend geometry comes from the factor $\sin^2(\Phi_0 + \theta_s)$. The optic mode in very thin films of 8OSI is too fast to measure the decay with an autocorrelator, although it can be measured[29] in the related material 8SI, so it can be detected only through its effect on the intensity, shown in Fig. 11 by the change in curvature of the splay data at large $q_y$ for decreasing temperature. Quantitatively, these observations suggest that the coupling parameter $H$ may be 10 - 100 times greater in a two-layer film of 8OSI than in two-layer 8SI. For 8OSI, the splay geometry is more strongly influenced by the optic mode, as the constant $K_6^{\parallel}$ is about four times $K_6^{\perp}$, which makes the acoustic and optic mode scattering somewhat closer in intensity.

In fact, the ratio of the splay to bend elastic constants, essentially $K_6^{\perp}/K_6^{\parallel}$, contains useful information[24] about the melting process of the crystalline SmJ (or SmG) phase into the hexatic liquid crystalline SmI (or SmF) phase. Because the molecular tilt slightly breaks the six-fold

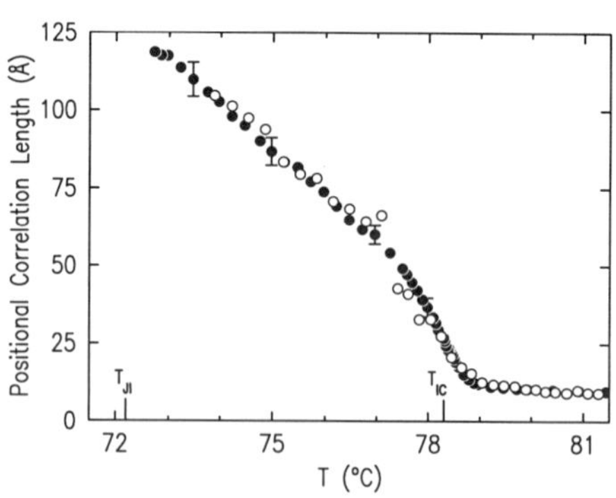


Fig. 10. Bond orientational stiffness constant (filled circles) obtained from light scattering in the SmI phase of a 3 μm thick film of 8OSI and scaled to compare with X-ray scattering measurements of the positional correlation length, $\xi$, which are shown as open circles.

rotational symmetry, there are two different types of dislocations possible in the crystalline phase. For example, in the SmJ phase, these are distinguished by whether the Burgers' vector lies along the tilt direction (type I) or at 60° to it (type II). The melting is predicted[31] to be different, depending which kind of dislocation pairs unbind first; in particular, for the hexatic phase,

$$\frac{K_6^{\perp}}{K_6^{\parallel}} = \left[\frac{a_0^2}{d^2}\left(\frac{n_I}{n_{II}} + \frac{1}{4}\right)\right]^{\pm 1} \tag{35}$$

where $n_I$, $n_{II}$ are the number densities of free type I, II dislocations, $a_0$, $d$ are the lattice constants along and perpendicular to a type I Burgers' vector ($d \approx \sqrt{3}/2a_0$), the "+" sign is taken for a SmI phase, and the "-" for SmF. If type I pairs unbind first (type I melting), a highly anisotropic situation arises, with $n_I >> n_{II}$. For type II melting, however, $n_I \approx n_{II}$. Thus, type II unbinding predicts

$$K_6^{\perp}/K_6^{\parallel} \approx 0.6 \text{ (SmF) or } 1.6 \text{ (SmI)},$$

and type I melting gives

$$K_6^{\perp}/K_6^{\parallel} \ll 0.6 \text{ (SmF)} \quad \text{or} \quad >> 1.6 \text{ (SmI)}.$$

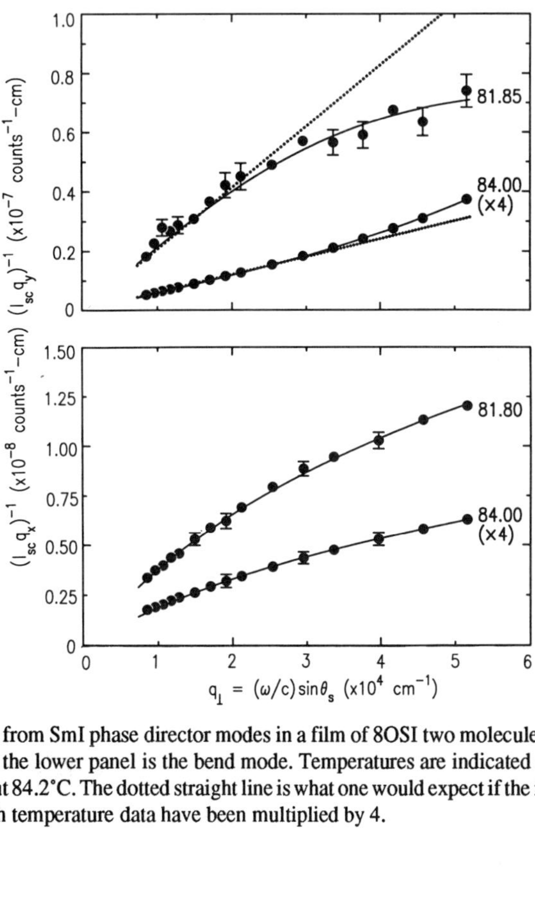


Fig. 11. Light scattering from SmI phase director modes in a film of 8OSI two molecules thick. The upper panel is for splay geometry, and the lower panel is the bend mode. Temperatures are indicated on the figure, and the SmI to SmC transition is about 84.2°C. The dotted straight line is what one would expect if the intensity were proportional to $q_y^{-2}$. Note that the high temperature data have been multiplied by 4.

The experimental values are

$$K_6^{\perp}/K_6^{\parallel} \approx 4 \quad \text{and} \quad K_6^{\perp}/K_6^{\parallel} \approx 0.6$$

for a hexatic two-layer film of 8OSI[30] and for a two-layer 8SI film[29], respectively.

These results *suggest* a type I melting process in 8OSI (with SmI hexatic phase) and a type II process in 8SI (with SmF hexatic phase). Although the hexatic phase in both bulk 8OSI and 8SI is purely SmI, its nature in two-layer films has not been definitively established. X-ray scattering experiments[26] suggest that at least near the SmC phase, two-layer 8OSI films are SmI. On the other hand, a substantial SmF region has been shown[29] to exist in five-layer films of 8SI. Since recent theory[32] suggests that a SmI - SmF transition should be driven by a softening of the optic mode (i.e., $H \to 0$) and experimentally $H(8SI) \ll H(8OSI)$, 8SI thin films may lie intrinsically closer to a SmF phase in a generalized phase diagram for hexatic materials. Before firm conclusions can be drawn about the melting process, the precise nature of hexatic order in thin films of the two substances must be confirmed by *direct* methods - e.g., X-ray[33] or electron diffraction[34]. Nevertheless, the striking difference in the magnitude of the bond-tilt coupling and the opposite anisotropies of the elasticities $K_6^{\perp}$ and $K_6^{\parallel}$ in 8OSI and 8SI are quite fascinating results, as the only difference between the two molecules is that 8SI has the $C_8H_{17}O$ at the left end in Fig. 1 replaced by $C_8H_{17}$.

## 7. Summary

In this Chapter, we have presented in as tutorial a way as we could, enough theoretical details to calculate the spectrum and intensity of light scattered by thermotropic liquid crystal phases. This is sufficient to enable a quantitative analysis of experiments, and we have illustrated its use with a number of examples from isotropic, nematic, and smectic phases.

ACKNOWLEDGEMENTS. This article is based upon work supported by the National Science Foundation under grants DMR-8619234, DMR-8719217, DMR-8813164, DMR-9014886. We appreciate helpful discussions with our colleague Mark Spector.

### References

1. M. Born, and E. Wolf, Principles of Optics, Pergamon Press, New York, 1965.
2. P.G. de Gennes, Phys. Letters A **30**, 454 (1969).
3. T.W. Stinson, J.D. Litster, and N.A. Clark, J. Phys. (Paris) **33**, C1-69 (1972).
4. L.D. Landau, and E.M. Lifshitz, Electrodynamics of Continuous Media, Addison-Wesley, Reading, Mass. (1960).
5. T.W. Stinson, and J.D. Litster, Phys. Rev. Lett. **30**, 688 (1973).
6. Groupe d'Etude des Cristaux Liquides (Orsay), J. Chem. Phys. **51**, 816 (1969).
7. P.G. de Gennes, The Physics of Liquid Crystals, Oxford University Press, 1974.
8. F.C. Frank, Disc. Faraday Soc. **25**, 19 (1958).
9. P.G. de Gennes. Solid. State. Comm. **10**, 753 (1972).
10. W.L. McMillan, Phys. Rev. A **4**, 1238 (1971).
11. B. Ocko, Ph.D. Thesis, Department of Physics, M.I.T. (1984).
12. F. Jänig and F. Brochard, J. Phys. (Paris) **34**, 411 (1973).
13. H. Birecki and J.D. Litster, Mol. Cryst. Liq. Cryst. **42**, 33 (1977).
14. S. Sprunt, L. Solomon, and J.D. Litster, Phys. Rev. Lett. **53**, 1923 (1984).
15. J.D. Litster, Phil. Trans. R. Soc. Lond. A **309**, 145-153 (1983), and references therein.
16. C. Dasgupta and B.I. Halperin, Phys. Rev. Lett. **47**, 1556 (1981).
17. A.R. Kortan, H. von Kanel, R.J. Birgeneau, and J.D. Litster, J. Phys (Paris) **45**, 529 (1984).

18. T.C. Lubensky, J. Chim. Phys. (Paris) **80**, 31 (1981).
19. C.W. Garland, G. Nounesis, and K.J. Stine, Phys. Rev. A **39**, 4919 (1989); G. Nounesis, C.W. Garland, G. Sigaud, and N.H. Tinh, Phys. Rev. A **43**, 1849 (1991); C.W. Garland , Chapter 11 in this volume.
20. J.D. Brock, R.J. Birgeneau, J.D. Litster, and A. Aharony, Physics Today, July 1989, p 52.
21. L.D. Landau, Phys. Z. Sowjet. 11, 545 (1937); N.D. Mermin, Phys. Rev. **176**, 250 (1968).
22. B.I. Halperin, D.R. Nelson, Phys. Rev. Lett. **41**, 121 (1978); D.R. Nelson, B.I. Halperin, Phys. Rev. B **19**, 2457 (1979); A.P. Young, Phys. Rev. B **19**, 1855 (1979).
23. H. Pleiner, H.R. Brand, Phys. Rev. A **29**, 911 (1984).
24. S. Sprunt, M.S. Spector, J.D. Litster, to be published.
25. A. Zippelius, B.I. Halperin, D.R. Nelson, Phys. Rev. B **22**, 2514 (1980).
26. J.D. Brock, D.Y. Noh, B.R. McClain, J.D. Litster, R.J. Birgeneau, A. Aharony, P.M. Horn, and J.C. Liang, Z. für Physik B **74**, 197 (1989).
27. R. Bruinsma, D.R. Nelson, Phys. Rev. B **23**, 402 (1981).
28. C.Y. Young, R. Pindak, N.A. Clark, and R.B. Meyer, Phys. Rev. Lett. **40**, 773 (1978).
29. S.B. Dierker, R. Pindak, Phys. Rev. Lett. **59**, 1002 (1987).
30. S. Sprunt, J.D. Litster, Phys. Rev. Lett. **59**, 2682 (1987).
31. S. Ostlund, B.I. Halperin, Phys. Rev. B **23**, 335 (1984).
32. J.V. Selinger, D.R. Nelson, Phys. Rev. Lett. **61**, 416 (1988).
33. S. Gierlotka, J. Przedmojski, B. Pura, Liquid Crystals **3**, 1535 (1988).
34. M. Cheng, J.T. Ho, S.W. Hui, R. Pindak, Phys. Rev. Lett. **59**, 1112 (1987).

Chapter 10

# CALORIMETRIC STUDIES OF LIQUID CRYSTAL PHASE TRANSITIONS: STEADY STATE ADIABATIC TECHNIQUES

J. THOEN

## 1. Introduction

During the last couple of decades calorimetric investigations have contributed substantially to the better understanding of phase transitions in general and of phase transitions in liquid crystals in particular. Although differential scanning calorimetry (DSC) is an important general characterisation tool, detailed thermal information near liquid crystal phase transitions is, however, mainly obtained with high-resolution adiabatic and ac calorimetric techniques. High-resolution ac calorimetry is described in Ref.1. Here we concentrate on results obtained by means of adiabatic methods. Special emphasis will be given to the possibility to distinguish, with some adiabatic scanning calorimetric techniques, between first-order and second-order phase transitions and obtain quantitative results for the true latent heat, when present.

The systems and phase transitions discussed and the results presented here focus on (1) the transition from the isotropic to liquid crystalline phases, (2) the nematic-smectic A transition for monolayer and partially bilayer smectic $A_d$ systems, (3) the nematic-smectic A-smectic C multicritical point. These choices are complementary to the subjects discussed in Ref.1, and taken together, give quite a broad review of high-resolution calorimetric investigations on thermotropic liquid crystals.

## 2. Steady State Adiabatic Calorimetry

Traditionally heat capacity measurements are carried out adiabatically by means of a heat pulse method. A known amount of heat $\Delta Q$ is supplied to the sample and its corresponding temperature rise $\Delta T$ is measured. This method has been used for detailed studies of the liquid gas critical points[2-4] and critical demixing in mixtures of ordinary liquids[5,6]. It was also applied by Anisimov and coworkers [7,8] to liquid crystal phase transitions, in particular for nematic-isotropic (NI) transitions. Recently Sorai et al.[9] used this method to measure the heat capacity $C_p$ of a discotic mesogen over a very large temperature range (17 to 397 K).

---

J.Thoen - Laboratorium voor Akoestiek en Warmtegeleiding, Departement Natuurkunde, Katholieke Universiteit Leuven, Celestijnenlaan 200D, B-3001 Leuven, Belgium

The above method is rather time consuming and not very suitable for automatic data collection. In this respect adiabatic scanning calorimetry has distinct advantages. It has been successfully used in the 1970's for high-resolution work near liquid-gas[10,11] and liquid-liquid critical points[12]. About ten years ago we[13], and later on also Anisimov et al.[14], started using it for liquid crystal phase transitions. The different possible modes of operation of a adiabatic scanning type calorimeter can be illustrated by means of the schematic representation of such a calorimeter in Fig. 1.

Instead of supplying heat to the sample (e.g. via the electric heater $H^1$ in stage 1) during a certain time, it is, however, also possible to operate in dynamic modes by continuously applying (or extracting for cooling runs) heating power to stage 1. The total heat capacity $C_1$, sample ($C_p$) plus holder ($C_h$), is then given by

$$C_1 = C_h + C_p = P_1/\dot{T}_1 = (P_1^m + P_1^l)/\dot{T}_1 \tag{1}$$

Besides the applied power $P_1^m$, the total power $P_1$ contains also a term $P_1^l$, representing heat leaks to the thermal shielding (stage 2). The term $P_1^l$, which can be positive or negative, should be negligible compared to $P_1^m$ or known. By keeping $P_1$ or $\dot{T}_1$ constant, combined with increasing or decreasing the temperature of stage 1, four practical modes of operation are obtained. In Table 1 the ideal working conditions for the four modes are summarized.

No conditions for stage 4 are given because it is usually kept at a constant temperature below the temperature of stage 3. If desired it can also be kept at a constant temperature difference with stage 3. Let us concentrate on the two heating modes 1 and 2, cooling modes are quite similar. For both heating modes one wants zero heat transfer between the stages 1, 2 and 3, and a constant heating power $P_1^m$ or $P_2^m$ is applied to stage 1 or stage 2.

Table 1. Modes and working conditions for the scanning calorimeter

| | mode 1 (heating) | mode 2 (heating) | mode 3 (cooling) | mode 4 (cooling) |
|---|---|---|---|---|
| $\Delta T_{12} = T_1 - T_2$ | 0 | 0 | $K^{(a)} > 0$ | $K > 0$ |
| $\Delta T_{23} = T_2 - T_3$ | 0 | 0 | $K > 0$ | $K > 0$ |
| $\dot{T}_1$ | $P_1^m/C_1$ | $=\dot{T}_2$ | $P_1^l/C_1$ | $=\dot{T}_2$ |
| $\dot{T}_2$ | $=\dot{T}_1$ | $P_2^m/C_2$ | $=\dot{T}_1$ | $(P_2^l - P_1^l)/C_2$ |
| $P_1^m$ | $K > 0$ | $P_2^m C_1/C_2$ | 0 | $-P_1^l - (P_1^l - P_2^l)C_1/C_2$ |
| $P_1^l$ | 0 | 0 | $K < 0$ | $K < 0$ |
| $P_2^m$ | $P_1^m C_2/C_1$ | $K > 0$ | $P_1^l(1 + C_2/C_1) - P_2^l$ | 0 |
| $P_2^{l(b)}$ | 0 | 0 | $K < 0$ | $K < 0$ |

(a) : The symbol K represents a constant value for the corresponding quantity in the first column.
(b) : Stands only for input from or output to the outside of stage 2.

In the former case one keeps the numerator of Eq. 1 constant, in the latter case the denominator is kept constant if the heat capacity $\dot{C}_2$ of the (reference) stage 2 is constant. In order to obtain $C_1$ with mode 1 it is necessary to obtain $\dot{T}_1$ from a careful measurement of the temperature versus time evolution of stage 1. In mode 2 one has to measure accurately the changing power $P_1^m$. Provided that $C_2(T)$ and $P_2^m$ are known one arrives at $C_1(T)$ from

$$C_1/C_2 = P_1^m/P_2^m \tag{2}$$

In the cooling modes 3 and 4 the calorimeter is operated in such a way that in the stationary state stages 1 to 3 cool down with a constant rate. This requires a temperature setting of stage 4 below the final temperature of the run. It is also essential that the thermal resistances between stages are sufficiently large and constant. In mode 3 the power $P_1^m$ is set equal to zero and the cooling rate $\dot{T}_1$ becomes inversely proportional to $C_1$ provided the heat loss $P_1^l$ is constant. In mode 4 the temperature differences are set similarly, but $P_2^m$ is now zero. The sample cools down with a constant rate provided $C_2$ and the thermal resistances remain constant. The power $P_1^m$ applied to stage 1 in order to keep $\Delta T_{12}$ constant has to be measured to arrive at $C_1$.

Although high-resolution cooling runs are possible and sometimes necessary for checking hysteresis effects, the heating runs are much more convenient to operate. In order to satisfy the conditions of zero temperature difference between stages 1, 2 and 3 in mode 1 and 2, sensitive servosystems are needed. This is particularly important for $\Delta T_{12}$. This might, however, become very difficult for mode 2 near a second-order or nearly second-order phase transition, where, because of the sharp increase of the heat capacity $C_1$, the power needed to keep a constant rate ($T_1 = T_2$ ) increases very rapidly. Moreover, if the phase transition is first-order it will not be possible to deliver instantaneously the finite amount of latent heat to the sample, and $T_1$ will lag behind $T_2$. This basic problem also occurs in differential scanning calorimeters because they essentially operate similarly to mode 2 (and at quite large scanning rates). The opposite situation occurs in the first mode. The servo system will have to deliver less power to stage 2 because $\dot{T}_1$

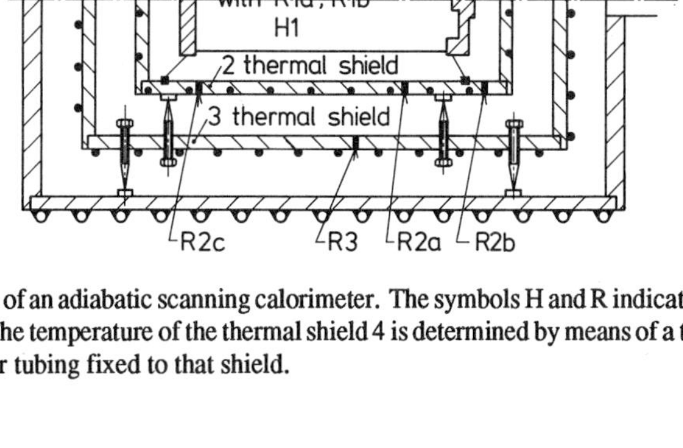


Fig. 1. Schematic diagram of an adiabatic scanning calorimeter. The symbols H and R indicate respectively, electric heaters and thermistors. The temperature of the thermal shield 4 is determined by means of a temperature controlled water flow through copper tubing fixed to that shield.

will slow down near the transition. First-order transition also do not pose a problem because in principle the rate $\dot{T}_1$ becomes zero for a time interval give by

$$\Delta t \equiv t_f - t_i = \Delta H_L / P_1 \tag{3}$$

with $\Delta H_L$ the latent heat of the transition. In fact, the direct experimental result T(t) immediately gives the enthalpy as a function of temperature by

$$H_1(T) = H_1(T_S) + P_1(t - t_s) \tag{4}$$

with $T_s$ the starting temperature of the scanning run at the time $t_s$.

Essential in the operation of a scanning calorimeter is the high-resolution control and measurement of the temperature and of the temperature difference between stage 1 and 2. For this purpose each stage containes thermistors (R) and heaters (H). In the past high-sensitivity servosystems for controlling the temperature difference between two stages, were based on ac bridges containing a thermistor of each stage[15]. In such a scheme a possible source of error resides in the unavoidable mismatch of the thermistors. This problem could be minimized by periodical (manual) readjustment of the bridge, or by compensating resistor networks[15]. This problem can largely be solved with today's sensitivity of digital multimeters, programmable power sources and flexibility of microcomputers in a software approach[16]. More details, also on the mechanical construction of an adiabatic scanning calorimeter, can be founds elsewhere[15].

## 3. Isotropic to Mesomorphic Transitions

### *3.1. The Isotropic to Nematic and Isotropic to Smectic Transitions*

A common difference between an isotropic liquid and all liquid crystalline states is the orientational order of the molecules. In fact this is the only difference between the nematic and the isotropic phase. A proper description of the orientational order requires the introduction of a tensor of the second rank[17,18]. This tensor can be diagonalized, and for anisotropic liquids with uniaxial symmetry one can arrive at a description by only one order parameter. For biaxial symmetry a second independent order parameter is required. The thermodynamic behavior in the vicinity of the NI transition is usually described in terms of the mean-field Landau-de Gennes theory[17]. For the uniaxial nematic case one can obtain the expansion of the free energy F in terms of the modulus of an order parameter Q.

$$F - F_0 = \frac{1}{2}AQ^2 - \frac{1}{3}BQ^3 + \frac{1}{4}CQ^4 + \frac{1}{6}DQ^6 + \ldots \tag{5}$$

In the isotropic phase one has $Q = 0$ and in the nematic phase $Q \neq 0$. In Eq. 5 one has $A=a(T - T^*)/T_{NI}$ and $B > 0$ for uniaxial phases with positive anisotropy. The presence of the cubic term, which does not disappear at $T_{NI}$, leads to a first order transition with a finite discontinuity in the order parameter ($Q_{NI} = 2B/3C$). $T^*$ is the stability limit of the isotropic phase. For B = 0 one would have a normal second-order transition at $T_{NI}$, because one has $T_{NI} = T^*/(1-2B^2/9aC)$. The specific heat excess in the nematic phase is given by[19]

$$C_P = aQ\left(\frac{\partial Q}{\partial T}\right)_P = \frac{a^2}{2CT_{NI}}\left[1+\frac{B}{2(aC)^{1/2}}\left(\frac{T^{**}-T}{T_{NI}}\right)^{-1/2}\right] \tag{6}$$

with $T^{**}$ the stability limit of the nematic phase. One thus has a singular contribution with an exponent $\alpha = 1/2$ in the nematic phase, resulting in a jump in $C_p$ at $T^{NI}$ equal to $\Delta C_p = 2a^2/CT_{NI}$. In the mean field second- order (B = 0 and C > 0) case the singular contribution is absent and only a jump $\Delta C_p = a^2/2CT_{NI}$ is present. The singular contribution follows from the mean-field behavior (with the critical exponent $\beta = 1/2$) of the order parameter, given by

$$Q = \frac{B}{2C}\left[1+\frac{2(aC)^{1/2}}{B}\left(\frac{T^{**}-T}{T_{NI}}\right)^{1/2}\right] \tag{7}$$

For the enthalpy discontinuity at $T_{NI}$ one obtains

$$H_I - H_N = 2aB^2/9C^2 \tag{8}$$

For B = 0 there is no enthalpy jump or latent heat and one has a critical point, or Landau point, on an otherwise first-order line as a function of B.

The Landau-de Gennes theory is a mean-field theory and classical critical exponents are thus obtained. Since, however, the molecular interactions are also short range in liquid crystals, one should expect nonclassical exponents to apply, and one should at best find qualitative agreement between experimental results and predictions by the Landau-de Gennes theory. The prediction that, because of the presence of the (small) cubic term in Eq. 5, the transition should be weakly first-order is in agreement with experimental observations. In Fig.2 parts of the enthalpy curves near $T_{NI}$ are given for three different compounds of the alkylcyanobiphenyl homologous series[20,21]. In that figure the dotted parts of the curves represent the latent heat jumps at the transitions. Numerical values for $\Delta H_L$ are also given in Table 2. These latent heat values

Table 2. Latent heats (kJ/mol) for alkylcyanobiphenyls

| compound | N - I | $S_A$ - I |
|---|---|---|
| 5CB | $0.390 \pm 0.003$ | - |
| 6CB | $0.293 \pm 0.003$ | - |
| 7CB | $0.578 \pm 0.004$ | - |
| 8CB | $0.612 \pm 0.005$ | - |
| 9CB | $1.20 \pm 0.03$ | - |
| 10CB | - | $2.83 \pm 0.05$ |
| 11CB | - | $3.8 \pm 0.1$ |

are indeed quite small, e.g. about two orders of magnitude smaller than the latent heat at the melting transition in these cyanobiphenyl compounds[13,22]. The values in Table 2 are typical for many liquid crystals. Since the latent heat at the NI transition is very small, one might expect many physical properties to display critical-like behavior, which could possibly be described by power laws with the appropriate critical exponents. These fluctuation effects are, indeed, observed in many properties including the specific heat capacity $C_p$. Fig. 3 gives the temperature dependence of $C_p$ near the NI transition for the same three substances as in Fig. 2. Power law fits can be made for these and similar data with an expression of the form

$$C_p/R = (A\,|\varepsilon|^{-\alpha} + B + E\,|\varepsilon| + ..) \qquad (9)$$

where $\varepsilon = (T - T_c)/T_c$ with $T_c = T^*$ for $T > T_{NI}$ and $T_c = T^{**}$ for $T < T_{NI}$. Fits to $C_p$ data for a series of compounds[19-23] with expressions of the type of Eq. 9 yield effective $\alpha$ values, which are strongly dependent on the fitting range. For $T < T_{NI}$, $\alpha_{eff}$ values fall usually in the range between 0.3 and 0.4. Because in the isotropic phase the pretransitional increase in $C_p$ is substantially smaller, the range for $\alpha_{eff}$ is even broader and values between the Ising ($\alpha = 0.11$) and tricritical ($\alpha = 0.5$) one are possible.

Because several other exponents, in particular the order parameter exponent $\beta$, seem to be quite close to the tricritical value (0.25 for $\beta$), it was argued by Keyes[24] and also by Anisimov et al[25,26] that indeed tricritical values should be expected. According to the conjectures of Keyes, two competing nematic order parameters (the uniaxial one and the biaxial one) are expected to exhibit diverging fluctuations near $T_{NI}$ even though the biaxial order parameter is zero in the nematic phase.It is, however, expected that the tricritical point can be reached in the presence of a suitable external field[27].

Anisimov et al. argued that the NI behavior looks like that near a tricritical point because of the smallness of the coefficient C in the Landau-de Gennes expansion Eq. 5.

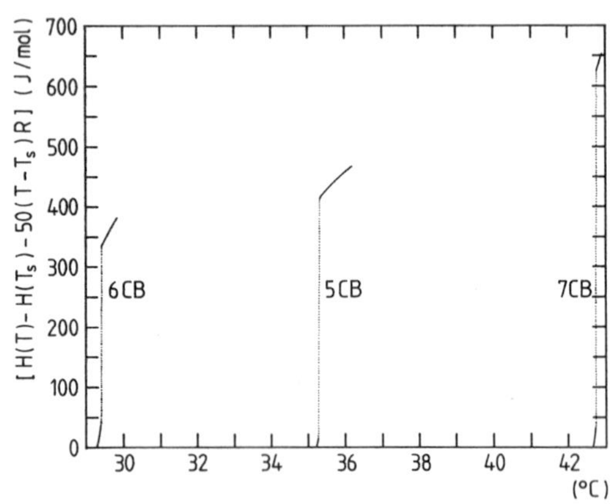


Fig. 2. Temperature dependence of the enthalpy near the nematic to isotropic (NI) transition for three different compounds of the alkylcyanobiphenyl (nCB) homologous series. Note that a linear background $50R(T - T_s)$ has been substracted for clarity from the direct enthalpy data (R is the gas constant).

Numerical estimates for MBBA gave a C value almost as small the B coefficient[19]. A specific crossover expression between tricritical and Ising-critical was also proposed for the specific heat capacity[26].

Good fits to data can be obtained with that expression (with several adjustable parameters)[19,20]. Adding up all experimental results there is still no unambiguous proof of what type of critical behavior is relevant for the NI transition. Several conflicting results are obtained, in particular when comparing values for the same parameters deduced from different physical properties. From the specific heat capacity e.g. one gets values for $(T^* - T_{NI})$ and $(T^{**} - T_{NI})$ which are consistently about one order of magnetude smaller than from several other quantities[19].

At the direct transition from the isotropic phase to smectic phases the additional positional order appears as a modulation superimposed on the overall (nematic-like) orientational order and also these transitions are first-order. In fact typical latent heat values for isotropic-smectic transitions are substantially larger than for NI transitions. This is illustrated in table 2 for the alkylcyanobiphenyl homologous series.

Because of the larger $\Delta H_L$ one can approach $T^*$ and $T^{**}$ even less closely and thus it is more difficult to arrive at meaningful conclusions about pretransitional behavior and critical exponents. In addition to the pretransitional effects arrising from the orientational order parameter fluctuations, one expects, on the basis of coupling between the nematic and the smectic order parameters, also a noticeable influence from smectic fluctuations[28,29]. As the $S_A$ phase approaches the NI transition for different compounds of an homologous series one expects e.g. a change in the temperature dependence of the susceptibility. Experimental evidence presented by Gohin et al.[28] and Anisimov et al.[30] seems to support this expectation.

This is, however, not in agreement with light scattering results by Zink and de Jeu[31] for the nCB series which indicate a very similar behavior from n = 5 up to n = 12. On the basis of the same theoretical arguments one also expects[19] an increase in the "background" part of the specific heat capacity when the width of the nematic range decreases. Evidence in this direction was presented in a paper by V.P. Voronov and F. Kholmurodov[32].

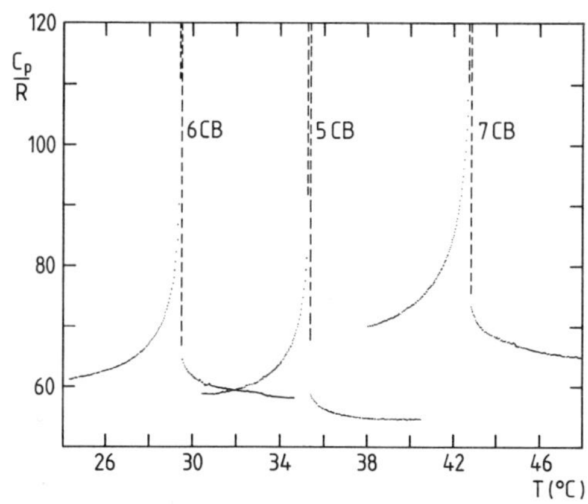


Fig. 3. Temperature dependence of the reduced specific heat capacity per mole (R is the gas constant) near the nematic to isotropic (NI) transition for three compounds of the alkylcyanobiphenyl (nCB) homologous series. The two vertical dashed lines at the transitions indicate the width of the two-phase region.

### *3.2. The Isotropic to Chiral Nematic Blue Phases*

In the case of optically active molecules one arrives at a special situation for the nematic phase. In addition to the long range orientational order there is a spatial variation of the director leading to a helical structure. The local preferred direction of alignment of the molecules is slightly rotated in adjacent planes perpendicular to the pitch axis. This phase is usually called chiral nematic ($N^*$) or cholesteric phase. The helical pitch in the $N^*$ phase is substantially larger than molecular dimensions and varies with the type of molecules. For a long pitch one has a direct first-order transition between the isotropic phase and the chiral nematic phase with characteristics very similar to the normal NI transition in nonchiral compounds. For cholesterics with sufficiently short pitch (typically less than 0.5 μm) there appear new types of liquid crystalline phases, the so called blue phases (BP's), in a small temperature span just below the $IN^*$ clearing point. During the last years it has been established that cholesteric liquid crystals of short pitch may form up to three distinct blue phases[33]. In order of increasing temperature they are named BPI to BPIII. Although blue phases have been noticed for years and observed by optical microscopy, not until recently have they been accepted as legitimate thermodynamic phases. The first experimental evidence for thermodynamic stability of at least two blue phases was obtained some ten to fifteen years ago by DSC and dilatometry[33-35]. Convincing evidence for the stability of the BPIII or "fog phase" was obtained from optical rotary dispersion measurements by Collings[36,37] and from high-precision measurements of the shear modulus, viscosity, and specific heat by Kleiman et al.[38]. All the phase transitions involving blue phase are theoretically expected to be first-order. This also seems to be the conclusion reached in most experimental investigations. In many cases this conclusion was reached on the basis of DSC measurements, which can, however, not discriminate between pretransitional heat-capacity increases and small latent heats[21,23]. Kleiman et al.[38] reported for cholesteryl nonanoate (CN) (which has all three blue phases) high-resolution results for the shear modulus, the viscosity and the specific heat capacity. For the measurements of $C_p$ they used an adiabatic calorimeter operated in a scanning mode in which the power required to keep the scanning rate dT/dt constant was monitored. Although much lower scanning rates were used, this is essentially the same as a DSC operation mode. From the discussion in section II it should be clear that it is generally not possible to discriminate between retransitional effects and latent heats with this mode of operation. On the basis of their results Kleiman et al. concluded the cholesteric-BPI, the BPI-BPII and BPII-BPIII transitions to be first-order. For the BPIII-isotropic transition, no firm conclusion could be reached and the possibility of a second-order transition was left open. In an effort to clarify the situation we also carried out a scanning calorimetric investigation of CN by means of the constant heating (of the sample) mode (see Section 2), allowing the separation of pretransitional and latent heat contributions to the enthalpy. Fig. 4 gives a general overview of our specific heat capacity results for a temperature range covering all phase transitions involving blue phases in cholesteryl nonanoate[39]. It is clear from Fig. 4 that the BPIII to isotropic phase transition is by far the most dominant peak. The other transitions appear as small, narrow features on a substantial specific heat capacity peak (extending well into the CH phase) related with the clearing transition to the isotropic phase. From the inspection of the enthalpy behavior it was found that these small features correspond with first-order transitions with very small latent heats[39]. The dominance of the BPIII to isotropic transition is an indication that a large amount of the heat supplied to the sample is used to reduce the local nematic order in the liquid crystalline phases. One may thus expect the transition to the isotropic phase to be quite similar to a normal NI transition and to be first-order. That the transition from BPIII to the isotropic phase is indeed a first-order one can be seen in Fig. 5 where the enthalpy versus temperature is given for a T-range of 0.2°C below and 0.2°C above the transition.

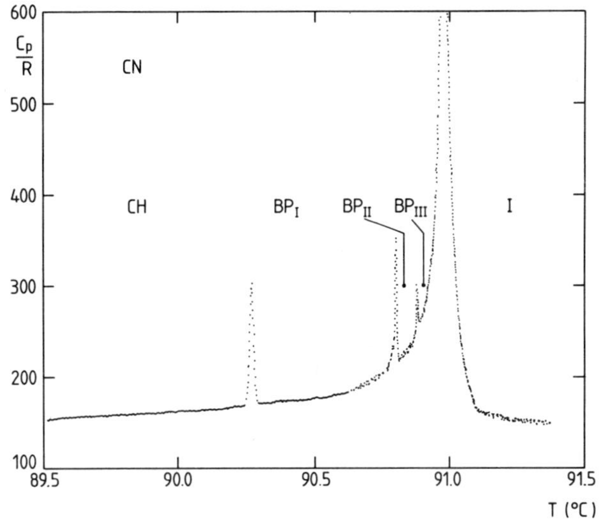


Fig. 4. General overview for the reduced specific heat capacity per mole $C_p$ /R (with R the gas constant) for a temperature range covering all phase transitions involving the blue phases in cholesteryl nonanoate (CN).

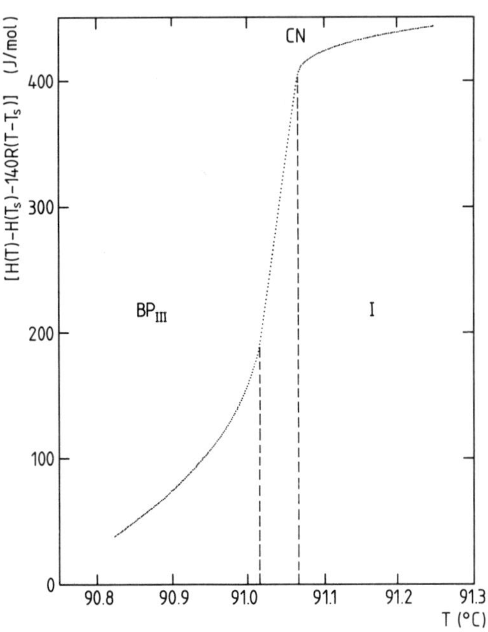


Fig. 5. Detailed plot of the enthalpy near the phase transition from the blue phase III to the isotropic phase in cholesteryl nonanoate (CN). Note that for clarity a large linear background 140R (T - $T_s$ ) with R the gas constant and $T_s$ = 90.73°C, has been substracted from the direct data. The two vertical dashed lines indicate the width of the two-phase region.

We believe that by now it is well established that the observed blue phases are genuine stable thermodynamic phases and the transitions are first-order ones, as expected theoretically. For many other interesting aspects of blue phases we refer the reader to recent review articles[40-42].

# 4. The Nematic to Smectic A Transition

## *4.1. Theoretical Background*

The study of the nematic to smectic A (NA) transition is an active area of research and has been vigorously investigated in an effort to determine the order of the transition and the universality class for critical behavior. The $S_A$ liquid crystal phase results from the development of a one-dimensional density wave in an orientationally ordered nematic phase. The smectic wave vector q for the $S_A$ phase is parallel to the nematic director (along the z-axis) and the $S_A$ order parameter $\Psi = |\Psi| e^{i\varphi}$ can be introduced by

$$\rho(r) = \rho_0[1 + R_e \Psi e^{iqz}] \tag{10}$$

The order parameter $\Psi$ has two components: magnitude $|\Psi|$ and phase $\varphi = qu(r)$, with u the displacement of a layer in the z direction away from the equilibrium position. The $S_A$ phase shows in itself a rich polymophism depending on the type of molecules involved. For nonpolar molecules, there is only one type of A phase with layer spacing $d = 2\pi/q = L$, with L the molecular length. This phase is often called the smectic $A_m$ (for monomeric) phase. For polar materials several distinct, thermodynamically stable, smectic A phases can exist. The most common cases are the monolayer $S_{A1}$ with $d \cong L$, the partial bilayer $S_{Ad}$ for which $L < d < 2L$ and the bilayer $S_{A2}$ with $d \cong 2L$.In addition there are also phases with in plane (x,y) modulation. This complex polymorphism is observed in "frustrated" liquid crystals with long aromatic cores with polar head groups[43]. Phase transitions for frustrated smectics will be discussed by Garland[1]. Here we limit ourselves to the $A_m$ case and to some $A_d$ smectics for shorter polar compounds with relatively narrow nematic ranges (if present). Typical examples are e.g. the alkylcyanobiphenyls with two aromatic rings.

The first description of the NA transition was given on the basis of molecular model considerations by McMillan[44] and Kobayashi[45]. By introducing a suitable intermolecular interaction potential in a mean-field model phase transitions between N and $S_A$ could be obtained. The predictions by the model strongly depended on the ratio between the length of the aromatic core and the total length of the molecule ($\cong$ layer spacing). For low values of this ratio the transition is second-order, for larger values it becomes first-order at a tricritical point for a McMillan ratio $r = T_{NA}/T_{NI} = 0.87$. The basic content of these theories was expressed by de Gennes[46] in terms of a Landau free energy expansion including a coupling between the nematic order parameter Q and the amplitude $|\Psi|$ of the $S_A$ density wave

$$\Delta F = a|\Psi|^2 + b_0|\Psi|^4 + \frac{(\delta Q)^2}{2\chi} - C|\Psi|^2 \delta Q \tag{11}$$

where $a = a_0(T-T_{NA})/T_{NA} = a_0\varepsilon$; $b_0, C > 0$; and $\delta Q = Q - Q_0(T)$ the change in nematic order induced by the formation of $S_A$ layering, and $\chi$ the temperature dependent nematic susceptibility.

This quantity $\chi$ is a function of the degree of saturation of the nematic order, therefore $\chi(T_{NA})$ depends on the width of the nematic range. Minimizing $\Delta F$ with respect to $\delta Q$ and eliminating it from $\Delta F$, one has

$$\Delta F = a\,|\Psi|^2 + \left(b_0^2 - \frac{1}{2}C^2\chi\right)|\Psi|^2 = a\,|\Psi|^2 + b\,|\Psi|^4 \tag{12}$$

For a wide nematic range $\chi(T_{NA})$ is small and $b > 0$, the transition is second-order. For narrow ranges $b < 0$ because $\chi(T_{NA})$ is large, resulting in a first-order transition. For stability reasons one then has to supplement Eq. 12 with a positive sixth order term in $|\Psi|$. A tricritical point occurs for $b = 0$ or $\chi = 2b_0/C^2$. As far as critical behavior is concerned the McMillan and this de Gennes expansion predict the classical mean field critical exponents. In order to make further progress smectic fluctuations have to be accounted for properly. The fact that the smectic A order parameter has two components led de Gennes[47] and McMillan[48] to conclude that the NA transition should be similar to that of superfluid helium, with a $He^3 + He^4$ - like tricritical point[48]. The NA transition should thus belong to the XY universality class[50,51]. Subsequent theoretical progress has mainly been based a Landau-Ginzberg free energy functional introduced by de Gennes[47]. That expression is closely related to that of a superconductor. The fact that the nematic director fluctuations couple to smectic order fluctuations in a complex way, makes, however, the analogy not complete, and leaves several potentially important differences, and makes it a difficult problem to solve[51]. There have been several theories based on the de Gennes model which have attempted to describe the observed critical behavior, including the anisotropic rate of~growth of the critical correlation length along and transverse to the layering direction. These theories either place the NA transition in the inverted three-dimensional XY universality class or in an anisotropic class[50,52]. Halperin et al.[53] also showed that, to the first-order in an expansion near four dimensions, the coupling between the director fluctuations and the smectic order parameter makes the transition always weakly first-order. Dasgupta and Halperin[54] questioned the validity of this result for the three dimensional case. They found that a lattice model for a superconductor had a phase transition asymptotically equivalent to that of a superfluid with an inverted temperature axis and showed no first-order effect.

## *4.2. Calorimetric Results*

### *4.2.1. Enthalpy and Latent Heat*

On the basis of DSC measurements it was generally believed at the beginning of the 1970's that the largest part, if not all, of the investigated NA transitions were first-order, and also that the tricritical McMillan ratio r = 0.87 was not far from being correct. But, as already pointed out, DSC can not distinguish between small latent heats and pretransitional effects. This was pointed out dramatically for 8CB by Kasting et al.[55] by showing that the pretransitional enthalpy change $\delta H_{AN}$, derived from ac calorimetry, was already larger than the DSC "latent heat". In an experiment[21] with our adiabatic scanning calorimeter we also did not find evidence for a latent heat in the direct enthalpy data for 8CB. As an upper limit we arrived at a value of 0.4 J/mol, which is more than three orders of magnitude smaller than the latent heat for the NI transition of the same substance. Subsequent measurements[22] in 9CB, which has a nematic range of only 2°C, turned out to be also nearly second-order. An upper limit of 5 J/mol was obtained for a possible latent heat.

Measurements in mixtures of 9CB and 10 CB, with even smaller N ranges, definitely had latent heat discontinuities in the enthalpy curves, thus proving the first-order character of part of the NA transition line[56]. In fact, all 9CB + 10CB mixtures investigated turned out to be first-order with a latent heat shrinking to zero at a tricritical point very near the NA transition of 9CB. A similar investigation[57] of 8CB + 10CB mixtures with N ranges from 7.0 K (8CB) to zero (mixtures with a mole fraction $x_{10CB} = 0.65$) resulted in a partly first and partly second-order NA-line with a tricritical point at $x_{10CB} = 0.314$. In Fig. 6 the enthalpy curves near $T_{NA}$ are given for several mixtures of 8CB + 10CB showing a firstorder transition. The latent heats decrease drastically with decreasing $x_{10CB}$, becoming zero at the tricritical point (mixture 3). Also using adiabatic scanning calorimetry Anisimov et al[58] made similar observations in mixtures of two nonpolar monomeric compounds. In a recent experiment, combining DSC and ac calorimetry, Stine and Garland[59] were able to determine true latent heat values and locate a tricritical point in mixtures of 4O.8 and 6O.8 (butyl and hexyloxybenzylidene octylaniline), also two nonpolar compounds.

### *4.2.2. Specific Heat Capacities*

During the last 10 to 15 years a substantial amount of high-resolution specific heat capacity data has been obtained for the NA transition in several systems. Initially, various measurements[60,23,21,61] of the $C_p$ variation near $T_{NA}$ have yielded rather larger positive values for the critical exponent $\alpha$ (usually in the range 0.15-0.30) with the exception of $\overline{8}$S5 (a member of the $\overline{n}$-thiosulfate liquid crystal systems) and 4O.7, for which $\alpha \cong 0$ values have been reported[62,63]. Brisbin et al[61] found that for the $\overline{n}$S5 series $\alpha$ depends on the nematic range, and they made the important observation that these varying exponents might be connected with a crossover from 3D-XY behavior ($\alpha = -0.007$) to tricritical behavior ($\alpha = 0.5$) with decreasing nematic range. The results for 8S5, 4O.7 and for mixtures[56] 7CB + 8CB are consistent with XY behavior. This general picture was confirmed by more elaborate studies[56-59] on the dependence of the effective critical exponent $\alpha$ on the McMillan ratio $r = T_{NA}/T_{NI}$.

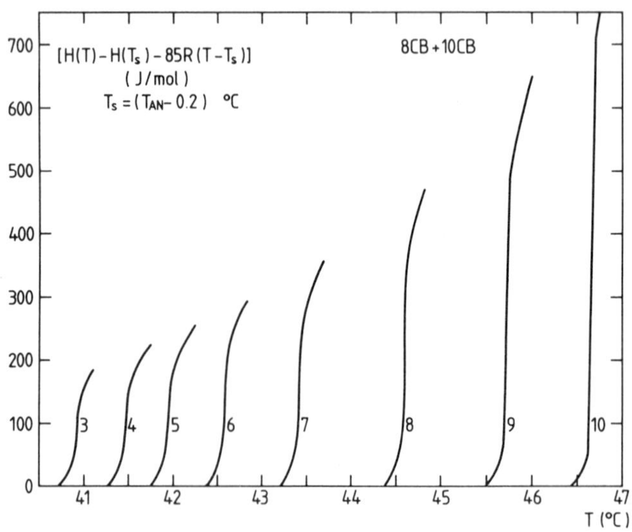


Fig. 6. Enthalpy versus temperature near $T_{NA}$ for tricritical mixture (N° 3) and seven mixtures of 8CB + 10CB with a first-order NA transition. A regular part $85R(T - T_s)$ is substracted for display reasons (R is the gas constant).

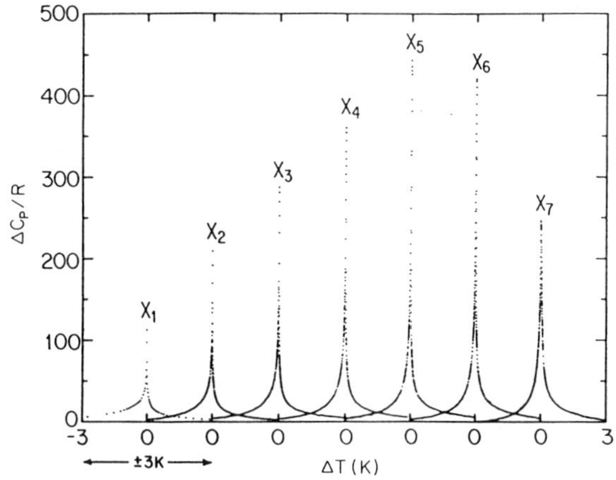


Fig. 7. Reduced excess heat capacity $\Delta C_p/R$ (with R the gas constant) for the NA transition in mixtures of 4O.8 + 6O.8. In each case data are shown over the range $-3K < T - T_c < 3K$. The values of the mole fraction X of 6O.8 are $X_1 = 0$ (pure 4O.8), $X_2 = 0.10$, $X_3 = 0.20$, $X_4 = 0.30$, $X_5 = 0.35$, $X_6 = 0.40$, and $X_7 = 0.50$ (Ref. 59).

The increase of $\alpha$ on approaching the tricritical point is accompanied by a dramatic increase of the heat capacity anomaly. This can be seen in Fig. 7 where the results of Stine and Garland[59] for binary mixtures of 4O.8 and 6O.8 are given. The former compound ($X_1$ in Fig. 7) has a continuous NA transition (with $\alpha \cong 0.13$), and the latter is first-order with a latent heat of 3.6 kJ/mol. For this binary system the tricritical concentration is near $X_5 = 0.35$. A summary of the dependence of $\alpha$ on the width of the nematic range is given in Fig. 8 for this, and some other systems. One clearly sees an evolution form the XY value of $\alpha$ for small r values (large N ranges) to the tricritical $\alpha$ value for larger r values (small N ranges).

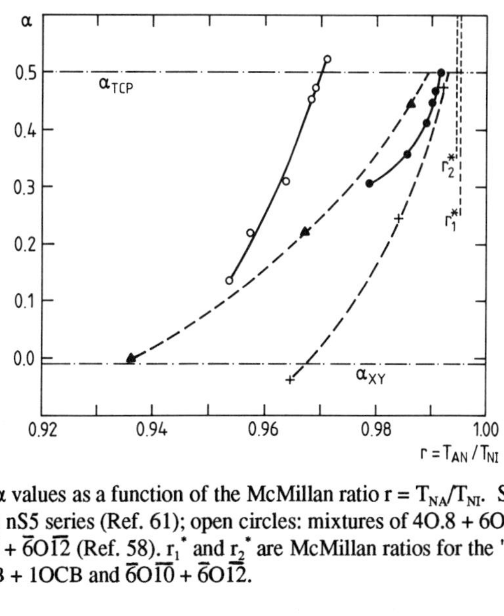


Fig. 8. Critical exponent $\alpha$ values as a function of the McMillan ratio $r = T_{NA}/T_{NI}$. Solid dots: mixtures of 8CB + 10CB ( Ref.57), triangles: nS5 series (Ref. 61); open circles: mixtures of 4O.8 + 6O.8 (Ref. 59); pluses: mixtures of $\bar{6}O\bar{8}$ + $\bar{6}O\overline{10}$ and $\bar{6}O\overline{10}$ + $\bar{6}O\overline{12}$ (Ref. 58). $r_1^*$ and $r_2^*$ are McMillan ratios for the "Landau tricritical point" (see text), respectively for 8CB + 1OCB and $\bar{6}O\overline{10}$ + $\bar{6}O\overline{12}$.

It should also be clear from this figure that the tricritical McMillan ratio r is different for each system. On the basis of the calorimetric data alone one would thus be inclined to condude that there is crossover from XY to tricritical behavior with changing width of the nematic range.However, high-resolution data for the correlation length of critical fluctuations $\xi$ show that the picture is substantially more complicated. Indeed, it is observed[64] that critical fluctuations have anisotropic rate of growth of the critical correlation lengths along the layering direction, $\xi_{//}$, and transverse to the layering, $\xi_{\perp}$, with different critical exponents: $\nu_{//} \neq \nu_{\perp}$. This is not expected for XY behavior ($\nu_{//} = \nu_{\perp} = 0.66$) or tricritical behavior ($\nu_{//} = \nu_{\perp} = 0.5$). On the other hand average $\nu$ values also seem to evolve from nearly XY to tricritical as a function of $r = T_{NA}/T_{NI}$. Furthermore, hyperscaling,$\alpha + \nu_{//} + 2\nu_{\perp} = 2$, and two-scale-factor universality, $A\xi_{//0}\,\xi^{2}_{\perp 0}$ = universal constant, between heat capacity and correlation length amplitudes, are also satisfied within experimental uncertainty[63].

### *4.2.3. Fisher Renormalization*

The anomalous fluctuation induced contribution to heat capacity near a second-order NA transition is usually described in terms of power law expressions of the form of Eq. 9. One uses this kind of expressions also to analyze $C_{px}$ data in mixtures along lines of constant concentration, although in principle one should apply expressions like Eq. 9 to data for $C_{p\varphi}$ as a function of $\varepsilon_{\varphi'}$, where $\varphi$ designates a path of constant chemical potential difference. This distinction becomes crucial when full Fisher renormalization[65] is realized because then one also has a renormalized critical exponent $\alpha_R = \alpha/(1-\alpha)$, which results in a non divergent heat capacity for $\alpha > 0$. The distinction between a measuring path of constant x or $\varphi$ is not very important as long as $dT_{NA}/dx$ is not too large, which is the case for the systems discussed so far.

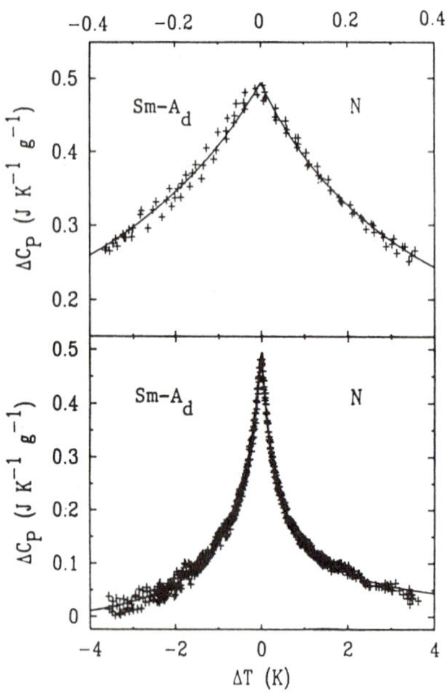


Fig. 9. The NA heat capacity anomaly for the tricritical mixture of TBBA + 8OCB. The solid line represents a fit with a fully Fisher renormalized tricritical exponent $\alpha_R = -1$ (Ref. 67).

However, some pathological situations, with dramatic modifications of the heat capacity behavior, do exist. Two binary systems showing full Fisher renormalized NA transitions have recently been studied in detail by ac calorimetry.

Huster et al.[66] found a renormalized NA tricritical point in a mixture of nonpolar heptyloxypentylphenylthiolbenzoate ($\bar{7}S5$) and polar octyloxycyanobiphenyl (8OCB). The mixture contained only a few mole percent of the polar component 8OCB. Stine and Garland[67] showed that the same effect occurs in a mixture containing a small mole fraction of the nonpolar compound terephtal-bis-butylaniline (TBBA) and a large mole fraction of the polar compound 8OCB. Both systems had very steep NA transition lines as a function of mole fraction. In the first case $dT_{NA}/dx = 289$ K and in the second one $dT_{NA}/dx = 230$ K at the tricritical point. For both cases a fully renormalized tricritical exponent $\alpha_R = -\alpha/(1-\alpha) = -1$ was obtained, resulting in a finite heat capacity cusp at the NA transition.

This is clearly visible in Fig. 9 where results for the tricritical concentration in TBBA + 8OCB are reproduced[67].

### *4.2.4. Fluctuation Induced First-Order*

It has been observed experimentally that along the first-order part of the NA line the latent heat variation passes from a nearly quadratic dependence on concentration close to the tricritical point to a nearly linear one far away from that point[57,58]. This in not in agreement with mean-field Landau theory without a cubic term, where it is predicted that the latent heat along the first-order side of the transition line should be a linear function of the distance to the tricritical point, i.e. in a mixture one should have: $\Delta H_L \sim (x{-}{-}x^*)$, with $x^*$ the tricritical concentration. Anisimov et al.[68] suggested that the origin of this discrepancy could be the presence of a small cubic term in the vicinity of the Landau tricritical point(LTP). This cubic term is expected to be the result of the coupling between director fluctuations and the smectic order parameter, as predicted in 1974 by Halperin et al[53]. In Ref. 68 it was found that the assumption of a cubic term in the Landau expression was consistent with the observed $\Delta H_L$ behavior. Recently a more detailed analysis of $\Delta H_L$ data for three different systems (9CB + 10CB, 8CB + 10CB, 6O10 + 6O12) was presented[69].

The data could be described well in terms of a mean-field scaling expression assuming the presence of the cubic term predicted by Halperin, Lubensky and Ma[53]. On the basis of this analysis,it is expected that the NA transition in 8CB should also be first-order. The expected latent heat value should apparently be lower than the upper limit of 0.4 J/mol obtained from adiabatic scanning calorimetry[3].

Evidence in support of a first-order NA transition for 8CB was also presented on the basis of observations of the dynamics of interface movements in undercooled and overheated samples[69,70]. Both the latent heat analysis and the dynamic measurements lend support to the presence of a cubic term in the mean field free energy, implying the NA transition to be at least weakly first-order.

The authors of Ref. 69 pointed out, there still remain some questions which have to be addressed. One additional point of concern has to do with the location of the "true" Landau tricritical point (LTP) in Fig. 8. From the results for the effective $\alpha$ values as a function of $r = T_{NA}/T_{NI}$ for the 8CB + 10CB and the 6 O m systems, it seems that the tricritical value $\alpha = 0.5$ is reached at the **r** value where the experimentally measured latent heats becomes zero and not at the value $\mathbf{r}^*$ for the LTP location obtained by Anisimov et al[69]. For the 4O.8 + 6O.8 system[59] a $\mathbf{r}^*$ value has not yet been determined.

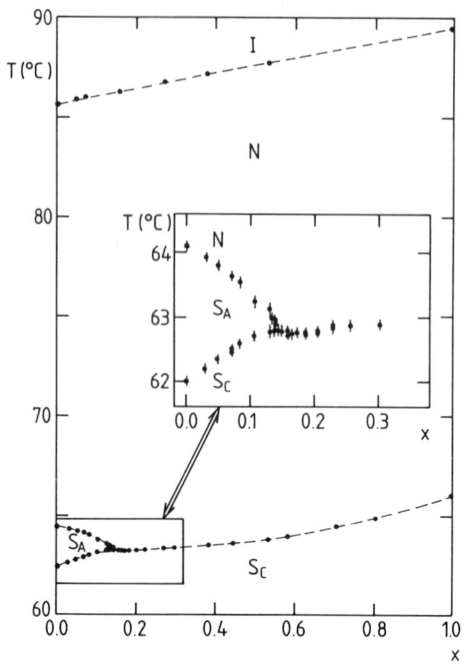


Fig.10. Phase diagram for the system 5O8 (left) + 6O8 (right). x is the weight fraction of 6O8 in the mixtures.

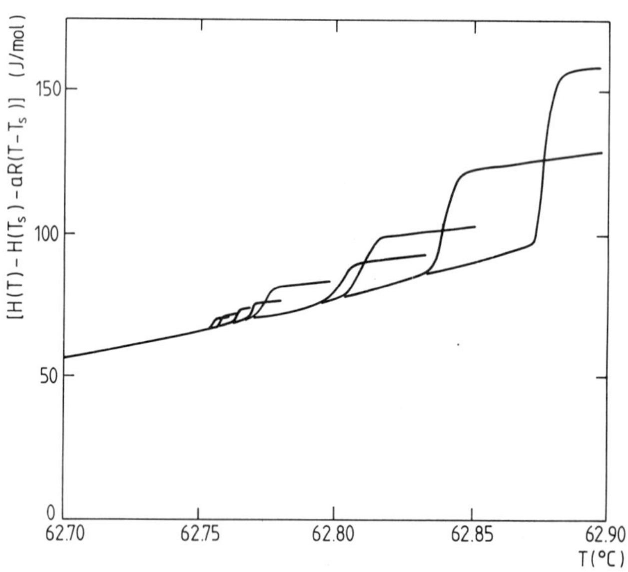


Fig.11. Temperature dependence of the enthalpy for several mixtures of $\bar{5}O\bar{8}$ + $\bar{6}O\bar{8}$ with direct nematic to smectic C transitions. Note that for clarity a large linear background $C_p^0$ = aR, has been subtracted from the direct experimental data. The curve with the smallest steplike increase is closest to the NAC point, and has a latent heat of (1.7 ± 0.3) J/mol.

# 5. The Nematic - Smectic A - Smectic C Multicritical Point

## 5.1. Theoretical Background

When a liquid crystal which has a nematic to smectic A (NA) and a smectic A to smectic C (AC) transition is mixed with one which has a direct nematic to smectic C (NC) transition, a phase diagram with a NAC multicritical point, where the NA, AC and NC phase transition lines merge, is obtained. Fig.10 gives the phase diagram with a NAC point for mixtures of $\bar{5}O\bar{8}$ + $\bar{6}O\bar{8}$, two compounds of the alkyloxyphenyl-octylbenzoate series[16]. The NAC point has been the subject of extensive theoretical and experimental studies during the past decade. The nature of the NAC point is, however, still not clearly established.

The many theoretical approaches to describe various aspects of the multicritical behavior fall into two categories, a basic difference being the number of order parameters in the model. The first class, introduced by Chen and Lubensky[71], is based on a single infinite dimensional order parameter for the smectic density wave. In a mean field analysis of their model these authors predicted that the NAC point should be a ($d = 3$, $n = 2$, $m = 2$) Lifshitz point. The second class of models involves two order parameters, one for the smectic density modulation and another for the smectic C tilt angle of the director with respect to the normal to the smectic layers[72-76]. On the basis of their dislocation loop theory, Grinstein and Toner[77] have shown that in this case the NAC point is, in fact, a decoupled tetracritical point and that in the immediate vicinity of the NAC point a biaxial nematic (N') phase should intervene between the nematic phase and the smectic C phase. This type of phase diagram near the NAC point is also consistent with the Chen--Lubensky model when fluctuations are included[77].

## 5.2. Calorimetric Results

Since the first discovery[78,79] of NAC multicritical points in 1977, several systems have been investigated experimentally with different techniques. The first system which has been studied extensively, also calorimetrically[80], was the binary system $\bar{7}S5$ + $\bar{8}S5$. Detailed calorimetric studies near the NAC point were also carried for $\bar{7}S5$ + 8OCB[81], $\bar{6}O\bar{8}$ + $\bar{6}O\overline{10}$[14] and $\bar{5}O\bar{8}$ + $\bar{6}O\bar{8}$[16]. Recently also a NAC* multicritical point in binary mixtures of two chiral compounds was investigated by means of ac calorimetry[82]. All these calorimetric data show essentially similar pictures of the thermal behavior near the NAC multicritical point. The NC transitions are first-order with a latent heat becoming zero at or very near the NAC point. This is illustrated in Fig. 11 where for different concentrations of $\bar{5}O\bar{8}$ + $\bar{6}O\bar{8}$ parts of the enthalpy curves near the NC transitions are given. This first-order character is expected theoretically[83], and also found in other systems[80,14]. AC transitions and NA transitions are found to be second-order. The heat capacity anomalies along the AC line as well as along the NC line become larger and sharper on approaching the NAC point, indicating the presence of a (mean field) tricritical point at or near the NAC point. On the other hand high resolution x-ray scattering experiments[84,85] are found to be in general agreement with the NAC point being a Lifshitz point. The NAC multicritical point seems to exhibit simultaneously the characteristics of a Lifshitz point and of a tricritical point. Sofar no experimental evidence has been presented proving the existence of the biaxial nematic phase. However, Xin Wen et al[82] observed a broad heat capacity hump in the N region for samples with near NAC* composition, possibly providing evidence for some biaxial or cholesteric modulation. Also in the data of Anisimov et al[14] there seems to be an indication of a similar anomalous $C_p$ behavior in the N phase near the NAC point in $\bar{6}O\bar{8}$ + $\bar{6}O\overline{10}$ mixtures.

## 6. Conclusions

During the past ten to fifteen years a great deal of progress has been made in the experimental characterization and theoretical understanding of several phase transitions in liquid crystals. This is not only so for the phase transitions and multicritical points considered here, but also for other cases and systems e.g. the frustrated smectics, discussed in Ref.1.

The general theoretical picture for the NI transition, as being weakly first-order, is certainly consistent with experimental observations. Concerning the influence of order parameter fluctuations and the type of critical behavior, the situation is less clear. The mean-field Landau-de Gennes approach is certainly not sufficient because deviating behavior is observed for many properties. Several other theoretical possibilities have been proposed, but no unambiguous choice for the type of critical behavior on the basis of experimental data has been possible sofar. The experimental situation is very unfavorable because the first-order discontinuities at $T_{NI}$ cause a lot of uncertainty in the fitting of pretransitional behavior. Probably high-resolution results very near or at a second-order NI Landau point could bring some progress here.

The NA transition is one of the liquid crystal phase transitions for which a vast amount of high quality experimental data have been accumulated. The change from first-order to second-order at a tricritical point, depending on the width of the nematic range (or the coupling between the N and $S_A$ order parameters), has been found experimentally. Several of the theoretical predictions for the critical behavior on the basis of the Landau-Ginzburg free energy have been found in accordance with experiment, provided crossover from (XY) critical to tricritical and Fisher renormalization is brought into the picture for some cases. Nevertheless, some basic issues remain to be explained, most importantly the anisotropic correlation length behavior, even at the tricritical point.

Also for the NAC multicritical point substantial progress has been made during the last decade. Several systems have been investigated with high-resolution calorimetric and other techniques. The thermal data point to a tricritical point near or at the NAC point. X-ray data on the other hand are better explained on the basis of the NAC point being a Lifshitz point. Experimental evidence for the theoretically predicted biaxial nematic phase near the NAC point, has not been obtained so far.

### References

1. C.W. Garland, Chapter 11 of this Volume (1991).
2. M.E. Bagatskii, A.V. Voronel' and V.G. Gusak, Sov. Phys. JETP, **16**: 517 (1963).
3. A.V. Voronel', Yu R. Chaskin, V.A. Popov and V.G. Simkin, Sov. Phys. JETP, **18**: 568 (1964).
4. A.V. Voronel', V.G. Gorbunova, Yu R. Chaskin and V.V. Shechekochikhina, Sov. Phys. JETP, **23**: 597 (1966).
5. M.A. Anisimov, A.V. Voronel' and T.M. Ovodova, Sov. Phys. JETP, **35**: 536 (1972).
6. J. Thoen, E. Bloemen and W. Van Dael, J. Chem. Phys. **68**: 735 (1978).
7. M.A. Anisimov, S.R. Garber, V.S. Esipov, V.M. Mamnitskii, G.I. Ovodov, L.A. Smolenko and E.L. Sorkin, Sov. Phys., JETP **45**: 1042 (1977).
8. M.A. Anisimov, E.L. Sorkin and V.M. Mamnitskii, J. Eng. Phys. (USSR) **39**: 1385 (1981).
9. M. Sorai, S. Asahina, C. Destrade and N.H. Tinh, Liq.Cryst. **7**: 163 (1990).
10. J.A. Lipa, C. Edwards and M.J. Buchingham, Phys. Rev. Lett. **25**: 1088 1970).
11. J.A. Lipa, C. Edwards and M.J. Buchingham, Phys. Rev. A **15**: 778 (1977).
12. E. Bloemen, J. Thoen and W. Van Dael, J. Chem. Phys. **73**: 4628 (1980).
13. J. Thoen, H. Marijnissen and W. Van Dael, Phys. Rev. A **26**: 2886 (1982).
14. M.A. Anisimov, V.P. Voronov, A.O. Kulkov and F. Kholmurodov, J. de Physique **46**: 2137 (1985).
15. J. Thoen, E. Bloemen, H. Marynissen and W. Van Dael, in:"Proceedings of the 8th Symposium on Thermophysical Properties, Nat. Bur. Stand. Maryland, 1981", Am. Soc. Mech. Eng., New York (1982).
16. J. Thoen and R. Parret, Liq. Cryst. **5**: 479 (1989).
17. P.G. de Gennes, "The Physics of Liquid Crystals", Clarendon Press, Oxford (1974).

18. G. Vertogen and W. de Jeu, "Thermotropic Liquid Crystals, Fundamentals", Springer-Verlag, Berlin (1988).
19. M.A. Anisimov, Mol. Cryst. Liq. Cryst. **162 A**: 1 (1988); and references therein.
20. J. Thoen and F. Detremerie, unpublished.
21. F. Detremerie, Lic. Thesis, Leuven (1984).
22. H. Marynissen, J. Thoen and W. Van Dael, Mol. Cryst. Liq. Cryst. **97**: 149 (1983).
23. G.B. Kasting, K.J. Lushington, and C.W. Garland, Phys. Rev. B **22**: 321 (1980).
24. P.J. Keyes, Phys. Lett. A **67**: 132 (1978).
25. M.A. Anisimov, V.M. Zaprudskii, V.M. Mamnitskii and E.L. Sorkin, JETP Lett. **30**: 491 (1979).
26. M.A. Anisimov, E.E. Gorodetskii and V.M. Zaprudskii, Sov. Phys. Usp. **24**: 57 (1981).
27. E. Grambergen, L. Longa, and W.H. de Jeu, Phys. Reports **135**: 195 (1986).
28. A. Gohin, C. Destrade, H. Gasparoux and J. Prost, J. de Physique **44**: 427 (1983).
29. M.A. Anisimov, E.E. Gorodetskii and V.E. Podneks, JETP Lett. **37**: 414 (1983).
30. M.A. Anisimov, V.I. Labko, G.L. Nikolaenko and I.K. Yudin, JETP Lett. **45**: 111 (1987).
31. H. Zink and W.H. de Jeu, Mol. Cryst. Liq. Cryst. **124**: 287 (1985).
32. V.P. Voronov and F. Kholmurodov, Sov. Phys. JETP **68**: 763 (1989).
33. H. Stegemeyer, Th. Blümel, K. Hiltrop, H. Onusseit and F. Porsch, Liq. Cryst. **1**: 1 (1986); and references therein.
34. D. Armitage and F.P. Price, J. Appl.Phys. **47**: 2735 (1976).
35. D. Armitage and F.P. Price, J.Chem. Phys. **66**: 3414 (1977).
36. P.J. Collings, Phys. Rev. A **30**: 1990 (1984).
37. P.J. Collings, Mol. Cryst. Liq. Cryst. **113**: 277 (1984).
38. R.N. Kleinman, D.J. Bischop, R. Pindak and P. Taborek, Phys. Rev. Lett. **53**: 2137 (1984).
39. J. Thoen, Phys. Rev. A **37**: 1754 (1988).
40. P.P. Crooker, Liq. Cryst. **5**: 751(1989).
41. D.C. Wright and N.D. Mermin, Rev. Mod. Phys. **61**: 385 (1989).
42. T. Seideman, Rep. Prog. Phys. **53**: 659 (1990).
43. F. Hardouin, A.M. Levelut, M.F. Achard and G. Sigaud, J. Chim. Phys. **80**: 53 (1983).
44. W.L. McMillan, Phys. Rev. A **4**: 1238 (1971).
45. K. Kobayashi, Phys. Lett A **31**: 125 (1970).
46. P.G. de Gennes, Mol. Cryst. Liq. Cryst. **21**: 49 (1973).
47. P.G. de Gennes, Sol. St. Commun. **10**: 753 (1972).
48. W.L. McMillan, Phys. Rev. A **6**: 936 (1972).
49. R. Alben, Sol. St. Commun. **13**: 1783 (1973).
50. T.C. Lubensky, J. Chim. Phys. **80**: 31 (1983).
51. J. Prost, Adv. Phys. **33**: 1 (1984).
52. D.R. Nelson and J. Toner, Phys. Rev. B **23**: 363 (1981).
53. B.I. Halperin, T.C. Lubensky and S.K. Ma, Phys. Rev. Lett. **32**: 292 (1974).
54. C. Dasgupta and B.I. Halperin, Phys. Rev. Lett. **47**: 1556 (1981).
55. G.B. Kasting, C.W. Garland and K.J. Lushington, J. de Physique **4**: 879 (1980).
56. J. Thoen, H. Marynissen and W. Van Dael, Phys. Rev. Lett. **52**: 204 (1984).
57. H. Marynissen, J. Thoen and W. Van Dael, Mol. Cryst. Liq. Cryst. **124**: 195 (1985).
58. M.A. Anisimov, V.P. Voronov, A.O. Kulkov, V.N. Petrukhov and F. Kholmurodov, Mol. Cryst. Liq. Cryst. **150b**: 399 (1987).
59. K.J. Stine and C.W. Garland, Phys. Rev. A **39**: 3148 (1989).
60. R.J. Birgeneau, C.W. Garland, G.B. Kasting and B.M. Ocko, Phys. Rev. A. **24**: 2624 (1981).
61. P. Brisbin, R. DeHoff, T.K. Lockhart and D.L. Johnson, Phys. Rev. Lett. **43**: 1171 (1979).
62. C.A. Schantz and D.L. Johnson, Phys. Rev. A **17**: 1504 (1978).
63. C.W. Garland, M. Meickle, B.M. Ocko, A.R. Kortan, C.R. Safinya, L.T. Yu, J.D. Litster and R.B. Birgeneau, Phys. Rev. A **27**:3234 (1983).
64. B.M. Ocko, R.J. Birgeneau and J.D. Litster, Z. Phys. B **62**: 487 (1986).
65. M.E. Fisher, Phys. Rev. **176**: 257 (1968).
66. M.E. Huster, K.J. Stine and C.W. Garland, Phys. Rev. A. **36**: 2364 (1987).
67. K.J. Stine and C.W. Garland, Phys. Rev. A. **39**: 1482 (1989).
68. M.A. Anisimov, V.P. Voronov, E.E. Gorodetskii, V.E. Podneks and F. Kholmurodov, JETP Lett. **45**: 425 (1987).
69. M.A. Anisimov, P.E. Cladis, E.E. Gorodetskii, D.A. Huse, V.E. Podneks, V.G. Taratuta, W. van Saarloos and V. Voronov, Phys. Rev. A **41**: 6749 (1990).
70. P.E. Cladis, W. van Saarloos, D.A. Huse, J.S. Patel, J.W. Goodby and P.L. Finn, Phys. Rev. Lett. **62**: 1764 (1989).

71. J. Chen and T.C. Lubensky, Phys. Rev. A **14**: 1202 (1976).
72. K.C. Chu and W.L. McMillan, Phys. Rev. A **15**: 1181 (1977).
73. L. Benguigi, J. de Physique, Colloq. **40**: C3-419 (1979).
74. C.C. Huang and S.C. Lien, Phys. Rev. Lett. **47**: 1917 (1981).
75. G. Grinstein and J. Toner, Phys. Rev. Lett. **51**: 2386 (1983).
76. B.S. Andereck and B.R. Patton, J. de Physique **48**: 1241 (1987).
77. G. Grinstein, T.C. Lubensky and J. Toner, Phys. Rev. B **33**: 3306 (1986).
78. G. Sigaud, F. Hardouin and M.F. Achard, Sol. St. Commun. **23**: 35 (1977).
79. D.L. Johnson, D. Allender, R. DeHoff, C. Maze, E. Oppenheim and R. Reynolds, Phys. Rev. B. **16**: 470 (1977).
80. R. De Hoff, R. Biggers, D. Brisbin and D.L. Johnson, Phys. Rev. A **25**: 472 (1982).
81. C.W. Garland and M.E. Huster, Phys. Rev. A **35**: 2365 (1987).
82. X. Wen, C.W. Garland and M.D. Wand, Phys. Rev. A **42**: 6087 (1990).
83. J. Swift, Phys. Rev. A **14**: 2274 (1976).
84. L.J. Martinez-Miranda, A.R. Kortan and R.J. Birgeneau, Phys. Rev. Lett. **56**: 2264 (1986).
85. L.J. Martinez-Miranda, A.R. Kortan and R.J. Birgeneau, Phys. Rev. A **36**: 2372 (1987).

Chapter 11

# CALORIMETRIC STUDIES OF LIQUID CRYSTAL PHASE TRANSITIONS: AC TECHNIQUES

C. W. GARLAND

## 1. Introduction

Liquid crystals exhibit a rich variety of second-order phase transitions and multicritical points. In order to characterize fully the nature of the critical fluctuations at such transitions, it is necessary to employ a variety of experimental techniques. Among these are high-resolution calorimetric methods, which provide direct information about the energy fluctuations. Such heat capacity data in combination with scattering results can establish the universality class of a given transition and test various theories of liquid crystal behavior.

High-resolution adiabatic calorimetric techniques have been described by Thoen.[1] The material presented here will focus on results obtained by high-resolution ac calorimetry. In this ac technique, a very small oscillatory heat input is supplied to a sample that is loosely coupled to a thermostat bath and the heat capacity $C_p$ is determined from the amplitude of the resulting temperature oscillations. The ac technique has been used extensively on liquid crystals and has the advantage of requiring only a small amount of sample. Indeed, the technique has been extended to the study of free-standing smectic films only three or four layers thick. Furthermore, it is possible to measure $C_p$ at hydrostatic pressures up to ~3000 bar. It should be noted that this method measures $C_p$ directly and not the enthalpy H. The liquid crystal phase transitions to be discussed in this Chapter include: (1) smectic-A−smectic-C (SmA-SmC) transitions and the analogous SmA−SmC* transitions in chiral materials, (2) smectic-hexatic transitions and (3) a variety of transitions occurring in "frustrated" smectic liquid crystals. These topics are complementary to those discussed by Thoen.[1] Taken together, these two Chapters provide a broad review of high-resolution calorimetric results for thermotropic smectic liquid crystals.

## 2. Method and Analysis

The most extensive development and use of ac calorimetry for the investigation of liquid crystals has been carried out by Johnson,[2] Huang,[3] and Garland.[4] In the simplest

C. W. Garland - Department of Chemistry, Massachusetts Institute of Technology, Cambridge, MA 02139 USA

*Phase Transitions in Liquid Crystals*, Edited by S. Martellucci and
A.N. Chester, Plenum Press, New York, 1992

designs for an ac calorimeter,[2-4] a sample of heat capacity $C_p$ is loosely coupled to a heat bath at temperature $T_0$. For an oscillatory power input $P_0 + P_1 \exp(i\omega t)$ to the sample, its temperature is given by $T(t) = T_0 + \Delta T_{dc} + \Delta T_{ac} \exp(i\omega t + i\phi)$, where $\phi$ is the phase shift between P(t) and T(t). For a bulk sample (typically 20-100 mg) at 1 atm, the sample thickness and operating frequency $\omega$ can be chosen so that the simple expression

$$C_p = P_1/\omega\Delta T_{ac} \tag{1}$$

is an excellent approximation.[4] Depending on the individual apparatus design, thickness and frequency can vary[3,4] from 0.1 mm and $\omega = 6$ to 1 mm and $\omega = 0.2$. Since $\Delta T_{ac}$ is small (typically 4 to 10 mK peak to peak), high-resolution measurements can be made close to critical points where $C_p$ exhibits singular behavior. Although latent heats $\Delta H$ cannot be determined, the existence of two-phase coexistence at a first-order transition can be detected qualitatively by abrupt anomalous changes[5] in the phase shift $\phi$.

The ac calorimetric technique is adaptable to high-pressure studies. In this case the heat capacity of the thermal link (dense argon gas) can no longer be neglected and the thermal conductivity of the sample plays a more important role due to the smaller thermal impedance mismatch with the surroundings. As a result, a more sophisticated analysis is required, the details of which are given elsewhere.[6-7]

An ac calorimeter design has also been developed for the study at 1 atm of free-standing liquid crystal films.[8] The phenomenological theory for this apparatus is closely related to the high-pressure case since even at 1atm exchange gas makes a significant contribution to the observed heat capacity relative to the very small $C_p$ of a thin film. Films ranging from three to a few hundred smectic layers have been investigated.[8]

The goal of ac calorimetry is to obtain $C_p$ data of sufficiently high resolution that one can analyze the detailed heat capacity behavior near a second-order phase transition or tricritical point. The $C_p$ behavior can range from mean-field, where fluctuations do not play a significant role, to any one of several critical universality classes where fluctuations play a dominant role (such as Ising and XY models).

In the mean-field case, the free energy can be described by the Landau expression $G = G_0 + at\psi^2 + b\psi^4 + c\psi^6$, where $\psi$ is the order parameter and $t \equiv (T - T_c)/T_c$ is the reduced temperature. For this model, the excess heat capacity $\Delta C_p = C_p - C_p^0$ is given by

$$\Delta C_p = A\frac{T}{T_c}\left(\frac{T_k - T_c}{T_k - T}\right)^{1/2} = A^*\frac{T}{T_c}(T_k - T)^{-1/2} \quad \text{for } T < T_c \tag{2}$$

where $A = a^2/2bT_c$, $T_k = T_c(1 + t_0/3)$ represents the metastability limit, $t_0 = b^2/ac$ and $A^* = (a^3/12cT_c)^{1/2}$. The quantity $C_p^0$ is the noncritical "background" heat capacity arising from the regular term $G_0$. $\Delta C_p = 0$ above the transition. Note that A is the size of the $C_p$ jump at $T_c$ for a second-order transition and $t_0$ is the full width at half maximum on a reduced temperature scale and thus characterizes the sharpness of the peak. At a tricritical point, $b = 0$ and $C_p$ diverges at $T_k$ with a power-law exponent of 0.5.

For critical fluctuation models, renormalization-group theory gives the following form for the excess heat capacity associated with the transition

$$\Delta C_p^{\pm} = A^{\pm}|t|^{-\alpha}\left(1 + D_1^{\pm}|t|^{\Delta_1} + D_2^{\pm}|t|^{2\Delta_1}\right) + B_c \tag{3}$$

where $\pm$ indicates above and below $T_C$. The corrections-to-scaling exponent $\Delta_1$ is approximately 0.5, $B_C$ is a constant term that arises from the critical free energy and $\alpha$ is the heat capacity critical exponent for a given universality class ($\alpha$ = +0.11 for the 3D Ising model and $\alpha = -0.007$ for the 3D XY model). In most cases, the second correction term $D_2^{\pm}|t|^{2\Delta_1}$ is small and can be neglected in the fitting of experimental data.

## 3. Smectic-A–Smectic-C Transitions

The SmC order parameter is a two-component vector defined by the tilt angle $\theta$ between the director and the layer normal and an azimuthal angle $\phi$. Thus, one expects that second-order SmA-SmC transitions should be in the 3D-XY universality class (i.e., exhibit critical behavior like that seen at the lambda transition in helium). Extensive measurements have been made on nonpolar liquid crystals and it is well established that all these SmA-SmC transitions are second-order transitions well described by a mean-field Landau model with an anomalously large sixth-order term.[9-11] An example of this behavior in heptyloxybenzylidene butylaniline (7O.4) is shown in Fig. 1. The values of the fitting parameters in Eq. (2) are[10] $T_C$ = 337.559 K, A = 47.8 R and $t_0 = 0.8 \times 10^{-3}$. The values of A and $t_0$ vary considerably for different samples and there is no microscopic theory that predicts these parameters. For example, A = 7.7 R and $t_0 = 1.3 \times 10^{-3}$ for 4O.7, values that are surprisingly different[10] from those for 7O.4. The observation of mean-field behavior can be explained[9] in terms of the Ginzburg criterion, which indicates that the SmA-SmC critical fluctuation regime lies within an experimentally inaccessible reduced temperature range[9] of $\pm 10^{-5}$. The large sixth-order coefficient c is consistent with this since it implies that the liquid crystal resists large tilt fluctuations, but a physical explanation of such behavior is still lacking.

For chiral liquid crystal molecules, the SmA phase undergoes a transition into a chiral smectic-C (SmC*) phase when the chiral center is coupled to a polar group perpendicular to the long molecular axis. As a result, the spontaneous polarization and molecular tilt direction vary helicoidally from layer to layer. Both first- and second-order SmA-SmC* transitions have been observed in pure materials and mixtures can exhibit a tricritical point. Both tricritical and second-order transitions can be quite well described by a Landau free energy with large sixth-order terms.[5,12,13]

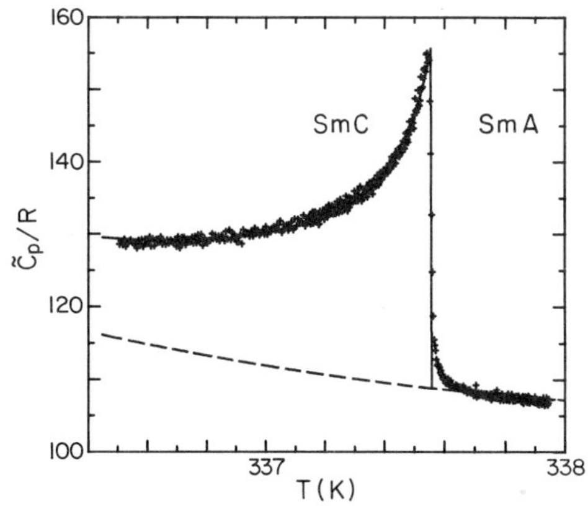


Fig. 1. Molar heat capacity $\tilde{C}_p$ of 7O.4 in the SmA-SmC transition region. Dashed curve represents the background $\tilde{C}_p$ variation and solid line is a Landau fit. (Ref. 10)

Frustrated smectic liquid crystals are polar compounds that exhibit a rich polymorphism of SmA and SmC phases (see paragraph 5), and SmA2-SmC2 transitions associated with the tilting of bilayer smectics have been observed. High-resolution calorimetric data are available on two materials that undergo this SmA2-SmC2 transition.[14,15] In both cases, the excess heat capacity $\Delta C_p$ increases in a steplike fashion from zero in the SmA2 phase to an essentially constant value in the SmC2 phase (see Figs. 2 and 7). The octylphenyl-chloro-cyanobenzoyloxy benzoate (DB8ClCN) data shown in Fig. 2 illustrate the striking contrast between SmA2-SmC2 behavior and that observed for SmA-SmC transitions in nonpolar compounds. A Landau fit to $\Delta C_p$ data for a SmA2-SmC2 transition would require a sixth-order coefficient c that is zero or very small (which means a large $t_0$ value). The challenging theoretical problem is to develop a microscopic model that explains this dramatic difference in $C_p$ behavior at SmA2-SmC2 and nonpolar SmA-SmC transitions.

## 4. Smectic-Hexatic Transitions

In some liquid crystals, a 3D hexatic phase can be observed between the smectic and crystalline (long-range 3D translational order) phases. These stacked hexatic phases, which exhibit long-range six-fold bond-orientational order but only short-range translational order within the smectic layers, have been extensively studied with scattering techniques.[16] The scaling of order parameter data away from the immediate vicinity of the smectic-hexatic transition is consistent with three-dimensional XY behavior, as expected from the symmetry of the hexatic order parameter.[17] However, $C_p$ data near the transition are inconsistent with such XY behavior. The transition sequence most commonly observed is SmA-HexB-CrB (or CrE) and systems of this type have been investigated extensively by Huang and coworkers.[18] The analogous transitions in tilted smectics are SmC-SmF-CrG and SmC-SmI-CrJ, where SmF and SmI are hexatic phases in which SmC-like order is supplemented by bond-orientational order. The tilt direction lies between two nearest-neighbor molecules in SmF and toward a nearest neighbor in SmI.

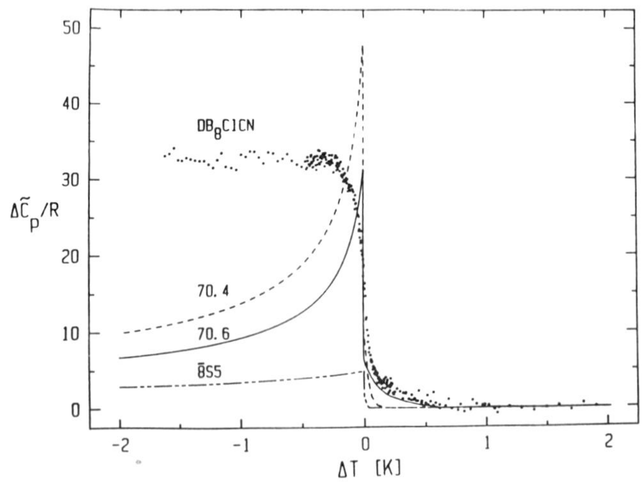


Fig. 2. Excess heat capacity of DB8ClCN associated with the SmA2-SmC2 transition. Also shown are $\Delta C_p$ variations for SmA-SmC transitions in several non- polar compounds. (Ref. 14)

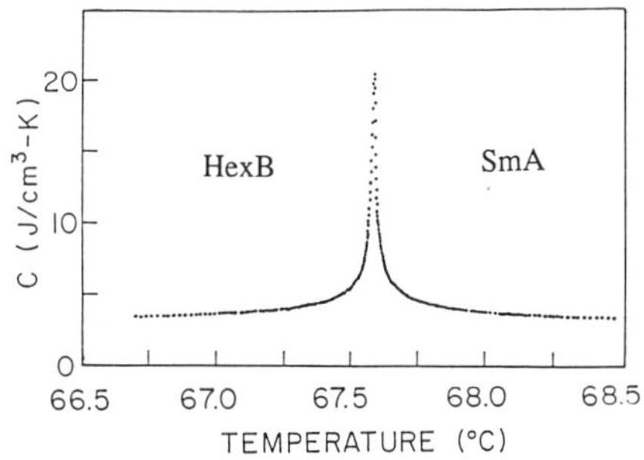


Fig. 3. Heat capacity per unit volume associated with the SmA-HexB transition in 65OBC. (Ref. 19)

The first high-resolution heat capacity study of a SmA-HexB transition was made on hexylpentyloxybiphenyl carboxylate (65OBC),[19] and the results shown in Fig. 3 are typical of all such transitions. These 65OBC data were fit with a power-law expression like Eq. (3) with $D_2^{\pm} = 0$ and $\Delta_1$ somewhat arbitrarily set equal to 0.75. The least-squares value of the critical exponent is[18-19] $\alpha = 0.60 \pm 0.03$. Heat capacity data on a wide variety of pure compounds and several binary mixtures yield large $\alpha$ values in the range 0.56 to 0.65 (except for 45OBC with $\alpha = 0.48$).[18] A critical exponent $\alpha \approx 0.6$ is completely inconsistent with XY behavior, for which $\alpha = -0.007$. One proposed explanation for such large $\alpha$ values is the assumption that the transition is in the vicinity of a tricritical point, for which $\alpha = 0.5$. On the basis of this assumption, several theoretical efforts have been made in order to explore the possible coupling of the hexatic order parameter to localized herring-bone order,[20] crystalline order,[17] and spatial fluctuations of the smectic layers.[21] However, $C_p$ measurements[22,23] along the SmA-HexB transition lines in liquid-crystal binary mixtures seem to rule out the possibility that either localized herring-bone or translational ordering might have an effect on the character of this transition.

The heat capacity associated with the SmC-SmI-CrJ sequence of transitions is shown in Fig. 4 for methylbutylphenyloctyloxybiphenyl carboxylate (8OSI).[24] In the case of tilted

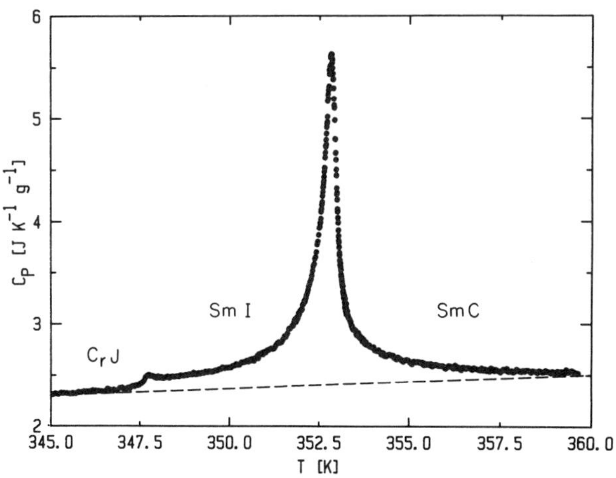


Fig. 4. Specific heat of 8OSI associated with the SmC-SmI and the first-order SmI-CrJ transitions. The dashed line represents $C_p^0$, the regular background contribution. (Ref. 24)

hexatics, there is a temperature independent field H (due to the tilt). This field induces bond orientational order in the SmC phase, which then grows rapidly on cooling into the SmI or SmF phase.

As a result of this field, SmC and SmI are not thermodynamically distinct phases and

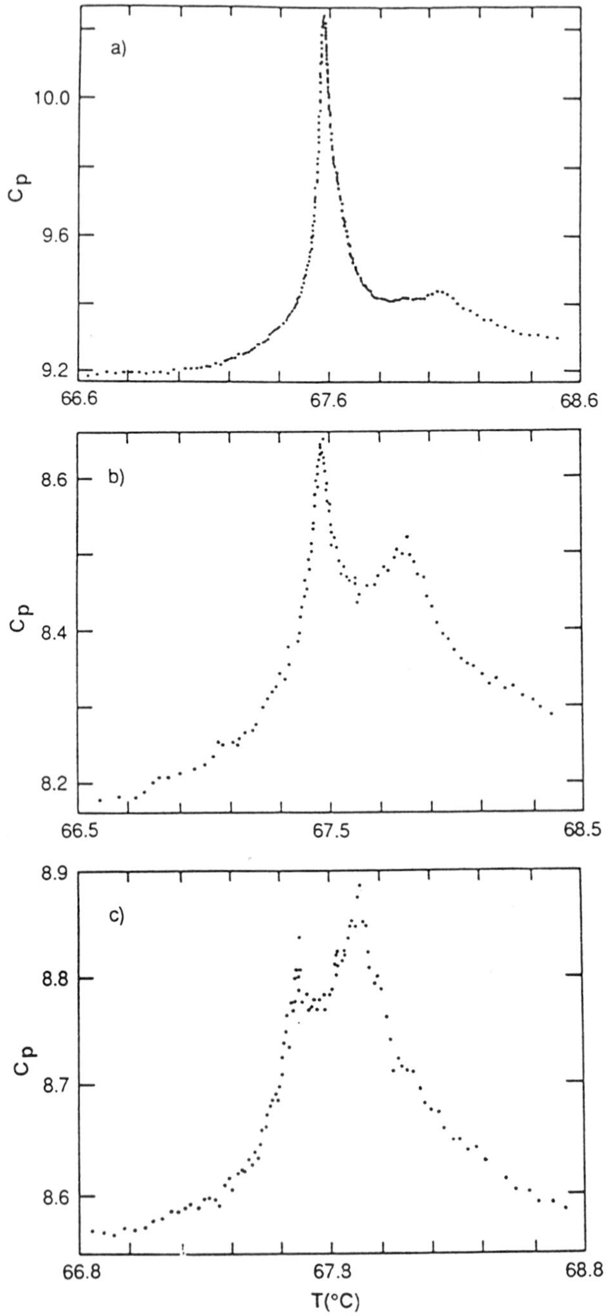


Fig. 5. Heat capacity associated with the SmA-HexB transition in 65OBC films with thickness of (a) ten layers, (b) six layers and (c) five layers. (Ref. 26)

the $C_p$ peak is clearly rounded over a 300 mK range. Both $\Delta C_p$ and the bond-orientational order parameter were fit simultaneously using a parametric equation of state with a finite H. The effective critical exponents ($\alpha = 0.47$ and $\beta = 0.077$) suggest that the SmC-SmI transition in 8OSI would be[24] weakly first order in the limit $H \rightarrow 0$. A study of the SmC-SmF transition in three homologs of terephtal-bis-alkylaniline (TBnA with n = 5,6,7) showed that this transition is strongly first order with a two-phase coexistence region[25] of ~3 K. It should also be noted in Fig. 4 that the SmI-CrJ freezing transition involves a much smaller enthalpy or entropy change than the SmC-SmI transition. The same conclusion is valid for the SmF-CrG versus SmC-SmF transitions and the HexB-CrB versus SmA-HexB.

Recently, Huang and coworkers have pioneered heat capacity studies of free-standing smectic films and have applied this technique to the SmA-HexB transition.[8,26] Results for 650BC samples of various thickness are shown in Fig. 5 and should be compared to the bulk sample result given in Fig. 3. The data in Fig. 5 show two distinct peaks associated with bond-orientational ordering of surface and interior layers. For the 5-layer film the high-temperature surface feature is larger than the bulk feature and a 4-layer film exhibits only one $C_p$ peak.[26] It appears likely that 3-layer and 4-layer data may provide a good approximation to the 2D limiting behavior that has been extensively studied theoretically.

## 5. Frustrated Smectic Phases Trasitions

Extensive smectic-A and smectic-C polymorphism has been observed in liquid crystal molecules with a long (three-ring) aromatic core, a strongly polar head group and an alkyl or alkoxy tail. Two competing smectic order parameters exist in these molecules and in general the two characteristic wavevectors are incommensurate. As a result of coupling between the order parameters, such "frustrated" smectics can exhibit a surprising variety of smectic-A phases: a monolayer smectic $SmA_1$, a partial bilayer $SmA_d$, a bilayer $SmA_2$, laterally modulated $Sm\tilde{A}$ (fluid antiphase) and $SmA_{cren}$ (crenelated phase), and the incommensurate $SmA_{ic}$. There are also many tilted smectic analogs: $SmC_1$, $SmC_d$, $SmC_2$ and $Sm\tilde{C}$. The theory of phase transitions in frustrated smectics has been extensively developed by Prost and coworkers.[27] A general review of the experimental behavior in frustrated smectics has been given elsewhere.[28] The results to be presented here include the N-$SmA_1$ and N-$SmA_2$ transitions (where N denotes the nematic phase), the $SmA_d$-$N_{reentrant}$-$SmA_1$ region, the $SmA_1$-$SmA_2$ and $SmA_1$-$Sm\tilde{A}$ transitions and the region of the $SmA_d$-$SmA_2$ critical point. In several cases there is good agreement between experiment and current theory, while in other cases there are puzzling features that require new experimental and theoretical efforts.

The N-SmA transition has been predicted theoretically to belong to the inverted XY universality class (XY critical behavior with an inverted $C_p$ amplitude ratio).[29] An extensive variety of investigations show that nonpolar N-SmA transitions and N-$SmA_d$ transitions in polar compounds do not conform to XY behavior.[1] However, recent studies of the N-$SmA_1$ transition in several different materials show that $C_p$ data for this transition are very well described by the XY model but do not show the expected amplitude inversion.[29-32] Critical heat capacity behavior at a typical N-$SmA_1$ transition is shown in Fig. 6. The theoretical fit with Eq. (3) to these octyloxyphenyl-cyanobenzloxy benzoate (8OPCBOB) data yields $\alpha = \alpha_{XY} = -0.007$, $A^-/A^+ = 0.987 \pm 0.002$ (XY value $0.9714 \pm 0.0126$, inverted XY value $1.0294 \pm 0.0134$), $D_1^-/D_1^+ = 1.03$ (theory $\cong 1.0$) and the dimensionless ratio $R^+ \equiv A^+ |D_1^+|^{\alpha/\Delta_1} B_c^{-1} = -1.074 \pm 0.025$ (XY value $-1.057 \pm 0.022$).[29] Each of the universal critical parameters that can be calculated with renormalization group theory show excellent agreement between experimental and the normal (noninverted) 3D-XY theoretical values for all N-$SmA_1$ transitions studied to date.

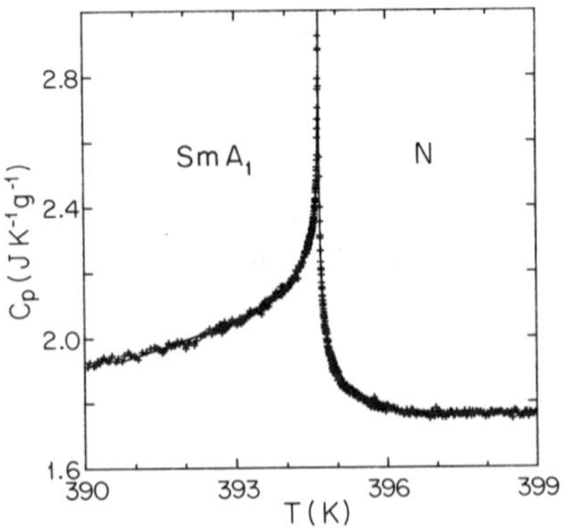


Fig. 6. Heat capacity data near the N-SmA$_1$ transition in 8OPCBOB and the least- squares fit with $\alpha = \alpha_{XY} = -0.007$. (Ref. 29)

Most of the N-SmA$_2$ transitions that have been studied so far are strongly first-order,[31] but a second-order transition has been observed recently in heptyloxycarbonylphenyl-cyanobenzoyloxy benzoate (7APCBB).[15] As shown in Fig. 7, this N-SmA$_2$ transition is also well described by the 3D-XY model: $\alpha$ = -0.007, $A^+ = -19.14$, $A^-/A^+ = 0.98$, $D_1^+ = 0.22$, $D_1^-/D_1^+ = 0.95$, $B_C$ = =18.80 are the fit parameters.

The existence of both first-order and second-order N-SmA$_2$ transitions indicates that crossover from XY to first order via a tricritical point should be observable in some appropriate mixture, but such experiments have not yet been carried out. This crossover behavior has, however, been observed for the N-SmA$_1$ transition in mixtures of octyl- and decyloxyphenyl-nitrobenzoyloxy benzoate (DB$_8$ONO$_2$ + DB$_{10}$ONO$_2$).[32,33] The phase diagram for this mixture involves multiple N and SmA$_d$ reentrance as well as SmA$_1$, SmA$_2$, Sm$\tilde{C}$ and SmC$_2$ phases.[33] A small part of this phase diagram is given in Fig. 8, where X represents the mole percent concentration of DB$_{10}$ONO$_2$. The second-order N-SmA$_1$

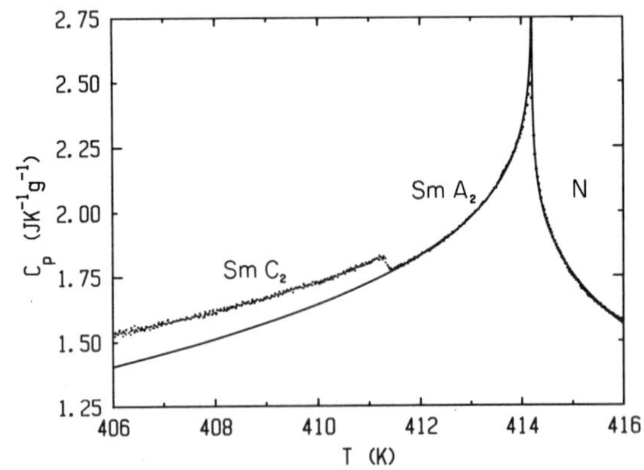


Fig. 7. Heat capacity of 7APCBB near the N-SmA$_2$ and SmA$_2$-SmC$_2$ transitions. The theoretical curve represents an XY fit to the N-SmA$_2$ peak using Eq. (3). Note also that the $C_p$ variation associated with the SmA$_2$-SmC$_2$ transition agrees well with that shown in Fig. 2. (Ref. 15)

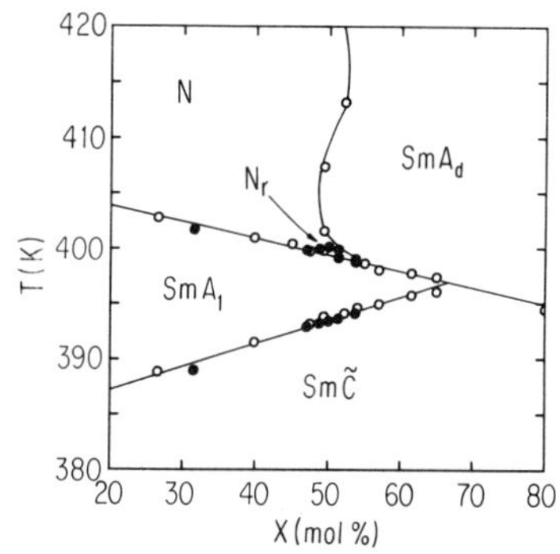


Fig. 8. Partial phase diagram for $DB_8ONO_2$ + $DB_{10}ONO_2$ mixtures. The open circles were obtained from optical microscopy and the solid points were obtained from $C_p$ measurements. (Ref. 32)

transition in pure $DB_8ONO_2$ and mixtures with $X \leq 47$ exhibit XY critical behavior. The transitions for mixtures with X = 48.7 and 50.0 are not XY-like but can be described by Eq. (3) using effective critical exponents $\alpha = 0.30$ and 0.42, respectively.[32] The $C_p$ variation for a near-tricritical mixture with X = 51.33 is shown in Fig. 9. The fit to these N-$SmA_1$ data with Eq. (3) yields[32,33] $\alpha = 0.5$ in agreement with the theoretical tricritical exponent and an amplitude ratio $A^-/A^+ = 1.66$. Very similar amplitude ratios have been reported for other N-SmA tricritical systems.[1,34] The dramatic difference in the size of the $SmA_d$-$N_r$ and $N_r$-$SmA_1$ peaks in Fig. 9 should also be noted. The relative size of these two peaks agrees qualitatively with theoretical expectations based on the frustrated spin-gas model.[35]

The generally accepted explanation for a second-order N-SmA transition becoming first order (via a tricritical point) is coupling between the nematic and smectic order parameters in a region where the nematic susceptibility $\chi_S$ is large. Tricritical points in previously studied systems occur near the SmA-N-I (isotropic) triple point,[1] whereas the N-I transition in $DB_8ONO_2$ + $DB_{10}ONO_2$ lies at ~500 K, far above the $SmA_d$-$N_r$-$SmA_1$ region shown in

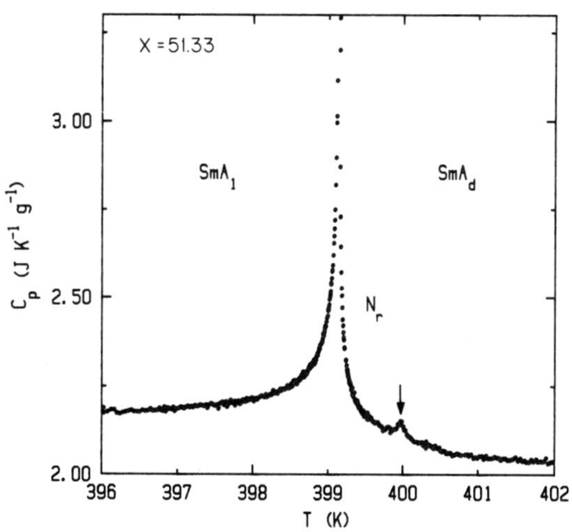


Fig. 9. The heat capacity variation in the $SmA_d$-$N_r$-$SmA_1$ transition region for a $DB_8ONO_2$ + $DB_{10}ONO_2$ mixture with X = 51.33. Note the very small $SmA_d$-$N_r$ peak. (Ref. 33)

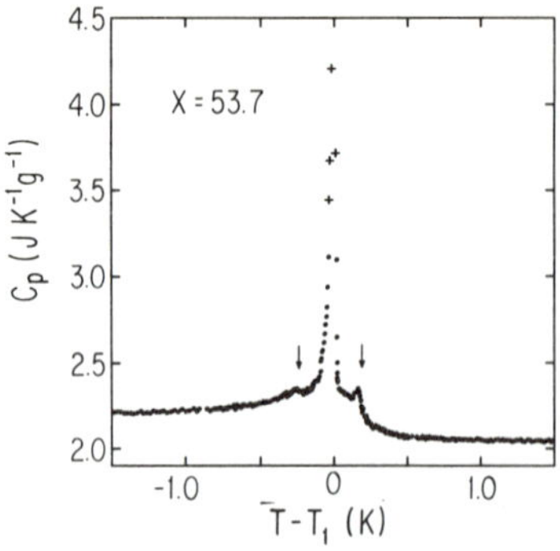


Fig. 10. Heat capacity measurements for a DB$_8$ONO$_2$ + DB$_{10}$ONO$_2$ mixture with X = 53.7. Crosses represent apparent $C_p$ values in a two-phase coexistence region. (Ref. 32)

Fig. 8. However, the tricritical point in DB$_8$ONO$_2$ + DB$_{10}$ONO$_2$ can be explained by a large $\chi_S$ value that arises due to the theoretically predicted existence of a nearby N-N critical point. A dislocation-loop melting model[36] shows that a SmA$_d$-N-SmA$_1$ "bicritical" region can be transformed into a SmA$_d$-N$_d$-N$_1$-SmA$_1$ sequence with a short first-order N$_d$-N$_1$ line which ends in a N$_d$-N$_1$ critical point. The N$_d$ phase is characterized by strong SmA$_d$-like fluctuations and the N$_1$ phase by strong SmA$_1$-like fluctuations. There is also a discontinuous jump in the nematic order parameter S at a first-order N$_d$-N$_1$ transition and a divergent $\chi_S$ at the N$_d$-N$_1$ critical point. The first experimental evidence of such phase behavior is shown in Fig. 10. The large central $C_p$ peak at $T_1$ = 398.86K is clearly first order with a coexistence region[32] of 35 mK and microscopic observations[37] show that the sample is nematic between the two arrows at 398.75K and 399.02K. Thus we can assign the central peak for this X = 53.7 sample to a N$_d$-N$_1$ transition and the small subsidiary peaks to SmA$_d$-N$_d$ and N$_1$-SmA$_1$ transitions.

Since frustrated smectics exhibit a rich SmA polymorphism there are several types of SmA-SmA transitions. Let us consider first the second-order SmA$_1$-SmA$_2$ transition which involves a doubling of the layer thickness from d = L to d = 2L, where L is the molecular length. Both the frustrated smectics model of Prost[27] and a renormalization group theory of Wang and Lubensky[38] place the SmA$_1$-SmA$_2$ transition in the 3D-Ising universality class,

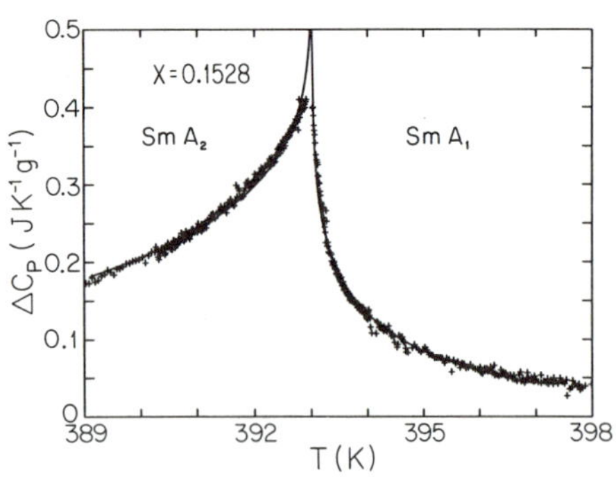


Fig. 11. Excess heat capacity associated with the SmA$_1$-SmA$_2$ transition in a DB$_5$CN + TBBA mixture. The fitting curve was obtained with Eq. (3) using $\alpha = -0.124$. (Ref. 31)

for which the critical exponents are $\gamma_I = 1.24$, $\nu_I = 0.63$ and $\alpha_I = 0.11$. This transition has been studied in mixtures of alkylphenyl cyanobenzoyloxy benzoate (DB$_n$CN) and terephthal-bis-butylaniline (TBBA),[31,39,40] which exhibit a N-SmA$_1$-SmA$_2$ phase sequence for TBBA mole fraction X in the range ~0.1 to ~0.2. The excess heat capacity associated with the SmA$_1$-SmA$_2$ transition is shown in Fig. 11 for a DB$_5$CN + TBBA mixture. A least-squares fit to $\Delta C_p$ with Eq. (3) letting $\alpha$ be a free parameter yielded $\alpha = -0.17$ for this sample. Comparable negative $\alpha$ values were obtained for other DB$_5$CN + TBBA mixtures,[31] and $\alpha = -0.15$ was obtained for a DB$_6$CN + TBBA sample with X = 0.15.[39] Furthermore, a high-resolution x-ray study of DB$_6$CN + TBBA mixtures gave[40] SmA$_1$-SmA$_2$ susceptibility and correlation length critical exponents $\gamma = 1.46 \pm 0.05$ and $\nu_{||} = \nu_{\perp} = 0.74 \pm 0.03$. The apparent disagreement between these experimental exponents and the Ising values can be explained in terms of Fisher renormalization of the critical behavior since the quantity $(dT_c/dX)^2$ has a large value.[1] The Fisher-renormalized Ising values are $\gamma_R = \gamma/(1-\alpha) = 1.39$, $\nu_R = \nu/(1-\alpha) = 0.71$ and $2-\alpha_R = (2-\alpha)/(1-\alpha)$ yielding $\alpha_R = -0.124$. The heat capacity data are all consistent with this negative $\alpha_R$ value, as indicated by the fit shown in Fig. 11.

Some liquid-crystal mixtures do not undergo the direct SmA$_1$-SmA$_2$ transition described above but exhibit the phase sequence SmA$_1$-Sm$\tilde{\text{A}}$-SmA$_{cren}$-SmA$_2$. The fluid antiphase Sm$\tilde{\text{A}}$ and the crenelated phase SmA$_{cren}$ are intermediate smectic phases with periodic long-range modulations of the dipole orientations within each layer. Such modulated phases have been most extensively studied in mixtures of DB$_n$CN and C$_5$stilbene (cyanobenzoyloxy pentylstilbene).[41] An overview of the heat capacity behavior for a DB$_5$CN + C$_5$stilbene mixture is shown in Fig. 12. The N-SmA$_1$ transition exhibits XY critical behavior as expected.[29,41] The SmA$_1$-Sm$\tilde{\text{A}}$ transition is predicted to be first order,[27] and one observes very weakly first order behavior with a narrow coexistence range. However, there is substantial pretransitional heat capacity over a considerable range, as indicated by the dashed curve representing the background heat capacity in the vicinity of the SmA$_1$-Sm$\tilde{\text{A}}$ transition. A detailed view of $\Delta C_p$ near the SmA$_1$-Sm$\tilde{\text{A}}$ transition is presented in Fig. 13 for two DB$_n$CN + C$_5$stilbene samples. The artificially high points associated with two-phase coexistence are not shown. Note that the pretransitional $\Delta C_p$ values in the SmA$_1$ phase

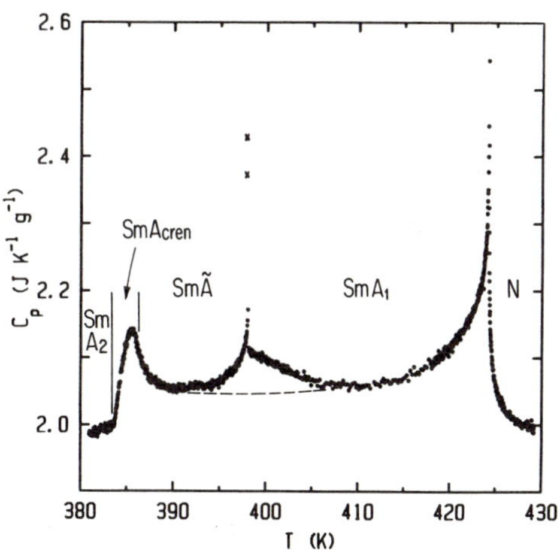


Fig. 12. Heat capacity of a DB$_5$CN + C$_5$stilbene mixture containing 49.2 mole percent C$_5$stilbene. The two points denoted by crosses are apparent $C_p$ values in a two-phase coexistence region at the weakly first-order SmA$_1$-Sm$\tilde{\text{A}}$ transition. (Ref. 41)

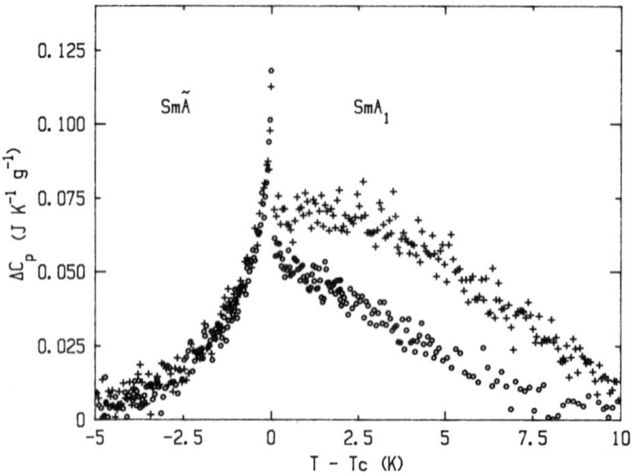


Fig. 13. Excess heat capacity associated with the $SmA_1$-Sm$\tilde{A}$ transition. Open circles are used for $DB_5CN$ + $C_5$stilbene and plus signs for $DB_6CN$ + $C_5$stilbene. (Ref. 41)

exhibit an anomalous temperature dependence. This excess heat capacity is associated with strong Sm$\tilde{A}$ fluctuations in the $SmA_1$ phase observed in x-ray studies. It can also be correlated with a very large increase in the shear viscosity $\eta$ observed on cooling toward the $SmA_1$-Sm$\tilde{A}$ transition.[41] Both $\Delta\eta$ and $\Delta C_p$ are consistent with substantial short-range Sm$\tilde{A}$ antiphase ordering in the $SmA_1$ phase, but no theory of such effects yet exists. Finally, it should be noted that the $C_p$ variation in Fig. 12 associated with the Sm$\tilde{A}$-$SmA_{cren}$-$SmA_2$ transition sequence is similar to the inverted Landau behavior observed for the Sm$\tilde{C}$-$SmC_2$ transition in other systems,[33] but with the peak truncated due to the presence of the $SmA_{cren}$ phase.

Since the $SmA_d$ and $SmA_2$ phases have the same macroscopic symmetry, there is no symmetry breaking and there cannot be a second-order $SmA_d$-$SmA_2$ transition. Thus the $SmA_d$ and $SmA_2$ phases must be separated by a first-order transition line, which can terminate at a critical point. It is predicted theoretically that this critical point belongs to a new universality class.[42] Experimental measurements,[43] although complicated by a distortion of the $C_p$ peak close to $T_C$, support the idea of a new critical behavior at the $SmA_d$-$SmA_2$ critical point.

ACKNOWLEDGEMENT. Many of the calorimetric results reported here are due to the vigorous efforts of my research coworkers K. Ema, G. Nounesis, K. J. Stine and X. Wen. This research was supported in part by National Science Foundation Grants DMR 87-02052, DMR 87-19217 and DMR 90-07611.

## References

1. J. Thoen, Chapter 10 of this Volume.
2. C. A. Schantz and D. L. Johnson, Phys. Rev. **A17**:1504 (1978).
3. J. M. Viner, D. Lamey, C. C. Huang, R. Pindak and J. W. Goodby, Phys. Rev. **A28**:2433 (1983).
4. C. W. Garland, Thermochim. Acta **88**:127 (1985) and references cited therein.
5. X. Wen, C. W. Garland and M. D. Wand, Phys. Rev. **A42**:6087 (1990).
6. J. D. Baloga and C. W. Garland, Rev. Sci. Inst. **48**:105 (1977).
7. G. B. Kasting, K. J. Lushington and C. W. Garland, Phys. Rev. **B22**:321 (1980).

8. R. Geer, T. Stoebe, T. Pitchford and C. C. Huang, Rev. Sci. Instrum. **62**:415 (1991).
9. R. J. Birgeneau, C W. Garland, A. R. Kortan, J. D. Litster, M. Meichle, B. M. Ocko, C. Rosenblatt, L. J. Yu and J. Goodby, Phys. Rev. **A27**:1251 (1983).
10. M. Meichle and C. W. Garland, Phys. Rev. **A27**:1251 (1983).
11. C. C. Huang and J. M. Viner, Phys. Rev. **A25**:3385 (1982); S. C. Lien, C. C. Huang and J. W. Goodby, Phys. Rev. **A29**:1371 (1984) and references therein.
12. S. Dumrongrattana, G. Nounesis and C. C. Huang, Phys. Rev. **A33**:2181 (1986); H. Y. Lui, C. C. Huang, Ch. Bahr and G. Heppke, Phys. Rev. Lett. **61**:345 (1988).
13. J. Boerio-Goates, C. W. Garland and R. Shashidhar, Phys. Rev. **A41**:3192 (1990).
14. Y. H. Jeong, K. J. Stine, C. W. Garland and Nguyen Huu Tinh, Phys. Rev. **A37**:3465 (1988).
15. X. Wen, C. W. Garland and G. Heppke, Phys. Rev. A, **A44** :5064 (1991).
16. R. Pindak, D. E. Moncton, S. C. Davey and J. W. Goodby, Phys. Rev. Lett. **46**:1135 (1981); J. D. Brock, A. Aharony, R. J. Birgeneau, K. W. Evans-Lutterodt, J. D. Litster, P. M. Horn, G. B. Stephenson and A. R. Tajbakhsh, Phys. Rev. Lett. **57**:98 (1986).
17. A. Aharony, R. J. Birgeneau, J. D. Brock and J. D. Litster, Phys. Rev. Lett. **57**:1012 (1986).
18. T. Pitchford, G. Nounesis, S. Dumrongrattana, J. M. Viner, C. C. Huang and J. W. Goodby, Phys. Rev. **A32**:1938 (1985).
19. G. Nounesis, C. C. Huang and J. W. Goodby, Phys. Rev. Lett. **56**:1712 (1986) and references cited therein.
20. R. Bruinsma and G. Aeppli, Phys. Rev. Lett. **48**:1625 (1982).
21. J. V. Selinger, J. Phys. (France) **49**:1387 (1988).
22. C. C. Huang, G. Nounesis, R. Geer, J. W. Goodby and D. Guillon, Phys. Rev. **A39**:3741 (1989).
23. G. Nounesis, R. Geer, H. Y. Lin, C. C. Huang and J. W. Goodby, Phys. Rev. **A40**:5468 (1989).
24. C. W. Garland, J. D. Litster and K. J. Stine, Mol. Cryst. Liq. Cryst. **170**:71 (1989).
25. K. J. Stine and C. W. Garland, Mol. Cryst. Liq. Cryst. **188**:91 (1990).
26. R. Geer, C. C. Huang, R. Pindak and J. W. Goodby, Phys. Rev. Lett. **63**:540 (1989).
27. J. Prost and P. Barois, J. Chim. Phys. **80**:65 (1983); P. Barois, J. Pommier and J. Prost, chap. 6 in: "Solitons in Liquid Crystals," L. Lam and J. Prost, eds., Springer-Verlag, (1989).
28. C. W. Garland, (1990) p. 221 in: "Geometry and Thermodynamics," J.-C. Toledano, ed., NATO ASI Ser. B, Plenum, New York .
29. C. W. Garland, G. Nounesis and K. J. Stine, Phys. Rev. **A39**:4919 (1989).
30. C. W. Garland, G. Nounesis K. J. Stine and G. Heppke, J. Phys. (France) **50**;2291 (1989).
31. P. Das, G. Nounesis, C. W. Garland, G. Sigaud and Nguyen Huu Tinh, Liq. Cryst. **7**:883 (1990).
32. G. Nounesis C. W. Garland and R. Shashidhar, Phys. Rev. **A43**:1849 (1991).
33. K. Ema, G. Nounesis, C. W. Garland and R. Shashidhar, Phys. Rev. **A39**:2599 (1989).
34. K. J. Stine and C. W. Garland, Phys. Rev. **A39**:3148 (1989).
35. J. O. Indekeu, A. N. Berker, C. Chiang and C. W. Garland, Phys. Rev. **A35**:1371 (1987).
36. J. Prost and J. Toner, Phys. Rev. **A36**:5008 (1987).
37. G. Nounesis, C. W. Garland and R. Shashidhar, to be published.
38. J. Wang and T. C. Lubensky, Phys. Rev. **A29**:2210 (1984).
39. C. W. Garland, C. Chiang and F. Hardouin, Liq. Cryst. **1**:81 (1986).
40. K. K. Chan, P. S. Pershan, L. B. Sorensen and F. Hardouin, Phys. Rev. **A34**:1420 (1986).
41. K. Ema, C. W. Garland, G. Sigaud and Nguyen Huu Tinh, Phys. Rev. **A39**:1369 (1989).
42. Y. Park, T. C. Lubensky, P. Barois and J. Prost, Phys. Rev. **A37**:2197 (1988).
43. Y. H. Jeong, G. Nounesis, C. W. Garland and R. Shashidhar, Phys. Rev. **A40**:4022 (1989); X. Wen, C. W. Garland , R. Shashidhar and P. Barois, Phys. Rev A, **B45**, March (1992).

Chapter 12

# PHOTOTHERMAL CALORIMETRY: SIMULTANEOUS MEASUREMENTS OF SPECIFIC HEAT AND THERMAL CONDUCTIVITY

M.MARINELLI, U.ZAMMIT, R.PIZZOFERRATO, F.SCUDIERI, S.MARTELLUCCI and F.MERCURI

## 1. Introduction

The photothermal techniques are based on the periodic heating induced in an investigated sample by a modulated light source, the sample temperature oscillations being detected using various techniques, such as gas microphone[1], piezoelectric[2], photodeflection[3] and photopyroelectric[4]. The photothermal signal, which is characterized by amplitude and phase, depends on the sample thermal as well as optical properties and may thus be used for thermal parameter measurements of the investigated sample.

The photothermal techniques, are a.c. techniques and they allow an adequate signal to noise ratio to be achieved while introducing temperature oscillations on the sample surface of only a few mK. The d.c. temperature variations within the sample induced by the a.c. heating source are also very small. The technique is thus very attractive for the study the behaviour of the thermal parameters in thermotropic liquid crystals in the vicinity of their phase transitions, where the parameters may depend strongly on temperature. Moreover, since under particular experimental conditions one of the photothermal signal channels, the phase channel, depends on only the sample thermal transport properties, the technique may be used to measure simultaneously the specific heat and the thermal conductivity of the sample, thus enabling the investigation of both the static and dynamic critical behaviours associated with the thermal parameters in the vicinity of liquid crystal phase transitions. In this paper we shall review the results which may be obtained in this respect using the gas microphone and photopyroelectric configurations.

## 2. Gas Microphone Configuration

In the gas microphone configuration the sample is contained in an air tight volume where a microphone is also contained (Fig.1).

---

M.Marinelli, U.Zammit, R.Pizzoferrato, F.Scudieri, S.Martellucci and F.Mercuri - Dipartimento Ingegneria Meccanica - II Università di Roma "Tor Vergata"

*Phase Transitions in Liquid Crystals*, Edited by S. Martellucci and A.N. Chester, Plenum Press, New York, 1992

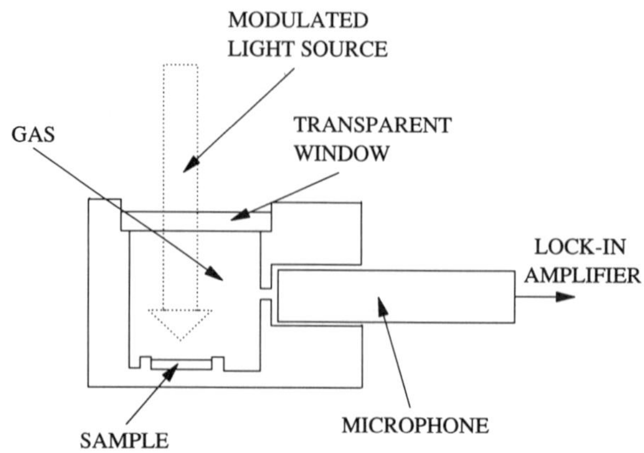


Fig.1. Photoacoustic cell configuration.

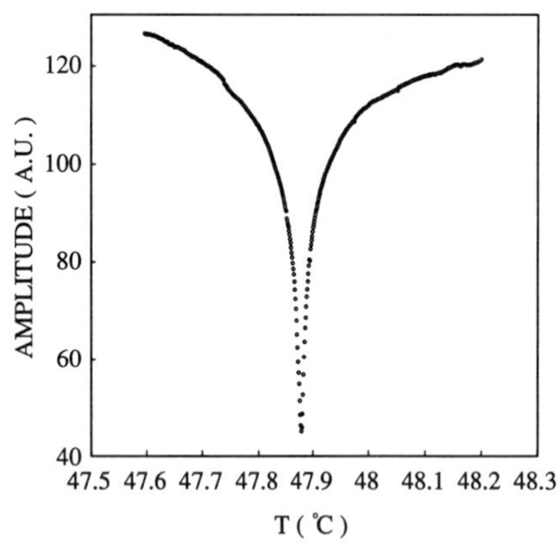


Fig.2. Photoacoustic signal amplitude over the smectic A - nematic phase transition in 9CB.

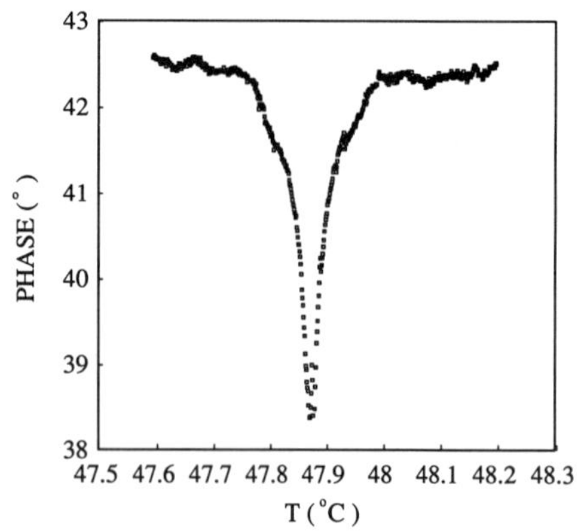


Fig.3. Photoacoustic signal phase over the smectic A -nematic phase transition in 9CB.

The light induced sample heating causes, by heat diffusion, the heating of the gas layer in the vicinity of the sample surface. This in turn causes a periodic overpressure in the gas which is detected by the microphone in a lock-in configuration. The photoacoustic signal amplitude and phase can thus be measured.

The expressions for the signal amplitude and phase have been developed in a one dimensional geometry case[1]. When the sample thickness $l$ is much larger than the thermal diffusion length $\mu = (2\alpha/\rho c)^{1/2}$ and optical absorption depth $\mu_\beta = 1/\beta$ in the sample, where $\alpha = k/\rho c$ is the sample thermal diffusivity ($\rho, c, k$ and $\beta$ being the density, specific heat, thermal conductivity and the optical absorption coefficient respectively), the sample is said to be optically and thermally thick. Under such conditions, the expressions for the photoacoustic signal amplitude $A$ and phase $\Phi$ become[5]

$$A = \frac{P_o \gamma \mu_g}{2^{1/2} l_g T_o}\left(\frac{I_o p}{2 e_g \omega^{1/2}(p^2+2p+2)^{1/2}(1+q)}\right) \qquad \phi = -\frac{\pi}{4} - \tan^{-1}\left(1+\frac{2}{p}\right) \tag{1}$$

where $p = \mu_s/\mu_\beta$, $q = e_s/e_g = (k_s c_s \rho_s / k_g c_g \rho_g)^{1/2}$, $e$ is the thermal effusivity, subscripts $s$ and $g$ refer to sample and coupling gas respectively, $I_0$ is the laser intensity, $P_0$ is the initial cell gas pressure, $T_0$ is the d.c. ambient temperature, $\gamma$ is the ratio of the gas specific heats at constant pressure and at constant volume, and $\omega$ is the modulation frequency in rad/s.

The quantity in the large brackets in the amplitude is the temperature oscillation induced on the sample surface. From the phase data the sample diffusivity $\alpha = k/\rho c$ can be derived, provided that $\beta$ is known, while from the amplitude data the effusivity $e$ and therefore $c$ and $k$ can also be obtained. The three thermal parameters can thus be simultaneously determined from measurements performed at one value of the modulation frequency.

We have applied the photoacoustic technique to the study of the anomalies observed in the behaviour of the thermal parameters in the vicinity of the smectic A - nematic (A-N) phase transitions in several alkylcyanobiphenyl samples with different nematic ranges. The second order A-N transition in liquid crystals was placed in the same universality class as a superfluid, namely the three dimensional XY-like transition[6]. Earlier observations concerning the behaviour of the specific heat alone[7] carried out in alkylcyanobiphenyl compounds and mixtures having different nematic phase temperature range showed that the critical exponent $\alpha$ of the anomaly varied between the tricritical value, 0.5, and -0.03, a value close to that predicted by the XY model (-0.007). This was interpreted as due to a crossover between tricritical and XY-like critical behaviour as the nematic range increased.

The measurements were performed on 5 mm diameter 0.4 mm deep samples contained in a gold plated copper sample holder. The samples were illuminated with a He-Ne laser operating at 3.39 μm wavelength, at which there is strong absorption by the C-H bonds of the liquid crystal compounds. The value of $\beta$ at such a wavelength is 1500 cm$^{-1}$ ($\mu_\beta$=7 μm). The modulation frequency was 30 Hz. With a typical $\alpha$ value for liquid crystals of $5\times10^{-4}$ cm$^2$s$^{-1}$, $\mu_s$=25 μm. The conditions for a thermally and optically thick sample are thus fulfilled. The laser intensity was 1 mW cm$^{-2}$ and the temperature oscillations induced on the sample surface, as estimated from the term within brackets in the expression for $A$, was only 0.5 mK. It is worthwhile pointing out that from Eq.1 it is clear that $p$ must not be much larger than unity or the phase would then tend to $-\pi/2$ and would become insensitive to $\alpha$. In our experimental conditions $p$=3.5, and thus useful for the measurement of $\alpha$. Finally, the d.c. temperature variation over the probed depth ($\mu_s$) for an optically thick sample is given by

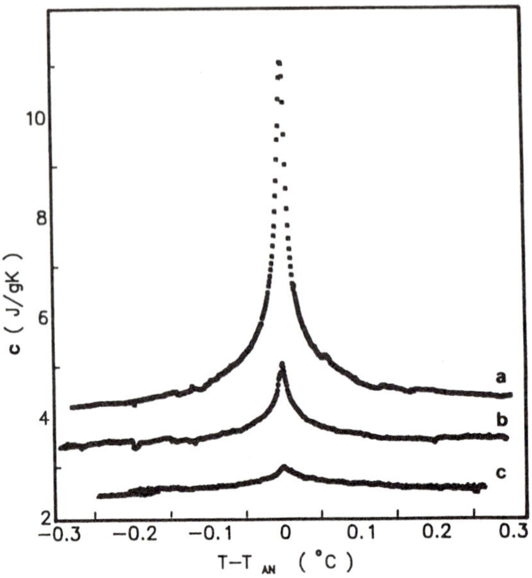


Fig.4. Specific heat data over the A-N transition in 9CB (a), 8CB (b) and 7CB+8CB mixture (c).Some data sets were shifted vertically to avoid overlapping.

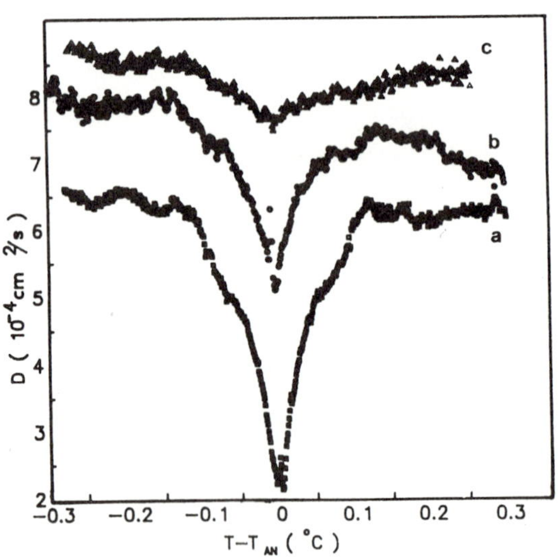


Fig.5. Thermal diffusivity heat data over the A-N transition in 9CB (a), 8CB (b) and 7CB+8CB mixture (c).Some data sets were shifted vertically to avoid overlapping.

$$\Delta T = \frac{I}{2k_s\beta_s}\left(\mu_s\left(1+\beta l_g\frac{k_s}{k_g}\right)\Big/\left(l_g\frac{k_s}{k_g}+l_b\frac{k_s}{k_b}+l_s\right)-1\right) \tag{2}$$

subscript $b$ refers to the sample backing material. Under our experimental conditions $\Delta T$=1.4 mK, which is negligible. The investigations were carried out on samples of 9CB, 8CB and a 0.76 8CB mole fraction mixture of 8CB and 7CB. The respective values of $T_{AN}/T_{NI}$ were 0.994, 0.967 and 0.962 respectively, where $T_{AN}$ and $T_{NI}$ are the smectic A - nematic and nematic - isotropic transition temperatures respectively. The temperature rate change during the measurements was less than 1 mK/min.

The amplitude and phase results for the 9CB sample are shown in Figs.2 and 3. Both feature very sharp dips over the transition region . The behaviour of $c$, $\alpha$ and $k$ are reported for all three samples in Figs.4, 5 and 6 respectively. In all the samples investigated, the heat capacity and the thermal conductivity show peaks over the transition region. The thermal diffusivity data show a dip with a small peak structure in the vicinity of the transition temperature. The peak is due to a sharper rise in the thermal conductivity than in the heat capacity in the region very near to the transition temperature. The slower rise in $c$ may be due to rounding of the data near the peak, as shown in the double log plot of the singular term contribution to the $c$ data, $\Delta c$, vs $t$ (Fig.7), which may affect the heat capacity more than the thermal conductivity. $\Delta c$ was obtained by fitting the $c$ data with the power law expression $c = A\,|t|^{-\alpha} + B + Et$ and then subtracting from the data the constant and linear terms: $\Delta c = c-(Et+B)$. $t=(T/T_{AN}-1)$ is the reduced temperature.

The critical exponents for the thermal parameters were obtained by fitting the data of each parameter with expressions containing singular and background terms. The results are reported in Table I. The points closest to the transition temperature considered in the fit procedure were those for which $|t| > 1.2\times10^{-5}$. To within experimental error the critical exponents calculated for $T>T_{AN}$ and $T<T_{AN}$ were equal, and their values are reported in Table I. In the following page together with the samples' respective $T_{AN}/T_{NI}$ values. The absolute values of the critical exponents of each thermal parameter progressively increase as $T_{AN}/T_{NI}$ increases.

Regarding the heat capacity data, the critical exponent values for 9CB and 8CB are in agreement with those already reported (0.5 and 0.3 respectively[7]). As for the mixture, the reported value is $\alpha=0.1$ for $T_{AN}/T_{NI}=0.961$. In the present case, the value of $\alpha$ is slightly larger, consistent with the trend of increasing values of $\alpha$ as $T_{AN}/T_{NI}$ increases. The good agreement of our $\alpha$ values with those obtained in literature confirms the reliability of the photoacoustic technique for the measurements of the thermal parameters in the vicinity of phase transitions.

Employing the hyperscaling relation[8] $d\nu = 2-\alpha$, where $d$=3 dimensions, $\nu$ may be determined from the experimental values of $\alpha$. Using the dynamic scaling law for an asymmetric planar magnet model which describes the superfluid-like transitions[9], $a=\nu/2(\varepsilon+\tilde{\alpha}/\nu)$ where $\varepsilon=4-d$ and $\tilde{\alpha}=\max(\alpha,0)$, the predicted values of the critical exponent $a$ of the thermal conductivity in the case of a 3 dimensional XY model can be obtained. The calculated values for 9CB, 8CB and the mixture are $a$=0.5, 0.43 and 0.38 respectively. Although the same trend is observed as that obtained experimentally for the $a$ values as a function of $T_{AN}/T_{NI}$, the values predicted by the scaling laws are substantially larger than the experimental ones and they would lead, in all the three cases, to a stronger divergence in the thermal conductivity than in the heat capacity over the phase transition. This would cause a divergence of the thermal diffusivity as well, in contrast with the reported data. This behaviour would also be found for a value of $\alpha$ consistent with the three dimensional XY model, as $a$ would be 0.34. It is most surprising that the discrepancy between predicted and experimental values increases as the $T_{AN}/T_{NI}$ value progressively decreases, and that is where the value of $\alpha$ tends to approach the $\alpha_{XY}$ value. The best agreement is in fact found at the TCP.

A similar result was obtained[10] for the Hexatic B - Smectic A transition in n-hexyl-4-n pentyloxybiphenyl-4-carboxylate (65OBC), another transition whose critical behaviour had been predicted to be XY-like. There was good agreement between the predicted and experimental values of $a$ even though a value of $\alpha$=0.6, much larger than $\alpha_{XY}$, had been obtained.

It has thus been shown that the critical exponents of the heat capacity, thermal conductivity and thermal diffusivity obtained at the smectic A - nematic transition in samples with different nematic ranges are inconsistent with the dynamic scaling law predicted by a model which applies to three dimensional XY-like phase transitions.

## 3. Photopyroelectric Configuration

Though useful for some thermal parameter studies near liquid crystal phase transitions, the gas microphone configuration has some limitations. Besides the microphone-limited frequency range of operation, and acoustic resonance effects which alter the ideal vs frequency behaviour of the photoacoustic signal, a major restriction is the limited sensitivity to thermal parameter changes, which is found particularly when the value of $p$ becomes much larger or much smaller than unity as discussed earlier. Even when the $p$ value is within a suitable range, the change which is found between the signal phase value found in one of the mesophases and the value found at the transition temperature is only about 4° in 9CB, as shown in Fig.3b, 1.5° in 8CB and 0.7° in the mixture.

It may therefore be very difficult to appreciate phase changes in transitions where small changes in the thermal diffusivity also occur, due to noise induced phase fluctuations. This result is typical for all the photothermal techniques which employ the front detection configuration, where detection is carried out on the same side of the sample as that which is illuminated by the light source.

The phase sensitivity from sample thermal diffusivity can be greatly enhanced by using a technique which employs a back detection configuration such as the photopyroelectric one sketched in Fig.8. The temperature changes induced in the sample are detected by means of a pyroelectric transducer placed on the opposite side of the sample with respect to the side where the light is incident. In this way, the thermal wave has to pass through the entire sample thickness before it is detected by the transducer, and the sensitivity of the signal from the sample thermal parameters changes is increased.

When the sample is optically and thermally thick, the expression for the signal amplitude and phase become[4]

$$A = \frac{K \exp[-(l_s/\mu_s)]}{(e_s + e_m)(e_p/e_s + 1)} \quad \text{and} \quad \Phi = -\frac{\pi}{2} - l_s/\mu_s \tag{3}$$

where $K$ is a constant which contains the transducer pyroelectric coefficient, and subscripts $p$ and $m$ refer to the pyroelectric transducer and the medium in front of the sample, respectively. The expressions show that once again the signal phase depends only on the sample thermal diffusivity and the amplitude also on the sample effusivity. However, in this case

$$\frac{d\Phi}{d\alpha_s} = \frac{1}{2\alpha_s}\frac{l_s}{\mu_s} \tag{4}$$

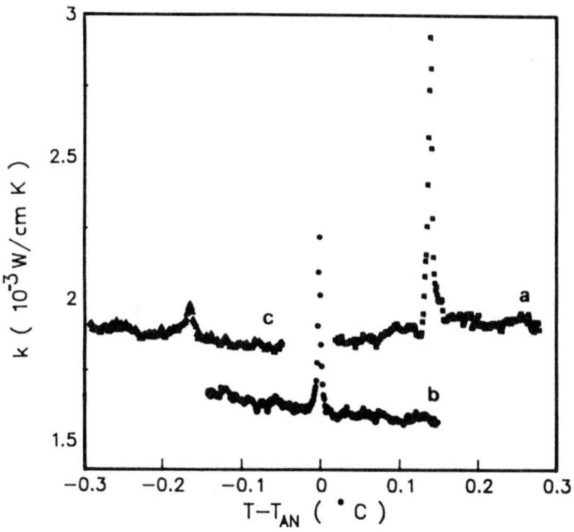


Fig.6. Thermal conductivity data over the A-N transition in 9CB (a), 8CB (b) and 7CB+8CB mixture (c).Some data sets were shifted horizontally to avoid overlapping.

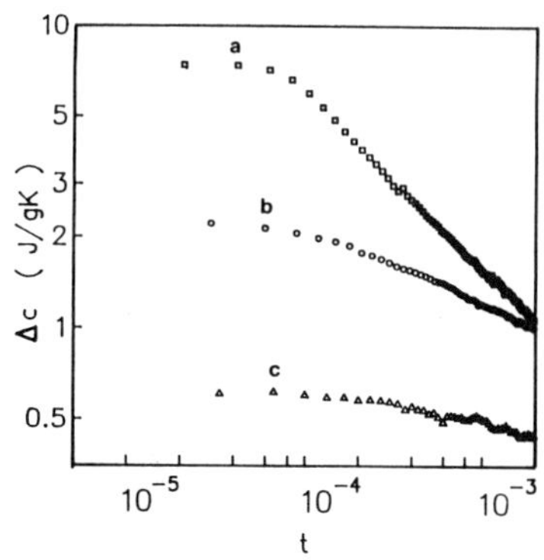


Fig.7. $\Delta c$ vs $t$ data over the A-N transition for 9CB (a), 8CB (b) and 7CB+8CB mixture (c).

Table I. Smectic A-Nematic to Nematic- Isotropic transition temperature ratios and critical exponents of heat capacity $\alpha$, thermal diffusivity $b$, and thermal conductivity $a$ of the samples investigated.

| sample | .24 7CB+.76 8CB mole fraction | 8CB | 9CB |
|---|---|---|---|
| $T_{AN}/T_{NI}$ | .962 | .967 | .994 |
| $\alpha(c)$ | .14±.03 | .28±.03 | .5±.04 |
| b (D) | -.01±.03 | -.04±.03 | -.08±.04 |
| a (k) | .13±.06 | .24±.06 | .44±.08 |
| a(th.) | 0.38 | 0.43 | 0.5 |

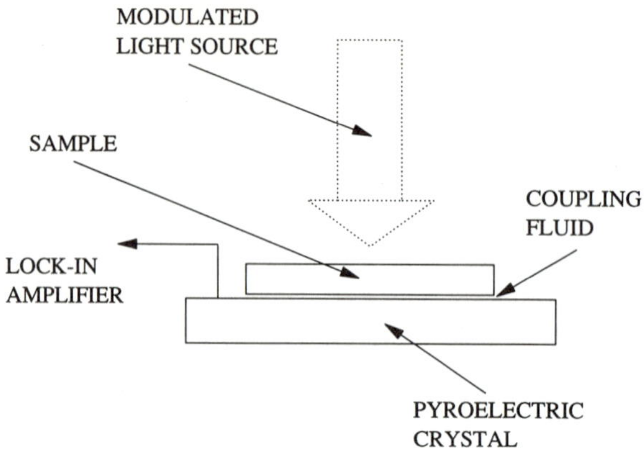


Fig.8 Photopyroelectric cell configuration.

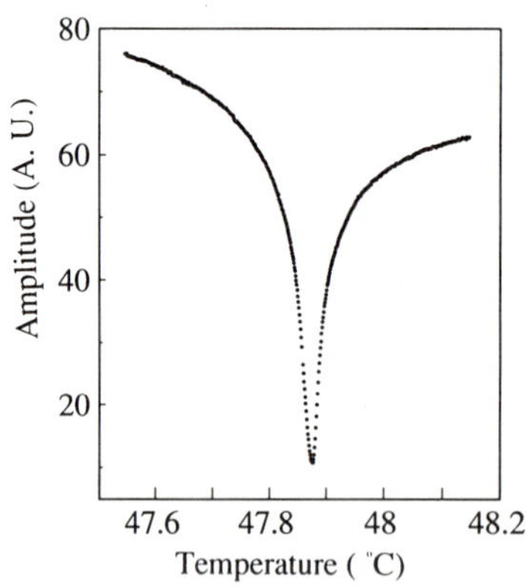


Fig.9. Photopyroelectric signal amplitude over the smectic A - nematic phase transition in 9CB.

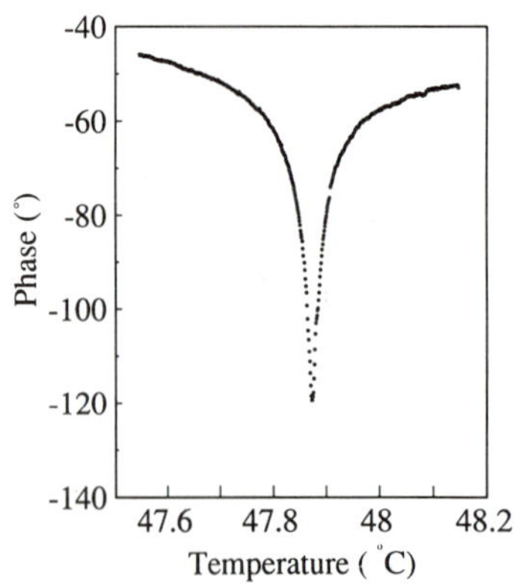


Fig.10. Photopyroelectric signal phase over the smectic A -nematic phase transition in 9CB.

which shows that the signal phase sensitivity from the sample thermal diffusivity changes increases as the ratio sample thickness to sample thermal diffusion length increases, which can be accomplished merely by increasing the modulation frequency. In contrast with the gas microphone the greatest sensitivity of the phase due to thermal diffusivity changes corresponded to $p=1$, and was therefore limited by the value of $\beta$. Moreover from the signal amplitude expression it can be noticed that the factor $l_s/\mu_s$ appears in the exponential function, which therefore acts as an amplification factor for changes in the sample thermal diffusivity.

The photopyroelectric signal amplitude and phase obtained at 23 Hz for a 50 μm thick for 9CB sample are shown in Figs.9 and 10. As shown, the relative change in the phase between the value in one mesophase and at the transition temperature is now about 80°, about 20 times greater than in the photoacoustic configuration. Moreover, the relative amplitude decrease through the transition, which was about 65% in the previous case, has now increased to about 90%.

The predicted behaviour of the photopyroelectric technique with respect to the gas microphone one is thus confirmed. A drawback of the photopyroelectric technique is that the sample acts as a low pass filter for the thermal waves passing through its thickness, and therefore it is not possible to work at large frequencies ( $l_s/\mu_s > 3$ ) without using a powerful light source and therefore introducing large temperature perturbations across the sample thickness.

## References


1. A.Rosencwaig and A.Gersho, J.Appl.Phys. **47**, 64 (1976)
2. W.B.Jackson and N.M.Amer, J.Appl.Phys.**51**, 3343 (1980)
3. W.B.Jackson, N.M.Amer, A.C.Boccara and D.Fournier, Appl.Opt. **20**, 1333 (1981)
4. A.Mandelis and M. Zver, Jour.Appl.Phys. **57**, 4421 (1985)
5. M.Marinelli, U.Zammit, F.Scudieri, S.Martellucci, F.Bloisi, L.Vicari and J.Quartieri, Nuovo Cimento D **9**, 557 (1987)
6. P.G.de Gennes, Solid State Commun., **10**, 753 (1972)
7. J.Thoen, H.Marynissen and W.Van Dael, Phys.Rev.Lett., **52**, 204 (1984)
8. L.P.Kadanoff, W.Götze, D.Hamblen, R.Hecht, E.A.S.Lewis, V.V.Palciauskas, M.Rayl, J.Swift, D.Aspnes and J.Kane, Rev.Mod.Phys. **39**, 395 (1967)
9. P.C.Hohenberg and B.I.Halperin, Rev.Mod.Phys., **49**, 435 (1977)
10. G.Nounesis, C.C.Huang and J.W.Goodby, Phys.Rev.Lett., **56**, 1712 (1986)

Chapter 13

# THREE DIMENSIONAL PHASE-SENSITIVE AC VOLTAMETRY: A NEW EXPERIMENTAL APPROACH FOR THE STUDY OF PHASE TRANSITIONS

N. PAPADOPOULOS, S. SOTIROPOULOS and P. NIKITAS

## 1. Introduction

A new methodology, which we name Three Dimensional Phase-Sensitive AC Voltametry, is proposed for the study of phase transitions in lyotropic liquid crystals. In this technique, we monitor the capacitance changes during adsorption of surface active compounds on mercury, as a function of both applied potential and time. The application method is applied to the case of a submicellar electrolytic solution of sodium dodecyl sulphate. It is seen that by means of the variation of the reconstructed C-E curves in time, not only short-time diffusion controlled processes but also long-time changes can be followed. The latter measurement is very useful in cases where phase changes proceed slowly.

Experimental observations have shown that a number of interesting phase transitions analogous to those appearing in multicomponent liquid systems can take place during adsorption on charged interfaces[1]. The presence in solution of neutral organic compounds can markedly affect the properties of the electrical double layer and the shape of the differential capacity curve. Differential capacity curves provide an important means for studying the electrical double layer at ideal polarized electrodes, because the differential capacitance measured with a lock in amplifier can be determined very accurately.

A feature of adsorption interactions, distinguishing them from the interactions between molecules in solution, is the very close approach of the molecules of the adsorbate to the atoms, ions or molecules forming the surface of the adsorbent. As a result, the interaction between particles of the adsorbate and the adsorbent is similar to interactions in condensed media, where distances between particles are also very small. This is why the phenomena of adsorption have many features in common with molecular association in liquids. The time dependence of these processes may reveal details of the first steps of molecular aggregation.

The formation of compact layers and, in general, all processes leading to a phase change

N.Papadopoulos, S.Sotiropoulos, P.Nikitas - Laboratory of Physical Chemistry, Department of Chemistry, Aristotle University of Thessaloniki, Thessaloniki 54006 Greece

*Phase Transitions in Liquid Crystals*, Edited by S. Martellucci and A.N. Chester, Plenum Press, New York, 1992

at the polarized mercury electrode, have usually been studied by: (1) differential capacitance measurements on a hanging mercury drop electrode (HMDE), either by a potential sweep (at a rather slow scan rate, for equilibrium to be established) or (2) when the kinetics of film formation are to be studied, by AC phase-sensitive chronoamperometry at fixed potentials[2-7].

The principal shortcoming of the first methodology in studying the equilibrium state of the adsorbate on the electrode is its semistatic nature: i.e. the electrode potential varies in time, and this can present a problem, even at very low scan rates, when the attainment of equilibrium is slow, as is the case for many phase changes[4,5,7-9] on Hg. Moreover, capacitance transients recorded at constant potentials are difficult to interpret as regards the potential dependence of the adsorption process. Sweep voltametry, for the study of adsorption and phase transitions that occur on the absorbate, has the drawback that the experimental variables are changing with time and potential, and many interesting phenomena are completely masked. When a potential step pertubation is applied, there is only the element of time to consider since all potential dependent quantities will be constant. Therefore, it should be informative to construct the entire C vs E curve at different times, by plotting isochronous C data (sampled from the capacitance transients) against the final potential values of the corresponding step experiments. In this way we can assemble a full picture of the dependence of the adsorption process both on time and applied potential. This picture resembles what we would obtain from phase sensitive AC polarography with an extended and fine-resolved time window. Furthermore, if we fix the birth potential of each drop of mercury outside the adsorption region, then the drop posesses no "past history" and the adsorption phenomenon can be studied from its very begining, under strictly potentiostatic conditions. Finally, the C vs E curves may represent the equilibrium state if no further change with time is observed, which is a prerequisitive for any thermodynamic treatment. It is probably evident that such a set of measurements can only be realized with an electrochemistry system interfaced to a computer.

The aim of this work is to obtain full (i.e. for closely spaced potentials) C-E curves at various times, and to introduce Three Dimensional Phase-Sensitive AC Voltametry as an alternative method for the study of time-dependent adsorption processes, one which is especially useful in the case of phase transitions. We chose sodium dodecyl sulphate (SDS) as the adsorbate because numerous studies have been published[8,11,15] of its adsorption on Hg, and phase changes have been proposed to take place both in the region around the e.c.m. and in the regions cathodic and anodic to it. Although an accepted explanation for these changes is presently lacking, the time dependence of this phase changes prompted us to use SDS as a convenient adsorbate for the application of our new methodology.

## 2. Experimental Procedure

Sodium Dodecyl Sulphate (SDS) from Fluka (purity > 99%) was twice recrystallised from absolute ethanol. Sodium sulphate, which served as the supporting electrolyte, was obtained from Aldrich. Pure nitrogen was used for de-aeration and all measurements were carried out at 25°C.

The working electrode was a Metrohm HMDE. A saturated (sat. NaCl) calomel electrode was used as the reference electrode, while a platinum hemisperical electrode served as the auxiliary electrode. All measurements were carried out with the help of a PAR model 170 electrochemistry system, interfaced to an AT compatible computer (VIP 200 12 MHz) via a 14 bit A-D/D-A card. A potential step was applied to the working electrode and the capacitive current of the electrical double layer, formed between the polarised Hg electrode and an aqueous .1 mmol SDS and .16 M sodium sulphate solution, was recorded.

It is evident that carrying out a large number of phase-sensitive AC chronoamperometric experiments (in our case up to 170), each one lasting more than a minute, becomes an exhausting task unless it is automated. Therefore, a synchronous motor (5 revolutions per minute) was installed on the top of the HMDE and its rotation was controlled by the computer via a relay. Because the rotation time was fixed, a reproducible drop was always formed. The birth potential for each drop was -1.6 V vs SCE. Preliminary potential sweep experiments on the HMDE showed that at such a cathodic potential the C vs E curves of the SDS coincide with those of the supporting electrolyte, indicating the absence of adsorption in that region. After allowing a 3 s resting time for the solution in the neighborhood of the drop, a potential step with a 5mV p-p modulation at 370 Hz was applied. The potential steps covered the entire +0.2 to -1.5 V vs SCE range in 10 mV intervals. The sampling rate was set to 5 msec/sample, and 15000 data points were stored in digital form. For the timing control of the experiment the internal timer and microtimer of the computer were used. The dislodging of each drop was achieved by instantaneous intense de-aeration which was activated by the computer at the end of each phase-sensitive AC chronoamperometric measurement. A schematic for the experimental set up is given in Fig. 1.

## 3. Results and Discussion

The differential capacitance vs potential curves for the system .1 mmol SDS +.16 sodium sulphate, which were from C(t) data, are given in Fig. 2 for two groups of selected time values.

Concerning the nature of the adsorbate states corresponding to the various portions of the capacitance curve, among other explanations we believe that a combination of Damaskin's[10] and Eda's[9] views is the most reasonable. That is, referring to the results obtained at longer times (Fig.2b), which are very likely to represent the equilibrium state, the wide cathodic peak can be attributed to an adsorption process leading to the formation of a compact layer followed by desorption of the adsorbate. The orientation of the SDS molecules in the cathodic and in the center pit regions is supposed to be with polar groups directed to the solution, thus making

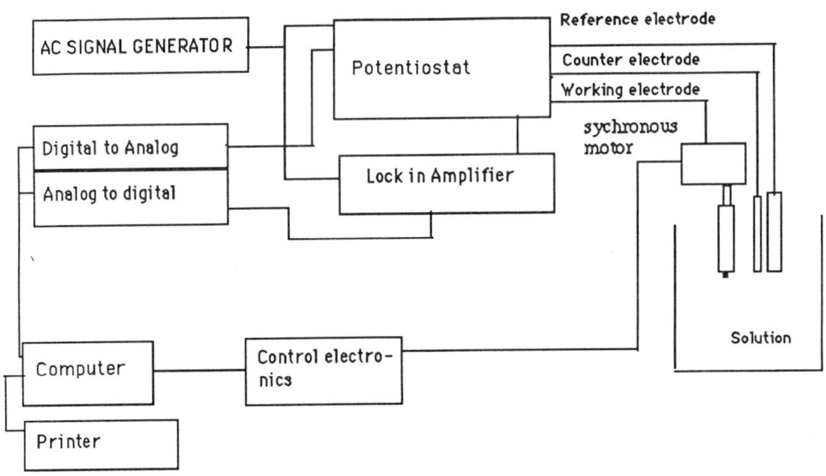


Fig. 1. Schematic of the experimental set-up.

the electrode hydrophilic. The compact film in the central pit extends from about -.25 to about. -1V vs. SCE, at which value it is destructed. However, this destruction is possibly followed by a reorientation of SDS molecules, leading to the formation of a type of "hemi-micelles" similar to those proposed by Fuerstenau and coworkers[15]. Micelles must dominate in the region from about -.25 V vs. SCE to the far anodic peak at about +.1 V. This peak can then be attributed to the destruction of the micellar film and a simultaneous desorption.

From Figs. 2a and 2b it is seen that the capacitance curve does not attain its final shape even at times as long as 60 s. Time effects are more pronounced in the anodic region. The time dependence of the capacitance curve may be due to a combination of diffusion and phase change kinetics. In order to estimate the diffusion effect we applied Koryta's equation[16], which gives the time needed for a diffusion adsortion process to reach equilibrium, and we obtained t=2.9 sec. This means that the continuing slight change of the capacitive current of the center pit, and the more pronounced change in the case of the micellar film supposed to be present at anodic potentials, should be attributed to the slow kinetics of changes in the adsorbate state.

In the reconstructed C-E curves, at longer time values (Fig. 2b), we see that the hump

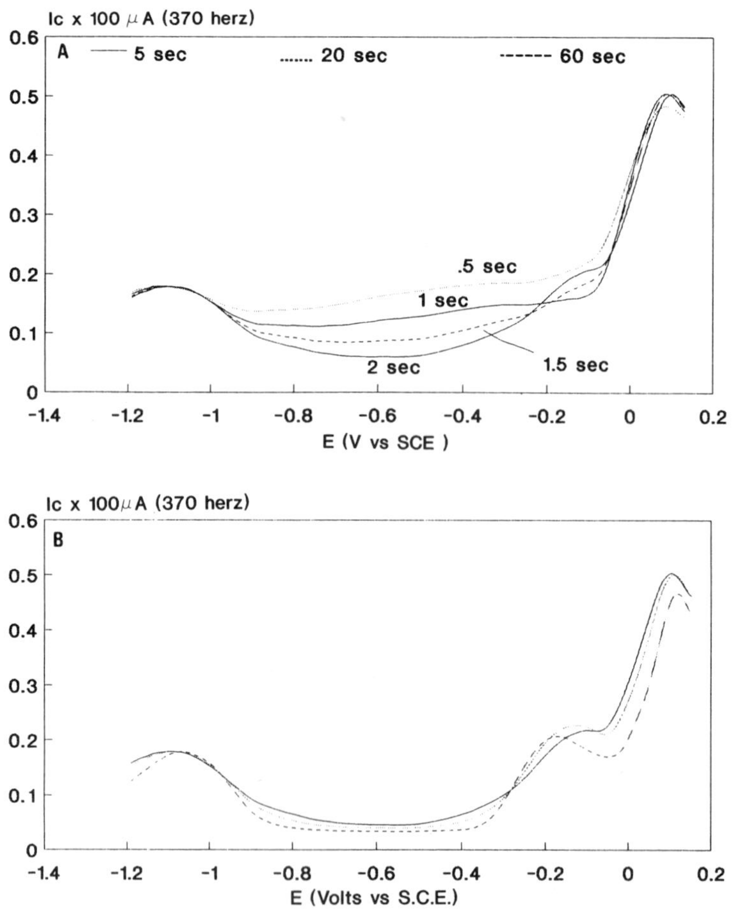


Fig. 2. Differential capacity vs. applied potential curves, constructed from chronoamperometric experiments for (a) short times and (b) long times.

assigned to the formation of the anodic plateau varies with time: we observe with time evolution a gradual steepening of the part of the curve which is associated with compact layer destruction at anodic potentials, together with a capacitance decrease of the anodic plateau corresponding to the micellar surface state. This implies that even at times as long as 60 sec the surface micellization is not completed.

In the case of the adsorption of SDS on Hg, the study of the variation of C-E curves at long periods has revealed a slow development of the micellar phase which is suspected to be present in the region anodic to the center pit. In a liquid bulk mixture of two components A and S an order-disorder phase transition, the homogeneous solution separates into two liquid phases, each one saturated with one of the two components. Therefore we must conclude that in an order-disorder phase transition at the electrode surface a separation of the interface into two new phases, each one saturated with either adsorbate or solvent molecules, occurs.

It is expected that Three-Dimensional Phase-Sensitive AC Voltametry will give even more useful results for slow adsorption processes, and current studies of some other micelle-forming surfactants are being conducted in our laboratory. Phase transitions are hindered processes and therefore are time dependent in their very nature. We believe that Three-Dimensional Phase-Sensitive AC Voltametry will facilitate answering questions such as:

a) What are the requirements for a phase transition to occur on the electrode surface;

b) What type of interfacial states may exhibit discontinuous transitions;

c) Under what conditions are such transitions controlled by nucleation and growth processes; and,

d) What are the criteria which characterize possible mechanisms.

## References

1. P. Nikitas, Electr. Act. **36**: 447 (1991).
2. R. Sridharan and R. de Levie, J. Electroanal. Chem. **201**: 133-143 (1986).
3. R. Sridharan and R. de Levie, J. Electroanal. Chem. **201**: 230 (1987).
4. L. Pospisil, J. Electroanal. Chem. **206**: 269 (1986).
5. T. Wandlowski and L. Pospisil, J. Electroanal. Chem. **258**: 179 (1989).
6. T. Wandlowski, J. Electroanal. Chem. **293**: 219 (1990).
7. L. Benedetti, M. Borsari, C. Fontanesi and E. Battistuzzi Gavioli, J. Chim. Phys. **87**: 1597-1607 (1990).
8. N. Batina, Z. Kozarac and B. Cocovic, J. Electroanal. Chem. **188**: 153 (1985).
9. Y. M. Temerk, M. M. Kamal, P. V. Valenta, Bioelectr. and Bioenerg. **24**: 165 (1990); **24**: 179 (1990).
10. K. Eda, J. Chem Soc. Jap. **80**: 343 (1959); **81**: 689 (1960).
11. B. Damaskin, N. V. Nikolaeva Fedorovich and R. V. Ivanova, Russ. J. Phys. Chem. **34**: 894 (1960).
12. D. Vollhardt, Colloid Polym. Scie. **254**: 64 (1976).
13. D. Vollhardt, U. Modrow, V. Retter M. Jehring and J. Siegler, J. Electroanal. Chem. **125**: 149 (1981); **125**: 157 (1981).
14. M. K. Kaisheva and V. K. Kaishev, Electrochimiya **22**: 804 (1986).
15. P. Somasundaran and D. W. Fuerstenau, J. Phys. Chem **70**: 90 (1966).
16. J. Koryta, Collect. Czech. Chem Commun **18**: 206 (1953).

Chapter 14

# ELECTROHYDRODYNAMIC (EHD) FLOW IN A SMECTIC A LIQUID CRYSTAL

E. SKJETNE, J. SAMSETH and A. JACKSON

## 1. Introduction

In EHDs one studies a liquid flow induced by an electric field. The liquid must contain a sufficient amount of ions which interacts with the external field. If the electric field is below a certain threshold the liquid remains stable, but conducts an ionic current. Above the threshold for EHD instability (EHDI), a liquid flow pattern develops. EHDIs are found in isotropic liquids as well as LCs. The book of Blinov[1] is a good introduction to the field. Here, we consider the EHD flow in a Sm-A LC doped with ions. The material is confined between two glass plates which are coated with transparent electrode patterns. The plates are parallel to the x-y plane and the molecules are initially oriented with their long-axis perpendicular to the plates, i.e. parallel to the z-axis (homeotropic orientation). The appearance of a EHDI in a homeotropic Sm-A was predicted theoreticly in 1972[2] and investigated experimentally from 1977[3-8]. The theory, is developed from the 1D Helfrich-Orsay model for a planar nematic LC (the molecular long-axis parallel to the plates limiting the sample), but takes into account some properties specific to the homeotropic Sm-A phase. The only velocity component in the 1D model is $v_z(x)$ which is modulated with a half period equal to the sample thickness. Experiments on Sm-A, report a circular flow in the x-y plane. It is obvious that the 1D model is insufficient to explain the observed flow pattern. It should be noted however, that in these experiments we are dealing with the situation occurring slightly above the threshold, whereas the theory predicts the situation only below the threshold. Firstly, we review in a simple way, some aspects of the Sm-A phase which are relevant for the observations presented in the experimental part. At the end we briefly discuss the observations.

## 2. EHDI and the Properties of the SM-A Phase

The structure of a Sm-A LCs may be described as a stack of liquid layers. Each layer is composed of molecules oriented with their molecular long-axis (director) perpendicular to the layer. The layers move freely with respect to each other. Due to the layered structure the ion

---

E. Skjetne and J. Samseth - Institutt for energiteknikk P.O.B. 40, N-2007 Kjeller, Norway
A. Jackson - Autodisplay a.s., c/o SINTEF, N-7034 Trondheim, Norway

*Phase Transitions in Liquid Crystals*, Edited by S. Martellucci and A.N. Chester, Plenum Press, New York, 1992

conductivity $\sigma_\perp$ perpendicular is larger than $\sigma_\parallel$ the conductivity parallel to the director $\sigma_a = \sigma_\parallel - \sigma_\perp < 0$. The molecules considered here have an positive dielectric anisotropy, i.e. $\varepsilon_a = \varepsilon_\parallel - \varepsilon_\perp > 0$.

Flow in a homeotropic Sm-A is visible through a polarizing microscope if it is accompanied with layer deformations and defects. The elastic bending energy of the layers is very small compared to the compression energy. Therefore the layers can be easily curved, as long as the layer thickness is kept nearly constant. In a thin sample with homeotropic orientation and flat layers at the limiting plates (which are assumed to be smooth), the undulations (thermal or forced) in the middle of the sample are clamped by the requirement of having nearly constant layer thickness. Permeation effects [10,11] (i.e. flow of molecules from one layer to the next) becomes important in a thin boundary layer near each plate. A velocity component tangential to the plates vanishes over the same boundary layer. The main structural defects allowed in a Sm-A LC are dislocations and focal conics (FCs)[12] . Edge dislocations correspond to the existence of one (or more) limited layer(s) in the bulk. The most general type of FCs consists of line defects which occur as pairs of confocal ellipses and hyperbolae. Another class of FCs which is observed in homeotropic samples, is the parabolic FC (PFC)[13]. In a PFC the line defects form a pair of confocal parabolae. The layers are strongly bent near the singularity lines. PFCs can be arranged in a 1D or 2D array or in a more disordered polygonal texture.

To understand why the flow is accompanied by a visible pattern, we have to consider the balance of angular momentum of the molecules. We consider the total torque

$$\vec{M} = \vec{M}_{elast} + \vec{M}_{visc} + \vec{M}_{elect} + \ldots = 0 \tag{1}$$

In our case the elastic torque $\vec{M}_{elast}$ is stabilizing due to the suppression of undulations. The viscous torque $\vec{M}_{visc}$ is assumed to be destabilizing and its components are proportional to the velocity gradients, i.e. $\vec{M} \sim \ddot{\alpha}\nabla\vec{v}$. $\ddot{\alpha}$ is a friction constant tensor that takes into account the geometry of the experiment. Velocity gradients arise from the electrostatic force acting on spatial accumulation of ions, i.e. space charges. The physical mechanisms which cause space charges formation in isotropic liquids may also lead to space charges in LCs. Other mechanisms takes only place in LCs due to the anisotropy of electric properties ($\varepsilon_a$, $\sigma_a$).

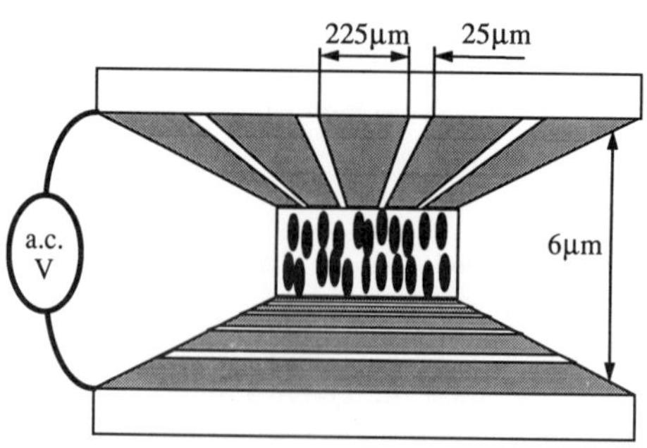


Fig. 1. A side view, in perspective, of a small part of the sample cell (relative dimensions are changed). The LC is oriented homeotropically between two glass plates which are coated by 225 μm wide electrode stripes which are separated by 25 μm. The stripes at the upper and lower glass are crossed so that a voltage (50 Hz and amplitude about 100V) applied across one upper and one lower stripe induces EHDI in the volume of 225 x 225 x 6 μm³.

Therefore, EHDIs are divided into the two main groups which reflects the nature of the space charge formation; *isotropic* and *anisotropic*. One isotropic mechanism, which is rather fundamental is the electrolytic separation of charges, i.e. an accumulation of positive and negative ions near the negative and positive electrode respectively. It leads to instability in the isotropic, nematic[14] and also suggested for the homeotropic Sm-A[15] phase. This effect occur in an a.c. electric field if the ions are able to drift along the z-axis to the opposite electrode during one half period. The liquid near the electrodes will be destabilized at a sufficient high electric field. In the case of a homeotropic Sm-A, the layers are stuck to the electrodes and can only be destabilized by large velocity gradients. In this case the electric torque $\vec{M}_{elect}$ is stabilizing ( $\varepsilon_a > 0$). Both the isotropic phase and the homeotropic nematic (or Sm-A) phase exhibit a circular flow in the x-y plane. This may indicate that the circular flow is induced by the same isotropic mechanism, even if the flow mechanism is not fully understood.

The 1D model referred above is based on the anisotropic Helfrich mechanism[16]. In this model (when applied to Sm-A), space charges are forming if the layers are undulated. An applied electric field in the z-direction causes positive and negative space charges to be separated in regions with largest negative and positive layer curvatures respectively. These space charges interact with the field and flow in opposite directions along the z-axis. The flow is accompanied with a shear which exerts a destabilizing torque on the layers. In addition, a small destabilizing electric torque $\vec{M}_{elect}$ arises from the electric field generated by the space charges in the x-y plane. However, the suppression of undulations in the Sm-A phase may be so strong that this mechanism is not the most important here.

In addition to the balance of the angular momentum discussed above, one should use the balance of linear momentum, i.e. hydrodynamic equations of the Sm-A phase including effects of boundary layers, and the electrostatic force acting on the space charge density $q$ from the total electric field E. The charge conservation must also be fulfilled

$$\frac{\partial q}{\partial t} + \nabla \vec{J} = 0 \qquad (2)$$

The current density $\vec{J}$ includes the ohmic term $\vec{J}_{ohm} = \vec{\vec{\sigma}} \cdot \vec{E}$ , convective term $\vec{J}_{conv}$ = q **v**( **r**, t) due to charge transfer with the moving fluid of velocity **v** ( **r**, t), and a term due to diffusion, so that

$$\vec{J} = \vec{J}_{ohm} + \vec{J}_{\mathrm{conv}} + \vec{J}_{diff}. \qquad (3)$$

The equations couple together.

This show that EHDI problems are in general complex. A complete analysis of the EHD flow in Sm-A has yet to be carried out.

## 3. Experimental Results

The LC is a mixture of alkyl and alkoxy cyanobiphenyls doped with positive and negative ions (about 0.2 weight %). It has a Sm-A phase at room temperature. The sample is confined homeotropically between two clean glass plates coated with transparent electrode stripes and separated 6 μm apart by glass fiber spacers as is seen in Fig. 1. Each stripe is 225 μm wide and separated 25 μm from its nearest neighbor. The direction of the stripes on the lower and upper plate is perpendicular to each other. The active area with electrode stripes is $12 \cdot 16\ \mathrm{cm}^2$. A voltage can be applied across the cell of individual stripes at the lower and upper plate, so that EHD flow is induced in separate volumes, each of size 6 · 225 · 225 $\mu m^3$. The voltage is given by a 50 Hz sinus signal generator tunable in the range 0-400 V. The cell is studied through a polarizing microscope using either photo or video equipment.

In all the patterns shown here we have used 50 Hz sinus signal with an amplitude in the range 100-120 V. Fig. 2a,b shows two typical flow patterns. The squares are given by the crossing electrode stripes. Dark areas correspond to a structure of nearly homeotropic orientation and light areas are light scattered from defects. The loops are visible trajectories of the defect flow. Here, we first present the static flow patterns then we show how the patterns are related to the observed velocity field.

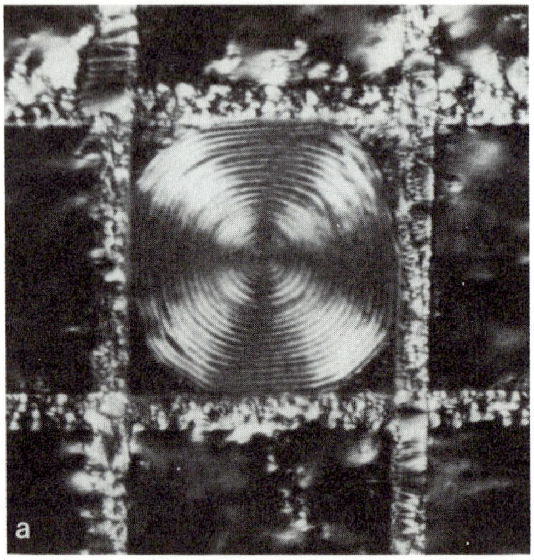


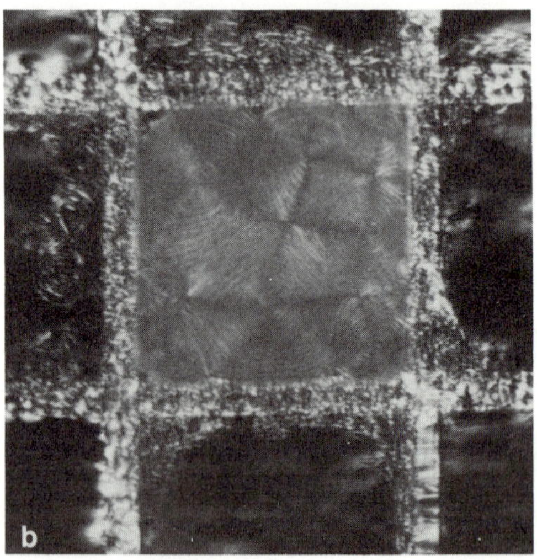


Fig. 2. Two typical flow patterns induced by a 50 Hz and 120 V (amplitude) signal in squares of dimension 225 x 225 $\mu m^2$. The LC layer is 6 micrometer thick. The simple one roll pattern (a) is most frequently observed. The flow is circular and the defect trajectories are closed loops. In the complex pattern of polygonal rolls (b), two neighboring rolls with a common interface turn in the opposite direction. The flow directions are illustrated in Fig. 4.

The simple flow pattern (Fig. 2 a) looks like a large roll. The roll consists of one core situated in the square center, which is enveloped by mainly closed loops, of nearly constant thickness. Near the core the shape of each loop is close to a circle. Closer to the square boundaries they are more square shaped. Loop dislocations are observed in the roll patterns, i.e. not closed loops. Some pixels show a similar pattern except of having minor rolls which fill the corners of the square and distort the outer envelop loop of the central core. An example of this is given in Fig. 3 a.

In the complex patterns there are more rolls as is seen in Fig. 2 b. The outermost loop which envelopes one core, defines the roll size. Two neighboring rolls can touch each other

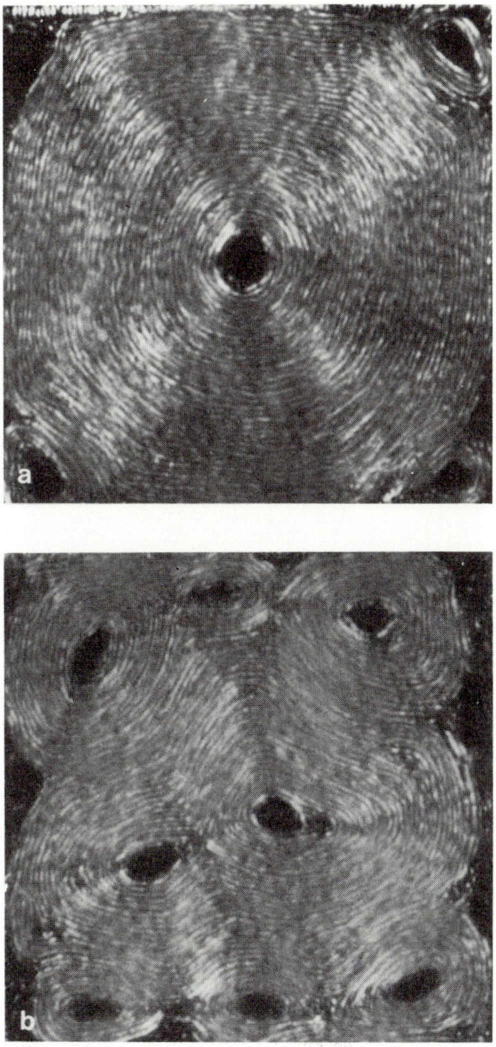


Fig. 3. Two flow patterns induced by a 50 Hz and 100 V (amplitude) signal in squares of dimension 225 x 225 $\mu m^2$. The black spots have nearly homeotropic orientation (regions with low shear velocity). In the simplest pattern (a), three small rolls fills the corners. The fourth corner might have a roll of lower velocity which is not vicible (no defects). In the complex pattern (b), the two central rolls are enveloped by a larger one. The three rolls turn in the same direction.

either along an almost straight line or in a single point. Therefore, a typical complex roll pattern looks like a square filled with polygons. In Fig. 3 b we see that two small rolls are inside one large roll. The complex pattern is most commonly observed in squares where there are glass fibre spacers, but it can also be found in squares without any spacers.

The velocity is always tangential to the loops. For one roll the circular flow is either clock-wise (here termed "- rolls" as the rotational axis points, into the paper, along the negative z-axis), or counter clock-wise ("+ rolls"). In 272 squares (one large roll in each), there was found 138 + and 134 - rolls. In the range from about 90 V to 200 V we observe a roll pattern which remains stable in the long time limit (10 minutes). Below 90 V the flow of defects do not form coherent patterns but follow locally the defect tracks in the roll structure. Above 200 V there is a breakdown of the uniform roll pattern. It is replaced by a more complex flow pattern. The flow velocity midway between the core and the square boundaries of a simple pattern, is of the order of 0.5 mm/s. The velocity profile out from the core shows a maximum before it decreases at the boundary. The slow velocity regions with homeotropic orientation (black spots), can be seen in Fig. 3a,b. If we turn off the voltage abruptly the flow stops instantanously.

The flow directions of the complex pattern must be evaluated in more detail. Fig. 4 illustrate schematically the flow pattern of Fig. 2 b. Each roll is indicated by a closed loop with a grey tone. Arrows, sign and grey tone indicate the flow direction. Complex flow patterns (in spacer free squares) follow a set of rules :

1. The loops in the vicinity of the square boundary are almost parallel to the boundary.
2. Two neighboring rolls of opposite sign touch each other along a line (often a straight line).
3. Two neighboring rolls of equal sign touch each other in a point.
4. If two or more rolls are enveloped by a larger roll they all have the same sign.
5. A large roll enveloping smaller rolls interact with the boundary and its neighbors as described by 1, 2, and 3.

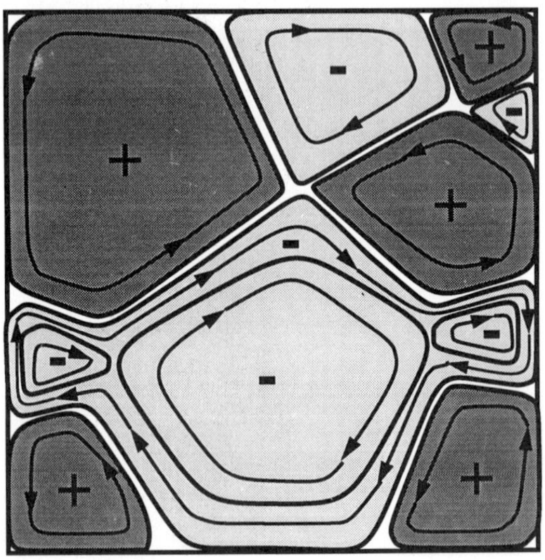


Fig. 4. A schematic illustration of the flow pattern of Fig. 2 b. Each roll is indicated by a closed loop with a grey tone. Arrows, sign and grey tone indicate the flow direction. The 6 rules in the text can be recognized in this Figure.

6. In a polygon pattern every polygon is surrounded by rolls of opposite sign or by the outer boundary. Only the corners of the polygon may touch a roll of same sign.

A complex pattern may undergo a transition towards a simple one (one large roll). The sequence in Fig. 5 a-i show the growth of a roll pattern. The smaller rolls are defeated by the larger one which grow from the boundaries into the center. One roll defeating the others can also be seen in patterns where the rolls fill the whole square. Typically, the growth of a

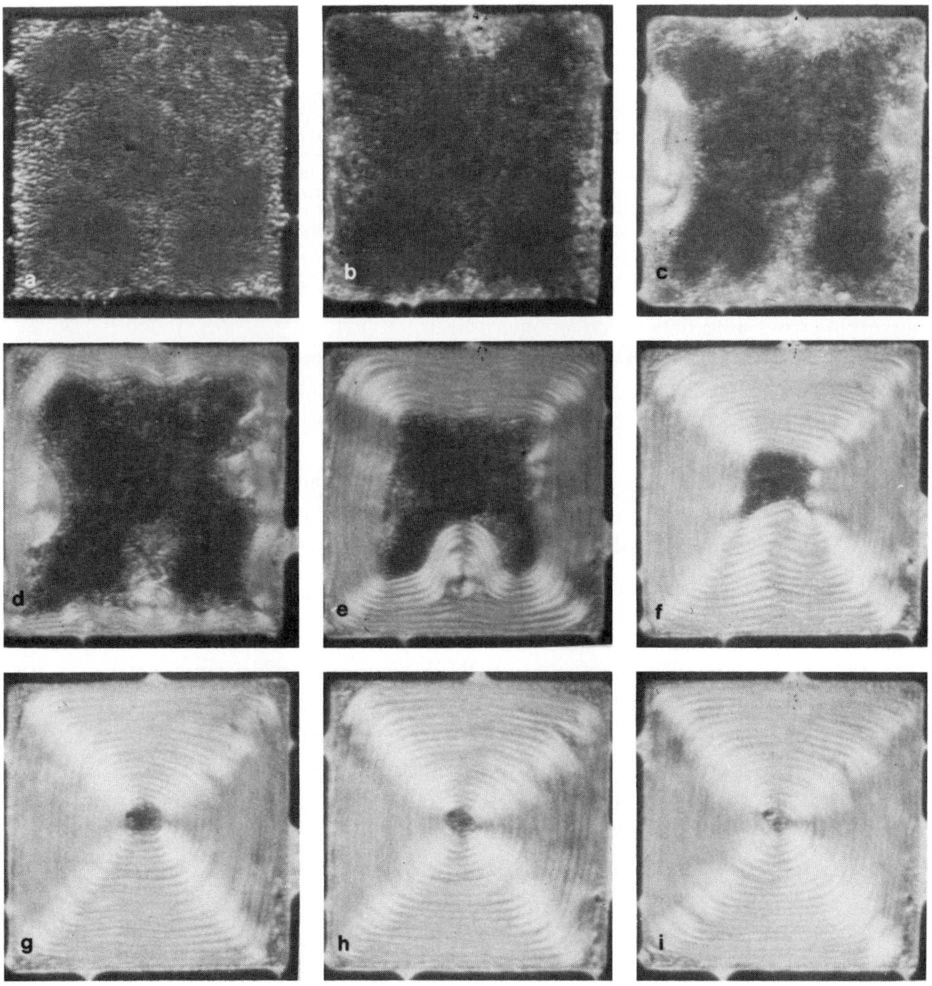


Fig. 5. The growth of a simple one roll pattern (induced by a 50 Hz and 120 V (amplitude) signal). There are 10 seconds between each photo. At the onset (a) there is no flow pattern. 10 and 20 seconds after onset (b and c); small rolls develope along the boundary. 30 seconds (d); one roll is along the whole square boundary. 40 seconds (e); this roll has defeated the small roll (at the left in d) and trapped a roll of opposite sign. 50 seconds (f); the trapped roll is defeated, but has deformed the large roll. 60 seconds (g); the roll have almost got its final shape. 70 and 80 seconds (h and i); the one large roll is stable except from the shift of a low velocity region along the up right diagonal.

pattern is seen as loops being formed around cores which nucleate simultaneously at different places.

## 4. Discussion

The fully developed EHD flow in Sm-A has been termed turbulent flow[8] because of the complex flow pattern. Here, we use well defined lateral boundaries to obtain simple, stable (one roll, closed trajectory) flow patterns. Therefore we do not find it appropriate to call the flow turbulent. The flow is highly viscous, as it damps out quickly when the voltage is turned off abruptly.

The nearly equal number of + and - rolls might be explained by the symmetry in the problem. As in an isotropic liquid, the homeotropic Sm-A LC has no preferred directions in the x-y plane. Then, there is no reason for having a larger number of one sort of flow (+ or -). The smaller rolls in the corners which are often seen in simple patterns but also in the complex one, is probably due to a combination of geometric constraints from the boundary and the preferred circular shape of the rolls.

The rules 1 to 5 of the complex pattern can not be understood fully as we do not know the growth mechanism. However, we give a few qualitative arguments which support the rules :

1. The rolls grows until they touch an other roll or the boundary. If they touch the boundary they cannot grow further. That is because there is no input energy (voltage) outside the square. The other part of the roll, which is not touching the boundary, may grow further. If the loops have a nearly constant thickness, this can only happen by introducing loop dislocations in the roll.
2. In the region where two neighboring rolls of different sign touch, they flow in the same direction, i.e. the flows are not hindered, but rather stimulated by each other. The further growth may therefore take place in the nearby homeotropic region, so that they get a line interface.
3. In the region where two neighboring rolls of same sign touch, they flow in opposite directions, i.e. the flows hinder each other and oppose a further growth in this region.
4. The pattern of a large roll enveloping smaller rolls, where all have the same sign, can be understood by a combination of 2 and 3. A loop enveloping all the rolls will follow the local flow direction along the whole envelope, except in the zero velocity point where the rolls touch. The loop outside is farther from the singular point and will be less affected by the singular point. The large roll may then keep on growing until it touches other rolls.
5. If it touches other rolls, it will interact with them as a single core roll.
6. The polygon pattern can be understood by applying 1 - 5 to more neighbor rolls.

The early stage of the growth of a pattern is characterized by nucleation of circular flow. The flow growing from the boundary seems to be exceptional and induced by the boundary condition. The circular flow, basic to all the patterns, cannot be understood without knowing the flow mechanism. However, the formation of trajectories of nearly constant thickness is not specific to the EHDI, but also observed in ultrasound and Poiseulle flow instabilities in a Sm-A[17]. We believe that an analysis of the hydrodynamic shear flow instabilities in Sm-A applied to the actual flow mechanism could explain our observations.

The aim of this chapter has been to serve as an introduction to the EHDI in a homeotropic Sm-A and the use of lateral boundary conditions to get more ordered flow patterns. The flow patterns are investigated qualitatively. It is surprising to find out how little is actually known about the EHDI in Sm-A. The EHDI mechanism is not fully understood. An understanding of the mechanism requires both more experiments and more precise models.

## References


1. L.M. Blinov, "Electro-Optical and Magneto-Optical Properties of Liquid Crystals," John Wiley & Sons, Chichester (1983).
2. Guerst and W.J.A. Goossens, Phys. Lett. **41A**:396 (1972).
3. V. N. Chirkov, D. F. Aliev, G. M. Radzhabov, and A. Kh. Zeinally, Sov. Phys. JETP **47**:5 (1978).
4. D. Coates, W.A. Crossland, J. H. Morrisy, and B Needham, J. Phys. D: Appl. Phys. **11**:2025 (1978).
5. V. N. Chirkov, D. F. Aliev, G. M. Radshabov, and A. Kh. Zeinally, Mol. Cryst. Liq. Cryst. **49**:293 (1979).
6. V. N. Chirkov, D. F. Aliev, and A. Kh. Zeinally, Sov.Phys.Crystallogr. **26**:1 (1979).
7. A. A. Razumov and E. A. Kirsanov, Sov.Phys. Tech. Phys. **27**:5 (1982).
8. P.-A. V. Kazlauskas, L. S. Klenevskis, and V. V. Skinderis, Sov.Phys. Tech. Phys. **28**:9 (1983).
9. E. Dubois-Violette, P.G. de Gennes and O.Parodi, J. Physique **32**:305 (1971).
10. P.G De Gennes, Phys. Fluids **17**:1645 (1974).
11. R. Ribotta and G. Durand, J. Physique **38**:179 (1977).
12. M. Kleman,"Points, Lines and Walls. In liquid crystals, magnetic systems and various ordered media", John Wiley & Sons, Chichester (1983).
13. Ch. S. Rosenblatt, R. Pindak, N. A. Clark and R. B. Meyer, J. Physique **38**:1105 (1977).
14. L. M. Blinov, A. N. Trufanov, V. G. Chigrinov and M. I. Barnik, Mol. Cryst. Liq. Cryst. **74**:1 (1981).
15. L. M. Blinov, M. I. Barnik, V. T. Lazareva and A. N. Trufanov, J. Physique, **40** C3:263 (1979).
16. W. Helfrich, J. of Chem. Phys. **51**:4092 (1969).
17. F. Scudieri, A. Ferrari and E. Gunduz, J. Physique **40**:C3-90 (1979).

Chapter 15

# INTRODUCTION TO LIQUID CRYSTALS AND THEIR MOLECULAR ORDER

R. SHASHIDHAR

## 1. Introduction

Liquid crystals constitute a state of matter in which the degree of molecular order is intermediate between the crystalline solid and the isotropic liquid. A liquid crystalline phase or mesomorphic phase can be attained either by varying temperature (thermotropic) or by adding a solvent (lyotropic). There are a variety of liquid crystalline phases which can be classified on the basis of the molecular order. Although different techniques like miscibility and optical observation of textures have been used to identify the different types of liquid crystalline phases, the most comprehensive understanding of the nature of the molecular order can be arrived at only by X-ray diffraction studies on oriented samples. The experimental techniques for X-ray investigations on liquid crystals vary depending on the nature of information that one is seeking.

The purpose of this Chapter is to introduce the reader to the details of the experimental techniques that are generally used for precise X-ray diffraction studies of both thermotropic and lyotropic liquid crystals. Before describing these details we shall first discuss the different types of liquid crystalline phases and the nature of molecular order in them.

Fig. 1 illustrates the progression of order through a sequence of mesomorphic phases exhibited by rod-like molecules[1]. The progression from the completely symmetric isotropic liquid through the mesomorphic phases into the crystalline phases can be described in terms of three separate types of order: orientational order, positional order and bond orientational order. The orientational order describes the situation wherein the molecules have some preferential orientation analogous to the spin orientational order of ferromagnetic materials. All liquid crystals have long range orientational order. In fact, this is the only type of long range order in the nematic phase, and, as a consequence, its physical properties are those of an anisotropic fluid. Fig 2 is a schematic representation of the nematic order, the molecules being represented by oblong ellipses. The average orientation of the ellipses is aligned; however, there is no long range order in the relative positions of the ellipses. Nematic phases are also observed for disk shaped molecules and for clusters of molecules that form micelles. They all share the common properties of being optically anisotropic and fluid-like, without any long range positional order.

---

R. Shashidhar - Center for Bio/Molecular Science & Engineering, Naval Research Laboratory, Code 6090, Washington, DC 20375-5000, USA; and Geo-Centers, Inc., 10903 Indian Head Highway, Fort Washington, MD 20744, USA.

*Phase Transitions in Liquid Crystals*, Edited by S. Martellucci and A.N. Chester, Plenum Press, New York, 1992

The second type of order is the positional order. In view of the fact that all smectic mesomophases have a layered structure, it is convenient to distinguish between positional order along the layer normal and perpendicular to it (within the layers). We can have three types of positional order:

1) The short range positional (SRO) order wherein the order is only correlated over a finite distance as in a simple fluid;
2) Long range positional order (LRO) as in a three dimensional crystal or spin orientation of a ferromagnet; and
3) Quasi-long range order (QLR), a concept which we will discuss in detail in another Chapter[2].

All smectics have one or more of types of positional order.

In the smectic-A and smectic-C phases, the molecules organize themselves into layers. One might describe them as forming a one dimensional periodic lattice in which the individual layers are two dimensional liquids. In the smectic-A phase (Fig. 3), the average molecular axis $<n>$ is normal to the smectic layers while for the smectic-C (Fig. 4) it makes a finite angle. It follows from this that the smectic-C phase has a lower symmetry than the smectic-A, and the phase transition from the smectic-A to smectic-C can be considered as the ordering of a two component order parameter, i.e. the two components of the projection of the molecular axis on the smectic layers[3]. All the higher order smectics, i.e., those which possess long range positional order have been labelled as crystal-B, crystal-C, etc. in Fig. 1. (For recent books on the structure on higher order smectics see Pershan[1] and Gray and Goodby[4]). They are better classified as three dimensional crystals and not as liquid crystals in the conventional sense.

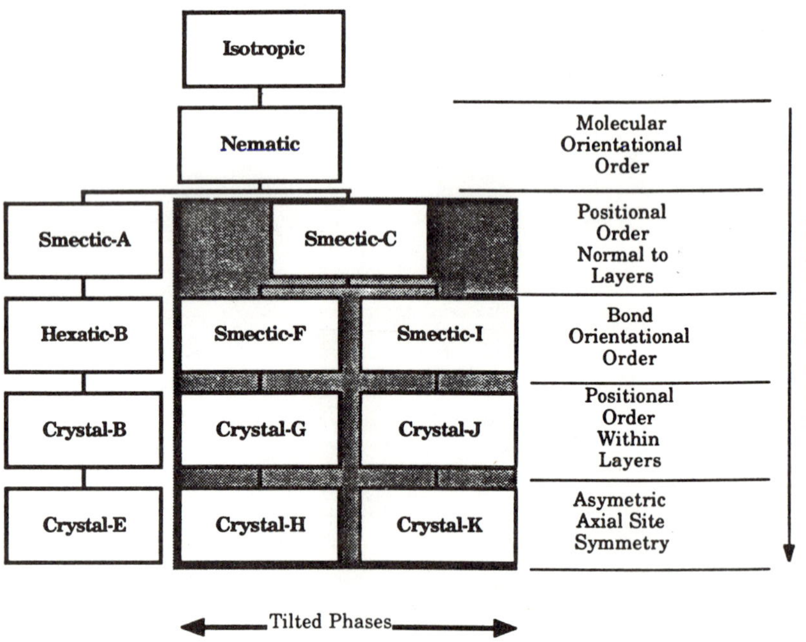


Fig. 1. Progression of order through different types of liquid crystalline phases (or mesophases) formed by rod-like molecules. The shaded areas indicate tilted smectics, i.e., the molecules in the layer are tilted with respect to the layer normal. [Ref. 1]

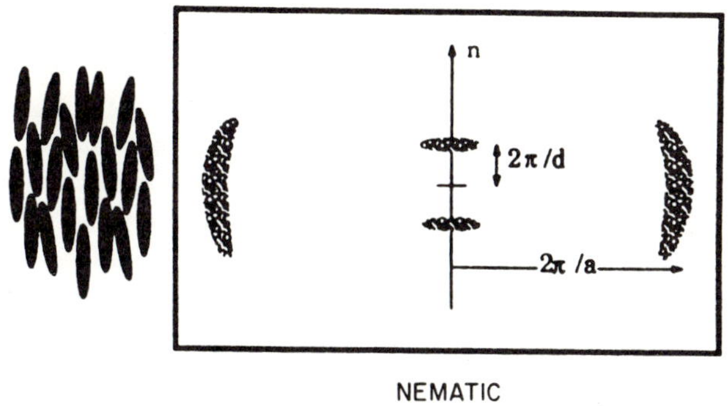


Fig. 2. Schematic representation of molecular order in the nematic phase and the X-ray scattering cross section in reciprocal space for this phase.

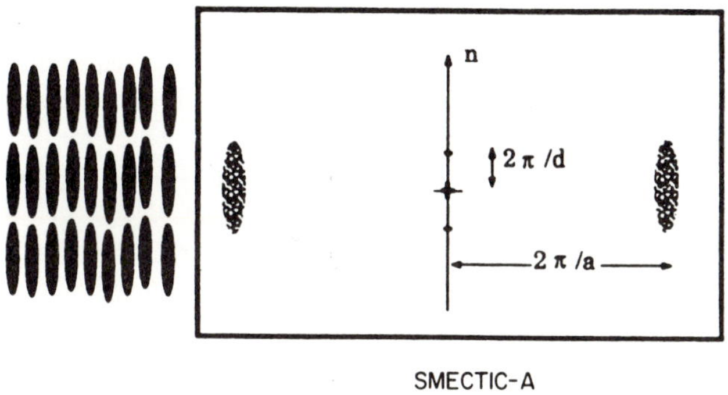


Fig. 3. Schematic representation of molecular order in the smectic A phase and the X-ray scattering cross section in reciprocal space for this phase.

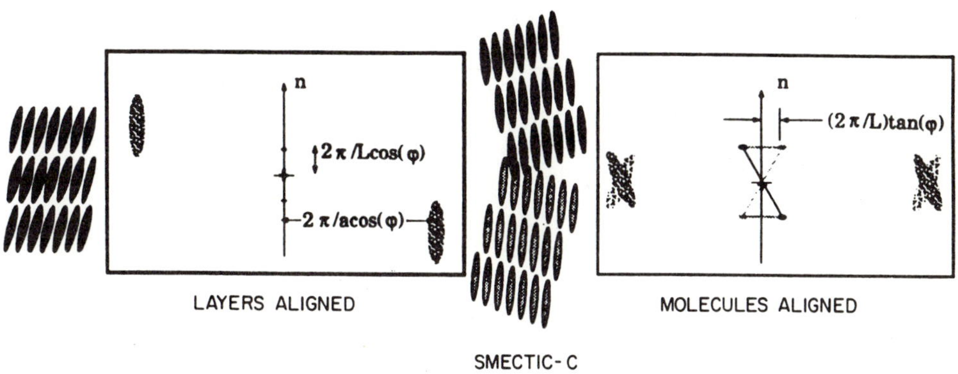


Fig. 4. Schematic representation of molecular order in the smectic C phase when the layers are aligned (left) and when the molecules are aligned (right). The corresponding X-ray scattering cross sections in reciprocal space for these two cases are also shown.

The third type of order is "bond orientational order"[5]. Let us imagine a set of six molecules represented by points which are so arranged that six nearest neighbors are distributed hexagonally in a plane. If we construct a perfect two-dimensional lattice from these points, each hexagon of six points would have the same orientation of its sides even though it is not in registry with the lattice. This situation, referred to as bond orientational order, is the order which characterizes the hexatic liquid crystalline phases: hexatic-B, smectic-F and smectic-I. In all the three hexatic phases the packing of the molecules within the layer is hexagonal. In the case of hexatic B the molecules are normal to the layer while in the case of smectic-F and smectic-I they are tilted, towards an edge of the hexagon in smectic-F and towards the apex for smectic-I. All the three hexatic phases are characterized by long range bond orientational order. The in-plane positional order in hexatic B is short range order while that in smectic-F and smectic-I is quasi long range order. Also, hexatic B has no interlayer correlations[6] while the tilted hexatics do have interlayer correlations[7,8].

We shall now discuss in some detail the reciprocal space picture of the different liquid crystalline phases. The X-ray scattering cross section of an oriented monodomain sample of the nematic phase with rod-like molecules usually exhibits a set of diffuse spots as illustrated in Fig. 2. The maximum of the cross section is along the average molecular axis $<n>$ at a value of $|q| = 2\pi/d$, where d is of the order of the molecular length L. This is a precursor to the smectic-A order that develops at lower temperatures for many materials. In addition, there is a diffuse ring along directions normal to $<n>$ at $|q| \approx 2\pi/a$, where a = 4.0 Å. This value of a is comparable to the average radius of the molecule. In some nematic systems, the near neighbor correlations favor antiparallel alignment and molecular centers tend to form pairs such that the peak of the scattering cross section can actually have values anywhere in the range from $2\pi/L$ to $2\pi/2L$[9,10].

There are also other cases where there are two diffuse peaks, corresponding to both $|q_1| = = 2\pi/L$ and $|q_2| = |q_1|/2$ that are precursors of a richer smectic-A morphology[9-11]. In addition, in all smectic A phases, the x-ray scattering has a diffuse ring along directions normal to $<n>$ at $|q| \approx 2\pi/a$ where a = 4.0 Å. This value of a is comparable to the average radius of the molecule.

On cooling, many nematic systems undergo a second order phase transition to a smectic-A phase and as the temperature approaches the nematic to smectic-A transition, the widths of these diffuse peaks become infinitesimally small. In 1972, de Gennes[12] demonstrated that this phenomena could be understood in analogy with either the transitions from normal fluidity to superfluidity in liquid helium, or normal conductivity to superconductivity in metals.

The X-ray cross section for the prototypical aligned monodomain smectic-A sample is shown in Fig. 3. It consists of a single set of sharp spots along the molecular axis at q ($=2\pi/d$) representing the QLRO along the layer normal, and a diffuse ring in the perpendicular direction at $|q| = 2\pi/a$ due to the SRO within the layer. The scattering cross section for an aligned smectic-C phase is similar to that of the smectic-A except that the molecular tilt alters the intensity distribution of the diffuse ring. This is illustrated in Fig. 4(1eft) for a monodomain sample. Fig. 4(right) illustrates the scattering pattern for a polydomain smectic-C sample in which the molecular axis remained fixed, but where the smectic layers are randomly distributed azymuthally around the molecular axis. A description of smectic-A and smectic-C as periodic stacks of two-dimensional liquids is an over simplification - the sharp spot along the molecular axis has a distinct temperature dependent shape indicative of QLRO that distinguishes it from the Bragg peaks due to true LRO as known in conventional three dimensional crystals. This aspect will be discussed separately[2].

An oriented domain of hexatic B phase, in which the molecules are normal within the layer, yields a set of six diffuse spots in reciprocal space[6]. The scattering intensities in reciprocal space from three dimensional tilted hexatic phases of smectic-I and smectic-F are shown schematically in Fig 5a and 5b respectively. The variation of the intensity along the $Q_L$-direction indicates the presence of interlayer correlations. The peak widths $Q_{L1,2}$ and $Q_{H1,2}$ correspond to the four inequivalent widths in the smectic-F phase. Similar inequivalent widths exist for the smectic-I phase also. The dark circle in (a) indicates the reciprocal space scan that directly measures hexatic order. A similar scan in the smectic-C phase would have intensity independent of $\chi$.

## 2. Experimental Setup for High Resolution X-ray Diffraction Studies of Liquid Crystals

A prerequisite for any X-ray diffraction studies of liquid crystals is a well aligned sample. Such an alignment is normally achieved using a magnetic field. The strength of the field required to align the sample in the nematic field is a few kilogauss for a typical sample thickness of 1-3 mm (thicker the sample, lesser will be the wall effects which compete with the magnetic field and hence better will be the alignment). Another important requirement is a good temperature controlled oven in which the sample, sealed in a flat cell or a glass capillary, will be located. Generally, whenever a temperature accuracy of a few mk is required, a double-stage heating system is used. For certain experiments, for instance, to probe the in-plane fluidity of a smectic or for a quick identification of the type of liquid crystalline phase or even for a routine measurement of the smectic layer spacing, low resolution experiments are adequate. Such experiments can be carried out using a regular (sealed-tube) x-ray generator and a photographic film or a linear position sensitive detector to record the diffraction maxima. For certain other cases which require a precise determination of the smectic-A layer spacing, a Guinier diffractometer has been. In this case, a medium resolution of $\approx 10^{-3}Å^{-1}$ has been obtained[13] by using a single focussing monochromator which separates out the $K_{\alpha 1}$ and $K_{\alpha 2}$ lines. However none of these experimental set-ups are adequate for certain other experiments. Whenever large correlation lengths are to be determined in the nematic phase or an accurate line profile analysis of the smectic diffraction peak is to be carried out, high resolution experiments are needed. For high resolution studies which permit a reciprocal resolution of $10^{-4}Å^{-1}$, the experimental setup is quite complex. Such high resolution studies are now being carried out in many different laboratories and they all have essentially the same basic features although there will be differences with regards to some experimental details. We shall describe, in the following, a "typical" experimental set-up used for high resolution X-ray studies of liquid crystals[14].

A schematic diagram of the scattering set-up is shown in Fig. 6. The spectrometer essentially consists of 1 ) a rotating anode X-ray source, 2) the primary beam monochromator, and 3) a 4-circle goniometer with the sample mounted on the Φ-arm and the analyzer/detector mounted on the 2 Φ-arm.

The high flux available from a rotating anode allows for high angular resolution of the experiments. The monochromator crystal (which can be Si (111) or Ge (220)) is rotated to Bragg reflect the strongest $K_\alpha$ line. The overall mosaic spread and the natural Darwin width of the crystal provide the horizontal in plane resolution of the reflected beam $k_i$, which is incident on the sample. The sample then scatters the incident beam into different directions and the scattering process is quasi-elastic. The analyzer crystal determines the in-plane resolution of the scattered X-ray beam $k_{f*}$. In this process only those X-rays that are scattered within an angular spread (determined by the overall quality of the analyzer crystal, mosaic spread and Darwin width) can Bragg reflect and reach the detector.

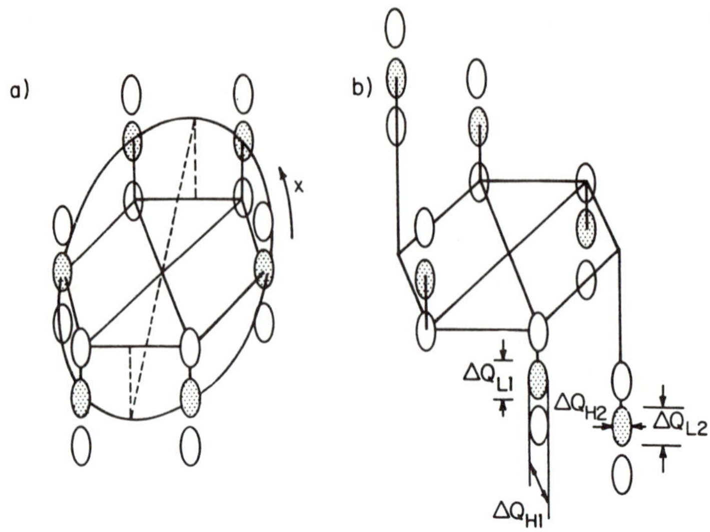


Fig.5. Scattering intensities in reciprocal space from three dimensional tilted hexatic phases: a) smectic-I and b) smectic-F. [Ref. 1]

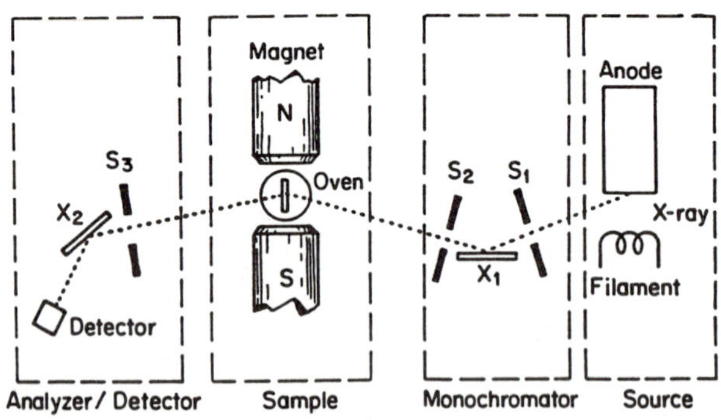


Fig.6. Schematic representation of a typical X-ray high resolution spectrometer. $S_1$, $S_2$ and $S_3$ are vertical slits. $X_1$, $X_2$ are Bragg reflecting crystals.

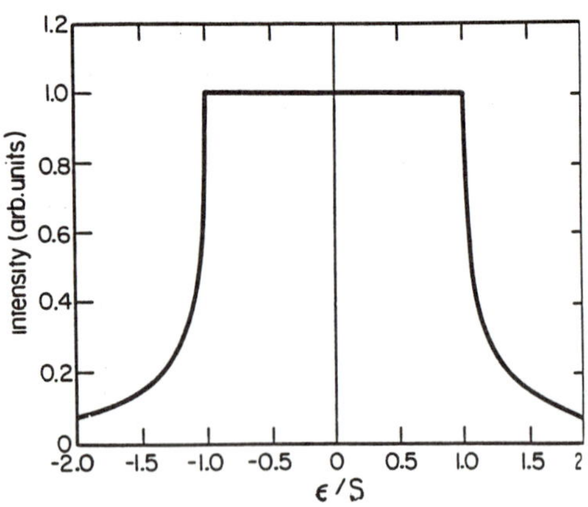


Fig.7. Darwin width of a non-absorbing perfect crystal with no mosaic spread.

We shall now discuss the components and resolution function of the spectrometer. Although the anode generates X-rays in all directions, the entrance slit $S_1$ accepts only X-rays coming from a take-off angle of 6° tangent to the anode. This makes the effective source size incident on the monochromater very small, 0.5 mm high and 1.0 mm wide, and allows for use of narrow collimation slits $S_1$ and $S_2$.

The monochromator consists of an entrance vertical slit ($S_1$), a Bragg reflection crystal ($X_1$) and an exit vertical slit ($S_2$). The crystal is rotated to Bragg reflect the desired X-ray line ($K_\alpha$) while $S_1$ and $S_2$ are then narrowed down to eliminate further scattering from ends of nearby X-ray lines and also to cut down on the Bremsstrahlung background. The crystal $X_1$ provides angular collimation which is limited only by the natural Darwin width and the mosaicity of $X_1$. The Darwin width is a finite size effect, i.e., the X-ray beam is attenuated as it penetrates the crystal. Thus, since X-rays measure the Fourier transform of the electron density correlation, a finite penetration depth gives a Bragg reflection of nonzero width, viz., the Darwin width. Fig. 7 shows the Darwin width of a non-absorbing perfect crystal with no mosaic spread. The line profile has long Lorentzian tails that fall off as $1/\varepsilon^2$, where $\varepsilon = \Phi - \Phi_0$, gives the mismatch in the Bragg angle. A proper characterization of the Lorentzian tails is essential to the data convolution process.

The sample cell containing the liquid crystal is placed in a two-stage automatically controlled temperature oven (accuracy ≈ 1 mK). The oven is placed in a magnetic field ( 4-6 kGauss) which aligns the molecules and gives a sample mosaicity of better than 0.5 degrees. The oven and the magnet are mounted on a goniometer head which has two degrees of tilt motion as well as the translational x-y motion needed to precisely center the sample in the X-ray beam. This goniometer assembly is then mounted on a Huber goniometer with two independent rotating precision stages, referred to as $\Phi$ and 2 $\Phi$. The $\Phi$-stage provides for rotation of the magnetsample assembly, while the analyzer crystal, detector and all the other components of the spectrometer after the sample are mounted on the 2 $\Phi$- stage. The rotation of the goniometer stages can be controlled to 0.0005°.

The analyzer functions as a collimator and provides angular resolution for the 2 $\Phi$- stage. Similar to the monochromator crystal $X_1$, the analyzer crystal $X_2$ collimates by Bragg reflection. The detector consists of NaI scintillation crystal mounted on a photomultiplier. Magnetic shielding around the detector prevents interference from the magnet. The detector signal is fed into a pulse height analyzer to minimize the Bremsstrahlung. We shall now discuss the spectrometer resolution in more detail.

## 3. Spectrometer Resolution

The two-crystal spectrometer discussed so far provides only in-plane (horizontal) collimation. In these experiments, vertical collimation is obtained by use of horizontal slits and Soller slits. Horizontal slits placed between the monochromator and the sample provide the vertical resolution of the X-rays incident on the sample ($k_i$). The vertical collimation of scattered X-rays ($k_f$) is determined by a horizontal set of Soller slits located between the sample and the analyzer. The X-ray beam incident on and scattered from the sample has a non-zero height. In the scattering studies of the nematic-smectic-A transition the interesting pretransitional fluctuations along the vertical direction are identical to the fluctuations along one of the in-plane (k) directions. For this reason the vertical resolution needs to be reduced by using horizontal slits (before and after the sample) that match the detector opening. On the other hand, the smectic-C and smectic-A experiment requires better vertical resolution and accordingly Soller slits are used after the sample.

The resolution along the vertical direction is obtained by geometric methods (slits) and so has no tails. The profile is well approximated by a Gaussian; this is probably because all of the imperfections, such as kinks in the Soller slit plates and nonparallel sides in a slit, tend to have a rounding effect on a simple triangular profile that results from geometric considerations of slits and Soller slits.

The collimation of X-rays in the vertical direction is usually about 0.1° (Fig. 8, for $k_i$ and $k_f$). The vertical resolution is then equal to about $dq_v \approx 2k \sin 0.1° = 0.014$ Å$^{-1}$ (FWHM) with $k = 2\pi/\lambda = 4.08$ Å$^{-1}$. In the nematic-smectic-A experiments it is necessary to reduce the vertical resolution by about an order of magnitude. We shall now consider the in-plane (horizontal) resolution function.

The in-plane resolution is obtained by two flat, perfect silicon (111) single crystals. The Bragg reflections have a width within a factor of two of that predicted by the Darwin theory for perfect non-absorbing crystals. The widths are measured by rotating the 2 Φ- stage of the goniometer so that the analyzer crystal rotates with respect to the monochromator crystal. The resulting rocking curve which we refer to as an "arm-zero" is shown in Fig. 8. The profile shown in this figure has $1/\Phi^2$ tails which is expected from the Darwin theory. The solid line is the result of a best fit to a Lorentzian which underestimates the center while overestimating the wings.

Since the scattering takes place in reciprocal space, we also need to know the resolution of the spectrometer in reciprocal space. Fig. 9 shows the in-plane (h-k) scattering geometry for reciprocal space. The X-ray incident on the sample is labeled $k_i$ and $k_f$ is the X-ray leaving the sample. The scattered X-ray is $q \equiv k_i - k_f$. The in-plane spectrometer resolution involves the convolution of the resolution of $k_i$ with the resolution of $k_f$. The resolution along the h and k axes (Fig. 9) are:

res (H) = cos Φ x convolution [res($k_f$)- res($k_i$)]
res (k) = sin Φ x convolution [res($k_f$)- res($k_i$)]

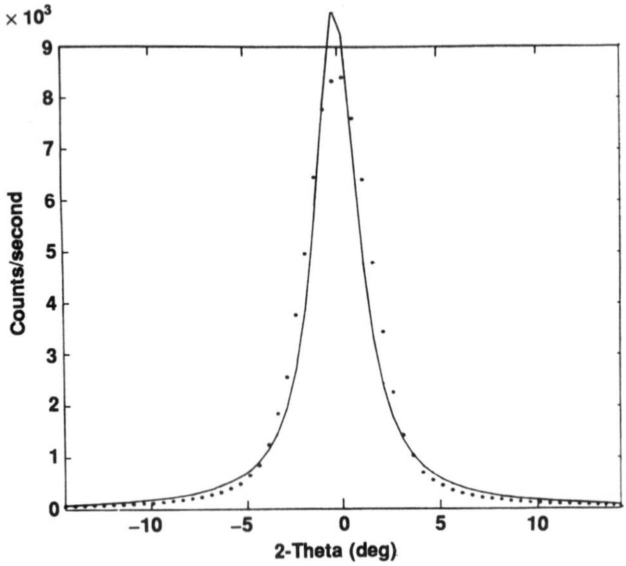


Fig.8. Measured "arm-zero" profile along the h-axis for a two-silicon crystal spectrometer. The solid line is the best fit to a Lorentzian.

Usually $|q| << |k_i| \approx |k_f|$, so the scattering angle is small. Typically $\Phi \approx 0.5°$, so that

$$\text{res}(k) \approx 0.03\ \text{res}(h) \text{ and } \text{res}(h) \approx \text{convolution } [\text{res}(k_f) \cdot \text{res}(k_i)].$$

Thus the resolution along h is different from the arm-zero profile. Typical arm-zero measurements give full width at half maxima (FWHM) values of 0.0035° for flat silicon. The different wavelengths for $K_{\alpha 1}$ and $K_{\alpha 2}$ scatter X-rays at angles differing by $\approx 0.007°$. Hence, using a silicon monochromator and analyzer [with an "arm-zero" width of 0.0035° (FWHM)], we obtain a double

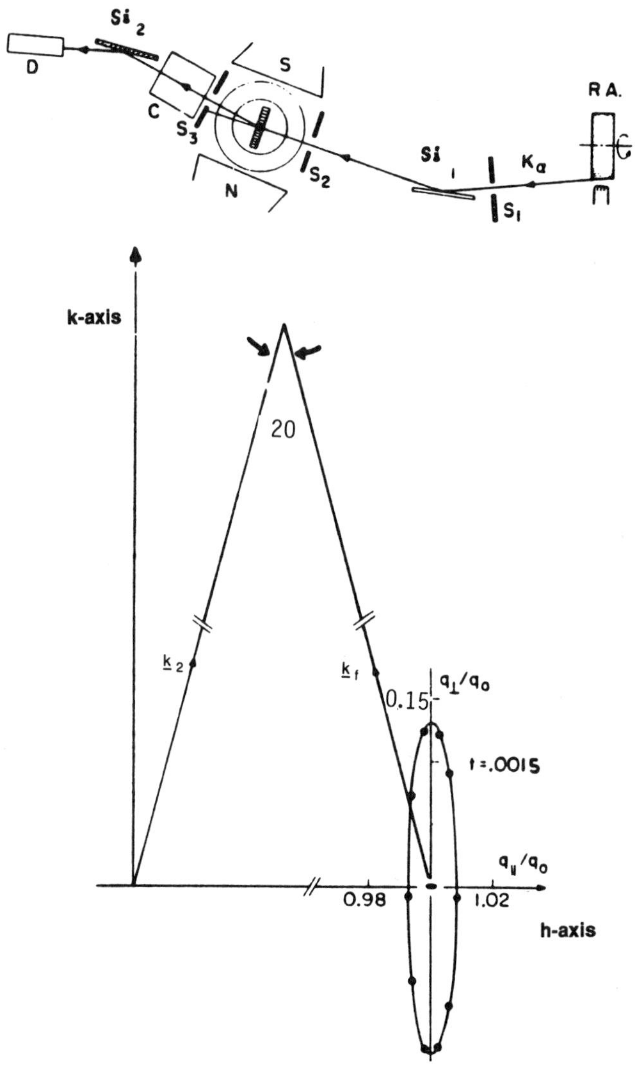


Fig.9. Top: Schematic diagram of the X-ray spectrometer with slits S1, S2 and S3: two silicon crystals: optional soller slits (C) and NaI detector. Bottom: Scattering diagram in the reciprocal (h-k) plane. Also shown is the scattering profile half intensity contour about 0.5 degree above the nematic-smectic A transition. The solid ellipse in the center is the in-plane instrumental resolution function.

peak resolution function consisting of the sum of two displaced (by 0.007°) peaks, one for $K_{\alpha1}$ and the other for $K_{\alpha2}$. This double peaked profile is hidden in the "arm-zero" scan because the scattering is centered at $\Phi=0$ so that $d\Phi = \tan\Phi(d\lambda/\lambda) = 0$.

From Fig. 9 we have $h \approx 2k \sin\Phi$; this leads to in-plane resolution along the h direction of $dh = k \cos\Phi \, d(2\Phi)$. For flat silicon $d(2\Phi) = 0.0035°$ (FWHM), and $k = 4.08$ Å$^{-1}$, so that $dh \approx 2.5 \times 10^{-4}$Å$^{-1}$. The resolution along k is $\tan 3° \approx 0.05$ narrower than along the h direction and in fact much better than what is required for most of the experiments. The scattering width along k is in fact limited by a property of the sample, namely, the mosaic spread. Typically, this spread is about 0.5° and yields an effective resolution of about $2\times10^{-3}$Å$^{-1}$ (FWHM) in the reciprocal scattering plane.

We shall now summarize the in-plane resolution: The double peaked resolution function along h is given by the sum of two peaks displaced by an amount determined by the $K_{\alpha1}$ - $K_{\alpha2}$ splitting for a scattering angle of $2\Phi \approx 3°$. Additionally, the natural line width broadens each peak by about 30% of the "arm-zero" width. The functional form for each peak in the resolution function is determined by the scattering profile of the arm-zero. From Fig. 9 we see that a Lorentzian lineshape does not adequately represent the intensity profile, even though, the tails ultimately fall off as $1/\Phi^2$. It is found that a modified lineshape given by:

$$I_{ARM-ZERO}(h) = \frac{a_0}{1+(h/K)^2+c(h/k)^4} \frac{h_c^2}{h^2+h_c^2} \tag{1}$$

with $h = 2k \sin\Phi$, characterized the armzero. It can be seen that for $h >> h_c^2$, $I(h) \approx 1/h^2$, while for $h < h_c$, the functional form is significantly affected by the presence of the $h^4$ term. The result of least-squares fits of the arm-zero to Eq.1 is shown in Fig. 10. The instrumental resolution

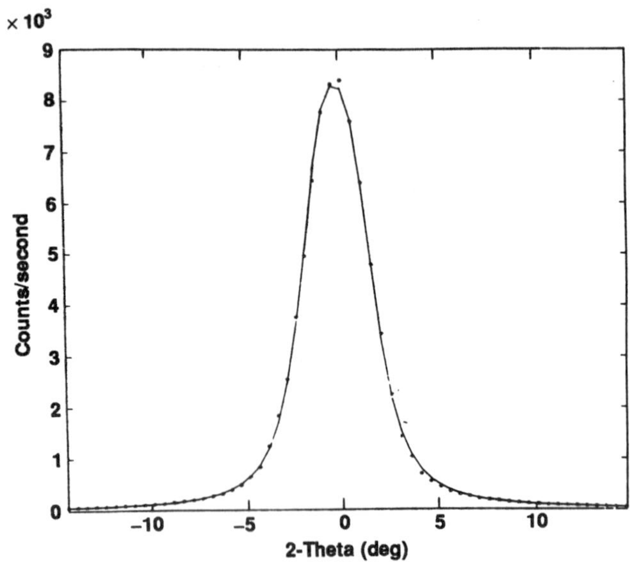


Fig. 10. Measured "arm-zero" profile along the h-axis. The solid line is the result of a least squared fit to a modified Lorentzian used in the data convolution analysis. The fitting is significantly better than the simple Lorentzian fitting shown in Fig. 8.

along k is typically two orders of magnitude narrower than the scattering scan widths along k and is treated as a delta function resolution in the data convolution analysis.

This type of experimental set-up has been used for many important X-ray studies on liquid crystals which have yielded several significantly new results. Some of these will be discussed in Chapter 16 of this volume.

ACKNOWLEDGEMENT. Technical editing by Ms. Shelley DeLozier is greatly appreciated.

## References

1. See, eg: P.S. Pershan, "Structure of Liquid Crystal Phases", World Scientific, Singapore (1988).
2. R. Shashidhar, Chapter 16 of this Volume (1991).
3. P.G. De Gennes, Mol. Cryst. Liq. Cryst. **21**: 49 (1973).
4. G.W. Gray and J. W. Goodby, "Smectic Liquid Crystals", Leonard Hill, Glasgow, (1984).
5. R.J. Birgeneau and J.D. Litster. J. de Phys. Lett. **39**: L399 (1978).
6. R. Pindak, D. Moncton, S.C. Davey and J.W. Goodby, Phys. Rev. Lett. **46**: 1135 (1981).
7. P.A.C. Gane and A.J. Leadbetter, Mol. Cryst. Liq. Cryst. **78**: 183 (1981).
8. J.J. Benattar, F. Moussa and M. Lambert, J. Chim. Phys. **80**: 99 (1983).
9. G. Sigaud, F. Hardouin, M.F. Achard and H. Gasparoux, J. de Phys. **40**: C3356 (1979).
10. A.M. Levelut, R.J. Tarento, F. Hardouin, M.F. Achard and G. Sigaud, Phys. Rev. A **24**: 2180 (1981).
11. C.R. Safinya, W.A. Varady, L.Y. Chiang and P. Dimon, Phys. Rev. Lett. **57**: 432 (1986).
12. P.G. de Gennes, Solid. State Commun. **10**: 753 (1972).
13. V.N. Raja, "High Pressure and X-ray Studies of Liquid Crystals", Ph.D. Thesis, University of Mysore (1988).
14. C.R. Safinya, " Xray Scattering Study of The Critical Behavior of Binary Liquid Crystal Mixtures", PhD. Thesis, M.I.T. (1981).

Chapter 16

# X-RAY STUDIES OF THERMOTROPIC LIQUID CRYSTALS

R. SHASHIDHAR

## 1. Introduction

Thermotropic liquid crystals are characterized by three types of order: orientational order, positional order and bond orientational order[1]. These can be short range order, quasi long range order or long range order. X-ray diffraction studies serve to probe the types and extent of the order. Such studies not only enable us to understand the structure of the mesophase, but also permits us to investigate how fluctuations associated with shortrange smectic-like order develop close to phase transitions. With the advent and extensive development of the high resolution X-ray scattering techniques, we are now in a position to measure correlation lengths extending to a few micrometers. This resolution, coupled with the ability to achieve a millikelvin temperature control, have led to a variety of new results which are fundamentally important in our understanding of condensed matter. In this chapter, we shall describe in some detail three important results: 1) observation of algebraic decay of positional order in smectic-A; 2) critical divergence of anisotropic correlation lengths near the nematic-smectic-A transition and 3) a new type of critical point involving two smectic A phases of the same symmetry.

## 2. Algebraic Decay of Positional Order in Smectic A: Landau-Peierls Instability

It has been known theoretically for some time that translational order as it occurs in a solid, cannot exist in two dimensions because it is destroyed by thermally excited fluctuations[2,3]. Thus, for crystalline solids the lower marginal dimensionality d° is two.

This is the spatial dimension at which thermal fluctuations prevent the establishment of the long-range order. For many systems d° is two, and this is one of the reasons for the intense current experimental and theoretical interest in two-dimensional materials. For these systems it is predicted that a transition occurs to a state of quasi-long range order in which the positional correlation functions do not extend to infinity, but decay algebraically as some power of the

R. Shashidhar - Center for Bio/Molecular Science & Engineering, Naval Research Laboratory, Code 6090, Washington, DC 20375-5000, USA; and Geo-Centers, Inc., 10903 Indian Head Highway, Fort Washington, MD 20744, USA.

*Phase Transitions in Liquid Crystals*, Edited by S. Martellucci and A.N. Chester, Plenum Press, New York, 1992

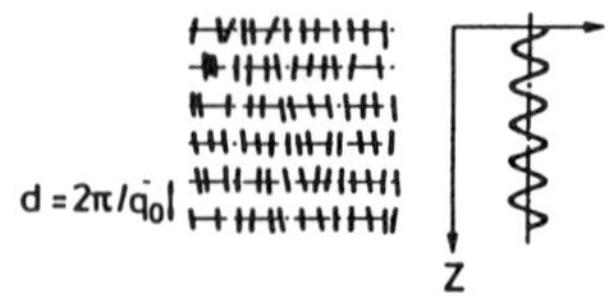


Fig. 1. Schematic representation of smectic A. The picture on the left is a classical picture. The figure on the right is a more correct representation of the smectic A: a density modulation which is sinusoidal. [Ref. 9]

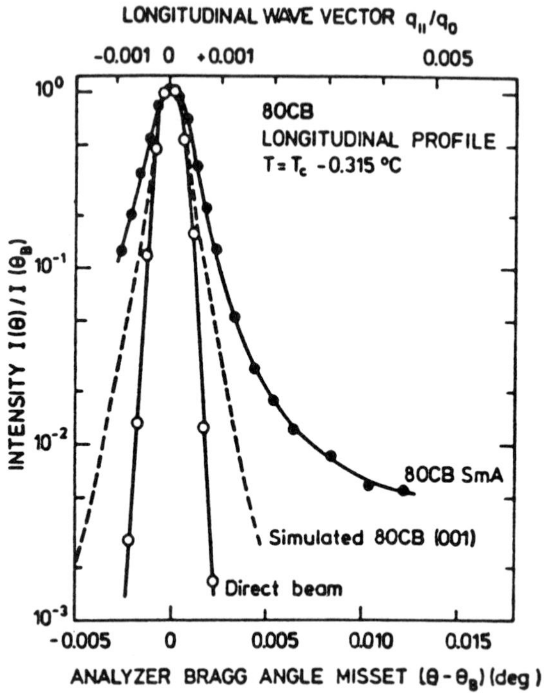


Fig. 2. Longitudinal line profile for the smectic A phase of 8OCB, direct beam and simulated profile of smectic A if it had true long range order. [Ref. 9]

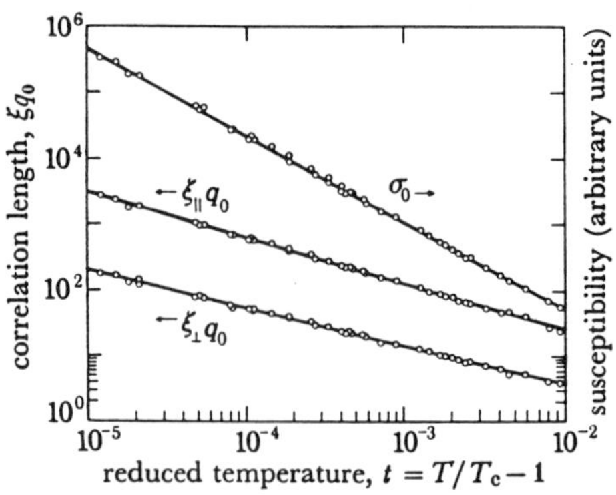


Fig. 3. Top: Susceptibility $\sigma_0$ and the longitudinal and transverse correlation lengths above the smectic A phase of 40.7. The value of $q_0$ is 0.222 $Å^{-1}$. Bottom: Comparison of the longitudinal correlations data for 8OCB obtained for X-ray diffraction (solid circles) and light scattering (open circles). [Ref. 11]

distance[4]. A similar algebraic decay of correlation functions occurs at an ordinary critical point and in both cases the power-law decay of correlations is accompanied by an infinite susceptibility. For various experimental reasons, this predicted behavior has not yet been observed directly in any two-dimensional system[5].

A closely related phenomenon is predicted to occur in smectic-A (SmA) and smectic-C (SmC) liquid crystals[6]. These are phases which possess both orientational long range order (LRO) of the anisotropic molecules and translational order intermediate between that of a liquid and a solid[7]. Explicitly, in the SmA and SmC phases there is a one-dimensional mass-density wave in a three-dimensional liquid. The density wave may be either along (SmA phase) or at an angle (SmC phase) to the nematic director. As we shall discuss below, it may be readily shown that the positional fluctuations diverge logarithmically at large distances at all temperatures in exact analogy with 2D crystals. Thus the smectic-A to nematic phase transition is not only the simplest example of melting, but liquid crystals also provide a three dimensional system in which the effects of divergent long-wavelength acoustic fluctuations can be studied experimentally.

A classical picture of smectic A is given in the left portion of Fig. 1. The molecular centers are arranged in layers with a defined layer spacing d but with a liquid-like order within each layer. The average orientation of the molecules is perpendicular to the layers in the SmA phase. If this picture was literally correct, Bragg scattering of X-rays should occur whenever the difference between incident and scattered wave vectors equals a multiple of $q_0 = 2\pi/d$. But it is known that in the smectic-A, the higher-order Bragg peaks are absent, or precisely, their intensity is at least a factor of $10^4$ less than that of the fundamental reflection. A more correct description of the SmA phase is therefore that it has sinusoidal density modulation[7] as shown in the right-hand part of Fig. 1. The dot-dashed lines through the molecules should not be interpreted as lattice planes but rather as planes of a certain phase of the density wave. The question is now whether this density wave has true long-range order.

The Landau-Peierls argument extended to liquid crystals by Caille[8], predicts that the smectic A liquid crystal in three dimensions cannot have true long range order and that the layer correlations should exhibit an algebraic decay. Experimentally, it is not trivial to distinguish between the "Landau-Peierls" scattering and a combination of Bragg scattering plus thermal diffuse scattering from the acoustic modes. The instrumental resolution must be very good to distinguish between a δ function (Bragg) and a cusp (Landau-Peierls). High angular resolution can be obtained by using perfect crystals together with an essentially monochromatic X-ray line source; only X-rays collimated within a few milliradians will fulfill the Bragg condition. The solution to this problem is to use multiple Bragg reflections in the collimating crystals, since the resolution after m Bragg reflections is that for a single reflection raised to the power of m.

Using a high resolution set-up based on this principle, Als-Nielsen et al[9] were able to prove unambiguously that the SmA line profile is consistent with the Landau-Peierls scattering. By investigating the resolution functions in great detail, they were able to analyze the line profiles of the diffracted peaks and showed that they are not delta functions (Fig. 2) Hence, the X-ray diffraction spots of a smectic-A are not true Bragg spots but quasi-Bragg spots and the smectic A phase does indeed lack true long-range positional order as expected from the Landau-Peierls argument.

## 3. Critical Divergence of Correlation Lengths Near Smectic-A Nematic (SmA-N) Transition

In several systems, when the nematic phase is cooled it undergoes a second order transition to the smectic-A phase. As the temperature approaches the nematic smectic-A transition, the width of diffuse X-ray scattering peaks of the nematic becomes infinitesimally small. The physical interpretation of the half width of the diffuse peak in the nematic is that the smectic fluctuations in the nematic phase are correlated over lengths $\xi$. In the mean field limit should vary as $(T-T_{NA})^{-1/2}$.

One of the major shortcomings of all mean field theories is that they do not take into account the difference between the average value of the order parameter $<\Psi>$ and the instantaneous value $\Psi = <\Psi> + \delta\Psi$, where $\delta\Psi$ represents the thermal fluctuations[10]. The usual effect expected from theories for this type of critical phenomena is a "renormalization" of the various terms in the free energy such that the temperature dependence of correlation length has the form

$$\xi(t) \propto t^{-\nu}$$

where $t \equiv (T-T^*)/T^*$, $T^* \equiv T_{NA}$ is the 2nd order transition temperature and $\nu$ is expected to have some universal value that is generally not equal to 0.5. One of the major unsolved problems of the nematic to smectic-A phase transition is the anisotropy of correlation lengths[11], i.e., the width along the scattering vector q varies as

$$\xi_{\parallel}^{-1} \propto t^{\nu_{\parallel}}$$

with a different temperature dependence than the width perpendicular to q, (Fig. 3)

$$\xi_{\perp}^{-1} \propto t^{\nu_{\perp}}.$$

The correlation lengths are measured by fitting the differential scattering cross sections to the empirical form:

$$\frac{d\sigma}{d\Omega} = \frac{\sigma}{1 + (Q_z - |q|)^{2\xi_{\parallel}^2} + Q_{\perp}^{2\xi_{\perp}^2} + c\left(Q_{\perp}^{2\xi_{\perp}^2}\right)^2}$$

The amplitude

$$\sigma \propto t^{-\gamma}$$

where the measured values of $\gamma$ are empirically found to be very close to the measured values for the sum

$$\nu_{\parallel} + \nu_{\perp}$$

Most of the systems that have been measured to date have values for

$$\nu_{\parallel} > 0.66 > \nu_{\perp} \quad \text{and} \quad \nu_{\parallel} - \nu_{\perp} = 0.1 \text{ to } 0.2$$

Table 1 lists some of the observed values for $\gamma$, $\nu_{\parallel}$ and $\nu_{\perp}$.

The anisotropy of correlation lengths seen experimentally for all materials listed in Table 1 poses a problem. Most theoretical treatments[12] predict isotropic critical behavior

Table 1. Critical Exponents from X-ray Scattering Studies of the Smectic A Nematic Transition.

| Molecule | $\gamma$ | $\nu_{\parallel}$ | $\nu_{\perp}$ |
|---|---|---|---|
| | | | |
| 40.7 | 1.46 | 0.78 | 0.65 |
| 85.5 | 1.53 | 0.83 | 0.68 |
| CBOOA | 1.30 | 0.70 | 0.62 |
| 40.8 | 1.31 | 0.70 | 0.57 |
| 8OCB | 1.32 | 0.71 | 0.58 |
| 9S5 | 1.31 | 0.71 | 0.57 |
| 8CB | 1.26 | 0.67 | 0.51 |
| 10S5 | 1.10 | 0.61 | 0.51 |
| 9CB | 1.10 | 0.57 | 0.39 |

$$\nu_{\parallel} = \nu_{\perp}$$

and some indicate that the A-N transition must always be first order which is contradicted by experiments. Theories based on the dislocation mechanism for the A-N transition[13-15] suggest

$$\nu_{\parallel} = 2\nu_{\perp}.$$

None of these predictions are supported by the experiments (see Table 1). Despite considerable theoretical and experimental efforts, the two different temperature dependences of the correlation length has remained one of the major unsolved questions in equilibrium statistical physics.

## 4. Polymorphic Smectic A Phases

The results described in the previous section required very high resolution ($10^{-4}$ Å$^{-1}$). There are several phenomena that can be studied when a lower resolution ($10^{-3}$ Å$^{-1}$) is adequate. A typical example is the transition between different fluid smectic phases. In this case, the relevant parameter to study is the smectic layer thickness or the periodicity of the density modulation. The wave-vector associated with the density modulation has been precisely determined at small intervals of temperature close to the transition. An area in which the lower resolution X-ray diffraction experiments have been remarkably successful is Polymorphic Smectic A.

Sigaud et al.[16] observed for the first time a smectic A-smectic A transition. X-ray studies[17] revealed that the higher and lower temperature A phases are respectively monolayer (or $A_1$) with the layer thickness (d) being approximately equal to the molecular length (l), and, bilayer ($A_2$) with d≈2 l. The existence of a partially bilayer smectic-A or $A_d$ phase with $1 < d < 2$ l being already known, it became clear that the Smectic A phase can exist in several polymorphic forms. This was soon followed by the discovery of several other types of new phases, viz, the antiphase or $\tilde{A}$, the crenelated phase or $A_{cre}$ and the ribbon phase or $\tilde{C}$[18,19]. So far, these phases have been observed only in strongly polar systems, i.e., materials in which the molecules possess a strongly polar cyano or nitro group at one end. (In the case of binary mixtures, at least one of the constituent compounds belong to this category).

The schematic representation of the arrangement of the molecules in these smectic A phases is given in Fig. 4(a) - (f). In the $A_1$, phase the dipolar heads are assumed to be arranged randomly within each layer. In the partially bilayer ($A_d$) phase, the preferential antiferroelectric ordering of the adjacent dipolar heads results in an interdigitated layer structure. The bilayer ($A_2$) phase has an almost perfect head-to-head arrangement of the dipolar heads. In the antiphase or $\tilde{A}$, the local order is the same as that in the $A_2$ phase but the direction of the dipolar arrangement alternates periodically within a layer and this results in a rectangular symmetry. The periodicity of the transverse modulations was found to be of the order of a few a hundred angstroms[19] and mostly temperature independent. Just like $\tilde{A}$, the crenelated smectic A ($A_{cre}$) phase also shows alternation of the dipolar arrangement between two neighboring domains but with the difference that these antiphase domains have unequal widths. It has been observed that with decrease in temperature the width of one domain increases at the expense of the neighboring one. In the $\tilde{C}$ (or smectic ribbon phase) the layers consist of pairs of molecules showing a bilayer type of arrangement which are seen to be broken periodically by defect walls oblique with respect to the director. Thus unlike the $\tilde{A}$ or $A_{cre}$ phases, the $\tilde{C}$ phase has an oblique lattice.

The types of diffraction patterns to be expected from the different smectic phases are also shown schematically in Fig. 4:

a) The $A_1$ phase shows a condensed peak at $2q_0$ ( $= 2\pi/l$) corresponding to the monolayer ordering and, generally, a diffuse scattering centered around $q_0 = 2\pi/l$) which is incommensurate with respect to $2q_0$.

b) The partially bilayer smectic A phase ($A_d$) shows a condensed peak at $q_0$, a weak second harmonic at $2q_0$ and a diffuse peak centered around $2q_0$.

c) The smectic $A_2$ phase shows condensed peaks at both $q_0$ and $2q_0$.

d) For the $\tilde{A}$ phase, there is a sharp peak at $2q_0$ and two condensed spots which are split out of the z-axis in a direction normal to it. The off-axis spots are centered around the $q_0$ position.

e) In the $A_{cre}$ phase all the spots of the $\tilde{A}$ are seen in addition to a condensed spot at $2q_0$.

f) In the $\tilde{C}$ phase, the diffraction pattern consists of spots split out of z-axis, but unlike in the A, these spots are not centered around $q_0$. Also these spots have highly asymmetric intensity. Since the layers are tilted with respect to the z-axis, the $2q_0$ spots are seen off z-axis.

A phenomenological theory which attempts to explain the origin of the different forms of A phase was proposed by Prost[20] almost concomitantly with the first observation of the A-A transition. This theory, which was enlarged in its scope by Prost and Barois[21] defines the free energy of a strongly polar smectic-A in terms of two order parameters, viz, the dipolar order parameter ($\Psi_1$) associated with an antiparallel pair of dipolar molecules, and the usual density order parameter ($\Psi_2$). The competition between these two types of order trying to condense at two incommensurate length scales leads to different types of A phases. (The same competition also leads to the occurrence of the reentrant nematic behavior[20,22] but we shall not be concerned with this aspect in the present article). In this definition, the nematic phase is identified as that in which $\Psi_1 = 0$ and $\Psi_2 = 0$, while $\Psi_1 = 0$ and $\Psi_2 \neq 0$ defines the $A_1$, phase. In the case of both $A_d$ and $A_2$ phases $\Psi_1$ as well as $\Psi_2$ have non-zero values, but $\Psi >> \Psi_2$ in the case of the $A_d$ phase while $\Psi_1 \sim \Psi_2$ in the case of the $A_2$ phase. This theory also predicts under certain conditions the existence of an incommensurate phase, i.e., a phase in which two collinear incommensurate density modulations exist independent of each other and the strongly coupled incommensurate phase where in the two periodicities are modulated in one-dimension leading to a structure consisting of defect walls or solitons[23]. The theory of Prost and Barois[21] has been extended by Barois, Prost and Lubensky[24] within the frame work of the mean-field (MF) theory to calculate the various phase diagrams that are possible involving the different types of A phases. These

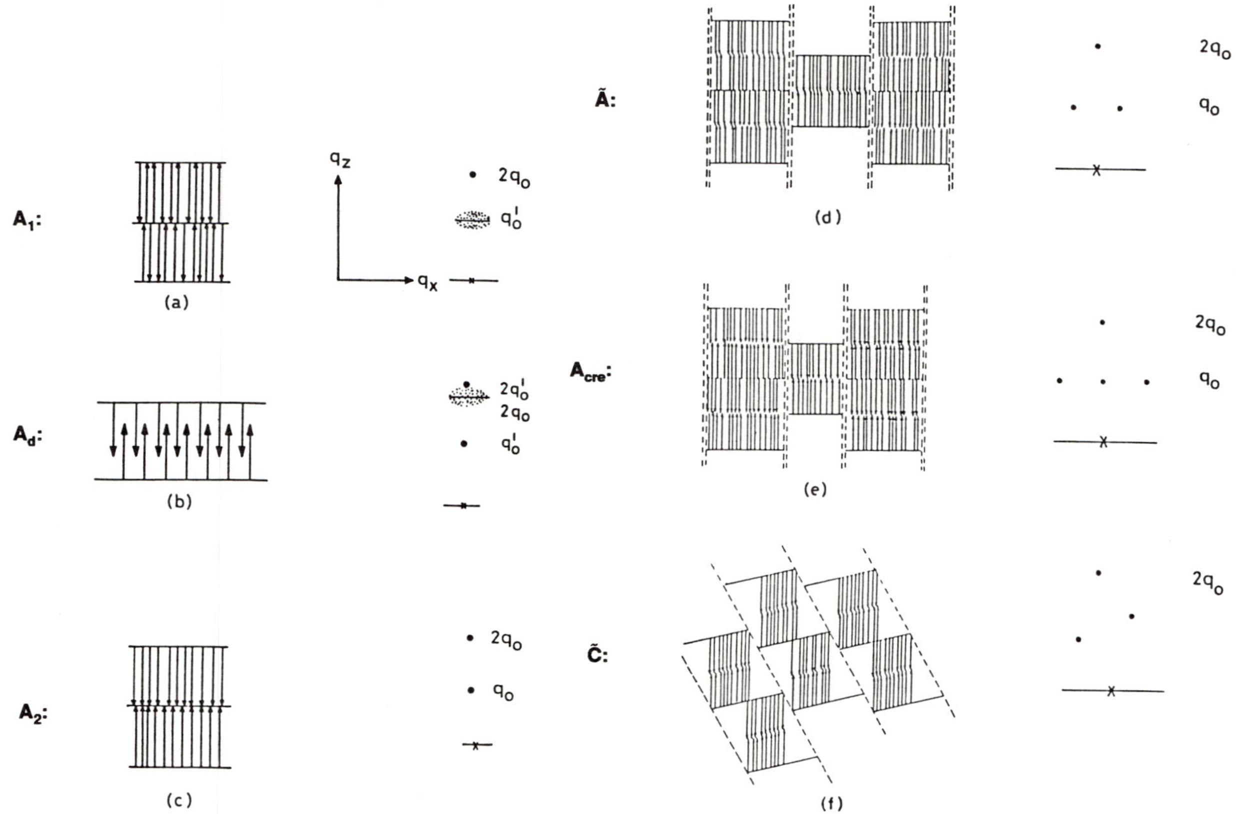


Fig. 4 (a) - (f). Schematic representation of the molecular arrangement along with their characteristic diffraction patterns for: monolayer phase ($A_1$); the partially bilayer phase ($A_d$), the bilayer phase ($A_2$); the smectic antiphase (Ã); the crenelated phase ($A_{cre}$); and, the ribbon phase (C̃). The dark circles represent condensed spots and the shaded ovals represent the diffuse spots due to fluctuations.

theoretical phase diagrams also predict several types of critical/multicritical points: (a) the N-$A_2$ tricritical point, (b) $A_1$-$A_2$ tricritical point, (c) the $A_d$-$A_2$ critical point and (d) and the $A_1$-$N_{re}$-$A_d$ bicritical point.On the experimental side, investigations carried out during the last few years have resulted in the observation of both types of incommensurate phase[25-27] as well as some of the critical or multicritical points mentioned above[27]. There have also been more rigorous theoretical treatments which take into account fluctuations. These fluctuation-corrected theories have made some important predictions regarding the universality of the $A_d$-$A_2$ critical point[28,29] and also concerning the different possible ways in which the $A_1$-$A_d$ transition boundary can terminate[30]. Experiments to verify these predictions have begun only very recently.

Phase transitions between the different forms of the smectic A phase are of considerable current interest because they are transitions between two phases with quasi-long range order (QLRO). The phases $A_1$, $A_d$ and $A_2$ have the same macroscopic symmetry differing from each other only by the wavelength of their periodic modulations. Hence it should be possible to go from $A_1$ to $A_d$ or from $A_d$ to $A_2$ by only varying the layer spacing (d). There can also be a first order transition between these phases at which the wave vector characterizing the phase should show a discontinuous jump. A line of such first order transitions can terminate at a critical point when the difference between the wave vectors in the A phases goes to zero, this critical point providing a continuous path between $A_1$/$A_2$ and $A_d$ phases. On the other hand, the $A_1$-$A_2$ transition can be either first or second order because of the exact double of the layer periodicity, i.e., the layer spacing (d) doubles as a result of the continuous vanishing of a spatial subharmonic. Experimentally, a tricritical point for the $A_1$-$A_2$ transition[31,32] as well as a $A_d$-$A_2$ critical point[33,34] have been observed. We shall briefly discuss the results concerning the latter.

## 5. $A_d$-$A_2$ Critical Point

As already mentioned, since the $A_d$ and $A_2$ phases have the same macroscopic symmetry and since the layer spacing ratio is not exactly equal to two, there can only be a first order transition between them. The theory of Barois, Prost and Lubensky[24], makes several interesting predictions concerning the $A_d$-$A_2$ transitions:

(1) The signature of the $A_d$-$A_2$ transition should be a discontinuous jump in the wave vector associated with the two phases. There should also be an accompanying discontinuity in the ratio of the 001 and 002 intensities at the transition.

(2) The first order $A_d$-$A_2$ phase boundary can terminate at a critical point (CP) similar to the gas-liquid Isinglike critical point.

(3) As the CP is approached the difference in the wave vectors $q_0$ and $q_0$ goes to zero and the distinction between $A_d$ and $A_2$ phases disappears. Beyond the CP and $A_d$ phase can change into the $A_2$ phase. These predictions have been verified experimentally as we shall see in the following.

The phase diagram of 11OPCBOB/90BCB[33] is shown in Fig. 5. For mole fractions of 110PCBOB (X) less than 0.52, the ribbon phase $\tilde{C}$ intervenes between the $A_d$ and $A_2$ phases. However, with increasing X, $\tilde{C}$ is suppressed, leading to a direct $A_d$-$A_2$ transition. This first order $A_d$-$A_2$ boundary then terminates at a critical point (CP) in the temperature - concentration plane. All these features have been deduced from the results of accurate X-ray experiments for a series of binary mixtures. For all $X < 0.642$, an $A_d$-$A_2$ transition is seen whose signature is the existence of a two-phase region in which the diffraction pattern consists of both $q_0$ and $q_0'$ (as well as their second harmonics which are not shown) as seen in Fig. 6. This is exactly as expected for a first order transition between two smectic phases with different layer periodicities. These figures also

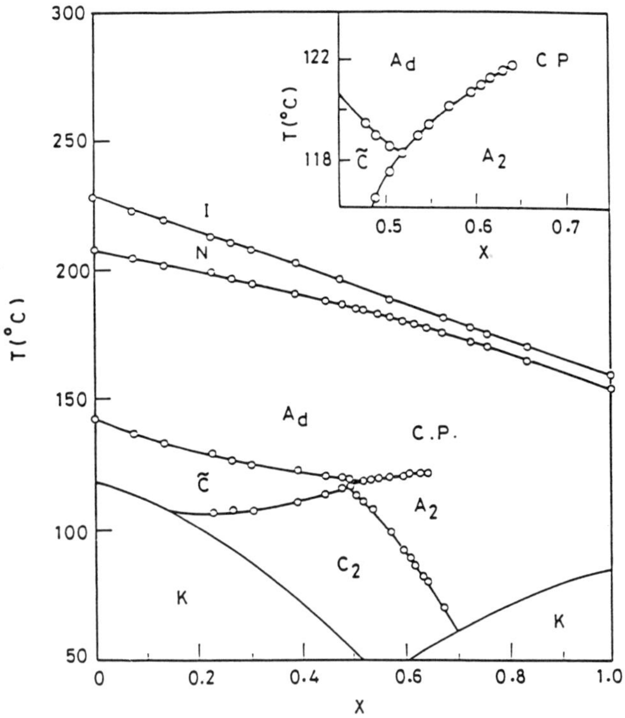


Fig. 5. Temperature - Concentration (T-X) diagram for binary mixtures of 11 OPCBOB and 9OBCB. The $A_d$-$A_2$ transition terminates at a critical point CP. [Ref. 33]

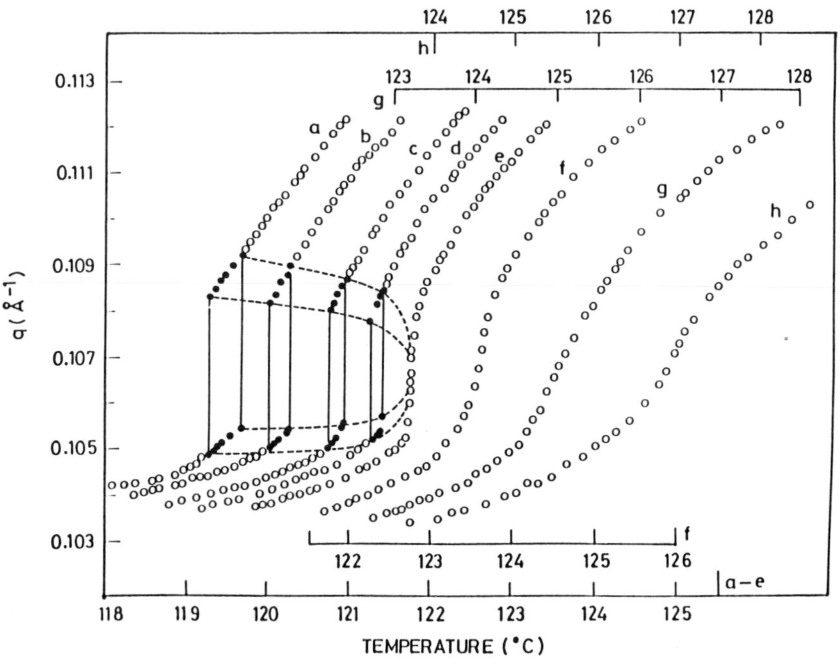


Fig. 6. Plots of wavevectors $q_0$ and $q_0'$ near the $A_d$-$A_2$ critical point. [Ref. 33]

show that the variations of $q_0$ and $q_0'$ in the coexistence region are nothing but continuations of the trends in the variation of the corresponding wave vectors in the $A_d$ and $A_2$ phases respectively. We can also associate the observed jump in the wave vector with the $A_d$-$A_2$ transition. With increasing X, the magnitude of the wave vector jump decreases with an accompanying decrease in the width of the two-phase region until at X = 0.642 no jump is observed. At this point, the wave vector variation is almost vertical and there is a point of inflection. All these features the shrinking of the two-phase region associated with a first order transition and the accompanying decrease in the difference between $q_0$ and $q_0$ at the transition, the extremely steep variation of the wave vector with an inflection point, are clearly indicative of the existence of a critical point at X = 0.642. For higher X values, i.e., on moving away from CP in the supercritical region, the point of inflection becomes progressively less pronounced and $A_2$ evolves continuously from $A_d$ to $A_2$ without a phase transition. Thus it has been established that the two smectic A phases $A_d$ and $A_2$ have the same macroscopic symmetry. More recently another instance of an $A_d$-$A_2$ critical point has been reported in binary mixtures involving members of the same homologous series[34].

As remarked earlier, the mean-field theory of Barois, Prost and Toner[24] the $A_d$-$A_2$ critical point or CP should have Ising symmetry, i.e., should be similar to the liquid - gas critical point. However, as pointed out by these authors themselves, the situation may be more complicated. Park, Lubensky, Barois and Prost[28] have carried out a more extensive theoretical investigation of the CP. They have developed a nonlinear elastic model to describe this critical point and have analyzed its critical properties using the $\varepsilon$ expansion. The theory predicts that the $A_d$-$A_2$ critical point should be of a new universality class with $d_c$, the upper critical dimension, being 6. (The upper critical dimension is defined as that above which the system should exhibit mean-field behavior). Furthermore, a recent extension of the theory[29] shows that $d_c$ drops to 2.5 in the presence of symmetry-breaking fields. In view of the unique and very complex character of the theoretical predictions, it is of considerable interest to characterize the $A_d$-$A_2$ critical point experimentally. Recent high resolution calorimetric studies have made significant advances in this direction[35,36].

ACKNOWLEDGEMENT. Technical editing by Ms. Shelley DeLozier is thankfully acknowledged.

## References


1. See eg: P.S. Pershan "Structures of Liquid Crystals", World Scientific (1988).
2. R.E. Peierls, Helv. Phys. Acta Suppl. **7**: 81 (1934).
3. L.D. Landau, in "Collected Papers of L.D. Landau", edited by D. ter Haar (Gordon and Breach, New York, 1965), 209.
4. J.M. Kosterlitz and D.J. Thouless, J. Phys. C **6**: 118 (1973).
5. Very interesting related phenomena are, however, seen in helium films. See, for example, D.J. Bishop and J.D. Reppy, Phys. Rev. Lett. **40**: 1727 (1978). Also recent experiments by D.E. Moncton and R. Pindak [Phys. Rev. Lett. **43**: 701 (1979)] indicate that powerlaw decay of correlation functions may be observable in smectic B films.
6. P.G. de Gennes," The Physics of Liquid Crystals" (Clarendon, Oxford, 1974).
7. R. Schaetzing and J.D. Litster, Advances in Liquid Crystals, **4**: 147 (1979).
8. A. Caille, C.R. Acad. Sci. Paris **274B**: 891 (1972).
9. J. Als-Nielsen, J.D. Litster, R.J. Birgeneau, M. Kaplan C.R. Safinya, A. Lindegaard-Andersen and S. Mathiesen, Phys. Rev. B, **22**: 312 (1980).
10. S.K. Ma, "Modern Theory of Critical Phenomena", W.A. Benjamin Inc. (1976).
11. R.J. Birgeneau, C.W. Garland, G.B. Kasting and B.M. Ocko, Phys. Rev. A **24**: 2624 (1981).
12. T.C. Lubensky, J. Chem. Phys. **80**: 31 (1983).
13. B.A. Huberman, D.M. Lubkin and S. Doniach, Solid St. Commun., **17**: 485 (1975).

14. W. Helfrich, J. de Phys, **39**: 1199 (1978).
15. D.R. Nelson and J. Toner, Phys. Rev. B **23**: 363 (1981).
16. G. Sigaud, F. Hardouin, M.F. Achard and H. Gasparoux, J. Physique, **40**: C3356 (1979).
17. F. Hardouin, A.M. Levelut, J.J. Benattar and G. Sigaud, J. Physique, **33**: 337 (1980).
18. F. Hardouin, A.M. Levelut, M.F. Archard and G. Sigaud, J. Chim Phys., **80**: 53 (1983).
19. A.M. Levelut, J. Physique Lett., **45**: L603 (1985).
20. J. Prost, in "Liquid Crystals of One and Two Dimensional Order", Eds. Helfrich, W. and Heppke, G. (SpringerVerlag, Berlin), 159 (1980).
21. J. Prost and P. Barois, J. Chim. Phys., **80**: 65 (1983).
22. P.E. Cladis, Phys. Rev. Lett., **35**: 48 (1975).
23. P. Barois, J. Pommier, and J. Prost, in "Solitons in Liquid Crystals", Eds. L. Lam and J. Prost (Springer Verlag), to be published.
24. P. Barois, J. Prost and T.C. Lubensky, J. Physique, **46**: 391 (1985).
25. B.R. Ratna, R. Shashidhar and V.N. Raja, Phys. Rev. Lett., **46**: 391 (1985).
26. B.R. Ratna, R. Shashidhar and V.N. Raja, in "Incommensurate Crystals, Liquid Crystals and Quasicrystals", Eds. Scott, J.F. and Clark, N.A., (Plenum), 259 (1987).
27. R. Shashidhar and B.R. Ratna, Liquid Crystals, **5**: 421 (1989).
28. J. Park, T.C. Lubensky, P. Barois and J. Prost, Phys. Rev. A **37**: 2197 (1988).
29. Y. Park, T.C. Lubensky and J. Prost, Liquid Crystals, **4**: 435 (1989).
30. J. Prost and J. Toner, Phys. Rev. A **36**: 5008 (1987).
31. K.K. Chan, P. Pershan, L.B. Sorensen and F. Hardouin, Phys. Rev. Lett. **54**: 1694 (1985).
32. K.K. Chan, P. Perhsan, L.B. Sorensen and F. Hardouin, Phys. Rev. A **34**: 1420 (1986).
33. R. Shashidhar, B.R. Ratna, S.K. Prasad, S. Somasekhara and G. Heppke, Phys. Rev. Lett., **59**: 1209 (1987).
34. S. Quentel, G. Heppke, S.K. Prasad, S. Pfeiffer and R. Shashidhar, Mol. Cryst. Liq. Cryst. Lett., **7**: 85 (1990).
35. Y.H. Jeong, G. Nounesis, C.W. Garland and R. Shashidhar, Phys. Rev. A **40**: 4022 (1989).
36. Xin-Wen, C.W. Garland and R. Shashidhar, to be published.

Chapter 17

# CRITICAL FLUCTUATIONS IN FREE-STANDING SMECTIC FILMS: AN X-RAY REFLECTIVITY STUDY

P. LAMBOOY, S. GIERLOTKA, I. W. HAMLEY and W. H. DE JEU

We report on the x-ray reflectivity of free-standing smectic films of 8CB. At temperatures well below the second-order nematic-to-smectic transition temperature $T_{NA}$ we find that the fluctuations of the smectic layers in the middle and at the surfaces have the same amplitude. However, on approaching $T_{NA}$ the fluctuations of the middle layers of the film are found to diverge while those at the surfaces are still quenched. The results can be understood qualitatively by including an additional surface term in the well-known Landau-De Gennes free energy for the smectic-A phase.

Much of the interest in smectic liquid crystals is related to the fact that these systems are at their lower marginal dimensionality. As a consequence the positional ordering of the layers is not truely long-range: the thermal fluctuations diverge logarithmically with the size of the sample. If u(**r**) is the layer displacement from its equilibrium position **r**, the well-known expression for the mean-squared fluctuations of u(**r**) is[1]:

$$\langle u^2(\mathbf{r}) \rangle = k_B T \ln(L/a) \Big/ \left(4\pi\sqrt{KB}\right) \tag{1}$$

where K and B are the elastic constants associated with layer bending and layer compression, respectively, L is the relevant macrosopic length, a the molecular dimension in the direction of L and $k_BT$ is the thermal energy. If the phase transition to the nematic phase is second-order, B can be expected to vanish at $T_{NA}$. In that case the layer fluctuations will diverge also for finite samples.

Here we report on preliminary x-ray observations of the smectic layer fluctuations on approaching $T_{NA}$ in free standing smectic films (FSSF). These films can be obtained over a hole in a substrate and have their layers (varying in number from two to very large) parallel to the surfaces[2]. Recently FSSF have been made large enough for x-ray reflectivity

P. Lambooy, S. Gierlotka, I. W. Hamley and W. H. de Jeu - FOM-Institute for Atomic and Molecular Physics, Kruislaan 407, 1098 SJ Amsterdam, Netherlands

S. Gierlotka - Permanent address: High Pressure Research Center, Polish Academy of Sciences, Sokolowska 29/37, 01-142 Warsawa, Poland.

W. H. de Jeu - Also at: The Open University, P.O. Box 2960, 6401 DL Heerlen, Netherlands.

*Phase Transitions in Liquid Crystals*, Edited by S. Martellucci and A.N. Chester, Plenum Press, New York, 1992

measurements to be performed[3,4]. While conventional transmission x-ray scattering of FSSF provides information about the in-plane structure, x-ray reflectivity gives an accurate determination of the electron density profile normal to the smectic layers[5]. This allows for a direct study of the layer fluctuation profile. At temperatures well below $T_{NA}$ we find that the fluctuations in the bulk and at the surfaces have the same amplitude. However, on approaching $T_{NA}$ we observe a divergence of the fluctuations in the bulk of the films while the fluctuations near the surfaces are quenched and remain finite.

The compound investigated was 4-*n*-octyl-4'-cyanobiphenyl (8CB), which has a smectic $A_d$ phase[6] with a bulk periodicity d=31.8 Å leading to $Q_0=2\pi/d=0.20$ Å$^{-1}$. 8CB has a second-order $SA_d$-N phase transition[7] at $T_{NA}\approx 33.5$ °C. The substance was obtained from BDH (Poole, Dorset, UK) and used without further purification. FSSF of 8CB were drawn over a hole of 10x20 mm$^2$ defined by four razor blades. The mosaic spread of 0.04° FWHM of the layers was due to the non-perfect planarity of the edges of the razor blades. The films were kept in a two-stage oven which was regulated to within 0.05 °C. The reflectivity data were obtained using films positioned vertically on a standard triple axis spectrometer as described in Ref.3.

Results for a 43 layer film of 8CB at several temperatures are shown in Fig. 1, together with the best fits to a model to be described below. A first qualitative interpretation of the data can be given by considering the two main contributions to the scattered intensity. In the first place there is the regular pattern of Kiessig fringes arising from the reflections against the two film/air interfaces and which extends over the whole range of Q-values. Secondly there is the smectic layering within the film leading to a finite size Bragg peak at $Q_0$ (equal to the bulk value) with its subsidary maxima. It is the interference between these two contributions that makes the method extremely sensitive to the electron density profile. The Bragg peak which is clearly present at 22.0 °C vanishes at temperatures close to $T_{NA}$ indicating the disappearance of the smectic layering in the middle of the film. However, the Kiessig fringes which depend on the roughness of the film/air interfaces do not show any appreciable change with temperature, indicating almost constant layering at the surfaces. This can be directly seen from the remaining interference minimum around Q = 0.19 Å$^{-1}$ in Fig. 1d, where the Bragg peak itself is already absent.

Starting point for a quantitative data analysis is the standard expression for the reflectivity[8]

$$R(Q) = R_F(Q)\left|\int dz \frac{\partial \rho(z)}{\partial z} \exp(iQz)\right|^2 = R_F(Q)|F(Q)|^2 \tag{2}$$

$R_F(Q)$ is the Fresnel reflectivity of a sharp flat interface and $\rho(z)$ is the projection of the electron density of the film on the z-axis normal to the film surface. We will restrict ourselves to a phenomenological form for $\rho(z)$. The simplest model for a smectic film with layer spacing d is a cosine modulation superposed on the average electron density $\rho_0$:

$$\rho(z) = \rho_0\,[1 - A\cos(Q_0 z)] \tag{3}$$

In order to incorporate different amplitudes at the surfaces as compared to the bulk we allow A to vary with z as a hyperbolic cosine:

$$A(z) = A_b + (A_s - A_b)\{\exp[\alpha(z-D/2)] + \exp[\alpha(-z-D/2)]\} \tag{4}$$

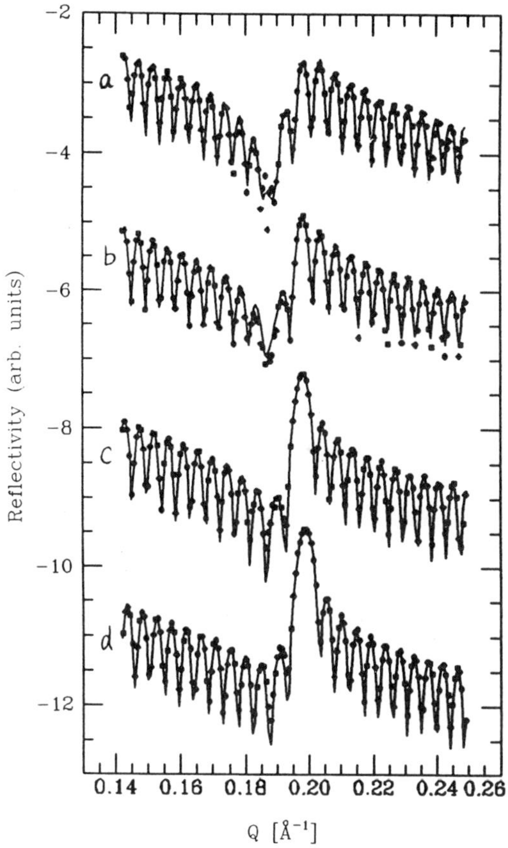


Fig. 1. Reflectivity curves of a 43 layer FSSF of 8CB at 32.8 °C, 32.0 °C, 31.0 °C and 22.0 °C, indicated as a, b, c and d, respectively. The dots are data points, the solid lines are fits using the model described in the text. The curves have been shifted for clarity.

A(z) represents the amplitude of the density modulation, varying from the surface value $A_s$ at $z=\pm\ 1/2\ D$ to the bulk value $A_b$. F(Q) is now given by:

$$F(Q)=2\sin\left(\frac{1}{2}QD\right)-(Q/Q_+)A_b\sin\left(\frac{1}{2}Q_+D\right)-(Q/Q_-)A_b\sin\left(\frac{1}{2}Q_-D\right)$$
$$-(A_s-A_b)\exp\left(-\frac{1}{2}\alpha D\right)\left[\frac{Q}{\alpha^2+Q_+^2}R_+ + \frac{Q}{\alpha^2+Q_-^2}R_-\right] \quad (5)$$

where

$$R_\pm = \alpha\sin\left(\frac{1}{2}Q_\pm D\right)\sinh\left(\frac{1}{2}\alpha D\right)+Q_\pm\sin\left(\frac{1}{2}Q_\pm D\right)\cosh\left(\frac{1}{2}\alpha D\right)$$

and $Q\pm = Q \pm Q_0$.

For each temperature $A_s$, $A_b$ and $\alpha$ were varied to obtain best fits to the experimental data. An additional structureless overlayer with a thickness of about 12 Å was needed in the model to describe the data correctly. The resulting electron-density profiles are shown in Fig.2. We conclude that the toplayers remain essentially constant over the whole temperature range. In Fig. 3 the amplitude of the middle layer and the top layers of the film is shown as a function of temperature. From this figure and the data at the next higher temperature (not shown), we conclude that the highest temperature run given in Fig.1d and 2d is within our accuracy at $T_{NA}$-T = 0.1 K.

A(z) as defined above still has to be related to the thermal fluctuations of the layers $\langle u^2(r)\rangle$. This can be done by noting that the layer fluctuations will smear out the projected density and thus diminish the amplitude of the modulation. This leads to

$$A(z) = S(Q_0)\exp(-Q_0^2\langle u^2(r)\rangle/2) \qquad (6)$$

where $S(Q_0)$ is the structure factor of a perfectly ordered smectic layer. We shall assume $S(Q_0)$ to be constant with temperature while $\langle u^2(r)\rangle$ is taken to be independent of x and y. The amplitude of the middle layer of the film is anticipated to be well described by Eqs.1 and 6. The behaviour of the surface layers will require an additional surface term to be added to the Landau-De Gennes free energy that is at the basis of Eq.(1). An explicit (but complicated) expression for $\langle u^2(r)\rangle$ taking such a term into account was published recently by Holyst *et al.*[9]. In Fig. 3 the predictions for the surface layer and the middle layer of the sample are

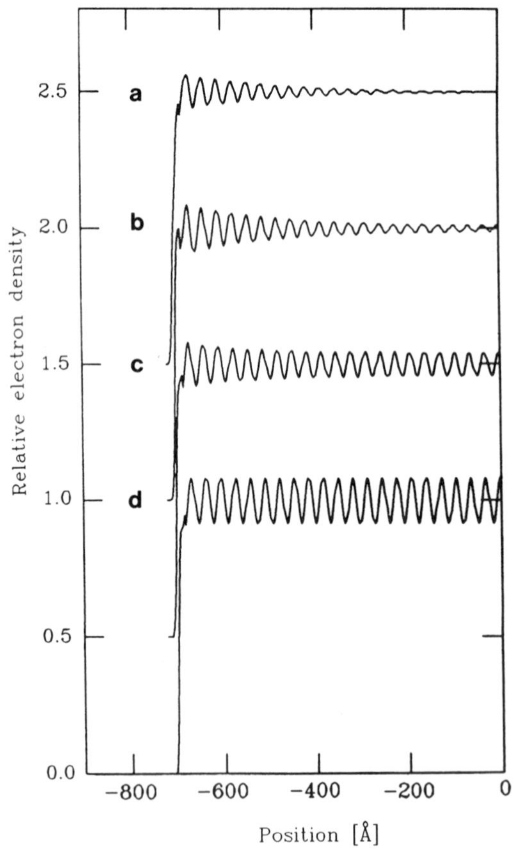


Fig. 2. Model electron densities corresponding to the curves in Fig. 1.

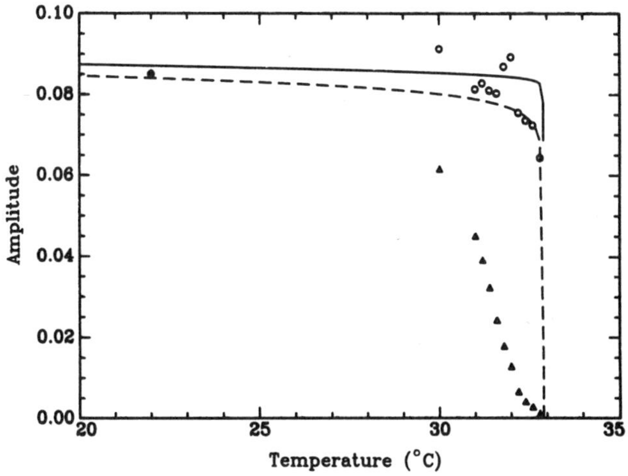


Fig. 3. Amplitude of the middle layer (triangles) and of a surface layer (dots). The lines are predictions of the Landau-de Gennes theory with an additional surface term; broken line: middle layer, solid line: surface layer.

shown. For B (see Ref.10), K (see Ref.11) and for the surface tension $\gamma$ (see Ref.12) experimental values of bulk 8CB are taken; L/a = 43 is the number of layers. A value $S(Q_0) \approx 0.12$ was chosen to get approximately agreement with the data for low temperatures. The amplitude of the density modulation of the surface layers is predicted to be almost constant in temperature in agreement with the experiment. The general behavior of the fluctuations in the middle of the film with increasing temperature is also well described, though the functional form is less satisfactory. The predicted behavior is essentially the same as for a layer in the bulk.

Several uncertainties remain. The rounding off of the fluctuation profile of the middle layer in Fig. 3 could be accounted for by a temperature gradient of the order of 2°C. This seems rather unprobable, and in fact later attempts to measure the temperature gradients over the sample, indicate values on order of magnitude smaller. Alternatively it could be an artefact due to the specific form chosen for A(z). Furthermore, the nature of the observed overlayer is not clear in spite of its reproducibility in the present series of experiments.

In conclusion we have observed in a 43 layer free standing film of 8CB a divergence of the layer fluctuations in the middle of the film on approaching $T_{NA}$ and quenching of the fluctuations of the surface layers.

ACKNOWLEDGEMENT. This work is part of the research programme of the Stichting voor Fundamenteel Onderzoek der Materie (Foundation for the Fundamental Research of Matter, FOM) and was made possible by financial support from the Nederlandse Organisatie voor Wetenschappelijk Onderzoek (Netherlands Organization for the Advancement of Research, NWO). I.W.H. thanks the Royal Society (London) for the award of a postdoctoral fellowship as part of the European Science Exchange Programme.

## References

1. See, for example, G. Vertogen and W. H. de Jeu, *Thermotropic Liquid Crystals, Fundamentals*, Springer Series in Chemical Physics vol. 45 (Springer, Heidelberg, 1988).

2. C.Y. Young, R. Pindak, N.A. Clark and R.B. Meyer, Phys. Rev. Lett. **40**, 773 (1978).
3. S. Gierlotka, P. Lambooy and W.H. de Jeu, Europhys. Lett. **12**, 341 (1990).
4. D.J. Tweet, R. Holyst, B.D. Swanson, H. Stragier and L.B. Sorensen, Phys. Rev. Lett. **65**, 2157 (1990).
5. For a review see: P.S. Pershan, *Proc. International Conference on X-ray and Neutron Scattering from Surfaces and Thin Films*, J.Phys.Colloque (Paris) **50**, C7-1 (1989).
6. S. Shashidar, Chapter 15 of this volume.
7. (a) J. Thoen, H. Marijnissen and W. van Dael, Phys.Rev.Lett. **52**, 204 (1984); (b) B.M. Ocko, R.J. Birgeneau and J.D. Litster, Phys.Rev.Lett. **52**, 208 (1984).
8. J.Als-Nielsen, *Handbook of Synchotron Radiation*, Vol. 3, North-Holland (Amsterdam, 1990).
9. R.Holyst, D.J. Tweet and L.B. Sorensen, Phys. Rev. Lett. **65**, 2153 (1990).
10. M. Benzekri, J.P. Marcerou, H.T. Nguyen and J.C. Rouillon, Phys. Rev. B **41**, 9032, (1990).
11. N.V. Madhusudana and R. Pratibha, Mol. Cryst. Liq. Cryst. **89**, 249 (1982).
12. A. Böttger and J.G.H. Joosten, Europhys. Lett. **4**, 1297 (1987).

Chapter 18

# DIELECTRIC AND HIGH PRESSURE STUDIES OF LIQUID CRYSTALS

R. SHASHIDHAR

## 1. Dielectric Studies

Dielectric studies probe the response of molecules to the application of an external electric field. In liquid crystals, the situation is quite complex: The materials exhibit anisotropy of dielectric permittivity while maintaining their fluidity. Also, the frequency response of the dielectric constants depends strongly on the relative direction of the electric field with respect to the long axis of the molecules. An investigation of the static permittivities gives important information on the dielectric anisotropy in relation to the molecular structure and dipole-moment. On the other hand, a study of the dielectric dispersion enables us to understand the different relaxation processes associated with the dipolar molecules of liquid crystals. However the interpretation of the dielectric dispersion data is difficult in view of the fact that the effects due to the nematic potential and viscosity are difficult to separate. Also, the activation energy calculated by the Arrehnius plots of the dielectric dispersion frequencies appear to be strongly influenced by short range order effects. This is particularly so when the range of the liquid crystalline phase is narrow and it is somewhat difficult to define a unique activation energy for a given liquid crystalline phase. In the present chapter, we shall discuss the behavior of static dielectric permittivity for different types of liquid crystals composed of nonpolar, weakly polar and strongly polar molecules. We shall also discuss how dielectric properties enable us to probe the nature of dipole-dipole interactions in liquid crystals.

### *1.1. Dielectric Anisotropy and Dipole Moment*

When an electric field is applied to materials consisting of non-polar molecules, only an induced polarization occurs. For static fields this consists of two parts: the electronic polarization and the ionic polarization. In materials composed of polar molecules, the dipoles orient parallel to the field and this results in orientational polarization in addition to the induced polarization. In liquids the dielectric permittivity is isotropic while in solids it is usually anisotropic. However, in solids the orientational polarization does not normally contribute significantly to the

R. Shashidhar - Center for Bio/Molecular Science & Engineering, Naval Research Laboratory, Code 6090, Washington, DC 20375-5000, USA; and Geo-Centers, Inc., 10903 Indian Head Highway, Fort Washington, MD 20744, USA.

*Phase Transitions in Liquid Crystals*, Edited by S. Martellucci and A.N. Chester, Plenum Press, New York, 1992

permittivity due to the lack of mobility of the molecules. On the other hand, in liquid crystals, the molecules retain their fluidity and hence the orientational polarization contributes significantly to the dielectric permittivity[1,2]. For a uniaxial liquid crystalline phase, i.e., a nematic phase or a smectic A phase, the principal dielectric constants are $\varepsilon_{\parallel}$ and $\varepsilon_{\perp}$, these being as the permittivities for measuring fields parallel and perpendicular to the uniaxial direction or director. The dielectric anisotropy $\Delta\varepsilon$ is defined as $\varepsilon_{\parallel}$- $\varepsilon_{\perp}$. In the case of a liquid crystal composed of non-polar molecules, the situation is relatively simple. The dielectric anisotropy is small (Fig. 1a). When a dipole moment is introduced into the molecule, it leads to rather large values of the dielectric anisotropy. The value of the dielectric anisotropy depends on the magnitude of the total dipole moment and on the angle between the dipole moment and the long axis of the molecule. In the case of a system which has a strong dipole moment like C ≡ N as a terminal group attached to one end of the molecule, $\Delta\varepsilon$ is large and positive[3,4] (Fig. 1b). On the other hand, if C ≡ N is attached as a side group, the net transverse dipole moment will be larger then the longitudinal component and the system will exhibit a large negative anisotropy[2] (see Fig. 1c). The anisotropy would also be affected if the bridging dipole moments are reversed, keeping the large terminal dipole group as the same. This effect has recently been demonstrated by a study of the dielectric permittivity of two isomeric cyano ester homologous series of compounds[5]. Both have the same terminal cyano group, but the ester linkages in the two series are oppositely disposed relative to the terminal end group.

For CPnOCB in which the bridging ester group is disposed additively relative to the C ≡ N group[6], the dielectric anisotropy is larger (by over a factor of 2) than that for nOPCB in which

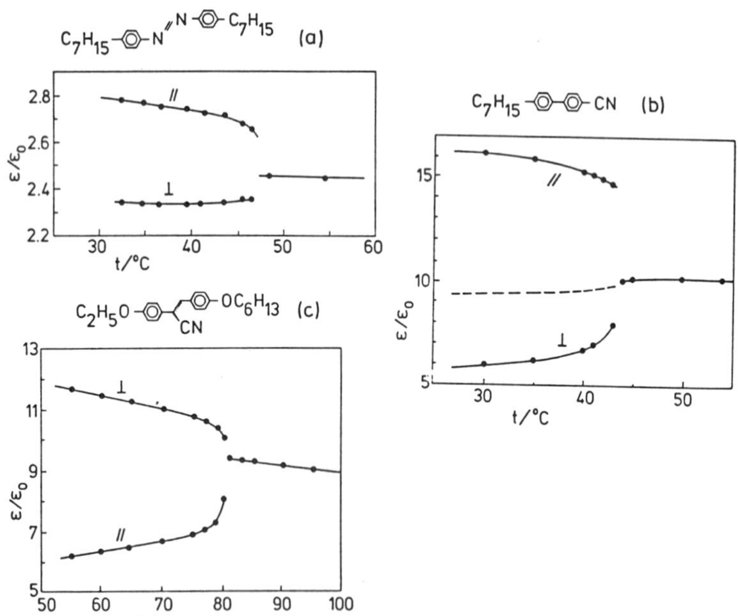


Fig. 1. Static dielectric permittivity in the nematic and isotropic phases of liquid crystals consisting of a) non-polar molecules b) molecules with a strongly polar group at on end and c) molecules with a strong transverse dipole. (Ref. 1)

the direction of the ester group opposes that of the end group[7] (see Figs. 2 and 3). Also, ε decreases with increasing chain length (n). This is essentially due to a decrease in $\varepsilon_{\parallel}$ ( $\varepsilon_{\perp}$ is not so prominently affected) indicating an increase of dipole-dipole correlations with increasing chain lengths.

The relation between the dielectric permittivity and the orientational order (S) of the medium has been discussed by Maier and Meier[8,9], who extended the Onsager theory to nematic liquid crystals. Their model satisfactorily accounts for many essential features of the permittivity of nematic liquid crystals. In the case of liquid crystals with a large asymmetry with regards to the disposition of the dipoles in the molecule, the situation gets more complex due to the internal field problem. This has been discussed extensively in literature by de Jeu[10] and by Kresse[11] and we shall not be discussing this aspect here.

## *1.2. Dipole-Dipole Interactions*

We shall now see how a study of dielectric properties enables us to probe the nature of dipole-dipole interactions in liquid crystals. The concept of near-neighbor antiferroelectric short range order was theoretically proposed[12] for materials with a large positive dielectric anisotropy,

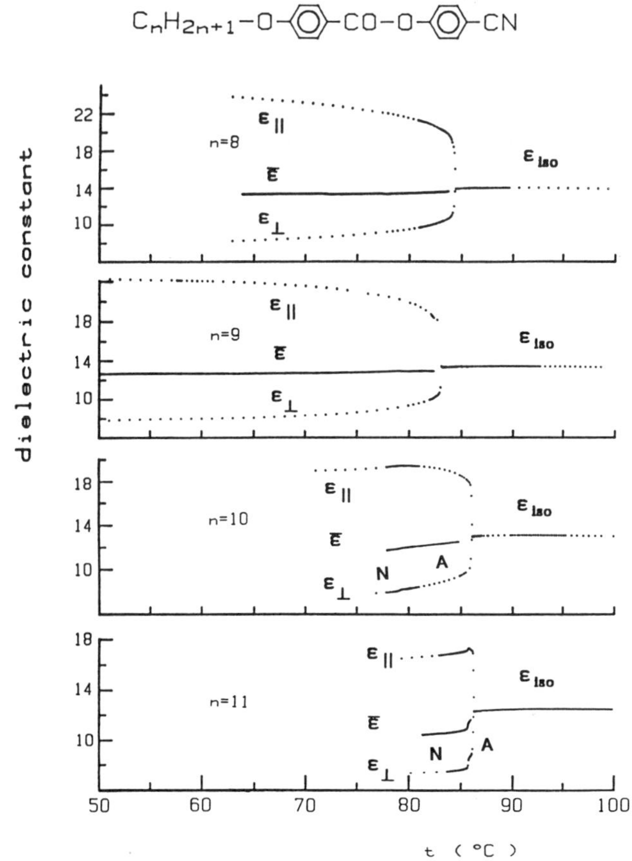


Fig. 2. Temperature dependence of the static dielectric permittivity of CPnOB series of materials. The molecular structure is also given. (Ref. 5)

i.e., for materials whose constituent molecules possess a strong terminal dipole moment (of the type shown in Fig. 1b). According to this model, two adjacent molecules prefer to have a configuration wherein the terminal dipoles overlap. (X-ray studies do confirm this molecular arrangement.) An important prediction of the theory is that throughout the nematic phase, the mean dielectric constant $\bar{\varepsilon} = (\varepsilon_{\parallel} + 2\,\varepsilon_{\perp}) / 3$ is less than the extrapolated isotropic value. Also, the theory predicts a sudden increase in $\bar{\varepsilon}$ at the nematic-isotropic transition. Both these effects were indeed observed soon after the theoretical model was proposed and have been confirmed in a wide variety of materials. A closer look at Figs. 2 and 3 illustrates this point. For all the materials, the $\bar{\varepsilon}$ value is less than the extrapolated $\varepsilon_{is}$ value. This difference in fact increases with decreasing temperature and is more pronounced in the smectic A phase as seen for the n = 10 and 11 homologs. Figs. 4a and 4b show the jump in $\bar{\varepsilon}$ at $T_{N\text{-}I}$ Clearly this jump is seen in all cases, the magnitude of the jump increasing with decreasing nematic range. Since it is known from independent studies that the decrease in the nematic range of these materials is accompanied by a pronounced increase in the smectic A-like local ordering of the molecules, we can infer that

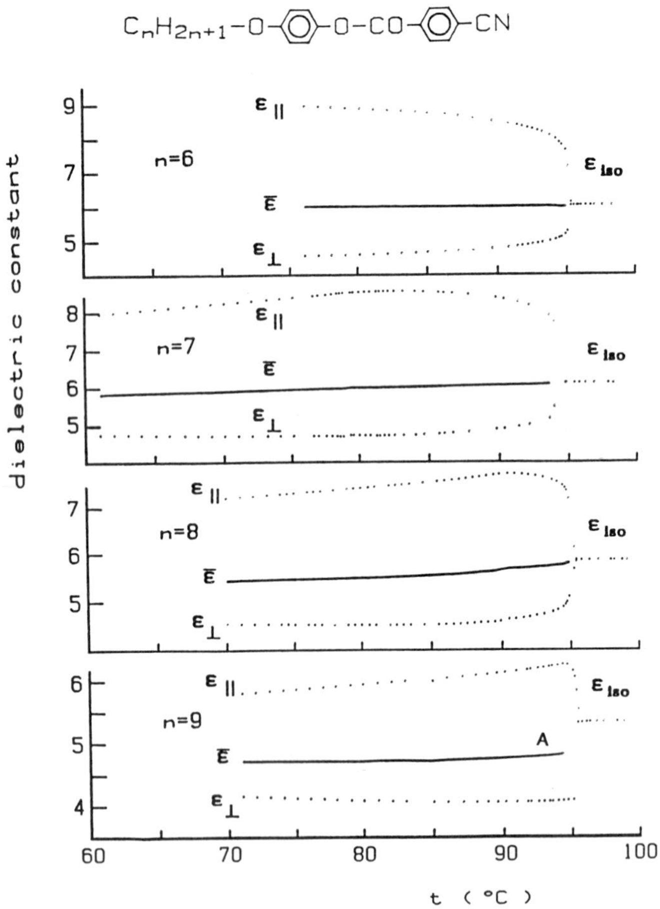


Fig. 3. Temperature dependence of the static dielectric permittivity of nOPCB series of compounds. The structure of the molecules is also shown. (Ref. 5)

the near-neighbor antiparallel correlations increase with increasing smectic ordering. This is also substantialized by the larger jump in $\bar{\varepsilon}$ observed for 9OPCB (Fig. 4b) at the direct smectic A-isotropic transition.

Another manifestation of the antiparallel dipole-dipole correlations in strongly polar groups was first demonstrated by Bradshaw and Raynes[14] and later confirmed by Thoen and Menu[15]. The dielectric permittivity in the isotropic phase, $\varepsilon_{is}$, was monitored as a function of temperature. It was found that on decreasing the temperature in the isotropic phase, $\varepsilon_{is}$ initially shows an increase which is in conformity with the behavior for dipolar liquids. However, as $T_{NI}$ is approached, $\varepsilon_{is}$ attains a maximum value and then decreases. This decrease in $\varepsilon_{is}$ is due to the building up of antiparallel correlations on approaching the nematic phase. Hence, dielectric studies can serve as a tool to probe pretransitional dipole-dipole correlation in the isotropic phase as well. This effect is shown in Figs 5a and 5b for both CPnOCB and nOPCB series of compounds. It is seen that although both the series show pre-transitional effects in the isotropic phase, the effect in nOPCB is much less pronounced than in CPnOB (note the differences in the scales in the Figs 5a and 5b). Hence the pre-transitional increase in $\bar{\varepsilon}$ in the nematic phase, as well as the pre-transitional decrease in $\varepsilon_{is}$ in the isotropic phase near $T_{NI}$, are much more pronounced for nOPCB, i.e., the materials for which the bridging dipole is additive with the end group. We can therefore infer that in the case of materials with a strong dipolar end group, larger the effective longitudinal dipole moment stronger are the antiparallel dipole dipole correlations both in the nematic and isotropic phases.

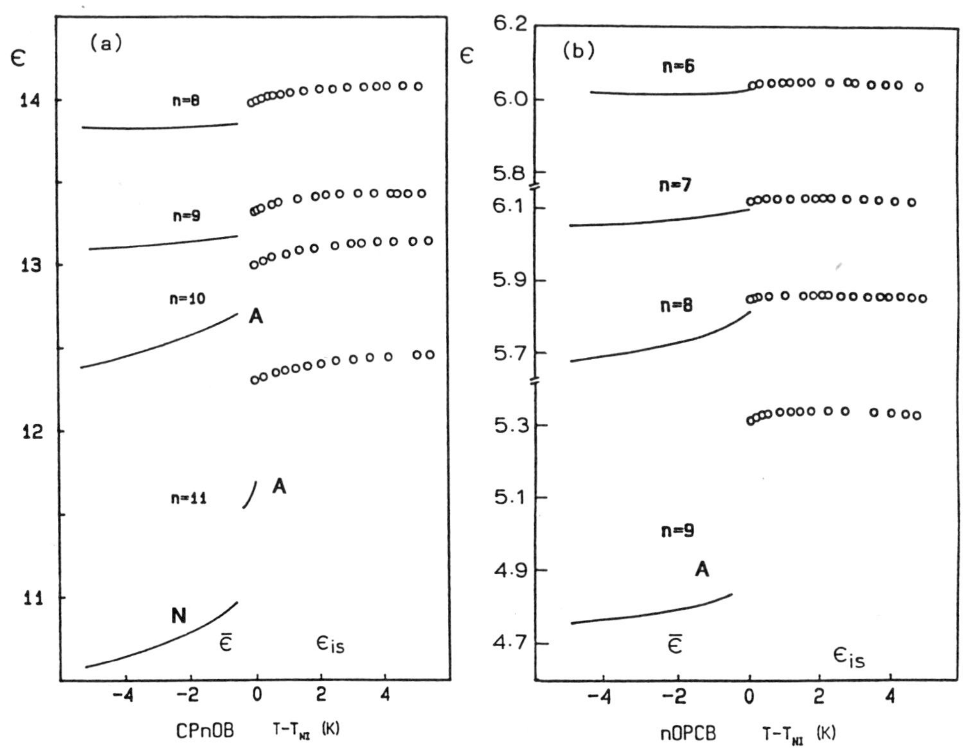


Fig. 4. Temperature variation of the average dielectric constant $\bar{\varepsilon} = (\varepsilon_{\parallel} + 2\varepsilon_{\perp})/3$ in the nematic phase and of the isotropic dielectric constant $\varepsilon_{is}$ for (a) CPnOB series and for (b) nOPCB series. (Ref. 5)

Recently similar pre-transitional effects due to antiparallel correlations of neighboring molecules have been observed[11] even in materials which do not possess a strongly polar terminal group. In this case, the compounds (called the "swallow-tail" compounds[17]) have two long "branched" chains at one end of the molecule. In the nematic phase, $\varepsilon$ is found to be significantly less than the extrapolated isotropic value and shows a jump at $T_{NI}$ Also, $\bar{\varepsilon}$ shows a pronounced decrease in the N phase. In fact, these antiparallel correlations are quite strong and are not affected easily by the addition of compounds with single chains.[18] The dipole-dipole correlations however are strongly reduced by the addition of compounds with lateral chain substituents.[19] It should be remarked that although these swallow-tailed compounds do not have a terminal cyano group, they do posses dipoles, both in the bridging group and in the branched chain. It is therefore not surprising that pronounced antiparallel dipole-dipole correlations should exist for these systems as well. It should also be recalled that Bradshaw and Raynes[14] had shown earlier that the pretransitional effects practically vanish for materials which have no effective molecular dipole moment. Thus it appears that antiparallel dipole correlations exist to some extent in all liquid crystal systems in which the molecules possess some net longitudinal dipole moment. Dielectric studies are useful to probe these dipole-dipole correlations in detail.

## 2. High Pressure Studies

Liquid crystals exhibit a rich variety of phase transitions.[20] A study of these transitions and the associated critical phenomena provides an insight into the understanding of other areas of condensed matter physics. High pressure is a very useful parameter in such a study since it permits us to probe (in single component systems) several types of critical points, some of which are rarely encountered experimentally in any other condensed systems. The first high pressure studies on liquid crystals were at the end of the last century[21], i.e., within a decade of the discovery of liquid crystals. However, during the next 75 years or so, only a few more pressure studies were conducted.[22] This is perhaps due to the difficulties associated with the design of the high pressure cells for liquid crystals. In fact, the design of the experimental set-up for high pressure investigations of liquid crystals presents certain difficulties which are not normally encountered in the study of solids under pressure. Liquid crystals are highly chemically active and react with most of the known pressure-transmitted media. It is therefore important to completely encapsulate the sample under study and at the same time ensure that the sample experiences hydrostatic pressure without any leaks. Also, the heat associated with the most of the phase transitions in liquid crystals are very small and hard to detect by conventional high pressure DTA apparatus. Finally, the transition temperatures are very sensitive to pressure and therefore a very precise control of pressure as well as temperature is required to monitor the phase transitions accurately. Two types of high pressure cells, viz., DTA[23] and optical[24], have been specially designed to overcome these problems. Such cells, capable of working up to pressures of about 7 kbar, have been successfully used to study a variety of liquid crystalline phase transitions at high pressure. These investigations have led to a number of new results of fundamental significance.[25] We shall briefly mention some of the early results and then discuss in some detail two of the most recent discoveries.

Perhaps the most significant of the early results is the observation of pressure-induced mesomorphism[26], i.e., the induction of liquid crystalline behavior at high pressure in materials which do not form liquid crystals at atmospheric pressure. This led to the first observation of triple points in single component systems. It was also found that in some other cases, pressure can suppress mesophases and this again leads to important consequences. If all the phase transitions involved are first order, a triple point results. On the other hand, if one or more of the transitions are second order, a variety of critical points can be observed. The potentialities for multicritical phenomena involving liquid crystalline transitions was realized with the observation of a tricritical point for the smectic A - nematic (or cholesteric) transition, the order of the transition changing from first to second at this point[27]. Since then, a number of other types of critical points have been found. In the next section, we shall describe the recent results on one of these, viz., and the nematic-smectic A-smectic C multicritical point.

## *2.1. The Nematic - Smectic A - Smectic C Multicritical Point*

The nematic-smectic A-smectic C multicritical point (or NAC point for short) is, by definition, a point at which three second order phase boundaries, viz., N - A, N - C and A - C, intersect. At this point all the phase transitions should be second order and the three phases become indistinguishable. (In contrast, at a triple point, the three phases merely coexist.) The existence of the NAC point was predicted theoretically by Chen and Lubensky[28] who expected it to be a type of Lifshitz point[29] defined earlier for magnetic systems. Soon afterwards, the NAC point was discovered experimentally by Johnson et al. and independently by Sigaud et al. in temperature-concentration (T-X) diagrams of binary liquid crystal systems[30]. This was followed by high resolution X-ray, calorimetric and light scattering studies which permitted explicit

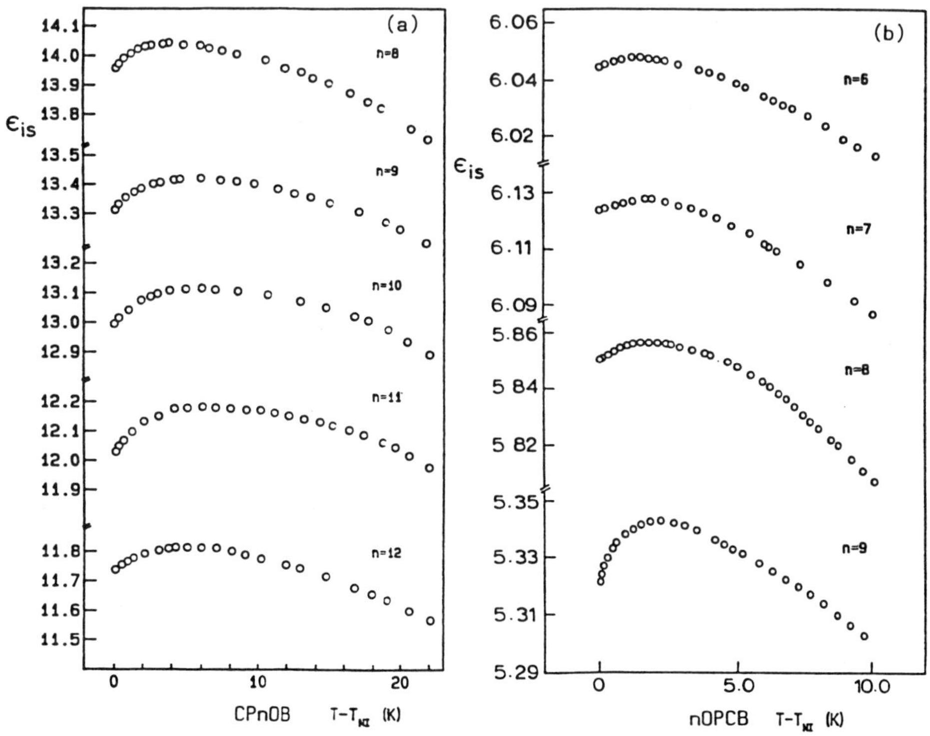


Fig. 5. An enlarged plot showing the variation of $\varepsilon_{is}$. Pretransitional effects are clearly seen in the isotropic phase. (Ref. 5)

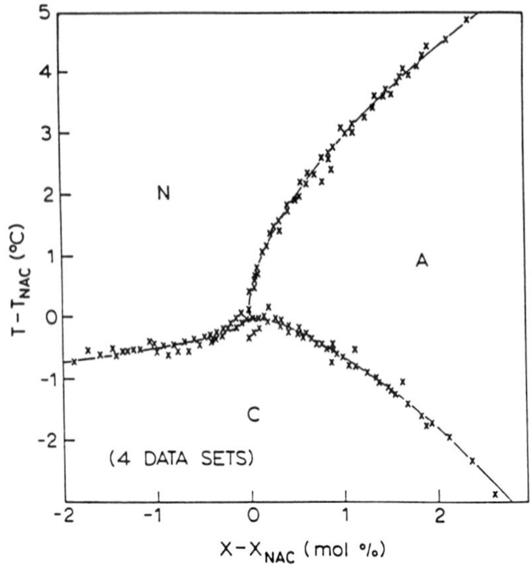


Fig. 6. The universal temperature concentration (T-X) plot of Johnson showing the data for 4 binary liquid crystal systems. (Ref. 31)

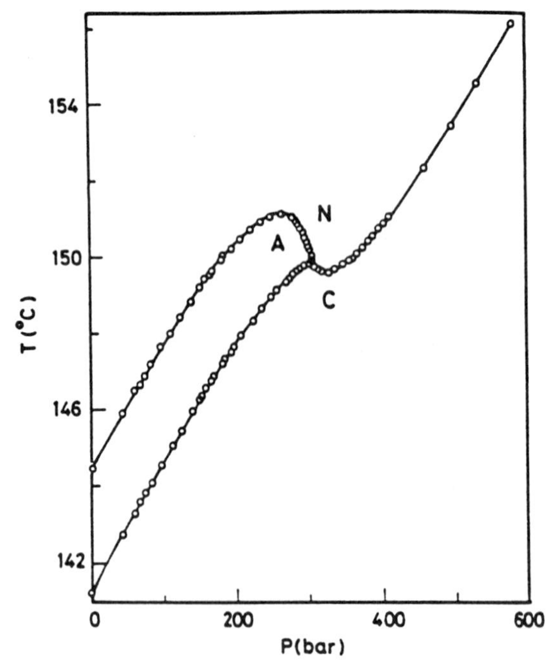


Fig. 7. The P-T diagram of 7APCBB showing the NAC multicritical point at 300 bar, 149.9°C. (Ref. 32)

comparison with theory. Johnson and his collaborators[30] obtained high resolution T-X diagrams for 4 binary systems exhibiting the NAC point. By analyzing the data in the close vicinity of the NAC point, they made an extremely important obseNation, viz., that the topology of the phase diagram near the NAC point exhibits universal behavior.[31] Fig. 6 gives the universal plot showing the data for all 4 binary systems. Since all these studies were on binary mixtures, the need arose to find the NAC point in a single component system (wherein concentration fluctuations do not play a role) so that the concept of universality can be proved unequivocally. This has been achieved[32] by the pressure studies on 7APCBB, i.e. 4-n-heptacylphenyl-4′-(4″cyanobenzoyloxybenzoate), a material which exhibits N, A and C phases at 1 bar.

The complete P-T diagram of 7APCBB is shown in Fig. 7. The NAC point is seen at 304 bar, 149.9°C. A feature of the P-T diagram is that the A-N and A-C boundaries are initially linear, but curl up dramatically as the NAC point is approached. The N-C boundary also shows the same behavior and hence pronounced singularities are seen for all the phase boundaries near the NAC point.

In order to analyze the shape of these singularities, high resolution P-T data were collected in the immediate vicinity of the NAC point. These data, obtained with a precision of ≈0.1 bar and ≈10mK, are shown in Fig. 8. It is clear that this diagram is strikingly similar to the universal T-X diagram of Fig. 6. It has also been shown that the exponents characterizing the phase boundaries of Figs. 6 and 8 are identical (see Refs. 31 and 32 for the expressions used in this analysis). Thus it has been established that the NAC multicritical point exhibits universal behavior.

Finally, it must also be mentioned that although the occurrence of a Lifshitz point in magnetic systems was theoretically predicted by Hornreich, Luban and Shtrikman as early as in 1975, most of the subsequent experimental efforts to observe such a magnetic Lifshitz point proved to be unsuccessful. So far, the existence of a Lifshitz point appears to have been established[33] in only a single instance, viz., MnP. The observation of the NAC point is therefore extremely important and studies on systems exhibiting such a point should lead to a better understanding of the Lifshitz point.

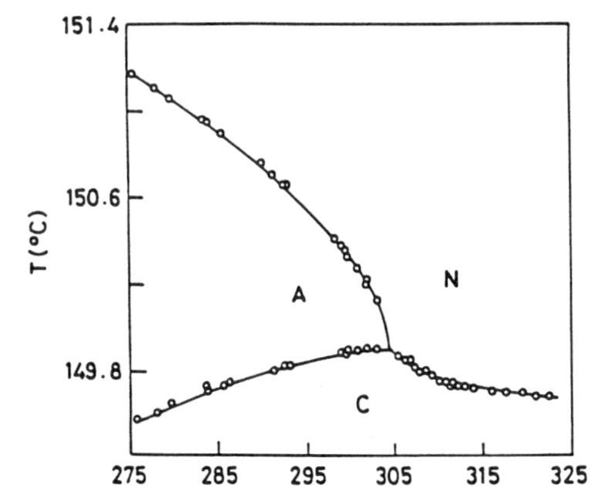


Fig. 8. High resolution P-T diagram in the vicinity of the NAC point in 7APCBB. The topology of this diagram is exactly similar to that of the T-X diagram shown in Fig. 5. (Ref. 32)

## *2.2. The Reentrant Nematic Phenomenon*

The binary mixtures of p-[p-hexyloxy-benzylidene)amino]benzonitrile (HBAB) and N-p-cyanobenzylidene-p-n-octoyloxyaniline (CBOOA) have been found[34] to show the following unusual sequence of transitions on cooling from the isotropic (liquid phase):isotropic - nematic - smectic A - nematic - solid.The second or lower temperature nematic phase has been designated as the "reentrant nematic" phase. Such instances of reentrance of a phase with high symmetry (or lower order) at temperatures lower than that at which a phase of lower symmetry (or more order) occurs are known in other areas of condensed matter physics, e.g., pressure dependence of one of the melting transitions in $He^3$ and the reentrant normal phase in superconductors doped with Kondo impurities[35,36]. The latter example is of particular interest since the analogy between smectic A and a charged superconductor is well known[37]. In the case of superconductors, the experimental observation of the reentrant normal phase followed a theoretical prediction[38] made earlier. In the case of liquid crystals, however, the occurrence of the reentrant nematic phase was not anticipated theoretically.

Soon after the observation of the $N_{re}$ phase in binary mixtures at 1 bar, Cladis et al.[39] also found a similar behavior in a single component system at high pressures. The P-T diagram of this material (8OCB) reproduced in Fig. 9 shows that in the pressure range of 1.6-1.8 kbar, the reentrant nematic phase is obtained on cooling the A phase. Since then, $N_{re}$ has been seen in single component systems at 1 bar as well[40] and this has led to a variety of experimental investigations including high pressure studies. The PT diagrams of a number of substances

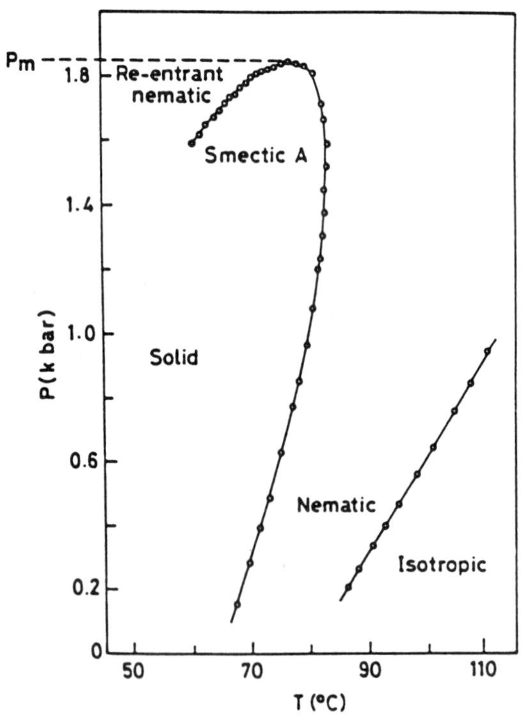


Fig. 9. P-T diagram of 80CB showing the reentrant nematic phase. $P_m$ is the maximum pressure up to which the smectic A phase exists. (Ref. 34)

exhibiting the $N_{re}$ phase at 1 bar have been studied and the following features, which are common to all the phase diagrams, have been observed[41,42].

i) the shape of the A-N phase boundary is elliptic,

ii) there is a maximum pressure ($P_m$) beyond which the A phase does not exist,

iii) $P_m$ is uniquely related to the temperature range (R) of the nematic phase at 1 bar by the relation $P_m = P_0\exp(-mR)$, where $P_0$ and m are constants which depend only on the number of benzene rings in the molecular structure, and,

iv) the tilt of the major axis of the elliptical A-N boundary is exactly the same as that of the linear nematic-isotropic phase boundary. This has been quantitatively verified for widely different tilts (see Figs. 10 and 11 ) showing thereby that coupling between smectic and nematic ordering manifests dramatically in the phase diagrams of reentrant systems.

As mentioned earlier, the occurrence of the reentrant normal phase in superconductors was first theoretically predicted soon observed experimentally. The possibility of a second reentrant phase, viz., a reentrant superconducting phase, occurring at lower temperatures was theoretically predicted. But such a phase does not seem to have been observed so far. In the case of liquid crystals however, not only doubly reentrant behavior[43], but even multiple reentrant phenomena have been found[44,45] recently. The material DB9.O.$NO_2$ exhibits (see Refs. 44 and 45) perhaps the richest variety of liquid crystalline phase transitions known to-date:

$$I - N - A_d - N_{re} - A_{d(re)} - N_{re} - A_1 - \tilde{C} - A_2 - C_2.$$

Here $A_d$, $A_1$ and $A_2$ are three polymorphic forms of the A phase which are characterized by a layer spacing (d) in relation to the molecular length ($l$); $d \cong l$ for the monolayer ($A_1$) phase, $d \cong 2l$ for the bilayer ($A_2$) phase while $l < d < 2\,l$ for the partially bilayer $A_d$ phase. ($\tilde{C}$ and $C_2$ are two forms of the C phase whose structures are described in Ref. 46) It is seen that the nematic phase appears thrice while the $A_d$ phase occurs twice and hence DB9.O.$NO_2$ is referred to as a multireentrant system. The phase diagram of this compound is shown in Fig. 12. Both the $A_d$ phases are completely counted in the P-T plane, the A-N boundaries being elliptically shaped as in singly reentrant systems discussed earlier. The delicate role that pressure plays in the occurrence of reentrant phase is demonstrated b the P-T diagram o DB1O.O.$NO_2$ (Fig. 13).

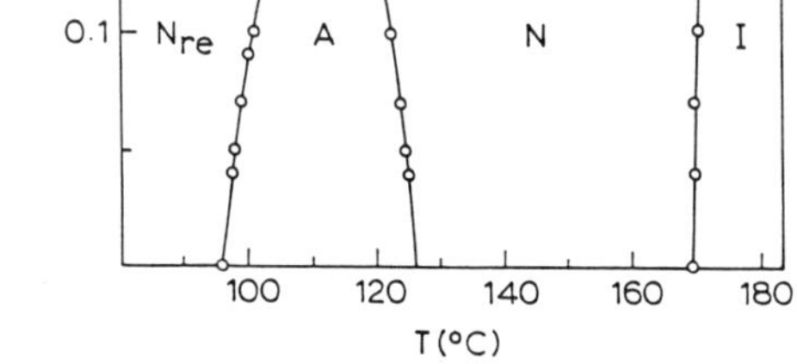


Fig. 10. P-T diagram of 11CPMαMCB (from Ref. 41). The solid lines are computer fits of the A-N and N-I data to equations to an ellipse and a straight line respectively. The major axis of the elliptic A-I boundary is seen to be parallel to the N-I line.

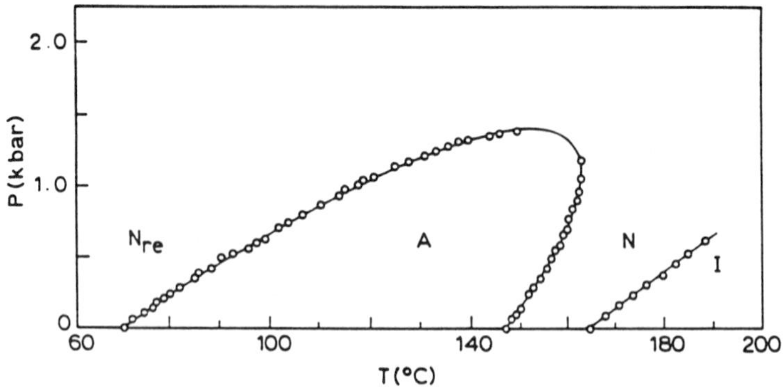


Fig. 11. P-T diagram of 12CPMαMCB (from Ref. 41). See also legend of Fig. 10a.

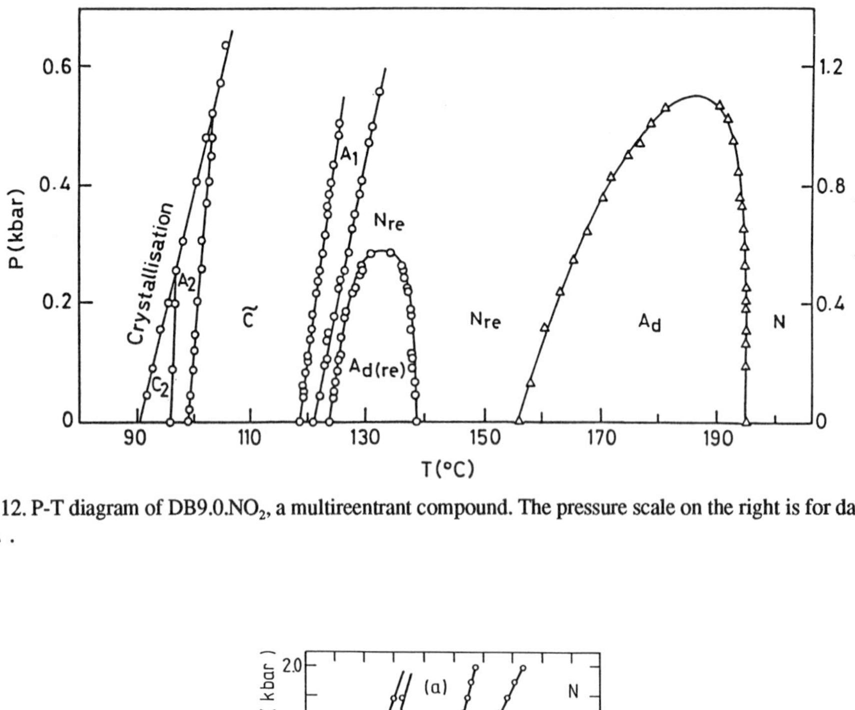


Fig. 12. P-T diagram of DB9.0.$NO_2$, a multireentrant compound. The pressure scale on the right is for data shown as △ .

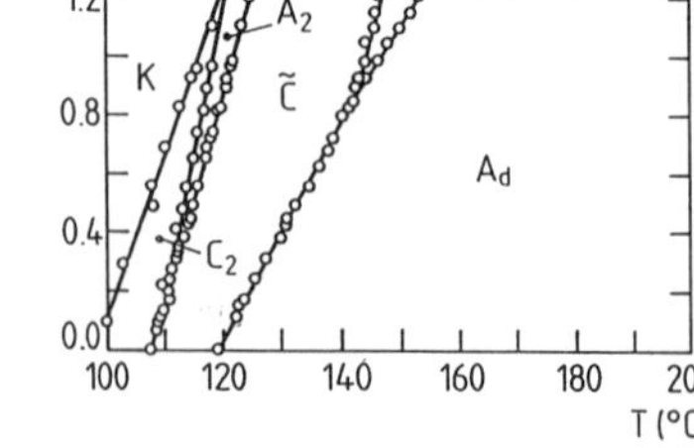


Fig. 13. P-T diagram of DB10.0.NO2, showing a multireentrant behavior in the pressure range 1.26-1.38 kbar.

This material, which is the next higher homolog of DB9.O.$NO_2$ exhibits only one $A_d$ phase at 1 bar. But over a small pressure range, viz., between 1.26 and 1.38 kbar, multireentrant behavior is seen.

Why does the nematic phase reenter? The origin of reentrant behavior has been the subject of several theoretical studies - phenomenological as well as molecular approaches. The thermodynamic approaches[47,48] yielded good fits of the experimental P-T data (for the A-N transition) to an ellipse. However, these treatments assume the A-N transition to be first order throughout the P-T plane while experimentally the transition has been found to be always second order[49]. The phenomenological theory due to Pershan and Prost[50] assumes that the smectic A phase can exist only near an "optimum density". Assuming that there are no pre-transitional anomalies near the A-N transition, the model predicts that the axis of the parabolic shaped A-N boundary should be parallel to lines of constant density or isochores. To test this prediction, a detailed PVT study of 8OCB was taken up[51]. It was seen that while there are pronounced pretransitional effects as well as a volume discontinuity at the N-I transition, no such effects are seen near the A-N transition. The lack of density anomalies in the vicinity of the A-N transition permits us to test the optimum density theory. Using the data obtained from different isotherms, lines of constant density have been evaluated (Fig. 14). It was found that these isochores are all parallel to each other, but not parallel to the axis of the A-N boundary. Hence, the optimum density model does not quantitatively describe reentrant behavior.

Amongst the molecular approaches to explain reentrant behavior, the most successful one appears to be the frustrated spin-gas model of polar liquid crystals[52,53]. According to this model, microscopic frustration of dipoles together with competing short-range antiferroelectric and ferroelectric orders can account for the observed polymorphism of reentrance in polar compounds. The model predicts phase diagrams involving one or more types of reentrant behavior and these diagrams are in good agreement with the experimental diagrams. It also predicts that the multiple reentrant phenomena should be very sensitive to the number of carbon atoms in the alkyl chain of the molecule. This is indeed the case experimentally also.

ACKNOWLEDGEMENT. Technical editing by Ms. Shelley DeLozier is greatly appreciated.

## References


1. F. Vertogen and W.H. de Jeu, in "Thermotropic Liquid Crystals", Springer-Verlag, p. 190 (1988).
2. W.H. de Jeu, in "Physical Properties of Liquid Crystalline Materials", Gordon and Breach, New York, (1980).
3. B.R. Ratna, M.S. Vijaya, R. Shashidhar and B.K. Sadashiva, Pramana Suppl. **1**: 69 (1973).
4. D. Lippens, J.P. Parneix and J.P. Chapoton, J. de Physique, **38**: 1665 (173).
5. G. Heppke, S. Pfeiffer, C. Nagabhushan and R. Shashidhar, Mol. Cryst. Liq. Cryst., **170**: 89 (1989).
6. A. Gobl-Wernsch, F. Heppke and R. Kopf, Z. Naturforsch., **36a**: 213 (1981).
7. J.W. Goodby, T.M. Leslie, P.E. Cladis, and P.L. Finn in "Liquid Crystals and Ordered Fluids", vol. 4, Gordon and Breach, New York, p. 203 (1982).
8. W. Maier and G. Meier, Z. Naturforsch., **16a**: 262 (1961).
9. W. Maier and G. Meier, Z. Naturforsch., **16a**: 470 (1961).
10. W. de Jeu in "Liquid Crystals", Ed. L. Liebert, Solid State Phys. Suppl. 14, Academic Press, New York, p. 109 (1978).
11. H. Kresse, Forschritte der Physik, **30**: 507 (1982).
12. N.V. Madhusudana and S. Chandrasekhar, Pramana Suppl. **I**, 225 (1973).
13. See, e.g.: P.S. Pershan, in "Structure of Liquid Crystal Phases", World Scientific, p. 35 (1988).
14. M.J. Bradshaw and E.P. Raynes, RSRE Newsletters, Res. Rev., **5**: 26 (1981).
15. J. Thoen and G. Menu, Mol. Cryst. Liq. Cryst., **97**: 163 (1983).
16. C. Nagabhushana, B.R. Ratna, R. Shashidhar, S. Chandrasekhar, H. Kresse and W. Weissflog, Mol. Cryst. Liq. Cryst. Lett., **5**: 87 (1988).
17. W. Weissflog, A. Wiegeleben, S. Diele and D. Demus, Cryst. Res. and Tech., **19**: 583 (1984).

18. H. Kresse, W. Weissflog, C. Nagabhushan and R. Shashidhar, Phys. Stat. Sol. (a),**101**: K77 (1987).
19. H. Kresse and P. Rubenstein, Phys. Stat. Sol. (a), **100**: K83 (1987).
20. S. Chandrasekhar, "Liquid Crystals", Cambridge University Press (1977).
21. Hulett, G.A., Z. Phys. Chem., **28**: 629 (1989).
22. For a review of high pressure studies till 1978, see S. Chandrasekhar and R. Shashidhar, Advances in Liquid Crystals, **4**: 83 (1979).
23. A.S. Reshamwala and R. Shashidhar, J. Phys. E., **10**: 183 (1977).
24. A.N. Kalkura, R. Shashidhar and M.S. Urs, J. de Phys., **44**: 51 (1983).
25. S. Chandrasekhar and R. Shashidhar, Advances in Liquid Crystals, **4**: 83 (1979).
26. S. Chandrasekhar, S. Ramaseshan, A.S. Reshamwala, B.K. Sadashiva, R. Shashidhar and V. Surendranath, Pramana Suppl., **1**: 117 (1973).
27. P.H. Keyes, H.T. Weston, and W.B. Daniels, Phys. Rev. Lett., **31**: 628 (1973); R. Shashidhar and S. Chandrasekhar, J. Phys. (Paris), **36**: C149 (1975).
28. J. Chen and T.C. Lubensky, Phys. Rev., A**14**: 1202 (1976).
29. M. Hornreich, M. Luban, and S. Shtrikman, Phys. Rev. Lett., **35**: 1678 (1975).
30. For the early work on NAC Systems, see e.g.: D.L. Johnson, J. Chim. Phys., **80**: 45 (1983).
31. D. Brisbin, D.L. Johnson, H. Fellner and M.E. Neubert, Phys. Lett., **50**: 178 (1983); D.L. Johnson (private communication).
32. R. Shashidhar, B.R. Ratna and S. Krishna Prasad, Phys. Rev. Lett., **53**: 2141 (1984).
33. See, e.g.: Y. Shapira, in "Multicritical Phenomena", Ed: R. Pynn and A. Skjeltrop, NATO Advanced Science Institute, Series B: Physics, Vol. 6 (Plenum), p. 129 (1986).
34. Cladis, P.E., Phys. Rev. Lett., **35**: 48 (1975).
35. A.C. Anderson, W. Reese, and J.C. Wheatley, Phys. Rev., **130**: 1644 (1963).
36. G. Riblet, K. Winzer, Solid State Commun., **9**: 1663 (1971).
37. W.L. McMillan, Phys. Rev A**4**: 1238 (1971); P.G. de Gennes, Solid State Commun. **10**: 753 (1972).
38. E. Muller-Hartmann and J. Zittartz, Phys. Rev. Lett., **26**: 428 (1971).
39. P.E. Cladis, R.K. Bogardus, W.B. Daniels and G.N. Taylor, Phys. Rev. Lett., **39**: 720 (1977).
40. For a recent review on materials exhibiting the $N_{re}$ phase, see : Tinh, Nguyen Huu, J. Chim. Phys., **80**: 83 (1983).
41. A.N. Kalkura, R. Shashidhar and M.S. Urs, J. Phys. (Paris), **44**: 51 (1983).
42. R. Shashidhar, S. Somasekhar and B.R. Ratna, Mol. Cryst. Liq. Cryst., **133**: 19 (1986).
43. F. Hardouin, G. Sigaud, M.F. Archard and H. Gasparoux, Phys. Lett., **71A**: 347 (1979).
44. Tinh, Nguyen Huu, F. Hardouin and C. Destrade, J. Phys. (Paris), **43**: 1127 (1982).
45. R. Shashidhar, B.R. Ratna, V. Surendranath, V.N. Raja, S. Krishna Prasad and C. Nagabhushana, J. Phys. (Paris) Lett., **46**: L445 (1985).
46. F. Hardouin, A.M. Levelut, M.F. Achard and G. Siguad, J. Chim. Phys., **80**: 53 (1983).
47. N.A. Clark, J. Phys (Paris), **40**: C3345 (1979).
48. D.D. Klug and E. Whalley, J. Chem. Phys., **71**: 1874 (1979).
49. G.B. Kasting, K.J. Lushington and C.W. Garland, Phys. Rev., B**22**: 321 (1980).
50. P.S. Pershan and J. Prost, J. Phys. (Paris) Lett., **40**: L27 (1979).
51. R. Shashidhar, P.H. Keyes and W.B. Daniels, Mol. Cryst. Liq. Cryst. Lett., **3**: 169 (1986).
52. J.O. Indekeu and A.N. Berker, J. Phys. (Paris), **49**: 353 (1988) and references cited therein.
53. V.N. Raja, B.R. Ratna, R. Shashidhar, G. Heppke, ChBahr, J.F. Marko, J.O. Indekeu and A.N. Berker, Phys. Rev. A, **39**: 4341 (1989).

Chapter 19

# MOLECULAR DYNAMICS IN NEMATIC PHASES: RAMAN AND INFRARED SPECTROSCOPY

M.P. FONTANA

## 1. Introduction

In this Chapter we shall describe the use of vibrational spectroscopy to study molecular reorientational dynamics in liquid crystalline mesophases, with particular emphasis on the temperature dependence as the nematic-isotropic (N-I) phase transition is approached.

Reorientational Brownian motion results from the random torques which molecules exert on each other as they diffuse through the liquid, and it provides an additional channel for damping of the vibrational mode excited in Raman scattering (RS) or in Infrared (IR) absorption. Thus the vibrational bandshape will contain information not only on vibrational dephasing and anharmonicity, but also on reorientational relaxation.

This particular application of vibrational spectroscopy is part of what is called fluctuation spectroscopy, and has been extensively used to study molecular dynamics in simple molecular fluids[1]. In the case of complex fluids such as liquid crystals the application is rather difficult, since even the simplest mesomorphic molecules have molecular weights of at least 200. Thus the vibrational spectra are generally complex and difficult to interpret, and the relaxation mechanisms reflect the complexity and plurality of intra- and intermolecular interactions. Furthermore, whereas the isotropy of simple fluids allows many simplifications in the theoretical analysis of the experimental data, the large orientational anisotropy which characterizes the mesomorphic state requires the full use of irreducible tensor formalism[2] to connect the experimentally determined correlation functions or relaxation times to the dynamical parameters which characterize the stochastic reorientations of the molecules.

Thus the potential information obtainable by vibrational spectroscopy is restricted both by experimental difficulty and theoretical complexity. However, in favourable cases and using particular care much information can be obtained by fluctuation spectroscopy concerning molecular dynamics in aligned mesophases. This Chapter will be devoted mainly to the application of Raman and IR spectroscopy and will be based mainly on the work our group has done on this subject. We shall not discuss the fundamentals of vibrational spectroscopy nor shall we dwell on vibrational mode assignments; for this refer the reader to the specialized literature[3].

M.P. Fontana - Dipartimento di Fisica, Università di Parma, Italy.

*Phase Transitions in Liquid Crystals*, Edited by S. Martellucci and A.N. Chester, Plenum Press, New York, 1992

## 2. Determination of Correlation Functions from IR and Raman Lineshapes

### *2.1. Raman and IR Fluctuation Spectroscopy*

The fundamental quantity which is measured in a spectroscopic experiment is the spectral density I(ω) of the transmitted (IR) or scattered (Raman) radiation. Such quantity is readily obtained from the experimental data. For the IR case for instance, I(ω) is given by:

$$I(\omega) = \frac{\sigma(\omega)n(\omega)}{\omega[1-\exp(-\beta\omega)]} \tag{1}$$

where, $\beta = \hbar/kT$, $\sigma(\omega)$ is the absorption coefficient, $n(\omega)$ the index of refraction, and the term in the square brackets is the statistical occupation factor of the mode of frequency ω. $I(\omega)$ is directly proportional to the transition probability for the change $n \rightarrow m$ in the vibrational quantum number, due to the absorption of a photon of frequency $\omega=\omega_{nm}$. This transition probability may in turn be expressed in terms of the time autocorrelation function (ACF) of the transition dipole moment $\vec{M}(t)$. The time dependence in $\vec{M}$ *is* due not only to the vibrational motion, but also to the rotational and translational molecular brownian motion; the contributions of these latter effects to the time dependence of $\vec{M}$ justify the name "fluctuation spectroscopy" given to the particular applications discussed here. We have:

$$<\vec{M}(t)\vec{M}(0)> = \int_{-\infty}^{+\infty} d\omega e^{-i\omega t} I(\omega) \tag{2}$$

that is, the spectral density of a vibrational bandshape and the ACF of the fluctuating dynamical variable are mutual FT's (Fourier tranforms). This is a particular case of the very general Wiener-Khintchine theorem[4].

Expanding $\vec{M}(t) = \vec{M}[Q(t)]$ in terms of the normal coordinates $Q^{\nu}$ and considering only the linear term, we have, for absorption due to the ν-th vibrational mode:

$$I^{\nu}(\omega) = FT < \vec{m}^{\nu}(t)\vec{m}^{\nu}(0) >$$

where $m^{\nu} = (\partial\vec{M}/\partial Q^{\nu})_0$. Now the vector $\vec{m}^{\nu}(t)$ may be separated into an angle-dependent component - the unit vector $\vec{\mu}^{\nu}(t)$ and an amplitude factor which is proportional to the normal coordinate $Q^{\nu}$(t). Assuming the *statistical independence* of vibrational and reorientational relaxation, we have:

$$I_i^{\nu}(\omega) = FT < \vec{m}^{\nu}(t)\vec{m}^{\nu}(0) >< Q^{\nu}(t)Q^{\nu}(0) > \tag{3}$$

where i = x,y,z refer to the components of $\vec{\mu}^{\nu}$ in the molecular frame.

Thus in an isotropic fluid the FT of the absorption bandshape can yield directly only the product of the vibrational and reorientational ACF's. This is a great handicap of IR spectroscopy in the study of molecular reorientations.

In the case of light scattering, the spectral density is proportional to:

$$I_{ij}(\omega) = FT < \alpha_{ij}^{A}(t)\alpha_{ij}^{B}(0) > \tag{4}$$

where i,j are the polarization indices, and A and B refer to the molecules. Expanding $\{\alpha\}$ in terms of the normal coordinates we have:

$$< \alpha_{ij}^{A}(t)\alpha_{ij}^{B}(0) > = < \alpha_{ij}^{oA}(t)\alpha_{ij}^{oB}(0) > + < \alpha_{ij}^{v}(t)\alpha_{ij}^{v}(0) >< Q^{v}(t)Q^{v}(0) > \tag{5}$$

where the first term yields Rayleigh and Rayleigh wing scattering, connected to the fluctuations of the average value of the polarizability; note its sensitivity to pair correlations. Such sensitivity disappears in the second, Raman term. In this case (similar considerations apply to high frequency IR absorption) in fact the coherence of pair correlations between molecules is destroyed by the vibrational term, the phase of which varies randomly from molecule to molecule. Thus the $A \neq B$ terms average to zero; this means that Raman scattering, and IR absorption, are sensitive only to the self part of the correlation functions.

Using the Raman term in Eq.5 we have:

$$I_{ij}^{v}(\omega) = FT < \beta^{v}(t)\beta^{v}(0) + \frac{1}{9}\overline{\alpha}^{2} >< Q^{v}(t)Q^{v}(0) > \tag{6}$$

where we have used the standard decomposition of the symmetric polarizability tensor into an isotropic (Trace $\{\alpha\} = \overrightarrow{\alpha}$ and anisotropic ($\{\beta_{ij}\}$) components.

Taking the Raman spectrum of a given mode in the polarized (VV) and depolarized (VH) geometries, the isotropic and anisotropic contributions may be separated (see details in Ref.5a):

$$\begin{aligned} FTI_{iso}^{v}(\omega) &= < Q^{v}(t)Q^{v}(0) > \\ FTI_{VH}^{v}(\omega) &= < \beta_{ij}^{v}(t)\beta_{ij}^{v}(o) >< Q^{v}(t)Q^{v}(0) > \end{aligned} \tag{7}$$

where we have chosen to write the FT of the experimental quantity, to emphasize that from such treatment of the data the autocorrelation function may be obtained directly. It is now straightforward to write the reorientational ACF:

$$C^{or}(t) = < \beta_{ij}^{v}(t)\beta_{ij}^{v}(0) > = \frac{FTI_{VH}^{v}(\omega)}{FTI_{iso}^{v}(\omega)} \tag{8}$$

Let us recall that since the trace of the Raman tensor corresponding to mode $Q^{v}$ must be non-vanishing, only totally symmetric vibrations can be used for this type of application. Another condition implicit in Eq.8 is the separability of rotational and vibrational relaxation. In case of strong cross-correlations the isotropic component may actually relax *more rapidly* than the anisotropic one; this would lead to a ACF which would increase exponentially in time, and this is not possible of course.

We conclude by pointing out that Eq.8 can be reformulated in the frequency domain using the convolution theorem of FT's:

$$I_{VH} = \int L(\omega')I^{v}_{iso}(\omega - \omega')d\omega' \tag{9}$$

where $L(\omega) = FTC^{or}(t)$. Thus a simpler way of analysing the bandshape may be to fit the experimental VH lineshape with the convolution of the isotropic component, using an assumed, parametrized form of $C^{or}(t)$. For instance, in the commonest case,

$$C^{or} \approx \exp(-t/\tau_{or})$$

hence $L(\omega)$ would be a Lorentzian, and the parameters of the fit would be the amplitude and the halfwidth (FWHM = Full Width at Half Maximum). Naturally this simpler method restricts the available options since it is based upon an assumed form of $C^{or}(t)$.

## 3. Reorientational Dynamics in Liquid Cristal

### *3.1 Preliminary considerations*

In fluctuation RS and IR spectroscopy the vibrational line frequency acts as a carrier frequency, which may be modulated by the rotational motion. Since the motion is random, the result will be a broadening of the spectral distribution of the vibrational line. In order to visualize the basic mechanism, we shall start by discussing the simpler case of IR absorption.

Let us consider a totally symmetric, strongly longitudinally polarized ($R \approx 6$ for instance, where $R$ is the dichroic ratio) mode. This could be the benzene ring stretching vibration, located at $\approx 1170\ cm^{-1}$ in the tolane derivatives. Such a mode will generate an oscillating dipole polarized along the main molecular axis. In a strongly anisotropic fluid such as a liquid crystal, the molecular diffusivity tensor can be reduced to the *spinning* ($D_{\parallel}$) and *tumbling* ($D_{\perp}$) diffusion coefficients. It is easy to see that such a mode can only be influenced by the tumbling motion of the molecular axis: in fact only such motion will induce an extra time dependence on the induced dipole moment $\mu^{v}$ as seen from the laboratory frame of reference.

A weakly polarized totally symmetric mode, such as for instance the oxygen-alkyl tail stretching band at $1047\ cm^{-1}$ in the tolane derivatives, corresponds instead to an induced dipole moment pointing away from the molecular axis ($\theta \approx 55°$ for the $1047\ cm^{-1}$ mode). In this case both the spinning and the tumbling motions will modulate the oscillating dipole in the laboratory frame of reference. However, due to the large difference in the relaxation times (spinning being much faster than tumbling), the spinning contribution to the bandshape will dominate. Thus by choosing the appropriate vibration, it is possible to probe only tumbling or only spinning reorientational motion. It is important to remark here that this is possible only in aligned anisotropic fluids which may have dichroic spectra. In an isotropic fluid, $I_i = I_j$, and therefore the reorientational and vibrational correlations cannot be separated by taking the ratio of their FT's.

The case of Raman spectroscopy is more complex, because the quantity which is modulated by the vibrations is a tensor, and therefore there is no a-priori easy way of connecting the type of vibration with the direction of the resulting induced dipole moment. Considerable simplification may be achieved if the polarizability modulation is strongly anisotropic. In this case only one tensor element will be non-negligible (say, the $\alpha_{zz}$), and in this sense the situation will be similar to the IR case. Actually, so far only strongly uniaxial Raman modes have been identified and used for this application.

We refer in particular to the stretching C≡N (cyanobiphenyls) and C≡C (tolane derivatives) vibrations in the 2220 $cm^{-1}$ range[5].

Although Raman spectroscopy is more difficult to apply, both conceptually and technically, it is also more general and more informative than IR spectroscopy. More general, because its tensorial character allows the study of reorientational dynamics in any molecular fluid; and more informative, because from the study of the polarization of Raman lines both the second and fourth moment of the orientational distribution function may be obtained (the so-called $< P_2 >$ and $< P_4 >$ order parameters). IR spectroscopy can yield reorientational information only in very indirect ways; it works at the same level as Raman spectroscopy only for ordered, anisotropic fluids, such as aligned liquid crystals, which however cannot be studied in their isotropic phase; furthermore, from the value of the IR dichroism only the order parameter $< P_2 >$ may be obtained.

In both Raman and IR spectroscopies the basic problem is the isolation of the rotational contribution to the bandshape in the presence of a much stronger vibrational broadening. For instance, for spinning motion the reorientational broadening at most reaches 3 $cm^{-1}$, whereas the total bandwidth can be as high as 20 $cm^{-1}$. The situation is even worse for the slower tumbling motion, where the broadening is always smaller than $\approx 0.6$ $cm^{-1}$, for total widths of up to $\approx 8$ $cm^{-1}$. For this and more general reasons, which shall presently be discussed, it is always advisable to determine reorientational relaxation times through the determination of the corresponding time ACF's, which may be obtained by FT of the vibro-reorientational bandshape.

Experimentally, for both the IR and the Raman case, the major requirement is the availability of well shaped single bands which are sufficiently isolated in position from other bands (a separation corresponding to about 10 times the bandwidth is deemed necessary for a reasonably accurate FT). This may not be easy, considering the complexity of the spectra. The accuracy of the FT procedure also requires a very good signal-to-noise ratio, in order to have a good signal even in the band tails. This is important also because of the unavoidable arbitrariness of the background subtraction procedure in the presence of noise.

Very high spectral resolution is not as stringent a requirement. In fact, the experimentally observed band is a convolution of the true bandshape $I_i^0(\omega)$ with the resolution function $\eta(\omega)$ of the spectrometer. When taking the FT, this convolution reduces to an ordinary product of $FT[I_i^0(\omega)]$ with $FT[\eta(\omega)]$, and this latter function drops out when taking the dichroic (IR) or depolarization (Raman) ratio.

Since most IR and Raman bands of interest in liquid crystals have bandwidths of the order of 5 to 20 $cm^{-1}$, a spectral bandpass of 1 to 3 $cm^{-1}$ is generally adequate. Of course the spectral resolution defines the time scale probed by the spectroscopic experiment. Thus it is also advisable to operate at sufficiently high resolution to probe the behavior of the ACF over a significant time interval. In practice the upper limit is about 5 ps. The lower time limit is set by how far on the tails of the band the experimental points can be used to perform the FT. In general, even under the most favourable conditions it is difficult to obtain meaningful data for $t \leq .2$ ps. An example of a reasonably accurate reorientational ACF is reported in Ref. 6, for the case of spinning rotations in nematic PAAf[6], using the line at 1028 $cm^{-1}$.

Ideally from such data one should be able to extract not only the relaxation time, but also information on the appropriate stochastic model describing the reorientations, together with the values of the relevant parameters[7]. In fact, the ACF's are connected with the reorientational transition probability $P(\Omega_0;\Omega_t)$. This function can be determined by solving the diffusion equation which is appropriate to the chosen dynamical model for the reorientations. A popular model to set up the equation is the small step model[8]; another one, which however seems less adequate for liquid crystals, is the strong collision model[9]. From the fit of the ACF thus obtained to the experimental one, the basic model parameters should be obtained. In practice, however, only the spinning (IR) or tumbling (IR and Raman) diffusion coefficients may be obtained with relative ease, together with the static order parameters $< P_2 >$ (IR) or $< P_2 >$ and $< P_4 >$ (Raman).

### *3.2 Lineshape analysis and correlation functions*

As stated before, the direct separation of vibrational and rotational relaxation is not possible in an isotropic fluid using IR bandshape analysis. However, it is possible in a dichroic sample, such as an aligned mesomorphic phase[10]. More specifically, we have:

$$I^{v}_{z(x)} \approx FT\eta(\omega) < \mu^{v}_{z(x)}(t)\mu^{v}_{z(x)}(0) >< Q^{v}(t)Q^{v}(0) > \tag{10}$$

where

$$\eta(\omega) = \frac{4\pi^2\omega L(\omega)}{n(\omega)\hbar c V}\left[1 - \exp\left(-\frac{\hbar\omega}{kT}\right)\right]$$

and $L(\omega)$ is the internal field, $n(\omega)$ the index of refraction, V the sample volume, and z(x) describes the incident light polarization. In what follows we shall neglect the frequency dependence of $\eta(\omega)$, which is expected to be much weaker than that of the bandshape. This approximation would influence only the long time behavior of the ACF, which is not available experimentally anyway due to finite instrumental resolution.

Since the vibrational ACF may be assumed to be angle-independent, it will be the same for the z and x polarizations, and it will drop out of the ratio:

$$\frac{a_z(t)}{a_x(t)} = \frac{< \mu^{v}_{z}(t)\mu^{v}_{z}(0) >< \mu^{v}_{x}(0)\mu^{v}_{x}(0) >}{< \mu^{v}_{x}(t)\mu^{v}_{x}(0) >< \mu^{v}_{z}(0)\mu^{v}_{z}(0) >} = C^{or}(t) \tag{11}$$

where $a_{z(x)}(t)$ are the normalized FT's of the IR absorption bandshapes and $C^{or}(t)$ is the reorientational ACF related to the molecular fluctuations which may be probed by the vibrational mode $Q^v$.

Clearly there will be upper and lower time limits to the validity of the experimental ACF, as previously discussed. In general $C^{or}(t)$ may be obtained reliably in the range 0.3 ps to 3 ps. Furthermore, we must recall that many unproven assumptions are involved in obtaining $C^{or}(t)$. We refer in particular to the separation of rotational and vibrational relaxations and to the homogenous character of the line broadening. This latter assumption in particular might not be justified in strongly structured fluids, where molecular residence times might be longer than reorientational times, thus leading to a multiplicity of local environments.

With many possibly relevant ACF's and therefore many microscopic parameters, how can we decide which are the most relevant for the problem at hand? Lacking a fundamental theory, we must resort to models. The basic assumptions of a model act as an *a priori* selector and eliminate most of the parameters from the calculation. This of course is an advantage, but is also the fundamental limit of modeling: we do not really know if some of the quantities we neglect might not be important factors in the dynamics.

Be that as it may, modeling is essential to be able to arrive at usable theoretical predictions. In order to connect with the experimental data, essentially two steps are necessary:

a. The experiment is performed in the laboratory frame of reference. Thus we must transform to it from the molecular frame. This is done through the use of the irreducible tensor representation formalism;

b. A specific dynamical model must be chosen, from which a form of the diffusion equation will be derived. From its solution usually in the form of a truncated series expansion the ACF's will be obtained.

In liquid crystals, the strong orientational anisotropy allows some simplification of an otherwise very difficult problem. In particular the microscopic reorientation diffusivity tensor $\{D^r\}$, which in general is orientation dependent, $D = D(\Omega)$ may be reduced to two diagonal elements:

$$D^r_{km} = \delta_{km} D^r_{kk}$$

and more particularly

$$D^r_{11} = D^r_{22} = D^r_{\perp}, \qquad D^r_{33} = D_{\parallel}$$

Furthermore, most of the available experimental evidence favors, among several possibilities, the model of small step brownian diffusion, at least for fluctuations in the time domain probed by vibrational spectroscopy. The most general diffusion equation in this case is[11]:

$$\frac{d}{dt} P(\Omega_0;\omega_t) = -\vec{J} \cdot D^r(\Omega)\{\vec{J} + [\vec{J}U(\Omega)]\} P(\Omega_0;\Omega_t) \tag{12}$$

where $\vec{J}$ is the angular momentum operator, $P$ is the probability that if the molecule was oriented along $\Omega_0$ at time zero, it will be oriented along $\Omega_t$ at time t. $U(\Omega)$ is the orientational pseudopotential defined by:

$$F(\Omega) = \frac{\exp[-U(\Omega)]}{Z}$$

where Z is the partition function and $F(\Omega)$ *is* the orientational distribution function. Under the assumption of cylindrical symmetry, Eq.12 reduces to:

$$\frac{\partial P}{\partial t} = -D^r_{\perp}\left[J^2 + \left(\frac{D^r_{\parallel}}{D^r_{\perp}} - 1\right)J_3^2 + \frac{1}{\sin\beta}\frac{\partial}{\beta}\left(\sin\beta\frac{dU(\beta)}{d\beta}\right)\right]P \tag{13}$$

which has been solved numerically[12].

However, in the time scale probed by our experiments,

$$tD^r_i \ll 1$$

in this case good approximate analytical solutions may be obtained[13].

Within the framework of the one exponential approximation, it is useful to consider the time derivative of the ratio of the FT's $a_i(t)$, which ( since we are in the $\ln(1+x) \approx x$ limit) should directly yield the coefficient in the exponential from the theory.

We have for instance:

$$\frac{d}{dt}\frac{a_z(t)}{a_x(t)} = 9 < P_2 > P_2(\cos\theta) D_\perp^r + < P_2 > (1 - P_2(\cos\theta))(1 - 2P_2(\cos\theta)) \cdot (D_\parallel^r - D_\perp^r) \cdot$$
$$\cdot \frac{1}{(1 - < P_2 > P_2(\cos\theta))(1 + 2 < P_2 > P_2(\cos\theta))} \tag{14}$$

where

$$P_2(\cos\theta) = 1/2(3\cos^2\theta - 1)$$

Eq.14 may be simplified considerably in some interesting limiting cases; in fact, for a longitudinally polarized band, $\theta \approx 0$ and we have:

$$\frac{d}{dt}\frac{a_z(t)}{a_x(t)} = \frac{9 < P_2 >}{(1 - < P_2 >)(1 + 2 < P_2 >)} D_\perp^r \tag{15}$$

For weakly polarized bands, for which the induced dipole moment points in the neighborhood of the "magic angle" $\theta \approx 54°$, we have:

$$\frac{d}{dt}\frac{a_z(t)}{a_x(t)} = < P_2 > (D_\parallel^r - D_\perp^r) \approx < P_2 > D_\parallel^r \tag{16}$$

since generally $D_\parallel^r \gg D_\perp^r$.

Using Eqs.15 and 16 we obtain the spinning and tumbling reorientational diffusion coefficients, provided $< P_2 >$ is known; in the IR case $< P_2 >$ may be determined from the dichroic ratio of a strongly longitudinally polarized band: $< P_2 > = R\text{-}1/R + 2$.

In Raman spectroscopy so far only vibrations corresponding to a strongly uniaxial Raman tensor component have been clearly identified. This is the simplest case from the analytical point of view, but only allows information to be obtained on the tumbling fluctuations. In spite of extensive search, no suitable mode for spinning fluctuations was found.

The differential polarizability tensor ACF may be expressed in terms of the functions $\phi_{mn}^2(t)$ and of the coefficients $A^{l,m}$ $(l = 2)$, which are the irreducible components of the differential polarizability tensor in the molecular frame. For instance:

$$< \alpha_{zz}^v(t)\alpha_{zz}^v(0) > = (A^{0,0})^2 + 2(2/3)^{1/2} A^{0,0} A^{2,0} < P_2 > + \frac{2}{3} \sum_{m=-2}^{2} (A^{2,m}) \phi_{0,m}^2(t)$$

$$< \alpha_{xz}^v(t)\alpha_{xz}^v(0) > = \frac{1}{2} \sum_{m=-2}^{2} (A^{2,m})^2 \phi_{1,m}^2(t)$$

For a vibrational mode associated to a strongly uniaxial differential polarizability tensor (*ie* $\{\alpha^v\} = \delta_{ij}$, with $\delta_{11} \approx \delta_{22} \approx 0, \delta_{33} \approx 1$ in the molecular frame), in the short time approximation we have:

$$\frac{d}{dt}\frac{a_{zz}(t)}{a_{xx}(t)} = \frac{8}{3} D_{\perp}^{r} \left[ \frac{14-5<P_2>-9<P_4>}{14-20<P_2>+6<P_4>} - \frac{7+5<P_2>-12<P_4>}{7+20<P_2>+8<P_4>} \right]$$

$$\frac{d}{dt}\frac{a_{zz}(t)}{a_{xz}(t)} = D_{\perp}^{r} \left[ \frac{14+5<P_2>+8<P_4>}{14+10<P_2>-24<P_4>} - \frac{8}{3}\frac{7+5<P_2>-12<P_4>}{7+20<P_2>+8<P_4>} \right]$$

and similar expressions for the other components. Thus from the data taken in the polarized and depolarized configurations we may obtain simultaneously $D_{\perp}^{r}, <P_2>, <P_4>$, by performing the VV-VH experiment in three suitable geometries: in particular, zz-zx and xx-zx for homogeneous alignement and xx-xy for homeotropic alignement.

## 4. Experimental Results

### *4.1 Temperature dependence of spinning and tumbling diffusion coefficients*

For the determination of the diffusion coefficients we have used the lines listed in Table I. From the data obtained using for instance the 1028 cm$^{-1}$ line in PAA, the relaxation time $\tau_{\parallel}^{1}$ and the diffusion coefficient $D_{\parallel}^{r}$ may be determined. The temperature dependence of $D_{\parallel}^{r}$ for nematic OET (octylethoxytolane) has been observed[14] in all nematic phases we have studied (see Table I).

Table I Typical spinning and tumbling probe

| Spinning |
|---|
| 1028 (PAA) 1047 (PMT,HET,OET), 1048 (EBBA): oxygen-alkyl tail stetching (R=1.8) |
| 1743 (NPOOB), 1730 (HOBDOP): carbonyl group stretching (R=0.9) |
| 833 (PMT,OET): out-of-plane deformation of H atoms in benzene ring (R=0.35) |
| 935 (Toluidine): Benzene ring + CH aromatic out-of-plane deformation (R=1.5) |
| 723 (Toluidines): Alkyl chain in-plane deformation (CH2 rocking) |
| **Tumbling** |
| 1175 (OET,HET,Toluidines), 1178 (PMT): benzene ring in-plane deformation (R=6) |
| 758, 910 (PAA), 765 (NPOOB): benzene in-plane deformations (R=5) |

It is clearly anomalous, since in principle the diffusion coefficient should increase more or less exponentially with temperature. We shall discuss the possible causes of such behavior later.

When measuring[14] the temperature behavior of the tumbling diffusion coefficient, obtained from the analysis of the 1175 cm$^{-1}$ line in nematic OET, the expected Arrhenius behavior was observed, with the proper value of the activation energy (namely, 30 Kcal/mole). The same behavior was observed using the $C \equiv C$ Raman line at 2220 cm$^{-1}$ in PMT. Thus both Raman and

IR spectroscopy yield quite similar results for tumbling fluctuations, and this, together with the quantitatively correct temperature behavior of the tumbling coefficient, shows the reliability of our method.

### *4.2 Short time limit of the correlation functions*

In principle, from the form of the ACF one should be able to deduce which stochastic model is appropriate to describe molecular spinning and tumbling. In practice, experimental problems limit the useful time range, and all models predict a relaxation-type behavior at sufficiently long times. Thus the study of the ACF's in this range cannot discriminate among them. The ACF is most sensitive to the details of the molecular dynamics in the transition region between the very short time inertial limit (when intermolecular torques have not had time to act) and the "Debye" relaxation range.

As already stated, the experimental problem in studying this region is connected with the necessity of performing the correct FT in the high frequency tails of an IR or Raman band. The signal-to-noise and background subtraction problems may make it very difficult to obain a reliable FT. In such an experiment particular care must be given to obtain a good spectrum in the band tails. The spectra were obtained[6] from an average over 10 iterations in each polarization. The inertial region appears to be reasonable, and from a parabolic fit (a rough but indicative approximation), we obtain the proper value of the moment of inertia relative to spinning and tumbling rotations.

Having obtained ACF's which look reliable in the short time range, we proceeded to fit them with the results from the chosen stochastic model, namely extended J-diffusion. In this model the parameters are the angular momentum correlation time $\tau_J$, the diffusion coefficients $D^r_{\parallel,\perp}$, and the root mean square rotation angle beween one reorientation event and another $<\varepsilon^2>^{1/2}$. From the experimental data[6] we were able to conclude that:

- For tumbling fluctuations, the small value of $<\varepsilon^2>^{1/2} \approx 3°$ indicates that the diffusion proceeds by small steps: this to our knowledge is the first *direct* experimental proof of a given stochastic model (namely, small step diffusion);

- For spinning fluctuations, the value of $<\varepsilon^2>^{1/2}$ is relatively high ($\approx 25°$); thus the small step model might not be quite as adequate to describe such diffusion.

- The anomalous temperature behavior of the spinning diffusion coefficient stands confirmed; thus such an effect cannot be due to the particular dynamical model used to interpret the data.

### *4.3 Pre-transitional behavior and molecular flexibility*

The temperature behavior of $D^r_{\parallel}$ is clearly anomalous. In the preceding paragraphs we have presented experimental evidence which excludes the possibility that the effect be due to the basic assumptions made in interpreting the data. This, plus the apparent generality of the effect, indicate that the slowing down of spinning fluctuations may be a general property of the nematic phase.

We can now attempt to discuss the possible origins of such behavior. Clearly perfectly smooth, rigid cylindrical molecules should have no pre-transitional slowing down in their spinning motion, or even any kind of spinning dynamics. Therefore molecular biaxiality or molecular flexibility may be important factors.

Let us consider the former alternative; in this case the slowing down is likely due to biaxial fluctuations, which become "critical", i.e. as a pre-transitional effect, as the N-I point is approached. This approach would lend support to the view that this transition is not classical[15](i.e.

Landau-de Gennes) but tricritical[16]. In this sense biaxiality not only has the static effect of altering the pseudopotential, but also a dynamical one, which induces additional correlations in the spinning fluctuations, therefore slowing them down (a spinning cogwheel is a good pictorial representation of this). In this framework, the spinning relaxation time $\tau_{\|}$ should have a power law behavior in the nematic phase. Such behavior was indeed found[17]; the double logarithmic plot of

$$(\tau_{\|}D_{\perp}^{r})^{-1} \approx \frac{T_0 - T}{T_0} \tag{17}$$

yielded in fact a straight line over the whole nematic phase, with an exponent $\gamma \approx 1$ and $T_0 = T_c + 0.3$, where $T_0$ is the superheating temperature. In Eq.17 $1/\tau_{\|}$ was divided by $D_{\perp}^{r}$ to take into account the temperature behavior of the rotational viscosity[18]. This result however cannot discriminate between classical and tricritical behavior, since both predict $\gamma = 1$.

In principle, the scaling behavior of the order parameters should depend on the type of transition. In fact, if $< P_2 > \approx (T_0 - T/T_0)^{\beta}$, then $\beta = 0.5$ for a classical transition and, $\beta = 0.25$ for a tricritical one. Most experimental evidence tends to favor low values of $\beta$, including our own data[14]. However such evidence is not cogent, since in the so called "critical" plots of $< P_2 >$ there are *three* adjustable parameters, namely $\beta$, $T_0$ and $< P_2 >_0$, the last one being the value of $< P_2 >$ extrapolated to $T_0$ (let us recall that the N-I transition is still first order, and what should be plotted is the difference $< P_2 > - < P_2 >_0$). Some of the ambiguity in the exponents derived from such plots may be removed using our results. In fact, we can determine $T_0$ independently from the slowing down of spinning fluctuations (Eq.17), and then use it in the critical plot of $< P_2 >$. We have done so, obtaining $\beta \approx 0.25$ for OET and $\beta \approx 0.3$ for EBBA[14]. These values would favor the tricritical model for the N-I transition. However the temperature resolution of our data (± 0.2 K) was not sufficient to obtain a good and reliable extrapolation for $< P_2 >_0$. This leaves some ambiguity in the results, although it seems highly unlikely that better data, which are needed to clearly assign the transition, would yield the classical value of $\beta$.

Even assuming the transition to be tricritical, there is another, more fundamental possible ambiguity. So far, we have assumed that the second fluid that goes "critical" is due to biaxial fluctuations. As we stated, there is an additional possibility to explain the spinning behavior, namely molecular flexibility. It is well known that mesogen molecules are not rigid; in fact, the author has directly shown that alkyl tails spin with a shorter relaxation time than the rigid cores[19]. Thus different parts of the molecule could have different dynamical behavior. This by the way would account for some discrepancies in the relaxation behavior observed by NMR[20] as compared with our IR observations of spinning motion.

Before any detailed model of pre-transitional behavior before the N-I transition is proposed, the complexity of spinning behavior must be understood more clearly. A promising approach is to compare our results with neutron scattering data, on normal and D-substituted samples as described in Chapter 20 of this volume.

## 5. Conclusions

The principal purpose of this Chapter is to show that, in favourable cases, IR and Raman bandshape analysis can yield quite detailed information not only on order parameters but also on the reorientational dynamics of molecules in liquid crystalline phases. Traditionally such determinations have been made using NMR, or probe (EPR and Fluorescence) spectroscopy. IR and Raman spectroscopy, when applicable, have the advantage of determining the time autocorrelation function of the relevant dynamical variable, and not just its time integral. We have shown that the connection between $C^{or}(t)$ and the manifold of ACF's involved in the dynamics may be quite complex; however, under reasonable assumptions it is possible to isolate a few important ACF's. The ensuing parametrization allows a reasonably quantitative description of reorientational dynamics, from the inertial limit to the Debye relaxation range. Once the basic probe characteristics have been determined, IR and Raman bandshape analysis may be usefully applied to study many relevant dynamical properties of liquid crystals. We have shown some applications taken from our own work: the discrimination between possible dynamical models through the precise determination of the ACF's in the short time limit; pre-transitional effects in the nematic phase and molecular flexibility. The latter applications should be particularly relevant to more complex systems such as lyotropics or polymeric liquid crystals.

A fundamental problem in the study of microscopic dynamics has always been the comparison of results obtained on presumably the same parameter using different techniques. In our own work we have shown that IR and Raman spectroscopy, in which the probe is electromagnetic and local, yield similar results. However, the data sometimes disagree with those obtained by NMR, neutron scattering or depolarized fluorescence (which in turn disagree with each other). From this point of view it is essential to perform several types of experiments on the same system. Each technique in fact has a definite time window, within which it is sensitive to time correlations. In complex systems such as liquid crystals there are several possible dynamical regimes, and one or another of them will be probed by a given technique. For example, we have studied 5CB and EBBA by time-resolved fluorescent probe depolarization spectroscopy[21]. We found that the tumbling relaxation time differed by as much as one order of magnitude from the "same" time determined by IR and Raman spectroscopy. Since our overall values are in agreement with other depolarization spectroscopy measurements[22], the discrepancy must be fundamental in character. In fact, Raman and IR spectroscopy probe dynamics on the $\Delta t \leq 10$ ps range, whereas nanosecond spectroscopy is sensitive to the $\Delta t \geq 1$ ns range. The discrepancy between the measured $\tau_\perp$ then implies that tumbling diffusion takes place over at least two different time scales.

Such result would have been impossible to obtain if measurements of the same physical quantity by different techniques (different time windows) had not been performed.

ACKNOWLEDGEMENT. The work reported here was performed in collaboration with I. Dozov, N. Kirov and B. Rosi.

## References

1. In this review we shall not try to present an exhaustive bibliography; the references we cite however do contain most of the relevant literature on the subject. General reviews on fundamentals can be found in: R. Kubo, Rep. Progr. Physics **29**, 255 (1966); D. Forster, "Hydrodynamics, Fluctuations, Broken Symmetry and Correlation Functions", Benjamin, 1975; B. J. Berne and R. Pecora "Dynamic Light Scattering", Wyley, N.Y., 1976; J. P. Hansen and I. R. McDonald, "Theory of Simple Liquids", Academic Press, 1976; A. Barker, R. Loudon, Rev. Mod. Phys. **44**, 18 (1972); Wilson E.B., Decius, J.C., Cross, P.C., *Molecular Vibrations,* Mc GrawHill Publ. Co., New York; Sverdlov, L.M., Kovner, M.A., Krainov, E.P., *Vibrational Spectra of Polyatomic Molecule,* (in Russian), Moskow, Nauka, 1970; Steele, D., *Theory of Vibrational Spectroscopy,* W.B.Saunders, Philadelphia, 1971; Volkenstein, M.V., Gribov, L.A., Eljashevich, M.A., Stepanov, B.I., *Molecular Vibrations,* (in Russian), Moskow, Nauka, 1972; J.H.R Clarke, in "Advances in IR and Raman spectroscopy", edited by R.J.H.Clark and R.E.Hester, Heyden, London 1977, vol.4, p.109.
2. M.E. Rose *Elementary Theory of Angular Momentum,* Wiley, New York (1957).

3. See f.i., P. Simova, N. Kirov, M. P. Fontana, H. Ratajczak, *Atlas of Vibrational Spectra of Liquid CrystalJ,* World Scientific, Singapore (1988); a particularly exhaustive references source is N. Kirov, P. Simova, *Vibrational Spectroscopy of Liquid Crystals,* Publishing House of B.A.S., Sofia (1984).
4. See f.i., M.C. Wang, G.E. Uhlenbeck, Rev. Mod. Phys **17**, 323 (1945).
5. N.Kirov, I.Dozov, M.P.Fontana, J.Chem. Phys. **83**, 5267 (1985); K.Miyano, J.Chem. Phys. **69**,4807 (1978); S. Jen, N.A.Clark, P.S.Pershan, J.Chem. Phys. **66**, 4635 (1977).
6. N. Kirov, I. Dozov, M.P. Fontana, B. Rosi, J. de Phys. **48**, 457 (1987).
7. W.Rothschild, "Dynamics of Molecular Liquids" Wiley, N.Y., 1984; see also J.H. Freed in *Spin Labelling: Theory and Applications,* L.J. Berliner ed., Academic Press, N.Y., 1976, Ch. 3; and refs. 1.
8. C.F.Polaczek, G.F. Bruno, J.H. Freed, J. Chem. Phys. **58**, 3185 (1973); G.R. Luckhurst, M. Setaka, C. Zannoni, Mol. Phys. **28**, 49 (1974).
9. G.R. Luckhurst, A. Sanson, Mol. Phys. **24**, 1297 (1972).
10. I.Dozov, N.Kirov, M.P.Fontana, J.Chem. Phys **81**, 2585 (1984).
11. G. Moro, P.L. Nordio, Mol. Cryst. Liq. Cryst., **104**, 361(1984), and references therein.
12. P.L.Nordio, P.Busolin, J. Chem. Phys. **55**, 5485 (1971).
13. I. Dozov, I. Penchev, J. Lumin. **22**, 69 (1980).
14. M.P.Fontana, B.Rosi, N.Kirov, I.Dozov, Phys. Rev A **33**, 4132 (1986).
15. Lin Lei, Phys. Rev. Letters **43**, 1604 (1979); C. Rosenblatt, Phys. Rev. A25, 1239 (1982); *ibidem,* **27**, 1234 (1983).
16. P.H. Keyes, Phys. Lett. **67A**, 132 (1978); P.H. Keyes, J.R. Shane, Phys. Rev. Letters **42**, 722 (1979).
17. M.P. Fontana, B. Rosi, N. Kirov, Phys. Rev. Letters, **56**,1708 (1986).
18. Lin Lei, Mol. Cryst. Liq. Cryst. **112**, 233 (1984).
19. M.P. Fontana, N. Kirov, J. Phys.(Paris) Lettres **46**, L341 (1985).
20. Regitze R. Vold, Robert L. Vold, J. Chem Phys. **88**, 1443 (1988) and references therein; also, M. Riccò, Ph.D. Dissertation, University of Parma (1988); see also the following Chapter.
21. I.Dozov, N.Kirov, M.P.Fontana, M.Manfredi, B.Rosi, R.Cywinski, Liquid Crystals **4**, 241(1989); A. Arcioni, F. Berinelli, R. Tarroni, C. Zannoni, Molec. Phys., **61**, 1161(1987).

Chapter 20

# MOLECULAR DYNAMICS IN NEMATIC PHASES: QENS AND NMR SPECTROSCOPY

M. RICCÒ and M.P. FONTANA

## 1. Introduction

Molecular dynamics in liquid crystals (particularly smectic and to a lesser extent nematic phases) has been extensively studied by quasi elastic neutron scattering (QENS)[1-11] as well as by magnetic relaxation[12-17]. Nevertheless, these two different techniques have never been combined, probably due to the fact that, quite often, the results obtained did not agree. The situation is particularly difficult in the nematic phase, where translational and rotational time scales are not well separated. There is, however, much interest in this type of combined study, since it is by coupled measurements that we may have a chance to unravel the complexities of the dynamics in these interesting materials. Furthermore, the extensive work on reorientational dynamics performed by Raman and Infrared fluctuation spectroscopy (see Chapter 19) showed several intriguing anomalies: the decrease of the spinning diffusion coefficient (for rotations about the main molecular axis) with increasing temperature, the large discrepancy between the temperature behavior and values of the tumbling diffusion coefficient (for rotations of the main molecular axis) determined with IR and Raman spectroscopy and those found with fluorescence depolarization spectroscopy. Thus it is important to use QENS, which directly probes the motions of atomic nuclei, and $^{13}C$ NMR, which is directly sensitive to molecular reorientations, to resolve these discrepancies and thus to gain finally a quantitative picture of molecular dynamics in such complex fluids.

In the following paragraphs we shall discuss briefly the fundamentals of QENS and NMR relaxation as they pertain to the study at hand; then we shall report on the application of such techniques to the study of a specific system, namely the nematic phase of EBBA (ethoxybenzilidene-bis-butyl-aniline), drawing essentially from the work our group has performed on this subject.

## 2. Quasi Elastic Neutron Scattering

Quasi elastic scattering can be viewed as due to the Doppler shift of the energy of scattered neutrons due to the fact the the scattering nucleus is moving away from or towards the observer.

---

M. Riccò and M.P. Fontana - Dipartimento di Fisica, Università di Parma, Parma, Italy.

*Phase Transitions in Liquid Crystals*, Edited by S. Martellucci and A.N. Chester, Plenum Press, New York, 1992

Since the velocity distribution of the nuclei will be random both in modulus and direction, the Doppler shift will result in a broadening of the incoming neutron energy distribution, which for simplicity can be assumed to be monochromatic. The resulting line shape will be directly related to the velocity probability distribution function. For the common case of a Gaussian distribution, the line shape will have a Lorentzian form, as we shall shortly show.

The actual experiment thus will have to be able to determine this lineshape, which means that the spectral resolution must be adequate: the time scale of the motions which will be detected is in fact determined by the resolution, which defines the "time window" of the experiment. All motions on a slower time scale will appear as a static distribution, which contributes to the elastic part of the scattering. This general feature is particularly important to take into account when working with complex fluids, which may be characterized by a plurality of dynamical regimes and hence different diffusion coefficients, each appropriate to a specific time window.

Another factor to consider when studying liquid crystals is their strong anisotropy: thus, many simplifications which apply for isotropic fluids cannot be made. For instance, all diffusion coefficients now will be tensor quantities, and we expect that the scattered intensity will depend on the vector $\vec{Q}$ (exchanged momentum), and not just on its modulus.

The strong anisotropy however allows the diffusion tensors to be well approximated by two scalars, the spinning and tumbling diffusion coefficients $D_{\parallel}^{r}$ and $D_{\perp}^{r}$ for reorientations and $D_{\perp}^{t}, D_{\parallel}^{t}$ for translations in the plane perpendicular to the molecular axis and along the molecular axis respectively.

The $\vec{Q}$-dependence of the scattered intensity allows a further simplification, since if all the molecules are aligned, the QENS experiment in which $\vec{Q} \perp \vec{n}$, where $\vec{n}$ is the nematic axis, will probe only the spinning fluctuations, thus considerably simplifying the analysis. In reality of course the molecules are not perfectly aligned; they are distributed according to an orientational probability distribution. This implies that tumbling fluctuations may be observed at least in principle. It also implies that in analysing the data an average over such a distribution must be performed.

All these features should help to understand why analysing QENS in liquid crystals, and particularly in nematics, may be a difficult task. In fact, to date only few quantitative studies have been published, and no quantitative temperature dependence has been determined. We shall show that the availability of the new IRIS spectrometer at the ISIS source of Rutherford Appleton Laboratory has provided enough of an experimental improvement to make possible the solution of some of the problems we have been discussing.

## *2.1. Theory and Data Analysis*

The observed quantity in a neutron scattering experiment is the double differential cross section $d^2\sigma/d\Omega dE'$, where $\Omega$ is the solid angle and E′ is the final energy. It can be expressed as the time Fourier trasform of the intermediate scattering law S($Q$,t)

$$\frac{d^2\sigma}{d\Omega dE'} = \frac{k'}{k}\frac{1}{2\pi\hbar}\int_{-\infty}^{+\infty} dt \exp(-i\omega t) S(Q,t) \tag{1}$$

where $\hbar\omega$ is the exchanged energy $E' - E = \hbar\omega$, $Q$ is the exchanged momentum $k' - k = Q$. Due to the very high incoherent scattering length of hydrogen (referred to as $b$), the observed scattering can be assumed to be entirely due to this nucleus. As a consequence, the intermediate scattering law can be expressed as an ensemble average of the following self correlation function of the position of the scattering nucleus $R$:

$$S_{incoh}(Q,t) = \Sigma_l \mid b_l \mid^2 < \exp\{iQ \cdot \hat{R}_l(0)\} \exp\{iQ \cdot \hat{R}_l(t)\} > \tag{2}$$

which, in the classical limit, is proportional[20] to the space Fourier trasform of the self-correlation function of $R$. All the different motions of the molecules (translation, rotation, flexibility, vibrations) contribute in principle to the quasi elastic broadening. If we assume that all these movements are completely decoupled from one another, the intermediate scattering law can be expressed as the product of such contributions, so that the differential cross section results from the convolution of those portions originating from the different dynamical processes.

As is well known[20], the vibrations contribute to a scattered intensity in the inelastic region with a consequent decrease of the intensity in the quasi elastic region, whose $Q$ dependence follows a Debye-Waller behavior. Thus vibrations do not alter the quasi-elastic spectrum. However it is useful to study the integrated intensity of such spectrum *versus* Q, both to obtain a measurement of the mean square displacement of scattering nucleus, and to test the reliability of the data.

### *2.1.1. Translational Diffusion*

The translational dynamics can be expected to obey some sort of diffusion equation. In the case of continuous translational diffusion, or small step diffusion, the time correlation function of particle positions $G_s$ will obey the simple diffusion equation:

$$\frac{\partial G_s}{\partial t} = D_t \Delta G_s \tag{3}$$

The ensuing intermediate scattering function is an exponential. The corresponding scattering cross section is thus a single lorentzian whose width is proportional to $Q^2$:

$$L_t(\omega) = \pi^{-1} \frac{D_t Q^2}{(D_t Q^2)^2 + \omega^2} \tag{4}$$

Although on general grounds we expect that in the Q→0 limit all diffusion models should lead to Eq. 4, as Q increases and we probe the system over smaller and smaller distance scales, the details of the microscopic structure and molecular interaction should lead to deviations from the hydrodynamic limit; in complex fluids such as liquid crystals we expect such deviations to be strong. On the other hand, for the few cases which can be treated exactly (harmonic oscillator, ideal gas, free diffusion), the intermediate scattering function turns out to be exponential. Thus we expect that deviations from the hydrodynamic limit will appear not as a different functional form of the intermediate scattering law, but rather as a Q-dependence of the otherwise constant diffusion coefficient $D_t$. This essentially means that we can still use Eq. 3 to deal with translational dynamics, but we should be prepared to see deviations from the $D_t \propto Q^2$ law at higher Q's.

### *2.1.2. Rotational Diffusion*

A rotating particle will occupy a finite region of space even in the limit $t \to \infty$. Thus the intermediate scattering function corresponding to a purely reorientational motion of the molecules will be finite at all times. The constant term in it, when Fourier transformed to yield the scattering cross section, will yield a $\delta(\omega)$ term, i.e. a purely elastic contribution.

The rotational contribution to the scattering cross section then consists of the sum of such purely elastic part and a sum over an infinite number of Lorentzians. In fact, Fourier transforming Eq. 1, for the case of isotropic rotations, we obtain the following scattering cross-section:

$$S(Q,\omega) = j_0^2(Qr)\delta(\omega) + \sum_{l=1}^{\infty}(2l+1)j_l^2(Qr)\frac{1}{\pi}\frac{l(l+1)D_r}{1+l(l+1)\omega^2 D_r^2} \tag{5}$$

Here the $j_l(Qr)$ are spherical Bessel functions. It is clear from Eq. 5 that a fit to experimental data might be quite difficult and not unique. However, two features help simplify the analysis. The first is that the halfwidths of the Lorentzians, which are independent of $Q$ as should be expected for space-localized motion, increase with $l^2$. Thus, within the time window, and hence the spectral resolution of the experiment, the higher order Lorentzians will be very broad, and will be indistinguishable from the background. In practice we expect only three, perhaps four, Lorentzians to be relevant.

Another guide in the analysis, and an important quantity to be determined for its own sake, is the so-called Elastic Incoherent Structure Factor (EISF), which empirically may be defined as the ratio of the elastic intensity to the total scattered intensity. Formally, since from the properties of the Fourier transforms, the integral over all energies of the scattering cross section is proportional to the $t \to 0$ limit of the intermediate scattering function, the EISF will be connected with the static spatial correlations of the rotating nuclei, averaged over time. Thus the EISF, which is a function of $Q$, yields information on the large scale characteristics of the rotational diffusion, such as the geometry of the rotation and of the rotating particle. This means that the $Q$-dependence of the EISF may be used to test the validity of the stochastic model chosen to interpret the data.

In practice, the energy resolution of the spectrometer sets a lower limit to the rate of the observed processes, below which the motion has to be considered static from the neutron spectroscopy point of view. In the observable time window of our experiment we see both spinning and translational diffusion while, as shown by other techniques[21,16], the tumbling motion is too slow to be resolved by the spectrometer, so that it has to be considered as contributing to a static distribution.

Two different spinning mechanisms have been tested in our experiment: uniaxial continuous rotational diffusion on a circle and the 180° jump reorientation very often assumed for the reorientation of aromatic rings. In the former case the expected differential cross-section is[22]:

$$\frac{d^2\sigma_{inc}^{rot}}{d\Omega dE'} = \frac{k'}{k}\frac{N}{2\pi\hbar}|b|_{inc}^2\left\{J_0^2(Qr\sin\theta)\delta(\omega) + \frac{2}{\pi}\sum_{l=1}^{\infty}J_l^2(Qr\sin\theta)\frac{D_r l^2}{(D_r l^2)^2+\omega^2}\right\} \tag{6}$$

where $r$ is the rotation radius, $\theta$ is the angle between the $Q$ vector and the rotation axis, $J_n^2(x)$ are Bessel functions of the first kind of order $n$ and $D_r$ is the diffusion coefficient. In the latter case, only one quasi-elastic component is expected[22]:

$$\frac{d^2\sigma_{inc}^{rot}}{d\Omega dE'} = \frac{k'}{k}\frac{N}{2\pi\hbar}|b|_{inc}^2\left\{\frac{1}{2}[1+\cos Q\cdot r]\delta(\omega) + \frac{1}{2}[1-\cos Q\cdot r]\frac{1}{\pi}\frac{2\tau}{4+\omega^2\tau^2}\right\}. \tag{7}$$

These equations decribe the case in which the axis of rotation is fixed in space while in the case of the real system we must include a distribution (we chose a Mayer-Saupe form) of the rotation axis around the nematic director. The average for such a distribution has been performed numerically. The calculation of the elastic and quasielastic structure factors for the two rotational models was based on the molecular structure determined by X-ray diffraction on the solid fully protonated compound[23].

## 2.2. Quasi Elastic Neutron Scattering in Nematic EBBA

In a previous preliminary work[18], we used the high resolution, high momentum transfer IRIS spectrometer at RAL to study QENS in the nematic phase of EBBA (ethoxybenzylidene-bis-butyl-anyline). Although we succeeded in separating the translational and rotational (spinning) contributions to the quasi elastic broadening, the temperature behavior of the spinning diffusion coefficient derived from our data still disagreed with similar data obtained by NMR relaxation. Here we wish to show that agreement of the measured dynamical parameters can be obtained if all the relevant dynamical processes, including molecular flexibility, influencing the broadening of the quasi elastic line in QENS and the relaxation time in NMR are taken into account, adopting a model dependent global fit of the data for both the techniques. The comparison of the QENS and NMR results leads to a more precise picture of molecular diffusion in the nematic mesophase. Due to the 15 μeV energy resolution of the neutron spectrometer we expect that both the translational diffusion of the molecules and their rotational diffusion around the long axis (*spinning*) will contribute to the quasi-elastic broadening of the elastic line while the diffusion around the short axis (*tumbling*) is expected to be too slow to produce observable effects.

### 2.2.1. The Experiment

The inverse geometry, time of flight spectrometer IRIS[19] provides a resolution of 15 μeV to an exchanged momentum ($Q$) ranging from 0.25 $A^{-1}$ to 1.85 $A^{-1}$; the incident energy is 1.82 meV. EBBA was selectively deuterated along the alkyl chains in order to "blind" the measurement of their motion, and spectra were taken at four temperatures within the nematic phase (T= 308 K, 317 K, 327 K and 337 K) and in the isotropic phase (T= 346.5 K). The sample was contained in a slab-shaped cell (50 × 50 mm) made of thin boronless glass. Its inner surfaces were treated with surfactants in order to get a homogeneous alignment in the nematic phase, and this alignment was checked by IR dichroism measurements. In order to enhance the scattering due to spinning dynamics the nematic director was aligned perpendicular to the scattering plane. The sample thickness was 250 μm which yielded a transmission coefficient of 0.87. A total of 29 spectra were recorded, covering an angular range from 15° to 165°. Standard programs were used to correct the data for absorbtion, self shielding, sample holder absorbtion, detector efficiency, empty cell subtraction, monitor normalization and background subtraction. The energy integrated intensity was verified to follow a regular Debye-Waller behavior with respect to $Q$ in order to rule out possible coherent scattering from the sample.

### 2.2.2. Results

The 29 energy spectra were analyzed simultaneously using the Minuit (Cernlibrary) fitting program for all the temperatures investigated. Due to the presence of translational broadening, the identification of the elastic and quasielastic intensity is not straightforward. In order to identify the different contributions all the components have to be fitted and, in order to reduce the fitting parameters, the fit has to be model dependent.

The rotational elastic and quasi elastic lines were convoluted with a translational lorentzian broadening and with the experimental resolution line (obtained from a vanadium slab run). The parameters of the fit were the intensity of the elastic line, that of the quasi-elastic rotational part, the rotational broadening (the same for all the energy spectra), and the translational broadening. No reasonably good fit of the data can be obtained in the nematic phase by letting the translational broadening follow a $Q^2$ law which however was found to apply in the isotropic liquid phase. Letting this broadening vary with respect to $Q$ a good fit of the energy spectra at all the $Q$ values can be obtained as is shown in Fig. 1.

A self consistency model test can be performed by comparing the EISF of the assumed model with the one calculated from the fitted intensities. Fig. 2 displays the result of this comparison for the continuous rotational diffusion model and for the 180° jump model. The former seems to fit quite well the predicted behavior, while the latter seems to diverge from it; especially at the higher $Q$ values where the adoption of a single quasi-elastic component (as predicted by Eq. 7) is probably totally inadequate.

At low $Q$'s ($Q \leq 1A^{-1}$) the translational broadening is proportional to $Q^2$. From such behavior a translational diffusion coefficient $D_{tr}$ can be deduced; it is plotted in Fig. 3 as a function of reciprocal temperature. It does not follow an Arrhenius behavior, probably due to the different mixing of the parallel and perpendicular translational coefficients at different temperatures.

At higher $Q$'s the linear dependence on $Q^2$ saturates. This effect was observed also by Bee et al[24] in the nematic phase of 2-OAOB. A more dramatic departure from the $Q^2$ law is observed at still higher $Q$'s, where a dip at about 2 $A^{-2}$ appears. This anomalous behavior was previously reported by us for hydrogenated EBBA[18] and its detailed analysis will be published elsewhere[25].

The rotational diffusion coefficient for the spinning motion is also (as is evident from Eq. 6) a product of the fit. It is presented in Fig. 4 as a function of temperature and it seems to follow a regular Arrhenius behavior. This behavior is remarkably different from that found for $D_{\parallel}$ in the fully hydrogenated EBBA[18], which was instead similar to the results obtained by IR dichroism bandshape analysis. We believe that the origin of the discrepancy is the extra disordered motion of the alkyl tails, which dominated the QENS of h-EBBA and also the IR results.

## 3. Nuclear Magnetic Relaxation

When in an NMR experiment the nuclear spin system is driven out of equilibrium by means of an RF pulse, it tends to reestablish its initial equilibrium state in a time called the relaxation time. The relaxation time relative to the component of the magnetization parallel to the applied magnetic field is associated with an exchange of energy (usually Zeeman energy) between the spin system and the lattice, and it is referred to as the spin lattice relaxation time ($T_1$).

The spin system is usually subject to magnetic interactions which can be modulated by the lattice degrees of freedom. If the spectrum of these fluctuations has a non-vanishing component at the Larmor frequency of the nuclei of interest, an exchange of energy takes place between the two systems. Usually many different interactions cooperate to induce the relaxation of the nuclear magnetization, so that extracting precise quantitative information from relaxation experiments becomes quite difficult. More reliable information can be extracted from isolated simple systems. An example of such a system in which one interaction can be singled out is $^{13}C$ in a C-H group of a molecule, because the strongest interaction between the two nuclei is the

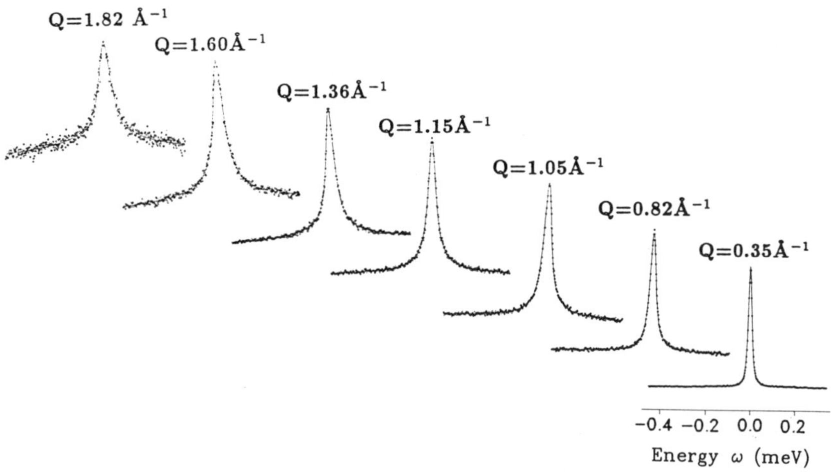


Fig.1. Fit of the measured spectra (dots) at different $Q$ values.

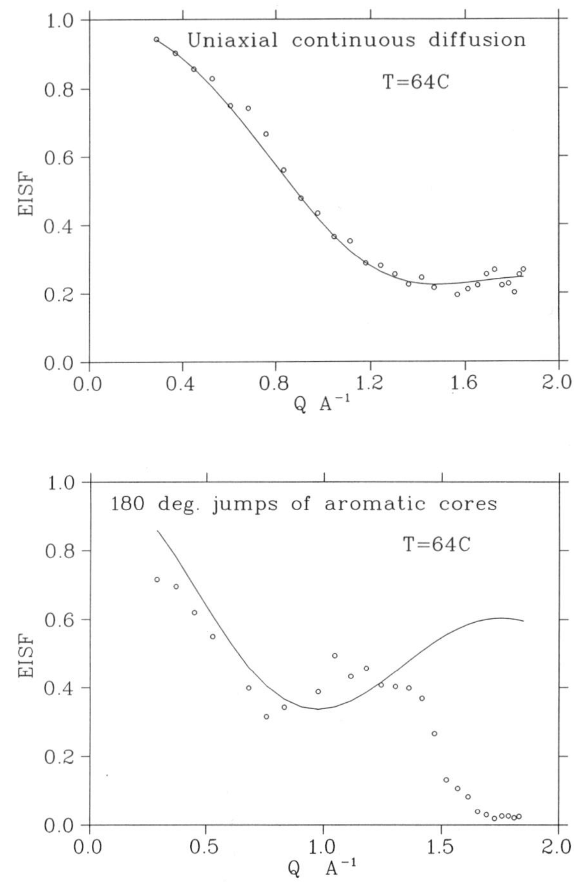


Fig.2. Comparison between fitted EISF values (dots) and theoretical predictions (continuous line) for two different diffusion models, namely: continuous diffusion on a circle (above) and 180° jumps (below).

dipolar one, which is completely known for a given geometry of the molecule. This system is not totally isolated from the rest of the molecule, but thanks to the low natural abundance of $^{13}C$ (≈ 1%) its homonuclear dipolar interaction can be totally neglected, and the corresponding interaction among hydrogens (which also affect the relaxation of carbon) can be removed by saturating these nuclei during the relaxation of the system. The reorientation of the molecule is the only dynamical process responsible for the relaxation of the $^{13}C$ nucleus, as no other process modulates the dipolar interaction. In this case, both the spinning and the tumbling motion are expected to contribute, while the slow collective fluctuations of the nematic director are supposed to have a negligible effect at the observed Larmor frequencies[17]. This fact, which does not hold true in the case of deuterium relaxation experiments, greatly simplifies the analysis of the $T_1$ data.

### 3.1. Experiment

$^{13}C$ relaxation measurements were performed using the commercial spectrometers Bruker AC100, CPX200 and AMX400. The fully protonated EBBA samples were purified by recrystallization. They were subsequently sealed under vacuum after several freeze-pump-thaw cycles.

The measurement of the spin lattice relaxation times ($T_1$) was done with a conventional full inversion recovery experiment under proton irradiation. The power of the decoupler was increased during the acquisition period (gated decoupling). In some instances, a composite pulse decoupling technique[26] (MLEV16) was adopted to avoid sample heating from the decoupler radio frequency. The Nuclear Overhauser Enhancement (NOE) factor was measured at different temperatures and Larmor frequencies by comparing the line intensities recorded under continuous proton irradiation with those obtained using the gated decoupling technique.

### 3.2. Theory

As discussed in the introductory paragraph, unlike the neutron scattering technique, magnetic relaxation is expected to be sensitive also to the slower reorientation around the short molecular axis (*tumbling*), so that the interpretation of the data has to be done on the basis of a more complex model which takes into account both dynamics.

As is suggested by neutron scattering results, a continuous diffusion model seems to be most suitable. We adopted a small step rotational diffusion model, according to which the conditional probability for the reorientation of the molecule (assumed to be a rigid round ellipsoid in an anisotropic restoring potential V(θ) is given by the following Smoluchowski equation[27]:

$$\frac{\partial P(\Omega_0,\Omega,t)}{\partial t}=-\left\{\boldsymbol{L}\cdot\boldsymbol{D}\cdot\left[\boldsymbol{L}+\boldsymbol{L}\frac{V(\theta)}{kT}\right]\right\}P(\Omega_0,\Omega,t) \tag{8}$$

In the case that heteronuclear dipolar relaxation with the bonded hydrogen is the only relaxation mechanism, the relaxation rate can be expressed in terms of the spectral densities $J_n(\omega)$ by the following[28]:

$$\frac{1}{T_{1D}}=\frac{1}{2}\left(\frac{\gamma_C\gamma_H\hbar}{r_{CH}^3}\right)^2[J_0(\omega_C-\omega_H)+3J_1(\omega_C)+6J_2(\omega_C+\omega_H)] \tag{9}$$

where the $\gamma$'s are the gyromagnetic ratios, $\hbar$ is Planck's constant divided by $2\pi$, the $\omega$'s are the resonance frequencies and $r_{CH}$ is the C-H bond length; while the spectral densities are the Fourier trasforms of the following orientational self-correlation functions:

$$J_m(\omega) = \sum_n \frac{1}{2}\int_{-\infty}^{+\infty} [< D_{m,n}^{(2)*}(\Omega(0)) D_{m,n}^{(2)}(\Omega(t)) > - |< D_{m,n}^{(2)} > \delta_{0,m}|^2] e^{i\omega\tau} d\tau \qquad (10)$$

where $D_{m,n}^{(l)}(\Omega)$ are Wigner functions and $\Omega$ represents the three Euler angles which describe the orientation of the molecular frame in the laboratory frame. The correlation function appearing in Eq. 10 can be calculated once the conditional probability, which is the solution of Eq. 8, is known:

$$< D_{m,n}^{(2)*}(0) D_{m,n}^{(2)}(t) > = \int\int D_{m,n}^{(2)*}(\Omega_0) D_{m,n}^{(2)}(\Omega) P(\Omega_0) P(\Omega_0, \Omega, t) d\Omega d\Omega_0 \qquad (11)$$

This equation represents the link between the model parameters $D_{\|}$ and $D_{\perp}$ (for spinning and tumbling diffusion respectively) and the relaxation rate. The solution of Eq. 8 in terms of diffusion coefficients has been carried out by P.L. Nordio et al.[27,29] and J.H. Freed[30].

Unfortunately, the heteronuclear dipolar mechanism is not the only one inducing the relaxation of the $^{13}C$ nucleus; other mechanisms, such as the chemical shift anisotropy, spin-rotation interaction direct coupling anisotropy, etc. are modulated by molecular reorientation. Nevertheless, the pure dipolar part of the relaxation can be isolated by measuring the Nuclear Overhauser Enhancement ($\eta_{sp}$) factor, as it is given by the following equation:

$$\frac{1}{T_{1D}} = \frac{\eta_{sp}}{\eta} \frac{1}{T_{1sp}} \qquad (12)$$

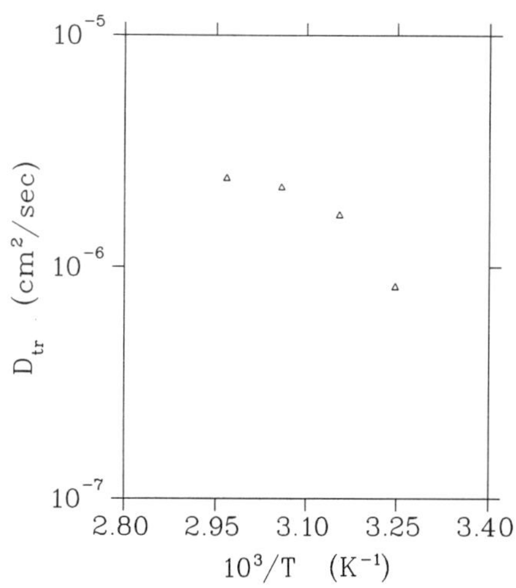


Fig.3. Translational diffusion coefficient at different temperatures within the nematic phase, as deduced from the low $Q$ behavior of the translational broadening.

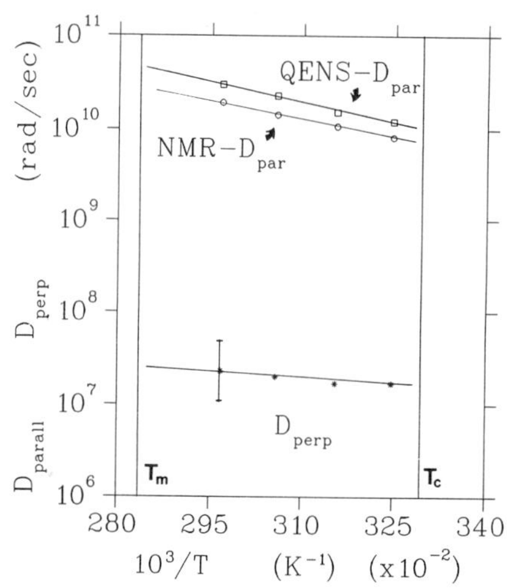


Fig.4. Rotational diffusion coefficients $D_{\|}$ (squares) from QENS and $D_{\|}$ (circles) and $D_{\perp}$ (stars) from NMR as a function of the inverse temperature in the nematic phase.

where $T_{1sp}$ is the measured relaxation time. Unlike the case of liquids composed of spherical molecules, in this case the calculated $\eta$ factor is not simply $1/2(\gamma_H/\gamma_C)$ but is expressed in term of spectral densities[31]:

$$\eta = \frac{\gamma_H}{\gamma_C} \frac{6J_2(\omega_H + \omega_C) - J_0(\omega_H - \omega_C)}{J_0(\omega_H - \omega_C) + 3J_1(\omega_C) + 6J_2(\omega_H + \omega_C)} \tag{13}$$

so that the extraction of the dynamical parameters from the measured relaxation times has to be performed using an iteration procedure, which has proven to be quickly convergent for the orthophenyl carbons in usual nematogens[31].

### *3.3. Results*

Fig. 5 represents the $^{13}C$ spectrum of EBBA in the nematic phase together with the assignment of the lines. We focused our attention on the relaxation of the orthophenylic carbons (the four highest peaks in the spectrum) because the collective director fluctuations can be proven to be quite ineffective in inducing relaxation of these carbons[32], thanks to the fact that the angle between the C-H bond and the molecular axis is very close to the magic angle (~55°). As a consequence, this contribution can be totally neglected in the analysis of the relaxation times, and this greatly simplifies the treatment of the data.

The relaxation time measured at a single Larmor frequency does not allow the determination of the two model parameters; to determine both parameters as well as to test the model, at least three Larmor frequency $T_1$ measurements are necessary. Our experiments have been performed at 2.35 T, 4.7 T and 9.4 T and the measured relaxation times within the nematic phase are displayed in Fig. 6 as a function of temperature.

As mentioned in the previous paragraph, the pure dipolar relaxation time can be deduced from the measured quantity through Eq. 12, which requires the measurement of the NOE factor and the use of an iteration procedure, due to the spectral density dependence of the $\eta$ factor (Eq. 13).

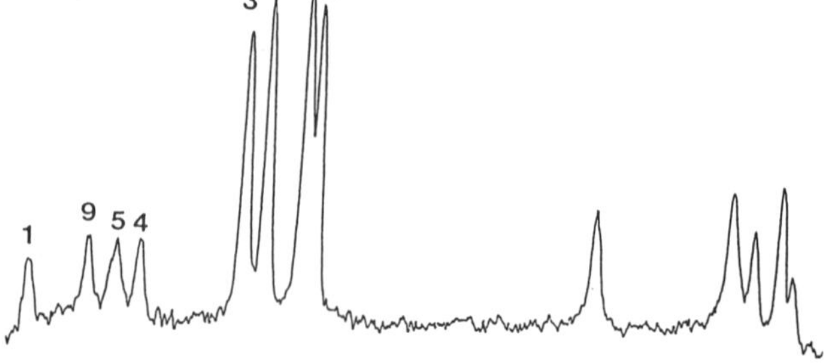


Fig.5. $^{13}C$ spectrum of EBBA in the nematic phase; the lines were assigned in agreement with Ref.16.

On the other hand, the values of the two diffusion coefficients are extracted from the relaxation times at the various Larmor frequencies by adopting a global fit.

The final values of the diffusion coefficients reproduce the values of the dipolar relaxation time within the experimental error, which confirms the goodness of the adopted model. These values are plotted versus inverse temperature in Fig. 4 together with the values of $D_{\parallel}$ measured in the QENS experiment. The agreement between the values extracted using the two different techniques appears quite good.

The very high ratio of the two different diffusion coefficients (~$10^3$), much higher than what is predicted by classical hydrodynamic theory (~10), suggests the presence of an internal rotation of the aromatic rings around their para axis. This internal rotation can be easily taken into account in the developement of the theory of the magnetic relaxation[33] but, as a consequence of the relatively small angle (6°)[23] between the phenyl para axis and the long axis of the molecular core, the fit cannot distinguish between internal and overall rotation. The investigation of the interphenyl carbon (referred to as nr. 5 in Fig. 5) will hopefully allow the distinction of the two different dynamical processes.

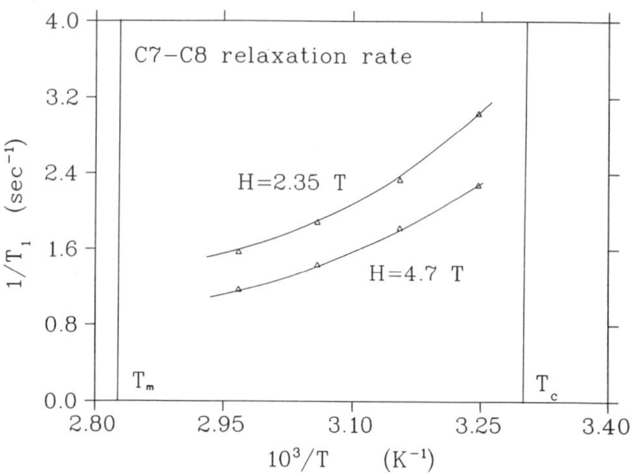


Fig.6. Measured $^{13}C$ spin lattice relaxation times at two different values of the applied field. The lines are simply guides for the eye.

## References


1. H. Ervet, F. Volino, A.J. Dianoux and R.E. Lechner, J. Physique Lett. **35** L-151 (1974).
2. F. Volino, A.J. Dianoux and H. Hervet, J. Physique Colloq. **37** C3-55 (1976).
3. M. Richardson, A.J. Leadbetter, C.J. Carlile and W.S. Howells, Molec. Phys. **35** 1697 (1978).
4. A.J. Leadbetter and M. Richardson, Molec. Phys. **35** 1191 (1978).
5. A.J. Leadbetter, M. Richardson and J.C. Frost, J. Physique Colloq. **40** C3-125 (1979).
6. A.J. Leadbetter, M. Richardson and J.C. Frost, Molec. Phys. **45** 1163 (1982).
7. D.H. Bonsor, A.J. Leadbetter and F.P. Temme, Molec. Phys. **36** 1805 (1978).
8. M. Richardson, A.J. Leadbetter, D.H. Bonsor and G.J. Kruger, Molec. Phys. **40** 741 (1980).
9. K. Chledowska, D. Chrusciel, J. Chrusciel, B. Janik, J.A. Janik, J.M. Janik, J. Krawczyk and K. Otnes, Liquid Crystals **1** 127 (1986).
10. P. Nguyen, J. Krawczyk, D. Chrusciel, J. Chrusciel, J.A. Janik, J.M. Janik, K. Otnes, H. Kresse, I. Natkaniec, S. Urban and S. Wrobel, Liquid Crystals **1** 561 (1986).

11. B. Cvikl, Liquid Crystals **2** 149 (1987).
12. T.M. Barbara, R.R. Vold and R.L. Vold, J. Chem. Phys. **79** 6338 (1983).
13. R.R. Vold, 1983 *Nuclear Magnetic Resonance of Liquid Crystals* edited by J.W. Emsley (D. Reidel Pub. Comp.) p.253.
14. R.Y. Dong and K.R. Sridharan, J. Chem. Phys. **82** 4838 (1985).
15. R.Y. Dong, J.Chem. Phys. **88** 3962 (1988).
16. J.S. Lewis, E. Tomchuk and E. Bock, Liquid Crystals **5** 1033 (1989).
17. F. Noak, M. Notter and W. Weiss, Liquid Crystals **3** 907 (1988).
18. M.P. Fontana, B. Rosi and M. Riccò, Physica B **156-157** 563 (1989).
19. C.J. Carlile, "*High Resolution Inelastic Spectrometer*" Proceeding of the 4$^{th}$ collaboration on advanced neutron sources (KENS report II) p.588 (1980).
20. M. Bèe, "*Quasielastic Neutron Scattering*" (1988) Adam Hilger Pub. Comp.
21. M.P.Fontana, B.Rosi, N.Kirov and I.Dozov, Phys.Rev. A **33** 4132 (1986).
22. A.J. Dianoux, F. Volino and H. Hervet, Molec. Phys. **30** 1181 (1975).
23. J. Howard, A.J. Leadbetter and M. Sherwood, Mol. Cryst. Liq. Cryst. **56** 271 (1980).
24. M. Bèe, A.J. Dianoux, J.A. Janik, J.M. Janik and R. Podsiadly, to be published in Liquid Crystal.
25. M.P. Fontana, M. Riccò and B. Rosi, to be published in Phys. Rev. Letters.
26. M.H. Levitt, R. Freeman and T. Frenkiel, J. Magn. Reson. **47** 328 (1982).
27. L. Nordio and P. Busolin, J. Chem. Phys. **55** 5485 (1971).
28. A. Abragam, *The principles of Nuclear Magnetism* (Oxford University Press), (1961).
29. P.L. Nordio, G. Rigatti and U. Segre, J. Chem. Phys. **56** 2117 (1972).
30. J.H. Freed, J.Chem. Phys. **66** 4183 (1977).
31. M. Riccò, Ph.D Thesis, University of Parma (A.A. 1987-88).
32. P.L. Nordio and U. Segre, Gazz. Chim. It. **106** 431 (1976).
33. R.Y. Dong, Mol. Cryst. Liq. Cryst. **141** 349 (1986).

# FERROELECTRIC LIQUID CRYSTALS

Chapter 21

# THE LAYER AND DIRECTOR STRUCTURES OF FERROELECTRIC LIQUID CRYSTALS

T. P. RIEKER and N. A. CLARK

## 1. Introduction

Liquid Crystals are a fascinating state of matter which have long been studied for their rich and diverse physical properties[1,2]. These materials have become of interest in a variety of applications, mainly due to their large optical anisotropy which is easily manipulated by applied fields and surface interactions. Twisted nematic (TN) devices have proved to be phenomenally useful and as a result, their response to applied fields and surface interactions is well understood.

Since Clark and Lagerwall[3] demonstrated the bistable Surface Stabilized Ferroelectric Liquid Crystal (SSFLC) device concept in 1980, a substantial effort has been made to understand in detail the structural and electro-optic characteristics of ferroelectric liquid crystals in thin cells[4]. The structures of such devices are inherently more complex than those found in TN devices due to the molecular layering of these phases. Additional complexities arise in the presence of surface interactions, which with ferroelectric liquid crystals (FLCs) influence both the molecular and layer orientations. This paper discusses the physics of layer and director organization in FLCs, as revealed by X-ray and optical studies[5,6].

Ferroelectric Liquid Crystal (FLC) phases are layered, having local orientation states described by the layer normal, $\hat{z}'$ the director, $\hat{n}$ (tilted from $\hat{z}'$ through the equilibrium angle, $\theta$) and the FLC polarization density[7], $\mathbf{P} = P_o\hat{z}' \, x \, \hat{n}$. The azimuthal orientation $\Phi$ of the $\hat{n}$ - $\mathbf{P}$ couple about $\hat{z}'$ is a nematic-like variable controlled by Frank elasticity electric fields and surface interactions. This basic architecture is shown in Fig.1. In a bulk FLC structure $\Phi$ varies linearly with position as one moves along $\hat{z}'$, with $\hat{n}$ and $\mathbf{P}$ precessing about $\hat{z}'$ to form a helical structure, a result of the FLC chirality.

Of interest in this paper are the layer and director structures found in the so-called SSFLC cells. For electro-optic applications FLCs are generally prepared between solid substrates as the dielectric in a transparent parallel plate capacitor. In the SSFLC geometry the layers are terminated by interfaces at which there are local energy barriers to change of $\Phi$. The resulting boundary condition on the $\Phi$ field (surface stabilization) leads to multiple stable orientation distributions $\Phi(r)$ (states) and the suppression of the bulk helix[3,4,8]. Application of

T. P. Rieker and N. A. Clark - Condensed Matter Laboratory, Department of Physics and Optoelectronic Computing Systems Center, University of Colorado, Boulder 80309 (U.S.A.)

*Phase Transitions in Liquid Crystals*, Edited by S. Martellucci and A.N. Chester, Plenum Press, New York, 1992

a voltage to the capacitor produces an electric field component in the layer planes and enables selection among the stable states.

To make a useful SSFLC device, the cell preparation step must achieve two distinct but related results. First an ordered layer structure (alignment) must be established, which generally means that the intersections of the layers with the plates are made to be linear and normal to a preselected direction. Secondly the boundary conditions on the $\Phi$ field must produce the desired electro-optic response.

For most device applications SSFLC alignment must be imposed by making the bounding plate surfaces suitably anisotropic, as is the case for nematic devices and allowing the FLC material to self-organize as it cools into the desired FLC phase. This process establishes the basic layer structure and the boundary conditions on $\Phi(r)$. This dual role of the alignment step along with an incomplete understanding of SSFLC cell physics has made the achievement of a completely satisfactory SSFLC alignment technology a significant challenge.

The simple model of the layer structure in SSFLC cells proposed by Clark and Lagerwall[3] has been successful in explaining the fundamental operation of such cells. However, this model doesn't adequately explain the occurrence of layering defects known as zig-zag walls which spontaneously form upon cooling into the smectic C phase. In addition, multistate switching and the nucleation and subsequent motion of internal disclination loops in the director field during switching[9] are left unexplained. Rondelez[10] and Lockhart and coworkers[11], while making independent determinations on the refractive index of thin $S_C^*$ samples, pointed out that the layer organization in such cells is uncertain.

In bulk samples, X-ray scattering experiments on smectic C samples aligned in a mag-

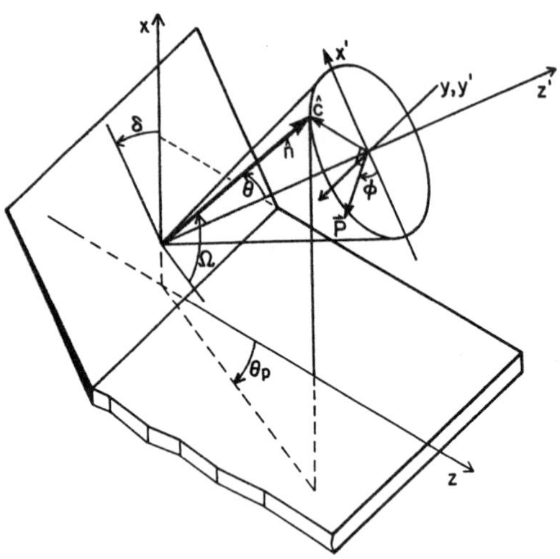


Fig.1. Geometry used to describe the smectic phases, in which we use one coordinate system affIxed to the bounding plates (x,y,z) and another fixed to the smectic layers ($x'$, $y'$, $z'$). $\delta$ is defined as the angle between the plane of the smectic layer and the bounding plate normal. For $\delta = 0$, $z$ and $\hat{z}'$ are coincident. In the $S_C$ phase, $\hat{n}$ is restricted to lie on the cone defined by the angle $\Theta$ centered on $z'$. Chiral molecules give rise to a ferroelectric polarization, $\mathbf{P} = P_0\,(\hat{z}' \times \hat{n})$. The azimuthal orientation of the $\hat{n}$ - $\mathbf{P}$ couplet is indicated by $\Phi$. $\Theta_p$ is the angle between the projections of $\hat{n}$ and $\hat{z}'$onto the substrate. The elevation angle, $\Omega$, is the angle between n and the plane of the substrate. For the $S_A$ phase, $\Theta$, $\Theta_p$ and $\Phi$ are zero and there is no ferroelectric polarization, $\mathbf{P} = 0$.

netic field ($\hat{n}$ // **B**) demonstrated that these samples self-organize into domains with conical layers with the cone axis and the director parallel to the applied **B** field[12]. Pelzl and coworkers proposed the same conical layer structure in thin smectic C samples aligned with an **E** field[13].

The experiments detailed here address the question of the equilibrium Local Layer Structure (LLS) in thin smectic C samples aligned in the absence of external fields. Based on

a

b

c

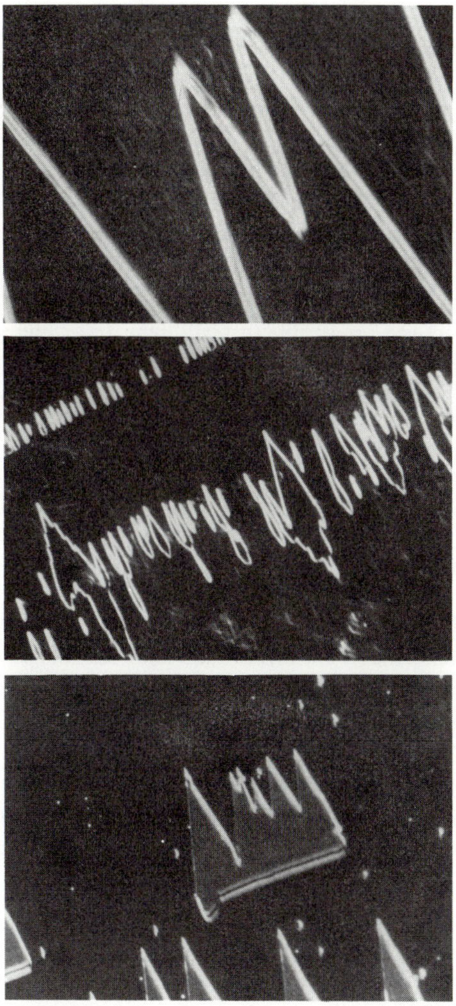

Fig.2. Typical zig-zag defects in SSFLC cells as they appear in polarized transmitted light microscopy. Arrows indicate the $S_C$ layer direction. Zig-zag defects are a characteristic feature of SSFLC cells with the chevron local layer structure. The zig-zag defect consists of narrow walls running nearly normal to the $S_C$ layers and broad walls running parallel to the layers, often forming loops as shown in (c). The detailed structure of the narrow and broad walls in a loop is shown in Fig.6. (a) A narrow zig-zag line in a 1.5 $\mu$m thick rubbed polymer *planar* aligned SSFLC cell. The cell is well-aligned giving excellent extinction between crossed polarizers except at the zig-zag line which transmits. (b) At lower magnification, arrays of zig-zag lines significantly degrade extinction. (c) Zig-zag loop in a 4 $\mu$m thick cell with zero applied electronic field. In a color picture color difference across the loop resulting from the chevron interface being displaced from the cell midplane in a direction toward the reader inside the loop and away from the reader outside the loop could be observed. See also the detailed 3d structure of a typical loop shown in Fig.19.

X-ray scattering and polarized light transmission microscopy experiments, we have been able to establish the local layer structure and the principles of layer organization in thin (~ 3 $\mu$m) smectic C cells. These investigations not only demonstrated that the origin of the zig-zag walls is layer geometry, but also explained many other features of SSFLC cells.

Consider, for example, Fig.2. This picture shows an SSFLC cell viewed in transmitted light through crossed polarizers and oriented so as to have the cell optic axis along the polarizer.

The FLC, aligned using a rubbed polymer coating on the plates, produces excellent extinction of the transmitted light, $\sim 10^{-3}$, everywhere except on the dramatic zig-zag line. There light leaks through and substantial numbers of such lines can seriously degrade the contrast or alter the optical appearance of of SSFLC cells, as Fig.2 b and c show. This spontaneously formed "zig-zag" defect is an example of an SSFLC cell structure which needs to be controlled if optimum devices are to be made.

Furthermore, such structures provide invaluable clues to understanding the principles of smectic C layer self-organization and electro-optic response, in the way that defects have been used to reveal the structures of LC phases since the time of Friedel[14].

In this paper we discuss the results of combined X-ray and optical experiments which show that the layer thermal contraction in the smectic C phase strongly influences the layer structure adopted and the boundary condition on $\Phi(r)$ and that the layer structure must be known if the director related electro-optic properties are to be at all understood.

## 2. Probing The Local Layer Structure In Ferroelectric Liquid

### *2.1. Experimental Introduction*

The experiments detailed here were undertaken for three main reasons. First, to understand the layer structure present in surface stabilized ferroelectric liquid crystal (SSFLC) cells. Second, to correlate the layer structure with optical and other observations. Third, to gain an understanding of the overall director layer interactions and the resultant structure.

To this end, SSFLC cells were prepared with various surface treatments for a given liquid crystal and studied by X-ray and optical techniques. The cells created for these experiments employed standard preparation methods with slight modifications for the X-ray experiments. This allows for a direct comparison of the results to known physical and electro-optic properties of such cells.

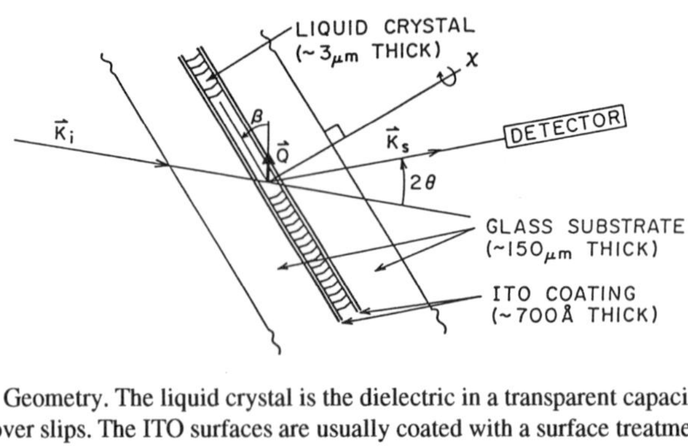


Fig.3. The Scattering Geometry. The liquid crystal is the dielectric in a transparent capacitor formed by a pair of (ITO) coated glass cover slips. The ITO surfaces are usually coated with a surface treatment.

## 2.2. General Cell Design Strategy

Surface Stabilized Ferroelectric Liquid Crystal (SSFLC) cells are prepared as shown in Fig.3 with the liquid crystal as the dielectric in a transparent capacitor formed by a pair of closely spaced ($t \approx 3\mu m$) indium-tin oxide (ITO) coated glass cover slips. The ITO coating allows for the application of an electric field **E** directly to the sample, perpendicular to the plates, without obstructing the transmission of light through the cell.

The SSFLC geometry requires parallel uniform spacing of the bounding plates for mechanical suppression of the bulk helix and for a constant phase shift of light which is transmitted through the cell. The most accurate spacing is accomplished by using photolithographic techniques in producing spacers from silicon monoxide, silicon dioxide and polyimide. Thin strips of Mylar or Teflon sheeting also work well and are often used for convenience.

The parallel planar bounding surfaces stabilize two orientation states of the director field in the smectic C phase. These surfaces however do not yield long range alignment of the director field. Aligned cells can be produced by shear flow, electromagnetic fields and anisotropic surface treatments. Surface treatments not only yield single layer-domains but also profoundly alter the liquid crystal-boundary surface interaction energy. These interactions orientationally constrain the director $\hat{n}$ at the surface, strongly influencing the orientation of the director field throughout the cell and can be tailored by the choice of surface treatment. Surface interactions can be strong enough to produce surface memory effects (SME)[15]. In such cells, optically observed textures of various phases can be imprinted on the surface. These textures are then visible in other, generally less ordered phases, when a thin surface layer of liquid crystal adheres to this imprint.

Those surface treatments which constrain the director field to lie in the plane of the bounding plates at the surface are said to produce planar alignment. Shear flow using clean or ITO coated glass and many buffed polymers on glass fall into this class. To produce a single layer-domain using shear flow, one of the bounding plates is moved relative to the other (usually done in the smectic A phase). This aligns the molecular long axis normal to the direction of motion and the layers along the direction of motion due to the anisotropic flow properties of the molecules. Extensive shearing of the plates causes homeotropic alignment of the director field ($\hat{n} \perp$ to the plates). Heating to the isotropic or cooling into a crystalline phase will destroy alignment.

Polymer surface treatments are applied on glass surface as thin layers of dilute solutions of polymer in an appropriate solvent. Anisotropic rubbing of these thin polymer layers results in orientation of the polymer chains. Polymer chain orientation and polymer crystallinity seem to be necessary to produce alignment of the liquid crystal director field[16]. In materials which exhibit the $S_A$ to $S_C$ phase sequence, the smectic layers form roughly normal to the buffing direction at the surface, independent of an applied **E** field. This is a consequence of layer formation in the smectic **A** phase where the molecules align approximately along the buffing direction. SSFLC cells prepared with such materials and with polymer alignment layers will in general maintain alignment when cycling the cell through the isotropic, liquid crystal and crystalline phases.

Patel and Goodby have demonstrated that compounds which exhibit the $N^*$ - $S_C^*$ phase transition in the SSFLC geometry can be aligned by the application of an electric field across a cell with a buffed polymer surface treatment[1]. With this method two domains of smectic layers form at large angles to the buffing direction normal.

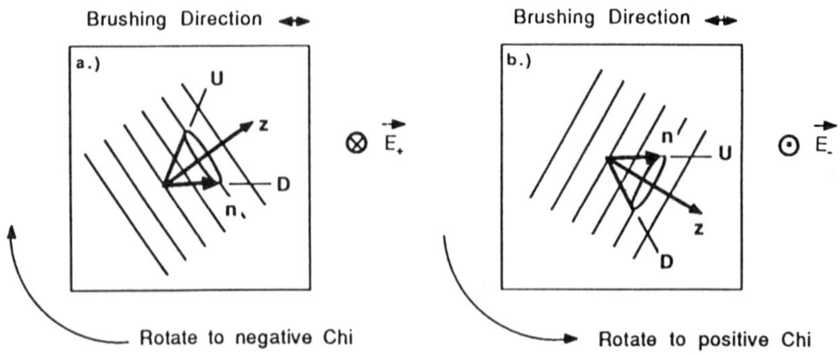


Fig.4. Smectic layer orientation as a consequence of an applied **E** field when cooling through the N*-$S_c^*$ phase transition. This is a result of the system simultaneously satisfying the imposed alignment direction, the large director tilt angle Θ, surface stabilization and the alignment of the polarization (**P**) along **E** . The sign conventions are for a material with a positive polarization (**P**=$P_0(\hat{z}'\mathrm{x}\hat{n})$), such as SCE-10 and are confirmed experimentally, see Fig.15 (a,b,c).

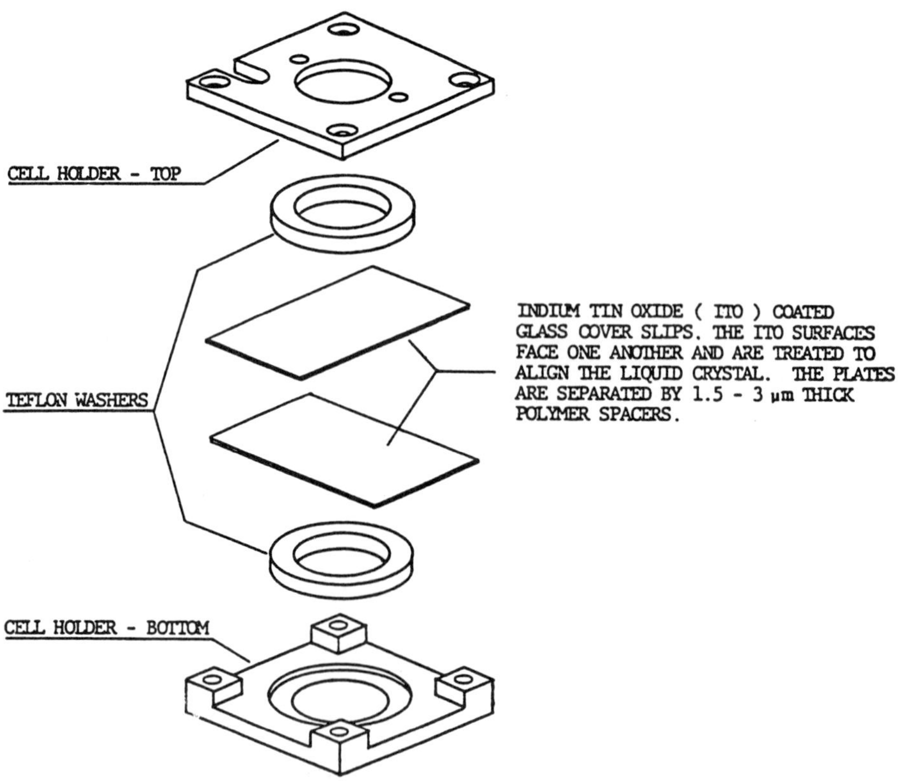


Fig.5. Scattering Cell Assembly.

This is a consequence of the system simultaneously satisfying the imposed alignment direction, the large director tilt angle Θ, surface stabilization and the alignment of the polarization (**P**) along **E**, see Fig.4. These cells do not maintain their layer organization when cycled through the various phases in the absence of an applied **E** field. Some anisotropic surface treatments produce uniform director alignment at the surface in which the director lies out of the plane of the bounding plates. This pretilt Ω of the director field relative to the bounding surface can vary from a few degrees to 90° (homeotropic alignment). Surface treatments which give a nonzero pretilt are lecithin, silanes, obliquely evaporated silicon monoxide (SiO) and some polymers.

Janning[17] demonstrated that nematic liquid crystals align very well on obliquely evaporated silicon monoxide (SiO) surfaces. Urbach, Boix and Guyon[18] were the first to demonstrate the alignment of smectics using this technique. Large, well aligned, defect free regions can be produced in the smectic phases by exploitation of a nematic phase above the smectic phases[19]. The surface induced pretilt, Q and the director orientation relative to the direction of evaporation are dependent on the angle of incidence of the deposited SiO layer[18]. The angle of incidence, $\theta_d$, is measured from the substrate normal to the line joining the substrate and the evaporation source (the direction of evaporation is defined as the direction of particle motion from the source at the substrate). For $\theta_d < 60°$, poor alignment is obtained. For $60° < \theta_d < 80°$ the director generally aligns perpendicular to the direction of evaporation. For larger angles the director is aligned along the direction of evaporation with Ω opening toward the direction of incidence. Smectic phases aligned with obliquely deposited SiO are of special interest to us because large defect free regions can readily be created. We wish to understand the difference in the local layer structure present in these cells in the light of our recent work in determining the origin of the zig-zag wall defect found in cells with planar alignment of the director field. The SiO treated surfaces are thought to cause alignment through steric interactions of the liquid crystal molecules with columnar SiO structures[20]. SiO treated surfaces maintain alignment through temperature cycling, but show a strong surface memory effect.

## 2.3. X-Ray Cell Design

The surface treatments used in these experiments were buffed polymers, clean ITO and obliquely evaporated SiO. The glass thickness was minimized to keep X-ray attenuation and stray scattering at negligible levels. In cells with a polymer alignment layer, one of the ITO coated plates had a 1 cm diameter, 3 $\mu$m thick polyimide ring fabricated upon it as a spacer. Both plates were subsequently coated with a $\approx 100$ Å thick nylon layer and were brushed or buffed to produce uniform planar alignment. For cells with an SiO alignment layer, both plates were coated with $\approx 300$ Å of SiO (measured normal to the plates) deposited at an 82° angle of incidence. In assembly of the cells the direction of evaporation on the two plates was aligned both parallel and antiparallel. In addition, a hybrid cell with an SiO and a nylon surface treatment was produced. To avoid contamination with a polymer, plates with obliquely evaporated SiO surfaces were separated by 1.5 $\mu$m thick Mylar strips. In these cells however, the sample thickness was $\approx 2$ $\mu$m, due to liquid crystal having wicked between the spacers and the plates.

The plates were clamped together in a cell holder as shown in Fig.5. To obtain the best alignment and avoid surface memory effect, the cells were filled in the isotropic phase then slowly cooled into the liquid crystalline phases. For samples which exhibit the chiral nematic to smectic C* phase transition a $\pm 2$ volt DC field was applied to enhance alignment. Wires were affixed to the ITO surfaces with conducting epoxy.

Table 1. SSFLC cells, filled with materials that exhibit the phase sequence $I \leftrightarrow S_A \leftrightarrow S_C^*$, studied by X-ray scattering. Brushed indicates thet the nylon surface treatment was brushed with an artisan's brush; buffed indicates, buffed with a buffing wheel. The arrows indicate the relative direction of brushing on the two plates. It was determined that the brushing or buffing direction had no effect on the LLS of these cells. HMDS is hexamethly-disilazane, a surfactant (see Refs. 21,22).

| Materials ⇒<br>Surfaces ↓ | W7-W82 [21]<br>Symb. Ref. to Fig. 9 | DOBAMBC [22] | W82 [21] |
|---|---|---|---|
| brushed nylon | ◆ both ↑↑, ↑↓ | Fig. 13 | Fig. 14 |
| buffed nylon | ◔, □, △, ▽, ■, ▼ | | |
| buffed PVA | ▲ | | |
| ITO (shear) | not plotted | | |
| brushed nylon (25 μm) | not plotted | | |
| nylon / HMDS | ● | | |

Table 2. SSFLC cells, filled with materials that exhibit the phase sequence $I \leftrightarrow N^* \leftrightarrow S_C^*$ studied by X-ray scattering. This sample was studied only on the rotating anode at EXXON (see Refs. 23).

| Materials ⇒<br>Surfaces ↓ | SCE-10 [23] |
|---|---|
| brushed nylon | Fig 15 a,b,c |

Table 3. SSFLC cells, filled with materials that exhibit the phase sequence $I \leftrightarrow N \leftrightarrow S_A \leftrightarrow S_C^*$ studied by X-ray scattering. The arrows indicate the orientation of the evaporation direction on the bounding plates (see Refs. 24,25,26).

| Materials ⇒<br>Surfaces ↓ | CS-1014 [24]<br>The symbols refer to Fig. 17 | ZLI-3654 [25] | SCE-3 [26] |
|---|---|---|---|
| brushed nylon | ○ (c) | □ (b) | |
| SiO ↑↑ | ■ (c) | △ (b) | |
| SiO ↑↓ | ◆ (a) | ○, □ (a) | ●, ■ (a) |
| SiO / brushed nylon | | ○ (b) | |

Table 4. Spectrometer comparison.The spectrometer at EXXON had a Rigaku RU-300 18 kW rotating copper anode source and a bent graphite monochrometer. The SSRL spectrometer uses synchrotron radiation and a double crystal (graphite) monochrometer. The resolution of a spectrometer is given by the FWHH of the arm zero scans; see Fig.6. The slit size are listed as in-plane / out-of-plane. Monitor counts are the number of counts wich the monitor reports before the number of counts at the detector is recorded, *i.e.*, a data point is generated. The intensity in the number of counts/sec per scattering volume per solid angle for a scattering peak in $\beta$ for the same cell at the same temperature, studied on the two spectrometers. The scattering volume is defined by the slits before the sample, wich determine the spot size and the sample thickness. The solid angle is that subtended by the detector as defined by the slits before the detector.

| | EXXON | SSRL |
|---|---|---|
| Wavelength($\AA$) | 1.54056 | 1.53978 |
| Slits before the sample | 1 / 2–3 (mm) | 100 / 100 ($\mu$m) |
| Slits before the detector | 1 / 4 (mm) | 380 / 500 ($\mu$m) |
| Resolution | 0.1° | 0.04° |
| Typical monitor counts | $5 \times 10^4$ | 65536 |
| Intensity ($\frac{Counts}{sec-\mu m^3}$) | 0.2 | $7 \times 10^3$ |

Tables 1,2,3 list the details of the SSFLC cells studied by X-ray scattering. These cells were evaluated optically by transmission polarized light microscopy, prior to X-ray scattering, to ensure uniform alignment and that the bulk director helix was suppressed. Homogeneous, defect free regions were identified for study by X-ray scattering by noting the position of the region with a calibrated reticle.

## *2.4. X-Ray Scattering*

The local layer structure of the cells detailed above were determined by X-ray experiments conducted at the EXXON Reaserch and Engineering Co. and at the Stanford Synchrotron Radiation Laboratory (SSRL), see Table 4 for a comparison. The 1 x 3 mm spot size of the EXXON spectrometer encompasses many zig-zag wall domains. At SSRL the beam was sufficiently small (100 x 100 $\mu$m) that the homogeneously structured regions separated by the zig-zag walls could be probed and the true local layer structure (LLS) observed. These areas were selected via polarized light microscopy to show no visible defects. The sample is mounted in a two stage oven, which provides temperature stability to $\pm$4mK. This oven has three mutually perpendicular linear translation stages for positioning the sample in the beam. The oven is in turn mounted to a four-circle goniometer. The sample is oriented normal to the incident X-ray beam and goniometer zeroes are defined.

The detector arm orientation ($2\theta$) was set to the approximate angle for smectic layer Bragg diffraction. The goniometer angles, $\chi$, $\theta$ and $2\theta$ were varied to locate and maximize the Bragg peaks. Different regions of a given sample were studied to probe the variation of the structure with position. Layer orientation data is presented as plots of scattering intensity I($\beta$) vs. $\beta$ and I($\chi$) vs. $\chi$, for a fixed orientation in $\chi$, $\beta$ temperature and applied voltage where $\beta$ is defined as $\theta$ - $2\theta/2$. Peak positions in the I($\beta$) vs. $\beta$ plots yield the tilt ($\delta$) of the smectic layers with respect to the normal of the bounding plates. Peak positions in the I($\chi$) vs. $\chi$ plots yield the rotation of the smectic layer normal $\hat{z}'$, away from the bufflng direction (Fig.6).

For our scattering geometry, $\chi$ scans give a measure of the 'straightness' of the layers along *y*. Typical peak widths, (FWHH), for $\chi$ scans are 10 - 20°, indicating a small curvature of the layers with respect to *y*. Typical step size was 1.0° for both spectrometers. see Fig.7 (a). For samples which exhibit the smectic A to smectic C* phase transition, the peak in $\chi$ was typically within $\approx$ 4° of the rubbing direction. Smectic layer Bragg scattering was observed at large values of $\chi$ for samples with a chiral nematic to smectic C* phase transition. The location and width of the peak in $\chi$ remained fixed as the temperature was varied in all samples, except for a cell prepared with brushed nylon surfaces and filled with ZLI-3654. In this sample, the $\chi$ peak split into two peaks (at the $S_A$ - $S_C^*$ transition symmetrically disposed about the peak position in the smectic A phase, which moved apart, to increasing $\pm\chi$ as the sample was cooled). This is believed to be due to the presence of large numbers of zig-zag walls.

Scans in $\theta$ measure $\delta$, the tilt of the layer with respect to the normal to the bounding plates. Results are reported as intensity as a function of $\beta$, where $\beta$ = 0 corresponds to smectic layers normal to the bounding plates. Typical peak widths (FWHH) in $\beta$ are 2°, indicating highly planar layers; see Fig.7 (b). Typical step sizes were 0.4° at EXXON and 0.2 - 0.4° at SSRL. In all samples, in the smectic C* phase, the peak position in $\beta$ moves to larger $|\beta|$ as the temperature is decreased.

Scans in $\theta/2\theta$ yield a measure of the interlayer spacing *d*, which can be calculated through Bragg's Law, $n\lambda = 2d\sin(2\theta/2)$. Both $\theta$ and $2\theta$ are varied with typical step sizes of 0.01°/ 0.02° at EXXON and 0.005°/0.01° at SSRL respectively. In all samples studied, the peak in this type of scan varied as a function of temperature in the smectic C* phase. Typical peak widths (FWHH) are 0.05°; see Fig.7 (c). These scans are not resolution limited (i.e., have a larger FWHH than the corresponding arm zero scan; see Fig.6) indicating some variance in the interlayer spacing.

## 3. X-Ray Scattering

### *3.1. Cells with Planar Alignment of $\hat{n}$*

Careful cooling through the $I$ - $S_A$ transition generally produces smectic A layers normal to the plates in the materials studied. However, the smectic A layers may also be tilted relative to the bounding plate normal at an angle $\delta_A$ that depends on the sample history. In the smectic C* phase, therefore, there are two local layer structure (LLS) scenarios, one for $\delta_A = 0$ and the other for $\delta_A \neq 0$. Typical data for $\delta_A = 0$ are shown in Fig.8 (a). In the smectic A phase a plot of scattered intensity I($\beta$) vs. $\beta$ shows a single peak at $\beta = 0$. As $T$ decreases through the $S_A$ – $S_C^*$ transition this peak splits to form a pair of peaks, symmetrically located at $\pm\delta_A(T)$.

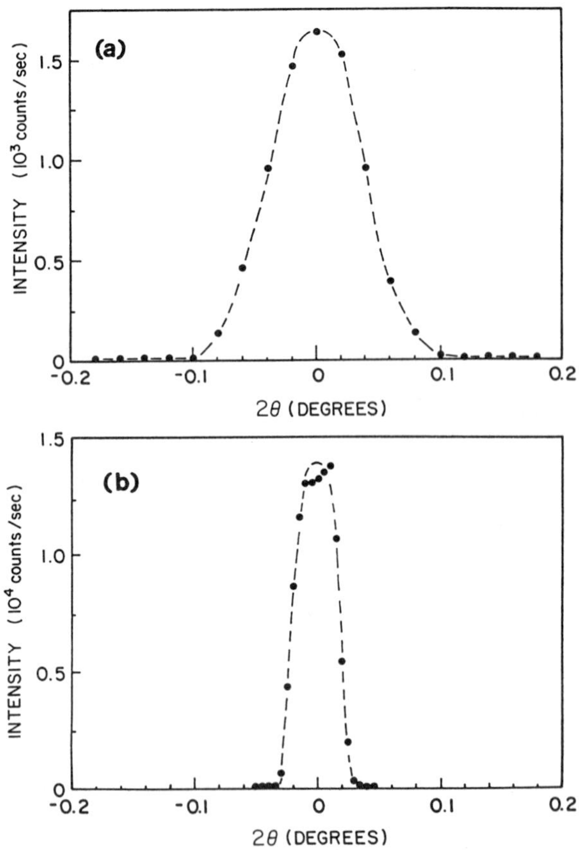


Fig.6. Typical arm zero scans, (a) EXXON, (b) SSRL. Note the high angular resolution of both spectrometers; 0.1° at EXXON and 0.04° at SSRL. The intensity of the two spectrometers can not be directly compared from these plots because the attenuators placed into the X-ray beam paths were not calibrated.

The splitting increases monotonically with decreasing temperature to saturate at about 17° in the W7-W82 mixture and 25° in DOBAMBC. At lower temperatures in the smectic C* phase, ≈ 10° C below $T_C$, the peaks in I($\beta$) become very sharp (FWHH $\Delta\beta$ = 1.2°) and contain more than 85% of the scattering intensity, indicating a LLS of highly planar tilted

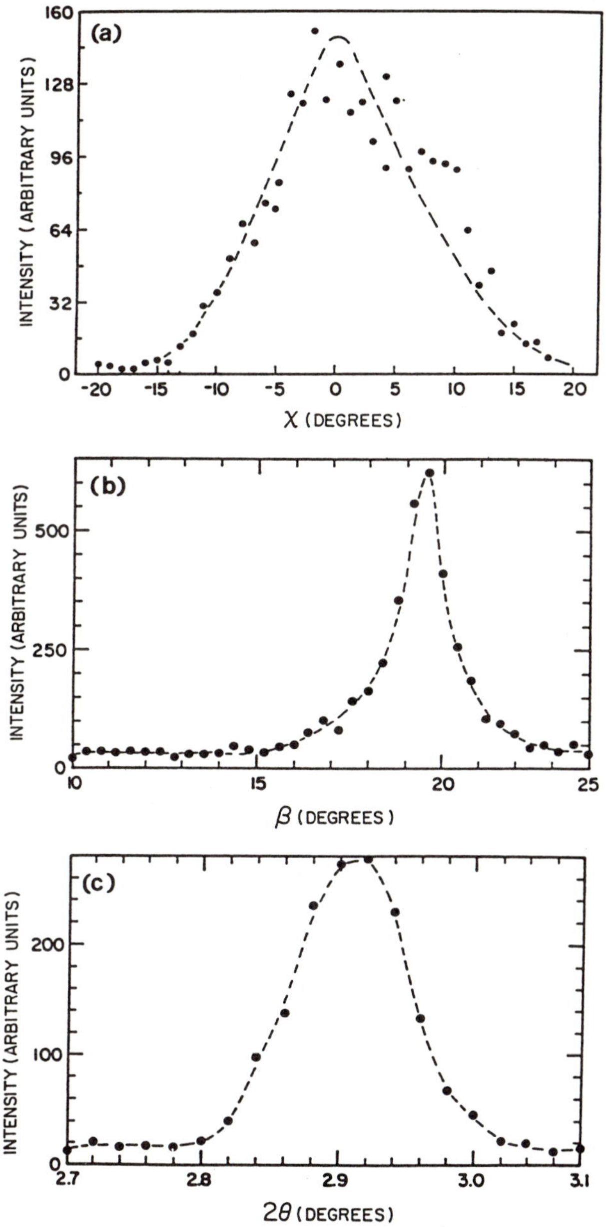


Fig.7. Typical $\chi$ (a), $\theta$ (b), $\theta/2\theta$ (c) scans[21]. These scans are for a cell prepared with a brushed nylon surface treatment and filled with W82.

layers, in nearly equal amounts. $\delta_C(T)$ is a *reversible* equilibrium function, monotonically increasing with decreasing temperature, which, as shown in Fig.9, is completely independent of surface treatment.

In some cases the two smectic C* peaks are of significantly different intensity, as shown in Fig.10 (a). In this case increasing the temperature causes the smaller peak to get weaker and disappear at a temperature $T_d$ in the smectic C* phase. For $T < T_d$ the peak positions $\delta_C(T)$ lie on the typical curve ($\nabla$ and in $\square$ Fig.9). For $T > T_d$ the location of the surviving peak becomes independent of $T$, maintaining the value $\delta_C(T)$ it had at the disappearance of the small peak, right up into the smectic A phase. This is the scenario for layers tilted in the smectic A phase ($\delta_A=\delta_C(T_d)$). It is also found to be *reversible.*

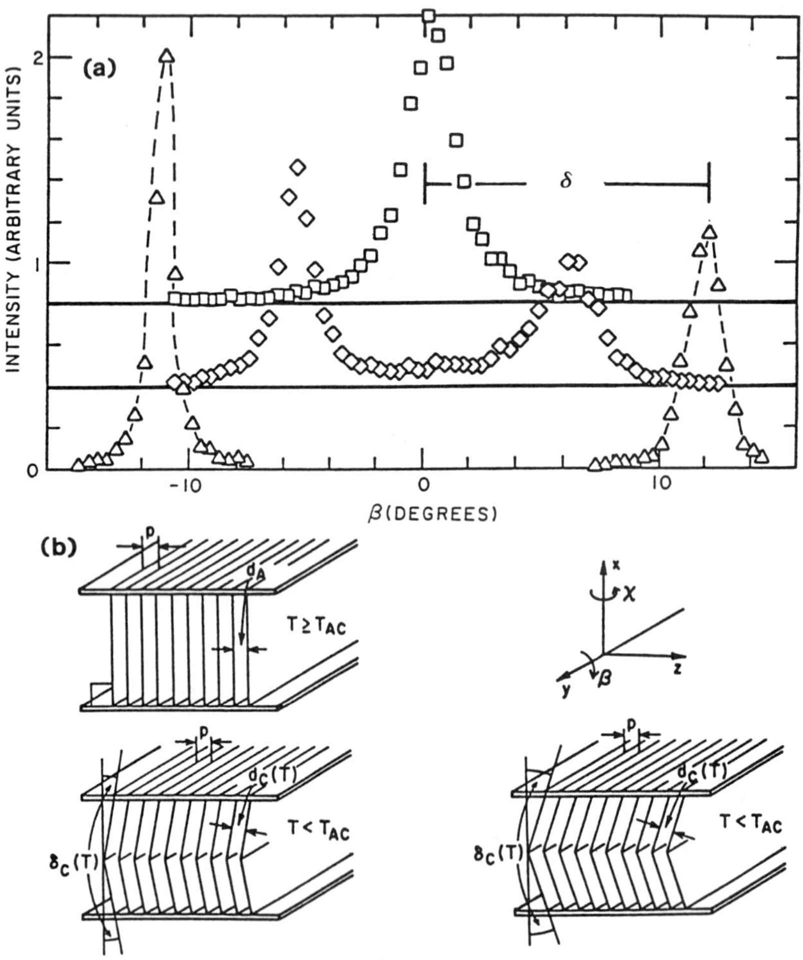


Fig.8. Plots of scattering intensity $I(\beta)$ vs. $\beta$, the angle between the substrate planes and the X-ray scattering vector **q**, from a 100 x 100 $\mu m^2$ zig-zag defect free area of a SSFLC cell. These plots indicate unambiguously that the local layer structure in our SSFLC cell is planar, but with two distinct tilts in the $S_C^*$ phase, with respect to the bounding plate normal. (a) $\delta_A = 0$ ($S_A$ layers normal to the bounding plates) ($\square$) $T = T_C + .2°C$, ($\diamond$) $T = T_C - .5°C$, ($\Delta$) $T = T_C - 2.2°C$. The dashed lines are a guide to the eye. (b) Local layer structure scenario for $\delta_A = 0$, showing the formation of symmetric chevrons for $T < T_C$

At each temperature studied a θ/2θ scan was done, and the corresponding interlayer spacing was determined using the Bragg equation;

$$d = \frac{\lambda}{2\sin(2\theta/2)}$$

This data are plotted for DOBAMBC (●) and the W7-W82 mixture (■) in Fig.11. The interlayer spacing has its maximum value, $d_A$, throughout the smectic A. The interlayer spacing in the smectic C* phase, $d_C$, has the value $d_A$ at the $S_A$ - $S_C$ phase transition and continuously decreases as the sample is cooled in the smectic C* phase. In the smectic A phase the interlayer spacings for W7-W82 and DOBAMBC are 31.4 Å and 33.2 Å respectively. As a comparison, the molecular lengths of these molecules were determined from Dreiding molecular models in their longest conformation. The lengths so determined are: W7- 30.2 Å, W82- 32 Å and DOBAMBC- 33.6 Å. At the $S_A$ - $S_C$ phase transition, no coexistence of different d-spacings is found in any sample studied. However, coexistence of different layer spacings was found in W82 at the $S_C^*$-$S_B$ transition, see below. The interlayer spacing as a function of temperature is also completely *reversible.*

The smectic C* layer tilt angle $\delta_C(T)$, determined from the peak location in I(β), is related to the smectic C* layer spacing $d_C(T)$ and the smectic A layer spacing $d_A$, (both obtained from the location of the scattering peak in 2θ as follows; see Fig.12:

$$d_C(T) = d_A \cos(\delta_C(T)) \qquad (1)$$

Electric fields of **E** < 5 V/μm applied to ferroelectric smectic C* samples had no effect on the layer tilt.

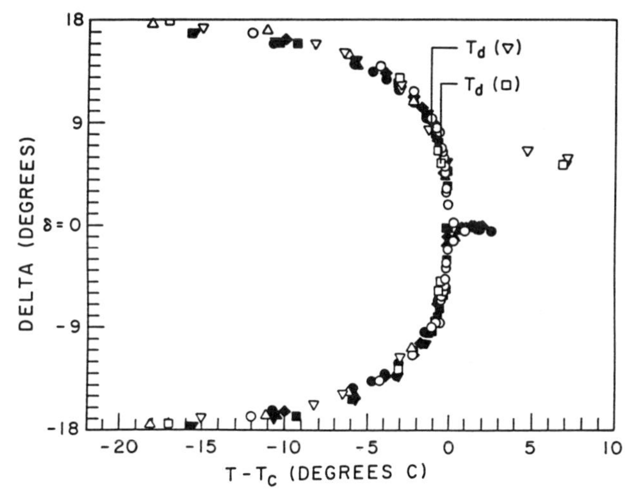


Fig.9. Layer Tilt δ as a function of $T$ - $T_C$, as obtained from the peak location in $I(\beta)$ data, showing upper and lower branches located symmetrically about δ = 0. (Solid points - rotating anode; open points - SSRL). The data yield no distinction among the various surface treatments used (see text and Table 1). With the exception of the ∇ and □ data, the points lie on a common curve with $\delta \rightarrow 0$ as $T \rightarrow T_C$ and $\delta_A = 0$ in the $S_A$ phase. For ∇ and □ the $S_A$ layers were tilted with $\delta_A \approx 7°$ in both the $S_A$ and $S_C^*$ phases for $T > T_d = T_C$ - 1°C, the temperature at which the peak on the lower branch appears with decreasing $T$. This and the fact that the ∇ and □ data lie on the common curve for $T < T_d$, indicates the formation of the asymmetric chevron for $T < T_d$.

In addition to the general behavior detailed above, the following specific experiments were performed: A thick, t $\approx$ 25$\mu$m, W7-W82 cell was prepared using a brushed nylon surface treatment to test the layering structure in the presence of the bulk FLC helix. This cell exhibited the same layer structure observed in other cells using this mixture. It is not plotted with the other data. A cell with cleaned ITO plates was prepared with the W7-W82 mixture and shear aligned. This cell showed the same layering structure as a function of temperature as all other cells prepared with this mixture. Again these data are not plotted.

A cell filled with DOBAMBC and aligned with a brushed nylon surface treatment was prepared, Fig.13. This cell exhibits a striking feature not seen in the cells filled with the W7-W82 mixture. In the smectic A phase, the layers are tilted with respect to the bounding plates

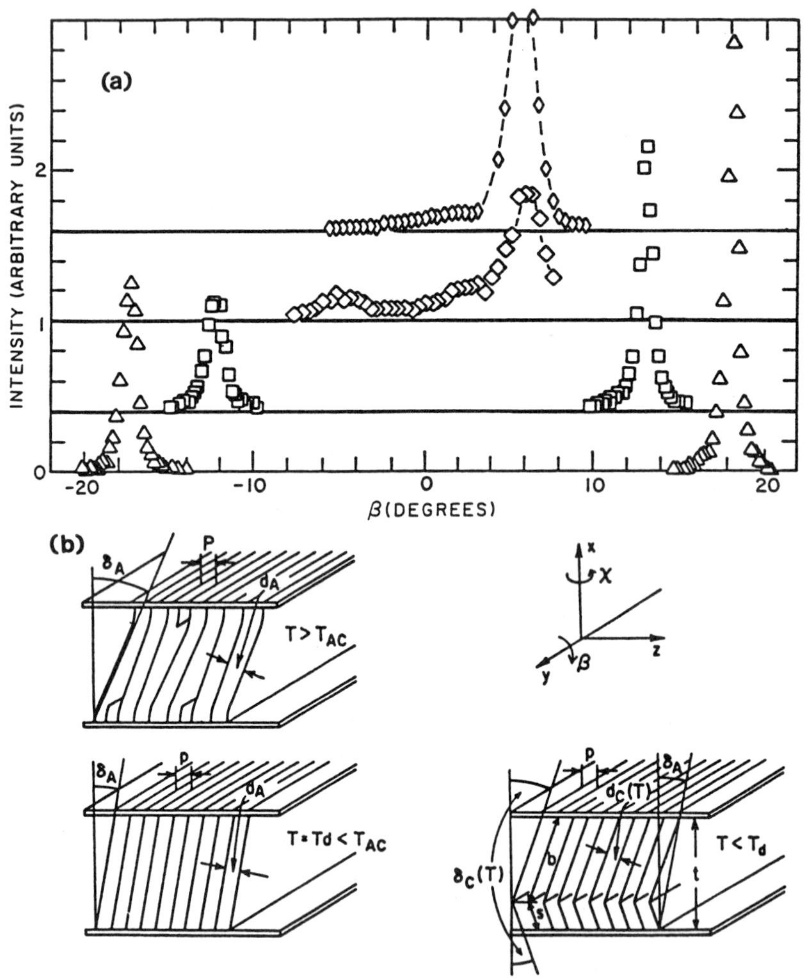


Fig.10. This plot is analogous to Fig.8, except that it details the formation of the asymmetric chevron. (a) $\delta_A$ = 7° ($S_A$ layers tilted) (◇) $T = T_C$ +69°C, (◇) $T = T_C$ - .5°C, (□) $T = T_C$ - 3.1°C, (Δ) $T = T_C$ - 17.0°C. The dashed lines have been added as a guide to the eye. (b) Local layer structure scenario for $\delta_A > O$, layer tilt and accompanying dislocations in the $S_A$ phase, tilted layers with no dislocations at $T = T_d$ and the formation of the asymmetric chevron for $T < T_d$.

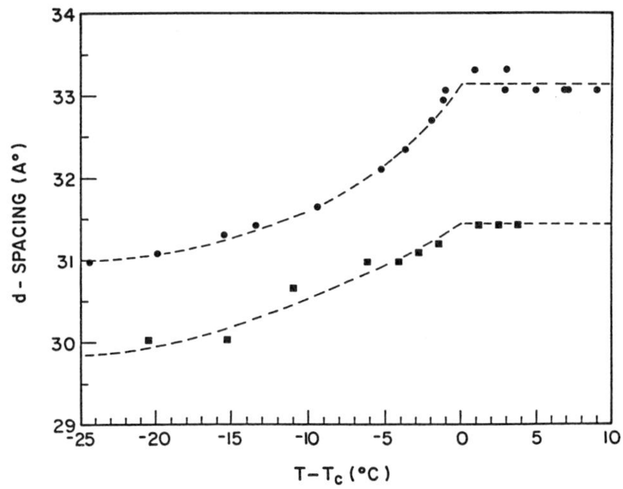


Fig.11. Interlayer spacing d, as a function of $T - T_C$; (●) DOBAMBC, (■) W7-W82. In the smectic A phase the interlayer spacings for W7-W82 and DOBAMBC are 31.4 A and 33.2 A respectively. As a comparison the molecular lengths of these molecules were determined from Dreiding molecular models in their longest conformation, as: W7- 30.2 A, W82- 32 A and DOBAMBC- 33.6 A. The dashed lines have been added as a guide to the eye.

as in the smectic C* phase. This effect persists through the cycling of the sample from the isotropic through the smectic C* phase and back to the isotropic phase. This tilt is evident from the presence of zig-zag walls into the smectic A phase as discussed with D.S. Parmar[27].

A recent paper by Takanishi et al.[28] clearly demonstrates the presence of tilted layers in the smectic A phase. This sample shows the same general behavior in the smectic C* phase, as do all materials tested with planar boundary conditions. The only differences lie in the saturation value of $\delta_C(T)$, the smectic C* temperature range and the rate at which $\delta_C(T)$

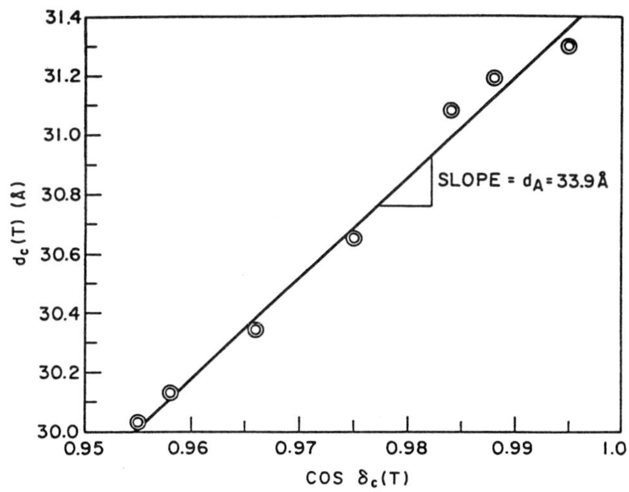


Fig.12. Plot of the smectic C* layer spacing vs. cos($\delta_C$(T)). The $S_A$ layer thickness from the slope is 33.9 A which compares well with the 2$\theta$ determined value of 31.4 Å.

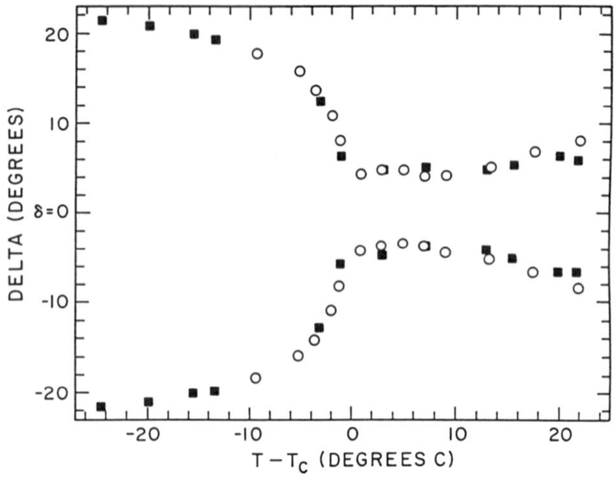


Fig.13. Plot of $\delta(T)$ vs. $T$ - $T_C$ for DOBAMBC. Both cooling (■) and heating (○) runs are indicated.

saturates with temperature. A cell prepared with brushed nylon surfaces was filled with W82. The experimental data are shown in Fig.14 and show several interesting features. Note the presence of two phase transitions. The $S_A$ - $S_C^*$ phase transition is at $T_C$ = 70°C. Upon cooling from $T_C$ the layers tilt continuously from being nearly upright, $\delta_A \approx 0$, in the smectic A phase, to saturate at $\delta \approx \pm 21.5°$, at 20°C below the transition. This is analogous to the behavior observed in other materials. The second transition, however, is unique in that an abrupt reorientation of the layers is observed at ≈38°C below $T_C$. The identity of this lower phase has been under dispute.

At the disputed phase transition the layers change from a large tilt angle to being

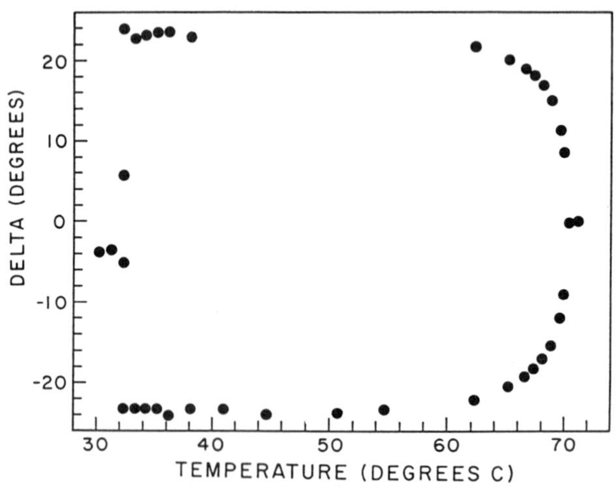


Fig.14. Layer tilt angle, $\delta(T)$ vs. $T$ - $T_C$ for W82. Note the presence of two phase transitions. The $S_A$ - $S_C^*$ transition is at 70°C and the $S_C^*$ - $S_B$ at 32°C. Note the near instantaneous change in the layer tilt at the $S_C^*$ - $S_B$ transition. The interlayer spacing changes from $d_C$ to $d_B$ = $d_A$ at this transition with a narrow region of coexistence.

upright on both heating and cooling across the transition. The change in layer tilt is accompanied by a change in the interlayer spacing. The layer spacing changes abruptly from that of the smectic C* phase to that found in the smectic A phase. This corroborates the view of P. Keller and coworkers [29], that the lower phase present is the smectic B phase. Note the coexistence of the two phases in a narrow temperature range at the phase transition. The two data points of large $\delta$ have an interlayer spacing identical to that for the smectic C phase just above the transition, while the two points with smaller $\delta$ have the smectic A layer spacing. The sharpness of the transition indicates that the $S_C^*$ - $S_B$ transition is first order.

### 3.2. *Materials which Exhibit a Chiral Nematic to Smectic C* Phase Transition*

In compounds which exhibit the smectic C* ($S_C^*$) to chiral nematic (N*) phase transition, the $S_C^*$ layers form directly from the N* phase. These materials demonstrate a large tilt of the director ($\hat{n}$) from the layer normal ($\hat{z}'$) immediately upon cooling into the $S_C^*$ phase[30].

Scattering intensity peaks are found only in the $S_C^*$ phase. Peaks in the I($\beta$) vs. $\beta$ scans are strongly dependent on the sample orientation in $\chi$ and the sample temperature. For samples grown with a positive (negative) field applied, maximum scattering was found at approximately -40° (+40°) in $\chi$ *immediately* upon entering the $S_C^*$ phase. The I($\beta$) vs. $\beta$ scattering data exhibits one or two peaks, or background dependent on the sample position in $\chi$ and the sample history, see below. Whenever two peaks are observed they appear symmetrically disposed about $\beta = 0$. This is most easily seen in Figs. 15 (a,b,c) which shows the peak positions in the I($\beta$) vs. $\beta$ data plotted as a function of temperature, applied voltage and sample position $\chi$. When characterizing a particular peak in the I($\beta$) vs. $\beta$ data only single scattering peaks in I($\chi$) vs. $\chi$ scans were observed. The peak location in $\chi$ does not vary with the sample temperature in the $S_C^*$ phase.

Samples prepared with a +1 V/$\mu$m field applied across the cell plates (Fig.15a) demonstrate the strongest scattering at $\chi = -36°$. Here two peaks are observed in the I($\beta$) vs. $\beta$ data symmetrically disposed about $\beta = 0$ with a Full Width at Half Height (FWHH) of 3 - 5°. A $\chi$ scan through each of these peaks in $\beta$ results in only one peak in I($\chi$) vs. X with a FWHH of $\approx$ 5 - 10°. I($\beta$) vs. $\beta$ scans at $\chi = 0$ reveal one or two scattering peaks, however these are about a factor of four weaker in intensity and broader (FWHH $\approx$ 5 - 8°) than those at $\chi = -36°$. I($\beta$) vs. $\beta$ scans at positive values of $\chi$ near 40°showed only background intensity.

Those samples prepared with a 1 V/$\mu$m applied field exhibit similar behavior except that the strongest scattering intensity is now found at $\chi = +42°$ and only background is observed at $\chi \approx -40°$, see Fig.15b.

Samples prepared in the absence of an applied field demonstrate the strongest scattering intensity at $\chi = 0$ with two peaks in the I($\beta$) vs. $\beta$ data. The scattering intensity here is comparable to that found at $\chi = 0$ in cells prepared with an applied field. Residual scattering intensity is found at both $\pm 40°$ in $\chi$, see Fig.15(c). However these peaks are not reproducible as the sample temperature is ramped up and down.

### 3.3. *Cells with Non-Planar Alignment of* $\hat{n}$

#### 3.3.1. *A-SiO Cells*

In cells with *antiparallel* alignment of the SiO evaporation directions (A-SiO cells), only one scattering peak is present in the $S_A$ and $S_C$ phases, as shown in the typical data in

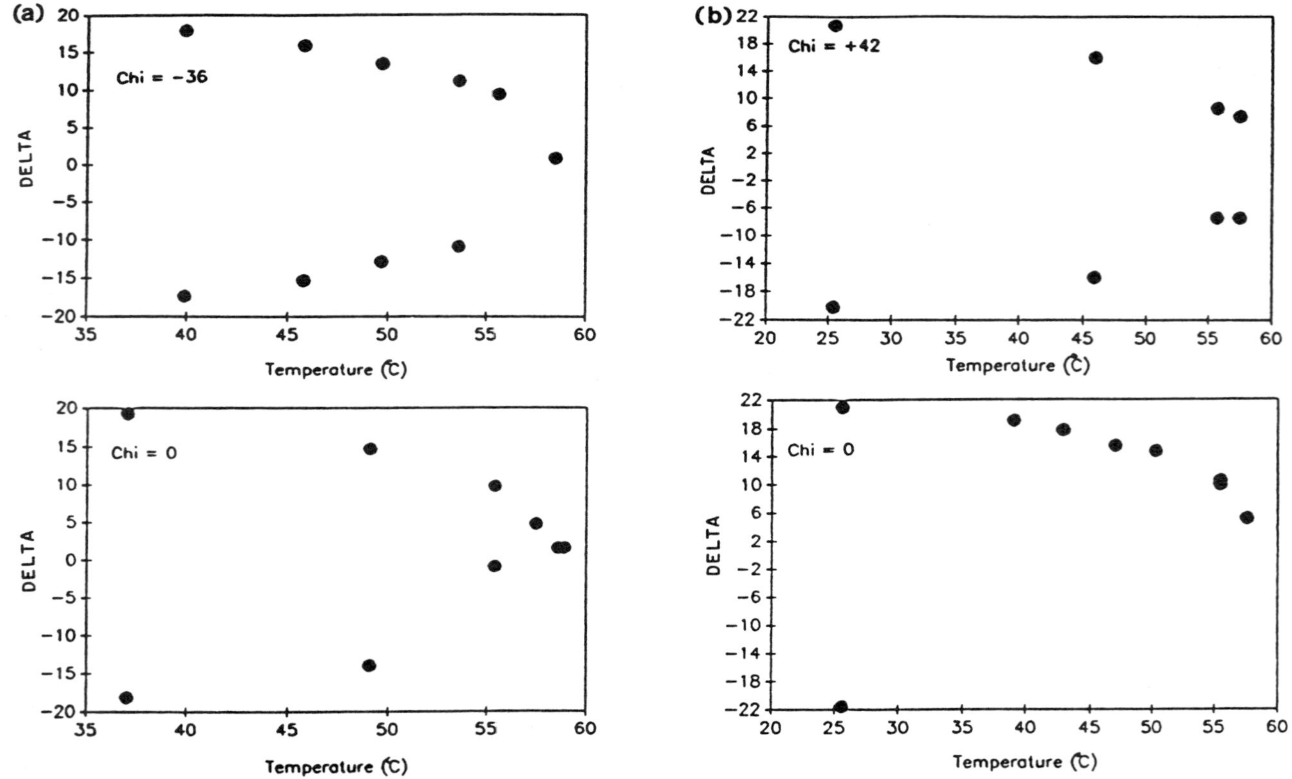


Fig.15 (a). Peak positions in the I($\beta$) vs. $\beta$ data for SCE-10 as a function of temperature, applied voltage and sample position $\chi$. (a) For +1 V/$\mu$m applied to the cell strong scattering is observed at $\chi$ = - 36°, weak scattering at $\chi$ = 0 and background for $\chi \approx$40°. (b). For -1 V/$\mu$m applied to the cell the situation is reversed with strong scattering observed at $\chi$ + 42° and background for $\chi \approx$- 40°.

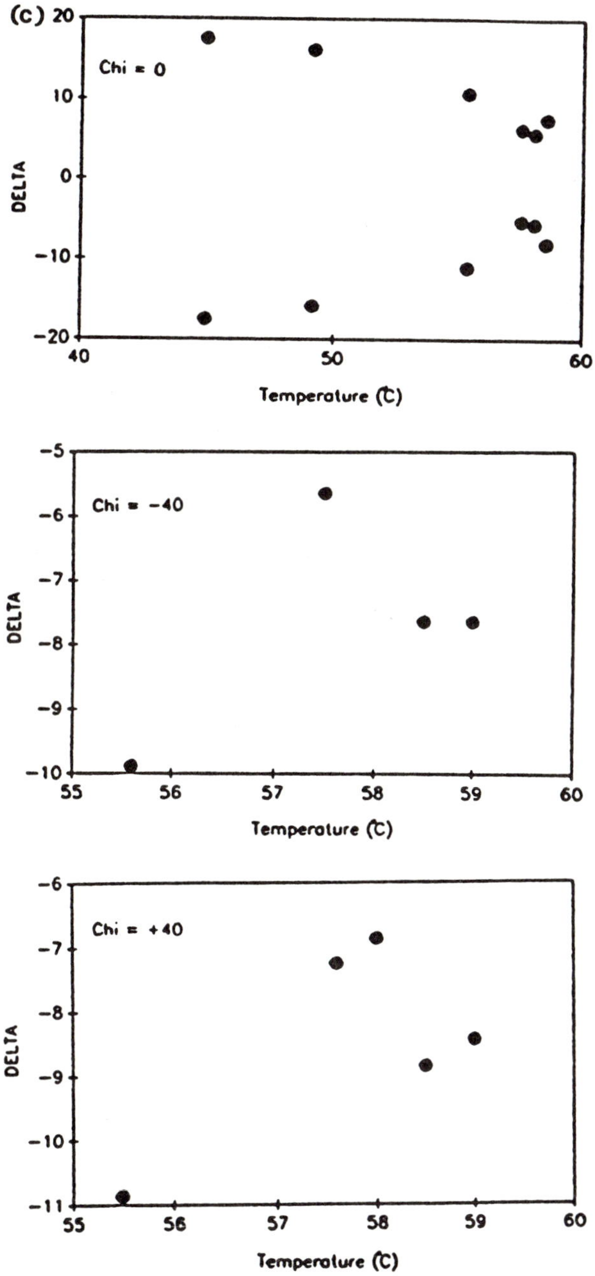


Fig.15 (c). For samples prepared in the absence of an applied field the strongest scattering is observed at $\chi = 0$ and weak scattering is observed at both large positive and negative values of $\chi$.

Fig.16 (a), in this case for ZLI-3654. This scattering is first detected as a single broad peak (Full Width at Half Height, FWHH = 17°) in I($\beta$) when the sample is cooled into the $S_A$ phase from the nematic. The corresponding layer tilt is large (> 15°) in comparison to cells with *planar* alignment.

This uniform tilt of the layers increases monotonically with decreasing temperature, with the I($\beta$) peak narrowing in the $S_C$ phase to a FWHH = 3° when the tilt has increased to 30°at 36°C below the $S_A$ - $S_C$ transition temperature.

A single sharp peak in I($\beta$) indicates a uniform tilt of planar layers, which is what is

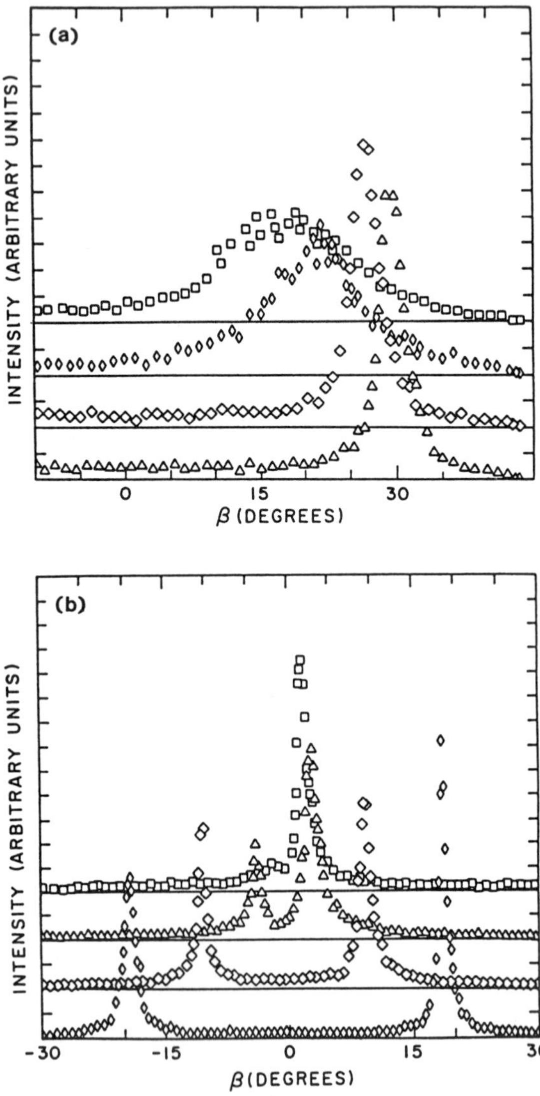


Fig.16. Typical data plotted as intensity I($\beta$) vs. $\beta$, the sample orientation. (a) ZLI-3654, A-SiO (corresponding to (□) in Fig.17 (a)): (□)$T = T_C + 10.9$°C, (◇)$T = T_C + .3$°C, (◇)$T = T_C - 15.1$°C, (Δ)$T = T_C - 36.1$°C. (b) ZLI- 3654, P-SiO (corresponding to (Δ) in Fig 17 (b)): (□)$T = T_C + 10.8$°C, (Δ)$T = T_C + 3.6$°C, (◇)$T = T_C - 3$°C, (◇)$T = T_C - 24.4$°C.

found at the lower temperatures, well down into the $S_C$ phase. Fig.17 (a) shows the temperature dependence of this layer tilt for several different commercial FLC mixtures, showing that in general δ(T) increases with decreasing temperature.

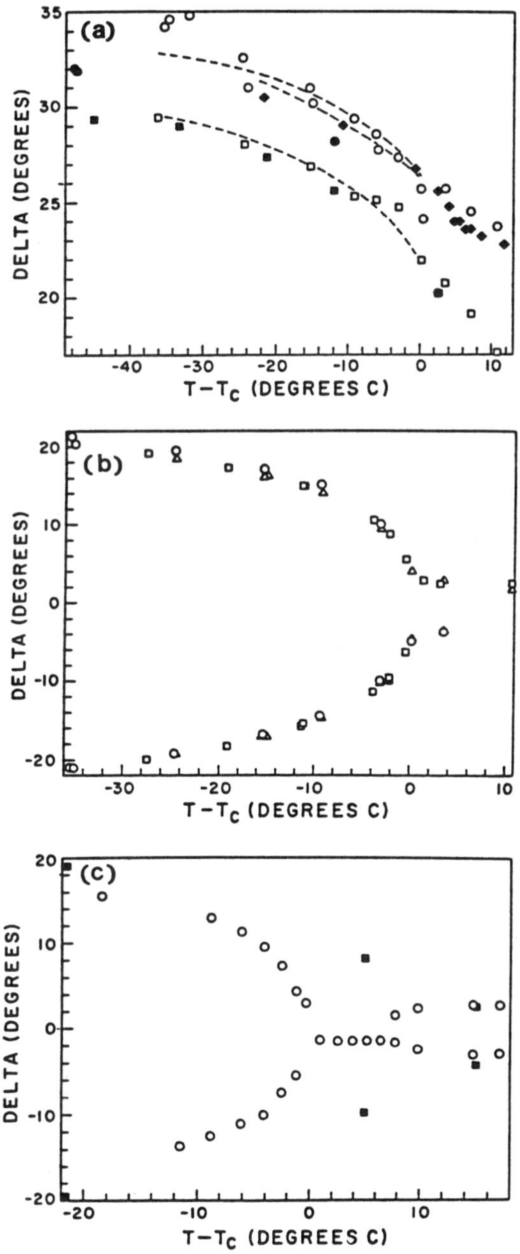


Fig.17. Layer tilt angle δ plotted as a function of $T - T_C$. (a) A-SiO cells, ZLI-3654 (□,○), SCE-3 (■,●) and CS-1014 (◆). (b) ZLI-3654; (□) brushed nylon surfaces (*planar* alignment), (Δ) P-SiO, (○) hybrid cell, one surface nylon the other oblique SiO. (c) CS-1014; (○) *planar* alignment, (■) P-SiO.

### *3.3.2. P-SiO Cells*

In cells with parallel alignment of the SiO surface treatments (P-SiO cells), two scattering peaks, disposed symmetrically in position about $\beta = 0$, but of different intensity, are observed in I(β) in the $S_A$ and $S_C$ phases as shown for ZLI-3654 in Fig.16 (b). These peaks are narrow, with a FWHH of 2° in the $S_A$ phase at $T_C + 10.8°C$ and decreasing to 1° in the $S_C$ phase at $T_C \sim 24.4°C$. The corresponding layer tilts increase monotonically as the sample is cooled through the $S_C$ phase. The appearance of symmetric peaks in I(β) is reminiscent of the case of *planar* alignment and Fig.17 (b) and (c) shows a comparison of P-SiO and *planar* layer tilts δ(T) for ZLI-3654 and CS-1014 respectively.

For the Merck sample, the peak position δ in the I(β) vs. β plots has the same temperature dependence in the smectic C* phase as the cell prepared with planar boundary conditions, see Fig.17 (b), while the Chisso sample has a completely different behavior; see Fig.17 (c). In the smectic A phase both the Merck and Chisso samples demonstrate some residual layer tilt.

A cell was prepared with one of the bounding plates coated with SiO and the other with nylon. This cell demonstrated the same behavior as the cell prepared with two nylon coated surfaces, see Fig.17 (b). This includes the presence of zig-zag walls when observed under crossed polarizers.

## 4. Layer Structures of SSFLC Cells

### *4.1. Introduction*

In the previous Section we detailed the X-ray results that gave a measure of the distribution of layer orientation in SSFLC cells. We also showed that the X-ray technique is very useful in establishing the layer orientation in such cells. In this Section we interpret these X-ray results to form, for the case of planar alignment of $\hat{n}$, a comprehensive picture of the effects which control the layer organization.

The understanding of these controlling effects has revealed that the origins of the heretofore puzzling zig-zag wall defect can be attributed to purely geometric consequences. We show that this defect mediates the change in chevron direction and give a geometric description which includes limiting cases.

The fact that the zig-zag wall defect formation lies in the layer geometry is further supported by our investigations of the layer structure in in cells with parallel and antiparallel SiO surface treatments. Such cells do not exhibit zig-zag defects in the smectic C* phase.

### *4.2. Layer Organization in Cells with Planar Alignment of* $\hat{n}$

The experimental facts outlined in the previous Section, have led us to formulate the following Anchored Layer Model (ALM) which accounts for all of the observed features of the LLS temperature dependence in cells with planar alignment of the director field. We begin by noting that the observed material dependent smectic C* layer tilt angle $\delta_c(T)$, given by Eq.1, is that which maintains the pitch $p$ of the layers at its smectic A value $p=d_A$, at all temperatures in the smectic C* phase ($p$ = length/layer in the $z$ direction; see Fig.8 (b)). *This suggests that in order to maintain p the layers tilt as a response to the shrinking* ($d_C < d_A$) *which occurs in the smectic C* phase.* In the smectic A phase the layers form a set of parallel, planar sheets. Fig.8 (b) shows the smectic A phase in which the layers have formed normal to the plates. *We assume that once the layers have formed and are anchored to the bounding surfaces, the anchoring positions are fixed, i.e.,* the intersections of the layers and the glass plates cannot move in the $z$ direction ($z$ is the layer normal for upright layers, $\delta = 0$). In this case the smectic A layers are continuous; there is no need for dislocations. If we further assume that *the layers remain continuous, i.e., dislocations do not form,* then any LLS must have $p=d_A$. The only smectic C* phase LLS having $p=d_A$ is that having a uniform tilt according to Eq.1. *Finally, the only kind of tilt which preserves the anchoring condition at both surfaces is the symmetric chevron.* There is no preference for the formation of chevroned layers pointing along $\pm z$. The X-ray scattering results do not differentiate between the two directions and we assume that both are present in a sample.

Now consider that the layers form with a tilt $\delta_A$ in the smectic A phase. At the surface the layers are normal to the bounding plates due to the director alignment parallel to $z$ imposed by the bounding surface and so at the surface $p_s = d_A$. In the middle $p_m=d_A\cos(\delta_A)$, so that dislocations must be present to make up for $p_s \neq p_m$, as discussed by Kleman and Williams[31] and shown in Fig.10 (b). As the sample is cooled through the $S_A$ - $S_C^*$ transition the layers begin to shrink and these dislocations move toward the sample center, reaching the center and completing whole layers when $\delta_C(T) = \delta_A$, *i.e.,* when the smectic C* pitch decreases to $p = d_A$, at the temperature $T = T_d$. At this point further shrinkage of the smectic C* layers is accommodated by tilting of the layers through the angle $\delta_C(T)$ and chevron formation. However, now the chevron is asymmetric; the sharp break in the layers initially forming at one surface are moving toward the middle as $T$ decreases. If $R$ gives the ratio of the lengths of the chevron sides, ($R = s/b$) then, referring to Fig.10 (b), the geometry gives:

$$(1+R)\Big/(1-R) = \frac{\tan(\delta_C)}{\tan(\delta_A)} \tag{2}$$

Using the integrated scattering intensities of the small-to-big peak $R = I_s/I_b$ to measure $R$, we plot the data in Fig.18 from the I(β)'s in Fig.10 (a). The fitted slope $(\tan(\delta_A))^{-1}$ yields $\delta_A = 9.6°$, which agrees favorably with the directly measured value of 7.2°. This scenario thus accounts for the behavior when the two peaks in I(β) are unequal in intensity and indicates

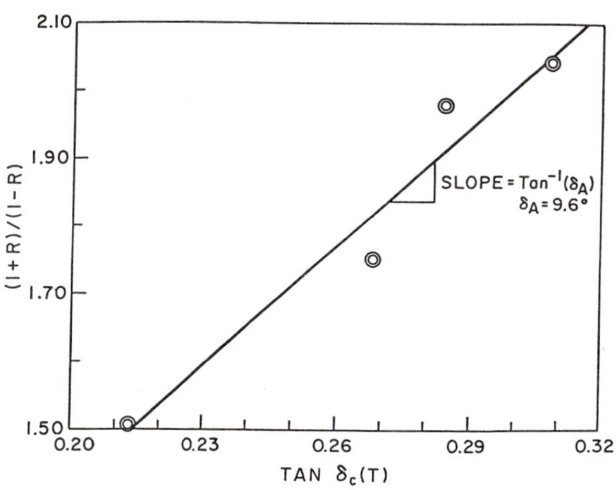


Fig.18. $(1+R)/(1-R)$ vs. $\tan(\delta_C(T))$, where $R = I_s/I_b$ is the intensity ratio of the two scattering peaks, as well as the ratio of the lengths of the sides of the chevrons, $R = s/b$.

that for the surface treatments used here the molecules are parallel to the plates in the smectic A phase.

The formation of symmetric and asymmetric chevrons in the same sample cell would require domains of different layer tilt in the smectic A phase. Such domains would require layer dislocations, which we have discounted on the scale of a few microns. On a larger scale, dislocations are introduced by surface defects such as dust particles. Therefore, we conclude that the LLS is homogeneous. However, on a larger scale the chevron 'tip' will vary in distance from the surfaces throughout the cell.

The formation of the chevron local layer structure (LLS) is not solely a feature of the smectic C phase. Layer shrinkage which drives the formation of the chevron (LLS) can arrise from a decrease in density. Recently Takanishi et al.[28] have reported the presence of tilted layer domains in the smectic A phase of 8CB[32]. These domains form as the sample is cooled from the nematic into the smectic A phase. The Japanese present layer tilt data in a 7°C temperature range with the layers starting in the bookshelf geometry at the N-$S_A$ transition. The maximum layer tilt obtained is 7°(average of both + and - layer tilts). Eq.1 for layer tilt due to layer shrinkage can also be applied to this case with $d_A$ is the smectic A layer spacing at the N-SA transition, $d_C$ is the temperature dependent interlayer spacing in the $S_A$ phase and δ(T) is the temperature dependent layer tilt angle. Using their data we calculate that the interlayer spacing need change by only 0.7% over the $S_A$ range to yield a layer tilt angle of 7°. Unfortunately, this change in interlayer spacing is just at the edge of resolution of our instrument.

Although the interlayer spacing can not be determined with high enough accuracy via X-ray scattering, precision density measurements will give a good estimate. A paper by Rao and Plslpatl[33] provides detailed density information on HBBA[34]. From their data we calculate a density change of 0.68% over a 6°C temperature range in the smectic A phase. We expect that nearly all of this density change is associated with layer shrinkage. Only a small portion is due to a decrease in intermolecular distance. If this assumption is correct then chevron formation can be driven by layer shrinkage resulting from a decrease in density. Indeed any sufficiently large change in density will cause the formation of chevron shaped layers in the smectic phases, provided that the layers are fixed at the bounding plates and that there are few defects to absorb the shrinkage.

## 4.3. The Zig-Zag Wall Defect

### 4.3.1. Introduction

SSFLC cells, when viewed in transmission polarized light microscopy, show a variety of characteristic features in their optical texture. The most striking and prevalent of these are the so-called zig-zag defects as shown in Fig.2, found in cells with planar alignment of the director field[8,35]. These defects can appear as tightly bunched rows of line defects, connected to one another in a zig-zag pattern, running nearly normal to $z$ with the rows running along the smectic layers, where $z$ is the layer normal for upright layers $\delta = 0$. Often they appear as isolated closed loops, with the narrow walls running at a small angle $\gamma$ to $z$, which come to a sharp point at one end and broad walls closing the loop normal to $z$ at the other end. Such closed loops will form the basis of our discussion, since they demonstrate all aspects of the zig-zag wall defect. Experiments using a 100 x 100 μm probe beam allowed us to confirm the presence of the chevron local layer structure inside such zig-zag defect loops. This discovery and subsequent microscopy have enabled us to build a detailed three dimensional structural model for the zig-zag defect walls.

### *4.3.2. Mediation of the Change in Chevron Direction*

The chevron formation in the smectic C* spontaneously breaks the smectic A mirror symmetry normal to $z$ and causes the smectic C* to be a mosaic of two kinds of domains, each having one of the two possible chevron directions. Zig-zag defects are the spontaneously formed boundaries between domains having taken opposite chevron directions, preserving the continuity of any single layer, as that layer buckles from the planar sheet of the smectic A phase to the two domains of chevron direction of the smectic C* phase. The appearance of the zig-zag defect is the major optical evidence of the formation of chevroned layers. From optical studies of dechiralization lines we have been able to ascertain that inside the loop the chevron points toward the broad wall of the defect, while outside the chevrons point in the opposite direction.

The narrow wall of the zig-zag defect is observed to be either straight ($\gamma$ fixed), or curved ($\gamma$ varied) as in the photomicrographs of Fig.2 and 21 respectively. The coordinate system formed by $z$ (ordinate) and y (abscissa) where $\hat{z}$ is normal to the layers, lies in the plane of the bounding plates and points from the broad walls toward the narrow walls and $\hat{y}$ lies along the layers is used to define the sign of $\gamma$. For a switch in chevron direction from along $\hat{z}(\uparrow)$ to against $\hat{z}(\downarrow)$ $(\downarrow$ to $\uparrow)$ for a positive (negative) slope of the local tangent to the narrow wall, $\gamma > 0$. For the tangent parallel to $z$, $\gamma = 0$. For a switch in chevron direction from $\uparrow$ to $\downarrow$ $(\downarrow$ to $\uparrow)$ for a negative (positive) slope of the tangent, $\gamma < 0$.

Moving along $y$ one encounters zig-zag walls at random intervals, indicating that the chevron direction is not nucleated in a regular fashion but rather is nucleated from defects in the smectic A layer structure. Moving along $z$ and crossing a narrow zig-zag wall the chevrons change direction, pointing out from the defect like so: <<<<<>>>>> for $\gamma > 0$ and into the defect like so: >>>>><<<<< for $\gamma < 0$. The broad wall of the defect running parallel to the layers mediates only the change: >>>>><<<<<. For those defects with straight narrow walls, this aspect of the structure results in the simple topology that as one moves in a cell parallel to $z$, one alternately encounters narrow, then broad lines. For those defects with curved narrow walls, one may encounter either narrow or broad walls, but one must always alternately encounter the two different chevron changes, >>>>><<<<< or <<<<<>>>>>. The chevrons may be symmetric or asymmetric. For the asymmetric case the height of the chevron tip is reflected through the midplane of the cell as you cross the zig-zag wall. We propose that the remarkable structure of these defects is a result of the same simple conditions that produce the chevron in the first place (the chevron golden rules):

(1)The planar and continuous smectic A layers are anchored in the smectic A phase with a pitch $p$ (length along $z$/layer) = $d_A$ and this anchoring persists into the smectic C* phase which, as a result, must also have $p = d_A$, but with thinner layers ($d_C < d_A$).

(2)There are no layer dislocations in either the smectic A or C phases.

(3)The interlayer spacing everywhere in the cell is the same and is temperature dependent in the smectic C* phase, with $d_C(T) \leq d_A$. The accommodation of $p = d_A$ is achieved via the tilt of continuous layers.

Fig.19, is a compilation of various observations on the layer and director structure found in cells with a chevron local layer structure (LLS). The three dimensional zig-zag wall model shown in Fig.19 (e) details a structure which mediates the switch in chevron direction while satisfying the above conditions. Fig. 19(g) shows an $x$ - $z$ section of the layering structure through the defect for the case of symmetric chevrons ($\delta_A = 0$). The chevrons achieve $p = d_A$ by rotating the layers about y through the layer tilt angle:

$$\delta_C(T) = \arccos\left(\frac{d_C(T)}{d_A}\right) \tag{3}$$

In making the narrow wall transition, the layers must be effectively thicker near the chevron plane in the transition region. This is achieved by rotating the layers away from being normal to $z$ through an angle $\alpha$. In order to mediate the chevron direction change this rotation must be about $x$ rather than $y$. The very simple continuous layer structure shown in Fig.19 (e) and (h), having a parallelogram-shaped planar element to mediate the chevron sign switch, results. Successive parallelograms are separated by the smectic C* layer spacing $d_C$. The limits of the

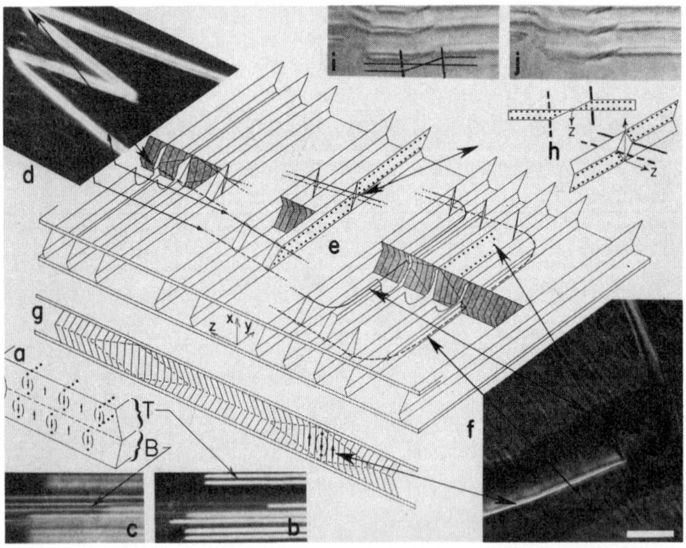


Fig.19. Schematic drawing of zig-zag wall defect and supporting evidence. (a) In sufflciently thick cells, a director constraint at the chevron should lead to independent dechiralization lines, (indicated throughout the figure by heavy dotted lines ...) appearing on either side of the chevron defect; (b,c) Photomicrographs obtained in polarized transmission microscopy of dechiralization lines in a 6$\mu$m thick SSFLC cell, focussing respectively near the top and bottom of the FLC cell, showing the two sets of lines; (d,f) Photomicrographs of respectively the narrow and broad walls of zig-zag defects in 3 to 4 $\mu$m thick cells; (e) The three dimensional structure of a closed zig-zag defect loop, the spontaneously formed defect separating one direction of chevron from the other in a smectic C formed between solid plates by cooling from the $S_A$ phase. The loop consists of narrow walls running nearly parallel to $z$ and a broad wall running parallel to $y$. These walls are delimited by the lines which focus sharply in (d) and (f), which are also indicated by the heavy solid and dashed lines in (e). In this example the mean layer tilt is nonzero [asymmetric chevron]. The defect structure is characterized by a construction of layers made up of planar elements, continuously connected to one another by the chevron discontinuities. Thus all of the solid lines on the layers are chevron discontinuities. The overall structure maintains everywhere the layer pitch $p = d_A$ established by layer anchoring in the $S_A$ phase.; (g) Section showing layer structure change upon passing through the narrow (<<<>>>, $\gamma > 0$) and broad (>>><<<) walls for the symmetric case. Upon reduction of an electric field dechiralization lines first appear in the center of the broad wall, where the layers arė chevron free (e,f,g); (h) Dechiralization lines dress the layer structure near a narrow zig-zag wall, outlining the limits of the wider chevron element; (i,j) Photos of dechiralization lines crossing zig-zag defects in a 5 $\mu$m thick cell, respectively focussed near the top and bottom of the SSFLC cell. The lines on opposite sides of the zig-zag wall focus at different $x$ values, as expected from (h). All photos (printed here in B&W) were made using a Zeiss Universal microscope with a UD condenser and an Epiplan 40X objective observing a 50% - 50% W7-W82 mixture[21]. The bar in (f) is 15 $\mu$m long.

zig-zag wall, which focus as sharp lines in the polarized light microscope Fig.19 (d), are the lines formed by the loci of points where the chevron planes intersect the parallelogram in successive smectic C layers, indicated by the heavy solid and dashed lines in Fig.19 (e). At the 'tips', where narrow walls meet the two defects merge in a continuous way.

In making the >>>>>:::<<<<< (broad wall) transition, the layers must be effectively thinner near the chevron plane in the transition region. This is achieved by rotating the layers through an angle less than $\delta_C(T)$, inside a right parallelogram cylinder running parallel to *y*. The limits of this cylinder, indicated by the heavy solid and dashed lines in Fig.19 (e), are the visible limits of the broad wall in Fig.19 (f). In the case of asymmetric chevrons, the lines on one side of a zig-zag wall focus near the top of the cell and those on the other side near the bottom, as would be expected from (e). At the center of the broad wall, the layers have their maximum thickness without a chevron (e), making this region the most favorable for forming dechiralization lines (f).

Further direct evidence for this model, obtained with the polarized light microscope is shown in Fig.19 (h,j). With a 5 to 10 $\mu$m thick cell it is possible to obtain zig-zag defects crossed by dechiralization lines. This is quite useful as the dechiralization lines dress the FLC layering structure, running along the boundaries of the wider planar element of an asymmetric chevron (h), enabling it to be directly observed. Figs.19 (i,j) show a typical result, with lines on opposite sides of a narrow wall focussing at different *x* values. Observations of such dechiralization lines have allowed us to identify the chevron directions as detailed above. The geometry of $\hat{n}$ at the chevron defect and solid bounding surfaces leads to a basic requirement for obtaining bistability in SSFLC cells: one must have[4] $\Theta(T) > \delta(T)$; see the following Section. Since, $\delta_C$ is determined solely by the bulk material property, $d_C(T)/d_A$ (see Eq.3), bistability requires that the bulk condition, $d_C(T) / d_A$, be satisfied by the material being employed. That is, the smectic C layer thickness must *be larger* than $d_A \cos(\Theta)$ that would be obtained in a model of sticklike molecules being tilted over. In fact most measurements in smectic C materials agree[12] with the above equation, with a few definite exceptions[36].

For the materials we studied in our experiments, $\Theta$ was always larger than $\delta$. The ratio of the layer tilt angle $\delta$ to the optically determined director tilt angle $\Theta$ is plotted in Fig.20, as a function of $T - T_C$ for the W7-W82 mixture. For this material $\delta = 0.87\ \Theta$. Since $\delta$ has the same temperature dependence as $\Theta$, it can be used substituted for $\Theta$ as an order parameter in thermodynamic calculations.

#### *4.3.3. Geometry of the Zig-Zag Wall Defect*

A complete understanding of the layer structure present in the zig-zag wall is obtained through a geometrical approach. We will consider only the case in which the layers start out normal to the plates in the smectic A phase, $\delta_A = 0$, giving rise to symmetric chevrons.

#### *4.3.4. Narrow Walls*

The narrow walls of the zig-zag wall defect are of the order of the cell thickness t in width, run nearly normal to the layers and are generally straight line segments, except where they curve to meet the broad walls. The angle $\gamma$ between *z* and the line tangent to the narrow wall is always small, typically less than 20°. The straight line sections of the narrow walls are of a uniform width *w*. The sections of the defect which curve to meet the broad walls widen as they curve to make a smooth transition to the broad walls. We have also observed cells in which the narrow section of the zig-zag wall is not straight, rather it curves along its length as in Fig.21. In this case the thickness of the wall is not uniform along its length and is seen to vary with $\gamma$.

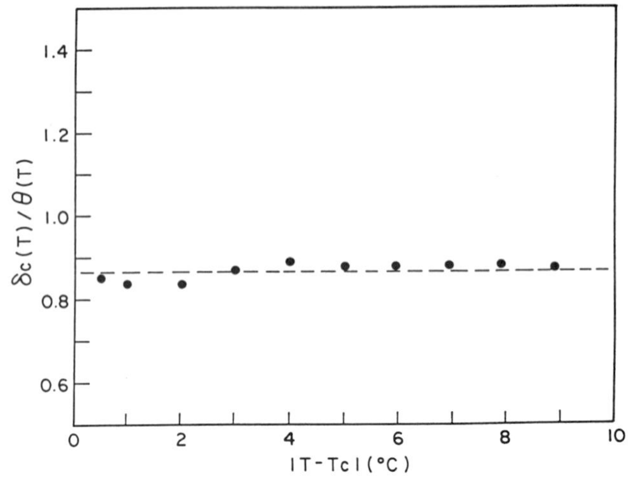


Fig.20. Plot of $\delta_C(T)$ / $\Theta(T)$ vs. $T$ - $T_C$ for the W7-W82 mixture. The dashed line indicates the average value of $\delta_C(T)$: $\delta_C(T) \approx 0.87\ \Theta(T)$. Since $\delta$ has the same temperature dependence as $\Theta$ it can be substituted for $\Theta$ as an order parameter.

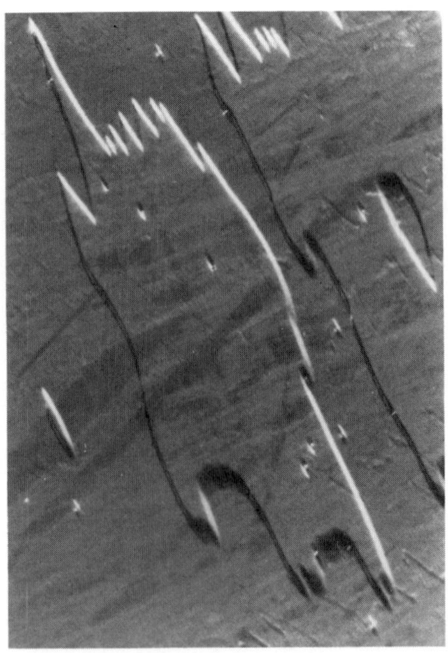

Fig.21. Photomicrograph of a zig-zag defect with curved walls. Note the variation of the wall thickness *w* with the curvature of the defect.

Fig.22 shows views of the smectic C* layering through a zig-zag wall defect and defines the relevant geometry. All coordinates are relative to the glass plates and the smectic A layer normal $z$ for upright layers. As depicted in Fig.22, $\gamma$ is the angle between the tangent to the narrow wall and $z$. The angle C* lies between $z$ and the normal to the upright parallelogram layer section, $\hat{u}$, which connects the two chevron domains. The angle between the layer normal $\hat{z}'$ of any of the chevron faces and $\hat{u}$ is $\pm\sigma$ is the angle of the smectic layer with respect to $x$. The parallelogram has width $h$; the zig-zag wall has width $w$; the SSFLC cell has fixed thickness $t$. All parameters, except $t$, are functions of temperature. All calculations will assume a fixed temperature. For the cells tested we have demonstrated that $\delta < \Theta$ and that $d_C = d_A \cos\delta$.

Curved narrow zig zag walls have the same general construction as shown in Fig.22. However, in narrow walls with both positive and negative $\gamma$ the chevrons point both out of and into the narrow wall respectively. This is illustrated schematically in Fig.23.

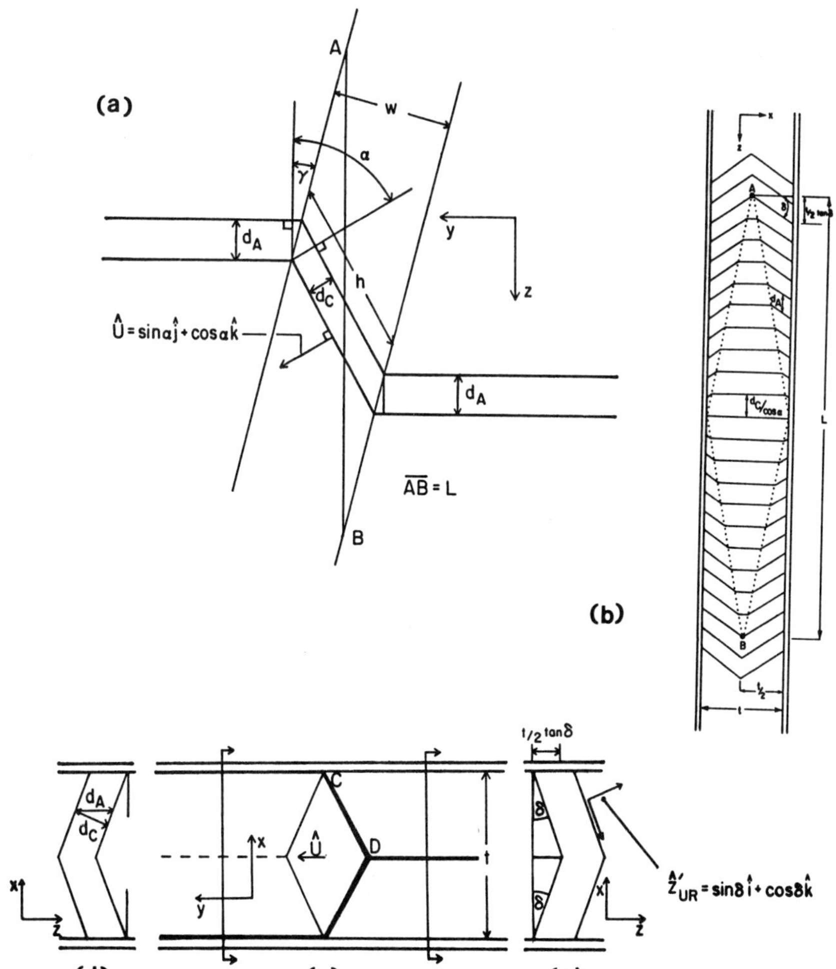


Fig.22. Schematic diagram of the layering structure of the zig-zag wall defect. (a) This is a top view, down $x$, of a slice through the mid plane of the cell which includes the apices of the chevron. (b) An $x$ - $z$ section through the zig zag wall indicated in (a) as the line segment $L$ from point A to B. (c) Frontal view of (a) along $z$. (d,e) Side views along $y$.

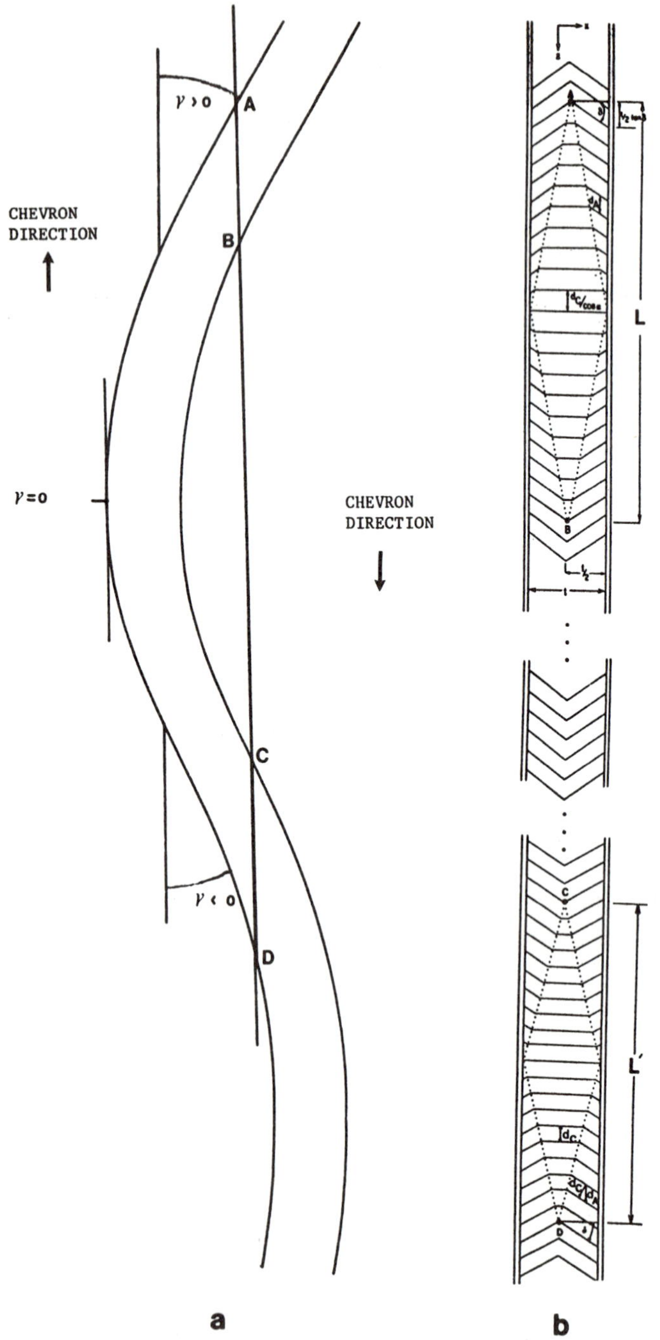


Fig.23. Schematic diagram of the layer structure in a curved narrow zig-zag wall such as that shown in Fig.21. (a) Line drawing of the curved narrow wall. (b) $x$ - $z$ sections through the narrow wall of (a) detailing the layer structure for the segments of $\pm\gamma$. The chevrons can point both into and out of the narrow wall defect.

From Fig.22 three pivotal geometrical relations are obtained.

(1) Consider the length 1, in Fig.22 (a). The two adjacent triangles can be used to eliminate l in terms of the adjacent angles and the interlayer spacing:

$$\frac{\cos(\alpha-\gamma)}{\cos(\gamma)}=\frac{d_C}{d_A}.$$

Combing with Eq.3, we obtain:

$$\frac{\cos(\alpha-\gamma)}{\cos(\gamma)}=\cos(\delta) \tag{4}$$

(2) The triangles bordering the perpendicular width of the zig-zag wall $w$, provide a relationship between $h$ and $L=\overline{AB}$:

$$w = h\cos(\alpha - \gamma)$$

$$w = L\sin(\gamma)$$

$$L\sin(\gamma) = h\cos(\alpha - \gamma)$$

The last equation can be rewritten using Eq.4 above as;

$$L\tan(\gamma) = h\cos(\delta) \tag{5}$$

(3) Equating the layer displacement at the midplane of the cell, due to the tilt of the layers by $\delta$, at the parallelogram with that at the chevron tips, we obtain;

$$h\sin(\alpha) = t\tan(\delta) \tag{6}$$

Which with Eq.5 above can be written as:

$$w=t\frac{\cos(\gamma)\sin(\delta)}{\sin(\alpha)} \tag{7}$$

We can verify the above geometric considerations from the conservation of layer number $N$. Consider an $x$- $z$ section* through the narrow wall, see Fig.22 (b), along the defect at the surface, where $p = d_A$ and at the midplane, where $p = d_C$, one obtains the following:

$$L=N\frac{d_C}{\cos(\alpha)} \qquad\qquad L=Nd_A+2\frac{t}{2}\tan(\delta)$$

Eliminating $N$ and substituting $\cos(\delta)$ for $d_C/d_A$:

$$L\left(1-\frac{\cos(\alpha)}{\cos(\delta)}\right)=t\tan(\delta) \tag{8}$$

Eq.8 can be obtained from the geometrically derived Eqs.4, 5 and 6. This implies that *layer conservation is an essential feature of the zig-zag wall defect.*

In general for planar alignment, in materials with the phase sequence $I$ - $S_A$ - $S_C^*$, β, δ, (α - γ) and Θ are all less than 30°. Therefore, the cosine of these angles can be expanded to order angle squared. Expansion of Eq.4, allows us to solve for one of the three angles in terms of the others.

$$\gamma \approx \frac{\alpha^2 - \delta^2}{2\alpha} \tag{9}$$

$$\alpha \approx \gamma + \sqrt{\gamma^2 + \delta^2} \tag{10}$$

Substitution of the above expression for α (Eq.10) into Eq.7 yields an expression for the thickness of the narrow wall $w$, in terms of the measurable quantities, γ and δ as follows:

$$w \approx t\frac{\sin(\delta)\cos(\gamma)}{\sin\left(\gamma + \sqrt{\gamma^2 + \delta^2}\right)}$$

This expression clearly shows the variation in the thickness of the narrow part of the zig-zag wall with the angle it makes with respect to $z$. The wall width $w$ is on the order of the sample thickness $t$, except where the wall makes a large excursions from running along $z$ ($\gamma << 0$) .

Several limiting cases for γ are of interest. First the trivial case. For the narrow walls running parallel to $z$, $\gamma = 0$, the Eqs.4 and 7, reduce to:

$\alpha = \delta$ ; and, $w = t$.

The fact that the narrow walls never run at a large angle to $z$ provides a second limit of interest. The upper limit on γ, is due to the desire of the system to minimize the splay deformation introduced by 'breaks' in the layers. The introduction of splay by layer breaks is discussed in much greater detail in the following Section. Any layer break can have a maximum angle of twice the smectic C* tilt angle, 2Θ. Beyond this, energetically unfavorable splay deformations are introduced. The relevant layer kink, which puts an upper limit on γ, is that which connects the chevron half planes to the mediating parallelogram. Consider the line segment $\overline{CD}$ in Fig.22 (c). By taking the dot product of the layer normal of any of the chevron half planes with that of the mediating parallelogram, we obtain;

$$\cos(\alpha)\cos(\delta) = \pm\cos(\sigma) \tag{11}$$

Where the ± sign depends on whether the layer bends away from the mediating parallelogram or toward it, respectively.

Small angle approximations are valid for only a few of the variables. To demonstrate the usefulness of the above discussion we present calculations of the important parameters using data obtained on the W7-W82 mixture. At saturation, δ = 18° and Θ = 20.7°. By the above argument, $\sigma_{max}$ = 41.4°. Eq.11 gives α = 37.9°. Now plugging these values into Eq.9 and Eq.7, we obtain the values $\gamma_{max}$ = 14.8° and $w$ = 0.49 $t$. This compares favorably with optical observations such as those in Fig.19 (d), for which γ = 11.4° and $w$ = 3.1 μm in a sample of thickness $t \approx 4$ μm.

#### 4.3.5. Broad Walls

As previously indicated the broad part of the zig-zag wall defect, of width $L'$, is seen to run along the layers, $\perp z$ and is observed to be many times wider than the cell thickness. Fig.24, is a schematic drawing of a cross-sectional view in the $x$ - $z$ plane, through the broad end of the zig-zag wall defect.

Preservation of layer number $N$ in the above figure, yields the following equations;

$$L' = Nd_C \qquad\qquad L' = Nd_A - 2\frac{t}{2}\tan(\delta)$$

eliminating $N$: $L'\left(\frac{1}{\cos(\delta)} - 1\right) = t\tan(\delta)$ ; and, expanding for small angles to order $\delta^2$:

$$L' = \frac{2t}{\delta} \tag{12}$$

$L'$ is in terms of experimentally determined quantities. We can compare the measured length of $L'$ with calculated values. Again, using our data for the W7-W82 mixture we calculate $L' = 25\ \mu$m, for $t = 4\ \mu$m and $\delta_C = 18°$. This compares favorably with our optical observations as in Fig.19 (f), which has a measured value of $L' = 31\ \mu$m.

### 4.4. Planar Cells with Nematic to Smectic C* Materials

Electric fields applied to samples which exhibit the N* - $S_C^*$ transition have a strong influence on the layering structure adopted by these materials in an SSFLC cell. This due to the additional constraint $\mathbf{P} \cdot \mathbf{E} = 0$ on the orientation of the $\hat{n}$ - $\mathbf{P}$ couplet and $\hat{z}'$ as the smectic layers form from the N* phase. The $\mathbf{E}$ field pins $\hat{n}$ at the extremes of the $S_C$ tilt cone ($\Phi = \pm\pi/2$) while n retains the orientation of the N* phase imposed by the anisotropic surface treatment, resulting in layer formation at large angles in $\chi$, see Fig.4.

The formation of the layers at large angles with respect to the normal of the buffing direction is an indication that the director tilt angle $\Theta_{N^*-S_C^*}$ indeed large right at the N*-$S_C^*$ transition. Once formed the layers become anchored ($\chi$ remains fixed) and layer shrinkage drives chevron formation. The angle $\chi$ is $\lesssim \Theta_{N^*-S_C^*}$ since the smectic layers are generally tilted ($\delta \neq 0$) as they form from the melt, see Fig.15 (a,b,c). Weak scattering in samples aligned with an applied $\mathbf{E}$ field which is found at $\chi = 0$ can be attributed to defect structures in the layering probably due to elastic deformation.

For samples prepared in the absence of an applied $\mathbf{E}$ field, the structure obtained depends on the preferred orientation of the $S_C^*$ tilt plane (containing $\hat{z}'$ and $\hat{n}$) at the surface. If the tilt plane lies normal to the surface, then uniformly tilted layers having $\chi = 0$ and a $\delta$ having a magnitude close to the director tilt angle, but of either sign, $\delta = \pm\Theta$, will form in a mosaic of domains. In the other extreme with the tilt plane parallel to the surface the two domains will have $\chi = \pm\Theta$ and $\delta = 0$, as is obtained for an applied AC field[1]. These two domains are also observed (see Fig.15 (c)). In these samples the predominant layer domain is found at $\chi = 0$. At $\chi = 0$ the layers form from the melt at $\delta \sim \pm 7°$ indicative of an equilibrium structure at the surface at the N* - $S_C^*$ transition which has n along the rubbing direction and $\Theta \sim 7°$. This due to either a small tilt angle $\Theta \sim 7°$ very near the transition or a hybrid surface-bulk LC structure giving an effectively large pretilt near the transition. Once the

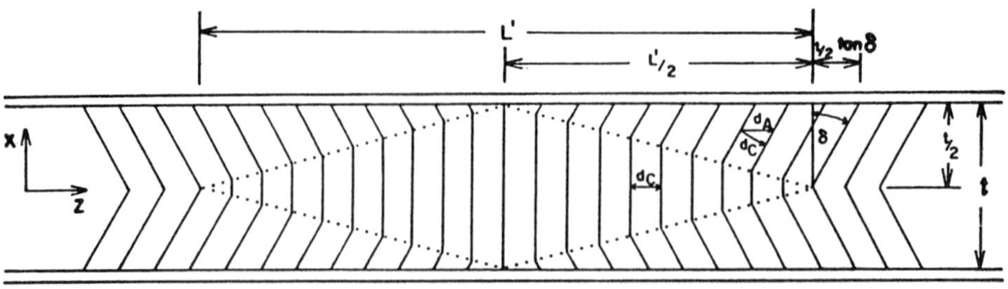


Fig.24. Cross-sectional (*x*- *z*) of the broad end of the zig-zag wall, such as that in Fig.19 (f), of width *L'*, detailing the layer structure within. The chevrons always point into such a defect.

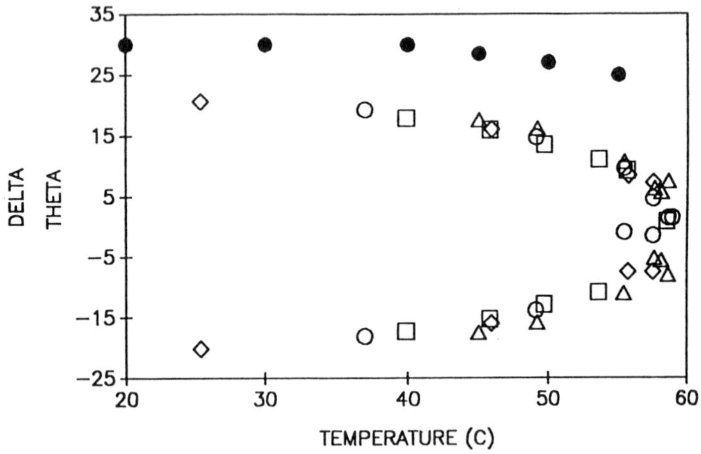


Fig.25. Compilation of data from Fig.15 (a,b,c) for which two values of $\delta$ at a given temperature are reported. Values of $\delta(T)$: ○ $\chi = 0$, +2V; Δ $\chi = 0$, 0V; □ $\chi =$ - 36°, +2V; ◇ $\chi =$ +42, - 2V. ● $\Theta(T)$, data supplied by BDH.

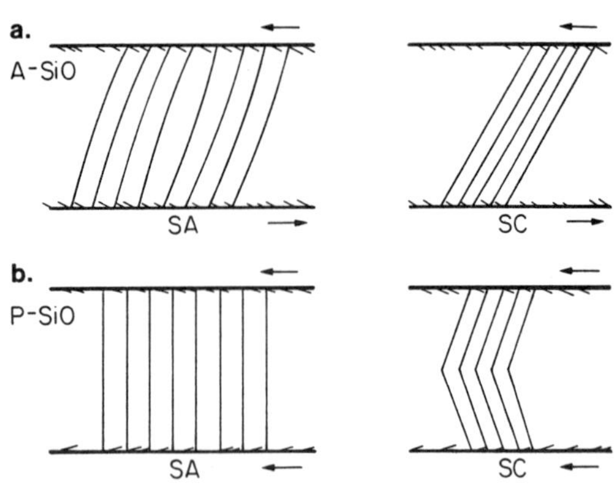


Fig.26. Sketches of the local layer structures found in (a) A-SiO cells and (b) P-SiO cells. At $T_C$ the layers are dislocation free and have a tilt $\delta_0$ equal to or less than the surface director tilt. Thermal contraction of the layers upon cooling in the $S_C$ phase results in the $S_C$ structures depicted.

layering is established it behaves in the usual way as the temperature is reduced and the layers shrink, i.e., $\hat{n}$ rotates away from the rubbing preferred surface orientation.

Fig. 25 is a compilation of the data from Fig.15 (a,b,c) for which two values of $\delta$ at a given temperature are reported. This shows that the layer tilt angle $\delta$ has the same temperature dependence irrespective of the applied field or sample orientation. This is as expected since layer shrinkage, which drives chevron formation is a material property. The director tilt angle $\Theta$ (from the BDH data sheet for SCE-10) is also plotted for comparison with $\delta$.

## 4.5. SiO Cells

For cells with oblique SiO surface treatment, the pretilt of the director field at the bounding plates strongly influences the layer structure present in the cell. These cells are of interest since they do not exhibit the common zig-zag wall defect in the smectic C* phase. Cells were prepared with both parallel (P-SiO cells) and antiparallel (A-SiO cells) alignment of the evaporation directions on the bounding plates. In addition cells were prepared with oriented polymer surface treatments for planar boundary conditions as a comparison. The two types of SiO cells show remarkably different behavior from oneanother and from cells with planar alignment filled with the same material, both in optical and X-ray observations. The liquid crystal materials used all exhibit the following phase sequence: $I \leftrightarrow N^* \leftrightarrow S_A \leftrightarrow S_C^*$.

As previously discussed, the nematic phase is required for good alignment of the director field. As the sample is cooled, alignment is preserved in the smectic phases. The materials used were all commercial multicomponent mixtures, chosen for their phase sequence and availability. Refs. 24,25,26 detail these liquid crystals.

### 4.5.1. A-SiO Cells: Layer Structure

In the A-SiO Cells the surface preferred director orientation on each plate is parallel to that on the other plate. This leads to a uniform director field in the nematic phase, pretilted by $\approx 20°$ from the surface plane and upon cooling produces layers having having a comparable average tilt $\delta$ in the $S_A$ phase; see Fig.26 (a). The tilt angle $\delta$ generally increases with decreasing temperature in the $S_A$ phase for reasons which are presently unclear. The peaks are generally quite broad in the $S_A$ phase, possibly a result of focal conic layer curvature.

The layer tilt $\delta(T)$ continues to increase with decreasing temperature in the $S_C$ phase (Fig.26 (a)), with a variation that can be attributed to the strongly temperature dependent layer thickness $d_C(T)$. In general, upon cooling through the $S_A$ phase, the layers achieve some nonzero history dependent tilt $\delta_0$ at $T_C$. Assuming that the layer pitch (# of layers / unit length along $z$) established at $T_C$ is to be maintained in the $S_C$ phase, then, as the sample is further cooled, the layers shrink and, neglecting dislocations, must increase their angle of tilt $\delta(T)$ according to the following equation:

$$\cos(\delta) = \frac{d_C(T)}{d_A}\cos(\delta_0) \tag{13}$$

Fig. 17 (a) presents data of layer tilt angle $\delta$ vs $T$ - $T_C$ for three materials along with curves calculated from Eq.13, using $d_C(T)$ and $d_A$ data from planar cells. *The agreement shows that layer shrinkage controls layer tilt in the $S_C$ phase. This is as in planar cells but is not accompanied by chevron formation as in planar cells.* Chevrons form in *planar* aligned cells as a consequence of three conditions: (1) anchoring of the layers, i.e., no displacement at the bounding plates; (2) absence of layer dislocations; (3) shrinkage of the layers with decreasing temperature in the $S_C$ phase. The absence of chevrons indicates the breakdown of at least one of these conditions. A tentative explanation of this difference is realized if we consider the geometry of chevron formation in a cell with tilted surface orientations shown in Fig.27. We assume that at $T_C$ the layers are dislocation free and have a tilt $\delta_0$ equal to or less than the surface director tilt $\Omega$. Thermal contraction upon cooling in the $S_C$ phase causes $\delta$ to increase, which with the above conditions in force requires chevron formation via layer reorientation through the angle $-2\delta_0$ at one of the plates. For $\delta_0$ sufficiently large this layer reorientation cannot take place while maintaining a director orientation at this surface which simultaneously satisfies the bulk $S_C$ tilt cone (constant $\hat{n} \cdot \hat{z}' = \cos \Theta(T)$ ) and surface preference constraints. Here $\Theta(T)$ is the equilibrium bulk $S_C$ director tilt angle. The range of initial layer tilt angle $\delta_0$ for which the constraints can be satisfied is given by:

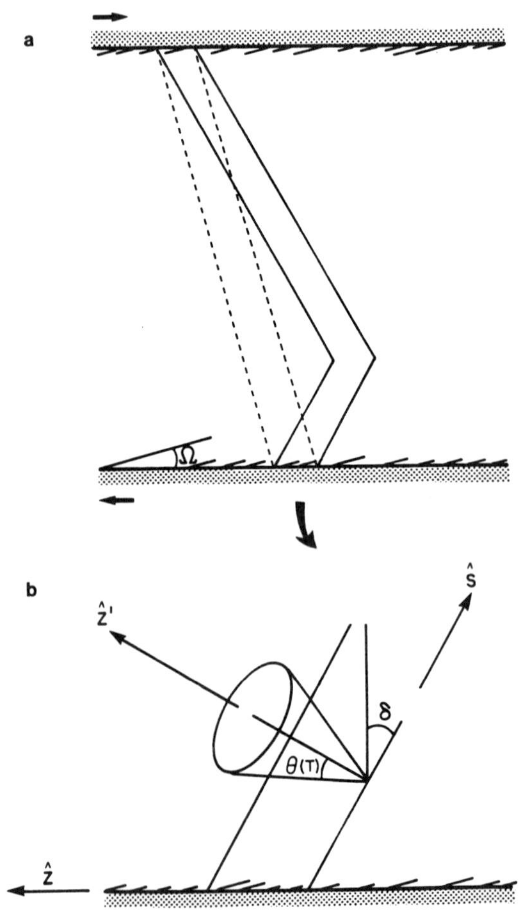


Fig.27. (a) Chevron formation in cells with antiparallel SiO or rubbed plates. The director in the nematic and $S_A$ phases is uniform and tilted out of the planes of the bounding plates. Upon cooling and entering the $S_C$ phase the layers are initially tilted through an angle $\delta_0$. To form a chevron the geometry requires a layer reorientation at one or the other of the surfaces through the angle $2\delta_0$;(b) If $\delta_0$ is too large (see Eq.14) at the surface the director cannot simultaneously lie on the bulk $S_C$ tilt cone and have the preferred orientation $\Omega = \delta_0$ relative to the surface.

$$\cos(\delta_0) > \cos(\delta_{0\max}) = \frac{d_A}{d_C(T)} \tag{14}$$

assuming a preferred surface tilt $\Omega = \delta_0$.

As $\delta_0$ increases beyond $\delta_{0max}$ the surface energy associated with the layer reorientation required for chevron formation increases, suppressing their formation. The result is a loss of one of the three above conditions. Since the X-ray data indicates that the layers have the expected temperature dependence of $d_C(T)$, i.e., that (3) still holds, it must be (1) or (2) that is lost. Which is not clear at this time.

Returning to Fig.16 (a), the narrowing of the I($\beta$) peak with increasing $\delta$ is quite striking. At $T_C$ a distribution of $\delta_0$'s over the range 15.6° - 26.4° is observed, narrowing to 3°, 36.1°C below $T_C$. One possible origin of this effect is that, as the layers shrink in the $S_C$ phase, the tilt angles $\delta$(T) obeying Eq.13 for various $\delta_0$'s collapse into a narrower range. However Fig.28, shows this not to be the case, with the net increase in $\delta$ being nearly independent of the starting $\delta_0$. This would appear to rule out a mosaic of regions of differently oriented planar layers as the origin of the broadening of I($\beta$) at higher temperatures, since from Fig.28, the peak width would hardly change if this were the case. Rather, this evidence points to locally curved layers in the $S_A$ phase and a relaxation to planar layers with decreasing temperature.

### *4.5.2. P-SiO Cells: Layer Structure*

Fig. 17 shows the layer tilt scenarios found in the P-SiO cells. In general the P-SiO treatment produces the chevron structure in the $S_C$ phase. In the $S_A$ phase the layer structure

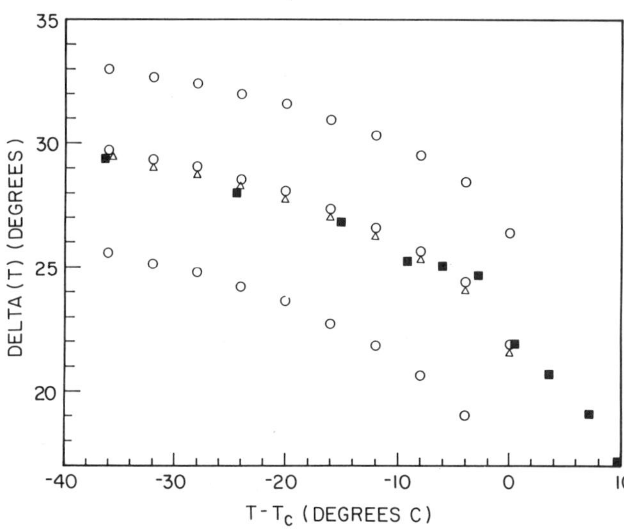


Fig.28. Plot of $\delta$ vs. $T$ - $T_C$. ■ data for A-SiO cell filled with ZLI-3654 (the same as in Fig.17 (a). All other curves are data calculated from Eq.13 with differing values of $\delta_0$. ○ points are for $\delta_0$'s calculated from the ratio of the cosine of $\delta_p(T)$, the layer tilt for *planar* alignment, to the cosine of $\delta_t(T)$, the layer tilt for an A-SiO cell. The Δ and both sets of ○ data points are for $\delta_0$'s taken as the peak, lower FWHM, upper FWHM, respectively from Fig.16 (a), (◇) curve.

varies from sample to sample, depending on sample history, with layer tilts ranging from zero to 15° in different samples. Thus, in contrast to the A-SiO surfaces, with P-SiO the $S_A$ layer tilts are not directly related to the surface tilt imposed by the boundary condition. This is presumably because both boundary conditions cannot be satisfied simultaneously with planar layers, as is possible with A-SiO. The layer structure adopted in P-SiO depends on the layer growth scenario at the N-$S_A$ transition. Fig. 17 (b) shows typical results when the layer tilt is small in the $S_A$ phase. The $\delta(T)$ curves for P-SiO, *planar* alignment (rubbed nylon) and a hybrid cell are essentially the same, indicating that the layer thickness controls $\delta(T)$ in all three cases. Fig. 17 (c) shows a case where the chevron structure is established with a large layer tilt. In this case the boundary conditions are simultaneously partially satisfied by chevron formation and the tilt angle increase with decreasing $T$ in the $S_C$ phase depends on the layer pitch established at the $S_A$ - $S_C$ transition, i.e., according to Eq.13.

Optically P-SiO cells are zig-zag defect free and exhibit domain mediated transitions at the two FLC-solid interfaces and at a third internal interface, as found in *planar* aligned cells having chevron layers and an internal chevron interface[6,37].

This observation, along with the X-ray results of symmetrically tilted layers indicates that the layers in the P-SiO cells have the chevron structure. Since zig-zags mediate the change in chevron direction, the absence of zig-zag defects indicates the presence of a single chevron direction in P-SiO cells, as might be expected since the surface pretilt condition breaks the energy degeneracy of the <<< and >>> chevron states.

## 5. Director Structures Of SSFLC Cells

### 5.1. Introduction

The determination of the layer structures present in SSFLC cells has given us new insight on the director structures of such cells.

In bulk samples the three dimensional director-polarization $\hat{n}$ - **P** structure is governed by the Frank elastic constants and electrostatic self-energy. Solid bounding plates introduce both steric and polar surface interactions. We demonstrate that the chevron local layer structure introduces an internal interface, which is a planar defect in the **P** field. This internal interface has a profound influence on the three dimensional director structure, both in static and dynamic cases.

In this Section we explain the various $\hat{n}$ - **P** structures of switching phenomena observed in cells with the chevron local layer structure (LLS) and in cells with obliquely evaporated SiO surface treatments. We demonstrate the formation of disclination loops and show the influence of the zig-zag wall on their formation.

### 5.2. Ideal SSFLC Cells

In all SSFLC cells, in the smectic C* phase, the director rotates about its cone of constraint to minimize the combination of elastic deformation energy, electrostatic self-energy, surface orientational energy and polar-surface interactions. First we consider the ideal SSFLC cell in which the director lies in the plane of the bounding plates, the planar smectic layers are normal or nearly normal to the plates and we neglect polar surface interactions. For such a cell the two FLC-bounding plate interfaces each stabilize two energetically degenerate states in the Smectic C phase. We depict this with the smectic C* cone of constraint, centered about the layer normal, with opening angle 2Θ, affixed to the bounding plate at its apex; see

Fig.29. Stable orientation states are those which simultaneously satisfy the surface boundary condition and allow the director to remain on the tilt cone.

For smectic layers normal to the bounding plates the tilt cone is bisected by the FLC-bounding plate interface; see Fig.29 (a). For planar boundary conditions the stable states are where the cone and bounding surface intersect ($\Phi = \pm\pi/2$). For this case the ferroelectric polarization **P** points into or out of the cell, normal to the plates. Thus the largest spread in the angle between the two orientation states and hence the optical axis is realized. For layers tilted from the normal by the angle $\delta$, with $\delta < \Theta$, the tilt cone rises from the surface by the same angle and the stable states in n are no longer at the extremes of the cone. At the surface, the polarization takes on a 'pretilt'[8]; see Fig.29 (b), that arises from two diflferent considerations. First, as the layers tilt back by the angle $\delta$, so does **P**, since it lies in the plane of the layers. Second and more importantly, as $\hat{n}$ reorients in $\Phi$ to meet the orientational constraints on the director field at the surface, the polarization *tilts from the cell normal.*

In general, in order to minimize the elastic energy due to deformations of the $\hat{n}$ - **P** field, the cell adopts a homogeneous orientation throughout. This results in two stable states, which we identify as U (up) and D (down) in terms of **P**. The reorientation from one stable state to the other is achieved with the application of an external electric field (two-state switching). For planar boundary conditions and upright layers, the director can reorient by moving either clockwise or counterclockwise on the tilt cone. However, the surface stabilized states are equilibrium positions since the torque $\mathbf{T} = \mathbf{P} \times \mathbf{E} = 0$ there, so the actual sense of switching could be determined by thermal fluctuations in the initial state. For layers tilted with $\delta < \Theta$, the direction of motion around the tilt cone for reorientation is the shortest path to the other stable state along the cone as determined by the pretilt of **P**. The limiting layer tilt is that with $\delta = \Theta$, which produces a monostable $\hat{n}$ - **P** structure, with **P** pinned at $\Phi = \pi/2$ see Fig.29 (c). For $\delta > \Theta$, the system can nolonger maintain the requirements of planar layers and the planar boundary condition.

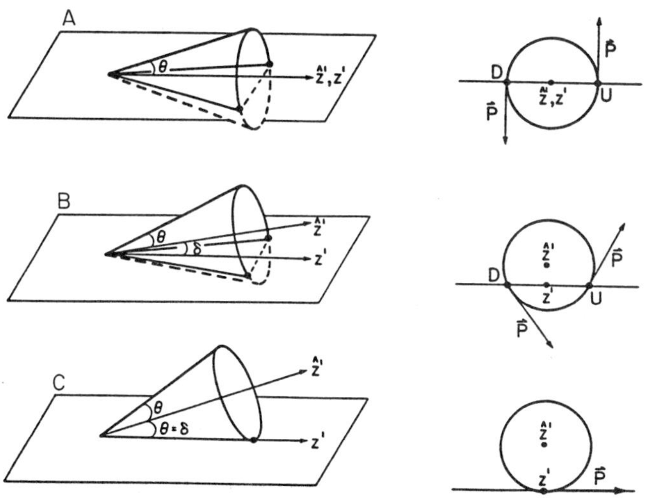


Fig.29. Schematic view of the $S_C^*$ tilt cone intersection with the bounding plates, showing both the stable director orientation states and the ferroelectric polarization **P**. The cone has opening angle $2\Theta$ and its apex is affixed to the bounding plate. (a) The longitudinal axis of the cone lies in the plane of the bounding plate for upright layers, $\delta = 0$. (b) The axis is tilted up by the layer tilt angle $\delta$, where $\delta < \Theta$. (c) $\Theta = \delta$.

Now we introduce polar surface interactions for which both the U and D states are local but nondegenerate minima in energy[8]. Recent experiments by Xue and Clark have demonstrated that SSFLC cells with nylon-treated ITO and clean ITO surfaces (both promote planar boundary conditions) prefer to have the ferroelectric polarization pointing into the sample[38]. In this case one observes three stable orientations of the $\hat{n}$ - **P** field (threestate switching). In the absence of an applied field, $\hat{n}$ undergoes a uniform twist from one of the plates to the other such that **P** is uniformly splayed, pointing into the sample at both surfaces. Homogeneous U and D regions of **P** are achieved with the application of an external electric field. As the field is removed the cell once again adopts the uniform splayed state. Reorientation takes place as described above except if the polar surface interactions are strong enough, the surfaces may not reorient at all. In this case the splay in **P** would collapse to a thin region at the surface whose depth is determined by a balance between the splay energy and **T**.

For nonplanar alignment of the director field at the bounding surfaces, the above discussion holds with the modification, $\delta = \delta_{NP} - \delta_0$ where $\delta_{NP}$ is the measured tilt of the layers and $\delta_0$ is the tilt of the layers in the smectic A phase as well as the tilt of $\hat{n}$ from the bounding surface. In an ideal SSFLC cell with uniformly tilted layers and uniform $\hat{n}$ - **P** structure (absence of bend, splay and twist) one can generate equations which relate the angles $\Theta, \Phi, \delta, \Theta_p$ and $\Psi$ shown in Fig.1. The following rotation matrix relates the primed coordinates which are affixed to the smectic layers to the unprimed ones affixed to the sample cell:

$$\begin{pmatrix} n_x \\ n_y \\ n_z \end{pmatrix} = \begin{pmatrix} \cos\delta & 0 & \sin\delta \\ 0 & 1 & 0 \\ -\sin\delta & 0 & \cos\delta \end{pmatrix} \begin{pmatrix} \sin\Theta\sin\Phi \\ \sin\Theta\cos\Phi \\ \cos\Theta \end{pmatrix}$$

From this matrix the projected molecular tilt angle $\Theta_p$ is given as:

$$\tan\Theta_p = n_y / n_z = \frac{\cos\phi}{\cot\Theta\cos\delta + \sin\Phi\sin\delta}$$

and the elevation angle of the director relative to the plane of the bounding plate as:

$$\sin\Psi = n_x / n = \cos\Theta\sin\delta - \sin\Theta\sin\phi\cos\delta.$$

These equations can be applied to either half of the chevron, with a change of sign in $\delta$ and a reorientation of $\Phi$ by 90°. For applied fields which saturate **P** ($\Phi = 0$ DOWN, $\Phi = \pi$ UP) the above equations reduce to:

$$\tan\Theta_{psat} = \frac{\tan\Theta}{\cos\delta} \tag{15}$$

and, $\sin\Psi = \cos\Theta\sin\delta$. Eq.15 indicates that the projected molecular tilt angle $\Theta_p$ which is measured optically is greater than or equal to the actual moleculat tilt angle $\Theta$.

Determination of the complete $\hat{n}$ - **P** structure in real SSFLC cells is difficult even though both the director tilt angle $\Theta(T)$ and the layer tilt angle $\delta(T)$ can both be measured for any given sample. This is due to the highly complex interactions of electrostatic, elastic, surface and polar surface energies. We can however model such structures and compare these models to spectroscopic data[39].

## 5.3. SSFLC Cells with a Chevron LLS

### 5.3.1. Introduction

SSFLC cells with a chevron LLS incorporate all the characteristics described above. The chevron tips define a plane parallel to the bounding plates, which acts as an internal interface adding additional constraints on the $\hat{n}$ - **P** field at this FLC-FLC interface. This plane is the first example to be found of the class of planar liquid crystal defects in which the order parameter discontinuities occur at a planar sheet, parallel to which there is full translational invariance. This internal chevron interface splits the cell into two nearly independent cells, optically and electrically in series. For the following discussion we neglect electrostatic interactions.

The chevron discontinuity significantly influences the orientational structure of the director field $\hat{n}(r)$ and polarization $\mathbf{P}(r) = P_0(\hat{z}' \times \hat{n})$ in its vicinity. Fig. 30 shows the tilt cones with directors $\hat{n}_+$ and $\hat{n}_-$, separated by the chevron reorientation angle $2\delta$, giving the orientations $\Phi_-$ and $\Phi_+$ immediately adjacent to the chevron in the upper and lower halves of the cell respectively. If $\delta$ satisfies the condition $\delta < \Theta$, the + and the - cones intersect at the open circles U (up) and D (down) in Fig.30. Adjusting $\Phi_-$ and $\Phi_+$ to either of these intersections *makes the orientation of* $\hat{n}$ *everywhere uniform and parallel to the chevron plane.* This condition and the layer tilt thus produce pretilt of the polarization in accord with the arguments of Clark and Handschy[8], with pretilt changing sign as the chevron is crossed (Fig.30). For $2\delta = 2\Theta$, the two cones overlap at only one point and **P** lies parallel to the glass plates, destroying bistability at the chevron interface. In these minimum energy states, the chevron elastic energy will arise from the Franklike elasticity resisting abrupt change of the smectic C* biaxial order parameter, i.e., reorientation of the ($S_C^*$ tilt) plane locally containing $\hat{z}'$ and $\hat{n}$ (dashed lines in Fig.30).

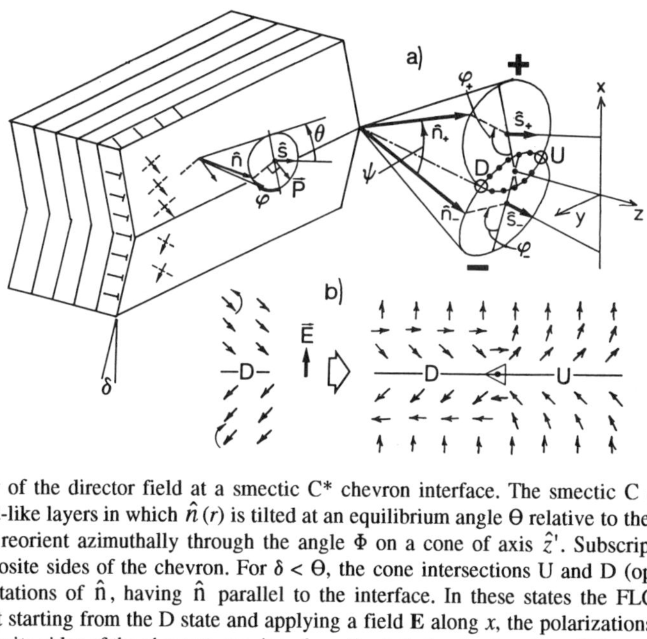


Fig.30. Geometry of the director field at a smectic C* chevron interface. The smectic C is a locally biaxial structure of liquid-like layers in which $\hat{n}(r)$ is tilted at an equilibrium angle $\Theta$ relative to the local layer normal $\hat{z}'$ and is free to reorient azimuthally through the angle $\Phi$ on a cone of axis $\hat{z}'$. Subscripts + and - refer to quantities on opposite sides of the chevron. For $\delta < \Theta$, the cone intersections U and D (open circles) are the equilibrium orientations of $\hat{n}$, having $\hat{n}$ parallel to the interface. In these states the FLC polarization **P** is pretilted such that starting from the D state and applying a field **E** along $x$, the polarizations rotate in opposite directions on opposite sides of the chevron, moving along the dotted paths from D to U for sufficiently large **E**.

Assuming an elastic constant $L$ for this reorientation and a layer compression elastic constant $B$, the width $dx$ in $x$ of the chevron tip will be determined by the competition of orientational and compressional forces to be $dx \approx L/(B\delta) \lesssim 100$ Å. We thus expect the chevron reorientation to be abrupt. The application of an electric field and/or the boundary conditions at the solid surfaces will in general lead to gradients in $\Phi$ and consequent torques tending to move $\Phi_+$ and $\Phi_-$ from U or D. The angle $\Psi$ between $\hat{n}_+$ and $\hat{n}_-$ across the chevron tip is given by:

$$\begin{aligned}\cos(\Psi(\Theta,\delta,\Phi_+,\Phi_-)) = \ & \cos(2\delta)\cos^2(\Theta)+\sin^2(\Theta)\sin(\Phi_+)\sin(\Phi_-)+ \\ & \sin(2\delta)\sin(\Theta)\cos(\Theta)\cos(\Phi_+)+ \\ & \cos(2\delta)\sin^2(\Theta)\cos(\Phi_-)\cos(\Phi_+)- \\ & \sin(2\delta)\sin(\Theta)\cos(\Theta)\cos(\Phi_-)\end{aligned} \tag{16}$$

For a given $\Phi_-$, the interface orientational elastic energy will be minimized by that $\Phi_+$ which satisfies the equation:

$$\begin{aligned}-\sin(\Psi)\left.\frac{\partial\Psi}{\partial\Phi_+}\right|_{\Phi_-} &= \sin(\Theta)\left[\sin(\Phi_-)\cos(\Phi_+)-\cos(2\delta)\cos(\Phi_-)\sin(\Phi_+)\right]- \\ & \sin(2\delta)\cos(\Theta)\sin(\Phi_+)=0\end{aligned} \tag{17}$$

obtained from minimizing the angle $\Psi$.

The key feature of this geometry is that, for $\delta$ comparable to $\Theta$, for the physically relevant condition, $|\Phi_+| < \Phi_{max}$, $\Phi_+$ varies over a finite range for a complete $2\pi$ rotation of $\Phi_-$; see Fig.31. In this case $\Phi_+$ can be decoupled from $\Phi_-$ with the application of torques large enough to overcome an orientational energy barrier of at most $\gamma \sim (K/dx)(\Psi_{max})^2 \sim (K/dx)(2\delta)^2 \sim 2\text{x}10$ ergs/cm$^2$, where $K$ is a typical nematic Frank elastic constant. This barrier is comparable to typical LC - solid surface interaction energies[40]. *Thus the chevron plane behaves as an internal FLC-FLC interface.*

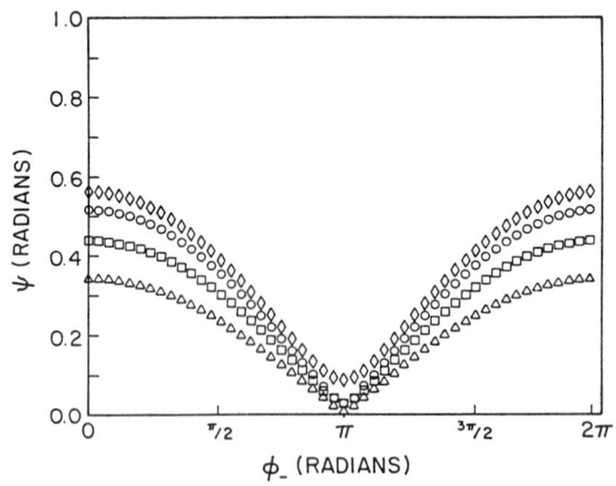


Fig.31. Plot of $\Psi$ vs. $\Phi_-$ from Eq.16 with $\Phi_+ = 0$ and the following values of $\delta$ and $\Theta$; $\Delta$, $\delta = 9.8°$, $\Theta = 10.0°$; □, $\delta = 12.6°$, $\Theta = 13.4°$; ○, $\delta = 14.8°$, $\Theta = 15.6°$; ◇, $\delta = 18.1°$, $\Theta = 18.6°$.

The three interfaces in a cell with the chevron LLS stabilize four $\hat{n}$ - **P** structures, Fig.32. On application of a moderate down **E** field the **P** field reorients as shown in (a), with all splay of **P** pressed to the interfaces. This traps a $\pi$ disclination sheet at the bottom surface, which will reorient via a surface transition for larger down **E** fields. As the applied **E** field is reduced to zero, the system relaxes to the structure depicted in (b). In the upper half of the cell, the $\hat{n}$ - **P** couplet adopts a uniform structure. In the lower half, the remaining splay in **P** (twist in $\hat{n}$) has been distributed throughout, in order to minimize the elastic energy. As the applied **E** field is increased in the up direction, the splay in **P** is again forced to the interfaces and reorientation of the director field at the chevron plane will occur as described below. For moderate up **E** fields, the **P** structure in (d) is observed. Again, for higher **E** fields, the $\pi$ disclination sheet is trapped at the top surface and will reorient via a surface transition. Reduction of the **E** field to zero results in the structure depicted in (c), which is analogous to the one in (b), except that the halves of the cell are interchanged and the sense of twist and splay of the $\hat{n}$ - **P** couplet are reversed. Thus the states (b) and (c) are optically distinguishable even in the case of the chevron plane bisecting the cell (symmetric chevron). This explains the observation of fourstate switching which occurs in cells with planar alignment of the director field[41]. Cells with the asymmetric chevron layer structure exibit different switching thresholds between the U and D states due to the two different compressions (pitches) of the splay deformation in the two halves of the cell. In cells with zig-zag wall defects (both chevron directions) we can directly observe this when the inside (outside) of a zig-zag wall switches before the outside (inside) area.

The energy barrier for the decoupling of $\Phi_+$ and $\Phi_-$ at the chevron plane is, in fact, much lower than described above in one situation of crucial importance, namely the response of the director to applied electric field **E**. The change of pretilt sign at the chevron leads to opposite signs of rotation in $\Phi$ on opposite sides of the chevron, Fig.30, producing in response to sufflciently high **E**, a transition from the D to U state in which $\hat{n}_+$ and $\hat{n}_-$ rotate on the dotted paths (...) from D to U in Fig.30. The surface energy maximum for this D to U

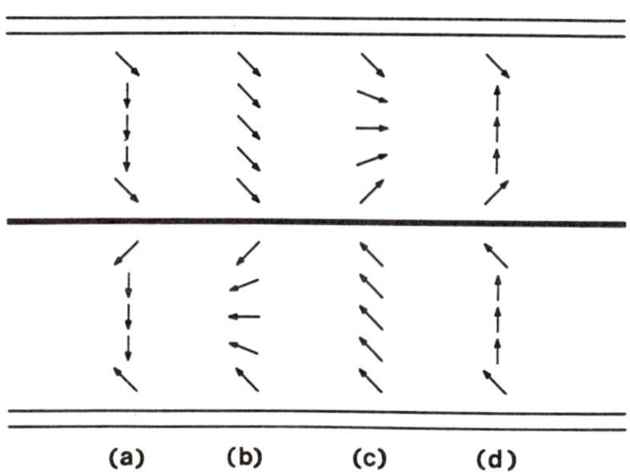


Fig.32. **P** structures stabilized in a cell with a chevron LLS. In this diagram the chevron is pointing out of the page toward the reader, as indicated by the heavy line at the chevron plane. The **P** structures drawn are for positive materials, $\mathbf{P} = P_0(\mathbf{z} \times \hat{n})$. (a) Moderate applied **E** field in the down direction. (b) Resultant structure as the field in (a) is reduced to **E** = 0. (d) Moderate applied **E** field in the up direction. (c) resultant structure as the field in (d) is reduced to **E** = 0. For larger applied **E** fields the $\pi$ disclinations trapped at the top and bottom of the cell, (d) and (a) respectively, would reorient as a surface transition. The reader is reminded that **P** lies in the plane of the tilted layers.

reorientation, $\gamma_{DU} \sim (K/dx)(\delta - \Phi)^2$, is much less than $\gamma$ when $\delta \sim \Theta$. For **E** fields exerting a torque density larger than $\gamma_{DU}$, $\hat{n}_+$ and $\hat{n}_-$ homogeneously decouple and the D to U transition occurs without domain walls [HIGH field limit[8]]. For lower **E** fields the opposite rotation sense above and below the chevron traps, at the chevron plane, a $2\pi$ disclination sheet in $\Phi$; see Fig.33. The chevron associated pretilt conditions thus account for the internal disclination identified by Ouchi and coworkers[42].

### *5.3.2. Boats*

Holes nucleating in the $2\pi$ sheet, shown in Fig.33, are the domain loops observed in strobe microscopy experiments[43,42]. These internal disclination loops are trapped in the chevron planes and move in these planes; see Fig.34. This can be observed directly by noting that in cells with asymmetric chevrons (see Fig.19 and Reference 44), the domain loops in a given region remain in fixed planes normal to $x$ and have the same plane of focus as the chevron plane, as discussed in Fig.19.

The disclination loops have been unambiguously identified only in cells with the chevron LLS. These loops have a unique shape arising from anisotropic growth rates. In general, for moderate applied **E** fields, they appear as well defined five sided objects with straight sides and sharp corners. Their shape resembles the outline of a 'boat,' hence this colloquial name has been adopted and will be used here for convenience. The faceted nature of these holes is due to a slow growth rate in the direction normal to the edges of the boat. The sides of the boat always lie along the smectic layers, perpendicular to $z$. The stern, which exists only for low voltages, is a straight line segment normal to the sides and the smectic layers. The bow is formed by two line segments which do not necessarily meet in a symmetric way, that is, the tip may be off center. For larger applied **E** fields (but less than that which produces uniform reorientation of the $\hat{n}$ - **P** field), the boats adopt a hexagonal shape, with bows on the front and back which grow out from the defect at nearly the same rates. As an aid to explain the $\hat{n}$ - **P** structures and correlate these to optical observations we once again define our geometry: $z$ lies in the plane of the bounding plates normal to upright smectic A layers; the U (up) direction is defined as pointing from the lower to the upper plate ($+x$); D (down) is the opposite of U; all arguments are for the case in which $\mathbf{P} = P_0(\hat{z}' \times \hat{n})$ is positive, where $\hat{z}'$ is the local layer normal. The chevron direction is as indicated in the various figures, with the heaviest lines indicating edges to the front of the figure.

Boats grow in an outward fashion with the bow of the boat and the sides moving faster than the stern. The direction of the boat is dependent on the switching sequence and the direction of the chevron. Fig.34 demonstrates that the trapping of the $2\pi$ disclination sheet at the chevron interface and the nucleation of a boat is just a consequence of the pretilt directions at the chevron (FLC-FLC) and FLC-bounding plate interfaces. In this figure the chevron points out of the page and shows the switching of a sample from the U to D state. The hole which nucleates in the $2\pi$ sheet has $\pm 2\pi$ disclination points at its ends, along the smectic layers, $y$. Switching from D to U interchanges the disclination points. This then demonstrates the change in direction of the boats with the switching sequence. Careful microscopic analysis of domain nucleation from zig-zag walls, as described in the following Section, has led us to assign the $+2\pi$ and $-2\pi$ disclinations to the bow and stern of the boats respectively which concurrs with Ouchi et al.[35]. The sides of the boat are $\pi$ twist walls of **P**.

### *5.3.3. Influence of the Zig-zag Wall on $\hat{n}$*

In Section 4.3 we demonstrated that the zig-zag wall mediates the change in chevron

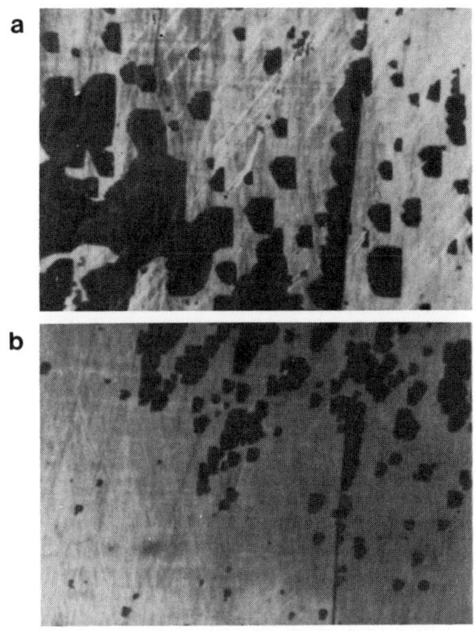

Fig.33. Photomicrographs showing the nucleation of holes ('boats') in the $2\pi$ disclination sheet trapped in the plane of the chevron interface; see also Fig.34. (a) Switching from the U to the D state. (b) Switching from the D to U state. Note that the boats change direction.

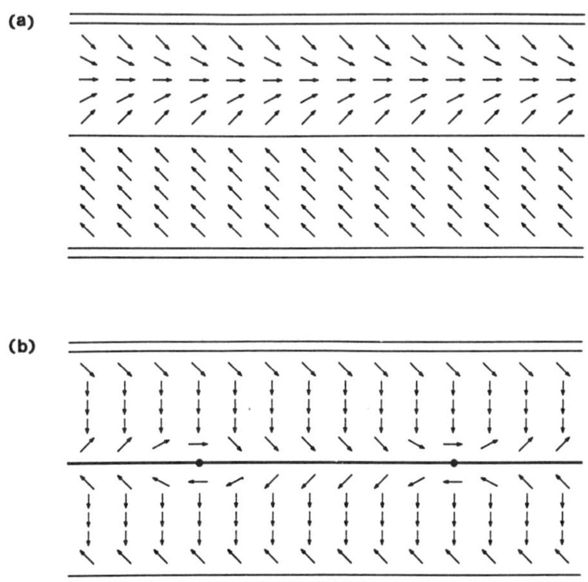

Fig.34. Formation of a $2\pi$ disclination sheet in cells with the chevron LLS. Schematic diagrams showing the **P** field for a positive material ($\mathbf{P} = P_0(\mathbf{z} \times \hat{\mathrm{n}})$) in a chevron cell where the chevron points out of the page and where polar surface interactions require that the polarization points into the cell. The **P** structure of (a) is that which results as an applied external **E** field in the UP direction is reduced to zero. As the applied **E** field is increased in the DOWN direction, the **P** field reorients, trapping a $2\pi$ disclination sheet at the chevron interface. Holes which nucleate in this sheet reorient the director field at the chevron interface by the motion of $\pm 2\pi$ walls (b) and have a boat shaped appearance as seen in Fig.33. The bow (stern) of the boat is assigned the $+2\pi$ ($-2\pi$) disclination from optical observations of switching in the vicinity of zig-zag walls, see text.

direction and that the structure which connects the two chevrons is a plane parallelogram; see Fig.22. Here we show that the zig-zag wall defect has a profound influence on the $\hat{n}$ - **P** structure in its vicinity and hence affect the switching characteristics of SSFLC cells. Two equations from Section 4.3 are pertinent to our discussion:

$$\cos(\alpha - \gamma) = \cos(\delta)\cos(\gamma) \tag{18}$$

$$\cos(\sigma) = \cos(\alpha)\cos(\delta) \tag{19}$$

where $\gamma$ is the angle that the narrow wall makes with $z$ and $\alpha$ is the angle between the normal of the mediating parallelogram and $z$. Each of the four layer breaks at the edges of the mediating parallelogram have the layer reorientation angle $\pm\sigma$, were the + and - signs are assigned for layers which bend away from, or toward the adjoining layer respectively; see Fig.35. The angle $\sigma$ is on the order of $2\delta$, the angle of layer reorientation across the chevron tip. *Thus each of the layer breaks at the zig-zag wall has an influence on the* $\hat{n}$ - **P** *structure in its icinity as does the chevron interface and all the arguments put forward above are in force.* The largest angle which $\sigma$ can adopt is $2\Theta$, beyond which energetically unfavorable splay deformations of the $\hat{n}$ field will occur. This is the argument used in the previous Section to limit the angle $\gamma$, the angle between the narrow section of the zig-zag wall defect and $z$.

Strobe microscopy of the switching dynamics in cells containing zig-zag defects show the nucleation of domain walls from the narrow walls of the defect. These domain walls are boats which move in the plane of the chevron tips as discussed in the last Section. For regions of the narrow wall with $\gamma > 0$, sterns of the boat disclinations emanate from *both sides of alternate* narrow walls, as in Fig 36. These domain loops continue to grow outward from alternate walls until they collapse onto adjacent narrow walls; see Fig 37. The nucleating wall

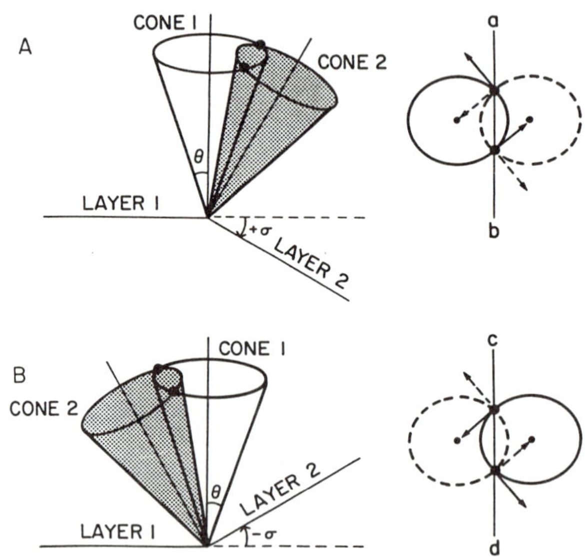


Fig.35. Schematic diagram of the director cones adjacent to the mediating parallelogram of the zig-zag wall defect. The stable $\hat{n}$ - **P** configurations are those where the two cones overlap (●). (a) Positive $\sigma$. (b) Negative $\sigma$.

Fig.36. Stroboscopic photomicrograph of boat nucleation from the narrow wall of the zig-zag wall defect. The bow of the boat shaped disclination loops emanate from both sides of alternate narrow walls of the zig-zag defect for regions with $\gamma > 0$. The wall which nucleates these domain walls alternates with a change in switching direction.

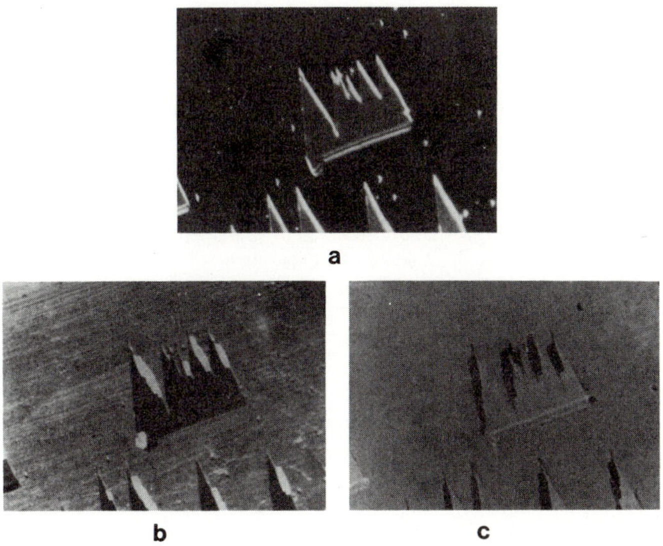

Fig.37. Photomicrographs showing a switching sequence in which domain walls emanate from alternate narrow walls of the zig-zag defect. Domain loops nucleate on both sides of alternate narrow walls (a) and collapse on the adjacent walls (c) for $\gamma > 0$. The wall which nucleates the domain loop alternates with a change in switching direction.

Table 5. Calculated zig-zag wall parameters for $\delta = 6°$. Angles are in degrees. *w/t* is the ratio of the zig-zag wall width to the cell thickness and is dimensionless. $\alpha$ is calculated by interation of Eq.18. $\sigma$ is calculated from Eq.19.

| $\gamma$ | $\alpha$ | $\lvert\sigma\rvert$ | $w/t$ |
|---|---|---|---|
| 15.0 | 31.1 | 31.6 | 0.2 |
| 12.0 | 25.4 | 26.1 | 0.2 |
| 9.0 | 19.8 | 20.7 | 0.3 |
| 6.0 | 14.5 | 15.6 | 0.4 |
| 3.0 | 9.7 | 11.4 | 0.6 |
| 0.0 | 6.0 | 8.5 | 1.0 |
| -3.0 | 3.7 | 7.0 | 1.6 |
| -6.0 | 2.5 | 6.5 | 2.4 |
| -9.0 | 1.8 | 6.3 | 3.3 |
| -12.0 | 1.4 | 6.2 | 4.2 |
| -15.0 | 1.1 | 6.1 | 5.1 |
| -18.0 | 0.9 | 6.1 | 6.0 |
| -21.0 | 0.8 | 6.1 | 7.0 |
| -24.0 | 0.7 | 6.0 | 7.9 |
| -27.0 | 0.6 | 6.0 | 8.7 |
| -30.0 | 0.5 | 6.0 | 9.6 |

Table 6. Calculated zig-zag wall parameters for $\delta = 12°$. Angles are in degrees. *w/t* is the ratio of the zig-zag wall width to the cell thickness and is dimensionless. $\alpha$ is calculated by interation of Eq.18. $\sigma$ is calculated from Eq.19.

| $\gamma$ | $\alpha$ | $\lvert\sigma\rvert$ | $w/t$ |
|---|---|---|---|
| 15.0 | 34.1 | 35.9 | 0.4 |
| 12.0 | 28.9 | 31.1 | 0.4 |
| 9.0 | 24.0 | 26.6 | 0.5 |
| 6.0 | 19.4 | 22.7 | 0.6 |
| 3.0 | 15.4 | 19.4 | 0.8 |
| 0.0 | 12.0 | 16.9 | 1.0 |
| -3.0 | 9.4 | 15.2 | 1.3 |
| -6.0 | 7.4 | 14.1 | 1.6 |
| -9.0 | 6.0 | 13.4 | 2.0 |
| -12.0 | 4.9 | 13.0 | 2.4 |
| -15.0 | 4.1 | 12.7 | 2.8 |
| -18.0 | 3.5 | 12.5 | 3.2 |
| -21.0 | 3.1 | 12.4 | 3.6 |
| -24.0 | 2.7 | 12.3 | 4.1 |
| -27.0 | 2.4 | 12.2 | 4.5 |
| -30.0 | 2.1 | 12.2 | 4.9 |

Table 7. Calculated zig-zag wall parameters for $\delta = 18°$. Angles are in degrees. *w/t* is the ratio of the zig-zag wall width to the cell thickness and is dimensionless. $\alpha$ is calculated by interation of Eq.18. $\sigma$ is calculated from Eq.19. This Table shows the widening of the narrow zig-zag walls as it connects with the broad wall. The calculated thickness of the broad wall is $\approx 25\ \mu m$.

| $\gamma$ | $\alpha$ | $\lvert\sigma\rvert$ | $w/t$ |
|---|---|---|---|
| 15.0 | 38.3 | 41.7 | 0.5 |
| 10.0 | 30.5 | 35.0 | 0.6 |
| 5.0 | 23.7 | 29.4 | 0.8 |
| 0.0 | 18.0 | 25.2 | 1.0 |
| -5.0 | 13.7 | 22.5 | 1.3 |
| -10.0 | 10.5 | 20.8 | 1.7 |
| -15.0 | 8.3 | 19.8 | 2.1 |
| -20.0 | 6.7 | 19.2 | 2.5 |
| -25.0 | 5.5 | 18.8 | 2.9 |
| -30.0 | 4.6 | 18.5 | 3.4 |
| -35.0 | 3.8 | 18.4 | 3.8 |
| -40.0 | 3.2 | 18.3 | 4.2 |
| -45.0 | 2.7 | 18.2 | 4.6 |
| -50.0 | 2.3 | 18.1 | 4.9 |
| -55.0 | 1.9 | 18.1 | 5.2 |
| -60.0 | 1.6 | 18.1 | 5.5 |
| -65.0 | 1.3 | 18.0 | 5.8 |
| -70.0 | 1.0 | 18.0 | 5.9 |
| -75.0 | 0.8 | 18.0 | 6.1 |
| -80.0 | 0.5 | 18.0 | 6.2 |
| -85.0 | 0.2 | 18.0 | 6.3 |

switches from one set of alternate walls to the other with reversal of the applied **E** field (change in switching direction).

For regions with $\gamma$ small or < 0, both the bow and stern of boat disclination loops are observed to emanate from both sides of the narrow walls. The domain loops spread out from the walls and generally coalesce with one another before encountering an adjacent wall since both walls nucleate the disclinations with either switching direction. Switching inside of the narrow walls is observed to occur as a nucleation of two point defects which race in opposite directions along the narrow wall. However the nucleation site of this switching process does not occur in as regular a fashion as that for regions external to the narrow walls.

These observations can be explained by considering the variation of $\alpha$ and $\sigma$ as $\gamma$ varies from $-90^\circ \leq \gamma \leq \gamma_{max}$: along the length of a curved narrow wall. This large range in $\gamma$ is accompanied by significant changes in $\alpha$ and more importantly in $\sigma$, as shown in Tables 5,6 and 7. These tables demonstrate the effects of changing $\gamma$ on $\alpha$ and $\sigma$ for various fixed values of $\delta$ (fixed temperatures). The values in these tables were calculated as follows: for a given $\delta$ and $\gamma$ Eq.18 was iterated to determine $\alpha$, which was in turn used in Eq.19 to determine the corresponding $\sigma$. As is evident from the tables, $\sigma$ is ~$\delta$ for most values of $\gamma < 0$, slowly

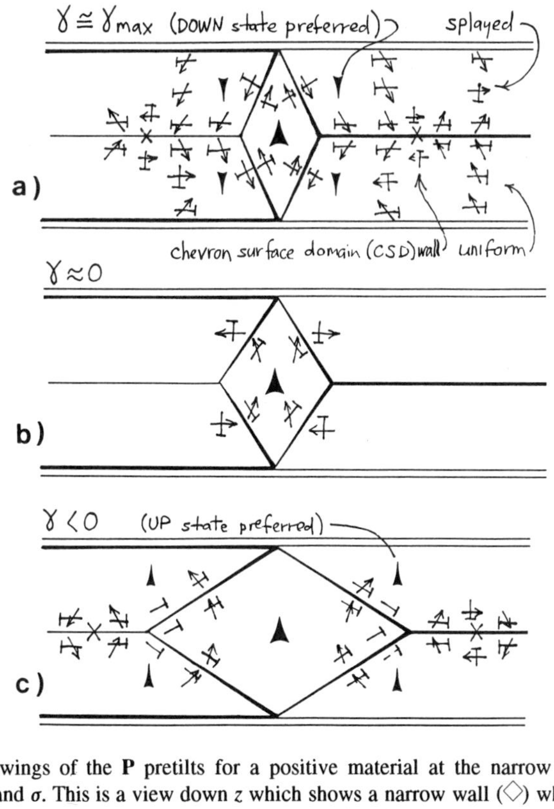


Fig.38. Schematic drawings of the **P** pretilts for a positive material at the narrow zig-zag wall defect for diflferent values of $\gamma$ and $\sigma$. This is a view down $z$ which shows a narrow wall (◇) which separates regions of different chevron directions. Left of the wall the chevron points into the paper; right, out of the paper. The strength of the pinning of the $\hat{n}$ - **P** fields at the edges of the narrow wall is dependent on $\sigma$ and $\gamma$. (a) For $\gamma > 0$; $\sigma > 2\delta$ the $\hat{n}$ - **P** fields are strongly pinned and sterns of boat disclinations are observed to nucleate. (b) $\gamma = 0$, $\sigma \sim \delta$. (c) For $\gamma < 0$; $\sigma < 2\delta$, the pretilt at the narrow walls have both U and D components and are not as strongly influenced by the narrow wall. In this regime both sterns and bows of boat disclinations can be nucleated.

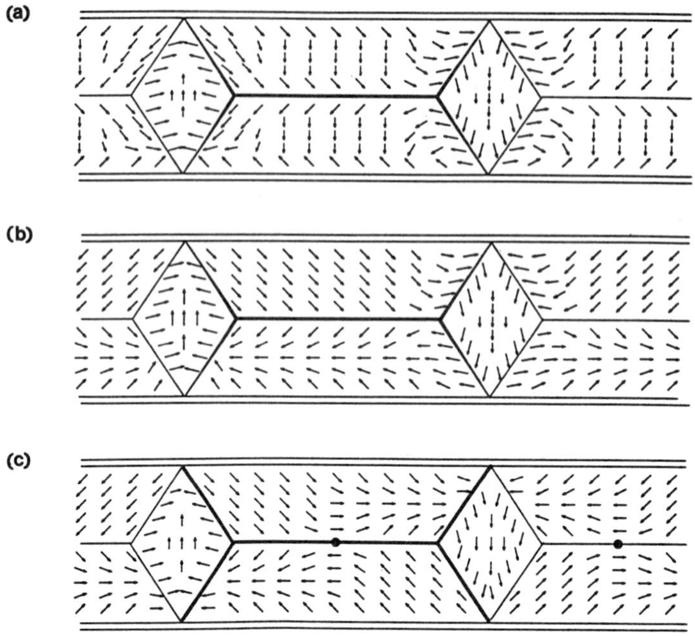


Fig.39. Schematic representation of the nucleation of a boat from a zig-zag wall. The arrows show the **P** field for a positive material. This diagram has the chevron pointing at the reader between two adjacent narrow zig-zag walls (◇ figures) and into the paper outside of them. For $\gamma > 0$; $\sigma > 2\delta$ (see Fig.38) the **P** fields stabilized external to the edges of the narrow walls (◇) alternately promote the U (right ◇) or D (left ◇) states. This demonstrates why switching domains nucleate from alternate walls, with the nucleating wall changing with switching direction; see Fig 37. (a) The resultant **P** structure with an externally applied DOWN **E** field. (b) **E** = 0. (c) As the **E** field is increased in the UP direction, **P** reorients trapping a $2\pi$ disclination sheet at the chevron interface. Since the **P** field already has a pretilt for the UP state at one of the walls, the stern of a boat is nucleated here, on both sides of the wall and expands throughout the cell to eventually collapses on an adjacent wall.

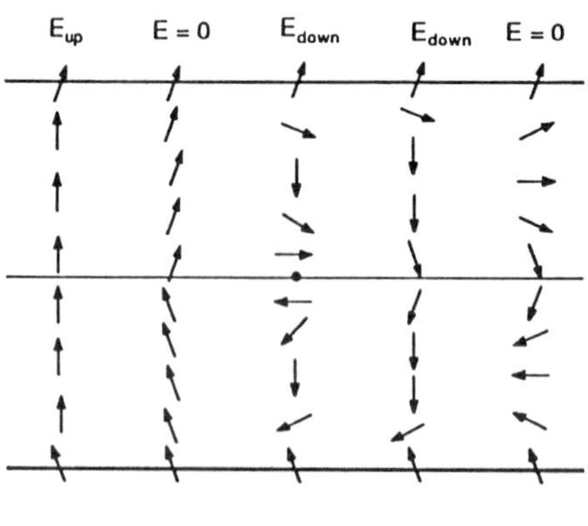


Fig.40. Polarization structures stabilized in an SSFLC cell with strong boundary conditions. Drawn for positive polarization.

increasing until it becomes slightly larger than $\delta$ for $\gamma \sim 0$. As $\gamma$ is further increased, $\sigma$ increases dramatically, quickly obtaining the value $2\Theta$.

As $|\sigma|$ increases so too does the pretilt of **P** stabilized by the the chevron parallelogram interface, with respect to this interface. For $\gamma > 0$ the layer breaks of the zig-zag wall have as large or a larger effect on the $\hat{n}$ - **P** field in their vicinity as the chevron tip (layer break $2\delta$). Fig. 38 (a) demonstrates that the **P** pretilts stabilized by these edges have only an U or D component for $\gamma > 0$ and that alternate narrow walls stabilize the U and D states. The inside of the narrow walls have the opposite pretilt sense from that external to them. These regions have a complex $\hat{n}$ - **P** structure due to the proximity of many layer breaks and is as yet not well understood. The intrinsic promotion of the U or D state by alternate narrow walls explains the nucleation of domain loops from alternate narrow walls and its dependence on switching direction. Fig 39 is a diagram of the $\hat{n}$ - **P** structure which demonstrates the nucleation of a domain wall (stern of a boat) from the narrow wall for $\gamma > 0$. This figure shows that the stern of a boat will emanate from both sides of a narrow wall and that the *stern is* a - $2\pi$ *disclination.*

### *5.3.4. Planar Boundary Conditions with Nematic to Smectic C* Materials*

The smectic layers in SSFLC cells with planar boundary conditions and filled with materials which exhibit the chiral nematic to smectic C* phase transition have been shown to form at approximately the molecular tilt angle $\Theta$ from the alignment direction imposed at the bounding surfaces. For these cells the orientational energy of the two surface stabilized states are no longer degenerate because they are not symmetrically disposed about the buffing direction. The orientation state which lies along the buffing direction must be energetically favored by construction. This is a case for strong pinning of the director at the surface which can give rise to the polarization states shown in Fig.40.

## 6. SSFLC Cells with Nonplanar Alignment of $\hat{n}$

### *6.1. Introduction*

As previously discussed, the director $\hat{n}$ has a large tilt angle $\Omega$ at the liquid crystal silicon monoxide (SiO) interface, which is dependent on the angle of evaporation with respect to the substrate normal. $\Omega$ is also material and temperature dependent. In nematic phases, for many materials studied, $\Omega$ increases with decreasing temperature from the $I$ - $N$ phase transition. $\Omega$ saturates near 20° at about 5°C below the transition, for cyano-biphenyls[45]. In smectic phases the director also orients at a large angle with respect to the plane of the substrate. However this tilt is difficult to measure due to the thinness of an SSFLC cell. We assume that once the director tilt has saturated in the nematic phase, it is frozen in as the sample is cooled into the smectic phases. This results in the formation of layers tilted by the angle $\Omega$ from the substrate normal in the smectic A phase. The layer structure at the surface may have a lot of defects (dislocations) at the $N$ - $S_A$ transition. These will anneal out as the sample is cooled further into the $S_A$ phase. The layer organization in SSFLC cells with SiO coated surfaces has been discussed in great detail. We have shown that cells with parallel alignment of the SiO evaporation directions, P-SiO cells, show a chevroned layer structure. While those cells with the evaporation directions aligned antiparallel, A-SiO cells, show a uniform tilting of the layers. In this Section we present optical evidence which corroborates our results from X-ray scattering.

## *6.2. A-SiO Cells: Optical Observations*

Observations in transmission polarized microscopy show that A-SiO cells lack zig-zag walls, consistent with the A-SiO cell uniform tilted layer structure and the requirement of chevron layers for zig-zags. Application of electric fields to A-SiO cells shows a splayed **n-P** structure stabilized by polar surface interactions[8,46] at low electric field, with surface domain mediated orientation transitions at one (the other) of the FLC-solid interfaces to uniform up (down) states at large + (-) electric fields.

The structure of the director-polarization **n-P** field in A-SiO cells with tilted planar layers is determined by conditions at the FLC-solid interfaces, namely by the geometry of the tilt cone and preferred surface orientation constraints. This geometry is shown in Fig.41 (a).

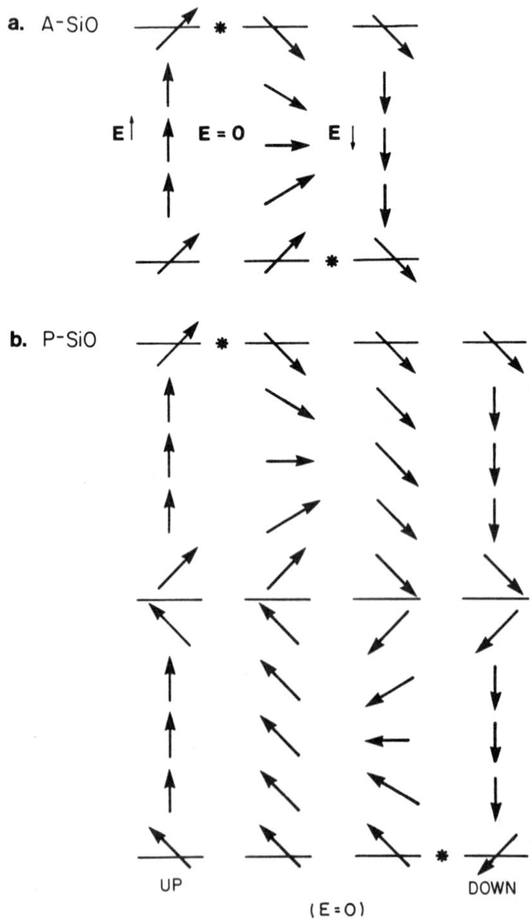


Fig.41. (a) A-SiO cells generally have a monostable equilibrium splayed state at E = 0 with **P** directed into or out of the surface (dependent on the material). Sufficiently strong applied fields produce surface domains (*) which mediate switching between nearly uniform UP and DOWN states. Possible variations of this scenario are discussed in Reference 4. (b) P-SiO cells exhibit the 4-state switching behavior characteristic of the chevron structure.

### 6.3. P-SiO Cells: Optical Obsevations and Director Field Structure

The switching characteristics in the **n-P** fields in the P-SiO cells are somewhat different from those of chevron cells having *planar* (low pretilt) surfaces, such as clean indium tin oxide or rubbed polymer. In the latter the chevron interface switches at low field with the characteristic boat shaped domains[35] shown in Fig.42 (a) and this switching[37] between states 2 and 3 in Fig.41 (b) produces significant contrast. The surface transitions occur via surface domains at higher fields (state 2→UP and state 3→DOWN) and produce much less contrast, since the surface disclinations which disappear via the domains are quite thin. The surface domains are generally irregular in shape as shown in Fig.42 (b). The chevron domains have much higher mobility than the surface domains.

In the P-SiO cells as in the low pretilt chevron cells three switching transitions are found, with the 2↔3 transition on the chevron interface (the one occurring at low voltage) producing rather little contrast and exhibiting irregularly shaped high mobility domains; see Fig.42 (c). On the other hand the (2→UP and 3→DOWN) surface transitions produce significant contrast in the P-SiO cells. Additionally, the surface domain has a characteristic triangular shape as shown in Fig.42 (d). Similar domain behavior has been reported for P-SiO cells by Yamada et al.[47].

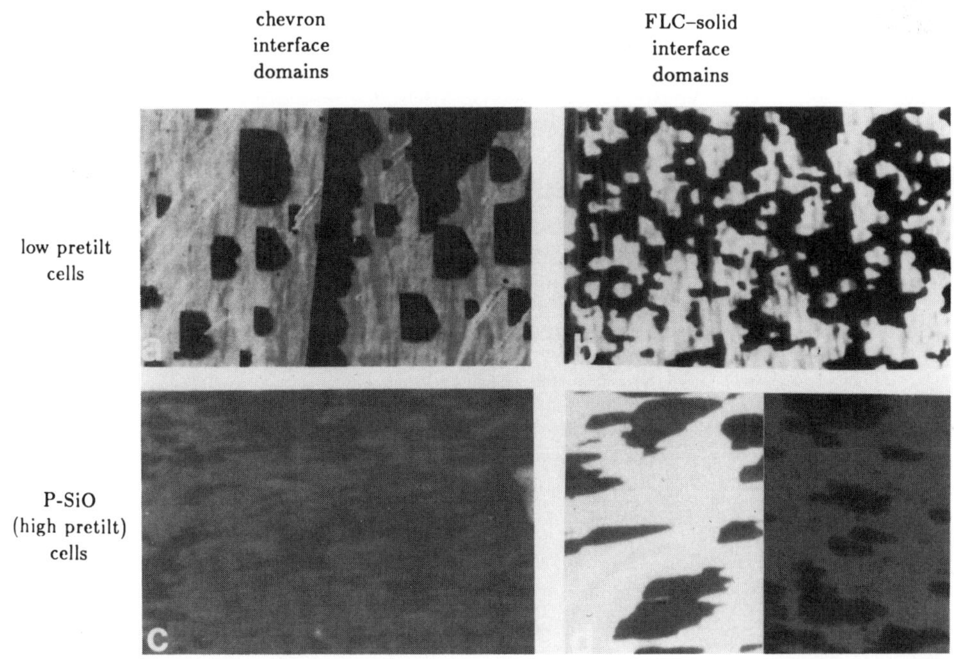


Fig.42. Comparison of domain walls at the chevron and FLC-solid interfaces in low pretilt and P-SiO SSFLC cells (reproduced here in B&W). In the low pretilt case the high contrast domains are at the chevron interface (switching from state 2 → 3 in Fig.41 (b)) and have a characteristic boat shape (a) and the FLC-solid surface transition domains (2 → UP in Fig.41 (b)) are irregularly shaped (b). In the P-SiO cells the chevron (2 → 3) domains have very little contrast and have irregular shapes (c), while the FLC-solid (2 → UP) domains have a characteristic triangular (shoe) shape (d). The photomicrographs of (d) show both the 2 → UP and the 3 → DOWN domains. The $S_C$ layers run horizontal in all photos.

The origin of these differences is not entirely clear at the present time. Since the layer tilt angles are the same for the two preparations the **n-P** boundary conditions at the chevron interface should be the same in the two cases. On the other hand the boundary conditions at the FLC-solid interfaces are likely to be quite different. Applying the arguments of Handschy and Clark[8] it would appear that qualitatively the nonpolar anchoring energy is weaker in the case of P-SiO than for the other low pretilt surface treatments. Thus in P-SiO the surface transition occurs at lower field with a correspondingly thicker surface disclination layer[8]. Additionally the higher surface domain contrast indicates surface switching through a larger angle in the P-SiO case. With typical low pretilt alignment the polarization pretilt $\Phi_p$ at the surface is $\Phi_p \sim 45°$ ($\Phi_p$ is the angle between **p** and $\hat{s}$ in Fig.27), using $\theta \sim 30°$, $\delta \sim 1.25\Theta$ and $\Omega = 0$ in

$$\sin\Phi_p = \frac{\tan(\delta - \Omega)}{\tan\Theta},$$

where $\Omega$ is the angle between the director and surface plane. The surface switching in this case produces a change ~ of about 90°. In the P-SiO case if the strongest surface attractor is conical with an angle of $\Omega \sim 25°$ between the director and surface as recently proposed by Durand[48], then $\delta - \Omega$ is small, $\Phi_p$ will be close to 0° and the change in $\Phi$ at the surface upon switching close to 180°. This should produce a larger optical effect upon switching as found.

The observations of Fig.42 show that the domain walls having significant contrast (i.e., a substantial change in the **n-P** field) have a particular domain wall shape - that is both interface and bulk factors determine the structure of the boat shaped domain on the chevron of Fig.42 (a) and the shoe shaped surface domain of Fig.42 (d).

ACKNOWLEDGEMENT. Work supported by NSF Solid State Chemistry Grant DMR 89-01657, ARO Grant DAAL03-90-G-0002 and by the Optoelectronic Computing Systems Center NSF Grant CDR 8622236.

## References

1. J.S. Patel and J.W. Goodby, J. Appl. Phys., **59** (7), (1986).
2. P.G. de Gennes, *The Physics of Liquid Crystals,* Oxford University Press (1974).
3. N.A. Clark and S.T. Lagerwall, Appl. Phys. Lett. **36**, 899 (1980).
4. N.A. Clark and S.T. Lagerwall, Ferroelectrics **59**, 25 (1984).
5. T.P. Rieker, N.A. Clark, G.S. Smith, D.S. Parmar, E.B. Sirota and C.R. Safinya, Phys. Rev. Lett. **59**, 2568 (1987).
6. N.A. Clark and T.P. Rieker, Phys Rev A **37**, 1053 (1988).
7. R.B. Meyer, 'On the Smectic C Phase of Liquid Crystals'; Presentation at the 5th International Liquid Crystal Conference, Stockholm (1974).
8. M.A. Handschy and N.A. Clark, Ferroelectrics **59**, 69 (1984).
9. Y. Ouchi, H. Takezoe and A. Fukuda, Jpn. J. Appl. Phys. Pt. 1 **26**, 1 (1987).
10. F. Rondelez, W. Urbach and H. Hervet, Phys. Rev. Lett., **41** (15), 1058-62, (1978); W. Urbach, H. Hervet and F. Rondelez, Mol. Cryst. Liq. Cryst., **46** (34), 20921, (1978).
11. T.E. Lockhart, D.W. Allender, W. Gelerinter and D.L. Johnson, Phys. Rev. A **20**, 1655 (1979).
12. C.R. Safinya, M. Kaplan, J. AlsNielsen, R.J. Birgenau, D. Davidov and J.D. Litster, Phys. Rev. B **21**, 4149 (1980); E.N. Keller, E. Nachaliel and D. Davidov, Phys. Rev. A **34**, 4363 (1986).
13. G. Pelzl, P. Kolbe, V. Preukschas, S. Diele und D. Demus, Mol. Cryst. Liq. Cryst. **53**, 167 (1979).
14. G. Freidel, Ann. Phys. (Paris) **18**, 277 (1922).
15. N.A. Clark, Phys. Rev. Lett. **55**, 292 (1985).
16. J.M. Geary, J.W. Goodby, A.R. Kmetz and J.S. Patel, J. Appl. Phys. **62**, 4100 (1987).
17. J.L. Janning, Appl. Phys. Lett. **21**, 173 (1972).

18. W. Urbach, M. Boix and E. Guyon, Appl. Phys. Lett. **25**, 479 (1974).
19. T. Uemura, N. Ohba, N. Wakita, H. Ohnishi, I. Ota, Japan Display, 464 (1986).
20. G. Hauck, Cryst. Res. Technol. **22**, 817 (1987).
21. W7-W82 is a 50:50 mixture of the compounds W7 and W82 obtained from Displaytech which exhibits the following phase sequence $I \xleftrightarrow{60^\circ C} S_A \xleftrightarrow{52^\circ C} S_C \xleftrightarrow{22^\circ C} X$. W7 is 4'-[(s)2methyl-3-oxa-1-pentyloxylphenyl 4-(decyloxy)benzoate (also known as $C_2C_{10}$). W82 is 4'-[(s)-1-hexyloxy-4-methyl.-phenyl 4-(decyloxy)benzoate (also known as 10.O7*).
22. The phase sequence $I \xleftrightarrow{118^\circ C} S_A \xleftrightarrow{92^\circ C} S_C \xleftrightarrow{60^\circ C} S_I^* \xleftrightarrow{70^\circ C} X$, DOBAMBC is decyloxybenzylidene-p'-amino- 2-methylbutylcinnamate.
23. The phase sequence: $I \xleftrightarrow{109^\circ C} Ch \xleftrightarrow{61^\circ C} S_C^* \xleftrightarrow{<-20^\circ C} S_I^*$, SCE-10 is a proprietary mixture, available from BDH.
24. The phase sequence $I \xleftrightarrow{81^\circ C} Ch \xleftrightarrow{69^\circ C} S_A \xleftrightarrow{54^\circ C} S_C$. Chisso CS-1014 is a proprietary mixture, available from Chisso.
25. The phase sequence: $I \xleftrightarrow{86^\circ C} Ch \xleftrightarrow{76^\circ C} S_A \xleftrightarrow{62^\circ C} S_C$. ZLI-3654 is a proprietary mixture, available from E. Merck.
26. The phase sequence: $I \xleftrightarrow{136^\circ C} Ch \xleftrightarrow{105^\circ C} S_A \xleftrightarrow{74^\circ C} S_C$. SCE-3 is a proprietary mixture, available from BDH.
27. D.S. Parmar (personal communications).
28. Y. Takanishi, Y. Ouchi, H. Takezoe and A. Fukuda, Jpn. J. Appl. Phys. Lett. **28**, 3 (1989).
29. P. Keller, P.E. Cladis, P.L. Finn and H.R. Brand, J. Physique **46**, 2203 (1985).
30. W. Helfrich and C.S. Oh, Mol. Cryst. Liq. Cryst., **14** (34), 28992.
31. C.E. Williams and M. Kleman, J. Physique Lett. **35**, L33 (1974).
32. 8CB is 4'-cyano-4-n-octylbiphenyl.
33. N.V.S. Rao and V.G.K.M. Plslpatl, J. Phys. Chem. **87**, 899902 (1983).
34. HBBA is N-(p-n-hexyloxybenzylidene)-p-n-butylaniline.
35. Y. Ouchi, H. Takezoe and A. Fukuda, Jpn. J. Appl. Phys. Pt. 1 **26**, 1 (1987).
36. M. Ocko, A.R. Kortan, R.J. Birgenau and J.W. Goodby, J. Phys. (Paris) **45**, 113 (1984) .
37. N.A. Clark, T.P. Rieker and J.E. Maclennan, Ferroelectrics, **85**, 79-97 (1988).
38. J. Xue, N.A. Clark and M.R. Meadows, Appl. Phys. Lett. **53**, 2397 (1988).
39. J.E. Maclennan 'Switching Dynamics and Structures of Ferroelectric Liquid Crystals in the SurfaceStabilized Geometry.' Ph.D. dissertation, University of Colorado, 1988.
40. J. Cognard, Mol. Cryst. Liq. Cryst., Suppl. Ser. **1**, 1 (1982).
41. M.A. Handschy 'Structures and Responses of Ferroelectric Liquid Crystals.' Ph.D. dissertation, University of Colorado, 1983.
42. Y. Ouchi, H. Takazoe and A. Fukuda, Jpn. J. Appl. Phys. **26**, 1 (1984).
43. M.A. Handschy and N.A. Clark, Appl Phys. Lett. **41**, 39 (1982).
44. M. Kleman, *Points, Lines and Walls* (Wiley, New York, 1981).
45. H.A. van Sprang and R.G. Aartsen, J. Appl. Phys. **56**, 251 (1984).
46. M.A. Handschy, N.A. Clark and S.T. Lagerwall, Phys. Rev. Lett., **51**, 471 (1983)
47. Y. Yamada, T. Tsuge, N. Yamamoto, M. Yamawachi, H. Orihara and Y. Ishibashi, Ferroelectrics, **85**, 123 (1988).
48. G. Durand, Second International Topical Meeting on Optics of Liquid Crystals, Torino, 1988.

Chapter 22

# MODELS FOR PHASE TRANSITIONS IN FERROELECTRIC LIQUID CRYSTALS: THEORY AND EXPERIMENTAL RESULTS

R. BLINC

## 1. Introduction

It has been shown[1] that the smectic $C^*$ (Sm $C^*$) liquid crystalline phase is ferroelectric if the molecules are chiral and have a permanent dipole moment transverse to their long molecular axes.

In the smectic $A$ (Sm $A$) phase the molecules rotate freely around their long axes. The director $\vec{n}$, describing the time averaged orientation of the long molecular axes, is oriented parallel to the smectic layer normals. Within the layers there is no positional ordering of the molecules, and thus each layer can be considered as a two-dimensional liquid. The spontaneous polarization is zero.

In the Sm $C^*$ phase the director tilts with respect to the layer normals by an angle $\Theta \neq 0$ and the rotation around the long molecular axes becomes biased. The in plane component of the spontaneous polarization $\vec{P}$ becomes non-zero in a direction perpendicular to the tilt:

$$\vec{\xi} = (\xi_1, \xi_2)$$

$$\xi_1 = n_x \, n_z = \Theta \cos\phi\,(z)$$

$$\xi_2 = n_y \, n_z = \Theta \sin\phi\,(z)$$

Thus we deal with an improper ferroelectric.

The tilt vector and the direction of the in-plane spontaneous polarization $\vec{P} = (P_x, P_y, 0)$ precess around the normal as one goes from one layer to the other thus forming a helical structure. This means that the azimuthal angle $\phi$ changes with $z$ as $\phi = 2\pi z/p$ where $p$ is the pitch of the helix, which is generally incommensurate to the spacing of the smectic layers $L$. The precession is slow on a molecular scale, since $L/p \sim 10^{-3}$. The helix can be unwound by an external field and the polarization becomes macroscopic.

The helix can be also unwound by surface effects in thin samples. We shall in the following treat only thick samples where surface effects can be neglected.

---

R.Blinc - J. Stefan Institute, University of Ljubljana, Jamova 39, Ljubljana, Yugoslavia

*Phase Transitions in Liquid Crystals*, Edited by S. Martellucci and A.N. Chester, Plenum Press, New York, 1992

## 2. Landau Theory: Experimental Facts and Theoretical Predictions

The first theoretical analysis of the dielectric properties and the dynamics of the tilt angle and polarization for the simplest case of an unwound helix has been given by Blinc[2].

The first complete theoretical description of the thermodynamic properties of the helical Sm $C^*$ phase has been given by Pikin and Indenbom[3] whereas the dynamics of the helical structure were described by Blinc and Žekš[4]. The analysis of their "classical" model[5,6] (Eq. 1a-b) shows a systematic and qualitative disagreement with the experimental data (Fig. 1).

Classical Landau expansion of the free-energy density:

$$g_o(z)=\frac{1}{2}a(\xi_1^2+\xi_1^2)+\frac{1}{4}b(\xi_1^2+\xi_2^2)^2-\Lambda\left(\xi_1\frac{d\xi_2}{dz}-\xi_2\frac{d\xi_1}{dz}\right)+ \\ +\frac{1}{2}K_3\left[\left(\frac{d\xi_1}{dz}\right)^2+\left(\frac{d\xi_2}{dz}\right)^2\right]+\frac{1}{2\varepsilon}(P_x^2+P_y^2)-\mu\left(P_x\frac{d\xi_1}{dz}+P_y\frac{d\xi_2}{dz}\right)+C(P_x\xi_2-P_y\xi_1) \tag{1a}$$

Generalized Landau model:

$$g_o(z)=\frac{1}{2}a(\xi_1^2+\xi_2^2)+\frac{1}{2}b(\xi_1^2+\xi_2^2)^2+\frac{1}{6}c(\xi_1^2+\xi_2^2)^3-\Lambda\left(\xi_1\frac{d\xi_2}{dz}-\xi_2\frac{d\xi_1}{dz}\right)+ \\ +\frac{1}{2}K_3\left[\left(\frac{d\xi_1}{dz}\right)^2+\left(\frac{d\xi_2}{dz}\right)^2\right]+\frac{1}{2\varepsilon}(P_x^2+P_y^2)-\mu\left(P_x\frac{d\xi_1}{dz}+P_y\frac{d\xi_2}{dz}\right)+C(P_x\xi_2-P_y\xi_1)- \\ -\frac{1}{2}\Omega(P_x\xi_2-P_y\xi_1)^2+\frac{1}{4}\eta(P_x^2+P_y^2)^2-d(\xi_1^2+\xi_2^2)\left(\xi_1\frac{d\xi_2}{dz}-\xi_2\frac{d\xi_1}{dz}\right) \tag{1b}$$

The pitch is not temperature independent, but exhibits a maximum[7] just below the Sm $A$ - Sm $C^*$ transition temperature $T_c$.

The ratio between the polarization and the tilt, $P/\Theta$, is not independent of $T$ but decreases to a low value[8] in a narrow $T$-interval close to $T_c$. The static dielectric susceptibility has a small anomaly at $T_c$ and increases well below $T_c$ to value which is much larger[9] than the maximum at $T_c$. The critical magnetic field for the unwinding of the helix shows[10] a minimum below $T_c$, thus allowing for a reentrant Sm $C^*$ phase and a "run-away" Lifshitz point.

The critical electric field as well exhibits a minimum close to $T_c$[11]. Motivated by this discrepancy an "extended" phenomenological model[5,6] has been introduced which can account qualitatively and quantitatively for the observed anomalous behaviour of ferroelectric liquid crystals close to $T_c$, except perhaps for some details of the phase diagram close to the Lifshitz point.

The most essential point of this model is that in addition to the "classical" chiral bilinear coupling between the polarization and the tilt, $P\Theta$, a large non-chiral biquadratic coupling, $P^2\,\Theta^2$, is introduced. Close to $T_c$, where the tilt is small, only the bilinear coupling term is important, while well below $T_c$, where $\Theta$ is large, the biquadratic term dominates.

The existence of the large biquadratic coupling term cannot be understood within the standard Landau expansion. We will show that it follows from the fact that the tilting of the molecules on going from the Sm $A$ to the Sm $C^*$ phase introduces not only a spontaneous polarization - which is a small chiral effect - but also a much larger quadrupolar ordering which is of non-chiral origin[5].

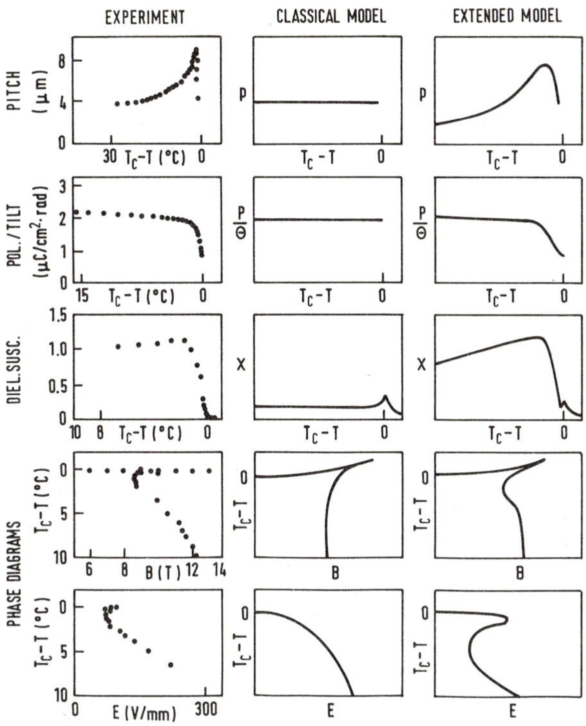


Fig.1. Comparison between the experiment, the predictions of the "classical" Landau model and the predictions of the "extended" Landau model for the temperature dependence of the pitch $P$, the ratio between the polarization and the tilt $P/\theta$, the temperature dependence of the static dielectric susceptibility and the phase diagrams in external magnetic and electric fields, applied transverse to the helicoidal axis.

## 3. A Microscopic Model for the Spontaneous Polarization

The influence of the molecular tilt on the ordering of the transverse dipoles can be thus described[5] by a single particle potential $V(\phi)$ for the rotation of the molecule around its long axis:

$$V(\phi) = -a_1\Theta\cos\phi - a_2\Theta^2\cos 2\phi \tag{2}$$

where $\phi$ determines the orientation of the transverse molecular dipole.

The first term is linear in the tilt. It induces polar ordering, $\langle\cos\phi\rangle \neq 0$, in the direction perpendicular to the tilt where $\phi = 0$. The second term is quadratic in $\Theta$ and is not of chiral character. It leads for $a_2 > 0$ to bipolar (quadrupolar) ordering, $\langle\cos 2\phi\rangle \neq 0$, in the direction perpendicular to the tilt.

The orientational free energy per molecule is now obtained as

$$F_{orient} = \langle V(\phi)\rangle + kT\langle\ln\rho\rangle \tag{3a}$$

where $\rho$ is the density matrix:

$$\rho = \frac{e^{-V(\phi)/kT}}{\int_0^{2\pi} e^{-V(\phi)/kT}\, d\phi} \tag{3b}$$

The above potential $V(\phi)$ leads to a free energy expansion in powers of the polar $\langle\cos\phi\rangle$ and quadrupolar $\langle\cos 2\phi\rangle$ order parameters:

$$F_1 = \frac{1}{2}C_1P^2 + \frac{1}{4}C_2P^4 + \frac{1}{2}\tilde{C}_2Q^2 + \frac{1}{4}\tilde{C}_2Q^4 - a_1\Theta P - a_2\Theta^2 Q - a_3P^2Q \tag{3c}$$

where

$$Q = K_1\langle\cos 2\phi\rangle \tag{4}$$

$$P = K_2\langle\cos\phi\rangle \tag{5}$$

By minimizing $F_1$ with respect to $Q$ and putting $\tilde{C}_2 = 0$ we find:

$$Q = K_1\langle\cos 2\phi\rangle = \frac{1}{C_1}[a_2\Theta^2 + a_3\langle\cos\phi\rangle^2] \tag{6}$$

The quadrupolar ordering is thus proportional to the square of the tilt and the square of the polar ordering. By inserting $Q$ from expression (6) into (3) one finds a biquadratic coupling between the tilt and the polarization, $P^2\Theta^2$, which is the result of quadrupolar ordering, i.e. the $\Theta^2\langle\cos 2\phi\rangle$ term. One also finds a $\Theta^4$ term. The above model predicts that the quadrupolar ordering is always much larger than the polar one:

$$|\langle\cos 2\phi\rangle| \gg \langle\cos\phi\rangle \tag{7}$$

This is indeed borne out by recent $^{13}C$ and $^{14}N$ NMR and NQR experiments.

Since $P = K_2\langle\cos\phi\rangle$ and

$$\langle\cos\phi\rangle = \frac{\int_0^{2\pi}\cos\phi\, e^{-V(\phi)/kT}d\phi}{\int_0^{2\pi}e^{-V(\phi)/kT}d\phi} \tag{8}$$

where $V(\phi) = -a_1\Theta\cos\phi - a_2\Theta^2\cos 2\phi$, it is clear that $\langle\cos\phi\rangle = P = 0$ in achiral systems where $a_1 = 0, a_2 \neq 0$.

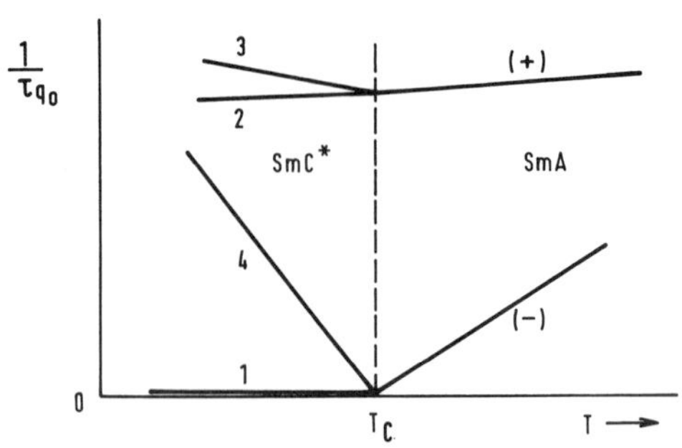


Fig.2. Schematic predicted temperature dependence of order parameter relaxation frequencies in ferroelectric liquid crystals at the SmA - SmC* transition. Mode 1 is the Goldstone (phason) mode and mode 4 the soft amplitudon mode of the SmC* phase.

The above free energy density, after adding the elastic and Lifshitz terms, has the same form as the extended Landau model proposed by Žekš[5] and Dumrongrattana and Huang[8], (Eq. 1b), where however the origin of the anomalously large biquadratic coupling term between the polarization and the tilt was not clear.

The polarization is here not simply proportional to the tilt as in the classical model. For the homogeneous case, the polarization is given by a solution of the equation

$$\left(\frac{1}{\varepsilon}-\Omega\Theta^2\right)P+\eta P^3-C\Theta=0 \tag{9}$$

The ratio $P/\Theta$ has the value $P/\Theta=\varepsilon/C$ close to $T_c$, where $\Theta$ and $P$ are small, and another value, $P/\Theta=(\Omega/\eta)^{1/2}$, far below $T_c$ where $\Theta$ is large.

For helicoidal systems the pitch of the helix $p=2\pi/q$ is obtained as:

$$q=\frac{1}{K_{33}}(\Lambda+d\Theta^2+\mu P/\Theta) \tag{10}$$

and is - in contrast to the classical case - temperature dependent.

## 4. The Soft Mode Nature of the Smectic A - Smectic C* Transition

The order parameter fluctuation spectrum consists of four coupled tilt polarization modes[4]: two "hard" modes describe out-of phase polarization-tilt angle fluctuations and have relaxation frequencies in the 100-500 MHz range whereas the two in-phase polarization-tilt angle fluctuation modes occur below 1 MHz (Fig. 2). They are commonly denoted as the soft (amplitudon) mode and the Goldstone (phason) mode (Fig. 3).

Denoting the amplitude changes by $\delta\Theta_1$ and $\delta P_1$ and the phase changes by $d\phi$ and $\delta P_2$ (Fig. 3) - where $\delta\vec{\Theta}_1 \parallel \vec{\xi}$, $\delta\vec{\phi}\perp\vec{\xi}$, $\delta\vec{P}_1 \parallel \vec{P}$, $\delta\vec{P}_2\perp\vec{P}$ - we can express the time dependent order parameters as:

$$\xi_1(t)=\Theta\cos(qz)+\delta\Theta_1(t)\cos(qz)-\delta\phi(t)\sin(qz) \tag{11a}$$

$$\xi_2(t)=\Theta\sin(qz)+\delta\Theta_1(t)\sin(qz)+\delta\phi(t)\cos(qz) \tag{11b}$$

$$P_x(t)=-P\sin(qz)-\delta P_1\sin(qz)-\delta P_2\cos(qz) \tag{11c}$$

$$P_y(t)=P\cos(qz)+\delta P_1\cos(qz)-\delta P_2\sin(qz) \tag{11d}$$

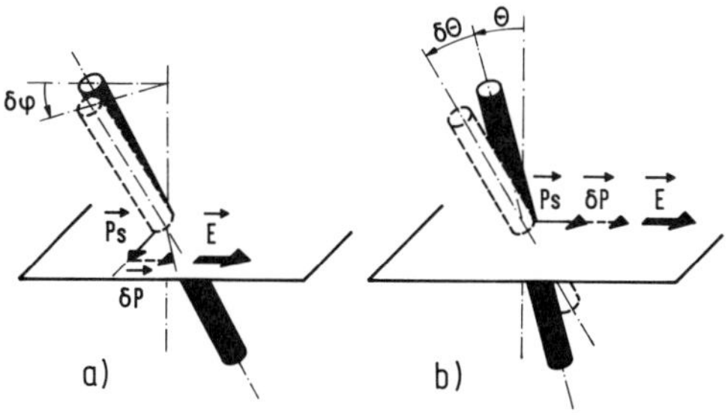


Fig.3. Schematic representation of the eigenvector of the phason (Goldstone) director mode and the amplitudon director mode.

The corresponding Landau-Khalatnikov relaxation equations are:

$$\frac{d\xi_1}{dt} = -\Gamma_1 \frac{\partial F}{\partial \xi_1}, \quad \frac{d\xi_2}{dt} = -\Gamma_2 \frac{\partial F}{\partial \xi_2} \tag{12a}$$

$$\frac{dP_x}{dt} = -\Gamma_3 \frac{\partial F}{\partial P_x}, \quad \frac{dP_y}{dt} = -\Gamma_4 \frac{\partial F}{\partial P_y} \tag{12b}$$

where

$$F = \frac{1}{L}\int_0^L g(z)dz \tag{12c}$$

and the $\Gamma_i$, $i = 1 - 4$, represent kinetic coefficients (i.e. inverse viscosities). When solving the dynamic equations one can use the fact that the eigenfrequencies of the director modes are well separated from those of the polarization modes. In the Sm *A* phase only amplitude fluctuations are present and the system exhibits a doubly degenerate soft director mode and a doubly degenerate high frequency polarization mode.

It is easy to show[12] that the relaxation rate of the doubly degenerate soft director mode above $T_c$, where $P = \Theta = 0$, is

$$\frac{1}{\tau_{soft}} = \gamma_S^{-1}[\alpha(T - T_c) + K_3(q_o \pm q_z)^2 + K_\perp q_\perp^2], \qquad T > T_c \tag{13a}$$

whereas below $T_c$ where $\Theta_o \neq 0$ the degeneracy of the soft mode is lifted and we have an optic-like amplitudon

$$\frac{1}{\tau_{ampl}} = \gamma_\Theta^{-1}[2\alpha(T_c - T) + K_3(q_o \pm q_z)^2 + K_\perp q_\perp^2] \tag{13b}$$

and a Goldstone mode-like gapless phason mode:

$$\frac{1}{\tau_{phason}} = \gamma_\Phi^{-1}[K_3(q_o \pm q_z)^2 + K_\perp q_\perp^2] \tag{13c}$$

Here $\gamma_S, \gamma_\Theta$ and $\gamma_\Phi$ are the viscosity coefficients for the soft, amplitudon and Goldstone modes respectively, $q_o = 2\pi/p$ where $p$ is the pitch of the static helix of the Sm $C^*$ phase, and $K_3$ and $K_\perp$ are elastic constants.

## 5. The Effects of External Electric and Magnetic Fields

It was predicted[13] some years ago that the phase diagram of ferroelectric liquid crystals in a magnetic field applied parallel to the smectic layers would exhibit a Lifshitz point ($H_L$, $T_L$) where the Sm *A*, Sm $C^*$ and Sm *C* phase transition lines meet. An experimental determination of the phase diagram has been performed for DOBAMBC[10] as well as for a mixture of racemic and chiral DOBAMBC[14]. Two unexpected results were obtained: a reentrant Sm $C^*$ phase was found and the Lifshitz point appeared to "run away" as $H \to H_L$ and $T \to T_L$.

The "classical" Landau model is unable to explain this behaviour. It is instructive to investigate the predicted phase diagram of the "classical" and "extended" Landau models of Eq.s 1a-b if a magnetic field term

$$g_H = -\frac{1}{2}\chi_a H^2 \xi_1^2 \tag{14}$$

is added to $g_0(z)$ in both cases. Here $\chi_a$ describes the diamagnetic anisotropy and H is the external magnetic field applied in the x-direction. $g_H$ provides for the "lock-in" term driving the transition from the modulated incommensurate Sm $C$* to the commensurate homogeneous Sm $C$ phase.

Using expression (1a) and (1b) and eliminating $P_x$ and $P_y$ by defining

$$P_x = \varepsilon\mu\xi_1' - \varepsilon C\xi_2 \tag{15a}$$

$$P_y = \varepsilon\mu\xi_2' + \varepsilon C\xi_1 \tag{15b}$$

one finds in the "classical" model the free energy density g(z):

$$g(z) = \frac{1}{2}(a - \varepsilon C^2)\Theta^2 + \frac{1}{4}b\Theta^4 - \tilde{\Lambda}\Theta^2\Phi' + \frac{1}{2}\tilde{K}_3(\Theta'^2 + \Theta^2\Phi'^2)$$

$$-\frac{1}{2}\chi_a H^2\Theta^2\cos^2\Phi, \quad a = \alpha(T - T_o) \tag{16}$$

Here

$$\tilde{\Lambda} = \Lambda + \varepsilon\mu C, \quad \tilde{K}_3 = K_3 - \varepsilon\mu^2, \quad T_c = T_o + \frac{1}{\alpha}[\varepsilon C^2 + \tilde{K}_3 q_o^2] \tag{17}$$

whereas

$$q_o = \frac{\tilde{\Lambda}}{\tilde{K}_3} \tag{18}$$

is T-independent.

The Lifshitz field $H_L$ is obtained as

$$H_L = 2\frac{\tilde{\Lambda}}{\sqrt{\tilde{K}_3\chi_a}} \tag{19}$$

and the Lifshitz temperature $T_L$ as:

$$T_L = T_0 + \frac{1}{\alpha}\varepsilon C^2 + 4\Delta T \tag{20}$$

Here

$$\Delta T = \frac{1}{\alpha}\frac{\tilde{\Lambda}^2}{\tilde{K}_3} \tag{21}$$

so that

$$T_L - T_c = 3\Delta T \tag{22}$$

The magnetic field dependence of the Sm $A$ - Sm $C^*$ transition line is given by:

$$T_{AC*} = T_0 + \frac{1}{\alpha}\varepsilon C^2 + \Delta T\left(1 + \frac{H^2}{H_L^2}\right)^2 \tag{23}$$

In the constant amplitude approximation $\Theta \neq \Theta(z)$ the critical field for the unwinding of the helix is obtained as:

$$H_c = \frac{\pi}{4} H_L \tag{24}$$

Here we minimized the free energy $F$ with respect to $\Phi(z)$ leading to the Sine-Gordon equation

$$\Phi'' = (\chi_a H^2 / \tilde{K}_3)\sin(2\Phi) \tag{25}$$

which admits $\pi$ phase soliton solutions. $H_c$ is $T$-independent and no reentrant Sm $C^*$ phase exists within this model, in contrast to the experimental data. In the "extended" Landau model, on the other hand, a reentrant Sm $C^*$ phase exists.

The critical field for the unwinding of the helix is now obtained as:

$$H_c = \frac{\pi}{2}\frac{\Lambda + d\Theta_0^2 + \mu P/\Theta_0}{\sqrt{K_3\chi_a}} \tag{26}$$

and is T- dependent. Expressions can be combined to relate $H_c$ and q:

$$H_c = \frac{\pi}{2}\sqrt{\frac{K_3}{\chi_a}}q = \pi^2\sqrt{\frac{K_3}{\chi_a}}\frac{1}{p} \tag{27}$$

The behaviour of a ferroelectric liquid crystal in an electric field is rather different from the magnetic case.

The reason for this difference lies in the fact that an in-plane electric field couples linearly to the spontaneous polarization, thus breaking the symmetry of the high temperature phase. This linear coupling takes place in addition to the quadratic coupling due to the anisotropy in the dielectric constant. The two terms which need to be added to the free energy density (Eq.s 1a-b) in the presence of an in-plane electric field are thus:

$$g_E = -\vec{P}\cdot\vec{E} - \frac{\varepsilon_a}{2}(\vec{\xi}\cdot\vec{E})^2 \tag{28}$$

where $\varepsilon_a = \varepsilon_\parallel - \varepsilon_\perp$. The second term is generally negligible compared with the first if the applied field is smaller than the critical field $E_c$ necessary for the unwinding of the helical structure. In the following we shall therefore consider only the linear coupling term.

The electric field tends to orient the polarization in the direction of the field.

If we apply the field in the x-direction perpendicular to the helical axis in the Sm $A$ phase, the field induces a polarization in the x-direction and a tilt in the y-direction. The Sm $A$ system thus transforms even for infinitesimal fields into the homogeneously tilted smectic $C$ phase.

When the transverse field is applied in the helicoidal Sm $C^*$ phase it distorts the helix and when the field is large enough the system transforms from the Sm $C^*$ into the homogeneously polarized Sm $C$ phase (Fig. 4).

In the constant amplitude approximation the electric field affects only the phase $\Phi = \Phi(z)$. Let us therefore consider only those terms in the free energy density which are $\Phi$-dependent:

$$g(\Phi) = \frac{1}{2}K_3\Theta^2\left(\frac{d\Phi}{dz}\right)^2 - K_3\Theta^2 q\left(\frac{d\Phi}{dz}\right) - EP\cos\Phi \tag{29}$$

Similarly to the magnetic case, minimization leads to the Sine-Gordon equation:

$$K_3\Theta^2\frac{d^2\Theta}{dz^2} = EP\sin\Phi \tag{30}$$

which gives as a solution a $2\pi$-soliton lattice. This is different from the magnetic, case where a $\pi$-soliton lattice appears because of the quadratic coupling to the order parameter.

With increasing $E$ the period of the $2\pi$-soliton lattice increases and goes to infinity at the critical field $E_c$, where the free energy difference between the single $2\pi$-soliton solution and the homogeneous solution becomes equal to zero. This limit of stability of the deformed helicoidal Sm $C^*$ phase is reached at

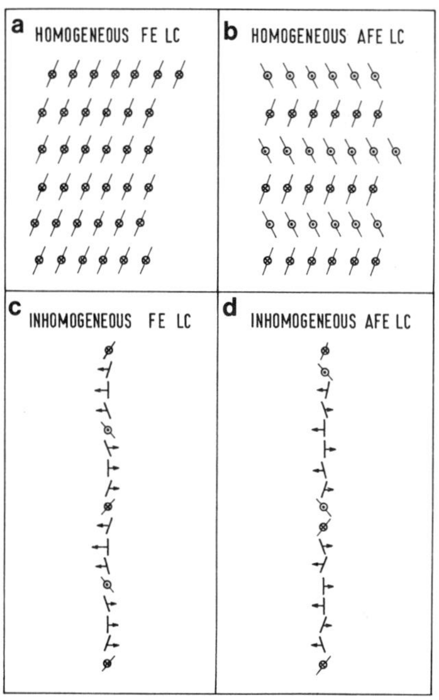


Fig.4. Schematic representation of the structure of: a) the homogeneous FE SmC$^*$ phase, b) the homogeneous AFE SmC$_A^*$ phase, c) the helicoidal FE SmC$^*$ phase, and d) the helicoidal AFE SmC$^*$ phase.

$$E_c = \frac{\pi^2}{16} K_3 \frac{q^2 \Theta^2}{P} \tag{31a}$$

Here $q$, $\Theta$ and $P$ are the zero field values of the wave vector, tilt and polarization of the Sm $C^*$ helix. $E_c$ can be also expressed as

$$E_c = \frac{\pi^2}{16 K_3} \left( \Lambda + d \cdot \Theta^2 + \mu \frac{P}{\Theta} \right)^2 \frac{\Theta^2}{P} \tag{31b}$$

The anomalous temperature dependence of $E_c$ thus reflects to a large extent the anomalous temperature dependence of the pitch $p = 2\pi/q$ of the helix:

$$q = \frac{1}{K_3} \left( \Lambda + d \cdot \Theta^2 + \mu \frac{P}{\Theta} \right) \tag{31c}$$

as $E_c$ is proportional to $\Theta^2/p^2 P$.

## 6. Antiferroelectric Liquid Crystals

In the antiferroelectric Sm $C^*_{AF}$ phase the molecules in neighbouring layers tilt in opposite direction so that the polarizations point in opposite directions similarly as in the unit cell of solid antiferroelectrics. The two primary order parameters here are the two tilt vectors $\vec{\xi}_1$ and $\vec{\xi}_2$, which represent the orientations of molecules in two adjacent layers and are defined[15] as

$$\vec{\xi}_i = (-n_{iy} n_{iz}, n_{iz} n_{iz}), \qquad i = 1,2 \tag{32}$$

where $\vec{n}_i$ is the director in the i-th layer and the z-axis is normal to the smectic layer.

It is convenient[15] to use linear combinations of the above two order parameters to represent antiferroelectric and ferroelectric order, respectively:

$$\vec{\xi}_a = (\vec{\xi}_1 - \vec{\xi}_2)/2 \tag{33a}$$

$$\vec{\xi}_f = (\vec{\xi}_1 + \vec{\xi}_2)/2 \tag{33b}$$

The two secondary order parameters are the in-plane polarizations $\vec{P}_1$ and $\vec{P}_2$ of two neighbouring layers. Similarly as with the tilt we introduce the antiferroelectric and ferroelectric in-plane polarization variables as:

$$\vec{P}_a = (\vec{P}_1 - \vec{P}_2)/2 \tag{34a}$$

$$\vec{P}_f = (\vec{P}_1 + \vec{P}_2)/2 \tag{34b}$$

All of the above variables transform like vectors under the rotations of the symmetry group $D_\infty$. However, the antiferroelectric and ferroelectric variables transform in different ways under translations. The former changes sign under translation parallel to the layer normal for a layer spacing while the latter does not change. The Landau free energy density[15] can hence be expressed within the classical model as:

$$g=\frac{\alpha_a'}{2}\xi_a^2+\frac{\alpha_f'}{2}\xi_f^2+\frac{\beta_a}{4}\xi_a^4+\frac{\beta_f}{4}\xi_f^4+\frac{\gamma_1}{2}\xi_a^2\xi_f^2+\frac{\gamma_2}{2}(\vec{\xi}_a\cdot\vec{\xi}_f)^2$$
$$+\lambda_a\vec{\xi}_a\cdot\vec{P}_a+\lambda_f\vec{\xi}_f\cdot\vec{P}_f+\frac{1}{2\chi_a}P_a^2+\frac{1}{2\chi_f}P_f^2-\vec{E}\cdot\vec{P}_f$$
$$+\delta_a\left(\xi_{ax}\frac{\partial\xi_{ay}}{\partial z}-\xi_{ay}\frac{\partial\xi_{ax}}{\delta z}\right)+\delta_f\left(\xi_{fx}\frac{\partial\xi_{fy}}{\partial z}-\xi_{fy}\frac{\partial\xi_{fx}}{\partial z}\right)$$
$$+\mu_a\left(P_{ax}\frac{\partial\xi_{ay}}{\partial z}-P_{ay}\frac{\partial\xi_{ay}}{\partial z}\right)+\mu_f\left(P_{fx}\frac{\partial\xi_{fy}}{\partial z}-P_{fy}\frac{\partial\xi_{fx}}{\partial z}\right)$$
$$+\frac{k_a}{2}\left(\frac{\partial\xi_a}{\partial z}\right)^2+\frac{k_f}{2}\left(\frac{\partial\xi_f}{\partial z}\right)^2 \qquad (35)$$

The above expression includes Lifshitz invariants with respect to ferroelectric as well as antiferroelectric variables. The Lifshitz invariants lead to a helical structure.

Eliminating $\vec{P}_a$ and $\vec{P}_f$ by minimization leads to a proportionality between $\vec{P}_a$ and $\xi_a$ and $\vec{P}_f$ and $\vec{\xi}_f$ if the external field $\vec{E}=0$. In this case one can show that in addition to the high temperature Sm$A$ paraelectric phase, three low temperature phases may exist. The phase transition sequence may thus be:

- paraelectric phase $(\vec{\xi}_a=\vec{\xi}_f)=0$,
- antiferroelectric phase $(\vec{\xi}_a\neq 0,\vec{\xi}_f=0)$,
- ferrielectric phase $(\vec{\xi}_a\neq 0,\vec{\xi}_f\neq 0)$,
- ferroelectric phase $(\vec{\xi}_a=0,\vec{\xi}_f\neq 0)$.

## 7. Spontaneous Polarization and Spontaneous Tilt

The temperature dependences[16] of the spontaneous polarization $P$ and the spontaneous tilt $\Theta$ for a) DOBA-1-MPC and b) DOBAMPC are shown in Fig. 5. The experimental values[8,16] of the ratio $P/\Theta$ are compared[17] with theoretical predictions of the microscopic model of Eq. 9, which leads via a power series expansion to the extended Landau model, in Fig. 6a. Here $P/\Theta$ is plotted versus $\Theta^2$ which varies roughly as $T_c - T$. We see that $P/\Theta$ is not constant as required by the simple classical Landau model but is in fact $T$-dependent as predicted by Eq. 9. The best fit[17] between the experimental and theoretical values is obtained for

$$\Theta^*=\frac{1}{\sqrt{\varepsilon\Omega}}=0.260rad \qquad (36a)$$

$$P^*=\frac{1}{\sqrt{\varepsilon\eta}}=7.01x10^{-6}As/m^2 \qquad (36b)$$

and

$$\beta=\frac{\varepsilon\eta^{1/2}(C+\mu q)}{\Omega^{1/2}}=0.04 \qquad (36c)$$

It should be stressed that the microscopic model[5,17] of expression (8) and the extended Landau model of Eq. 9 give identical results for small values of the tilt angle, i.e. close to $T_c$ where both predict the same $T$-dependence of $P/\Theta$. At lower temperatures the Landau model predicts values of $P/\Theta$ which are too large compared with the microscopic model.

One should also add a note on the electric field dependence of the spontaneous polarization[17]. The electric field not only orders the molecular dipole moments but also increases the pitch of the helix thus making the system softer and more susceptible to electric field changes: see Fig. 6b. For $E > E_c$ the polarization is field independent and equals its maximum possible value.

## 8. Dielectric Spectroscopy of the Soft and Goldstone Modes

Dielectric spectroscopy measures the response of the in-plane polarization of the system at zero wave-vector to an external homogeneous field. In the Sm $A$ phase, amplitude and phase fluctuations are degenerate. The system exhibits one doubly degenerate soft mode director and one doubly degenerate hard (polarization) mode. We shall consider only the soft director mode behaviour.

In the approximation where each mode is characterized by a single relaxation frequency the complex dielectric constant of the soft mode can be written as

$$\varepsilon^*(\omega) = \frac{\Delta\varepsilon_s}{1 + i\omega\tau_s} + \varepsilon_\infty, \qquad T > T_c \tag{37a}$$

where the dielectric strength is

$$\Delta\varepsilon_s = \frac{\varepsilon^2 C^2}{\alpha(T - T_c) + (K_3 - \varepsilon\mu^2)q^2} \tag{37b}$$

and the relaxation frequency

$$f = \frac{1}{2\pi\tau_s} = \frac{\Gamma}{2\pi}[\alpha(T - T_c) + (K_3 - \varepsilon\mu^2)q^2] \tag{37c}$$

while $q$ is the wave-vector of the pitch at $T_c$.

In the smectic $C^*$ phase the degeneracy of the soft mode is lifted and we find a contribution from the Goldstone mode and another from the amplitudon mode:

$$\varepsilon^*(\omega) = \frac{\Delta\varepsilon_G}{1 + i\omega\tau_G} + \frac{\Delta\varepsilon_A}{1 + i\omega\tau_A} + \varepsilon_\infty, \qquad T < T_c \tag{38a}$$

where

$$\Delta\varepsilon_G = \frac{1}{2K_3}\left(\frac{P_o}{q\Theta_o}\right)^2 \tag{38b}$$

and

$$f_G = \frac{1}{2\pi\tau_G} = \frac{\Gamma K_3}{2\pi} q^2 \tag{38c}$$

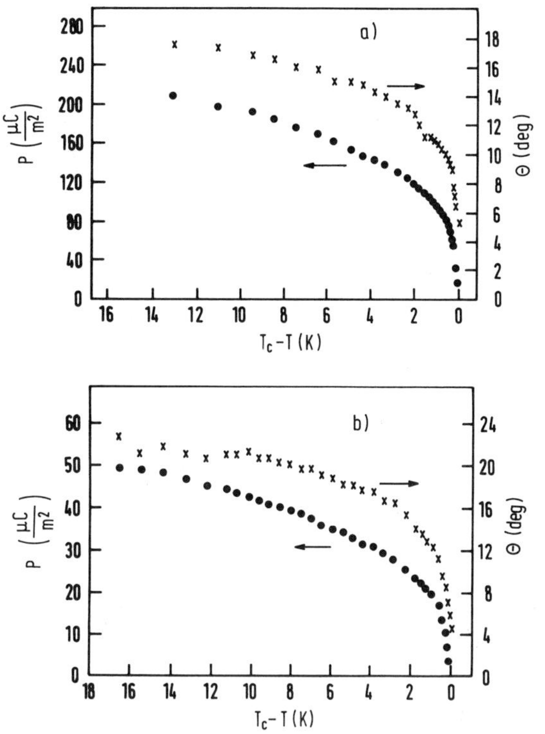


Fig.5. Temperature dependence[16] of the spontaneous tilt and the spontaneous polarization in: a) DOBA-1-MPC, and b) DOBAMBC.

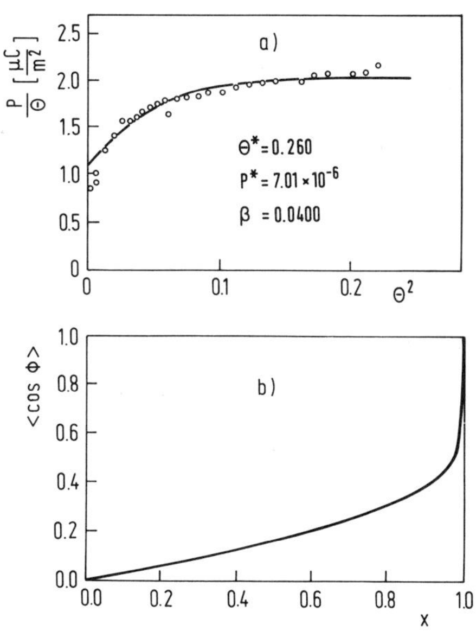


Fig.6. a) Comparison between the experimental and theoretical temperature dependences[17] of the ratio between the polarization and the tilt, $P/\theta$. b) Electric field dependence of the polar order parameter <cos $\phi$>. Here $\chi = E/E_c$.

and $\Delta\varepsilon_A$ and $f_A$ are given by expression (37b) and (37c) with $\alpha(T-T_c)$ replaced by $2\alpha(T_c-T)$.

Here $P$ is the spontaneous polarization, $\Theta$ the spontaneous tilt, $q = 2\pi/P$ the wave-vector of the pitch $p$ and $K_3$ the twist elastic constant while $\mu$ and $C$ are coefficients of the flexo- and piezo-electric bilinear coupling. $\varepsilon$ represents the dielectric strength of the polarization mode and $\Gamma$ is a kinetic coefficient. From the Cole-Cole plots the dielectric strength and relaxation frequencies (Fig. 7) of the soft, amplitudon and phason (Goldstone) modes of SCE 9 and DOBA-1-MPC have been determined[16] at the non-critical value of the wave-vector 0. It is clearly seen that:

i) The predicted splitting of the soft mode into an amplitudon and a phason (Goldstone) mode on crossing $T_c$ takes place.

ii) The relaxation frequency of the Goldstone mode $f_G = 1/(2\pi\tau_G)$ at $q = 0$ is for DOBA-1-MPC of the order of 400 Hz and nearly $T$-independent between $T_c - T = 16$ K and $T_c - T = 1$ K.

iii) Very close to the Sm $C^*$ - Sm $A$ transition but still below $T_c$ the Goldstone mode frequency sharply increases reflecting the decrease in the pitch of the helix.

iv) The maximum in the dielectric strength of the Goldstone mode occurs below $T_c$: the product of the dielectric strength and relaxation frequency

$$f_G \Delta_{\varepsilon G} = \frac{\Gamma}{4\pi}\left(\frac{P}{\Theta}\right)^2 \tag{39}$$

is indeed proportional to $(P/\Theta)^2$ as predicted, thus supporting the extended Landau model. It should be stressed that below $T_c$ the amplitudon (soft) mode is hard to observe by dielectric spectroscopy as it is partly covered by the Goldstone mode which has a much larger dielectric strength. This can be overcome by the application of a dc-bias field $E > E_c$ which suppresses[18] the Goldstone mode.

One should also mention that the high frequency "hard" modes, describing the "out of phase" fluctuations of the polarization and the tilt have been recently observed[19] by broad band dielectric spectroscopy in the 250-100 MHz region.

## 9. Light Mixing Spectroscopy of the Dispersion of the Soft Mode, the Goldstone (phason) Mode and the Amplitudon Mode

The order parameter components $\xi_1$ and $\xi_2$ transform[12] in the same way as the dielectric tensor components $\varepsilon_{xy}$ and $\varepsilon_{yz}$. Thus,

$$\varepsilon_{xz} = A\xi_1, \qquad \varepsilon_{yz} = A\xi_2 \tag{40}$$

and the spectral density of the spectral density of the scattered light in the vicinity of the first Bragg peak is simply[12]

$$I(\vec{q},\omega) = \langle |\Delta\varepsilon_{mn}(\vec{q},\omega)e_{Im}e_{sn}|^2\rangle \tag{41}$$

where $\vec{q} = \vec{k}_I - \vec{k}_s$, $\omega = \omega_I - \omega_s$ and $e_{Im}$ and $e_{sn}$ are components of the polarization vectors of the incoming and scattered light.

Samples of CE8 were prepared both in the bookshelf ($d = 75\mu$) and in the homeotropic ($d = 50\mu$) alignement geometry. A He-Ne 1 mW laser was used. The incident light was polarized as the extraordinary wave and the scattered light as the ordinary wave.

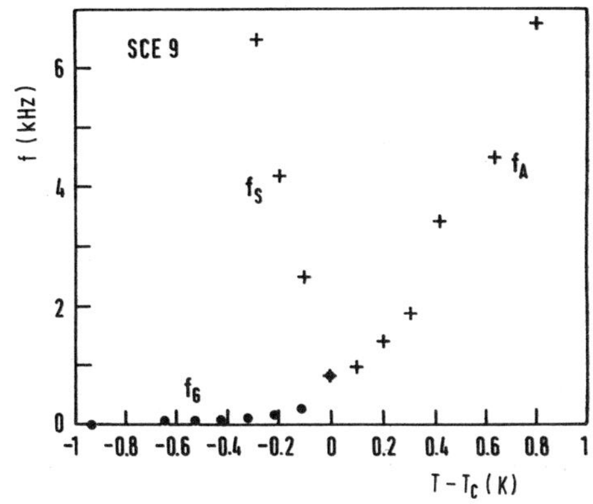


Fig.7. Temperature dependence of the soft, amplitudon and phason (Goldstone) mode relaxation rates - determined from dielectric data[16,25] - on passing from the SmA into the SmC* phase in SCE 9.

The temperature dependence of the order parameter relaxation rates above and below $T_c$ at constant scattering angle *(q)* is shown[20] in Fig. 8. Above $T_c$ we get a single soft mode relaxation frequency which linearly decreases towards $T_c$ with a slope

$$\left(\frac{d(1/\tau)}{dT}\right)_{soft} = 0.96 \times 10^6\, Hz/K\ (1 \pm 0.04), \qquad T > T_c$$

Two modes, an amplitudon and a phason are observed below $T_c$. The inverse relaxation time of the amplitudon is strongly temperature dependent. Its frequency increases linearly with ($T_c$) with a slope

$$\left(\frac{d(1/\tau)}{dT}\right)_{amplit} = -0.77 \times 10^6\, Hz/K\ (1 \pm 0.10), \qquad T < T_c$$

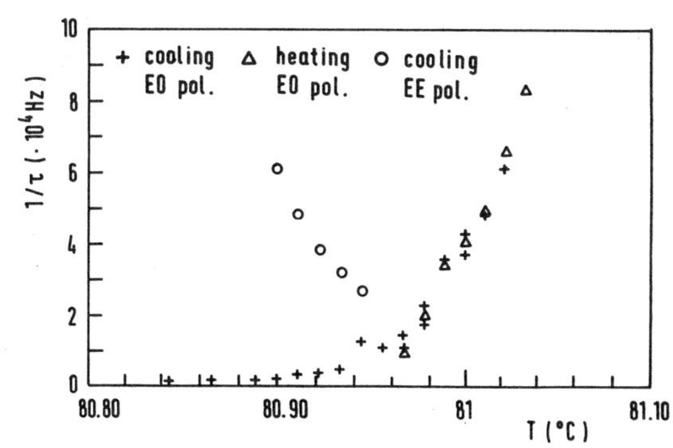


Fig.8. Temperature dependence of order parameter relaxation rates at the SmA - SmC* transition in CE 8 as determined from light scattering[20] measurements. The data were taken at constant exterior scattering angle. The scattering vector was not exactly equal to $q_0$ and was changing slightly with temperature because of the temperature dependence of the indices of refraction. This causes the measured phason relaxation rate to vary with temperature.

The ratio of the high and low $T$ slopes and Curie-Weiss constants is thus different from the one predicted by the Landau theory, which is $C^+/C^- = 1:2$, and different from the ratio found in solid crystalline ferroelectrics.

The phason relaxation rate is only weakly $T$-dependent. The dispersion[12,10] of the phason (Goldstone) relaxation rate is shown in Fig. 9.

## 10. $^{13}$C NMR, $^{14}$N NQR and Local Order Parameter Fluctuations

In view of the shielding of the nuclei by the surrounding electrons, the effective magnetic field at the nuclear sites is given by

$$\vec{B} = \vec{B}_o(1 - \underline{\sigma}) \tag{42}$$

where $\underline{\sigma}$ is the chemical shift tensor and $\vec{B}_o$ the external magnetic field density. The averaging of $\underline{\sigma}$ by order parameter fluctuations allows for a determination of these fluctuations at the various nuclear sites in the molecule, and thus a measurement of the non-rigidity of the molecule[21,22]. It should be stressed that the chemical shift tensor measures only local properties.
We can therefore avoid dealing with the precession of the molecules in the Sm $C^*$ phase when going from one layer to the other. This precession is very slow on a molecular level $L/p = 10^{-3}$ (where $L$ is the layer thickness) and changes the symmetry globally and not locally.

In the isotropic liquid phase the anisotropic part of the $^{13}$C chemical shift tensor is averaged out due to fast isotropic molecular rotations. Because of this averaging, one observes in this phase only the isotropic part of $\underline{\sigma}$:

$$\sigma_i = \frac{1}{3} Tr\ \underline{\sigma} \tag{43}$$

In the Sm $A$ phase the molecules order in such a way that the normals to the smectic layers are parallel to the direction of the external magnetic field. The molecular director $\vec{n}$ also points along the magnetic field. In view of the fast rotation of the molecules around their long axis, fluctuations of the long axes about the director and internal conformational motions, the component of the chemical shift tensor in the direction of the magnetic field is obtained[21,22] as:

$$\sigma = \sigma_i + \frac{2}{3} S(\sigma_\parallel - \sigma_\perp) \tag{44}$$

Here $S$ is the nematic order parameter, $\sigma_\parallel = \sigma_{zz}$ the component of $\underline{\sigma}$ along the direction of the director i.e. the average direction of the long molecular axis, and $\sigma_\perp = (\sigma_{xx} + \sigma_{yy})/2$ the average component in the plane perpendicular to the long molecular axes (Fig. 10).

In the Sm $C^*$ phase the director is tilted away from the layer normal. If this were the only change and the Sm $C^*$ phase were locally just a tilted Sm $A$ phase, one would expect that[21, 22]

$$\sigma = \sigma_i + \frac{2}{3} S(\sigma_\parallel - \sigma_\perp)\left(\frac{3}{2}\cos^2\Theta - \frac{1}{2}\right) + \Delta \tag{45}$$

with $\Delta = 0$ where $\Theta$ is the tilt angle. Since the average tilt angle $\Theta$ is known from independent X-ray and optical measurements, the above relation can be easily checked.

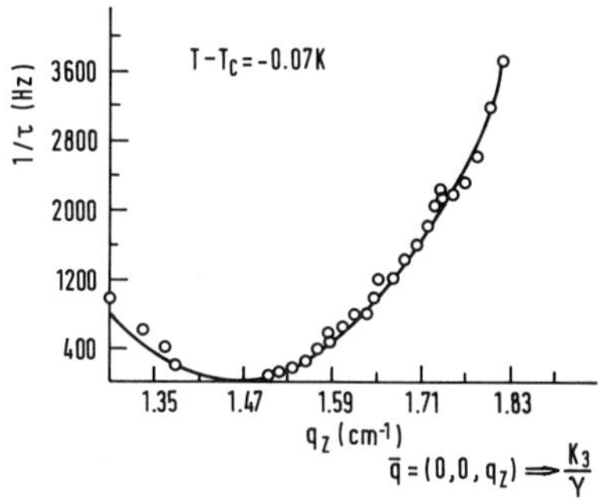


Fig.9. Phason dispersion[20] below $T_c$ in CE 8.

In fact, the tilt breaks the axial symmetry of the Sm *A* phase even for non-chiral systems and induces a quadrupolar (bipolar) ordering $\langle\cos 2\phi\rangle \neq 0$ of the transverse molecular dipoles while conserving the mirror plane. This leads to the biquadratic $P^2\Theta^2$ coupling term in the "extended" Landau model and to an additional term

$$\Delta = \Delta_{bp} = (\sigma_{xx} - \sigma_{yy}) \sin^2 \Theta_o \langle\cos 2\phi\rangle \qquad (46a)$$

where $\phi$ describes the rotation around the long molecular axis and is measured from the direction which is perpendicular to the tilt. In addition, for chiral systems the tilt breaks the *xz* mirror plane symmetry, thus allowing for a spontaneous polarization in the *y*-direction i.e. in a direction perpendicular to the tilt. This leads to a bilinear coupling between the polarization and the tilt, and to another additional term[21,22]

$$\Delta = \Delta_p = 2\sigma_{xz} \sin 2\Theta_o \langle\cos\phi\rangle \qquad (46b)$$

It should also be noted that the tilt induces even for non-chiral systems an anisotropy in the fluctuations around the director, leading to a term $\Delta_a$ the maximum value of which is

$$\Delta_{a,max} = \frac{2}{3}(1 - S)(\sigma_{\|} - \sigma_{\perp}) \sin^2 \Theta_o \qquad (46c)$$

This allows for a determination of $S$, $\langle\cos\phi\rangle$, $\langle\cos 2\phi\rangle$ and $\Theta$ over the molecule from the measured chemical shifts, if the chemical shift tensors $\underline{\sigma}$ are known for all nuclei in question[21,22].

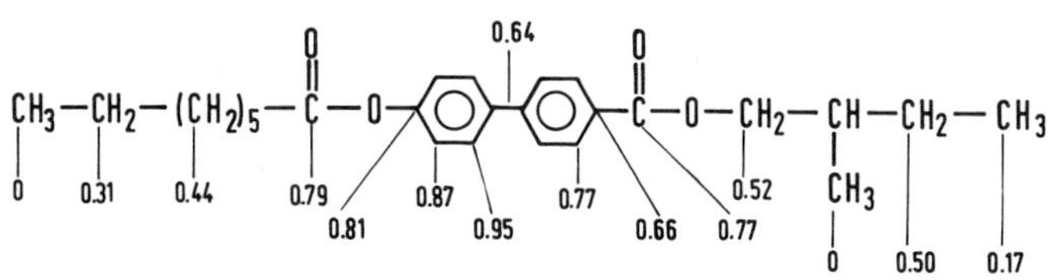


Fig.10. Variation of the nematic order parameter S along the IS-2424 molecule as determined from [13]C chemical shifts in the SmA phase at 60 °C.

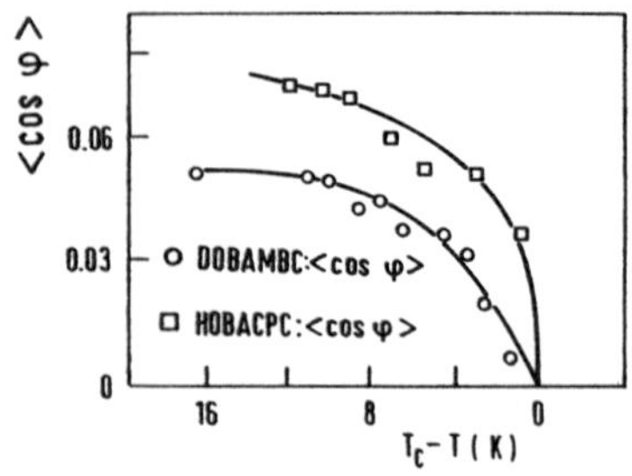


Fig.11. Temperature dependence of the polar order parameter <cos ϕ> of the *C*=0 group in DOBAMBC and HOBACPC.

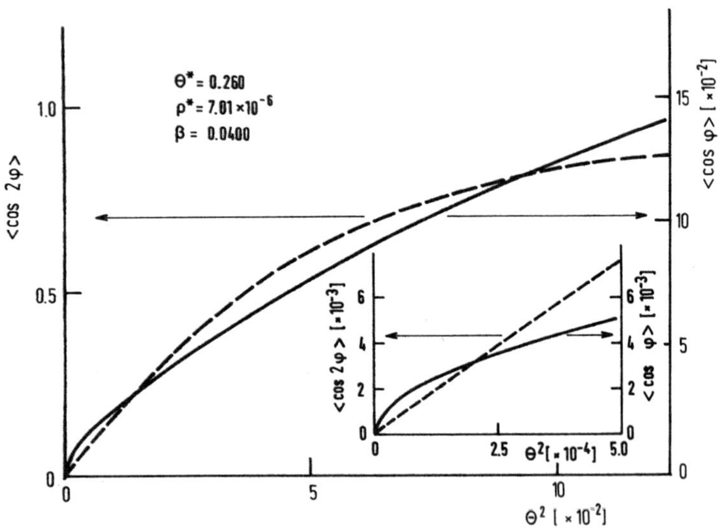


Fig.12. Polar and quadrupolar order parameters as functions[17] of the square of tilt angle $\theta^2$, which for small $\theta$ is roughly proportional to $T_c$ - T.

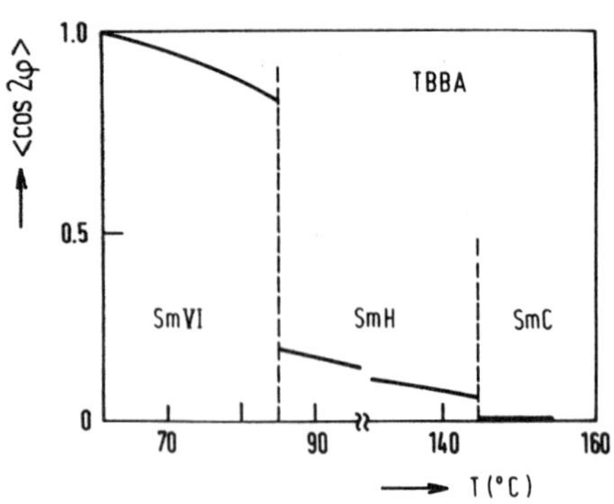


Fig.13. Temperature dependence of the quadrupolar order parameter <cos 2ϕ> in achiral TBBA[24].

In all cases the $\langle\cos\phi\rangle$ values determined (Fig. 11) from the NMR data agree with those obtained from macroscopic polarization measurements $P = N\mu\langle\cos\phi\rangle$.

It should be stressed that the obtained $\langle\cos\phi\rangle$ values for the $C = 0$ groups are rather small, $\langle\cos\phi\rangle \sim 10^{-2}$, the $\langle\cos 2\phi\rangle$ values obtained are significantly larger: whereas for aromatic carbons[22] e.g. $\langle\cos 2\phi\rangle \sim 0.11 - 0.12$. This fully agrees with the predictions of the microscopic model. The variations of $< \cos 2\phi >$ and $< \cos\phi >$ with $\Theta^2$, i.e. $T_c - T$, calculated from this model are shown in Fig. 12.

The biasing of the rotation of the transverse molecular dipoles around the long molecular axis has also been studied by $^{14}N$ nuclear quadrupole resonance (NQR)[21,22] in achiral TBBA and chiral TBACA. Here the possible biasing of the molecular rotation is ascertained via a determination of the asymmetry parameter

$$\eta = (V_{XX} - V_{YY})/V_{ZZ} \tag{47a}$$

of the electric field gradient (EFS) tensor at the $^{14}N$ sites. For free uniaxial rotation the largest principal axis $V_{ZZ}$ will point along the long molecular axis and the EFG tensor will be axially symmetric: $V_{XX} = V_{YY}$, $\eta = 0$. Any anisotropy in the rotation around the long molecular axis will destroy this axial symmetry, resulting in the appearance of three pure NQR $^{14}N$ lines

$$\nu_{1,2} = (3/4)(eQV_{ZZ}/h)(1 \pm \eta/3) \tag{47b}$$

$$\nu_3 = \nu_2 - \nu_1 \tag{47c}$$

instead of a single one for $\eta = 0$. For bipolar (i.e. quadrupolar) biasing of the uniaxial rotation one has $\langle\cos 2\phi\rangle \neq 0$, $\langle\cos\phi\rangle = 0$. For $\langle\cos 2\phi\rangle \leq 0,3$ one finds in TBBA

$$\eta = r\langle\cos 2\phi\rangle \tag{48a}$$

i.e. the asymmetry parameter is directly proportional to the quadrupolar ordering. For pure polar biasing $\langle\cos 2\phi\rangle = 0, \langle\cos\phi\rangle \neq 0$ one finds on the other hand

$$\eta = K\langle\cos\phi\rangle^2, \quad \langle\cos\phi\rangle \ll 1 \tag{48b}$$

Here $K$ and $r$ depend[23,24] on the five independent EFG tensor components in the molecular frames. The temperature dependence of the quadrupolar order parameter $\langle\cos 2\phi\rangle$ in achiral TBBA is shown in Fig. 13.

## 11. Phase Diagrams in External Fields

The *(T, E)*[25] and *(T, H)*[14] phase diagrams in external transverse fields are shown in Fig.s 14 and 15. The *(T, E)* phase diagram is shown for DOBA-1-MPC where the spontaneous polarization is about three times as large as in DOBAMBC. The experimental phase diagram showing the transition between the helicoidal and the homogeneously polarized Sm $C^*$ phase clearly shows the anomalous temperature dependence of the critical electric field $E_c$ predicted by the extended Landau model which includes polar as well as quadrupolar ordering.

The same is true for the phase diagram in a magnetic field of a mixture of five parts racemic

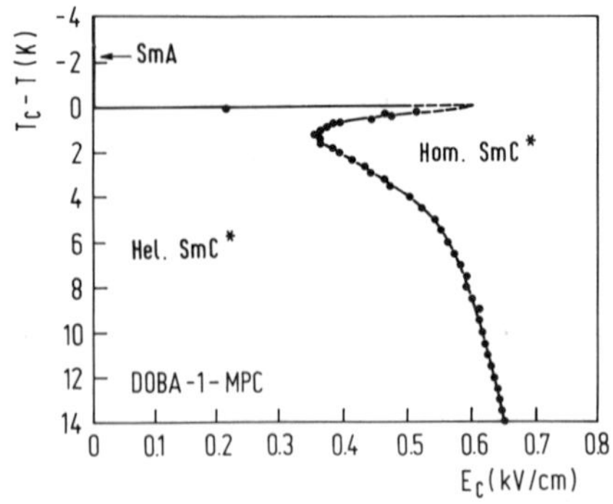


Fig.14. Phase diagram of DOBA-1-MPC in an external transverse electric field[16].

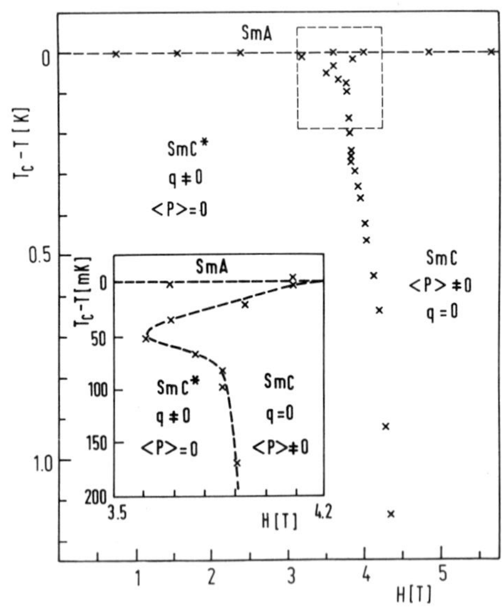


Fig.15. Phase diagram[14] of a mixture of chiral and racemic DOBAMBC in an external transverse magnetic field.

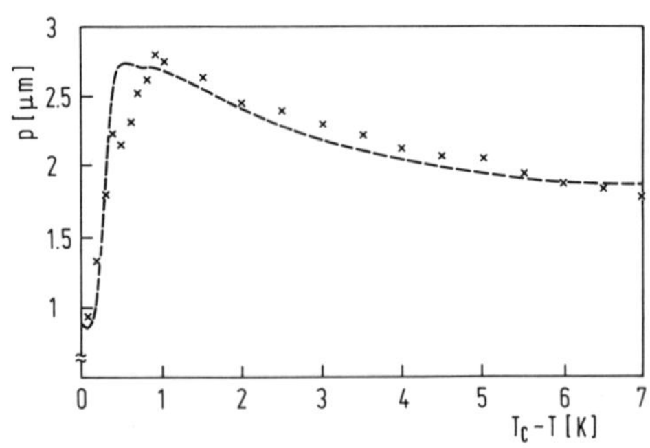


Fig.16. Comparison between the experimental and theoretical temperature dependences of the helical pitch in DOBAMBC[26].

and four parts chiral DOBAMBC[14]. In contrast to pure chiral DOBAMBC[10] magnetic field sweeps up to five Tesla were sufficient to unwind the Sm $C^*$ helix in the whole temperature range from $T_c - T$ = 1K to $T_c$. In contrast to DOBAMBC, in this mixture the Lifshitz point is experimentally accessible, with $H_L = 4.16/\mathrm{T}(1 \pm 0.01)$.

The anomalous forms of the above two phase diagrams showing a reentrant Sm $C^*$ phase can be related via expressions (31a) and (27) to the anomalous temperature dependence of the pitch of the helix[26], which can also be explained by the extended Landau model. As it can be seen from Fig. 16, the agreement between the experimental and theoretical (Eq. 10) values of the pitch is very satisfactory.

ACKNOWLEDGEMENT. The author would like to acknowledge the help of his colleagues B. Žekš, A. Levstik, M. Čopič, I. Muševič, A. Seppen, Z. Kutnjak and B. Urbanc-Kutnjak, who significantly contributed to this work.

## References

1. R.B. Meyer, L. Liebert, L. Strzelecki and P. Keller, J. Phys. (Paris) Lett. **36**, L69 (1975)
2. R. Blinc, Phys. Stat. Sol. **B 70**, K29 (1975)
3. S.A. Pikin and V.L. Indenboom, Uspekhi Fiz. Nauk **125**, 251 (1978)
4. R. Blinc and B. Žekš, Phys. Rev. **A 18**, 740 (1978)
5. B. Žekš, Mol. Cryst. Liqu. Cryst. **114**, 259 (1984). See also B. Urbanc and B.Žekš, Liq. Cryst. **5**, 1075 (1989)
6. I. Muševič, B. Žekš, R. Blinc, L. Jansen, A. Seppen and P. Wyder, Ferroelectrics **58**, 71 (1984). See also T. Carlsson, B. Zeks, A. Levstik, C. Filipic, I. Levstik, and R. Blinc, Phys. Rev. **A 36**, 1484 (1987)
7. B.I. Ostrovskii, A.Z. Robinovich, A.S. Sonin and V.A. Strukov, Zh. Eksp. Teor. Fiz. **74**, 1748 (1978)
8. S. Dumrograttana and C.C. Huang, Phys. Rev. Lett. **56**, 464 (1986)
9. A. Levstik, T. Carlsson, C. Filipič, I. Levstik and B. Žekš, Phys. Rev. **A 35**, 3527 (1987)
10. I. Muševič, B. Žekš, R. Blinc, Th. Rasing and P. Wyder, Phys. Rev. Lett. **48**, 192 (1982)
11. H. Takezoe, K. Kondo, K. Myasato, S. Abe, T. Tsuchiya, A. Fukuda and E. Kuze, Ferroelectrics **58**, 55 (1984)
12. I. Muševič, R. Blinc, B. Žekš, C. Filipič, M. Čopič, A. Seppen, P. Wyder and A. Levanyuk, Phys. Rev. Lett. **60**, 1530 (1988)
13. A. Michelson, Phys. Rev. Lett. **39**, 464 (1977)
14. A. Seppen, Ph.D. Thesis, University of Nijmegen (1988)
15. H. Orihara and Y. Ishibashi, Jpn. J. of Applied Phys. **29**, L115 (1990)
16. Z. Kutnjak, M.Sc. Thesis, University of Ljubljana (1991)
17. B. Urbanc-Kutnjak, M.Sc. Thesis, University of Ljubljana (1991)
18. S.U. Vallerien, F. Kremer, H. Kapitza, R. Zentel, W. Frank, Phys. Letters **A 138**, 219 (1989)
19. S.U. Vallerien, F. Dremer, T. Gelhar and A.E. Wachter, to be published
20. M. Čopič, I. Drevenšek, I. Muševič and R. Blinc to be published in the Proceedings of the Advanced Research Workshop on Quasi-Crystals, Liquid Crystals and Incommensurate Insulators, Preveza, Greece, September 1989
21. M. Luzar, V. Rutar, J. Seliger and R. Blinc, Ferroelectrics **58**, 115 (1984)
22. R. Blinc, J. Dolinšek, M. Luzar and J. Seliger, Liq. Cryst. **3**, 663 (1988)
23. J. Seliger, V. Žagar and R. Blinc, Phys. Rev. Lett. **38**, 411 (1977)
24. J. Seliger, V. Žagar and R. Blinc, Phys. Rev. **A 17**, 1149 (1978)
25. A. Levstik, Z. Kutnjak, B. Žekš, S. Dumrongrattana and C. C. Huang, J. Phys. II France 1, 797 (1991)
26. R. Blinc, I. Muševič and B. Žekš, Mol. Cryst. Liq. Crystals, 202, 117 (1991)

Chapter 23

# THE PHASE TRANSITION FROM THE SmC* TO THE SmC PHASE INDUCED BY AN EXTERNAL MAGNETIC FIELD

B. KUTNJAK-URBANC and B. ŽEKŠ

## 1. Introduction

Some chiral compounds exhibit a helicoidal smectic-C* (SmC*) phase which is of considerable interest due to its remarkable properties and potential applications. In the SmC* phase rod-like molecules are tilted with respect to the layer normal, and the director precesses from one layer to another so that the helicoidal structure is formed with a period of about $10^3$ layers thicknesses. The period of the helicoidal structure is strongly sensitive to the strength of the field, electric or magnetic, so that at some critical field the sample unwinds and the structure becomes similar to that in the SmC phase. The critical field values depend strongly on the temperature.

The phase diagram H-T of the SmC* phase in an external magnetic field applied parallel to the smectic layers was originally studied by Michelson[1]. The point at which three transition lines, SmA→ SmC*, SmA→ SmC and SmC*→ SmC, meet is called a Lifshitz point because of its very interesting properties. As the system proceeds along the SmA→ SmC* transition line, the wave vector of the modulated phase SmC* continuously approaches zero as the Lifshitz point is approached. It is also known that the phase transition SmC*→ SmC we will be interested in should be first order in the vicinity of the Lifshitz point. Usually this transition is described in the constant amplitude approximation (CAA) which is known to be valid deep in the SmC* phase, and leads to a continuous type of transition, similar to the transition in cholesterics[2]. Therefore, there exists a tricritical point (TCP) on the SmC*→ SmC transition line at which the order of the transition changes.

The experiments[3] show that the critical field is a nonmonotonic function of temperature. The phase sequence SmA→ SmC*→ SmC→ SmC* is observed upon lowering the temperature. The SmC* therefore exhibits re-entrant behavior. In some recent theoretical approaches[4,5] the phase diagram in magnetic field has been studied in a model which does not adequately describe the temperature dependences of some physical properties (the pitch of the helix, for example). The aim of our contribution is to study the phase transition SmC*→ SmC first in the classical model[6], and then in the extended Landau model[7]. The numerical treatment is described briefly; a detailed calculation will be given elsewhere. The comparison between

B. Kutnjak-Urbanc - J. Stefan Institute, Jamova 39, 61111 Ljubljana, Yugoslavia
B. Žekš - Institute of Biophysics, Medical Faculty, Lipičeva 2, 61105 Ljubljana, Yugoslavia

*Phase Transitions in Liquid Crystals*, Edited by S. Martellucci and A.N. Chester, Plenum Press, New York, 1992

the exact and the approximated (CAA) results is made in both the classical and the extended Landau model.

## 2. The Classical Landau Model

For a Landau description of the SmA, SmC* and SmC phases two two-component order parameters are mainly used, the primary order parameter is a tilt $\boldsymbol{\xi} = (\xi_1, \xi_2)$ and the secondary order parameter is a polarization $\mathbf{P}=(P_x,P_y)$. According to Pikin and Indenbom[8] the free energy density expansion, with an additional term due to the coupling between the tilt $\boldsymbol{\xi}$ and the external magnetic field **H,** can be written

$$f(z)=\frac{1}{2}a\left(\xi_1^2+\xi_2^2\right)+\frac{1}{4}b\left(\xi_1^2+\xi_2^2\right)^2+\frac{1}{2}K_3\left[\left(\frac{d\xi_1}{dz}\right)^2+\left(\frac{d\xi_2}{dz}\right)^2\right]$$
$$-\Lambda\left(\xi_1\frac{d\xi_2}{dz}-\xi_2\frac{d\xi_1}{dz}\right)+\frac{1}{2}\frac{1}{\varepsilon}\left(P_x^2+P_y^2\right)+C\left(P_x\xi_2-P_y\xi_1\right)$$
$$-\mu\left(P_x\frac{d\xi_1}{dz}+P_y\frac{d\xi_2}{dz}\right)-\frac{1}{2}\chi_a H^2\xi_1^2 \tag{1}$$

where $\chi_a = \chi_{//} - \chi_{\perp}$ denotes the magnetic anisotropy which is assumed to be positive, so that under the influence of the magnetic field the tilt $\boldsymbol{\xi}$ tends to align parallel to the field **H**. The $\Lambda$-term is the Lifshitz term, the $K_3$-term is the elastic term and there are also two bilinear coupling terms between both order parameters, the flexoelectric $\mu$-term and the piezoelectric C-term. The constant $\varepsilon$ denotes the high temperature dielectric constant. The only temperature dependent term is the one with $a = a_0\,(T - T_0)$, and b is a positive constant. The secondary order parameter **P** can be eliminated from the expansion (1) by partial minimization with respect to $P_x$ and $P_y$. Upon substituting the results into (1), one is left with only the first four terms and the last magnetic coupling term, however with slightly different coefficients $\tilde{a} = a - \varepsilon C^2$, $\tilde{\Lambda} = \Lambda + \varepsilon\mu C$ and $\tilde{K}_3 = K_3 - \varepsilon\mu^2$ instead of the original a, $\Lambda$ and $K_3$. The free energy density now depends only on the tilt $\boldsymbol{\xi}$, which has magnitude $\boldsymbol{\theta}$ and phase $\boldsymbol{\Phi}$, so that $\xi_1 + i\xi_2 = \theta\exp(i\Phi)$. The Lifshitz point in the H-T diagram is determined by the Lifshitz field $H_L$ and the Lifshitz temperature $T_L$[1]

$$H_L = 2\sqrt{\frac{\tilde{\Lambda}^2}{\chi_a\tilde{K}_3}} \qquad T_L = \tilde{T}_0 + 4\frac{\tilde{\Lambda}^2}{a_0\tilde{K}_3} \tag{2}$$

where $\tilde{a} = a_0\left(T - \tilde{T}_0\right)$ and $\tilde{T}_0 = T_0 + \varepsilon C^2/a_0$. One can now introduce a dimensionless field h and a dimensionless temperature t

$$h = \frac{H}{H_L} \qquad t = \frac{T_L - T}{T^*} \qquad \text{with} \quad T^* = \frac{\tilde{\Lambda}^2}{a_0\tilde{K}_3} \tag{3}$$

By appropriate choice of other scaling parameters

$$z^* = \frac{\tilde{K}_3}{\tilde{\Lambda}} \qquad \theta^* = \frac{\tilde{\Lambda}}{\sqrt{b\tilde{K}_3}} \qquad f^* = \frac{\tilde{\Lambda}^4}{b\tilde{K}_3^{\,2}} \tag{4}$$

and dimensionless quantities

$$\tilde{z} = \frac{z}{z^*} \qquad \tilde{\theta} = \frac{\theta}{\theta^*} \qquad \tilde{f} = \frac{f}{f^*} \tag{5}$$

one arrives at a dimensionless expression for the free energy density $\tilde{f}$:

$$\tilde{f}(\tilde{z}) = -\frac{1}{2}\tau\tilde{\theta}^2 + \frac{1}{4}\tilde{\theta}^4 + \frac{1}{2}\left(\frac{d\tilde{\theta}}{d\tilde{z}}\right)^2 + \frac{1}{2}\tilde{\theta}^2\left(\frac{d\Phi}{d\tilde{z}}\right)^2 - \tilde{\theta}^2\frac{d\Phi}{d\tilde{z}} + 2h^2\tilde{\theta}^2\sin^2\Phi \tag{6}$$

in which no other parameters appear except the dimensionless quantities field h and temperature t ($\tau$=t-4+4h$^2$).

The Euler-Lagrange equations that follow from the minimization of the free energy density (1) are two coupled nonlinear second order differential equations for functions $\tilde{\theta}$ and $\Phi$ respectively. In the CAA only the phase $\Phi$ of the order parameter is allowed to vary with $\tilde{z}$, and it obeys the sine-Gordon equation[3]. This approximation leads to the continuous phase transition from the SmC* to the SmC phase, which occurs at the critical field $H_C = \pi/4\ H_L$ or $h_c = \pi/4$. The critical field is temperature independent, and the pitch of the helicoidal structure diverges logarithmically as the transition line SmC*→SmC is approached from the SmC*-phase side[8]. In the vicinity of the Lifshitz point where the phase transition is first order there are three critical lines. Two of them are the stability limits of the SmC and SmC* phases, and between these two limit lines there is the critical field line at which these two phases have equal free energies. In the region of the continuous transition all three lines coincide. Our aim is to determine the stability limit of the SmC phase, which is located left of the critical transition line, at lower fields. We are therefore looking for single soliton solutions of Euler-Lagrange equations. In order to determine the other stability line, one should study multi-soliton solutions as well; this will be done elsewhere.

The Euler-Lagrange equations should be solved numerically. The procedure used is similar to that of the previous work on this field[4,5]. One linearizes the equations in the vicinity of the trial solutions, solves these linearized differential equations with appropriate boundary conditions, corrects the trial solutions and iterates until the desired accuracy is obtained. For the initial functions we take a constant tilt magnitude $\tilde{\theta}$ obtained by minimizing the free energy of the commensurate SmC phase and the tilt phase $\Phi$ which is a single soliton solution of the sine-Gordon equation. The described iteration converges for all values of $t \in [1, \infty)$. To approach closer to the Lifshitz point, instead of the above initial functions, we must use solutions from the near points of the phase diagram. The transition field line in Fig.1 is determined using the classical Landau model at a given temperature by changing the field h until the free energies of both phases are equal.

Next, one would like to find the position of the tricritical point. As predicted by Jacobs and Walker[9] (and also other authors) the order of the phase transition should affect the asymptotic behavior of the solutions $\tilde{\theta}$ and $\Phi$. A pure exponential asymptotic behavior would lead to a repulsive interaction between discommensurations, and an exponentially damped sinusoidal behavior would be characteristic of an attractive interaction between discommensurations. The former behavior leads to the continuous and the latter one to the discontinuous type of transition. To analyse the asymptotic behavior, one linearizes the Euler-Lagrange equations around the asymtotic values of $\tilde{\theta}$ and $\Phi$ and obtains linear differential

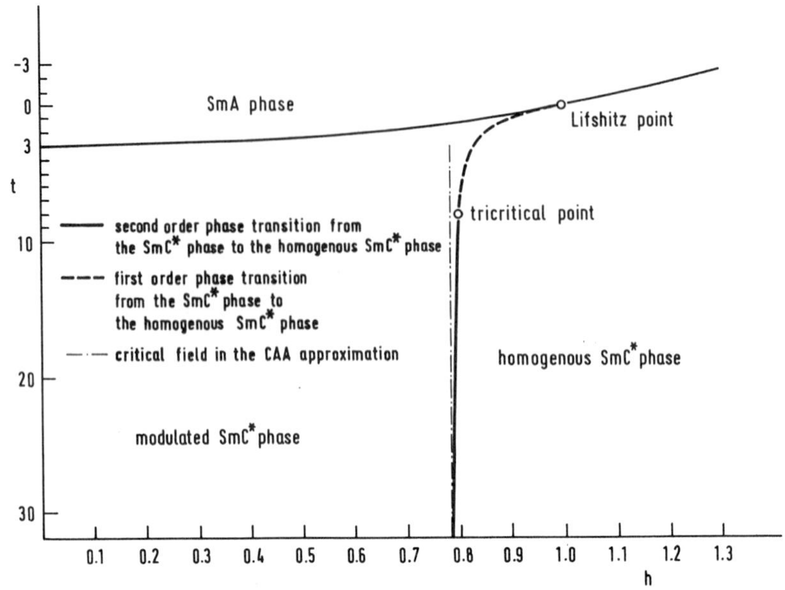


Fig.1. The phase diagram h-t in an external magnetic field, calculated using the classical Landau model. The phase transition SmC*⟶SmC is of first order between the Lifshitz and the tricritical point (dashed curve) whereas it is continuous at lower temperatures (solid curve).

equations with constant coefficients, which have solutions of the form const.exp($\lambda\tilde{z}$). One finds four solutions for the exponent $\lambda_{1,2,3,4} = \pm\sqrt{A \pm \sqrt{A^2 - B^2}}$, where $A = t - 6 + 6h^2$ and $B = \sqrt{8h^2(t - 4 + 4h^2)}$. The conditions $A^2 \geq B^2$ and $A \geq 0$ lead to four real roots and therefore to the continuous type of transition, whereas all other cases lead to the first order transition. One can see in Fig.1 that the critical field h lies in the interval [$\pi/4$, 1] at all temperatures. The lower field gives us an estimate $t_1 \approx 7.9$ and the higher field leads to $t_2 = 8$. The tricritical point therefore lies somewhere in the interval [$t_1$, $t_2$], its position is indicated in Fig.1.

On observing the numerical solutions of $\tilde{\theta}(\tilde{z})$ and $\Phi(\tilde{z})$ the minimum of $\tilde{\theta}$ is clearly seen in the middle of the discommensuration, where $d\Phi/d\tilde{z}$ has its maximum value (Fig.2). These solutions are therefore similar to those found by Jacobs and Walker[9]. The amplitude $\tilde{\theta}$ exhibits overshooting behavior in the whole temperature region, whereas the phase $\Phi$ shows such a behavior only for $0 \leq t \leq t_0$, where $t_0 \in (6,9)$.

To conclude this section we can compare the results for the critical field line $h_c(t)$ in the classical model with the constant value calculated in the CAA. We conclude that CAA gives good results for all $t \geq 30$ as seen from Fig.1. The comparison of the model predictions with the experimental results[11,3] shows that the classical model is not able to explain the temperature dependence of the ratio $P/\theta$, of the pitch, nor can it lead to the observed re-entrant behavior.

## 3. The Extended Landau Model

In order to explain the re-entrant behavior of the SmC* phase in an external magnetic field one must add to the free energy density expansion (1) the terms

$$+\frac{1}{6}c\left(\xi_1^2+\xi_2^2\right)^3$$
$$-d\left(\xi_1^2+\xi_2^2\right)\left(\xi_1\frac{d\xi_2}{d\tilde{z}}-\xi_2\frac{d\xi_1}{d\tilde{z}}\right)$$
$$+\frac{1}{4}\eta\left(P_x^2+P_y^2\right)^2$$
$$-\Omega\left(P_x\xi_2-P_y\xi_1\right)^2 \tag{7}$$

The full explanation of these terms is given for example by T. Carlsson et al[7]. The second d-term is of chiral character and it affects the temperature dependence of the pitch of the helix at zero field on a wide temperature range. The Ω-term describes the biquadratic coupling between the tilt and the polarization, and is essential for explaining the temperature dependence of the polarization ratio vs. the tilt and of the pitch at zero field[10]. A full discussion of the results for temperature dependence of some physical quantities comparing experiment with the classical and the extended model, is given elsewhere[11].

If one tries to proceed in the same way as in the case of the classical model, one finds that the elimination of the polarization order parameter $\mathbf{P}=(P_x,P_y)$ cannot be done analytically, in contrast to the classical model. In addition to the magnitude and the phase of the tilt, $\tilde{\theta}$ and $\Phi$, the magnitude and the phase of the polarization, P and $\tilde{\Phi}$ (where $P_x+iP_y$ = $P$=exp $(i\tilde{\Phi}+i\pi/2)$), have to be calculated. The two order parameters are therefore not necessarily perpendicular as in the case of zero field.

One now introduces a dimensionless field h and dimensionless temperature t in the same way as in the classical model (see Eqs. 2 and 3). There are several ways to introduce scaling parameters for the coordinate, the tilt, the polarization and the free energy density. In any case, one is left with six dimensionless parameters in the dimensionless free energy density. Next, we have to solve four Euler-Lagrange equations for four variables $\tilde{\theta}$, $\Phi$, $\tilde{P}$ and

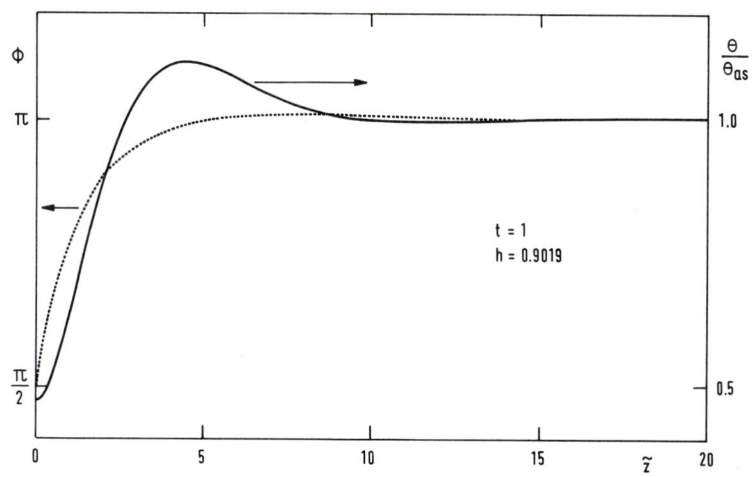


Fig.2. The typical shape of the single soliton solution for the magnitude $\tilde{\theta}$ (solid curve) and for the phase Φ (dashed curve) at t = 1 and h = 0.9019. The middle of the discommensuration is at $\tilde{z}$ = 0, where the minimum of θ occurs. The overshooting of θ can also be clearly seen.

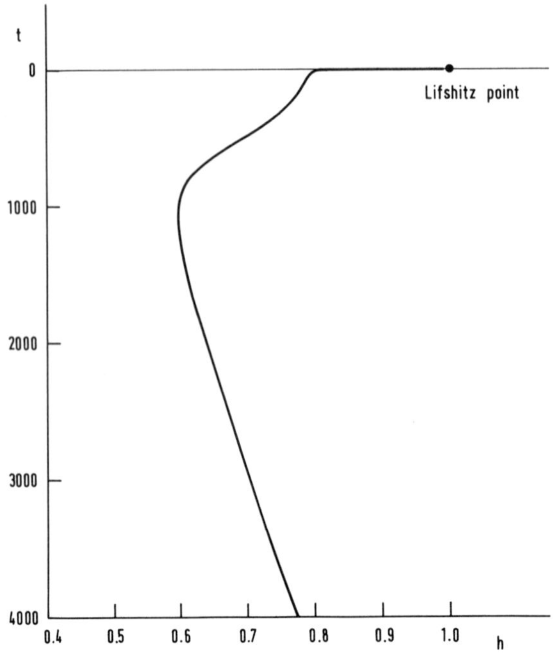


Fig.3. The phase diagram h-t in the external magnetic field calculated using the extended Landau model. As t increases, the re-entrant behavior with the phase sequence SmA⟶SmC*⟶SmC⟶SmC* occurs for fields h∈ (0.6, 1).

$\tilde{\Phi}$. All of these are coupled nonlinear equations: for $\tilde{\theta}$ and $\Phi$ one gets two differential equations of the second order, and for $\tilde{P}$ and $\tilde{\Phi}$ two algebraic equations. Again one has to find solutions numerically, by linearizing around the trial solutions, correcting the trial solutions and iterating until the desired accuracy is reached. For the initial functions $\tilde{\theta}$ and $\tilde{P}$ we take constants with values which minimize the free energy of the commensurate SmC phase; the initial phase $\tilde{\Phi}$ is equal to $\Phi$; and the initial phase $\Phi$ is a single soliton solution of the sine-Gordon equation. The six parameters that enter the calculation we determine by using values obtained from the best fit to some physical quantities using DOBAMBC[7].

The critical field transition line was calculated in the same way as in the classical model. The result is shown in Fig.3, where the phase sequence SmA→ SmC*→ SmC→ SmC* can be observed on lowering the temperature for the fields h∈ (0.6, 1.0). The critical field $h_c(t)$ reaches its minimum value at about $t \approx 1170$ and for larger t it increases. Two comparisons can be made here:

a) Between the rigorous results and the constant-amplitude approximation $\tilde{\theta}$ = const. and $\tilde{P}$=const. in the extended model; and,

b) Between the results of the extended and the classical model.

In the former case a) one calculates the critical field in the approximation of both constant amplitudes and gets the result $h_c \propto \tilde{q}_0$, where $\tilde{q}_0$ is a wave vector of the helicoidal structure at the zero field. When the SmA→ SmC* transition line is approached, $h_c$ tends to a constant value $\pi / 4$ which is different from the real value $h_c = 1$ at the Lifshitz point. Fig.4 shows that this approximation gives good values for $t > 20$, where it coincides with the extended-model values. Taking into account the values of Landau coefficients[7] for DOBAMBC one finds that t = 20 corresponds to $|T_C - T| \approx 10^{-2}K$. For smaller values of t this approximation breaks down. In the regime where $h_c(t)$ has a minimum, the extended model and this approximation give the same values for $h_c(t)$. One can therefore conclude that the minimum of $h_c(t)$ corresponds to a maximal value of the helical pitch at zero field.

The comparison in case b) is shown in the same Fig.4. One finds the classical-model and the extended-model values for $h_c(t)$ the same for $t < 200$ which corresponds to $|T_C - T| < 0.1\,K$ in DOBAMBC[7]. This result is expected because close to the SmA$\rightarrow$ SmC* transition higher order terms in the extended free energy density expansion can be neglected and one is left with the classical model. For higher values of t, deep in the SmC* phase, the difference between the two models is obvious from the results of the previous section.

Another comment should be made concerning the tricritical point. From Fig.4 it is clear that the predicted region of first order phase transition SmC*$\rightarrow$ SmC lies at temperatures at which the classical model is undoubtly valid. The results concerning the estimated position of the tricritical point are applicable in the extended model as well.

## 4. Conclusions

The phase transition from the SmC* to the SmC phase induced by an external magnetic field has been studied in the classical model, which is shown to be valid near the Lifshitz point, but fails to describe the re-entrant SmC* phase determined by measurement[3,12,13]. In order to predict such a behaviour one must include higher order terms in the free energy density expansion, and study this transition using the extended Landau model.

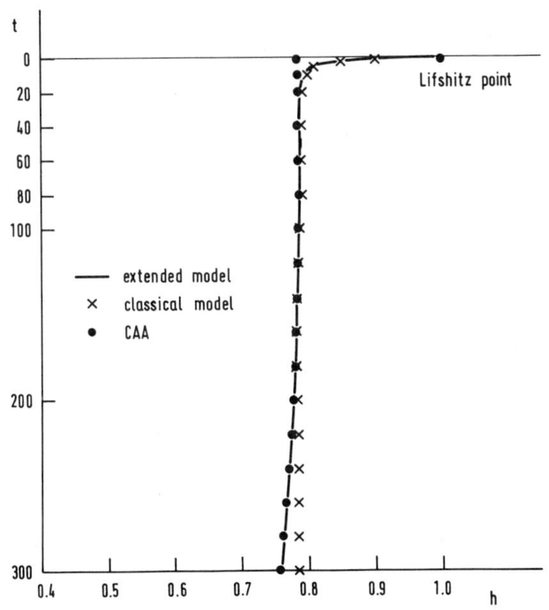


Fig.4. The phase diagram h-t, comparing the classical model to the exact and constant amplitude approximated (CAA) results of the extended model.

Experiments on DOBAMBC[3,12] show that the re-entrant phase sequence SmA→ SmC*→ SmC→ SmC* appears on lowering the temperature for fields higher than 8.5T. The critical field reaches its minimum value at about $|T_C - T| \approx 1\,K$, and this corresponds to the temperature at which the pitch of the helicoidal structure is the largest. The extended model gives results that agree very well with this data. On evaluating T* from the Landau expansion coefficients fitted to DOBAMBC[7], one finds that the temperature at which we obtain the smallest critical field actually corresponds to $|T_C - T| \approx 1\,K$.

Our conclusions differs from the calculation of Jacobs and Benguigui[4,5], where the authors used a model which did not give the observed maximum in the temperature dependence of the pitch at zero field just below the transition, and therefore no relation between the temperature dependences of the pitch and of the critical magnetic field was found.

Calculation of the estimated temperature regime in which the phase transition SmC*→ SmC would be discontinuous gives us $|T_C - T_{TCP}| \approx 10^{-3}\,K$. That is quite a narrow temperature regime where measurements are very difficult. Experiments[3,8] performed on samples in the "bookshelf" geometry show a first order transition at all temperatures, whereas experiments on samples in a homeotropic geometry show a continuous type of the transition. This geometry-dependent behavior can be explained by dynamical domain wall effects in finite samples. Due to these experimental difficulties the tricritical point has not been detected yet.

ACKNOWLEDGEMENTS. The authors are indebted to Dr. Rajko Krivec for stimulating discussions concerning the numerical treatment.

## References

1. A. Michelson, Phys. Rev. Lett. **39**, 464 (1977).
2. P.G.deGennes, Solid State Commun. **6**, 163 (1968).
3. I.Muševič, B. Žekš and R. Blinc, Th. Rasing, P. Wyder, Phys. Rev. A **48**, 192 (1982).
4. L. Benguigui and A. E. Jacobs, Ferroelectrics **84**, 379 (1988).
5. A. E. Jacobs and L. Benguigui, Phys. Rev. A **39**, 3622 (1989).
6. S. A. Pikin and V. L. Indenbom, Ferroelectrics **20**, 151 (1978).
7. T. Carlsson, B. Žekš, C. Filipič, A. Levstik and R. Blinc, Mol. Cryst. Liq. Cryst. **163**, 11 (1988).
8. R. Blinc, I. Muševič and B. Žekš, Phys. Scrip. **T35**, 38 (1991).
9. A. E. Jacobs and M. B. Walker, Phys. Rev. B **21**, 4132 (1980).
10. B. Žekš, Mol. Cryst. Liq. Cryst. **114**, 259 (1984).
11. B. Žekš, T. Carlsson, C. Filipič and B. Urbanc, Ferroelectrics **84**, 3 (1988).
12. I. Muševič, B. Žekš, R. Blinc, L. Jansen, A. Seppen and P. Wyder, Ferroelectrics **58**, 71 (1984).
13. A. Seppen, Ph. D. Thesis, University of Nijmegen (1988).

# POLYMERIC LIQUID CRYSTALS

Chapter 24

# PHASE TRANSITIONS AND PHASE DIAGRAMS IN LIQUID CRYSTALLINE POLYMERS

G. SIGAUD

## 1. Introduction

It is trivial to recall that the simplest and commonest model to represent a LMM is the anonymous uncharacteristic rigid rod shown in Fig. 1. Indeed liquid crystalline properties are readily associated to molecular or supramolecular anisotropy and therefore the first modelization of a macromolecular mesogen do not differ from this simple picture: Fig. 1 can serve as well provided one considers possible an infinite lengthening of this object. However this seemingly simple modification has non negligible outcomes, both theoretical and practical. First, the steric interactions in assemblies of such very long rods become a major parameter in the occurence of an anisotropic state. The great achievement of early theoretical works by Onsager and Flory is to have introduced the axial ratio x = L/d (Fig. 1) to account for these repulsive forces. Refined models including now attractive interactions[1] show that the isotropic-nematic transition is expected only within a given range of axial ratio limited by an upper critical value (≈6) (Fig. 2). In addition the order parameter at the transition for long rods comes out to be close to 1, significantly higher than the value from the Maier and Saupe's theory (≈0.43) which describes satisfactorily the behavior of short LMMs. What system seems more suitable than a sequence of phenyl rings to make up rigid-rod polymers ? In order to test this model and evaluate the meaning and the limitations of the phase diagram in Fig. 2, let us consider the Table 1 which presents some relevant characteristics of polyphenylenes: $H - \{ - \phi - \}_n - H$.

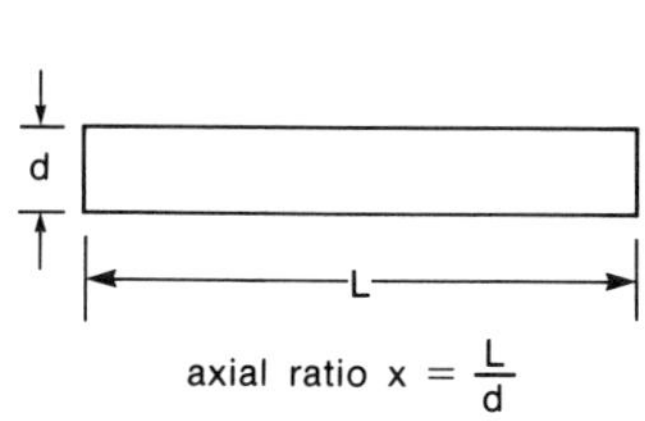


Fig. 1. Definition of the axial ratio in the model of rod-like mesogen.

G. Sigaud - IBM Research Division, Almaden Research Centre, 650 Harry Roas, San Jose, California 95120-6099, USA. Permanent address: Centre de Recherches Paul Pascal, C.N.R.S., Universite de Bordeaux I, Ave. A. Schweitzer, F-33600, Pessac, France.

*Phase Transitions in Liquid Crystals*, Edited by S. Martellucci and A.N. Chester, Plenum Press, New York, 1992

Table 1. Data for polyphenylenes H - { - $\phi$ - }$_n$ - H (Ref.s 1 and 2).

| n | axial ratio | $T_{N\text{-}I}$ (°C) estimated[(a)] | Polymorphism |
|---|---|---|---|
| 2 | ≈2 | | not nematic K-I 70 °C |
| 3 | ≈3 | | not nematic K-I 210 °C |
| 4 | ≈4 | ≈120 | not nematic K-I 320 °C |
| 5 | ≈5 | ≈450 | N-I ≈430 °C |
| 6 | ≈6 | ≈1900 | N-I ≈560 °C |
| 7 | ≈7 | not isotropic | K-N ≈550 °C and decomposition. |

a) Estimated $T_{N\text{-}I}$ has been normalized on quinquephenyl (n = 5).

We note with some concem that the critical axial ratio is reached for n as small as 6 or 7. The nematic phase cannot be observed on lowest terms owing to the stability of the crystalline phase and decomposition occurs around the N-I temperature for the higher oligomers. So, little hope would have been left to experimentalists to conduct extensive studies on strictly rigid rodlike macromolecules ! Moreover, the experimental study of solutions of rigid rods is often problematic as well since their solubility is very limited in most cases as illustrated in the phase diagram in Fig. 3. These introductory remarks are willingly grim but show that more realistic LCP materials were obviously needed to boost both theoretical and experimental researchs. From this standpoint the key advance was the input of some flexibility in these strictly rigid systems. Two different approaches have been successfully considered both based on the concept of flexible spacers: main-chain semi flexible polymers in Fig. 4a and polymers with mesogenic sidegroups in Fig. 4b[4]. Note that in the former case the degree of stiffness of the macromolecular sequence can still be very high and suffice to drive its overall rod-like character. In principle, anisotropy can be achieved regardless of the mesogenicity of the repeat unit. In contrast the rodlike units affixed to a highly flexible and non mesogenic backbone are necessarily mesogenic in the latter case. In turn, however, anisotropic phases can be expected despite no or very low anisotropy of shape of the overall macromolecule. These remarks stress the new shift in the interest for LCPs towards a better knowledge of the role of the flexible moieties in the anisotropic phase and at the transition(s). To what extent do they participate to the liquid

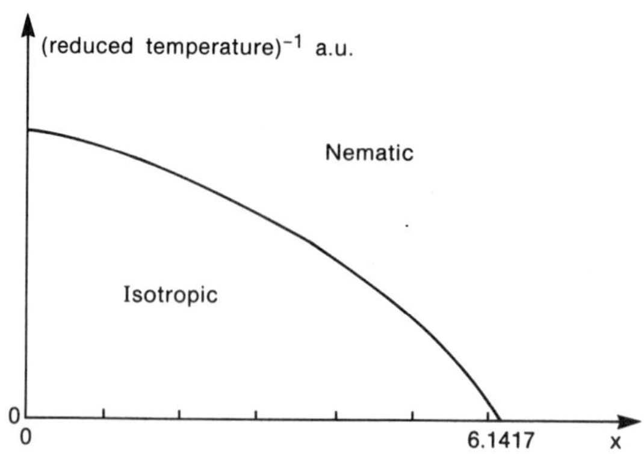


Fig. 2. Nematic-Isotropic phase line as a function of axial ratio and inverse temperature (Ref. 1).

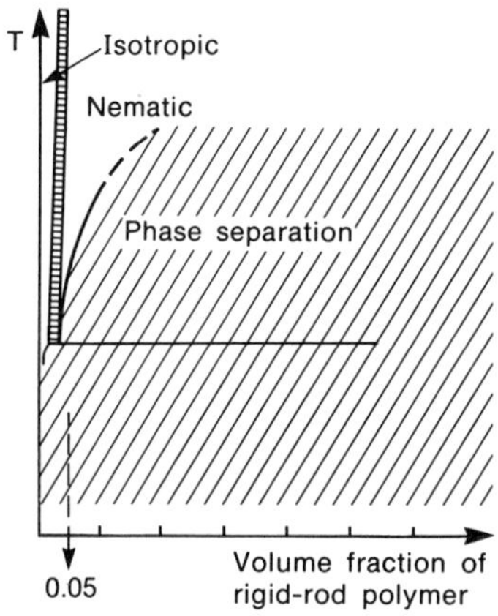


Fig. 3. Phase diagram for a rigid-rod polymer-solvent system. Note the very low concentration in polymer at which the I-N transition locates and the steepness of the lines.

crystalline structure ? The next Section summarizes some theoretical answers to this question in the case of the nematic phase.

## 2. Descriptive Theoretical Overview

In the model of partially flexible main-chain LCP, the relevant parameter is still a ratio of two characteristic lengths l/d, but 1 is now a renormalized length (the Kuhn's

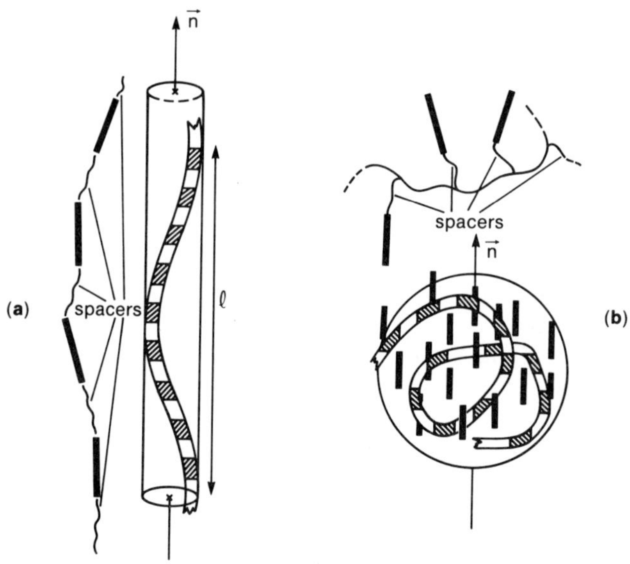


Fig. 4. a) Model for main-chain LCP; 1 defines the Kuhn's segment (Ref. 3). b) Model for side-chain LCP (Ref. 4).

segment, Fig. 4a). The higher 1, the higher the stiffness of the chain. Different types of flexibility can be considered but constant flexibility along the chain (as represented in Fig. 4a) is more likely to describe the experimental systems in which rod-like segments are regularly connected via flexible spacers. We consider here only the case of melts of thermotropic semi-flexible LCPs. The Fig. 5a presents the typical phase diagram for a semi flexible system[5]. N-I transitions are also expected below a limiting value of l/d with this model. It is satisfying to note that $\{l/d\}_c$ is much larger than $\{L/d\}_c$ for the system of rigid rods. The Fig. 5b shows the evolution of the orientational order parameter S at the transition as a function of l/d. S increases from a value comparable to the Maier and Saupe's theory up to the large value comparable to the order parameter of a system of rigid rods (≈0.85). Such high order parameters cannot be expected if not coupled with a strong conformational ordering of the chain in the N phase. The increase of $<R_z^2>$ should account for the elongation of the chain in the z direction defined by the director and the anisotropy of chain conformation to be theoretically achieved appears significant. The calculated values for $<R_z^2>/<R_x^2>$ range from 3 to more than 10 at the transition and this anisotropy is expected to be exponentially temperature dependent in the nematic phase. Comparison with neutron scattering and NMR results is made further.

In LCPs encompassing rigid rods and flexible spacers in the main chain different stages of orientational order are expected. The phase diagram in Fig. 6 shows two nematic phases. In $N_1$ the flexible moieties are weakly oriented while in $N_2$ they are highly oriented. Note that a 1st order transition is predicted between the two nematics and that a $N_1$-$N_2$ critical point is thus expected[6]. Basically, melts of side-chain LCPs can be described in the same way[5,7,8]. However the relevant parameters are modified :1 is the efficient length along the non mesogenic backbone (kind of Kuhn's length) and 1' is the distance between two mesogenic groups along the backbone (Fig. 7). Using this model, three main classes of nematic phases can then be described[7]. One with the mesogenic groups aligned along the "*easy axis*" and the backbone in the "*easy plane*" ($N_I$, Fig 8a). One with the mesogenic group in the "*easy plane*" and the backbone along the "*easy axis*" ($N_{II}$, Fig. 8b). Note that the $N_{II}$ phase can be biaxial ($N_{IIb}$) if the mesogenic groups are themselves oriented inside the "*easy plane*". One with both the mesogenic groups and the backbone aligned along the "*easy axis*" ($N_{III}$, Fig. 8c). A less ordered $N_{III}$ phase has been also proposed which is intermediate between the isotropic liquid and $N_{III}$. In $N_{III}$ the orientation of the mesogenic groups is high while the backbone remains poorly ordered. As the temperature decreases, the coupling with the nematic field induces the orientation of the backbone[7] (Fig. 9). Phase transitions among these different nematics are expected as a function of temperature and flexibility of the backbone (Fig. 10).

In summary, the interest of the phase behavior of LCPs as described by theories lies in the consequences of a more or less strong coupling of the conformation of the flexible or semiflexible backbone with the nematic order which drives a variety of transitions between nematic phases not possible in LMMs. At last one can remark that a few theoretical works[9,10] are aimed at the description of other mesophases than nematic although the sidechain LCPs are highly smectogenic as we will see further in the experimental Sections.

## 3. Characteristic Phase Behaviors of Neat Thermotropic LCPS

### *3.1. Role of intrinsic macromolecular parameters*

As we focus our attention on the differences in phase behavior between LCPs and LMMs, the influence of the intrinsic features of a polymer have to be primarily considered.

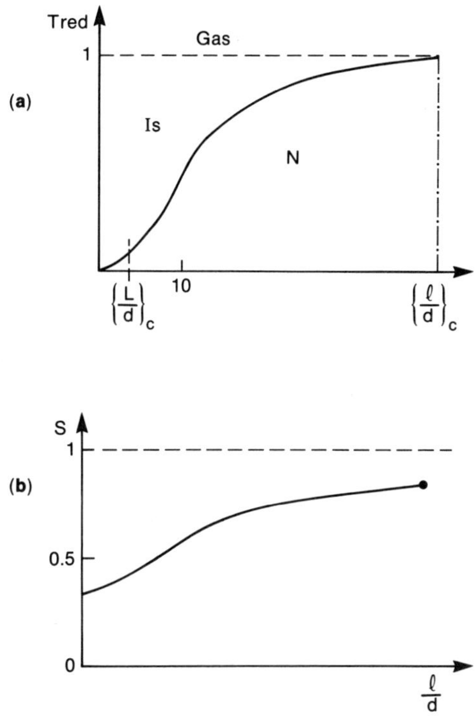


Fig. 5. a) Phase behavior and, b) evolution of the order pararneter as a function of the renormalized axial ratio l/d (Ref. 5).

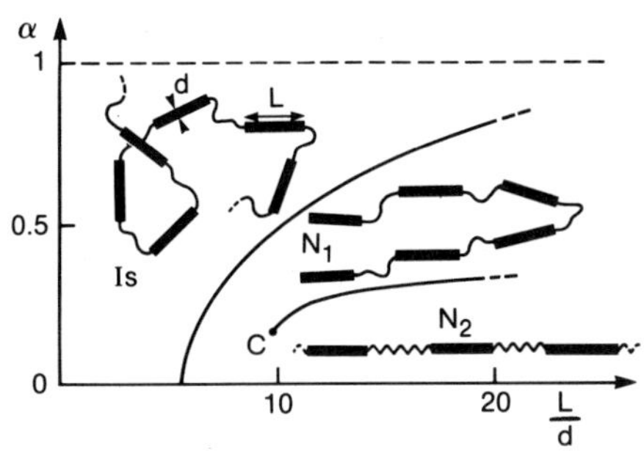


Fig. 6. Theoretical phase diagram for a LCP alternating rigid parts (with axial ratio L/d ) and flexible spacers (occupying a volume fraction α) (Ref. 6).

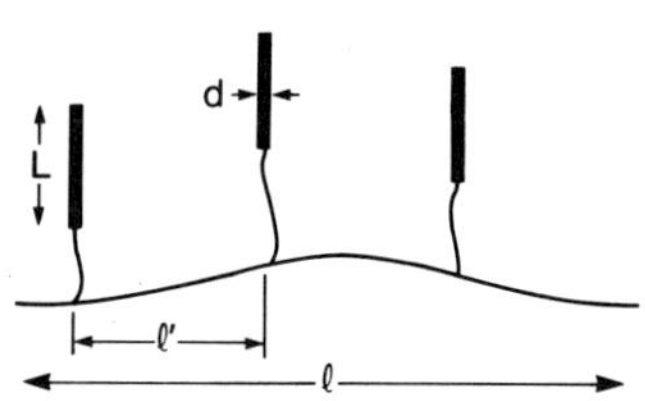


Fig. 7. Relevant parameters for the description of a side-chain LCP.

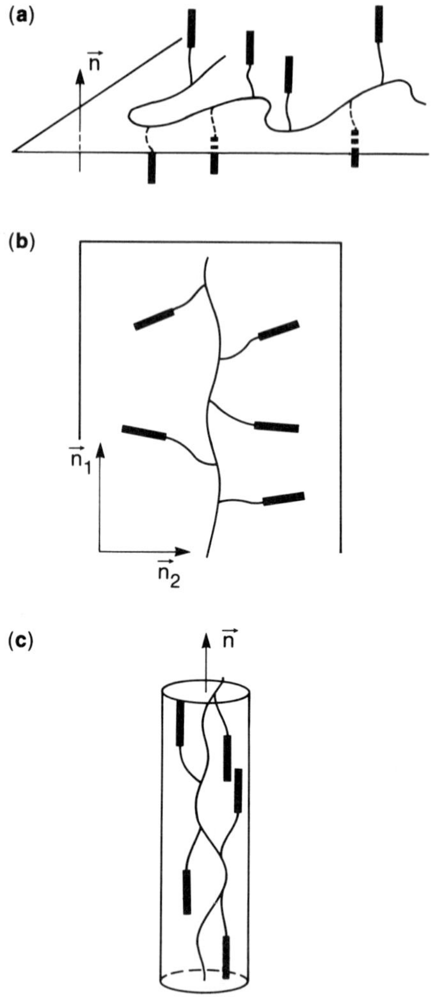


Fig. 8. Different types of nematic order expected with a side-chain LCP; a) $N_I$ ; b) $N_{II}$ and $N_{IIb}$; and c) $N_{III}$ (Ref. 7).

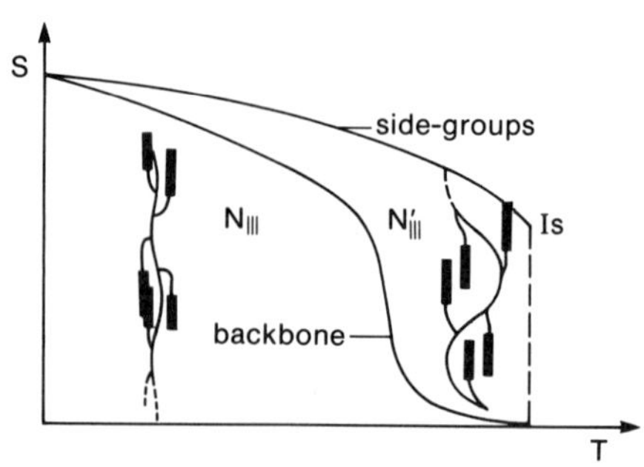


Fig. 9. Phases $N_{III}$ and $N_{III'}$ differ by the degree of order of the backbone.

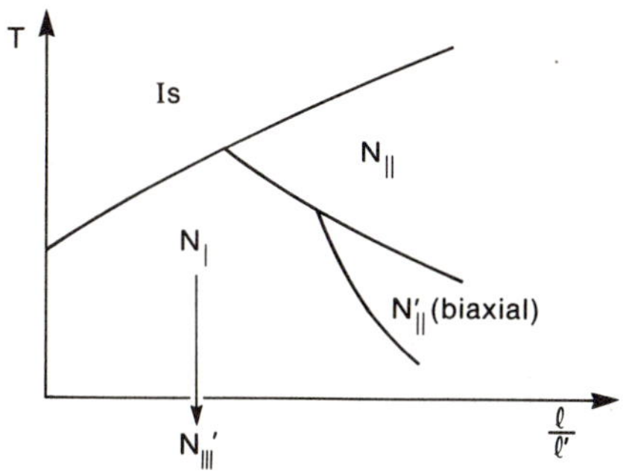


Fig. 10. Phase lines among the different nematic phases as a function of temperature and ratio l/l'; this ratio increases with decreasing flexibility of the backbone at constant interval between mesogens or with decreasing interval and constant flexibility (Ref. 5).

We discuss successively the role of the degree of polymerization, of the polydispersity and chemical heterogeneity, of the amorphous character of a polymer chain, of the flexibility of the backbone and of cross-linking.

### *3.1.1. Influence of the degree of polymerization $DP_n$*

A typical evolution of the clearing temperature with $DP_n$ is shown in the Fig. 11. $T_{MI}$ tends to an asymptotic value as the molar mass $\overline{Mw}$ increases. $T_m$, the melting temperature, and $T_g$, the glassy transition temperature, follow a similar evolution. Typically the value at plateau is about $10^4$ for linear main-chain LCP[11] and side-chain LCPs as well [12,13,14] corresponding to 10 to 20 repeat units. It must be noted that different mesophases can be stabilized differently by increasing $DP_n$. For instance, a N phase will transform gradually in a $S_C$ one as Mw increases in a linear LCP[15] or a N* phase will show up above a $S_A$ for a given degree of polymerization in a side-chain LCP[14]. This effect has been compared to the effect of pressure in a LMM[14]. A schematic example of such a behavior is given in Fig. 11. Systematic investigations in this domain are currently reported[16,17,18].

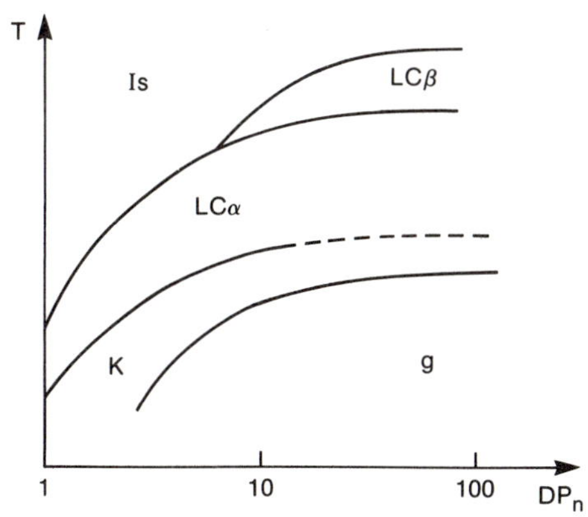


Fig. 11. Evolution of the transition temperatures with $DP_n$; note that a new polymorphism can result from a change in $DP_n$.

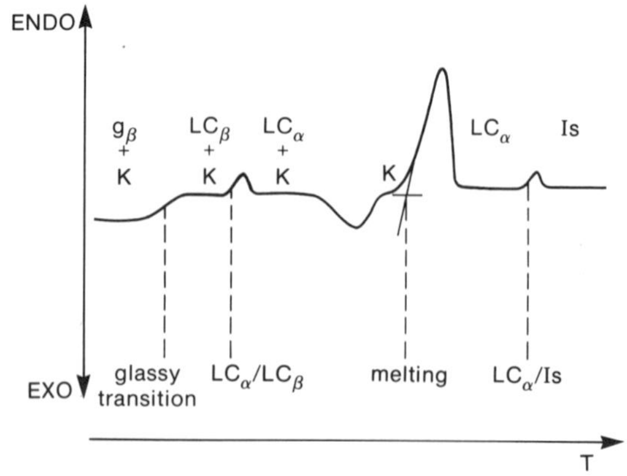


Fig. 12. Example of the thermal events likely to show up in the DSC recording upon heating an untreated semi-crystalline LCP with two mesophases $LC_\alpha$ (enantiotropic) and $LC_\beta$(monotropic).

### *3. 1. 2. Influence of the heterogeneity of a LCP material*

The very large biphasic domain observed through clarification is often a nightmare in studying LCPs since it makes the interpretation of physical measurements at the transition difficult. Two specific kinds of heterogeneity can account for this behavior: (i) dispersion of $\overline{Mw}$ (i.e. polydispersity) and (ii) chemical heterogeneity along the chain. As discussed previously, the clarification temperature is dependent on $\overline{Mw}$. So, different fractions in a same sample will become anisotropic at different temperatures. According to the shape of the curve in Fig. 11, it is quite understandable that, given a molar mass distribution, the biphasic gap is larger for lower $\overline{Mw}$s. This has been actually confirmed experimentally: in a biphasic equilibrium the isotropic and the nematic phases have different $\overline{Mw}$ distributions with higher masses in the anisotropic phase[19, 20].

The consequences of chemical heterogeneity are examplified by the following characteristics of chemically ordered and chemically disordered LCPs[21]:

$$-\{\text{-O-}\phi\text{-OOC-}\phi\text{-OOC-(CH}_2)_5\text{COO-}\phi\text{-COO-}\phi\text{-OOC-(CH}_2)_5\text{-CO-}\}_x\text{-}$$

( N + I ) range = 5 °C. $T_{NI}$ = 280 °C.

$$-\{\text{-CO-(CH}_2)_5\text{-COO-}\phi\text{-O-}\}_m\text{-CO(CH}_2)_5\text{-CO-}\{\text{-O-}\phi\text{-CO-}\}_n\text{-O-}\phi\text{-O-}\{\text{-CO-}\phi\text{-O-}\}_p\text{-}$$

(The three types of basic units m, n, p, are randomly connected.)

( N + I ) range = 119 °C. $T_{NI}$ = 388 °C.

Note the higher temperature of clarification of the " last droplet " in the case of the disordered system. The term polyflexibility is proposed to describe this structural heterogeneity in linear LCPs[21] since flexibility is likely to vary along the chain depending on the sequential stucture. One can thus expect copolymers to be especially sensitive to a non regular ...A-B-A-B... sequence. For example, the broadening of the ( S + I ) domain in side-chain copolysiloxanes[13] which occurs as the average number, y, of non mesogenic dimethylsiloxanes

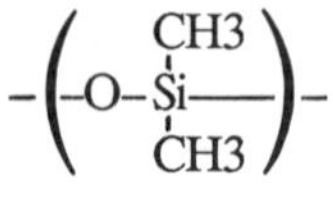

units separating two mesogenic groups is increased may originate from variations around this mean value from one segment to the other. The end groups participate also in chemical heterogeneity especially in the case of shorter chains.

### *3.1.3. Glassy transition in LCPs*

Surprisingly, this subject has deseNed little interest so far. The amorphous character of some LCPs has been mainly recognized as a way to tailor anisotropic glasses with supposedly future applications. For the experimentalist trained to the study of LMMs, however, the more or less amorphous character of a LCP introduces new ways to conduct experiments. To have a full understanding of the thermal behavior of a LCP, different runs are required with either annealed or at the opposite quenched samples. An example of some events which can be encountered on one DSC recording for a semi-crystalline LCP are shown in Fig. 12. The choice of the suitable conditions for X-ray experiments (which are performed only on powder samples in most cases owing to the diff1culty to prepare sizable monodomains) requires similar careful attention. As for the specific results reported in the literature one can mention the possibility of multiple glassy transitions in side-chain LCPs which can be assigned respectively to the mesogenic group at high temperature and to the backbone at low temperature[13] and the report of an opposite odd-even variation of Tg whith respect to the clarif1cation temperature in linear polyesters[22].

### *3. 1. 4. Flexibility of the backbone*

The influence of the flexibility of the backbone on the phase behavior can be analyzed in main-chain LCPs varying the length of the spacer or changing its nature ( i.e. using -CH2-CH2-O- segments instead of all methylenic chains[23] ). In all cases a higher flexibility means lower transition temperatures. With sidechain LCPs backbones of different chemical structures can be used such as:

PolyMethAcrylates :
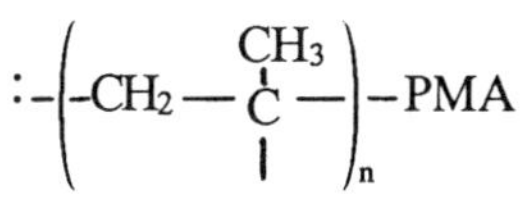


PolyAcrylates:
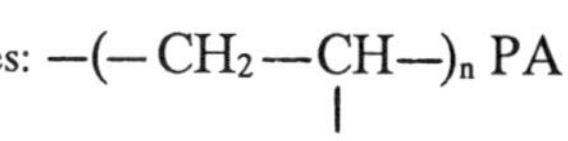


PolyMethylSiloxanes:
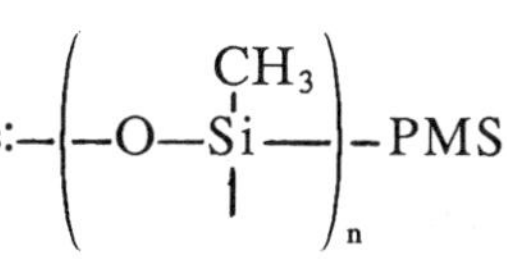


PolyPhosphAzenes:
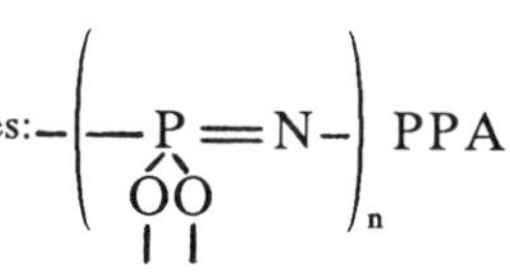


The classification is usually made considering Tg of the bare backbone and the above backbones rank in the following order: PMS > PPA > PA > PMA from the more flexible to the less flexible. However, the temperatures of transition do not always follow this sequence as shown in Table 2.

Connections between flexibility and steric hindrance along the backbone seem more relevant and are confirmed by the phase behavior of LCPs with partially substituted backbones (see the paragraph 3.4.). At last, the nature of the backbone seems to affect in a more consistent way the crystallinity with a higher tendency to crystallize for more flexible backbone[27].

Table 2. Temperatures of transition and variations of the backbone.

| Backbone | side-group | polymorphism | Ref. |
|---|---|---|---|
| PMS | $-(CH_2)_6-O-\phi-\phi-CN$ | S 166 I | 24 |
| PA | $-COO-(CH_2)_6-O-\phi-\phi-CN$ | S 125 I | 24 |
| PMA | id | S 100 I | 24 |
| PMS | $-(CH_2)_4-COO-\phi-OOC-\phi-OCH_3$ | S 124 N 128 I | 25 |
| PMS | $-(CH_2)_4-COO-\phi-OOC-\phi-OCH_3$ | N 125 I | 25 |
| PA | $-COO-(CH_2)_4-COO-\phi-OOC-\phi-OCH_3$ | N 165 I | 26 |
| PMA | id | N 150 I | 26 |
| $-(-CH_2-C-)-_n$ | id | N 155 I | 26 |
| $(-CH-C-)-_n$ | id | N 129 I | 26 |
| $-(-CH_2-CH(OCH_3)-CH-CH-)-_n$ | id | N 200 I | 26 |

### *3. 1. 5. Cross-linking*

Different types of cross-linkings can be imagined: the bifunctional agent can be either mesogen like or non mesogen. In this latter case it can be either stiff or flexible. The LC properties of a side-chain LCP will disappear above 10-15% of linked sites as a result of a simultaneous slight decline in $T_{MI}$ and strong increase of $T_g$. Linear LCPs seem more cross-linkable since LC phases are still observed for proportions in linking agent as high as 20%[28, 29]. Moreover a modification of the nature of the mesophase from smectic to nematic may occur at high degree of cross-linking[30]. A sharper reversible transition to the isotropic state characterizes also the behavior of elastomers. It must be noted that a low ordering into a state similar to the paranematic phase induced in LMMs by the application of a magnetic or electric field[31], is possible under the application of a mechanical strain as a result of the coupling between the network and the orientational order[32, 33, 34]. Consequently a critical point may exist with increasing stress.

## 3.2. Modifications in Chain Structure

Most kinds of mesophases described in LMMs have been already observed in LCPs owing to a huge effort in organic synthesis and recent reviews[35, 36] analyze the relationships between chemical constitution, molecular structure and LC properties of polymers. So, in this Section we would like again to sort only some special structures connected to the polymer character of the compound and which depart from the usual views of a nematic phase in which the director is parallel to the long axis of the mesogen and of a smectic phase in which the layer spacing is readily related to its longitudinal dimension.

### *3.2.1. Linear LCPs*

The polymorphism of linear LCPs seems deceivingly simple owing to their high crystalline character which most often rules out the observation of other mesophases than nematics. The specific nature of these nematics will be detailed further. However some main-chain LCPs present mesophases with very unusual structural features. Classical hexagonal columnar phases corresponding to the piling of disc-shaped mesogenic groups have been observed as expected in LCPs[37, 38]. But the structure of poly(di-n-alkylysilanes),

$$-\left(-\begin{array}{c} R_1 \\ | \\ Si \\ | \\ R_2 \end{array}-\right)-$$

(R1, R2 = $C_mH_{2m+1}$) is quite different (note that the repeat unit is not mesogenic at all). Here the hexagonal lattice is defined by the elongated polymer chains ( with no correlation in the position of the repeat units of adjacent chains ). A high degree of conformational order (all-gauche helix ) characterizes this state[39, 40]. In this regard, the following semi-flexible linear LC copolymer which looks like more conventional, also exhibits unexpectedly an hexagonal array of polymer chain columns[41]

$$-\{-O-\phi-(CH_2)_2-\phi-O\,(CH_2)_8\}_x-\{-O-\phi-(CH_2)_2-\phi-O\,(CH_2)_{12}-\}_y-$$

Note the "incommensurability" of the flexible parts; the "commensurable" homopolymers have only an ordered smectic B phase. Note also that the insert of a methyl group on one of the phenyls readily transforms the mesomorphism to a simple I-N. Rigid-rod polyesters and polyamides with lateral aliphatic and non-mesogenic chains (Fig. 13) form a non conventional layered phase in which the axes of the rigid rods are parallel to the smectic planes. The extended lateral flexible chains fill the interlayer spacing[42]. Other remarkable structures have been described in parent compounds[43]. The symmetries of the mesophases in these compounds correspond to the various manners to stack "segmented boards" in a more or less disordered way (Fig. 14b-d). These structures show the ability for the molecular biaxiality of these LCPs to be reflected in their mesophases. In LMMs such lattices are not observed owing to cylindrical symmetry (rod or disc) resulting from allowed rotation around a molecular axis. Consequently a new genus has been created: these sanidic mesophases are noted $\Sigma$. These systems display also a nematic phase which is reminiscent of the $N_{IIb}$ phase in Section 2.

### *3.2.2. Side-chain LCPs*

Their polymorphism appear classical at first (N, $S_A$, $S_C$). However, many differences characterize their phase behavior compared to LMMs. At this point we need now to differentiate two subclasses of LCPs with rod-like side-groups. LCPs with side-end fixed LC groups are strong smectogens as a result of their comb structure (seLCP, Fig. 15). In contrast side-on fixed groups lead to nematic LCPs (soLCPs, Fig. 15). In this latter class the existence of nematics with biaxial character has been claimed[44] (and will be discussed later).

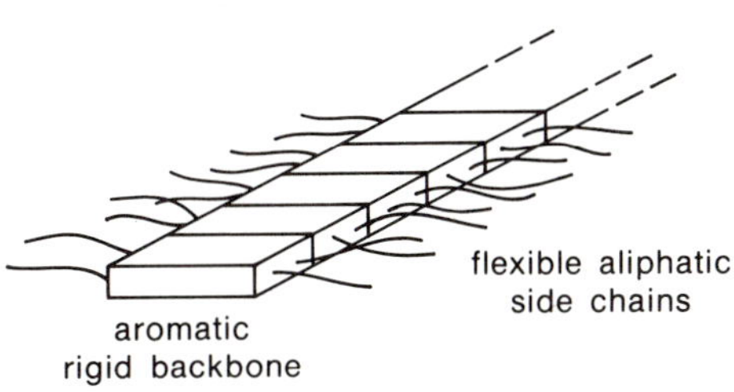


Fig. 13. Molecular architecture of LCPs likely to present board-like structures.

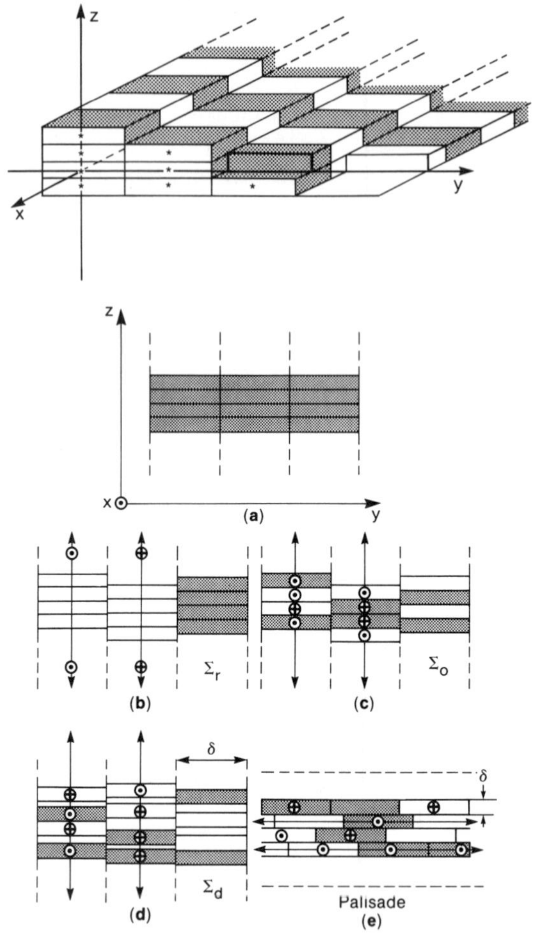


Fig. 14 Sanidic structures. Top: ordered stacking of macromolecular segmented boards. a): Cross section of the previous structure. b) c) and d): Sanidic phases described to date (Ref. 43); double arrows and connected pointing vectors mean that the whole stacks are allowed to slide up and down or back and forth, whilst individual pointing vectors indicate motions of single boards. e): The "palisade" structure (Ref. 49) can be also classified as sanidic. Note that structures d) and e) are smectic-like phases (δ layer spacing).

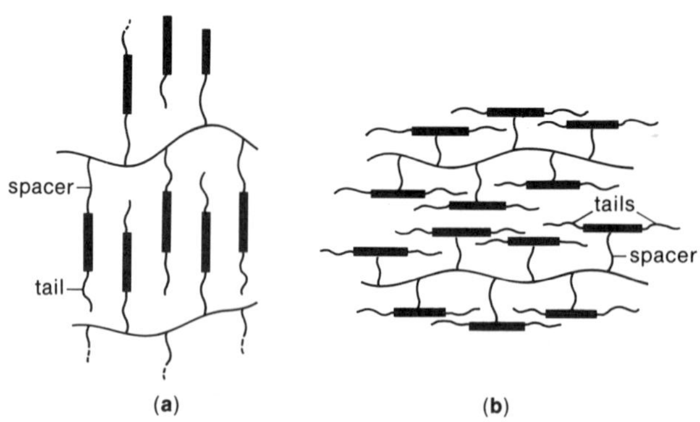


Fig. 15. Side-chain LCPs with: a) side-end (seLCP) and b) side-on fixed mesogens (soLCP).

The distinctive character of the smectic phases of seLCPs is the incommensurability of the layer spacing with the side-group: a variety of interdigitated smectic structures have been proposed to match both lengths. In LMMs this behavior is unusual and specific of unsymmetric higly polar mesogens[45]. In this regard there is a constitutive disymmetry in seLCPs as a result of the differentiation between spacer and tail. Therefore the variations of the polymorphism occuring as a function of the spacer and tail lengths differ (Table 3).

Table 3. Comparison of phase sequences in LCPs and LMMs.

| Tipically LC polymorphism | | | |
|---|---|---|---|
| in non polar seLCOs | | | in non polar LMMs |
| PMS | | PMA | |
| increasing tail length | increasing spacer length | increasing spacer length | increasing aliphatic tail length |
| | $S_A$ - I | $S_A$ - I | $S_B$ - N - I |
| $S_A$ - (N) -I | $S_A$ - N - I | $S_C$ - I | $S_C$ - $S_A$ - I |
| $S_B$ - $S_A$ - I | $S_A$ - I | $S_A$ - N - I | $S_C$ - $S_A$ - N - I |
| | $S_C$-$S_A$ - I | $S_C$-$S_A$ - I | ordered S - $S_A$ - I |

In addition these phase sequences are not as regular as they are in homologous series of LMMs. An example is given by non-polar PMA-LCPs: $S_{A2}$ - I, $S_{C2}$ - I, $S_{C1}$-I, $S_{A1}$ - I, $S_{A1}$ - N - I transitions are observed with increasing spacer length[45] (indexes$_2$ and$_1$ stand for layers with respectively bimolecular and monomolecular structures). Other unusual phase sequences, including nematic reentrance ($N_R$ - $S_A$ - N - I), are reported[46, 47, 48] in seLCPs with polar mesogenic groups. The $N_R$ phase in LCPs presents novel characteristics which are discussed in the next Section. At last the Fig. 16 represents the uncommon board-like structure proposed for a seLCP in order to account for the lattice parameters of its smectic phase[49] which could be classified as sanidic (Fig. 14e).

Compared with seLCPs the simple N-I "polymorphism" of soLCPs looks merely interesting. But in this case the special features of the nematic order deserves more attention (see next section). LC copolymers combining in the same material the characteristics of main-chain and side-chain LCPs present also an interesting phase behavior[50] with modulated smectic layers.

## 3.3. Coupling of Chain Conformation with Liquid Crystalline Order

Information on this coupling can be obtained from different sources. Transition entropies ($\Delta S$), diamagnetic anisotropy ($\Delta\chi$) measurements, NMR and DMR dipolar and

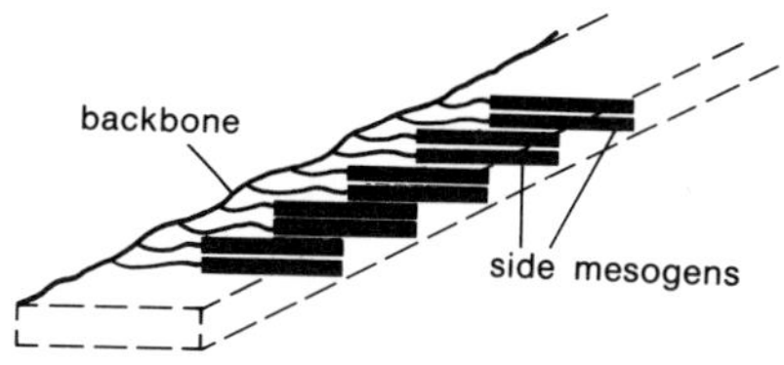


Fig. 16. Chain organization in palisade smectic (Ref. 49, see Fig. 14).

quadrupolar splittings are all well suited to provide an evaluation of the orientational order. A more direct insight to the chain conformation is given by Small Angle Neutron Scattering (SANS) and Small Angle X-ray Scattering (SAXS) which can be applied both in the nematic and in more ordered mesophases provided that oriented samples can be prepared.

The Fig. 17 is the result of an extensive compilation of $\Delta S_{NI}$ measurements reported in the literature. They are sorted as follows (for all terms used describing the classification of LMMs see Ref. 51):

(i) LMMs of all kinds (rod-like, disc-like, ...) on a wide range of molar masses (250 to 3000);
(ii) Twin compounds separated in two subclasses: (iia) siameses and (iib) dimer model compounds for main-chain LCPs;
(iii) Linear semiflexible LCPs with even number (n = 10) of spacer units;
(iv) Side-chain LCPs (both se and soLCPs) and
(v) Combined main-chain side-chain LCPs.

Average values are: LMMs: 0.006 $Jg^{-1}K^{-1}$; Dimers: 0.030 $Jg^{-1}K^{-1}$; and Linear LCPs: 0.029 $Jg^{-1}K^{-1}$.

The comment of these results is straightforward: the N-I entropy of transition is significantly different from LMMs for both the linear LCPs and their model dimers. It is striking to remark that the average value is the same for the last two classes suggesting comparable degrees of order. This can be assigned to some conformational order of the flexible spacer but the corresponding energy difference is far from what should be added for all trans spacers. All other mesogens (LMMs and side-chain LCPs) have comparable values. Despite the few data points, the combined LCPs seem to nicely classify either as semi-flexible main-chain LCPs or as side-chain LCPs depending on the combination. Keep also in mind for further discussion that the soLCPs have classical value of ΔS. The large differences recorded in ΔS between odd and even members in series of dimers or linear LCPs is another clear evidence of the conformational order of the flexible parts in the nematic phase[52]. The higher amplitudes of the odd-even variations of $T_{NI}$ in linear LCPs and dimers are also indicative of the different role of the spacers compared to the flexible ends in LMMs (Fig. 18). The even-odd effect can be more dramatic with an alternation of phases in the polyesters in Table 4.

Table 4. Odd-even alternation of phases in : $-\{-CO-\phi-OCO-\phi-COO-\phi-COO(CH2)_n O-\}_x-$ (Ref. 55)

| n | S | N | T in °C | I |
|---|---|---|---|---|
| 2 | | * | 365 | * |
| 3 | * | | 315 | * |
| 4 | | * | 345 | * |
| 5 | * | | 267 | * |
| 6 | | * | 290 | * |
| 7 | * | | 253 | * |
| 8 | * | | 220 | * |
| 9 | * | | 233 | * |
| 10 | * | | 266 | * |
| 11 | * | | 245 | * |

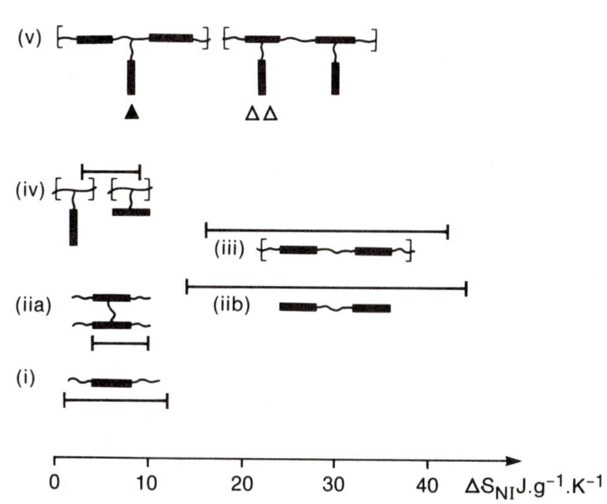


Fig. 17. N-I transition entropy (in $Jg^{-1}K^{-1}$) in different classes of mesogens: (i) LMMs; (ii) Twin compounds: (iia) Siameses and (iib) LCP dimers with even number ( n = 10 ) of spacer units; (iii) Linear semiflexible LCPs with even number ( n = 10 ) of spacer units; (iv) Side-chain LCPs; and (v) Combined main-chain side-chain LCPs.

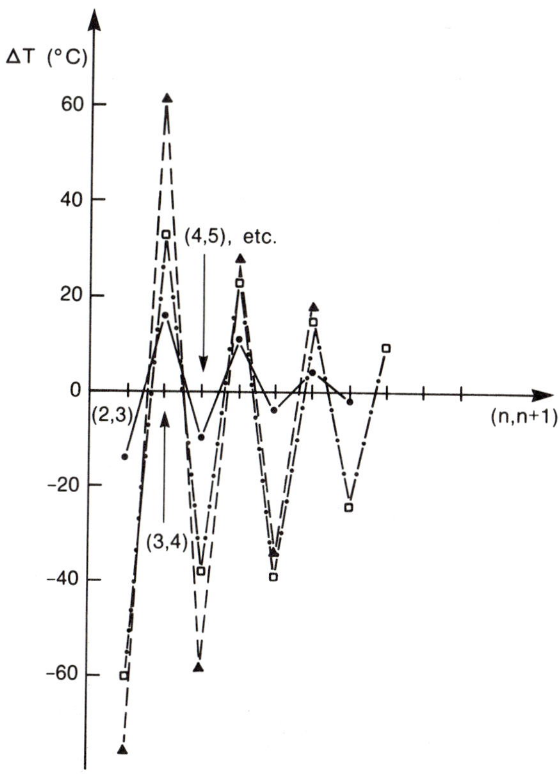


Fig. 18. Compared amplitudes of the odd-even variations of $T_{NI}$: ● LMMs: $C_nH_{2n+1}O - \phi - COO - \phi - OC_nH_{2n+1}$ (Ref. 53); ▲ Dimers: $CH_3O - \phi - N = N - \phi - O(CH_2)_n O - \phi - N = N - \phi - OCH_3$ (Ref. 53); and, □ LCPs: $\{CO - \phi - O(CH_2)_n O - \phi - COO - \phi - O(CH_2)_{10}O - \phi - O -\}_x$ (Ref. 54). ΔT is the difference between the clarification temperatures of two successive homologues.

The diamagnetic anisotropy change ($\Delta\chi$) at the N-I transition shows comparable differences between even and odd LCPs on one hand and between LCPs and model monomers on the other hand but also between LCPs and dimers[56, 57, 58]. In addition some unusual features in the thermal evolution of $\Delta\chi$ are sometimes observed which are discussed below. These results are consistent with the evaluation of the orientational order parameters for the overall material and for different segments in the spacer[59] reported from magnetic resonance techniques which all conclude to a higher value in the LCPs and model dimers with even number of $CH_2$ units. Though highly extended, the flexible parts have a variety of conformers to chose from and the more likely appear to be those with every other $CH_2$-$CH_2$ bond in trans conformation and each bond in between being free to adopt other conformations[60]. The large anisotropy of the chains has been also demonstrated through SANS experiments ($<R_z^2>/<R_x^2>$ 20!)[61], All in all, these results are consistent with the picture of the $N_2$ phase in the Fig. 6 . In the nematic phase of soLCPs the anisotropy of the backbone is also prolate and the ratio $<R_z^2>/<R_x^2>$ which tops 14 at about [$T_{NI}$- 20], suggests a highly extended backbone along the direction of the nematic director as in linear LCPs. The comparison can be pursued further since the uncommon thermal evolution of $\Delta\chi$ noticed above in linear LCPs is similarly observed in some soLCPs and has been now systematically described[62, 63]. A typical example of this phenomenon which occurs in the vicinity of the isotropic phase is shown in Fig. 19: the curve $\Delta\chi = f(T)$ presents a minimum. This seems to challenge the " universal " evolution of the order parameter in the nematic phase described in the Maier and Saupe's theory, i.e. a regular increase of S as T decreases actually as it does in all LMMs. Up to now this peculiar transient feature which seems not connected to a real phase transition is thus specific of these two classes of LCPs. Whether it reflects a non negligible biaxiality (sustained by conoscopic interference patterns on soLCPs[43] and $S_C$ fluctuations shown by X-rays in both soLCPs and linear LCPs[22, 62]) or a two steps ordering involving first the mesogenic parts and then the flexible moieties (respectively spacers for linear LCPs and backbone for soLCPs) as proposed for the $N_{III'}$-$N_{III}$ phase change in Fig. 9 and for the $N_1$-$N_2$ transition in Fig. 6, is still purely hypothetical. To conclude this comparison it is worth noting that $\Delta S_{NI}$ in soLCPs is closer to $\Delta S_{NI}$ in LMMs than to $\Delta S_{NI}$ in linear LCPs (see Fig. 17) which suggests that the orientational order parameter and/or the conformational order are/is not especially high despite the extension of the backbone.

The situation appears reversed in seLCPs. The anisotropy of the chain in the N phase is very low and consistent with the model $N_I$ ( Fig. 8a ), i.e. ($<R_z^2>/<R_x^2>$ )< 1; the backbone adopts a slightly oblate shape. As the temperature decreases the anisotropy increases and becomes important in the $S_A$ phase ($<R_z^2>/<R_x^2> \approx 0.4$) as a result of the higher confinement of the backbone between the mesogenic layers[64, 65]. However if a $N_R$ occurs a reversal of the chain anisotropy is surprisingly observed ($<R_z^2>/<R_x^2> \approx 4$), i.e. this $N_R$ phase is of type $N_{III}$ (Fig. 8)[66]. Here again $S_C$ fluctuations are detected by X-rays. NMR and DMR have provided evidence for a characteristic conformational order of the spacer while the free tail behaves as observed for alkyl chains in LMMs[67, 68].

## 3.4. Copolymers

The original behaviors generated by the various combinations tested in LC coPolymers ( LCcoPs ) are also worth describing. Two constitutions of LCcoPs can be considered: (i) these which associate two different mesogenic units and (ii) these which associate a mesogenic and a non-mesogenic unit.

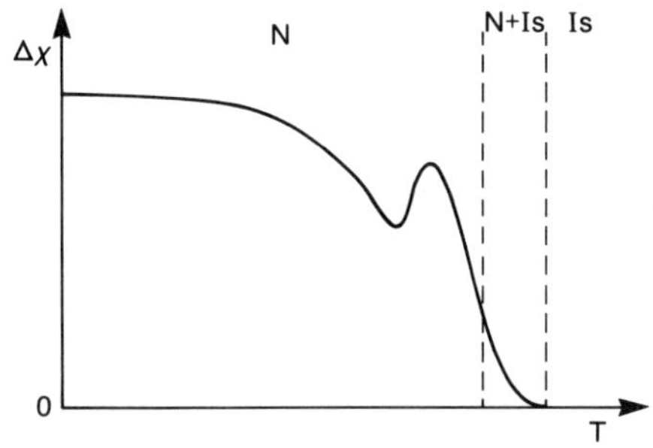


Fig. 19. Anomaly in the $\Delta\chi = f(T)$ curve observed at the N-I transition in some linear LCPs and soLCPs.

(i) Generally speaking, a random copolymer combining two counits each corresponding to an homopolymer with LC properties behaves as expected for a binary mixture in LMMs. This is valid for both main-chain and side-chain LCPs. Combinations of homologous counits show a nearly " ideal " phase behavior according to the fit of the phase lines by the Schröder-Van Laar equations[69, 70]. In contrast interesting " non ideal " phase behaviors can be created when chemically different side-groups are associated (i.e.: polar and non polar[69]). In this regard the combinations of so and se fixed side-groups currently investigated are likely to exhibit highly non-ideal properties[71].

(ii) In linear rigid LCcoPs, the use of non mesogenic, non linear counits has mainly served to decrease the high melting and isotropization temperatures by the introduction of kinks and chemical heterogeneity. Adding non mesogenic units in a side chain LCPs results in a decrease of the number of mesogenic units along the backbone (Fig. 20) which

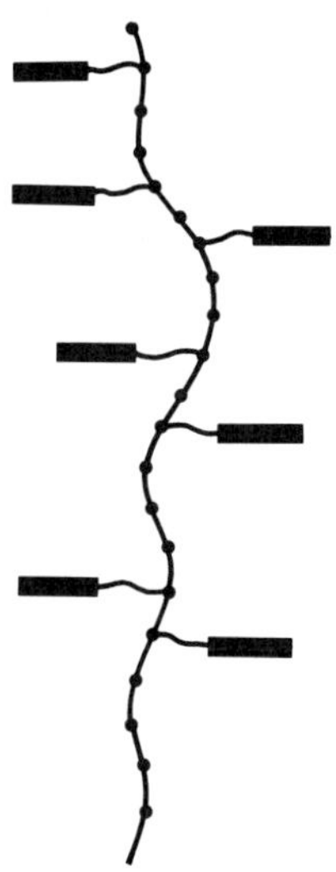

Fig. 20. "Diluted" side-chain LCcoP. Note the possible dispersion in the interval between consecutive mesogens around the average composition (≈1/3 occupied sites in this case).

allows to study the consequences of sort of "intramolecular dilution" upon the LC properties[72]. More precisely, one forces together in different proportions two moieties likely to behave in highly opposite ways if they could separate. A flexible backbone tends to an isotropic coil while the mesogens comply with the director orientation.

A second possible incompatibility is the chemical difference between both moieties (i.e.: a PolyDiMethylSiloxane would readily phase separate from a common LMM). As expected the mesophase, usually smectic, is destabilized upon reduction of the proportion of mesogenic groups. However, it is possible in some cases to preserve a smectic order at temperatures as high as 100 °C down to 1 over 10 site occupied with mesogenic groups (corresponding to a volume fraction in mesogen as low as 0.45)[73]. In these materials the number of mesogenic units per macromolecule may average only 3 with a degree of polymerization of about 30 ! A process of microphase separation seems responsible for the stability of these smectic phases[13, 74, 75]. X-rays measurement of the layer swelling as a function of the proportion of non mesogenic backbone suggests a surprisingly high degree of conformational order of the sections which separate two consecutive side-groups[76] (Fig. 21). This microsegregation is also likely to be important in more ordered phases which characterize the phase behavior of diluted LCcoPs at lower temperature. Either layered or hexagonal, they are reminiscent of the structures of di or triblockcopolymers[73, 77].

## 4. Phase Diagrams

As far as thermotropic systems are concerned, one can consider three types of binary systems involving LCPs: blend of two LCPs, solution of a LCP in a non liquid crystalline solvent and mixture of a LCP and a LMM. First, we emphazise that in such mixtures the presence of a macromolecule and the difference in size between the two components necessarily makes the solution non ideal. As a consequence gaps of miscibility are likely owing to a severe reduction of the entropy of mixing. The coefficient $\chi$ in the excess enthalpy of mixing defined in classical works by Flory and Huggins[78] actually controls the miscibility of a polymer:

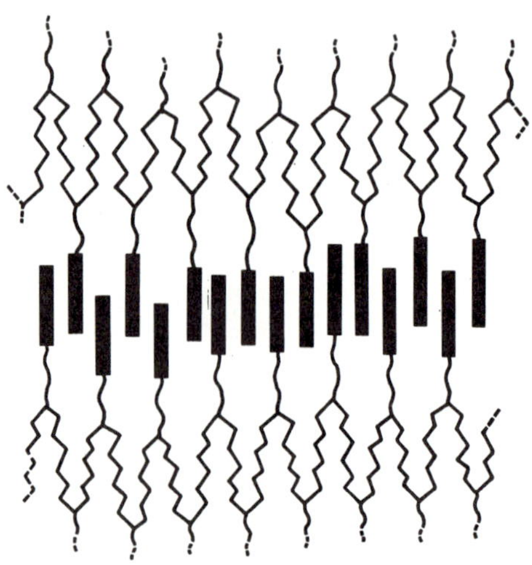

Fig. 21. Model of microphase separated smectic layer in a diluted LCcoP.

$$\chi(T) = \frac{1}{2} - \Psi \left\{ \frac{\Theta}{T} \right\}$$

In this regard, the experimental study of the first type of mixture is very difficult in practice because two polymers are hardly miscible. Theoretical models describe the phase lines of the isotropic-anisotropic phase equilibrium which characterizes the second case as a function of axial ratio and composition[79]. Indeed these solutions show a remarkable stability of the nematic phase even at large dilutions, but so far the agreement with theory is not satisfactory[80]. More interesting in view of the analysis of the differences between a LCP and a LMM is the third case which allows a direct comparison of the compatibility of their mesophases. A typical phase diagram for a LCP-LMM binary system[81, 82, 83, 84] is shown in Fig. 22. One notes immediately the gap of miscibility, very unusual in mixtures of LMMs, here induced by the polymer component as explained above. A comment needs first to define $T_c$ and $\phi_c$, respectively the critical temperature and the critical concentration at the consolute point. $\phi_c$($\phi$ is the volume fraction of polymer in the mixture) depends mainly on the degree of polymerization x: $\phi_c \approx 1/(1+x^{1/2})$. This explains the important shift of the two phase domain towards the LMM side since x is usually 100 or over $T_c$ depends on $\Theta$, $\Psi$ and the degree of polymerization x:

$$\frac{1}{T_c} = \frac{1}{\Theta}\{1 + \frac{1}{\Psi}\{\frac{1}{\sqrt{a}} + \frac{1}{2x}\}\} \tag{1}$$

$T_c$ is thus a convenient parameter to analyze the behavior of such a solution (i) at constant x which gives the influence of changing the nature of one of the component (which plays simultaneously on $\Theta$ and $\Psi$) and (ii) at constant $\Theta$ and $\Psi$ (i.e. for a given LCP-LMM pair) which gives the influence of the degree of polymerization. A simple contact experiment provides[84] a measure of $T_c$. The following summarizes then the results obtained so far from the analysis of the modifications of $T_c$ as a function of the parameters defined above[85, 86]. As

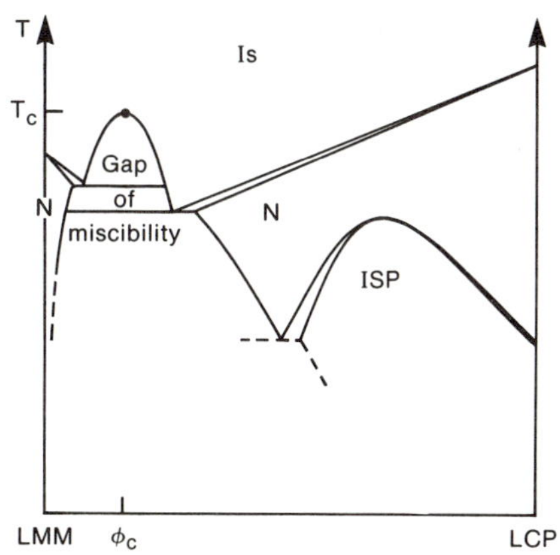


Fig. 22. Phase diagram for LCP-LMM binary mixtures. The gap of miscibility is characterized by its critical coordinates $T_c$ and $\phi_c$. In contrast a homogeneous smectic phase can be promoted in the same mixture at different compositions (ISP).

in usual polymer-solvent systems, the higher the degree of polymerization of the LCP, the lower the miscibility in a given LMM. Since, as it could be expected, the phase separation may occur either in the isotropic state as in usual polymer-solvent systems, or in the anisotropic state, it is possible to evaluate $\chi$ in one and the other phase from the measurements of $T_c$ as a function of x (Eq. 1). This means a way to test the coupling of the miscibility with the order parameter as theoretically suggested[87, 88]:

$$\chi^N(T) = \chi^I(T) + \tau S^2(T)\, T^{-1}$$

In side-chain LCP-LMM pairs $\tau$ is positive, i.e. the miscibility is less in the N phase than in the I phase. In contrast, in linear LCP-LMM systems $\tau$ seems negative which means a better miscibility in the N phase[89]! Generally speaking, a linear LCP seems to solubilize more easily than a side-chain LCP based also on the qualitative observation that few LMMs phase separate with the former while a variety phase separate with the latter. Following the above arguments on the respective roles of the entropy and enthalpy of mixing in polymer-solvent systems, a significant chemical difference, i.e. a high difference in polarity, is not a prerequisite to observe phase separation. Nevertheless the nature of each component modify the solvent-solute interaction. In side-chain LCP-LMM pairs it has been shown that miscibility is improved by less flexibility in the LMM (long stiff core and short aliphatic tails) and in contrast a higher overall flexibility of the LCP (more flexible backbone, longer spacer or tail(s), larger interval between mesogenic groups along the backbone). In linear LCP-LMM pairs the miscibility is improved by longer spacers as well. A greater rigidity of the backbone in soLCPs due to coupling (see the paragraph 3.3.) could also explain their lower solubility compared to seLCPs[84].

Other remarkable non-ideal behaviors are azeotropic features such as enhanced or induced smectic phases (ISPs) (Fig. 22). ISPs in LMM binary mixtures are observed mostly in mixtures of compounds with well defined chemical characteristics and the highest stability is usually reached for a 1:1 composition. These facts have led to interpret this phenomenon as the result of a weak donor-acceptor complexation. A minimum in layer spacing is also connected to the maximum stability of the smectic phase[90]. By comparison ISPs are more frequent in side-chain LCPs-LMMs mixtures[91]. As for phase separation no obvious difference in chemical nature is necessary and the stability of the ISP is similarly related to the flexibility of the components. In full contrast with the previous systems the maximum stability of the ISP corresponds to a maximum in layer spacing[44].

## 5. Conclusion and Perspectives

As shown above, a polymeric backbone whether highly rigid or at the opposite as flexible as possible is not only able to preserve most of the properties of its mesogenic moieties if any but it takes a most interesting part in distinctive liquid crystalline behaviors. In this regard, couplings of the chain configuration with LC order still deserve much attention. The study of new materials can help such as main chain LCPs made up with less rigid and less mesogenic repeat units (i.e.: - $\phi$ - $CH_2CH_2$ - $\phi$ -, Ref. 92) or model compounds such as dimers which are an interesting link between LCPs and LMMs[55, 56, 57, 93, 94, 95, 96]. The problem of $N_x$-$N_y$ transitions recently tackled in different works[61, 62, 97, 98] is still pending. The use of copolymers as pseudo binary phase diagrams seems a convenient way to investigate the interactions between the different parts of a

LCP. In particular it seems worth pursuing the study of the stuctural consequences in "mixing" unlike parts. A number of frustrated combinations can be imagined - repeat units with different length or nature, incompatible backbone and LC side-groups, antagonistic chain structures - among which a few examples have been described yet (for example combined LCPs, diluted systems, se versus so fixation). Phase transitions in LC elastomers is also an open field. At last, to temperate our forewords, the study of systems of strictly rigid rods could revive either with systematic comparison of the steric and dipolar characteristics of new model compounds with theories[99] or using totally novel systems, for example inorganic materials such as ionic infinite clusters and their solutions[100].

ACKNOWLEDGEMENT. The author acknowledge his support by I.B.M.-U.S.A. as visiting scientist at the Almaden Research Center while preparing this Chapter.

## References

1. P.J. Flory and G. Ronca , Mol. Cryst. Iiq. Cryst. **54**:311 (1979).
2. P.A. Irvine, Da Cheng Wu and P.J. Flory , J. Chem. Soc.. Faraday Trans I **80**:1795 (1984).
3. A. Roviello and A. Sirigu, J. Polym. Sci.. Polvm. Lett. Ed. **13**:455 (1975).
4. H. Finkelmann, H. Ringsdorf and J.H. Wendorff, Makromol. Chem. **179**:273 (1978).
5. A.N. Semenov and A.R. Khokhlkov, Sov. Phys. Usp. **31**:988 (1988).
6. S.V. Vasilenko, A.R. Khokhlov and V.P. Shibaev, Polymer Sci. USSR, 26, (1984).
7. X.J. Wang and M. Warner, J. Phys. A **20**:713 (1987); W. Renz and M. Warner, Proc. of the Royal Soc. of London, series A **417**:213 (1988).
8. F. Auriemma, P. Corradini and M. Vacatello , J. Chem. Phys. **93**:8314 (1990).
9. F. Dowell, Mol. Cryst. Liq. Cryst. **157**:203 (1988).
10. W. Renz, Mol. Cryst. Liq. cryst. **155**:549 (1988).
11. A. Blumstein, S. Vilasagar, S. Ponrathnam, S.B. Clough, R.B. Blumstein and G. Maret, J. of Polym. Sci. **20**:877 (1982).
12. P. Shibaev and N.A. Plate, Pure & Appl. Chem. **57**:1589 (1985).
13. V. Percec and B. Hahn, Macromolecules **22**:1588 (1989).
14. H. Stevens, G. Rehage and H. Finkelmann , Macromolecules **17**:851 (1984).
15. R.S. Kumar, S.B. Clough and A. Blumstein, Mol. Cryst. Liq. Cryst. **157**:387 (1988).
16. V. Percec and M. Lee, Polymer Bulletin **25**:123 (1991).
17. V. Percec, M. Lee and H. Jonsson, J. of Polymer Sci. **29**:327 (1991).
18. V. Percec and M. Lee, Macromolecules **24**:1017 (1991).
19. R.B. Blumstein and A. Blumstein, Mol. Cryst. Liq. Cryst. **165**:361(1988).
20. J.F. D'Allest, P.P. Wu, A. Blumstein and R.B. Blumstein, Mol. Cryst. Liq. Crst. Letters **3**:103 (1986).
21. J.S. Moore and S.I. Stupp , Macromolecules **21**:1217-1234 (1988).
22. A. Blumstein, Polymer J. **17**:277 (1985).
23. W. Volksen, D.Y. Yoon and P. Cotts, Macromolecules **22**:3846 (1989).
24. V. P. Shibaev and N.A. Plate, Pure & Appl. Chem. **57**:1589 (1985).
25. P. Keller, Mol. Cryst. Liq. Cryst. **157**:193 (1988).
26. P. Keller, Mol. Cryst. Liq. Cryst. **155**:37 (1988).
27. V. Percec, D. Tomazos and R.A. Willingharn, Polymer Bulletin **22**:199 (1989).
28. R. Zentel and G. Reckert, Makromol. Chem. **187**:1915 (1986).
29. G. Mitchell, F. Davis and A. Ashman, Polymer **28**:637 (1987).
30. C. Degert, M. Mauzac, H. Richard and G. Sigaud, 13th ILLC, Vancouver (Canada), July 22-27 (1990).
31. E. Gramsbergen, L. Longa and W. de Jeu, Phys. Rep. **135**:195 (1986).
32. P.G. de Gennes, C.R. Acad. Sci. Paris **281B**:101(1975).
33. K.P. Gelling and M. Warner, Mol. Cryst. Liq. Cryst. **155**:539 (1988).
34. J. Schatzle, W. Kaufhold and H. Finkelmann, Makromol. chem. **190**:3269 (1989).
35. H. Ringsdorf, B. Schlarb and J. Venzmer, Angew. Chem. Int. Ed. Engl. **27**:113 (1988).
36. W. Brostow, Polymer, **31**:979 (1990).
37. W. Kreuder and H. Ringsdorf, Makromol. Chem. Rapid Commun. **4**:807 (1983).
38. W. Kreuder, H. Ringsdorf and P. Tschirner, Makromol. Chem. Rapid Commun. **6**:987 (1985).
39. P. Weber, D. Guillon, A. Skoulios and R.D. Miller, J. Phys. France **50**:793 (1989).
40. P. Weber, D. Guillon, A. Skoulios and R.D. Miller, Liquid Crystals **8**:825 (1990).
41. G. Ungar, J.L. Feijoo, V. Percec and R. Yourd, Macromolecules, in press.
42. M. Ballauff and G.F. Schmidt, Mol. Cryst. Liq. cryst. **147**:163 (1987).

43. M. Ebert, O. Herrmann-Schonherr, J.H. Wendorff, H. Ringsdorf and P. Tschirner, Liquid Crystals **7**:63 (1990).
44. F. Hessel, RP. Herr and H. Finkelmann, Makromol. Chem. **188**:1597 (1987).
45. F. Hardouin, G. Sigaud, P. Keller, H. Richard, H.T. Nguyen, M. Mauzac and M.F. Achard, Liquid Crystals **5**:463 (1989) and references therein.
46. T.I. Gubina, S.G. Kostromin, R.V. Talroze, V.P. Shibaev and N.A. Plate, Vysokol. Soedin. **B28**:394 (1986).
47. N. Lacoudre, A. Le Borgne, N. Spassky, J.P. Vairon, P. Le Barny, J.C. Dubois, S. Esselin, C. Friedrich and C. Noel, Mol. Cryst. Liq.Cryst. **155**:113 (1988).
48. V. Percec, private communication.
49. R. Duran, D. Guillon and A. Skoulios, Makromol. Chem. Rapid Commun. **8**:321 (1987).
50. B.W. Endres, M. Ebert, J.H. Wendorff, B. Reck and H. Ringsdorf, Liquid Crystals **7**:217 (1990).
51. D. Demus, Liquid Crystals **5**:75 (1989).
52. A. Blumstein and O. Thomas, Macromolecules **15**:1264 (1982).
53. D. Demus and H. Zachse, in:"Fluessige Kristalle in Tabellen", VEB Deutscher Verlag fur Grundstoffindustrie, Leipzig.
54. A.C. Griffin and S.J. Havens, J.of Polymer Science. Pol. Phys. Ed. **19**:951(1981).
55. R.W. Lenz, J. of Polymer Sci.. Polymer Symp. **72**:1 (1985); R.W. Lenz, Pure & Appl. Chem. **57**:1537 (1985).
56. G. Sigaud, D.Y. Yoon and A.C. Griffin, Macromolecules **16**:875 (1983).
57. P. Esnault, D. Galland, F. Volino and R.B. Blumstein, Macromolecules **22**:2734 (1989).
58. H. Furuya, T. Dries, K. Fuhrmann, A. Abe, M. Ballauff and E. Fischer, Macromolecules **23**:4122 (1990).
59. K. Kohlammer, K. Muller and G. Kothe, Liquid Crystals **5**:1525 (1989) and references therein.
60. D.Y. Yoon and S. Bruckner, Macromolecules **18**:651 (1985).
61. J.F. D'Allest, P. Maissa, A. ten Bosch, P. Sixou, A. Blumstein, R.B. Blumstein, J. Teixera and L. Noirez, Phys. Rev. Lett. **61**:2562 (1988).
62. P. Keller, F. Hardouin, M. Mauzac and M.F. Achard, Mol. Cryst. Liq. Cryst. **155**:171 (1988).
63. F. Hardouin, S. Mery, M.F. Achard, M. Mauzac, P. Davidson and P. Keller, Liquid Crystals **8**:565 (1990); S. Mery, PhD thesis, Universite de Bordeaux I, # 510, (1990).
64. F. Moussa, J.P. Cotton, F. Hardouin, P. Keller, M. Lambert, G. Pepy, M. Mauzac and H. Richard, J. Phys. France **48**:1079 (1987).
65. F. Hardouin, L. Noirez, P. Keller, M. Lambert, F. Moussa and G. Pepy, Mol. cryst. Liq. Cryst. **155**:389 (1988).
66. L. Noirez, P. Keller, P. Davidson, F. Hardouin and J.P. Cotton, J. Phys. France **49**:1993 (1988).
67. M. Mauzac, H. Richard and L. Latie, Macromolecules **23**:753 (1990).
68. H. Oulyadi, F. Laupretre, L. Monnerie, M. Mauzac, H. Richard and H. Gasparoux, Macromolecules **23**:1965 (1990).
69. M.F. Achard, M. Mauzac, H. Richard, G. Sigaud and F. Hardouin, Eur. Polym. J. **25**:593 (1989).
70. see e.g. V. Percec and Y. Tsuda, Polymer Bulletin **23**:225 (1990).
71. G.W. Gray, J.S. Hill and D. Lacey, Angew. Chem. Int. Ed. Engl. Adv. Mater. **28**:1120 (1989).
72. S. Diele, S. Oelsner, F. Kuschel, B. Hisgen, H. Ringsdorf and R. Zentel, Makromol. Chem. **188**:1993 (1987).
73. M.F. Achard, H.T. Nguyen, H. Richard, M. Mauzac and F. Hardouin, Liquid Crystals **8**:533 (1990).
74. S. Diele, S. Oelsner, F. Hisgen and H. Ringsdorf, Mol. Cryst. Liq. Cryst. **155**:399 (1988).
75. H.T. Nguyen, M.F. Achard, F. Hardouin, M. Mauzac, H. Richard and G. Sigaud, Liquid Crystals **7**:385 (1990).
76. M.F. Achard, H.T. Nguyen, M. Mauzac, F. Hardouin, G. Sigaud and H. Richard, 13th ILLC, Vancouver (Canada), July 2227 (1990).
77. T. Hashimoto, M. Shibayama, M. Fujimura and H. Kawai, in :"Block Copolymers", Dale J. Meier, Ed., Harwood Academic Publishers, pp. 63-107 (1979).
78. see e.g. "Polymer Solutions", Cassassa E.F., Berry C.G., in: "Comprehensive Polymer Science", vol. 2, pp 71-120, Allen Bevington, ed., Pergamon (1989).
79. M. Warner and P.J. Flory, J. of Chemical Phys. **73**:6327 (1980).
80. C. Viney, D.Y. Yoon, B. Reck and H. Ringsdorf, Macromolecules **22**:4088 (1989).
81. C. Weill, C. Casagrande, M. Veyssie and H. Finkelmann, J. Phys. France **47**:887 (1982).
82. H. Benthack-Thoms and H. Finkelmann, Makromol. Chem. **186**:1895 (1985).
83. C. Casagrande, M.A. Guedeau and M. Veyssie, Mol. Cryst. liq. Cryst. Lett. **4**:107 (1987).
84. G. Sigaud, M.F. Achard, F. Hardouin, M. Mauzac, H. Richard and H. Gasparoux, Macromolecules **20**:578 (1987).
85. G. Sigaud, M.F. Achard, F. Hardouin, C. Coulon, H. Richard and M. Mauzac, Macromolecules **24**:565(1990).
86. F. Hardouin, G. Sigaud and M.F. Achard, in :"Liquid Crystalline and Mesomorphic Polymers", L.Lam and V.P. Shibaev, Eds, Springer-Verlag, (1991).
87. F. Brochard, C.R. Hebd. Acad. Sci. **289B**:229 (1979).

88. F. Brochard, J. Jouffroy and P. Levinson, J. Phys. France **45**:1125 (1984).
89. M.F. Achard, private communication.
90. W. H. de Jeu, L. Longa and D. Demus, J. Chem. Phys. **84**:6410, (1986).
91. M.F. Achard, G. Sigaud, P. Keller and F. Hardouin, Makromol. Chem. **189**:2159 (1988).
92. V. Percec and R. Yourd, Macromolecules **22**:3229 (1989).
93. J.A. Buglione, A. Roviello and A. Sirigu, Mol. Cryst. Liq. Cryst. **106**:169 (1984).
94. V. Krone and H. Ringsdorf, Liquid Crystals **2**:411 (1987).
95. J.W. Emsley, N.J. Heaton, G.R. Luckhurst and G.N. Shilstone, Mol. Phys. **64**:377 (1988).
96. G.M. Janini, N.T. Tilfil and G.M. Muschik, Liquid Crystals **7**:545 (1990).
97. G. Ungar, J.L. Feijoo, A. Keller, R. Yourd and V. Percec, Macromolecules **23**:3411 (1990).
98. V. Krone, H. Ringsdorf, M. Ebert, H. Hirschmann and J.H. Wendorff, Liquid Crystals **9**:165 (1991).
99. C. Viney, R.J. Twieg, C.M. Dannels and M.Y. Chang, Mol. Cryst. Liq. Cryst. Letters **7**:147 (1990).
100. F.J. di Salvo, Science **247**:649 (1990).

Chapter 25

# MOLECULARLY NON-HOMOGENEOUS NEMATIC POLYMERS

X. J. WANG and M. WARNER

## 1. Introduction

Most main chain liquid crystalline polymers are synthesized with the mesogenic groups (rods) linked by flexible spacers, such as repeated methylene groups. The theories presented so far deal either with rigid rods system (Flory[1] and Onsager[2]), or model semiflexible polymer liquid crystals by either the freely-jointed rod chain, or by the worm-like chain[3-6]. The rigid rod model does not allow any flexibility of polymer chain. However when the chain is not completely rigid, chain flexibility in fact has significant effect on the properties of polymers. The freely-jointed rod chain treats the polymer as repeated rods, but it fails to deduce the transition to rod behaviour expected when nematic field is strong. On the other hand, the worm-like chain theory has been successful and gives better agreement with experiments. More recently Yurasova and Semenov[7] presented a model which attacks the non-homogeneous chain mentioned above. But they relax the concept that chain tangent vector is a unit vector, so that the problem can be analysed in terms of the standard Wiener integral which is solvable mathematically. It is well known that the relaxation of this constraint can fail to give correct results even when dealing with the simple case, e.g. the long isotropic chain[8].

Actually, the flexible spacers are constrained by the mesogenic units to which they are linked to have some orientational order. In turn they affect the order of the mesogenic units - if the spacers are not very long and hence not perfect flexible, they can retain to some extent the orientations of the ends of spacer allowing the spacers to talk to each other via the mesogenic unit in between. They thus indirectly mediate nematic interactions. It is expected that spacers have an order differing from that of the mesogenic units.

In this work a model based on previous worm-like theory - the Spheroidal approach[4,6] is presented. It describes the influence of flexible spacers on the nematic - isotropic (N-I) phase transition and some other statistical properties, such as nematic order of two components, i.e. mesogenic units and spacers. Comparisons are made with the worm-like chain model and jointed rod models including the case of freely-jointed rods model.

The approach can be extended to monomers, dimers, trimers and so on. We will discuss those problems in separate paper. Although they have the same basic chemical formulae, the monomers, dimers, trimers and polymers whether linear or cyclic have their own

X. J. Wang and M. Warner - Cavendish Laboratory, Madingley Road, Cambridge CB3 0HE, UK

*Phase Transitions in Liquid Crystals*, Edited by S. Martellucci and
A.N. Chester, Plenum Press, New York, 1992

characteristics, including nematic - isotropic phase transition, nematic order, all available in the literature.

## 2. Non-Homogeneous Nematic Chain Model

Fig. 1 is the schematic of the main chain liquid crystal polymer with which we are concerned. The chain is composed alternatively of rod-like, mesogenic units, referred as A segments of length a each and flexible spacers, denoted as B worm-like sections of chain of total contour length b each, the degree of polymerization of the polymer being N. A segments are treated in terms of the Maier-Saupe theory[9], but the coupling constant diluted by a factor ρ, the volume fraction of rod component. Assuming the two parts have the same cross section, the volume fraction is thus

$$\rho = a/(a+b) \tag{1}$$

The Spheroidal approach[4] is exploited to describe the spacers, that is the section B of chain. B chains are so long that they can be considered to be perfect flexible. Their stiffness is characterized by the bend constant ε. This yields a persistence length $D^{-1} = 2\varepsilon/k_BT$. Perfect flexibility (and hence loss of memory of direction when traversing B) thus corresponds to $b >> D^{-1}$. The B chain also suffers an interaction from the mesogenic units, the coupling constant being $v_{ab}$. In the nematic phase, the mesogenic units A impose an orientational order, denoted $S_B$, on the B segments. $S_B$ is not necessarily the same as that of the A segments, $S_A$. Usually $S_B < S_A$, i.e. the flexible part is less ordered than the mesogenic part. Both are ensemble averaged over all the segments, i.e.

$$S_A = \frac{1}{N} < \sum_{i=1}^{N} P_2(\mathbf{u}^i \cdot \mathbf{n}) > \text{all A segments}$$

$$S_B = \frac{1}{N} < \sum_{i=1}^{N} P_2(\mathbf{u}(s_i) \cdot \mathbf{n}) > \text{all B segments} \tag{2}$$

where $P_2$ is the Legendre polynomial, $P_2(x) = (3x^2-1)/2$, $\mathbf{u}$ is either the tangent vector $\mathbf{u}(s_i)$ describing the i-th B chain at contour length $s_i$ or being the tangent vector $\mathbf{u}^i$ that describes the direction of the rod part (A) of the i-th monomer. $\mathbf{n}$ is the preferred direction, i.e. the director.

The order parameter of the system as a whole S is expressed by the weighted mean of

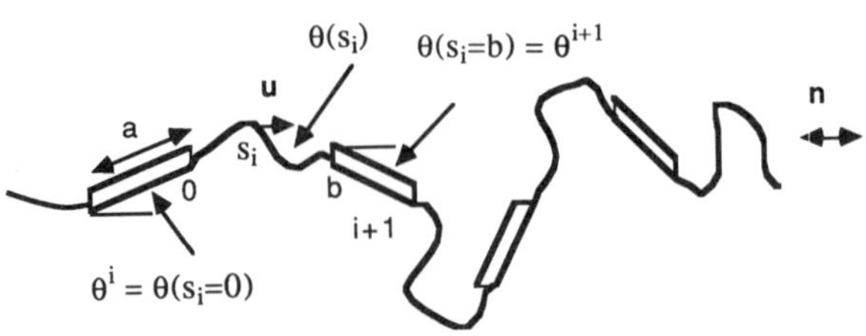


Fig. 1. Schematic of molecularly non-homogeneous nematic chain. a, b are the lengths of A segments and B sections of chain, respectively. $\mathbf{u}$ is the tangent vector of chain at contour length $s_i$ in i-th monomer along chain. $\theta(s_i)$ and $\theta^i$ are the angles the B section at $s_i$ and A segment make with respect to director n, respectively.

the orders of the two components

$$S = \rho S_A + (1 - \rho)S_B \tag{3}$$

The order parameter measured by birefringence will be weighted not only by volume fraction but also by relative dielectric anisotropy. By selective deuteration NMR can measure $S_A$ and $S_B$ separately, in fact being able to measure $S_B$ as a function of distance along the spacer.

Generally, there are three interactions in the system: the A-A, B-B and A-B interactions, represented by couplings per unit arc length $v_{aa}$, $v_{bb}$ and $v_{ab}$ ($v_{ab} = v_{ba}$). The hamiltonian of the i-th monomer consisting of an A segment linked with a B section, $H^i$, is

$$\begin{aligned} H^i &= -\left[v_{aa}\rho S_A + v_{ab}(1-\rho)S_B\right]\cdot aP_2\left(\mathbf{u}^i\cdot\mathbf{n}\right) + \\ &\int_0^b \left\{-\left[v_{ab}\rho S_A + v_{bb}(1-\rho)S_B\right]\cdot P_2\left(\mathbf{u}(s_i)\cdot\mathbf{n}\right) + \frac{\varepsilon}{2}\left(\frac{d\mathbf{u}}{ds_i}\right)^2\right\}ds_i \\ &\equiv -W_A aP_2\left(\mathbf{u}^i\cdot\mathbf{n}\right) + \int\left\{-W_B P_2(\mathbf{u}\cdot\mathbf{n}) + \frac{\varepsilon}{2}\left(\frac{d\mathbf{u}}{ds_i}\right)^2\right\}ds_i \end{aligned} \tag{4}$$

where the couplings $W_A$ and $W_B$ are respectively

$$\begin{aligned} W_A &= v_{aa}\rho S_A + v_{ab}(1-\rho)S_B \\ W_B &= v_{ab}\rho S_A + v_{bb}(1-\rho)S_B \end{aligned} \tag{5}$$

The subscripts A and B denote the A and B segments respectively.

The total hamiltonian H is the summation over all the N monomers

$$H = \sum_{i=1}^{N} H^i \tag{6}$$

The partition function Z of the polymer chain results from summing over all angles $\theta^i$ for the rod (A) segments and over all choices of the shapes of chains $\mathbf{u}(s_i)$ for the B sections:

$$\begin{aligned} Z = &\int\prod_{i=1}^{N} d\left(\cos\theta^i\right)\int\prod_{i=1}^{N} D\{\mathbf{u}(s_i)\}\exp\left[\sum_{i=1}^{N}\beta a W_A P_2\left(\cos\theta^i\right)\right] \\ &\exp\left[\sum_{i=1}^{N}\int_0^b\left[\beta W_B P_2\left(\cos\theta(s_i)\right) - \frac{\beta\varepsilon}{2}\left(\frac{d\mathbf{u}}{ds_i}\right)^2\right]ds_i\right] \\ &\prod_{i=1}^{N}\delta\left(\theta^i - \theta(s_i=0)\right)\prod_{i=1}^{N-1}\delta\left(\theta^{i+1} - \theta(s_i=b)\right) \end{aligned} \tag{7}$$

where the Dirac delta functions, $\delta$, ensure that the ends of spacers, B, have the same orientation as the conjugated A rods to which they are attached, see Fig. 1. $\beta$ is $1/k_BT$.

In terms of the Spheroidal model of worm-like chains[4,6], the part of Z concerned with each B spacer can be expressed by the Spheroidal wave functions. It is simply the conditional probability (Green function or propagator) that a chain starting with angle $\theta(s_i = 0)$ from one rod, ends with $\theta(s_i = b)$ at the next: $G(\theta(b), \theta(0); b) = \Sigma\, Sp_n(\theta_0)\, Sp_n(\theta_b) \exp(-\lambda_n bD)$. Here we assume $Sp_n$ is normalized to units. Performing the sums over the rod directions (the $\delta$-functions simply set them equal to the directions of the relevant chain ends):

$$Z = \int \prod_{i=1}^{N} \exp\left[\beta a W_A P_2\left(z_0^i\right)\right] \prod_{i=1}^{N-1} \left(z_b^i - z_0^{i+1}\right)$$
$$\prod_{i=1}^{N} \left[ Sp_0\left(z_0^i\right) Sp_0\left(z_b^i\right) \exp(-\lambda_0 Db)\, d\left(z_0^i\right) d\left(z_b^i\right) \right] \quad (8)$$

where $\lambda_n$ and $Sp_n$ are the n-th eigenvalue and eigenfunction of the Spheroidal wave equation. We have changed to the more natural variable, $z = \cos\theta$ and taken only the ground state $n = 0$ in G. With this restriction the partition function Z is simply given by

$$Z = \exp(-\lambda_0 DbN) \left\{ \int \exp[\beta a W_A P_2(z)] [Sp_0(z)]^2 dz \right\}^{N-1}$$
$$\left\{ \int Sp_0(z_N)\, dz_N \right\} \left\{ \int \exp[\beta a W_A P_2(z_1)] Sp_0(z_1)\, dz_1 \right\} \quad (9)$$

The next state to be involved is, for symmetry reasons, $n = 2$. We discuss this in detail in an appendix but at this stage note one requires $\exp[-(\lambda_2-\lambda_0)bD]$ be small. In the isotropic state $\lambda_2 - \lambda_0$ rapidly gets large and this requirement becomes more easily satisfied. When the degree of polymerization N is high, neglect the term without the N factor in the log(Z) and hence

$$Z \cong \exp(-\lambda_0 DbN) \left\{ \int \exp[\beta a W_A P_2(z)] [Sp_0(z)]^2 dz \right\}^{N} \quad (10)$$

The total free energy per chain has the form

$$F = -k_B T \ln Z - \frac{1}{2} < U > \quad (11)$$

where the first part arises from the partition function and the second term is from mean field theory to avoid double counting pair interactions. Putting Z and U into Eq. 11 yields

$$F/(Nk_B T) = \frac{\beta}{2}(aW_A S_A + bW_B S_B) + \lambda_0 Db$$
$$-\ln\left\{ \int \exp[\beta a W_A P_2(z)] [Sp_0(z)]^2 dz \right\} \quad (12)$$

The second and third terms are familiar as the mean field free energy of a worm chain of length Nb. The first term is the usual Maier-Saupe mean field term for an isolated rod. The last term would simply be that of Maier-Saupe if the chains were isotropic (the function $Sp_0(z) = 1$). The deviation from the simple rod form comes from additional nematic effects in the worm chain tied to each end of a rod segment - the two factors of $Sp_0(z)$ included in the rod partition integral. It is the term that makes the problem differ from simply that of rods mixed with nematogenic worm chains.

## 3. Nematic Order

Minimizing the free energy with respect to the S's gives the order parameters. For simplicity, we transfer the derivatives of free energy with respect to $S_A$ and $S_B$ into those with respect to the intermediate coupling parameters $W_A$ and $W_B$.

The relationship between ($W_A$, $W_B$) and ($S_A$,$S_B$) is

$$\begin{pmatrix} W_A \\ W_B \end{pmatrix} = \begin{pmatrix} v_{aa} & v_{ab} \\ v_{ab} & v_{bb} \end{pmatrix} \begin{pmatrix} \rho S_A \\ (1-\rho)S_B \end{pmatrix} = \mathbf{M} \begin{pmatrix} \rho S_A \\ (1-\rho)S_B \end{pmatrix} \tag{13}$$

Alternatively Eq. 13 is written as

$$\mathbf{W} = \mathbf{M}\,\mathbf{S} \tag{14}$$

where the column vectors **W** and **S** are $(W_A, W_B)^T$ and $(\rho S_A, (1-\rho)S_B)^T$ respectively, and $(...)^T$ denotes transpose.

Minimizing the free energy with respect to the order parameters becomes

$$\frac{dF}{d\mathbf{S}} = \mathbf{M}^{-1}\frac{dF}{d\mathbf{W}} \tag{15}$$

where $\mathbf{M}^{-1}$ is the inverse matrix well defined if the determinant det $\mathbf{M} = v_{aa}v_{bb} - v_{ab}^2$ is non zero.

If det $\mathbf{M} \neq 0$ the conditions $dF/d\mathbf{S} = 0$ are equivalent to the equilibrium conditions $dF/d\mathbf{W} = 0$. There is not any physical reason why the determinant should vanish except at accidental point. Minimizing the free energy with respect to $W_A$ and $W_B$ respectively yields the order parameters of both A segments and B sections. The order parameter of the A segments is given by

$$S_A = \frac{\int_{-1}^{1} \exp[\beta a W_A P_2(z)][Sp_0(z)]^2 P_2(z)\,dz}{\int_{-1}^{1} \exp[\beta a W_A P_2(z)][Sp_0(z)]^2\,dz} \tag{16}$$

The discussion of the new rod term in Eq. 12 for F is underlined by this result: when averaging $P_2$ the probability $\exp(\beta a W_A P_2)$ is enhanced by the nematic tendency of the worms to which the rod is attached. The order of B sections is

$$S_B = \frac{3}{2}\frac{\partial \lambda_0}{\partial \Delta^2} - \frac{3D^{-1}}{b}\frac{\int \exp[\beta a W_A P_2(z)]Sp_0(z)\frac{\partial Sp_0(z)}{\partial \Delta^2}dz}{\int \exp[\beta a W_A P_2(z)][Sp_0(z)]^2 dz} \tag{17}$$

The coupling constant $\Delta^2$ is an important parameter in the Spheroidal approach, $\Delta^2 = -3\beta^2\varepsilon W_B$. It is what determines the functions $Sp_n(z)$ and the $\lambda_n$ are purely function of $\Delta^2$.

This is why we transformed from the **S** to the **W** variables.The Pauli trick[6] gives the derivatives of $\lambda_0$ and $Sp_0$ with respect to $\Delta^2$ in terms of the matrix elements. Consequently, $S_B$ is

$$S_B = (P_2)_{0,0} + \frac{2D^{-1}}{b} \sum_{n=2}^{\text{even}} \frac{(P_2)_{0,0}}{\lambda_n - \lambda_0} \frac{\int \exp[\beta a W_A P_2(z)]\, Sp_0(z)\, Sp_n(z)\, dz}{\int \exp[\beta a W_A P_2(z)]\, [Sp_0(z)]^2\, dz} \tag{18}$$

where $(P_2)_{0,0} = \int Sp_0(z)\, P_2(z)\, Sp_0(z)\, dz$, the second term arises from the ends of spacers connected with rigid rods. When the chain is long enough compared with the persistence length $D^{-1}$, the second part is small and may be neglected[6]. Otherwise it becomes important.

Substituting $S_A$ and $S_B$ into Eq. 3 gives the resultant order parameter of the system. The Fig. 2 shows the dependence of nematic order of the mesogen, the spacer segments and of the whole polymer, $S_A$, $S_B$ and S. $S_A$ has typically a Maier-Saupe-like temperature dependence. $S_B$ is a distinctly different function of temperature varying enormously with the polymer parameters.

In temperature region where $S_A$ undergoes its transition the B sections are actually in the 'paranematic state' under the influence of the nematic field of the A segments. $S_B$ is therefore small. As temperature decreases the B sections order and thus $S_B$ becomes greater. $S_B$ is usually less than the order of the mesogenic units, A.

## 4. N - I Phase Transition

When the nematic - isotropic transition occurs the free energy difference of the phases is equal to zero, i.e.

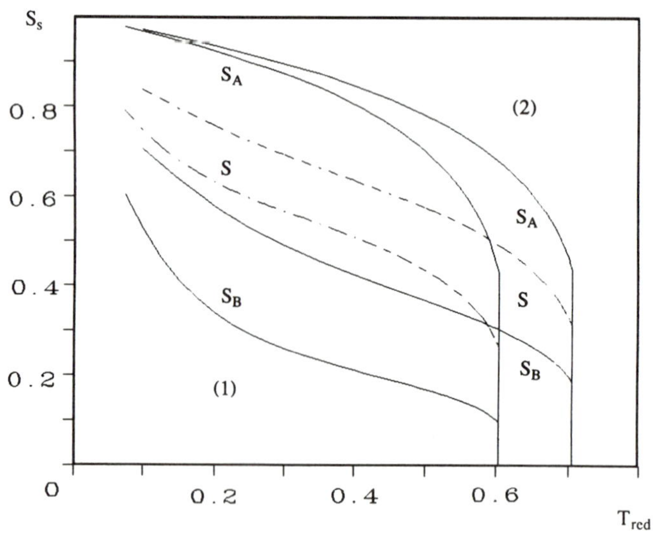


Fig. 2. Nematic order of whole system and its two individual parts, i.e. mesogen and spacer, as a function of temperature reduced by N-I transition temperature of Maier-Saupe model for pure rods. Two families of curves are corresponding to a: b: $D^{-1}$ = 2: 2: 1 (1) and 1: 1: 1 (2) respectively. The interaction constants $v_{aa}$: $v_{ab}$: $v_{bb}$ = 1: 0.5: 0.2.

$$\Delta F\big/(Nk_BT) = \frac{\beta}{2}\left\{aS_A\left[\rho v_{aa}S_A + (1-\rho)v_{ab}S_B\right] + bS_B\left[\rho v_{ab}S_A + (1-\rho)v_{bb}S_B\right]\right\}$$
$$+\lambda_0 bD - \ln\left(\int \exp\left\{\beta a\left[\rho v_{aa}S_A + (1-\rho)v_{ab}S_B\right]P_2(z)\right\}\left[Sp_0(z)\right]^2 dz\right) = 0 \qquad (19)$$

We normalize temperature by the Maier-Saupe transition temperature for pure rigid rods of length a, $T_0 = 0.2202\, av_{aa}/k_B$. Knowing the material parameters, such as the interaction ratio, $v_{aa}$:$v_{ab}$:$v_{bb}$ and the length ratio a:b:$D^{-1}$, i.e. the volume fraction ρ and the length of the spacer measured in the units of the persistence length b/$D^{-1}$, one can find the nematic-isotropic phase transition temperature of the liquid crystalline polymers compared with the Maier-Saupe value of N-I phase transition temperature.

The coupling constants $W_A$ and $W_B$ or $\Delta^2$ are expressed in terms of the material parameters mentioned above as

$$\beta a W_A = \frac{1}{0.2202}\frac{T_0}{T}\left[\rho S_A + (1-\rho)\frac{v_{ab}}{v_{aa}}S_B\right] \qquad (20)$$

and

$$\Delta^2 = -3\beta^2\varepsilon W_B = -\frac{3}{2}\frac{1}{0.2202}\frac{D^{-1}}{b}\frac{T_0}{T}\left[(1-\rho)\frac{v_{ab}}{v_{aa}}S_A + \frac{(1-\rho)^2}{\rho}\frac{v_{bb}}{v_{aa}}S_B\right] \qquad (21)$$

## 5. Latent Entropy

The entropy of a chain is given[6] by

$$S_N = -\, k_B < \ln f > \qquad (22)$$

where f is the probability, i.e.

$$S_N = k_B \ln Z - k_B\beta\frac{\partial}{\partial\beta}\ln Z \qquad (23)$$

Put Z in Eq. 10 into Eq. 25 and obtain the entropy $S_N$ in the nematic state

$$S_N\big/k_B = -\lambda_0 bDN + N\left\{\ln\int \exp[\beta a W_A P_2][Sp_0]^2 dz\right\} -$$
$$\beta\left\{-\lambda_0 bN\frac{\partial D}{\partial\beta} - bDN\frac{\partial\lambda_0}{\partial\beta} + N\frac{\frac{\partial}{\partial\beta}\left(\int \exp[\beta a W_A P_2][Sp_0]^2 dz\right)}{\int \exp[\beta a W_A P_2][Sp_0]^2 dz}\right\} \qquad (24)$$

The derivatives ∂/∂β in the last term are complicated but can be simplified with the aid of Eqs. 16 and 18. $S_N$ is given by

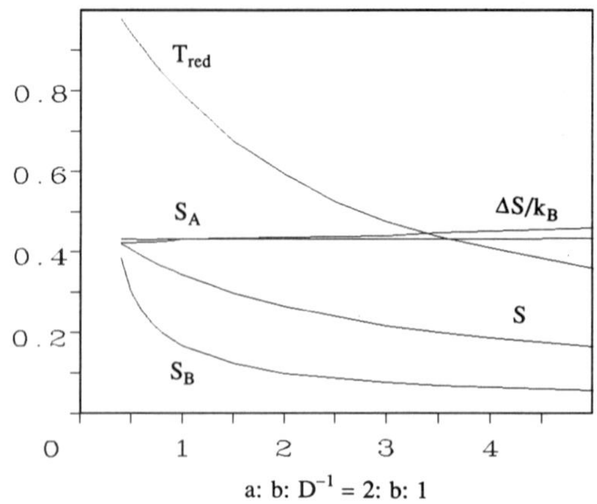


Fig. 3. Dependence of N-I transition temperature, order parameters and latent entropy on spacer length b.

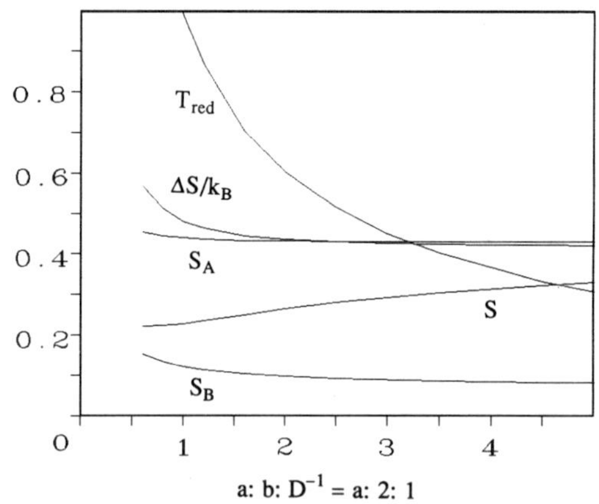


Fig. 4. Dependence of N-I transition temperature, relative to Maier-Saupe N-I transition temperature of pure rods with a = 2, order parameters and latent entropy on mesogen length a.

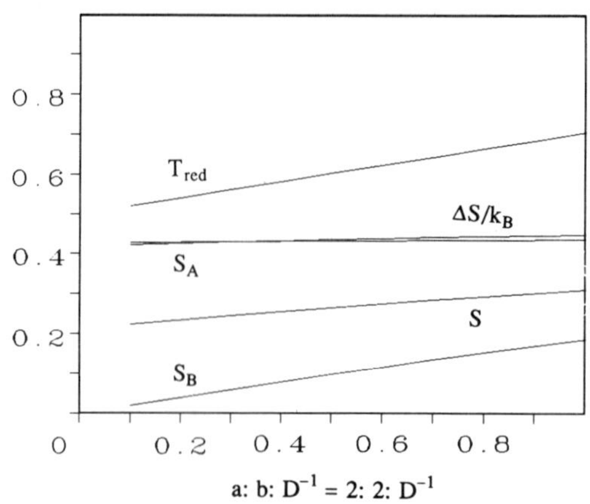


Fig. 5. Dependence of N-I transition temperature, order parameters and latent entropy on persistence length $D^{-1}$.

$$S_N/Nk_B = -\lambda_0 bD + \ln\left\{\int \exp[\beta a W_A P_2][Sp_0]^2 dz\right\} - \beta(bW_B S_B + aW_A S_A) \quad (25)$$

Consequently, the latent entropy per monomer at N - I transition is

$$\Delta S_N/(Nk_B) = \frac{\beta}{2}(bW_B S_B + aW_A S_A) \quad (26)$$

## 6. Results and Discussion

Experiments have found that most polymers with small N are not fusible[10] or do not show a nematic phase until decomposition. On the other hand the low molecular mass liquid crystals composed by the same mesogenic units can exhibit nematic phase.

The fact is agreement with our expectation, see Fig. 3, where the reduced N-I transition temperature is much greater than 1 and it may be far beyond the temperature of experiments in which the polymers may still be in the condensed phase.

The N-I transition, including aspects such as the transition temperature, order parameters and the latent entropy, depends on polymer parameters, such as lengths a, b, $D^{-1}$ and interactions $v_{aa}$, $v_{ab}$ and $v_{bb}$, shown in Figs. 3-7.

One of the important results examines the effect of the length of the flexible spacer on the transition, such as the transition temperature and the latent entropy, see Fig. 3. The theoretical conclusion is qualitatively in agreement with many experiments, for instance the Refs.11-13 which investigated the effect of both the number of carbon atoms in flexible spacer and the mesogenic groups on the N-I phase transition of polymers, the latter can be compared with the Figs. 4 and 6-7. The influence of the flexibility of spacer was also investigated[14], corresponding to Fig. 5.

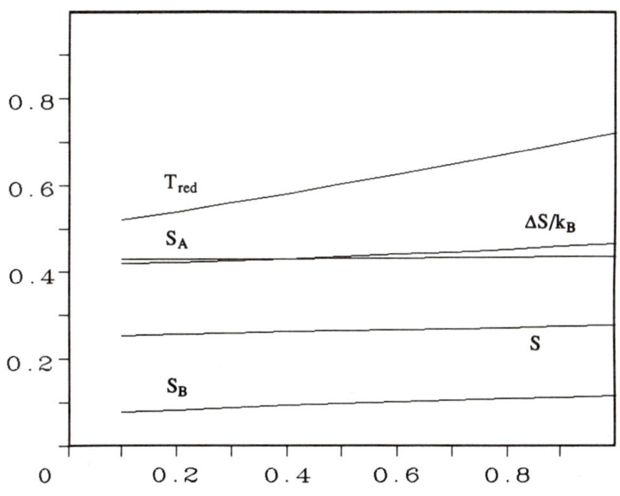


Fig. 6. Dependence of N-I transition temperature, order parameters and latent entropy on interaction ratio $v_{ab}/v_{aa}$.

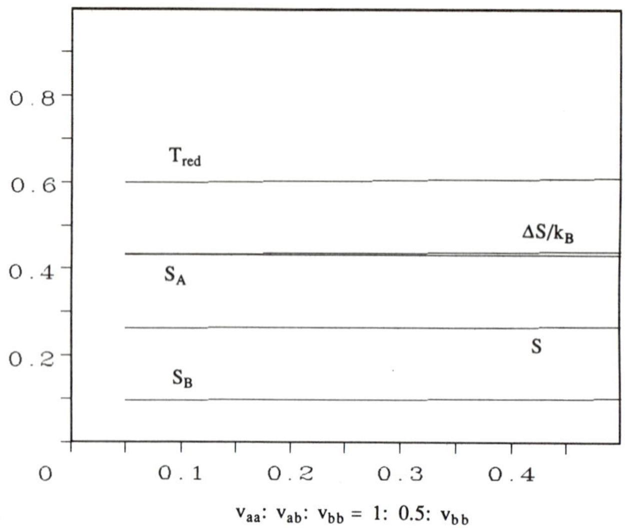


Fig. 7. Dependence of N-I transition temperature, order parameters and latent entropy on interaction ratio $v_{bb}/v_{aa}$.

The results also demonstrate that the order parameter of the mesogenic units at the N-I transition essentially does not vary, remaining about 0.434 despite variation in the order of the flexible spacer. The interaction ratio $v_{bb}/v_{aa}$ does not visibly affect the properties of the polymer. On the other hand N-I transition temperature is linearly proportional to the interaction ratio $v_{ab}/v_{aa}$.

One extreme is where the worm spacer flexibility decreases and the sections of rigid rod become short. This is the nematic worm chain limit. The other limit is where the worm spacer becomes so short and hence effectively the spacer as a whole is of limited flexibility, that successive rods have strong angular correlations induced between them. This then corresponds to the non - freely jointed nematic rods model, also treated elsewhere.

Experiments are being carried out on polymers simultaneously with deuteriated spacers and deuteriated mesogenic groups and deuteriated whole polymers as well, which would be good candidates with which to test the theory.

ACKNOWLEDGEMENT. One of the authors (XJW) acknowledges supports from SERC and RSRE, Malvern.

## Appendix

Only the first two states in Z, i.e. the ground state and the first excited state have been taken into account in the model so far. The reason for the approximation is based on the evidence that the eigenvalues, which appear in exponentials in the propagator G, coalesce in pairs as $\Delta^2$ becomes large. For instance $\lambda_1$ and $\lambda_0$ come together, as do the next eigenvalues $\lambda_2$ and $\lambda_3$ while at the same time becoming distant from $\lambda_0$ and $\lambda_1$, thus giving these terms exponentially smaller weight. Inserting the second two excited states into the partition function yields

$$Z = \int \ldots \prod_{i=1}^{N} \exp\left[\beta a W_A P_2\left(z_0^i\right)\right] \prod_{i=1}^{N-1}\left(z_b^i - z_0^{i+1}\right) \prod_{i=1}^{N} \{ [Sp_0\left(z_0^i\right) Sp_0\left(z_b^i\right) \exp(-\lambda_0 Db) +$$
$$Sp_1\left(z_0^i\right) Sp_1\left(z_b^i\right) \exp(-\lambda_1 Db) + Sp_2\left(z_0^i\right) Sp_2\left(z_b^i\right) \exp(-\lambda_2 Db)] d\left(z_0^i\right) d\left(z_b^i\right) \} \quad \text{(A1)}$$

The terms in $Sp_1$ give zero on symmetry grounds (they are dipolar in character and vanish in quadrupolar medium). Denoting the following integrals as matrix elements

$$G_{m,n} = \int \exp[\beta a W_A P_2(z)] Sp_m(z) Sp_n(z) dz \quad \text{(A2)}$$

Z reduces to

$$Z = \exp(-\lambda_0 DbN)(G_{0,0})^N \left\{1 + {}^{N-1}C_1 \exp\left[-(\lambda_2 - \lambda_0)Db\right]\left(\frac{G_{0,2}}{G_{0,0}}\right)^2\right.$$
$$\left. + {}^{N-1}C_2 \exp\left[-2(\lambda_2 - \lambda_0)Db\right]\left(\frac{G_{0,2}}{G_{0,0}}\right)^4 \left[1 + \frac{2}{N-2}\frac{G_{2,2}G_{0,0}}{(G_{0,2})^2}\right] + \ldots\right\} \quad \text{(A3)}$$

The ratio of the terms i-th and (i+1)-th is approximately

$$\frac{N-i}{i}\exp\left[-(\lambda_2 - \lambda_0)Db\right]\left(\frac{G_{0,2}}{G_{0,0}}\right)^2 \quad \text{(A4)}$$

The formulae from Eq. 10 are expected to be good provided the following condition is satisfied, i.e. the immediate next term in Z which is neglected in Eq. 10 must be less than unity

$$N \exp\left[-(\lambda_2 - \lambda_0)Db\right]\left(\frac{G_{0,2}}{G_{0,0}}\right)^2 << 1 \quad \text{(A5)}$$

or

$$b/D^{-1} > \frac{1}{\lambda_2 - \lambda_0} \ln\left[N\left(\frac{G_{0,2}}{G_{0,0}}\right)^2\right] \quad \text{(A6)}$$

Thus the higher terms as sketched in Eqs. A3 and A4 are even more negligible. The quadratic term of matrix elements ratio is always less than unity and the eigenvalue difference ($\lambda_2$-$\lambda_0$) is greater than 6, the non-nematic field value. When the nematic field becomes stronger the matrix element ratio becomes very small and $\lambda_2 - \lambda_0 >> 6$. Taking typical values N = 20, 100 or 400 yields the condition on spacer length: $b/D^{-1} > 0.499$, 0.768 or 0.998. The results do not change very much as N becomes even larger because of the logarithm in Eq. A6. The numerics confirm that for the typical values a: b: $D^{-1}$ = 2 : 2: 1 and $v_{aa}$: $v_{ab}$: $v_{bb}$ = 1: 0.5: 0.2, the left hand side of Eq. A6 is always below $8*10^{-4}$ N and is even less at N-I transition, $3.6*10^{-4}$N (where $\Delta^2 = -1.7$). It even demonstrates that Eq. A6 holds for a wide range of chain length. The above values for $b/D^{-1}$ are actually an under estimate.

## References


1. P. J. Flory, Proc. R. Soc. London, Ser. **A234**: 73 (1956).
2. L. Onsager, Ann. N. Y. Acad. Sci., **51**: 627 (1949).
3. A. ten Bosch, P. Maissa and P. Sixou, Phys. Letts., **94A**: 298 (1983).
4. M. Warner, J. M. F. Gunn and A. Baumgärter, J. Phys., **A18**: 3007 (1985).
5. V. V. Rusakov and M. I. Shliomis, J. Phys. Letts.(Paris), **46**: L1065 (1985).
6. X. J. Wang and M. Warner, J. Phys., **A19**: 2215 (1986).
7. T. A. Yurasova and A. N. Semenov, to appear in Mol. Cryst. Liq. Cryst.
8. K. Freed, in: "Advances in Chemical Physics", Vol. **XXII**, I. Prigogine and S. A. Rice, ed., Wiley-Interscience, N.Y. (1972).
9. W. Maier and A. Saupe, Teil I., Z. Naturf., **14a**: 882 (1959).
10. A. Roviello and A. Sirigu, Makromol. Chem., **183**: 895 (1982).
11. S. Antoun, R. W. Lenz and J. -I. Jin, J. Polym. Sci., Polym. Chem. Ed., **19**: 1901 (1981).
12. D. van Luyen and L. Strzelecki, Euro. Polym. J., **16**: 303 (1980).
13. C. K. Ober, J-I. Jin and R. W. Lenz, in:"Liquid Crystal Polymer I", M. Gordon and N. A. Plate, ed., Springer-Verlag, Berlin (1984).
14. V. Frosini, S. de Petris, G. Galli and E. Chiellini, in: "Recent Advances in Liquid Crystalline Polymers", L. Lawrence Chapoy, ed., Elsevier, London (1985).

# LYOTROPIC LIQUID CRYSTALS

Chapter 26

# INTRODUCTION AND GENERAL THEORY OF LYOTROPIC LIQUID CRYSTALS

D. ROUX

## 1. Introduction

Amphilic molecules incorporate two antagonistic chemical functions. One part of the molecule is very soluble in water (hydrophilic) and another part is very soluble in an organic solvent (hydrophobic). Due to these opposite effects the molecules in solution aggregate spontaneously into objects of very different size and shape, among which one principally observes spheres, cylinders and twodimensional objects (membranes). Many of the phases observed in a typical phase diagram are liquid crystals. For example, the lamellar phase, which consists of a stack of membranes with a long range order in the direction perpendicular to the plane, is a smectic A phase. At first these phases appeared relatively uninteresting, but now they have been extensively studied and they exhibit extremely interesting and peculiar behavior. In certain cases, close to a transition toward nematic or hexagonal phases, the lamellar phase cannot be described as a stack of flat membranes but exhibits many defects that can be observed experimentally.In principle, the lyotropic nature of the phase allows the repeating distance d between the membranes to vary. In many systems this variation remains small (typically ranging from 30 to 60 Å), however in certain cases it is possible to prepare lamellar phases with extremely large repeating distances (up to several 1000 Å). These systems are especially interesting because, as we will see, they permit us to study the interactions between membranes over a wide range of conditions. They are also a unique example of a colloidal smectic A phase.

The reason that extreme dilutions may exist is due to the existence of long range repulsive interactions between membranes. Two cases have been identified and studied either the interaction is electrostatic in origin, or it is due to the enhancement of thermally excited undulations of the membranes.

The description of this smectic phase can be carried out at two levels (corresponding to two length scales) without the necessity for a complete microscopic theory. For lengths larger than the repeating distance d the system behaves as a regular smectic phase, but for smaller lengths it is still possible to have a continuous description (at least for dilute phases). This description corresponds to a colloidal system and can be done in terms of interacting membranes. Consequently, it is possible to use a simple theory to link the macroscopic properties of the phase such as the elastic constants (compressibility B and bending constant $K_1$), to the microscopic

---

D. Roux - Centre de Recherche Paul Pascal, Ave du Dr. Schweitzer F33600 Pessac, France.

*Phase Transitions in Liquid Crystals*, Edited by S. Martellucci and A.N. Chester, Plenum Press, New York, 1992

properties of the membranes. Indeed, it can be shown that B is related to the interactions between membranes and $K_1$ to the flexibility of the membranes. Measurements of these macroscopic constants give direct information on membrane properties which are normally accessible only by much more detailed measurement.

In order to measure these elastic constants, several approaches have been used. Some experiments involve direct measurements of the compressibility; for example, the force machine of Israelachvili measures B directly. Other methods are based on analysis of the order parameter by EPR or NMR. However, our methods will be based on the analysis of fluctuations. Due to the one-dimensional nature of the order in the smectic state, the long wave length fluctuations exhibit the so-called Landau-Peierls instability. The X-ray structure factor can be analysed by using the Caille theory and elastic constants can be extracted quite easily. Another way of measuring the elastic constant is to use dynamic light scattering. It can be shown that the control of both the wave vector and the polarization allows one to determine all the elastic constants independently.

## 2. Liquid Membranes

The effect of thermal fluctuations on macroscopic interfaces separating two phases (liquid-gas, solid-liquid, liquid-liquid, etc.) has been extensively studied and it is well established that, except in a few rare cases, capillary waves may be understood as being controlled by the surface tension. Owing to a strong repulsive effect due to the surface tension, the fluctuations of the interface remain weak; indeed the mean square distance $<u^2>$ over which the interface fluctuates is much smaller than the interface size L ($<u^2> << L^2$).

In certain cases of interest, the surface tension term does not play a role in suppressing fluctuations. For example, membranes in a solvent at thermodynamical equilibrium fluctuate but their TOTAL area remains constant, and consequently these fluctuations are controlled by a higher order term in the film elastic energy, the curvature term. A membrane constituted by atoms or molecules free to move in the surface defined by the membrane is in a two-dimensional liquid state and its curvature energy is given by[1]

$$E_{(ela)} = \int_S dS\left[\frac{1}{2}\kappa(H - H_0)^2 + \bar{\kappa}K\right] \quad (1)$$

H and K are the mean ($1/R_1+1/R_2$) and gaussian ($1/R_1R_2$) curvatures ($R_i$ are the principal radii of curvature). $H_0$ is the spontaneous radius of curvature describing asymmetry between the two sides of the membrane; for a double layer this term is equal to zero. The elastic coefficients $\kappa$ and $\bar{\kappa}$ are respectively the mean bending constant and the gaussian bending constant.

In contrast to tethered membranes[2], liquid membranes do not have a fixed topology, and thus the topology of the surface may change. If we use the Gauss-Bonnet theorem, we can show that the integral over the entire surface of the gaussian curvature is a topological invariant

$$\int_S K dS = 4\pi(n_c - n_h) \quad (2)$$

where $n_c$ is the number of disjoint pieces or components and $n_h$ is the number of handles of the surface. Consequently, the gaussian curvature is only a function only of the topology, and the gaussian bending constant may be considered as a chemical potential for the genus of the surface.

The rigidities $\kappa$ and $\bar{\kappa}$ are both expressed in units of energy, and this allows us to define two separate regimes regions of approximation for the statistical physics of membranes. In order to understand the effect of thermal fluctuations, we need to compare $\kappa$ to $\kappa_B T$, ($\kappa_B$ is the Boltzmann constant and T the temperature) when $\kappa >> \kappa_B T$, the membrane is rigid and thermal fluctuations have little effects on the shape of the membrane; on the other hand, when $\kappa \approx k_B T$ (flexible films) thermal fluctuations renormalize the elastic constants $\kappa$ and $\bar{\kappa}$. By perturbation theory it can be shown that[3-5]

$$\kappa = \kappa_0 - \frac{\alpha}{4\pi} k_B T \ln\left(\frac{q_{max}}{q_{min}}\right) \tag{3a}$$

$$\bar{\kappa} = \bar{\kappa}_0 + \frac{B}{6\pi} k_B T \ln\left(\frac{q_{max}}{q_{min}}\right) \tag{3b}$$

where $\alpha$ and $\beta$ are numerical constants which can be calculated according to Ref. 5, $\alpha = 3$ and $\beta = 5$. In Eqs. 3, $q_{max}$ and $q_{min}$ are in-plane cutoffs corresponding respectively to the inverse of a molecular length (the smallest allowed length scale of fluctuations) and the inverse of the in-plane correlation length of the fluctuations which corresponds to the transverse size of the membrane in the case of a free membrane. This renormalization may be understood as the result of the shape dependence of the entropy of ripples; the logarithmic dependence on q is a well known characteristic of fluctuations in two-dimensional systems. Among the many consequences, the effect of thermal fluctuations leads to a loss of the orientational order for a free membrane. Eq. 3 allows us to define a persistence length[6] (which is the correlation length of the orientational order) as the length at which the effective bending constant goes to zero (with $a$ being a molecular length):

$$\xi_k = a e^{4\pi\kappa_0/\alpha k_B T} \tag{4}$$

## 3. Lamellar Phases

The experimental realization of liquid membranes is possible by using surfactants in solution. Indeed, amphiphilic molecules aggregate into objects of variable shapes and sizes. Among the most common structures are spheres (micelles), cylinders (hexagonal phases) and two-dimensional objects (membranes). These membranes are stacked either in a long range ordered phase (lamellar phase) or randomly distributed in space (sponge phase[7]).

One of the most interesting properties of the lamellar phase is its stability for a very wide range of dilution (that is, membrane over solvent ratio, see Fig. 1). The repeating distance of the lamellar ordering may vary continuously from 10 to several thousands of Å[8,9]. In order to explain how such a large variation in distance between membranes still maintains lamellar ordering, one needs to invoke repulsive interactions. There are two types of such interactions that lead to an extreme expansion of the lamellar phases electrostatic interactions and undulation forces.

## 4. Interaction Between Membranes

To analyze membrane interactions we must determine the interaction potential V(d) between two membranes of thickness $\delta$ separated by a distance d.

There are two cases leading to repulsive interactions which are long range enough to explain the large dilution range obtained in the experimental systems which have been described previously in the first case, when electrostatic interactions dominate, there are no excess charges except for counter-ions between two membranes swollen with water; in the second case, with neutral membranes, a steric interaction arises from the enhancement of thermally excited undulations, as recently proposed by Helfrich[10].

### *4.1. Electrostatic Interactions*

In the absence of layer undulations (we will come back to this point in the next paragraph), the long range repulsive interaction is due to electrostatic interactions, and occurs mainly for charged surfactants swollen with water. This interaction can be calculated from the one-dimensional Poisson-Boltzmann equation. Two limiting cases need to be considered (1) when the water between the membranes contains only the counter-ions of the charged surfactant, on the other hand salt has been added so that the Debye length is smaller than the distance d between membranes. In the first case an exact solution has been calculated by Parsegian[11], and its expansion for large d is[12]

$$V_{\text{elec}} = \frac{\pi k_B T}{4 L_B d}\left[1 - \frac{a}{L_B d} + \left[\frac{a}{L_B d}\right]^2 + \dots\right] \tag{5}$$

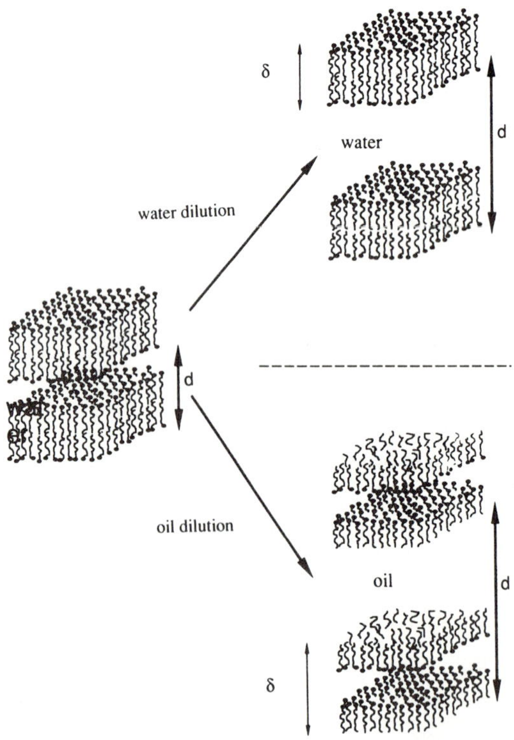


Fig.1. Schematic drawing of the membrane structure of the lyotropic lamellar phase in certain cases, the lamellar phase can be swollen either with water (hydrophylic solvent) or oil (hydrophobic solvent), leading to direct or inverted bilayers. δ is the membrane thickness and d is the repeati distance.

$L_B$ is the Bjerrum length ($L_B = (\pi e^2)/(\varepsilon k_B T) \approx 7$ Å at room temperature, for the dielectric constant of water $\varepsilon$=80). One should notice that the dominant term in Eq. (5) varies as 1/d (very long range) and is not dependent on the area *a* per charged polar head (this corresponds to the so-called ionic condensation). With added salt the interaction is screened and decreases exponentially [13].

## *4.2. Undulation Forces*

Besides this classical interaction, Helfrich[10] has proposed a novel interaction arising explicitly from thermal fluctuations. Let us consider a "free" membrane at finite temperature, which undulates due to thermal fluctuations. Now let us assume that this membrane is confined between two impenetrable walls (or even, other membranes) situated at a distance smaller than the persistence length $\xi_k$ [6]. The fluctuations are controlled by the first part of Eq. 1 (fluctuations of gaussian curvature are not considered), and if we rewrite this equation in terms of the displacement u of the membrane from its equilibrium position at T = 0, we obtain

$$F = \frac{1}{2}\int \kappa(\nabla^2 u)^2 dS \tag{6}$$

The equipartition theorem leads to the average value of the square of the displacement

$$< u_q^2 > \propto \frac{k_B T}{\kappa q^4} \tag{7}$$

Integrating over all the q modes we get

$$< u^2 > \approx \int_{q_{min}}^{q_{max}} < u_q^2 > q\,dq \propto \frac{k_B T}{\kappa q_{min}^2} \tag{8}$$

where $q_{max}$ is a large q cutoff (small distances) and corresponds to a molecular length (typically, the in-plane surfactant size). For a membrane restricted to fluctuate between two walls at a distance d, we expect that $< u^2 > \propto d^2$. This allows us to determine $q_{min}$, which corresponds to the inverse of the in-plane correlation length of the fluctuations

$$q_{min} \propto \sqrt{\frac{k_B T}{\kappa}}\frac{1}{d} \tag{9}$$

We can use this information to calculate first, the excess area due to thermal undulation, and second the interaction due to the confinement. Thermal undulations imply that the real area S of the membrane is larger than the projected area $S_0$, and simple geometrical arguments show that[14]

$$S = S_0\sqrt{1 + \nabla u^2} \tag{10}$$

Using Eq. 7 and 9 we obtain the excess area

$$\frac{S - S_0}{S_0} = \frac{\Delta S}{S_0} \propto \frac{k_B T}{\kappa}\ln\frac{q_{max}}{q_{min}} \approx \frac{k_B T}{\kappa}\ln\frac{d}{a} \tag{11}$$

where a is the in-plane cutoff length. The logarithmic correction due to the undulation is a characteristic feature of the two-dimensional nature of the fluctuations.

Using the same heuristic argument, we can calculate the interactions between membranes resulting from the pressure due to the undulations.

We have seen that the inplane correlation length $\xi_{\|}$ was defined as $1/q_{min}$ (Eq. 9). Assuming that the pressure against the walls is described as an ensemble of disks of size $\xi_{\|}$[2,14], we obtain

$$P = N\frac{k_BT}{V} = \frac{Nk_BT}{Nd\xi_{\|}^2} \propto \frac{k_BT}{\kappa}\frac{k_BT}{d^3} \tag{12}$$

This leads to a free energy per unit area proportional to $k_BT/\kappa$ and to $1/d^2$. A more exact calculation using the Landau-de Gennes elastic energy for a smectic leads to[10]

$$\frac{F}{S} = \frac{3\pi^2}{128}\frac{(k_BT)^2}{\kappa}\frac{1}{r^2} \tag{13}$$

## 5. Phase Diagrams

Fig. 2 presents two typical phase diagrams of binary systems which illustrate the two types of dilute lamellar phases. In these systems the lamellar phase (L) exists for a wide range of concentrations, including the very dilute region. The corresponding repeat distance can be very large since $d \propto 1/\phi$, where $\phi$ is the surfactant volume fraction. For the two systems, d varies continuously from 40 Å to 3000 Å (C12E5 water system, Fig. 2b)[15] and 50 Å to 200 Å (AOT water system, Fig. 2a)[16]. The reason for the existence of dilute lamellar phases is the presence of long range repulsive interactions between the surfactant bilayers. For the AOT-water system the interaction is electrostatic in origin, where as for the C12E5 system the interaction is due to undulation forces.

## 6. Measuring Interactions

Lamellar phases are especially useful for quantitative study, to measure membrane flexibility and interactions. Indeed, to study such systems, we can use certain recent developments in liquid crystal science. These lamellar phases correspond to smectic A phases, which have complex behavior due to the nature of their elastic energy[17,18]. Moreover, it is possible to prepare samples with warying repeat distance d, allowing measurements as a varying function of the membrane distance.

We will briefly summarize some of the results one can obtain from scattering experiments. Fig. 3 is a schematic drawing of a typical neutron or x-ray scattering profile. There are three types of information one can obtain; the peak position leads to a direct measurement of the repeat distance d; the small angle scattering is due to concentration fluctuations; and the shape of the peak is dominated by long wave length fluctuations of the layer position. The overall shape of the structure factor is then determined by the elastic constant of the smectic phase, which can be easily linked to the membrane properties using a microscopic model .

The physics of lamellar phases is described by the basic elasticity of binary smectic A. The free energy density f of twocomponent, isotherrnal and incompressible smectics A, up to second order in layer displacement u and concentration fluctuations $\delta c$ (mass fraction), is[18,19]

$$f = \frac{1}{2}B(\partial_z u)^2 + \frac{1}{2}K(\nabla_\perp^2 u)^2 + \frac{1}{2\chi}\delta c^2 + C_c \delta c \partial_z u \tag{14}$$

with B defined as the layer compression modulus at constant concentration, K as the bending modulus, $\chi$ as the osmotic compressibility at constant layer spacing and $C_c$ as the coupling constant between layer displacement and concentration fluctuations. This is the basic equation that we will extensively use to describe the scattering behavior of the phase.

Of particular significance is the combination

$$\overline{B} = B - C_c^2\chi$$

which is the layer compressibility modulus at *constant chemical potential.* It is straightforward to check that

$$f = 1/2\overline{B}(\delta_z u)^2 + \text{curvature terms}$$

keeping the chemical potential constant

$$\partial f / \partial \delta c = 0$$

In the case of lyotropic smectics, this elastic constant $\overline{B}$ is directly related to the interactions between membranes[19,20].

In the two limiting cases of systems dominated by electrostatic or undulation forces one can relate these macroscopic constants to the microscopic interactions between membranes[21,22]. One finds

i) Electrostatic interactions

$$\chi^{-1} \approx \frac{\overline{\rho}^2 v_s \varepsilon d}{m_s^2 \delta}, \quad B \approx \frac{\varepsilon\delta}{v_s d}, \quad C \approx \frac{\overline{\rho}\varepsilon}{m_s}, \quad \overline{B} = \frac{\pi k_B T}{2L_b d^2} \tag{15}$$

ii) Sterically stabilized colloidal smectics (quoting the results of Lubensky et al.[22] for crumpled membranes)

$$\chi^{-1} \approx \frac{12\pi^2 v_s^2 \overline{\rho}^2 \kappa}{\left(1-\frac{4}{\pi^2}\right) m_s^2 \delta^2 d}, \quad B \approx \frac{12\pi^2\kappa}{\left(1-\frac{4}{\pi^2}\right) d^3}, \quad C \approx \frac{12\pi^2 v_s \overline{\rho}\kappa}{\left(1-\frac{4}{\pi^2}\right) m_s \delta d^2}, \quad \overline{B} \approx \frac{9\pi^2 (k_B T)^2}{64\kappa d^3} \tag{16}$$

For both kind of systems, the smectic splay constant K is related to the membrane bending modulus $\kappa$ by

$$K = \kappa / d \tag{17}$$

However, in the electrostatic case there is still an open question about the effect of electrostatic terms on the bending rigidity[23], which could be a function of *d*.

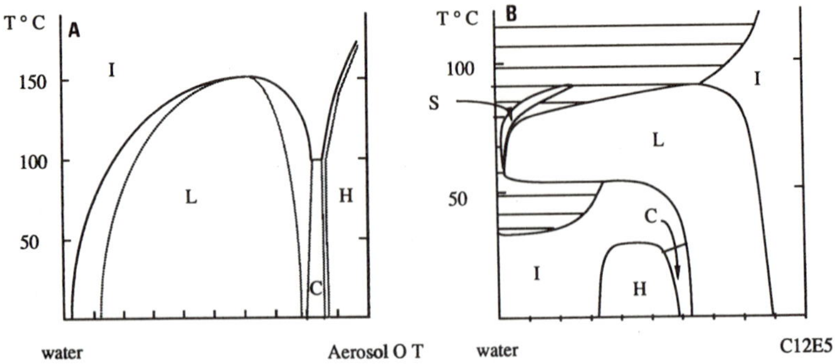


Fig.2. Phase diagrams of binary systems exhibiting dilute lamellar phases: 2a) gives the phase diagram of a charged surfactant (AOT) and water (Ref. 16); and 2b) the phase diagram of an nonionic surfactant (C12E5) and water (Ref. 15). Highly dilute lamellar phases appear because of the existence of long range repulsive interactions electrostatics (a) or undulations (b).

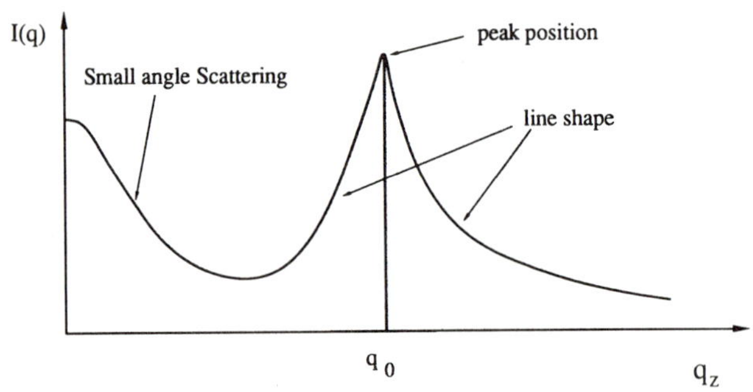


Fig.3. Schematic drawing of a typical scattering profile as obtained using neutron or x-ray scattering. Three types of information can be obtained from the analysis of this curve the peak position gives the repeat distance, the small angle scattering is due to concentration fluctuations, and the shape of the peak is related to longwave length fluctuations (Landau- Peielrs effect).

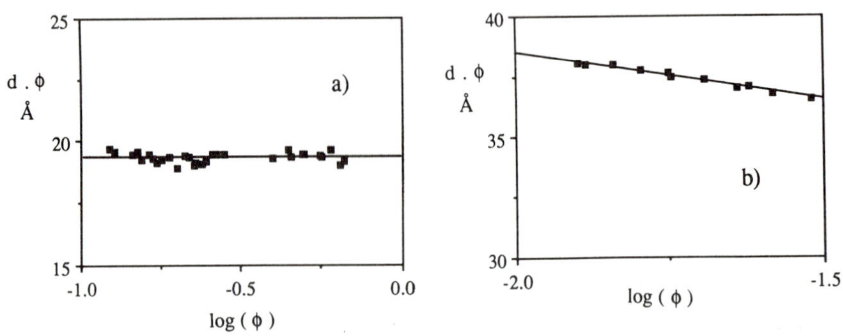


Fig.4. Deviation from the pure geometrical behavior for the repeati distance as a function of the membrane volume fraction ( $d = \delta/\varphi$). We have plotted the product $d.\varphi$ against the logarithm of $\varphi$. In the case of AOT the data are extracted from Ref. 16 and no noticable deviation is seen. In contrast, data from Ref. 15 exhibit a systematic deviation that is linear in $\ln \varphi$, corresponding to the excess area due to the undulations of the membranes.

## 7. Peak Position

In a perfectly flat ensemble of membranes in a smectic state it is easy to relate the peak position to the concentration of membranes. Let us describe the microscopic structure as a stack of surfactant bilayers of thickness $\delta$ and repeating distance d (Fig. 1). If $\Sigma$ is the surfactant area per polar head and $n_s$ is the number of surfactant molecules, the total membrane area and the total volume are, respectively, $S = n_s \Sigma$, $V = d S$, and the volume fraction of surfactant is

$$\phi = \frac{n_s v_s}{V} = \frac{v_s}{\Sigma d} = \frac{\delta}{d} \tag{18}$$

with $v_s$ the surfactant molecular volume.

This description is only valid for flat membranes, which occur when electrostatic interactions dominate. However, when the interaction between membranes is due to undulations, the membranes are no longer flat. We have seen that the excess area due to undulations leads to a logarithmic correction (see Eq. 11) and Eq. 18 becomes[15]

$$\phi \approx \frac{\delta}{d}\left(1 + \frac{k_B T}{4\pi\kappa}\ln\left(\frac{d}{a}\right)\right) \tag{19}$$

The case of electrostatic interactions, where undulations are relatively weak, corresponds to Eq.18 when the distances between membranes are large compared with the thickness of the membranes[16]. However, in the case of strongly fluctuating systems such as those leading to undulation interactions, we expect a deviation from the linear behavior of Eq 19. Fig. 4 presents data from two binary systems one (AOT-water[16]) is stabilized by electrostatic interactions and the other (C12E5-water[15]) by undulations. In order to emphasize the difference, only the systematic deviation from linear behavior has been shown ($d\varphi$ versus $\ln\varphi$). It is clear that the deviation is observable only in the system dominated by undulation forces, as expected from pertubation theory, and a value of $\kappa$ can be extracted from the data[15].

## 8. Small Angle Scattering

It can be shown[20] that small angle scattering is dominated by the term arising from the concentration fluctuations $<\delta c(0)\delta c(q)>$ when the shape of the peak is mainly due to the fluctuations of the layer displacement $<u(0)\, u(q)>$. Using the equipartition theorem and the free energy (Eq. 14), the correlation functions can be calculated

$$<u(q)u(-q)> = \frac{k_B T}{\overline{B} q_z^2 + K q_\perp^4} \tag{20a}$$

$$<\delta c(q)\delta c(-q)> = \frac{k_B T \chi (B q_z^2 + K q_\perp^4)}{\overline{B} q_z^2 + K q_\perp^4} \tag{20b}$$

The small angle scattering corresponds to the second term (Eq. 20b). One of the most obvious implications of this expression is that the intensity at q=0 is anisotropic depending on the direction of approach. Effectively, from Eq. 20b we have

$$< \delta c(q_z,0)\delta c(-q_z,0) > \rightarrow k_B T \frac{\chi^B}{\overline{B}}, \qquad q_z \rightarrow 0$$

$$< \delta c(0,\mathbf{q}_\perp)\delta c(0,-\mathbf{q}_\perp) > \rightarrow k_B T \chi, \qquad \mathbf{q}_\perp \rightarrow 0 \tag{21}$$

For reason of simplicity, we have not included gradient terms for the concentration fluctuations, nor gradient terms for the layer displacement which are higher order than usual for smectics[18,20]. These extra terms are required to represent the q-dependence of the (anisotropic) small-angle scattering which appears in the most dilute samples, the complete description of which is given in Ref 20.

Experiments exhibit effectively a very anisotropic small angle scattering as shown in Fig.5, which was obtained on oriented samples of a dilute lamellar phase using neutron scattering[20]. The analysis of this small angle scattering leads to information about the elastic constants. As expected from the theory, the scattering is more intense by a factor $B/\overline{B}$ along $q_z$ than it is along $q_\perp$ (note that by definition $\overline{B}$ is always smaller than B). Taking into account the gradient terms, we obtain a q dependence which is Orstein-Zernike like in the two principal directions[20].

## 9. Landau-Peierls Effect

Now let us examine the other part of the scattering, which leads to the peak arising from the long range order of the membranes.

From the elastic free energy Eq. 14, the direct space amplitude of the thermal fluctuations of the layer displacement can be calculated

$$< u^2 > \approx \frac{k_B T}{2\pi^2\sqrt{KB}} \ln\left(\frac{L}{d}\right) \tag{22}$$

where L corresponds to the sample size ($L \sim v^{1/3}$).

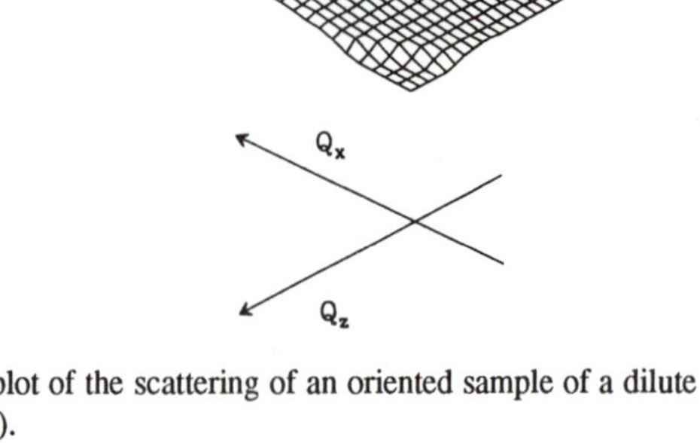


Fig.5. Three dimensional plot of the scattering of an oriented sample of a dilute lamellar phase, obtained using neutron scattering (Ref. 20).

As is clear from Eq. 22, the mean square displacement diverges as the sample size L. This effect of (long wave length) thermal fluctuations was recognized very early by Peierls[24] and Landau[25], and is equivalent to the divergence of thermal fluctuations in two dimensional solids. In both cases, the logarithmic divergence of the thermal fluctuations occurs because the smectic phase in three dimensions and the solid phase in two dimensions are each at their lowest marginal dimension, which is the dimension below which no long-range order could exist. Due to the (logarithmic) weakness of the divergence the systems are called quasi ordered; the finite size cutoff is sufficient to limit fluctuations, and solids at two dimensions or smectics at three dimensions can be observed for finite samples. Although thermal fluctuations do not completely destroy the ordering, they have consequences on the physics of smectics which are just beginning to be understood[17]. One of the first known consequences affects the X-ray structure factor[26].

The X-ray structure factor is proportional to the Fourier transform of the correlation function

$$G(\mathbf{r}) = \exp[-\frac{1}{2}q_0^2 < (u(\mathbf{r}) - u(0))^2 >] \tag{23}$$

This quantity converges (unlike $<u(\mathbf{r})^2>$), and after some algebra[26] we obtain

$$G(\mathbf{r}) = G(\rho, z) \propto \left[\frac{1}{\rho}\right]^{2\eta} e^{-\eta\left[2\gamma + E_1\left(\frac{\rho^2}{4\lambda z}\right)\right]} \tag{24}$$

where $\gamma$ is Euler's constant, $E_1(x)$ is the exponential integral $E_1(x) = \int_{-x}^{\infty} e^{-t}/t dt$ $\mathbf{r}^2 = z^2 + \rho^2$, and $\rho^2 = x^2 + y^2$.

$\eta$ and $\lambda$ are two parameters related to the elastic constant K, and $\overline{\mathbf{B}}$ as defined in Eq. 14

$$\eta_m = m^2 q_0^2 \frac{k_B T}{8\pi\sqrt{\overline{B}K}} \tag{25}$$

$$\lambda = \sqrt{\frac{K}{\overline{B}}} \tag{26}$$

m is the number of the harmonic of the structure factor.

In the special directions $\mathbf{r}_{/\!/} = (0,0,z)$ and $\mathbf{r}_{\perp} = (x,y,0)$ the correlation function given by Eq. 24 simplifies to a power law

$$G(r) \propto \begin{cases} \left[\frac{d}{z}\right]^{\eta}, \mathbf{r} = \mathbf{r}_{\parallel} = (0,0,\mathbf{z}) \\ \left[\frac{\lambda d}{\mathbf{r}^2}\right]^{\eta}, \mathbf{r} = \mathbf{r}_{\perp} = (\mathbf{x},\mathbf{y},0) \end{cases} \tag{27}$$

The Fourier transform of these power laws in the two directions $q_{/\!/}$ and $q_{\perp}$ gives X-ray peaks which are cusps instead of the delta functions which are expected for true solids

$$S(0,0,q_{//}) \propto \frac{1}{[q_{//}-q_m]^{2-\eta_m}} \tag{28a}$$

$$S(q_\perp,0,q_m) \propto \frac{1}{q_\perp^{4-2\eta_m}} \tag{28b}$$

This power law decay of the X-ray structure factor due to the lack of true long range order has been observed for thermotropic smectics[27,28] close to the nematic smectic phase transition and allows accurate determination of $\eta$. It should be emphasized that the very special nature of smectic phases in three dimensions permits, in principle, a direct measurement of K and $\overline{B}$ by an analysis of the X-ray structure factor.

X-ray studies of dilute lamellar phases[12,29,30] have shown that the Landau-Peierls effect may be even more important than in thermotropic smectic phases. Indeed, Synchrotron high resolution measurements allow precise measurements of $\eta$ and consequently give information on the behavior of the elastic constants of these systems as a function of the dilution. During the last few years, systematic studies of different systems of dilute lamellar phases have led to a better understanding of interactions between membranes, their relevance for biological systems, and the identification of molecular parameters responsible for the rigidity of the membranes.

## 10. Conclusion

We hope that we have demonstrated several important features of dilute lamellar phases. First, as illustrated by the strong Landau-Peierls effect observable in these systems, they are very useful for studying the properties of smectics. Moreover, the stability of these systems for a wide dilution range has allowed accurate and convincing measurements of interaction forces. The principal result of these studies is to demonstrate the important role of the flexibility (characterized by the bending constant $\kappa$) and its effect on the interactions between membranes by means of undulation interactions.

The experimental techniques which have been used are complementary. The high resolution X-ray measurements used on powder samples are obviously of great practical interest, since they permit the direct determination of the coefficient $\eta\{\propto(K\,\overline{B})^{-1/2}\}$. Using the Landau Peierls analysis to obtain the structure factor, there is no need for other models to obtain this coefficient from the experimental data. In the case of undulation interactions, this coefficient illustrates the remarkably universal behavior of these interactions. Indeed, $\eta$ is only a function of the ratio between the membrane thickness and the repeat distance, and it approaches a universal number (4/3) for very large distances. In principle, this approach allows also the determination of the length $\lambda$, which together with $\eta$ allows for a determination of the two elastic constants K and $\overline{B}$. However, the latter coefficient is obtained by analysis of the asymmetry of the peak, and consequently is much more subtle and less accurately determined than $\eta$. With this technique, another problem arises due to the appearance of small angle scattering for a d spacing larger than 200Å , as a result of concentration fluctuations[20].

We did not mention another technique which permits determination of the elastic constants. In fact, dynamic light scattering on oriented samples allows a separate determination of the flexibility (K) through the undulation mode and the interactions ($\overline{B}$) through the baroclinic mode[21]. The possibility of selecting the wave vector and the polarization of light, so as to separate concentration from order parameter fluctuations, and out-of-plane from in-plane scattering, makes this approach very promising.

Concerning the measurements, all approaches agree on the main features of these systems. When membrane interactions are dominated by electrostatic forces, the Poisson-Boltzman solution gives a very good representation of these interactions and confirms a well established fact using other techniques[11]. The most novel result, in the studies of these dilute lamellar phases is the existence of undulation forces. Indeed, even thought this interaction was proposed as a possible membranemembrane interaction by Helfrich ten years ago, its relevance has been unclear because of a lack of systems and techniques. Both the d and κ dependence of this interaction ($1/d^2\kappa$) have been confirmed. The principal reason for the dominance of this interaction is due to the large flexibility of certain membranes when cosurfactant molecules are added to regular surfactant (SDS or lipids). Measurements of κ indicate that $\kappa \approx k_B T$, and the choice of the cosurfactant is often crucial for obtaining such a small value.

The relevance of this mechanism for real cell membrane-membrane interactions is not clear. However, even if this system is still very far from a real membrane, the investigation of the effect of pentanol on the interactions between lipidic membrane of DMPC is a first step in assessing the biological relevance of undulation interactions[30].

ACKNOWLEDGEMENTS. I would like to aknowledge the help of my colleagues who have participated to the work presented here: S.T. Milner, F. Nallet, J. Prost, C.R. Safinya and R. Strey.

## References


1. W. Helfrich, Z. Naturforsh. **28a**: 693 (1973).
2. D. R. Nelson, in "Statistical mechanics of membranes and surfaces", eds. D.R. Nelson, T. Piran and S. Weinberg (World Scientific, Singapore, 1989).
3. W. Helfrich, J. Phys. France, **46**: 1263 (1985).
4. L. Peliti and S. Leibler, Phys. Rev. Lett. **54**: 1690 (1986).
5. F. David, in "Statistical mechanics of membranes and surfaces" eds. D.R. Nelson, T. Piran and S. Weinberg (World Scientific, Singapore, 1989).
6. P.G. de Gennes and C. Taupin, J. Phvs. Chem. **86**: 2294 (1982).
7. D. Roux and M. E. Cates, in "Proceedings of the 4th Nishinomayia Yukawa Symposium" Springer Verlag, to be published.
8. W.J. Benton and C.A. Miller, J. Phvs. Chem., **87**: 4981 (1983); M. Dvolaitzky, R. Ober, J. Billard, C. Taupin, J. Charvolin and Y. Hendrix, C. R. Acad. Sci. **45**: 295 (1981); D. Roux and A. M. Bellocq, in "Physics of Amphiphiles", ed. V. Degiorgio and M. Corti, North Holland Amsterdam, (1985); A. M. Bellocq and D. Roux in "Microemulsions Structure and Dynamics" eds. S. Friberg and P. Bothorel (CRC Press, Boca Baton, 1987) p. 33; and, J.M. Dimeglio, M. Dyolaitsky and C. Taupin, J. Phys. Chem. **89**: 871 (1985).
9. F. Larche, J. Appell, G. Porte, P. Bassereau and J. Marignan, Phys. Rev. Lett., **56**: 1700 (1985).
10. W. Helfrich, Z. Naturforsh, **33a**: 305 (1978).
11. A. Parsegian, N. Fuller and R.P. Rand, Proc. Natl. Acad. Sci. **76**: 2750 (1979).
12. D. Roux and C. Safinya, J. Physique, **49**: 307 (1988).
13. J.N. Israelachvili, in "Intermolecular and surface forces", Academic Press, Orlando, (1985).
14. W. Helfrich and R.M. Servuss, Il Nuovo Cimento **3**: 137 (1984).
15. R. Strey, R. Schomaecker, D. Roux, F. Nallet and U. Olsson J. Chem Soc. Faraday Trans., **86**: 2253 (1990).
16. P. Ekwall, L. Mandell and K. Fontell, J. Colloid Interface Sci., **33**: 215 (1970); and, K. Fontell, J. Colloid Interface Sci., **44**: 318 (1973).
17. J. Prost in "Advances in Physics", **33**: 146 (1984).
18. P.G. de Gennes in "The Physics of Liquid Crystals", (Clarendon Oxford 1974).
19. F. Brochard and P. G. de Gennes, Pramana. suppl., **1**: 1 (1975).
20. F. Nallet, D. Roux and S.T. Milner, J.Physique France, **51**: 2333 (1990) G. Porte, J. Marigna, P. Bassereau and R, May, Europh. Lett., **7**: 718 (1988).
21. F. Nallet, D. Roux and J. Prost, Phys. Rev. Lett., **62**: 276(1989); F. Nallet, D. Roux and J. Prost, Progr. Colloid Polvm. Sci. **79**: 313 (1989); and, F. Nallet, D. Roux and J. Prost, J. Phys. France, **50**: 3147 (1989).
22. T. C. Lubensky, J. Prost and S. Ramaswamy, J. Physique France, **51**: 93 (1990).

23. Y. Kantor and M. Kardar Europh. Lett., **9**: 53 (1989); J. Toner, Phys. Rev. Lett., **62**: 905 (1989); and, F. Pincus, J. F. Joanny and D. Andelman, Europhysics Lett., **11**: 76 (1990).
24. R. E. Peierls, Annls. Inst. Henri Poincare 5 177 (1935) Helv. Phys. Acta **7**, Suppl. **81** (1934).
25. L.D. Landau, Phys. Z. SowjUn., **2**: 26 (1937) or collected papers of L.D. Landau edited by D. Ter Haar (Gordon and Breach, NY 1965), p. 209.
26. A. Caille, C. R. Hebd. Acad. Scien. Ser. B, **274**: 891 (1972); and, L. Gunther, Y. Imry and J. Lajzerowics Phys. Rev. A, **22**: 1733 (1980).
27. J. Als-Nielsen, J.D. Litster, R.J. Birgeneau, M. Kaplan, C.R. Safinya, A. Lindegaard-Andersen and S. Mathiesen, Phys. Rev. B, **22**: 312 (1980).
28. J. Als-Nielsen in "Symmetries and broken Symmetries", edited by N. Bocarra (Paris IDSET), p. 107.
29. C. R. Safinya, D. Roux, G.S. Smith, S.K. Sinha, P. Dimon, N.A. Clark and A.M. Bellocq, Phys. Rev. Lett., **57**: 2718 (1986); and, P. Bassereau, J. Marignan, G. Porte, J. Physique, **48**: 673 (1987).
30. C.R.Safinya, E. Sirota, D. Roux and G.S. Smith, Phys. Rev. Lett. **62**: 1134 (1989).

Chapter 27

# PHASES AND PHASE TRANSITIONS IN NEMATIC LYOTROPICS

R. BARTOLINO, G. CHIDICHIMO, A. GOLEMME and F.P. NICOLETTA

## 1. Introduction

Mixtures of amphiphilic molecules and water may form nematic lyotropic liquid crystals characterized by aggregates of amphiphiles with finite shape and long range orientational order. The Liquid Crystal International Community refers to them as positive uniaxial, biaxial and negative uniaxial nematics[1,2,3]. Such systems occur in a concentration range between the diluted isotropic micellar phase, characterized by small spherical aggregates, and the more packed hexagonal, or lamellar, phases, in which the aggregates are infinite cylinders with bidimensional hexagonal symmetry or infinite lamellae in a unidimensional lattice respectively[4].

Such richness of mesophases has been a driving force for experimental and theoretical studies of phase transitions.

This chapter is divided into two parts. In the first part we present a brief review of experimental work on lyotropic nematics by treating the micellar shape and size in the three lyotropic nematic mesophases, and their mean field behaviour.

In the second part we report some more recent results obtained from Small Angle Neutron Scattering and from $^{2}$H-NMR measurements performed on the potassium laurate, decylammonium hydrochloride, water system.

Up to 1980 there were only two lyotropic nematic mesophases uniaxial positive and uniaxial negative, described respectively as cylinderlike nematics, Nc, and disk-like nematics, Nd. This distinction (which is not unique, because it is possible to obtain uniaxial positive disk-like nematics and uniaxial negative cylinder-like nematics, e.g. in fluorinated systems[5]), was due to the idea, later tested experimentally, that the micellar shape was an oblate or a prolate ellipsoid with a long range orientational order along a specific direction, called the director **n**[4,6].

The systems that give rise to nematic mesophases are binary, ternary or more-component solutions of amphiphiles. We should emphasize that even if it is possible to obtain a lyotropic nematic mesophase with a binary system, up to now all the richness of the mesophase varieties is exhibited only with at least three components. We report the sodium decylsulphate, decanol, water system phase diagram in Fig. 1, as an example[4].

---

R.Bartolino and F.P.Nicoletta - Dipartimento di Fisica, GNSM and INFM Unità di Cosenza, Università della Calabria, 87036 Arcavacata di Rende (Cosenza), Italy.
G.Chidichimo and A.Golemme - Dipartimento di Chimica, Università della Calabria, 87036 Arcavacata di Rende (Cosenza), Italy.

*Phase Transitions in Liquid Crystals*, Edited by S. Martellucci and A.N. Chester, Plenum Press, New York, 1992

Theoretical works[7] had predicted the existence of a third nematic phase, more ordered than the previous ones and characterized by a second, **m**, and a third, **l**, director, mutually perpendicular and perpendicular to the conventional uniaxial director **n**. This new phase, called biaxial, should have long range orientational order also along **m** and **l**, and it should be between the two uniaxial nematic phases in a concentration phase diagram. The phase transitions among these mesophases is expected to be second order and the biaxial mesophases should be a mixture of disks and cylinders, even if such a mixture should be thermodynamically unstable[8].

This discrepancy, and the experimental delays in finding such a phase, led to the idea that the biaxial nematic mesophase was only a nice theoretical exercise. In 1980 Yu and Saupe reported the first experimental evidence for the existence of a biaxial nematic phase, Nbx, in a lyotropic mixture of potassium laurate, decanol and water[1]. This phase was found in the middle between the disk and the cylindrical region and it gave rise to a reentrant phenomenon. In fact, it was possible to have the following sequence of mesophases as a function of temperature Nd-Nbx-Nc-Nbx-Nd. The phase transitions were second order as theoretically predicted and there was no experimental evidence, from $^2$H-NMR measurements, of a biaxial mesophase consisting merely of a mixture of the two uniaxial mesophases. Such findings suggested that it was possible to reach the cylinder-like shape by a continuous change of the shape of disk-like aggregates passing through biaxial micelles. These shape modifications were accomplished by the transferring of the amphiphiles from the micelles to the solution and viceversa.

The pioneering works of Yu and Saupe stimulated research on other lyotropic nematic systems with a biaxial phase. No more than two years later, birefringence measurements[2] on sodium decylsulphate, decanol, water showed that this system has a biaxial phase between the uniaxial ones. It seemed then that it was easy to find ternary systems with a biaxial nematic phase, but this was not the case. Up to now only one system besides the previous two presents all three nematic phases excepting systems in which only atom is chemically substituted as in the rubidium laurate, decanol, water system[9]. Of these, the potassium laurate, decylammonium hydrochloride, water[3] system has the advantage of being the most stable against deterioration since esterification of alcohols does not occur. The absence of biaxial nematics in binary systems of one amphiphile in water suggests that some particular effects are related to the presence in the solutions of two or more amphiphiles with different chain lengths and/or of different polar heads. In fact, it was found that in lyotropics segregational effects take place and the different amphiphiles are not uniformly distributed. Some molecules have higher concentration in the flat part of the aggregate, others in that part with a smaller curvature radius, as confirmed by $^2$H-NMR in the ribbon phase[10] and by neutron experiments with a contrast technique in nematics[11].

Each micelle is constituted by ~ 1000 amphiphiles packed by the hydrophobic effect, and their typical dimensions are ~ 60 Å × ~ 60 Å × ~30 Å as determined by X-ray diffraction experiments by Charvolin et al.[4]. From a macroscopic point of view, lyotropic nematics appear as homogeneous fluid solutions, more viscous than ordinary liquids and characterized by very low birefringence ($\Delta n \sim 10^{-3}$) as compared with thermotropics. Strong magnetic fields must be applied to obtain a well aligned sample, even if it is only intended as an orienting medium in NMR experiments[12]. It is possible to lower the required magnetic field intensity by doping the mesophases with ferrofluids (~ $10^{-4}$ %). Ferrofluids are colloidal suspensions of magnetic ($Fe_3O_4$) particles, coated with stabilizers. In the presence of a small magnetic field such particles give rise to a "grating" over the whole sample that produces a favored alignment direction of the micelles. This kind of doping is known not to introduce any change in the phase diagram of the system[13].

Lyotropic nematics follow a mean field behaviour, as is shown by the linear variation of the invariants of the order parameter as a function of temperature obtained from conoscopic measurements[14], by the values of some critical exponents at the phase transition to the isotropic phase[15], by the critical slowing down of the two pretransitional relaxation modes of the order parameter[16].

From these last measurements it is possible to obtain the correlation length, which is always larger than ~ 300 Å and of course larger than the typical micellar size. This might exclude the possibility of nematic phase transitions as driven by changes of their shapes[17].

Further experimental evidence which agrees with this hypothesis appears in the local biaxiality obtained by neutron scattering experiments[11] and the pseudo-lamellar structure observed by synchrotron X-ray measurements[18]. These experiments suggest that the intrinsic micellar shape is always biaxial and the different nematic mesophases are the result of overall fluctuations.

## 2. SANS and $^2$H-NMR Experiments

We have studied the potassium laurate (KL), decylammonium hydrochloride (DaCl), water system by Small Angle Neutron Scattering (SANS) and $^2$H-NMR experiments.

The choice of this particular system has been suggested by a recent work[3] in which the KL-DaCl-$H_2O$ system was found to exhibit all three nematic phases as a function of temperature and concentration.

### 2.1. SANS Experiments

SANS experiments on combinations of fully deuterated components were performed using the PAXY spectrometer at the Orphée reactor in Saclay.

In particular, we have prepared four different series:

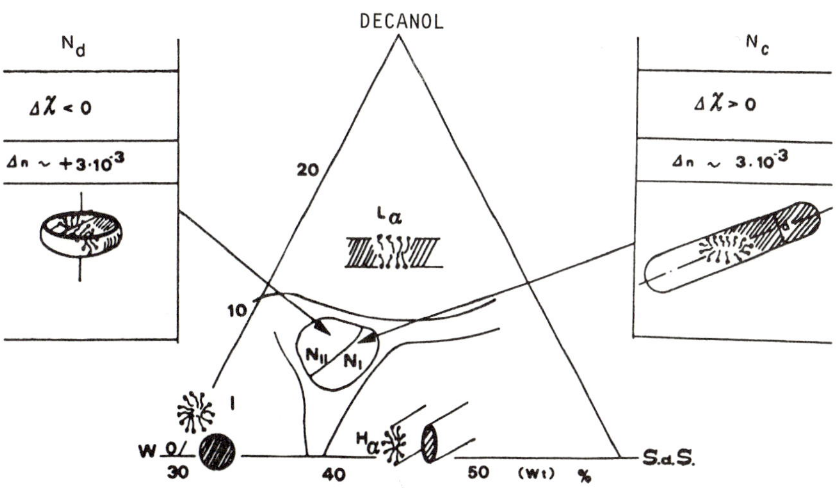


Fig.1. Phase diagram of Sodiumdecylsulphate-decanol-water with some features of the uniaxial mesophases, known up to 1980.

S1) $KL_{d23}$, DaCl, $D_2O$;
S2) KL, $DaCl_{d21}$, $D_2O$;
S3) $KL_{d23}$, $DaCl_{d21}$, $H_2O$;
S4) KL, DaCl, $D_2O$.

The molar ratio between amphiphiles and water is kept constant (0.05), whereas the relative molar of the two amphiles is varied within the range 0.14 to 0.15.

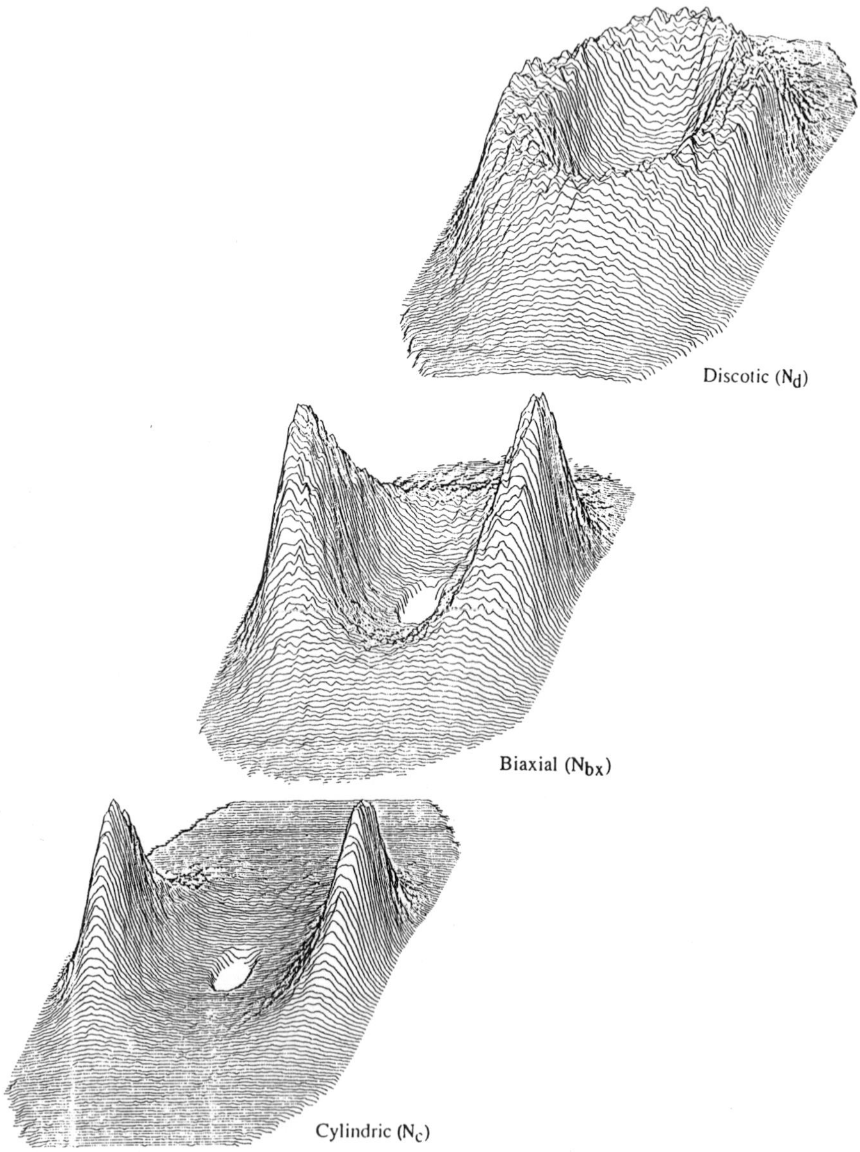


Fig.2. 3D neutron diffraction pattern from KL-DaCl-$D_2O$ system in the three nematic mesophases.

Samples were sealed in Hellma quartz cells, 1 mm in thickness, and an horizontal magnetic field, **H**, of 20 kG was used normal to the neutron beam. Some samples were also doped with a very small quantity of ferrofluid to obtain an easier alignment. The cells were located in an oven, the temperature of which was stabilized at ~± 0.02 °C. The scattered neutrons were detected by a 128 × 128 element multidetector placed ~ 2.075 m from the sample. The neutron wavelength was 6.00 Å.

The phases were identified by optical and NMR techniques[3]. For some samples of the different series the mesomorphism is that reported in Table 1. We emphasize that the isotopic substitutions can change the sequence of mesophases.

Table 1. Mesomorphism and phase transition temperatures for the samples reported in the text.

| | |
|---|---|
| 1) $I\text{-}^{8.0}\text{-}Nd\text{-}^{13.7}\text{-}Nbx\text{-}^{17.2}\text{-}Nc\text{-}^{34.0}\text{-}I$ | R = 0.140 |
| 2) $I\text{-}^{9.1}\text{-}Nd\text{-}^{14.2}\text{-}Nbx\text{-}^{17.8}\text{-}Nc\text{-}^{36.3}\text{-}Nbx\text{-}^{36.3}\text{-}Nd\text{-}^{41.0}\text{-}I$ | R = 0.150 |
| 3) $I\text{-}^{5.1}\text{-}Nd\text{-}^{7.4}\text{-}Nbx\text{-}^{14.2}\text{-}Nc\text{-}^{27.0}\text{-}Nbx\text{-}^{31.4}\text{-}Nd\text{-}^{45.1}\text{-}I$ | R = 0.145 |
| 4) $I\text{-}^{11.0}\text{-}Nd\text{-}^{14.6}\text{-}Nbx\text{-}^{21.2}\text{-}Nc\text{-}^{30.1}\text{-}Nbx\text{-}^{32.0}\text{-}Nd\text{-}^{46.0}\text{-}I$ | R = 0.145 |

If we assume that the micelles of our lyotropic system are bricks, the dimensions of which are A, B, C (with A > B > C), we expect that they will orient with the largest surface of the bricks perpendicular to the neutron beam, due to the magnetic field and to the cell wall interactions. In the Nc phase, the neutron diffraction patterns will be sensitive to the A dimension along the **H** axis, and to a convolution of the B and C dimensions along the direction perpendicular to **H**, according to the model of free rotations along the magnetic field.

In the Nd phase the axis of such free rotations will be along the neutron beam direction, and the diffraction patterns will be characterized by the convolution of A and B bands in the plane perpendicular to the beam. If the micelles are not well oriented, or if they exhibit some fluctuations out from that plane there will be a contribution of the C band as well. In the Nbx phase the rotations are replaced by small amplitude fluctuations and the diffraction is, in principle, due to a convolution of A, B, and C bands. The neutron diffraction patterns are very similar for all the different series and further studies of our experimental data are still in progress[20]. We report in Fig. 2 some 3D images of the reconstructed scattered neutron pattern of sample S4 in the three different mesophases, Nd, Nbx, Nc. It is possible to obtain the values of A, B, and C from vertical and horizontal cuts. The XY neutron spectra cuts were fitted by one or two dimension gaussian or lorentzian functions; thereafter the nature, position, width and relative intensity of the peaks were calculated. Such fits always reveal the presence of A, B and C bands in the three nematic mesophases. In Figs. 3 and 4 we report the positions of the peaks and their relative heights. We observe that the A, B and C dimensions are almost the same (even if there is a small increase in the A dimension) but the relative heights of the peaks change. The averaged volume of the micelles is ~ 98 Å × ~ 71 Å × ~ 53 Å as obtained from the positions of A, B, C bands, in agreement with X-ray measurements.

## 2.2. *$^2$H-NMR on Deuterium of the $D_2O$ Molecule*

It has been demonstrated[3] that $^2$H-NMR is a powerful tool for determining the phase boundaries and the asymmetry parameter of the averaged electrical field gradient tensor by

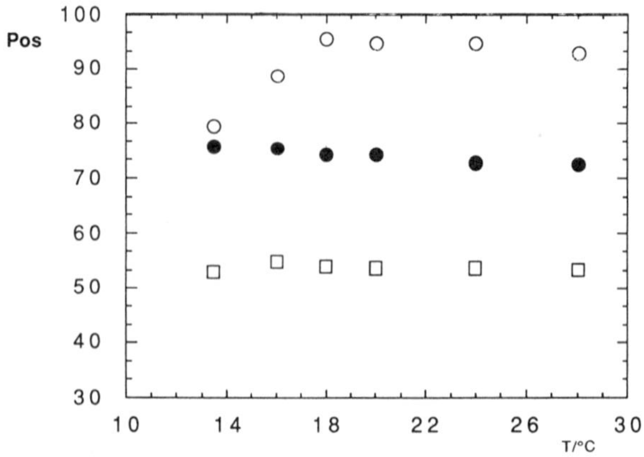


Fig.3. Positions in arbitrary units of A, B and C diffraction peaks as a function of temperature in KL-DaCl-$D_2O$ system. Dots: A peaks; Full dots: B peaks; Square: C peaks.

following the evolution of the $D_2O$ spectra as a function of temperature and of sample orientation with respect to the magnetic field. In particular we have performed for each temperature three different $^2$H-NMR measurements:

1) static measurements;

2) measurements after a $\pi/2$ rotation of the sample along an axis normal to the external field; and,

3) measurements with the sample spinning along an axis normal to the external field.

For these three different measurements the following formula for the reduced quadrupole resonance of the $D_2O$ holds[19]:

$$\nu_{\pm} = \pm\frac{3}{4}\nu\left[\frac{3\cos^2\theta_0 - 1}{2} + \frac{\eta\sin^2\theta_0\cos 2\phi_0}{2}\right] \tag{1}$$

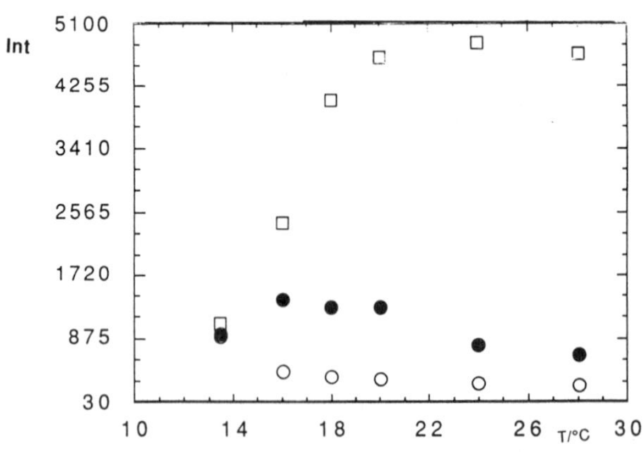


Fig.4. Experimental behaviour of the intensity of A, B and C diffraction peaks as a function of temperature in KL-DaCl-$D_2O$ system. A peaks open dots; B peaks full dots; C peaks square. There is an increase of A peak intensity respect to B and C intensities as the temperature increases.

where $\bar{\nu}$ is the averaged quadrupolar coupling constant, $\theta_0$ and $\varphi_0$ are the polar and azimuthal angles of the external magnetic field in the phase reference frame, and $\eta$ is the asymmetry parameter, which is given by $(V_{xx}-V_{yy})/V_{zz}$, (the $V_{ii}$ are the field gradient tensor components and in the usual convention $|V_{xx}| < |V_{yy}| < |V_{zz}|$). $^2$H-NMR experiments on deuterium of the $D_2O$ molecule were performed on a Varian XL100 spectrometer with a temperature control accuracy of $\pm 0.5$ °C. The intensity of the magnetic field is ~ 23 kGauss. The samples were sealed in glass tubes and their mesomorphism was tested by optical microscopy. In the Nd phase the magnetic field will orient the micelles with $\theta_0 = \pi/2$, as the Nd mesophase has a negative diamagnetic anisotropy. If we perform a $D_2O$ spectrum we will obtain a doublet if this value of $\theta_0$ is present throughout the sample. Each micellar frame is characterized by different values of $\varphi_0$ (which does not affect formula (1) since $\eta = 0$) as can be seen if we perform a $\pi/2$ rotation experiment. In this case, if the acquisition time is faster than the reorientation time, it will be possible to detect a typical "cylindrical" spectrum, i.e. with every $0 < \theta_0 < \pi$ equally probable. Spinning along a direction perpendicular to the **H** axis will force the director to align along such a direction which is the only unperturbed one. Thus the sample will be characterized by the same value of $\theta_0$ and, of course, by the same lineshape as in the static case even if the director distribution is quite different. In the Nc phase the director will be parallel to **H** ($\theta_0$=0) and, after a $\pi/2$ rotation, the splittings will be half as wide as in the static case. If we spin such a sample there will be two sets of singularities due to the static spectrum and to the $\pi/2$ rotated one. In the biaxial mesophase we expect an intermediate behaviour between the previous two. The values and/or distributions of the angles $\theta_0$ and $\varphi_0$ are reported in Table 2. according to the different measurements which were performed. These values, when put into Eq. 1, give rise to spectral lineshapes separated by $\Delta\nu = \nu_+ + \nu_-$ as reported in Table 3.

Table 2. Values of the angles $\theta_0$ and $\varphi_0$ for different phases. The two different biaxial cases, $N_{bxa}$ and $N_{bxb}$, are the two cases shown in Fig. 3 of Ref. 3.

| Mesophase | Static | $\pi/2$ Rotation | Spinning |
|---|---|---|---|
| $N_d$ | $\theta_0 = \pi/2, \eta = 0$ | $\theta_0 = \text{distr.}, \eta = 0$ | $\theta_0 = \pi/2, \eta = 0$ |
| $N_{bxa}$ | $\theta_0 = \pi/2, \phi_0 = \pi/2$ | $\theta_0 = \pi/2, \phi_0 = 0$ | $\theta_0 = \pi/2, \phi_0 = \text{distr.}$ |
| $N_{bxb}$ | $\theta_0 = 0$ | $\theta_0 = \pi/2, \phi_0 = 0$ | $\theta_0 = \text{distr.}, \phi_0 = 0$ |
| $N_c$ | $\theta_0 = 0, \eta = 0$ | $\theta_0 = \pi/2, \eta = 0$ | $\theta_0 = \text{distr.}, \eta = 0$ |

Table 3. Values of the quadrupolar splittings of the $D_2O$ molecule in different phases. $N_{bxa}$ and $N_{bxb}$ refer to the two cases shown in Fig. 3 of Ref. 3. In cases where the spectrum is not single doublet, the angle dependent reduced resonance frequency is given.

| Mesophase | Static | $\pi/2$ Rotation | Spinning |
|---|---|---|---|
| $N_d$ | $\frac{3}{4}\bar{\nu}$ | $\frac{3}{4}\bar{\nu}[3\cos^2\theta_0 - 1]$ | $\frac{3}{4}\bar{\nu}$ |
| $N_{bxa}$ | $\frac{3}{4}\bar{\nu}[1+\eta]$ | $\frac{3}{4}\bar{\nu}[1-\eta]$ | $\frac{3}{4}\bar{\nu}[-1+\eta\cos 2\phi_0]$ |
| $N_{bxb}$ | $\frac{3}{2}\bar{\nu}$ | $\frac{3}{4}\bar{\nu}[1-\eta]$ | $\frac{3}{4}\bar{\nu}[3\cos^2\theta_0 - 1 + \eta\sin^2\theta_0]$ |
| $N_c$ | $\frac{3}{2}\bar{\nu}$ | $\frac{3}{4}\bar{\nu}$ | $\frac{3}{4}\bar{\nu}[3\cos^2\theta_0 - 1]$ |

Experiments confirm that each nematic mesophase can be characterized by a well defined set of lineshapes as a function of the measurement parameters, as shown in Fig. 5. In such a way it is possible to draw the nematic phase boundaries of any lyotropic system.

In particular, we have determined in such a way the phase diagram of KL-DaCl-$D_2O$, which is reported in Fig. 6.

The presence of two Nbx rows in Tables 2 and 3 is due to the fact that there are two possible orientations of the magnetic field with respect to the phase symmetry axes, but at any given time only one agrees with the experiments.

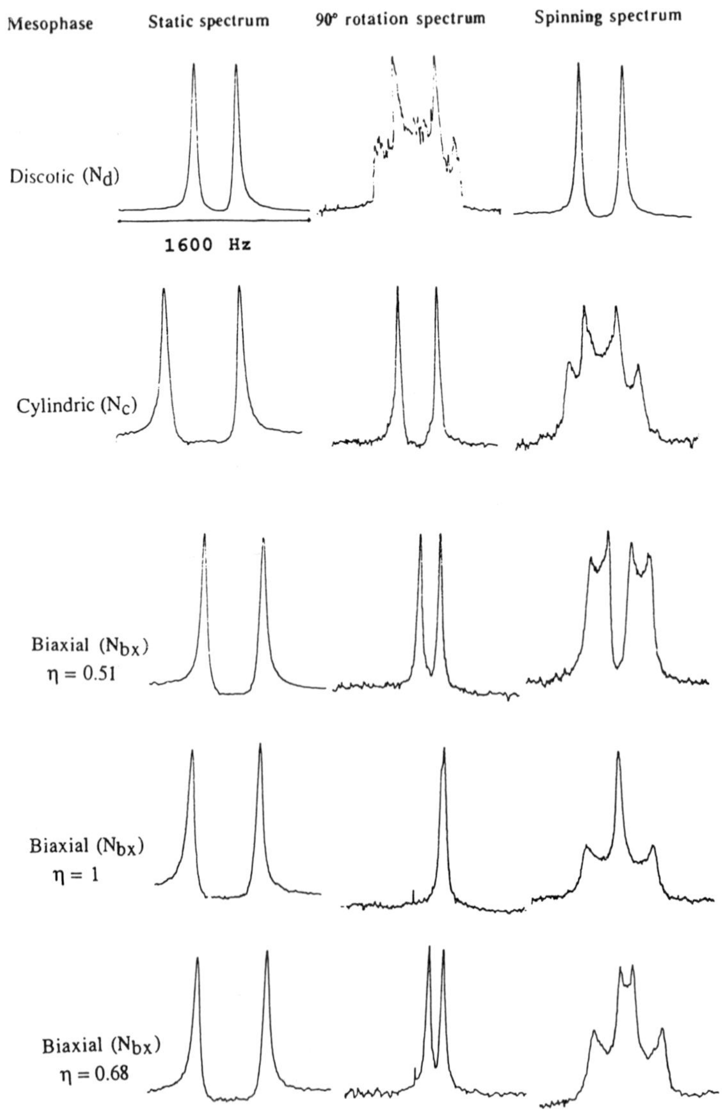


Fig.5. Different $^2$H-NMR lineshapes obtained in the different nematic mesophases as a function of the experimental conditions.

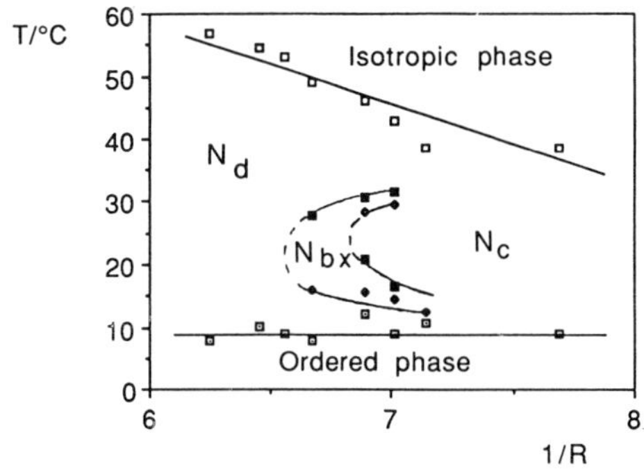


Fig.6. Phase diagram of KL-DaCl-$D_2O$ system as obtained by following the evolution of the $D_2O$ spectra as a function of the temperature and of the sample orientation with respect to the magnetic field.

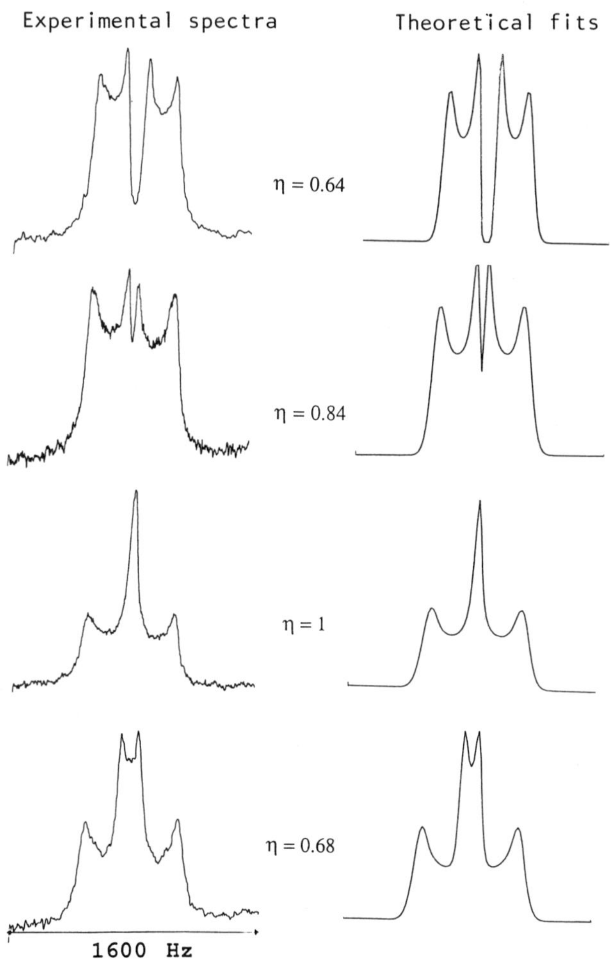


Fig.7. Experimental lineshapes and theoretical fits for different values of the asymmetry parameter, $\eta$.

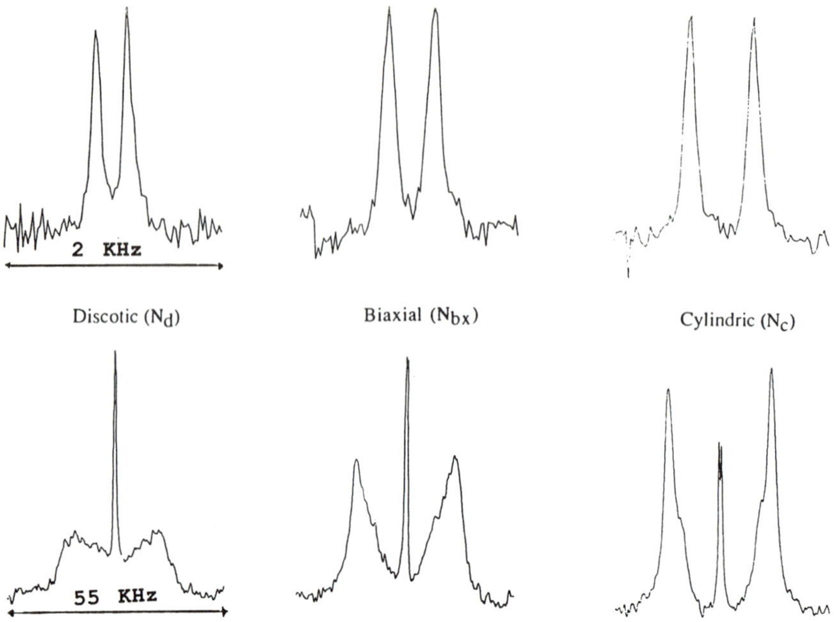


Fig.8. $^2$H-NMR spectra on deuterium in $D_2O$ (upper row), $KL\alpha d_2$ and $DaCl\alpha d_2$ (lower row) in the Nd, Nbx, Nc mesophases.

This finds its explanation in the usual convention for which $|V_{zz}| > |V_{yy}| > |V_{xx}|$ (see Ref. 3 for further details). Experimentally this can be easily seen in the spinning lineshapes of the biaxial nematics in the first and in the last row of Fig. 7, which are quite different even if they have almost the same value of $\eta$. Theoretical fits are reported in Fig. 8 and agree with the experimental lineshapes and the suggested explanations.

We emphasize that the value of the asymmetry parameter used in the fits is that obtained from the experimental lineashapes according to the formulas

$$\eta = \frac{\Delta\nu_s - \Delta\nu_{\pi/2}}{\Delta\nu_s + \Delta\nu_{\pi/2}} \quad \text{and} \quad \eta = \frac{\Delta\nu_s - 2\Delta\nu_{\pi/2}}{\Delta\nu_s} \tag{2}$$

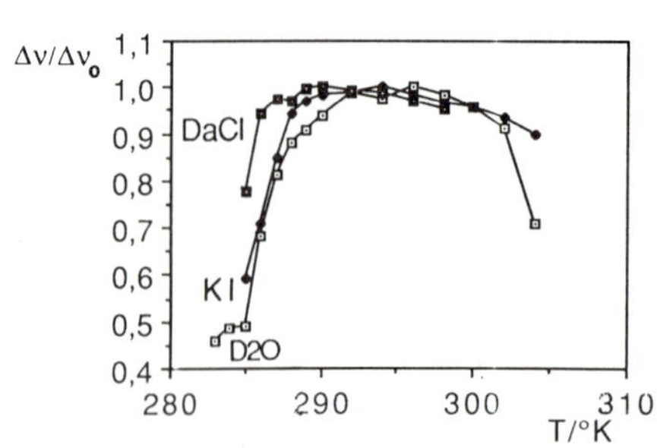


Fig.9. Normalized splittings, $\Delta\upsilon/\Delta\upsilon_0$, from $D_20$, $KL\alpha d_2$ and $DaCl\alpha d_2$ as a function of temperature. $\Delta\upsilon_0$ is the maximum value of the splittings.

which holds for the two possible orientations of H in the phase reference frame. $\Delta\nu_s$ and $\Delta\nu_{\pi/2}$ are respectively the splittings in the static and after $\pi/2$ rotation measurements.

The Eqs. 2 allow us to determine the asymmetry parameter in the Nbx phase as a function of temperature, also.

## *2.3. $^2$H-NMR on Deuterium of the Amphiphiles*

Further information may be achieved when $^2$H-NMR measurements are performed on the deuterium of selectively deuterated amphiphiles. We have deuterated in the alpha position the two amphiphiles of the potassium laurate, decylammonium hydrochloride, water system and determined their averaged electrical field gradient tensor as a function of temperature and concentration. We have performed experiments on samples which contained mostly protonic water in order to prevent the overlapping of the $D_2O$ signals with those from the deuterium on the amphiphilic molecules. The presence of $D_2O$ is important in order to check that no change in composition has occurred during the measurements. We have studied mixtures with only one deuterated amphiphile and the behaviour is similar to the mixtures with both of them deuterated. The use of the singly deuterated amphiphile has the disadvantage that it is not easy to have two identical samples in order to compare the evolution of both lipids.

Experiments were performed on a Bruker MSL300 spectrometer with a temperature accuracy of $\pm$ 0.5 °C and a magnetic field of ~ 80 kGauss.

If we assume the presence of a non uniform distribution of the two amphiphiles, due to segregational effects, we expect that $^2$H-NMR on the deuterium of the amphiphiles could be sensitive to differences of phase transitions driven by changes of full rotations and/or fluctuations or by changes in the amphiphile dynamics.

In Fig. 8 we report a sequence of spectra obtained from samples containing deuterium in the alpha position of each amphiphile and $D_2O$ in the Nd, Nbx and Nc phases. In the Nd phase the spectrum of $KL\alpha d_2$ and $DaCl\alpha d_2$ assumes a powder pattern profile, which suggests that the diffusion of the lipids in the aggregates is strongly reduced in such a mesophase.

If we look at the behaviour of the normalized splittings of the three components of our system as a function of temperature, shown in Fig. 9, we observe that the normalized $D_2O$ quadrupolar splitting changes faster than that of amphiphiles at the phase transition between the Nbx and Nc mesophases.

Further experiments are in progress in order to determine the behaviour of the asymmetry parameter of both amphiphiles in the three mesophases[21].

### References


1. L.J. Yu and A. Saupe, Phys. Rev. Lett., **45**, 1000, 1980.
2. R. Bartolino, T. Chiaranza, M. Meuti and R. Compagnoni, Phys. Rev. A, **26**, 1116, 1982.
3. E.A. Oliveira, L. Liebert and A.M. Figueiredo Neto, Liq. Crystals, **5**, 1669, 1989; and, F.P. Nicoletta, G. Chidichimo, A. Golemme and N. Picci, Liq. Crystals, **10**, 665, 1991.
4. J. Charvolin, E.T. Samulsky and A.M. Levelut, J. de Physique Lett., **40**, L587, 1979; and, Y. Hendrikx and J. Charvolin, J.de Physique, **42**, 1427, 1981.
5. N. Boden, P.H. Jackson, K. McMullen and M.C. Holmes, Chem. Phys. Lett., **65**, 476, 1979.
6. Y. Hendrikx, J. Charvolin, M. Rawiso, L. Liebert and M.C. Holmes, J.Phys.Chem., **87**, 3391, 1983; and, M.C. Holmes and J. Charvolin, J.Phys.Chem., **88**, 810, 1984.
7. C.S. Shih and R. Alben, J.Chem.Phys., **57**, 3035, 1972; R. Alben, Phys.Rev.Lett., **30**, 778, 1973; J.P. Straley, Phys.Rev. A., **10**, 1881 , 1974; R. Alben, J. Chem. Phys., **59**, 4299, 1973; A. Stroobants and H.N.W. Lekkerkerker, Mol. Cryst. Liq. Cryst., **89**, 67, 1982; R.G. Caflisch, Z.Y. Chen, A.N. Berker and J.M. Deutch, Phys. Rev. A., **30**, 2562, 1984; and, Z.Y. Chen and J.M. Deutch, J.Chem.Phys., **80**, 2151, 1984.

8. P. Palffy-Muhoray, J.R. de Bruyn and R. Alben, J. Chem. Phys., **82**, 5294, 1975.
9. J. Malthete, L. Liebert, A.M. Levelut and Y. Galerne, Compte R. Ac. Science, **303**, 1073, 1986.
10. G. Chidichimo, A. Golemme, J.W. Doane and P.W. Westermann, J. Phys. Chem., **82**, 536, 1985.
11. Y. Hendrikx, J. Charvolin and M. Rawiso, J. Colloid. Interface Sci., **100**, 597, 1984.
12. K.D. Lawson and T.J. Flautt, J. Am. Chem. Soc., **89**, 5489, 1967; K.D. Lawson and T.J. Flautt, Mol. Cryst. Liq. Cryst., **7**, 322, 1969; and, K. Radley and L.W. Reeves, Can. J. Chem., **53**, 2998, 1975.
13. L. Liebert and A. Martinet, J. de Phys. Lett., **40**, L363, 1979.
14. A.M. Figueiredo Neto, Y. Galerne, A.M. Levelut and L. Liebert, J. de Phys. Lett., **46**, L999, 1985; and, A.M. Figueiredo Neto, L. Liebert and Y. Galerne, J. Phys. Chem., **89**, 3737, 1985.
15. S. Kumar, J.D. Litster and C. Rosemblatt, Phys. Rev. A, **28**, 1890, 1983; S. Kumar, L.J. Yu and J.D. Litster, Phys. Rev. Lett., **50**, 1672, 1983; and, C. Rosemblatt, S. Kumar and J.D. Litster, Phys. Rev. A, **29**, 1010, 1984.
16. M.B. Lacerda Santos, Y. Galerne and G. Durand, Phys. Rev. Lett., **53**, 787, 1984; and, M.B. Lacerda Santos and G. Durand, J. de Physique, **47**, 529, 1986.
17. Y. Galerne, Mol. Cryst. Liq. Cryst., **165**, 131, 1988.
18. A.M. Figueiredo Neto, Y. Galerne, A.M. Levelut and L. Liebert, J. de Phys. Lett., **46**, L499, 1985; and, A.M. Figueiredo Neto, L. Liebert and Y. Galerne, J. Phys. Chem. **89**, 3737, 1985.
19. D.J. Photinos, P.J. Bos, J.W. Doane and M.E. Neubert, Phys. Rev. A, **20**, 2203, 1979.
20. R. Bartolino, V. Formoso, F.P. Nicoletta, Y. Galerne and G. Pepy, to be published.
21. F.P. Nicoletta, G. Chidichimo, A. Golemme and N. Picci, to be published.

Chapter 28

# X-RAY STUDIES OF LYOTROPIC LIQUID CRYSTALS

R. SHASHIDHAR

There are two broad classes of liquid crystalline systems, the thermotropic and the lyotropic. The historical difference between these two is that the lyotropic systems are always mixtures (or solutions) of unlike molecules in which one is a nonmesogenic liquid. Solutions of soap and water are typical examples of lyotropics. Their mesomorphic phases appear as a function of either concentration or temperature. In contrast, the thermotropic systems are usually formed from a single chemical component, and the mesomorphic phases appear primarily as a result of temperature changes. The molecular distinction between the two is that one of the molecules in the lyotropic solution always has a polar hydrophilic part, often called the "head group", and one or more hydrophobic alkyl chains called "tails". Although these molecules can form mesogenic phases as single component systems, they more easily do so in solution with either water or oil. This is due to a competition between the hydrophilic and hydrophobic interactions, as well as other things such as packing and steric constraints[1,2]. The molecules that form thermotropic liquid crystalline phases also have hydrophilic and hydrophobic parts, the disparity in the affinity of these parts for either water or oil is much less and most of these molecules are relatively insoluble in water. Nevertheless, it is interesting that different parts of typical thermotropic molecules do have some of the same features as the lyotropic molecules. For example, although the rod-like thermotropic molecules always have an acyl chain at one or both ends of a more rigid section, the chain lengths are rarely as long as those of the lyotropic molecules. Also, the solubility of different parts of the thermotropic molecules, when separated, are not as disparate as those of the lyotropic molecules, are definitely different.

A typical molecule that exhibits the lyotropic liquid crystalline phase is shown in Fig. 1. It is a phospholipid 1,2-dipalmitoyl-L-phosphatidylcholine (DPPC). The molecule of DPPC has a polar head group which has no net change (Zwitterion), a glycerol backbone and two long saturated hydrocarbon chains. An important feature of amphiphiles such as surfactants and lipids is that they self-assemble into different types of structures in presence of water. These structures are shown schematically in Fig. 2.

Micellar systems tend to form nematic phases. The lipid bilayer systems form liquid crystals which are in several respects similar to the smectic-A of thermotropic systems. The intermolecular forces that are responsible for the self-association of such structures are complex

---

R. Shashidhar - Center for Bio/Molecular Science & Engineering, Naval Research Laboratory, Code 6090, Washington, DC 20375-5000, USA; and Geo-Centers, Inc., 10903 Indian Head Highway, Fort Washington, MD 20744, USA.

*Phase Transitions in Liquid Crystals*, Edited by S. Martellucci and A.N. Chester, Plenum Press, New York, 1992

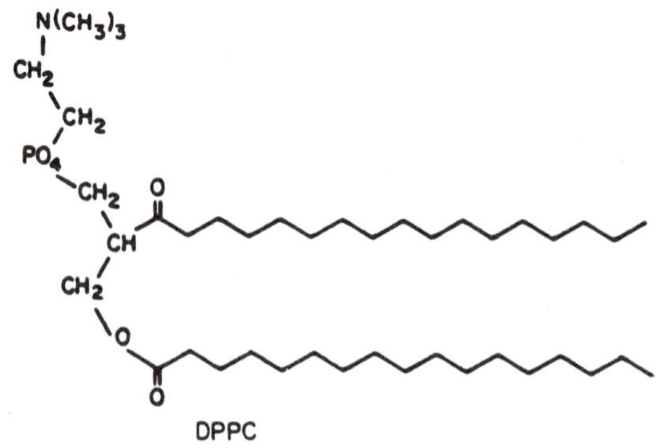


Fig. 1. Chemical formula of 1,2-dipalmitoyl-L-phosphatidylcholine (DPPC)

- a combination of van der Waals forces, hydrophobic and hydrophilic interactions and electrostatic interactions. [See Refs. 3 - 5 for reviews and books on this aspect]. In this Chapter, we will be concerned only with the bilayer structures (membranes) formed by phospholipid-water systems.

Stacked bilayer membranes in the phospholipid-water system are studied for both their

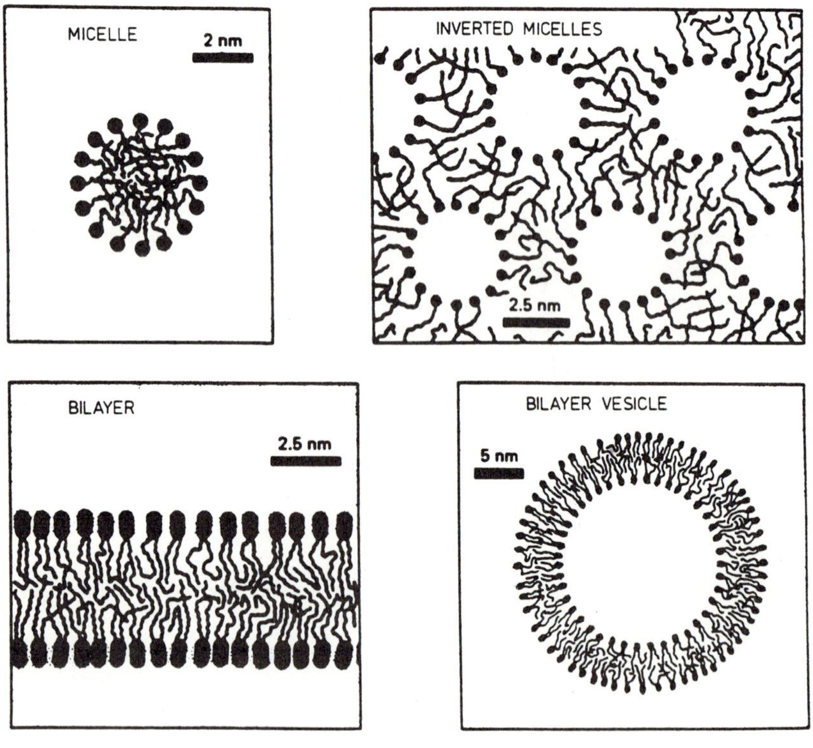


Fig. 2. Self-assembling structures of surfactant and lipid molecules - micelle, inverted micelle, bilayer and unilamellar vesicle.

biological and physical significance. Lipid bilayers are the basic building blocks of living cell membranes. Therefore, a knowledge of lipid-lipid and lipid-protein interactions is essential to understand endocytosis, fusion and other functions of the living cell. For example, previous work[5-9] has shown that these systems exhibit fascinating physical properties due to the complex interplay between attractive van der Waals and repulsive screened electrostatic, hydration and entropic forces that exist between two membrane sheets separated by an aqueous medium. Additionally, precisely because of the ability to change the intermembrane distance through solvent dilution, these systems are prototype models for elucidating the nature of phases and transitions in two dimensions, and the statistical behavior of fluid and ordered surfaces embedded in three dimensional space[9,10].

The structural nature of the phases and the phase diagrams of phospholipids and water has been the topic of many experimental investigations[11-14]. Yet, since the early work of Luzzati[11] and co-workers, essentially the same experimental procedures have been used: a sample of fixed lipid/water concentration (by wt. percent) is sealed in a thin walled sample tube and X-ray diffraction studies performed on these samples as a function of temperature. These experiments showed that the three main phases in the phospholipid-water system are the $L_\alpha$, $L_\beta$ and $P_\beta$ (Fig. 3). All three of these phases exhibit a stacked water-bilayer structure with a well defined repeat spacing, d. The distinguishing feature of the structures is the in-plane order of the hydrocarbon chains. The $L_\alpha$ phase is the highest temperature phase of the three for any given concentration. In this phase, the chains are liquid-like with no regular in-plane structure. In the lower temperature phases, the chains are frozen into a nearly all trans configuration. In the $L_\beta$ phase, the chains form a close packed lattice in which the bilayer planes are flat and the chains are tilted with respect to the bilayer normal. In the $P_\beta$ phase, the chains are still frozen and tilted, but the bilayer planes become "rippled" with a long wavelength modulation of periodicity ~120-200 Å.

Much of the detailed information about the nature of the in-plane structure and all of the information about the in-plane and out-of-plane correlations are lost in the X-ray diffraction patterns when studying polycrystalline powders. Recently, high resolution X-ray studies have been conducted on lipid bilayer systems - both bulk[15] and freely suspended layers[16]. A feature of the X-ray diffraction pattern from a stacked bilayer (lyotropic smectic-A) is that multiple order of diffraction maxima can be observed. This is because the layering is very well defined in the lyotropic smectic-A while it is not so in the case of thermotropic smectic-A. This permits a rigorous test of the Landau-Peierls theory. Also, the problem of sample mosaic which is difficult

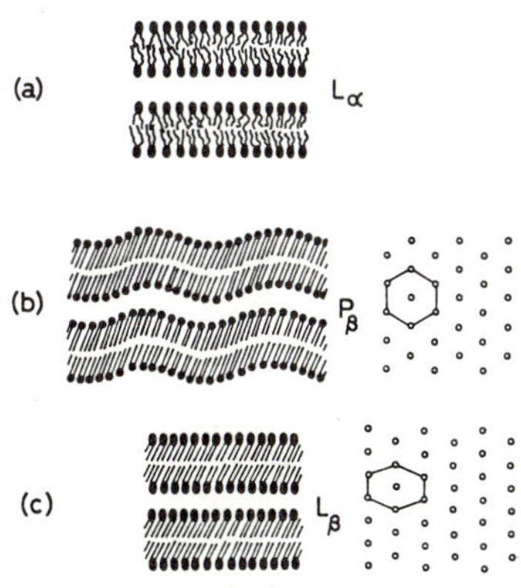


Fig. 3. The structures of $L_\alpha$, $P_\beta$ and $L_\beta$ phases exhibited by DMPC - water systems.

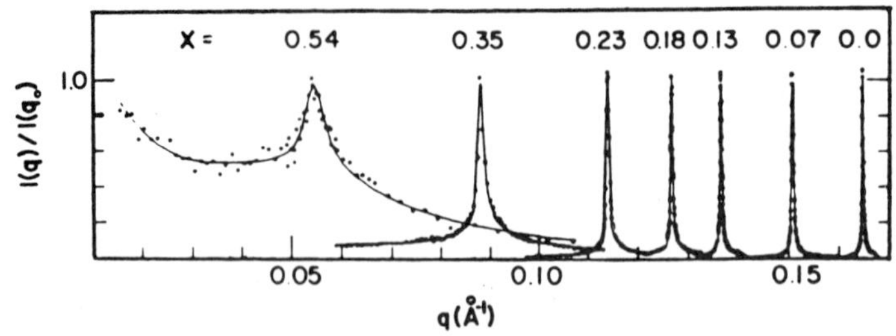


Fig. 4. Longitudinal profiles of the first harmonic of seven different mixtures along the dodecane dilution path. The percentage dodecane by weight of the mixtures (x) is indicated above each profile. All peak intensities are normalized to unity. The solid lines are fits by the Caille power law line shape [from Ref. 26].

to control in thermotropic smectic A has been resolved by using a three dimensional powder. Although the conditions on the analysis are delicate, Safinya et al.[15] demonstrated that for a peflect powder, for which the microcrystals are sufficiently large, the powder line shape does allow unambiguous determination of all of the parameters of the anisotropic line shape. In this Chapter, we shall discuss in some detail results of some of the recent high resolution X-ray studies on lyotropic smectic-A.

Landau and Peierls[20] first demonstrated that the mean square layer displacements of a system of stacked fluid layers diverge logarithmically $<u^2>$ In (L/a), with sample size L, destroying conventional long-range order (a is of order of the intermolecular distance). For the X-ray structure factor, the consequences are dramatic. Caille[24] has shown that for a smectic-A liquid-crystal phase, which has the identical elastic free energy as that for the $L_\alpha$ phase, conventional δ-function Bragg peaks at $(O,O,q_m=mq_0=m2\pi/d)$ (harmonic order m=1,2...) are replaced by singularities with asymptotic power-law behavior

$$S(O,O,q_z) \approx |q_z - q_m|^{-2+m} \quad \text{and} \quad S(O,O,q_z) \approx q^{-4+2m}$$

Here $q_m = m^2 q_0^2 k_B T/8\pi\, x(BK)^{1/2}$ and $q_z$ and $q_{//}$ are components of the wave vector normal and parallel to the layers. This power-law behavior has been confirmed[25] for the first harmonic $q_z = q_0$ in the smectic-A phase of liquid crystals.

Fig. 4 gives the longitudinal profiles of the first harmonic of seven different mixtures along the dodecane dilution path[26]. These data have been obtained using the Brookhaven Synchrotron source and a high resolution set-up with a longitudinal resolution of $8\times10^{-5}$ $A^{-1}$. The line profiles

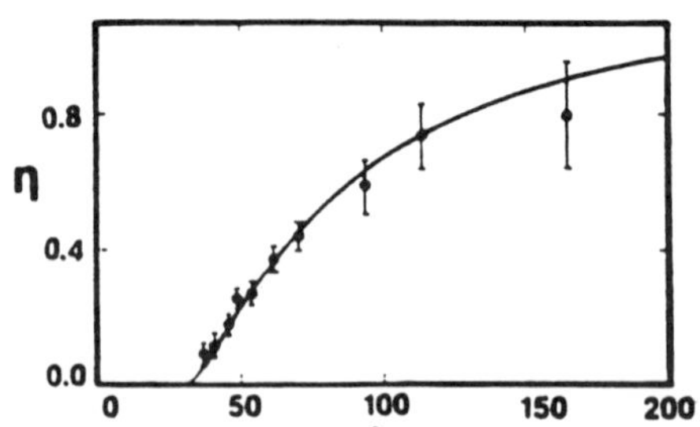


Fig. 5. Power-law exponent as a function of the intermembrane distance for mixtures along the dilution path. The solid line is the prediction of the model of Helfrich of entropically driven interactions [from Ref. 26].

have been carefully analyzed taking into account the instrumental resolutions finite size effects. The appropriately modified finite-size and powder averaged Caille cross-section gives a very good fit to the experimental data (solid lines of Fig. 4). Thus the Landau-Peierls arrangement is unambiguously proven experimentally in lyotropic smectic-A.

The basic forces between two parallel membrane sheets involve attractive van der Waals interactions, repulsive hydration and screened electrostatic forces. Helfrich[17] has proposed that large, thermally induced, out-of-plane layer fluctuations give rise to a repulsive interaction between membranes because of steric hindrance in multilayer systems. More generally, steric repulsions are known to be the dominant interactions associated with wandering walls of the incommensurate phases[18]. This interaction has been calculated for a multilayer system using the Landau-de Gennes[19,20] elastic theory of smectic-A liquid crystals.

The experimental relevance of this interaction has remained unclear. To identify unambiguously the origin of inter-membrane repulsion, it is crucial to carry out studies in uncharged systems. To date, studies of neutral lipids have been limited to systems which exhibit a narrow range of intermembrane separations between 1.0 and 2.0 nm, where the long-range van der Waals and steric interactions are both expected to fall off as $1/d^2$. Moreover, these studies have been associated with lipid systems in which the membrane curvature elasticity[21] $k_c > 20k_BT$, in which case van der Waals interactions are significantly larger. Thus, although the concept of this novel fluctuation-induced steric interaction has stimulated a large body of theoretical and experimental work, its significance in biological membranes remains largely controversial.

To elucidate the relevance of this interaction in multilayer systems, Safinya et al. have conducted a study of the synchrotron diffuse X-ray scattering study in the lamellar $L_\alpha$ phase of the quaternary mixture of sodium dodecyl sulfate (SDS, the surfactant), pentanol (cosurfactant), water, and dodecane as a function of dodecane dilution. An important result emerges from their studies[26]. The scattering exhibits a power law behavior given by

$$S(q) \approx | q - q_0 |^{-P}$$

where $P \approx 1 - \eta$, at $q = q_0$.

A plot of $\eta$ as a function of inter-membrane distance is given in Fig. 5 along with the theoretical prediction (solid line) of Helfrich based on a model of entropically driven steric interactions. The solid line agrees very well with experiments. This then provides compelling evidence that in this SDS multimembrane system, the intermembrane interactions are dominated by the Helfrich mechanism of entropically driven steric interactions.

Another important recent development on the X-ray studies of lyotropics is the study of oriented, freely suspended films of DMPC-water[27]. Upon drawing, such films spontaneously orient with the bilayer planes parallel to the film's surface. These films are drawn in a temperature controlled oven in which the chemical potential of water is controlled by an external humidity system. This technique permits a determination of the in-plane structure as well as the layer spacing (d) for any value of temperature (T) and relative humidity (RH) between O° and 80°C and 0% and 100%, respectively.

These methods have been used to study the detailed structure and correlations in the $L_\beta$ phase of DMPC[16]. It has been shown that the $L_\beta$ actually consists of 3 distinct phases, each distinguished by the direction of the chain tilt with respect to the position of its in-plane nearest neighbors. It is also found that there are no positional correlations across the water layers; hence these phases are shown not to be 3-dimensional in nature, but 2-dimensional structures. In addition, the in-plane correlations are found to be relatively short range ( 200 Å) which suggest a hexatic, but the in-plane lineshapes are also consistent with a 2-dimensional solid with finite size domains. We shall discuss these aspects in the following.

Table 1. The extension of the alphabetical nomenclature of tiled layered liquid crystal phases to include phases with the newly discovered intermediate tilt directions [from Ref. 16].

| $\phi$ | hexatic | crystalline | herring bone |
|---|---|---|---|
| 0° | F | G | H |
| 30° | I | J | K |
| $0° \leq \phi \leq 30°$ | L | M | N |

Using the X-ray data, the T-RH phase diagram has been constructed for the DMPC water system (Fig. 6). There are three new phases: $L_{\beta F}$, $L_{\beta L}$ and $L_{\beta I}$ The $L_\beta$ is retained as meaning a lamellar phase in which the molecules are tilted with respect to the bilayer normal. The labels F and L come from liquid crystal nomenclature where the F phase is a hexatic liquid crystalline phase with the molecules tilting between nearest neighbors and the I phase is a hexatic liquid crystalline phase where the molecules tilt toward nearest neighbors. Table 1 lists the known hexatic, crystalline and herringbone structures as previously labeled. A new row has been added to include those structures in which the molecules tilt in a common direction which is neither toward nor between nearest neighbors.

Continuing the alphabetical labeling as in thermotropics, the phases have been labelled, L, M and N phases. The results indicate that these are lower dimensional systems from the longitudinal scans through the in-plane peaks. Also, since the positional correlations do not extend beyond a single bilayer, it is clear that they are not 3-D crystals.

Thus X-ray scattering studies on lyotropics has been useful in enhancing our understanding of liquid crystals in general.

ACKNOWLEDGEMENTS. Technical editing by Ms. Shelley DeLozier is greatly appreciated.

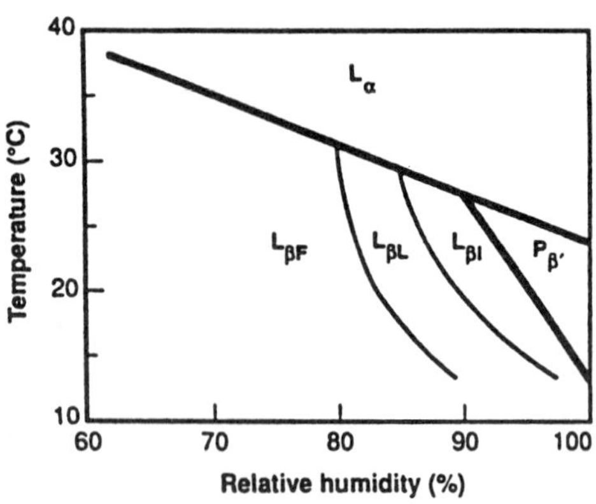


Fig. 6. The phase diagram of DMPC-water as a function of temperature and relative humidity, showing the $L_\alpha$, $P_\beta$ and the newly distinguished $L_{\beta F}$, $L_{\beta L}$ and $L_{\beta I}$ phases. The transitions among the different $L_\beta$ appear second order, while the other transitions are first order [from Ref. 27].

## References


1. P.S. Pershan, "Structure of Liquid Crystal Masses", World Scientific, Singapore (1988).
2. S.A. Safran and N.A. Clark, Editors, "Physics of Complex and Supramolecular Fluids", Wiley, New York (1987).
3. J.N. Israelachvili, "Intermolecular and Surface Forces", Academic, Orlando (1985).
4. J. Meunier, D. Langevin and N. Boccara, Editors, "Physics of Amphiphilic Layers", Springer Verlag, Berlin, Heidelberg, New York (1987).
5. V.A. Parsegian, N. Fuller and R.P. Rand, Proc. Natl. Acad. Sci., **76**: 2750 (1979).
6. A.C. Cowley, N.L. Fuller, R.P. Rand and V.A. Parsegian, Biochem. **17**: 3163 (1978).
7. J.N. Israelachvili, "Intermolecular and Surface Forces", Academic, Orlando (1985); J. Mahanty and B.W. Ninham, "Dispersion Forces", London (1971).
8. C.R. Safinya, D. Roux, B.S.Smith, S.K. Sinha, P. Dimon, N.A. Clark and A.M. Bellocq, Phys. Rev. Lett. **57**: 2718 (1986); D.Roux and C.R. Safinya, J. Phys. (Paris) **49**: 307 (1988); F. Nallet, D. Roux and J. Prost, Phys. Rev. Lett. **62**: 276 (1989).
9. R. Lipowsky and S. Leibler, Phys. Rev. Lett. **56**: 2541 (1986); R.E. Goldstein and. Leibler, Phys. Rev. Lett. **61**: 2213 (1988).
10. Y. Kantor, M. Kardar and D.R. Nelson, Phys. Rev. Lett. **57**: 791 (1986); D.R. Nelson and L. Peliti, J. Phys. (Paris) **48**: 1085 (1987); J.A. Aronovitz and T.C. Lubensky, Phys. Rev. Lett. **60**: 2634 (1988); For a broad discussion see "Statistical Mechanics of Membranes and Surfaces", Ed. D. Nelson, T. Piran and S. Weinberg, World Scientific (1988).
11. V. Luzzati, Biol. Membr. **1**: 71 (1968); A. Tardieu, V. Luzati and F.C. Reman, J. Mol. Biol. **75**: 711 (1973).
12. M.J. Janiak, D.M. Small and G.G. Shipley, J. Biol. Chem. **254**: 6068 (1979).
13. G.L. Kirk, S.M. Gruner and D.L. Stein, Biochemistry **23**: 1093 (1984); G.L. Kirk and S.M. Gruner, J. de Physique 46 (1985); M.B. Schneider and W.W. Webb, J. Phys. (Paris) **45**: 273 (1984).
14. S. Doniach, J. Chem. Phys. **70**: 10 (1979); W.K. Chan and W.W. Webb, Phys. Rev. Lett. **46**: 39 (1981).
15. C.R. Safinya, D. Rons, B.S. Smith, S.K. Sinha, P. Dimon, N.A. Clark and A.M. Bellocq, Phys. Rev. Lett. **57**: 2718 (1986).
16. G.S. Smith, E.B. Sirota, C.R. Safinya, R.J. Plano and N.A. Clark, J. Chem. Phys.(in press).
17. W. Helfrich, Z. Naturforsch, **33a**: 305 (1978).
18. S.G. J. Mochrie, A.R. Kortan, R.J. Birgeneau and P.M. Horn, Z. Phys. **B 62**: 79 (1985).
19. P.G. de Gennes. "The Physics of Liquid Crystals" (Clarendon, Oxford, 1974).
20. L.D. Landau, p. 209; in "Collected Papers of L.S. Landau" edited by D. ter Haar (Gordon and Breach, New York, 1965) R.E. Peierls, Helv. Phys. Acta, **7**, Suppl. 81 (1934).
21. M.B. Schneider, J.T. Jenkins and W.W. Webb, J. Phys. (Paris) **45**, 1457 (1984).
22. D. Roux and A.M. Bellocq, "Physics of Amphiphiles", edited by V. DeGiorgio and M. Corti (NorthHolland, Amsterdam, 1985).
23. J.M. di Meglio, M. Dvolaitsky and C. Taupin, J. Phys. Chem. **89**: 871 (1985).
24. A. Caille, C.R. Acad. Sci. Ser. B **274**: 891 (1972).
25. J. Als-Nielsen, J.D. Litster, R.J. Birgeneau, M. Kaplan, C.R. Safinya, A. Lindegaard-Andersen and S. Mathiesen, Phys. Rev. B **22**: 312 (1980).
26. C.R. Safinya, D. Roux, G.S. Smith, S.K. Sinha, P. Dimon, N.A. Clark and A.M. Bellocq, Phys. Rev. Lett. **57**: 2718 (1986).
27. G.S. Smith, C.R. Safinya, D. Roux and N.A. Clark, Mol. Cryst. Liq. Cryst. **144**: 235 (1987); G.S. Smith, E.B. Sirota, C.R. Safinya and N.A. Clark, Phys. Rev. Lett. **60**: 813 (1988); E.B. Sirota, G.S. Smith, C.R. Safinya, R.J. Plano and N.A. Clark, Science **242**: 1406 (1988).

Chapter 29

# PHASE TRANSITIONS IN SOME BIOMOLECULES

P. MARIANI and F. RUSTICHELLI

## 1. Introduction

Phase transitions occurring in some biomolecules organized in liquid crystalline phases will be considered.

In particular, three different cases will be discussed. The first case is the transition in oriented lecithins from $L\beta'$, to the ripple phase $P\beta'$, as recently studied by Scanning Tunnelling Microscopy and X-ray diffraction[1-4]. The second case consists of the analysis of phase behaviour of the cubic phases formed by lipid containing systems: such phases, which are the most sophisticated among those exhibited by lipids, have gained recent attention as regards their structures[5,6], their biological implications[7,8] and their theory[9,10]. The third case is related to the lyomesophases formed by polynucleotide and polynucleoside derivatives (see, for example, Ref. 11); in particular, we will consider the phase behaviour of some guanosine derivatives[12] which are interesting in connection with physical and physiological properties of DNA.

The first two systems are considered because they represent the most sophisticated supramolecular aggregations of phospholipids which have recently become well understood, thanks to new experimental techniques and theoretical treatments (see, for example, Ref.s 1-6, 9, 10, 13, 14). For a more detailed description of the properties and the rich polymorphism exhibited by aqueous dispersions of phospholipids (which are the principal components of biological membranes and which consist of a polar head, i.e. the hydrophilic part, and one or two hydrocarbon chains, i.e. the hydrophobic part), one should refer to Ref.s 15 and 16. It is useful to recall that by varying the water concentration or the temperature, one can obtain quite different phases, ranging from spherical micelles to more ordered structures having one, two or three-dimensional positional order.

## 2. Phase Transitions in Phospholipids from the Lamellar Gel $L\beta'$ to the Ripple Phase $P\beta'$

Fig. 1 represents the basic features of the $L\beta'$ and $P\beta'$ phases both of which are lamellar[17]. In both phases lipidic bilayers alternate with water layers, and the hydrocarbon chains are tilted relative to the normal to the layers.

---

P.Mariani and F.Rustichelli - Istituto di Fisica Medica, Facoltà di Medicina e Chirurgia, Università di Ancona, Italy.

*Phase Transitions in Liquid Crystals*, Edited by S. Martellucci and A.N. Chester, Plenum Press, New York, 1992

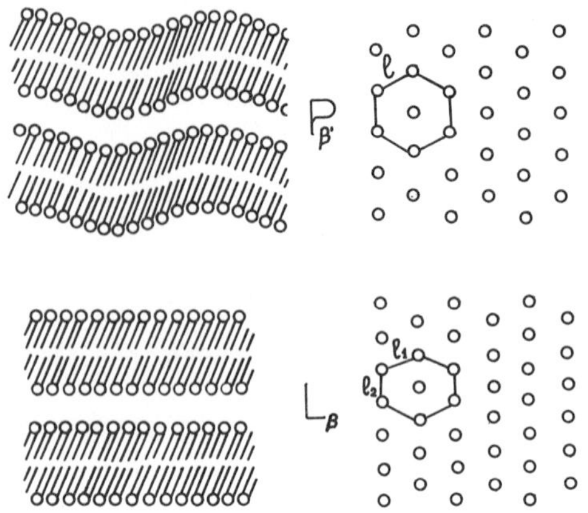


Fig.1. Schematic representation of the structural organization of the gel phase Lβ' and of the ripple phase Pβ' in aqueous dispersions of lipids.

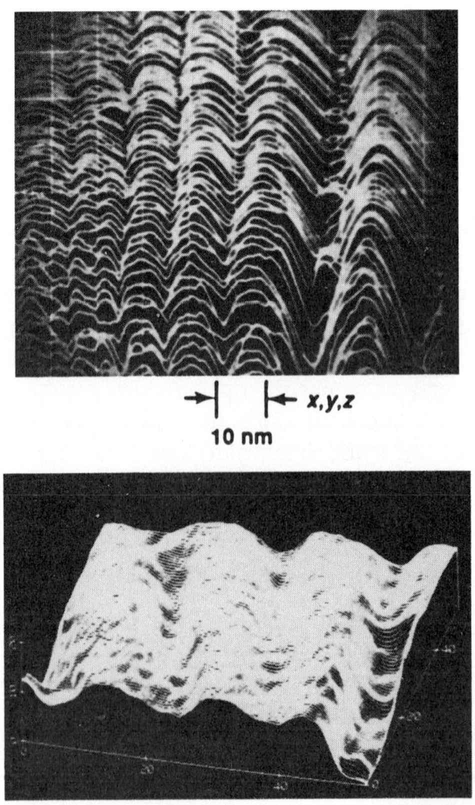


Fig.2. Scanning tunnelling microscope images of freeze-fracture replicas of ripples in the Pβ' phase of DMPC (a) Storage oscilloscope screen without y scan, the x axis being the horizontal straight line; (b) Digital image. Note the fine structure that crosses the ripples roughly orthogonal to the ripple direction. (From Ref. 1).

However, whereas in the $L\beta'$ phase the lipidic sheets are planar and the lattice is 1-dimensional, in the $P\beta'$ phase the lipidic sheets are distorted by a periodic ripple so that the lattice is 2-dimensional and the corresponding interlamellar distances are higher than in the $L\beta'$ phase. Moreover, as regards the in-plane intermolecular organization, in the $L\beta'$ phase the chain axes are ordered according to a distorted hexagonal packing, whereas a hexagonal packing occurs in the $P\beta'$ phase[17].

Among the large number of structural investigations of the lamellar lyotropic liquid-crystal phases of lipidic aqueous dispersions (see, for example, Ref.s 15-17), four noteworthy experiments should be mentioned which help elucidate the structural modifications occurring at the temperature-driven phase transition $L\beta' \rightarrow P\beta'$[1-4]. Moreover, a model was recently developed[18], using the mean-field approximation, which accounts for many of the observed physical properties of the lamellar lyotropic liquid-crystal phases of lipid aqueous dispersions.

The first experiment which will be discussed was performed by using scanning tunnelling microscopy (STM) on a freeze-fracture replica[1]; the three-dimensional contours of the ripple phase $P\beta'$ of dimyristoyl-phosphatidylcholine (DMPC) were imaged for the first time with very high spatial resolution, namely better than a nanometer. Fig. 2a and 2b show the obtained images. A ripple wavelength of 130 Å and an amplitude of 45 Å were obtained. This information constitutes an elegant confirmation and refinement of what was known before by other techniques[17]. In addition, however, a puzzling unforeseen result was obtained. A fine structure of apparently periodic nature crosses the ripples roughly orthogonally to the ripple direction. The authors conclude that it is difficult to say whether these additional ripples are related to the molecular structure of the $P\beta'$ phase or whether they are some unknown artifact. A partial solution to this question was found in an other experiment[3], which will be discussed below.

In the second experiment[2], which investigates by high resolution X-ray diffraction the same lipid system, it was shown that the $L\beta'$ phase consists in fact of three phases of 2-dimensional nature, each distinguished by the direction of the chain tilt with respect to the 2-D bond direction.

In the third experiment, the structural modifications occurring at the $L\beta' \rightarrow P\beta'$ phase transition in fully hydrated highly oriented samples of dipalmitoyl-phosphatidylcholine (DPPC) were investigated[3]. The principal result obtained was the observation, for the first time, that the $L\beta' \rightarrow P\beta'$ phase transition is characterized by a 30° (modulus 60°) change of the lattice tilt direction, in addition to the well-known lattice symmetry transition from orthorhombic to hexagonal. This observation appears to support the hypothesis that the secondary ripple structure in $P\beta'$ observed by STM[1] is more likely a real physical fact than an artifact. We would like to describe this work in more detail.

X-ray diffraction was carried out with a pinhole-type camera (Kiessig) resolving 0.3 to 30 nm reflections from Ni-filtered CuK$\alpha$ radiation. Small-angle diffraction patterns of the $P\beta'$ phase at 41°C reveal a two-dimensionally resolved ripple structure by several satellite spots about the meridian (Fig. 3a, with reduced brightness within the inset). The ripple period is 200 Å at a layer period of 71 Å. Fig. 3b indicates all detected reflections (solid circles).

Within the resolution of the spot sizes the two-dimensional lattice of the ripple structure has orthorhombic symmetry ($\gamma$=90°±2°).

Fig. 4a reports the wide-angle X-ray-diffraction patterns obtained from the $L\beta'$ and $P\beta'$ phases. For the first time, eight separated first-order wide-angle reflections have been observed from the $P\beta'$ phase of DPPC, four above and four below the equator. Six reflections, two of which lie on the equator, occur in the $L\beta'$ phase as reported earlier (only the symmetric left and right halves are shown). This clearly indicates that the related $L\beta' \rightarrow P\beta'$ phase transition is characterized by a change in the tilt direction of the hydrocarbon chains with respect to the lattice orientation in the membrane plane.

The superimposed parallel density traces along the meridional direction of the wide-angle diagram result in Fig.s 4b and 4c, where the maxima of the envelopes determine the positions of the separated reflections; in the Lβ′ phase all three reflections have equal weights, while in the Pβ′ case the two outer reflections have half the intensity of the two inner ones according to their multiplicity as assigned by indices. The splitting angles Θ are 13° and 26° in the Pβ′ phase at 41°C. The Lβ′ phase at 25°C gives $\Theta_1 = 0°$ and $\Theta_2 = 24.5°$. As discussed earlier,[19] the splitting angles relate to the hydrocarbon-chain inclination angle according to the rules of rotation patterns of misaligned single crystals. All the first-order reflections of the hydrocarbon lattice in the Pβ′ phase have an equal spacing of 4.2 Å, and thus hexagonal symmetry. In the Lβ′ phase the (200) reflections on the equator relate to 4.23 Å, while the other ones correspond to 4.10 Å. This clearly indicates an orthorhombic symmetry, which differs from hexagonal by 3%. Although the phase transition from Lβ′ to Pβ′ undergoes a change in lattice symmetry from orthorhombic to hexagonal, we keep the equivalent orthorhombic indexing for simplicity. The choice of indexing will become more evident in the following.

The c axis of the hydrocarbon lattice is parallel to the chain axis and the unit-cell length is two c bonds. The splitting geometry of the inclined hydrocarbon lattice depends on the direction of the tilt axis in the $a^*$ - $b^*$ plane of the reciprocal lattice (Fig. 3a). This axis spans an angle Ψ with $a^*$. The projections of the first-order reciprocal-lattice (hk0) points along the tilt axes result in different numbers and weights of reciprocal-lattice points and positions on the projection axis, depending on Ψ. They give the weights and number of observable reflections when the lattice is tilted with respect to the membrane normal and then rotated about it. The rotations of the lattice in real space and of the reciprocal lattice must be considered, as the in-plane organization of chains does not correspond to a single crystal but rather to a polycrystal, in spite of the fact that the sample is aligned and there is a regular periodicity of electron density along the normal to the layer.

The two projections along $a^*$ and $b^*$ of Fig. 5a correspond to six equal reflections of the Lβ′ phase (vertical line on the right side) and to eight Pβ′ reflections of different weights (top line). Fig. 6 represents in a clearer way the tilt directions of chains in Lβ′ and Pβ′ phases, respectively.

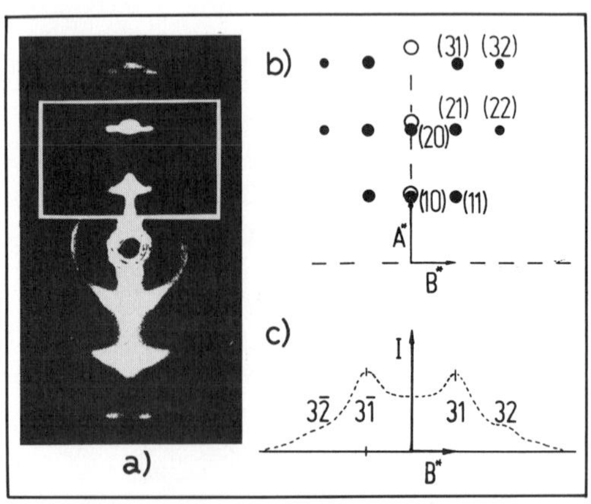


Fig.3. Small-angle X-ray diffraction of the Pβ′ phase of fully hydrated DPPC. (a) Oriented diffraction pattern of ripples of 20-nm period; (b) Detected reflections of two-dimensional orthorhombic symmetry (solid circles). Residual Lβ′ period is shown by open circles; (c) Densitometric trace along the third-layer line. (From Ref. 3).

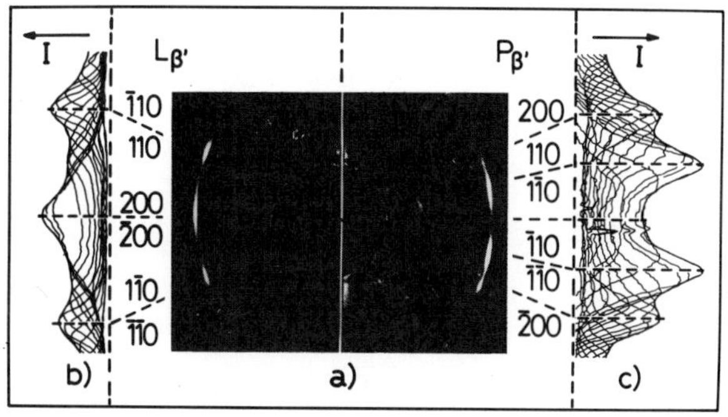


Fig.4. Wide-angle X-ray diffraction of oriented DPPC. (a) Left, Lβ′ phase with three arcs (of six); right, Pβ′ phase with four arcs (of eight). Only half of the symmetric patterns are shown; (b) Multiple vertical densitometric traces of separated $CH_2$-lattice reflections due to inclined chains in Lβ′; (c) Scans of Pβ′ with different weights in envelope. All traces are taken from the indicated halves of the patterns, but indices refer to the complete set. (From Ref. 3).

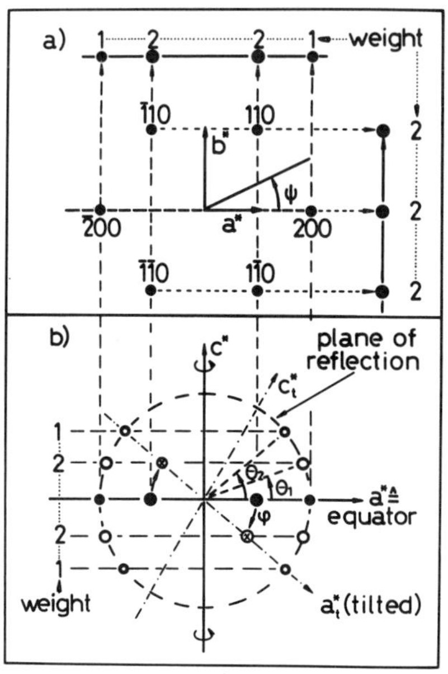


Fig.5. Reciprocal $CH_2$-lattice inclination geometry. (a) Orthorhombic (face-centered) lattice points of $a^*$-$b^*$ plane with projections along $a^*$ and $b^*$ (the corresponding tilt axes of Lβ′ and Pβ′) determine number and weights of expected reflections, depending on the angle Ψ between tilt axis and $a^*$; (b) Ewald plane of reflection with projection along $b^*$ axes on equator (solid circles, Pβ′) inclined by φ and finally rotated about $c^*$ gives splitting angles $\Theta_1$ $\Theta_2$ and positions of observed reflections (open circles). (From Ref. 3).

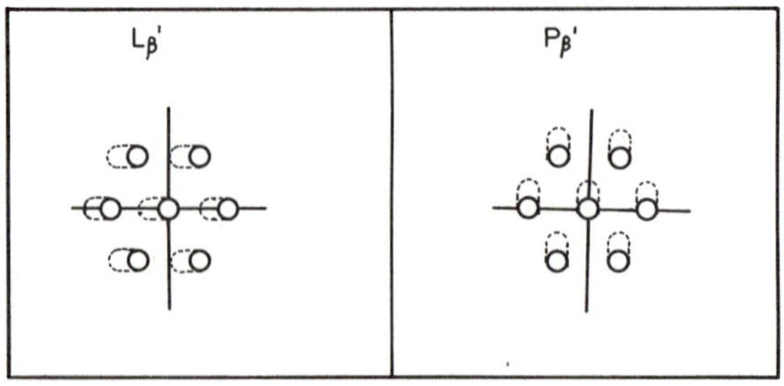


Fig.6. Tilt directions of aliphatic chains in the Lβ′ and Pβ′ phases.

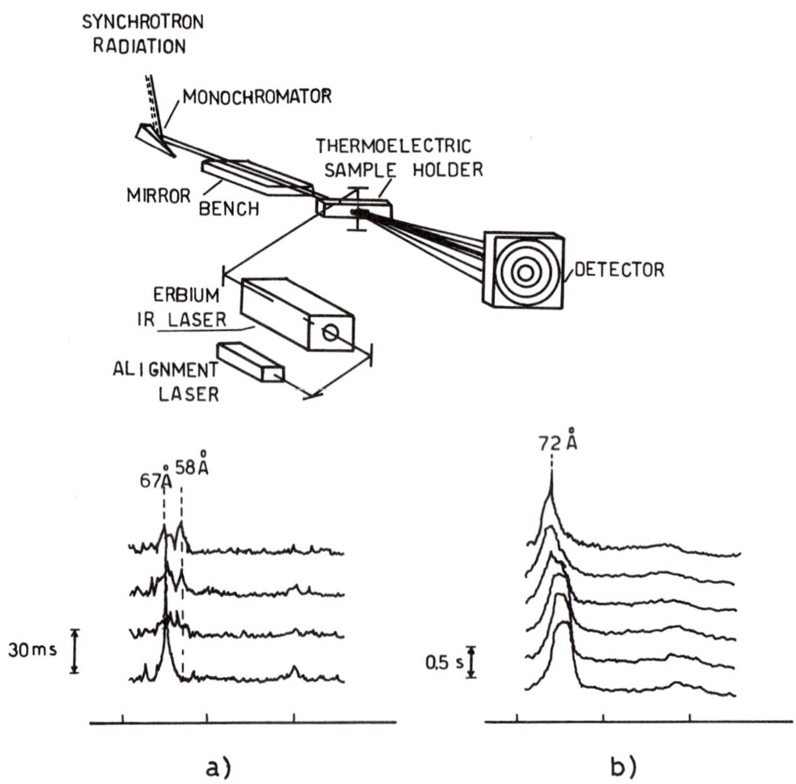


Fig.7. Top a schematic view of the instrumentation used to carry out phase-transition measurements by millisecond time-resolved X-ray diffraction with synchrotron radiation and temperature jumps by infrared laser pulses. Bottom: (a) Time resolved X-ray small-angle patterns of the relaxation process after a temperature jump through the Lβ′ → Pβ′ phase transition of DPPC, starting from 28°C. Time resolution 30 ms; (b) Time resolution 0.5 s. (From Ref. 4).

Their splitting geometry is given in Fig. 5b in the (Ewald) plane of reflection, where the equator is parallel to the average bilayer orientation. If the hydrocarbon lattice were untilted, the chain axis $c^*$ would be vertical and coincide with the membrane normal. The clockwise inclination of the lattice about the axis perpendicular to the drawing plane creates the tilted directions $c_t^*$, $a_t^*$ and the reciprocal-lattice points on the inclined dot-dashed line from the upper region on the left to the lower region on the right. The crosses are out of the reflection plane, but as the reciprocal lattice of this tilted arrangement is to be rotated about the (vertical) bilayer normal (as the sample has no preferential orientation about this axis, as mentioned above), this causes the crossed positions to meet the plane on the dashed circle (open symbols).

Clearly, the splitting angles $\Theta_n$ are related to the tilt angle $\varphi$ of the $CH_2$ lattice, depending on the orientation of the tilt axis $\Psi$. The reader should refer to Ref. 19 for more details. If $\Psi = 0$ one obtains six reflections and this is the case of the Lβ′ phase, for which a tilt angle $\varphi = 28°$ was derived from the splitting angle $\Theta_2 = 26°$ of the two symmetrical non equatorial reflections (see Fig. 5b). If $\Psi = 90°$ (or 30°) one obtains eight reflections and this is the case of the Pβ′ phase, for which a tilt angle $\varphi = 26°$ was obtained, coinciding with the $\Theta_2$ value. Concerning the ripple structure, from a series of considerations reported in Ref. 19 a model was derived according to which the tilt direction is practically perpendicular to the main ripple wave vector and conseguently parallel to the secondary undulations observed by STM[1], which supports the possible physical nature of these undulations.

The fourth experiment related to the Lβ′ → Pβ′ phase transition was performed by using a recently developed technique, namely millisecond time-resolved X-ray diffraction of synchrotron radiation[4]. By using this technique, while inducing temperature jumps by infrared laser pulses, it is possible to investigate the kinetics of lipid phase transitions. The results of the application of this technique through the transition considered to lead to ripple formation are reported in Fig. 7. Fig. 7a, which describes the kinetics for shorter times, in 30 ms steps, shows that the first step after the temperature jump, reveals a new diffraction peak, corresponding to a new lamellar periodicity of 58 Å; this periodicity coexists adiabatically with the original 67 Å periodicity associated with the Lβ′ phase for a time period of about 10 ms for DMPC and about 100 ms for DPPC. Moreover this period appears to increase when the final value of the temperature jump increases and when NaCl is present in the solution. Other measurements have shown that the 58 Å periodicity is already established after ≈ 1 ms and should be attributed to a disclination of the layers (see Fig. 7), characterized by the fact that the long axis of the hydrocarbon chains becomes perpendicular to the bilayer plane. This elimination of the tilt would facilitate rapid molecular rotation around the long axis, as is observed by spectroscopic techniques[20]. As a second slower step, a fusion of the two peaks occurs, leading to a broader peak which moves toward smaller q values, i.e. higher periodicities; this continues until the final value of 72 Å, characteristic of the monoclinic rippled Pβ′ phase, is reached after a time of about 4 s. However, the signal intensity continues to increase reaching the steady state maximum value after several minutes. Fig. 8 reports a schematic model of the Lβ′ → Pβ′ phase transition for DPPC, according to the present authors, for different times following the temperature jump. A possible interpretation of the 58 Å lamellar periodicity is the formation of Lα domains in the short time transient; in fact, this periodicity would correspond to the Lα phase in isothermal conditions at 50°C, i.e. 10°C above the Pβ′ → Lα main transition temperature. For other possible explanations of the occurrence of the thin periodicity, and for a more detailed discussion of the Lβ′ → Pβ′ phase transition, including the slow kinetic investigations by other authors[21], the reader should refer to Ref. 4. The conclusions of the analysis are in favour of a spinodal decomposition process, i.e. a nucleation and growth mechanism, for the Lβ′ → Pβ′ phase transition.

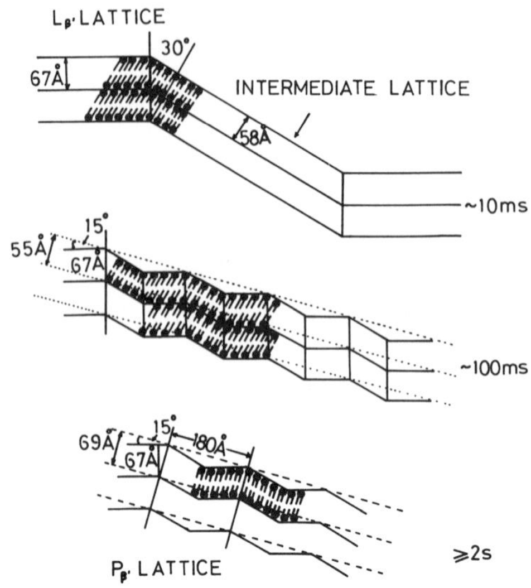


Fig.8. Schematic model of the mechanism of the Lβ′ → Pβ′ phase transition in DPPC, involving the rapid formation of an intermediate 58 Å lattice originating from the parent Lβ′ lattice through disclination by an angle of 30°. (From Ref. 4).

# 3. Polymorphism of Lipids: Non-Lamellar Phases

## 3.1. The Structure of Cubic Phases

The many studies that have been performed on lipid extracts, purified lipid components and synthetic lipids have established that, under suitable conditions of temperature and hydration, lipids may spontaneously form lamellar phases[7,22]. As reported above, such phases consist of one-dimensional periodic stacks of identical bilayers, each separated by a water layer of constant thickness. However, it has been shown that this lamellar organization is only one of the large variety of liquid crystalline structures which may be adopted by hydrated lipid systems[15,22,23]; to date, among non-lamellar lyotropic arrangements, the cubic phases appear as the most complex and "intriguing"[23,24].

Non-lamellar phases can be described by using a very pictorial type of criterion[6,25]. In fact, considering the molecular structure of an amphiphilic molecule, two different moieties can be visualized in all lyotropic phases, i.e., the regions occupied by the hydrocarbon chains and those occupied by the water, the interface being covered by the polar groups of the molecules. As general rule, it appears possible to identify the shape of the polar (or paraffinic) regions and then to describe with simple geometrical considerations such "structure elements"[6,15,25]. Therefore, the whole structure can be described by considering that the hydrocarbon or the polar medium forms a continuous matrix into which structure elements of opposite polarity are fitted. Moreover, it must be also considered that in the lyotropic phases, and with the exception of those formed by infinite sheets (namely the lamellar phases), a topological distinction could be made between the interior and the exterior volumes of the structure elements[15,22]. In particular, when the paraffinic chains fill the volume inside the structure element and the water the outside region, the structure is called of type I (oil-in-water, normal). On the contrary, the structure is type II (water-in-oil, inverted) when the structure element appears to be filled by the water and embedded in a paraffinic matrix (see, for example, Fig. 9).

It must be also noticed that, with the exclusion of the two-dimensional Pδ phase, in such phases the hydrocarbon chains show a liquid-like organization[6,26] , i.e. the so-called α conformation. This conformation can be visualized as highly disordered, like that of a liquid paraffin, even if, as a consequence of the fact that one chain end is anchored at the interface, the average chain orientation is perpendicular to the lipid-water interface[26]; this orientation is more and more pronounced as the area-per-chain ($S_{ch}$) decreases (when the water content decreases, for example). It must be also noticed that such a "liquid" may well display complex molecular movements[26].

Moreover, in the non-lamellar phases, simple geometric considerations indicate that the chains must be folded in fairly regular way in order to uniformly fill the oddly shaped volumes offered to them[22,26,27]. However, the conformation of the chains does not appear to be profoundly different in the lamellar and the non-lamellar phases, as assumed or proposed by some authors[22,26,27].

Regarding the structure of non-lamellar phases, we will limit ourselves to the description of the most prevalent hexagonal and cubic phases. In the hexagonal H phase (sketched in Fig. 9), the structure elements are rigid infinitely long rods, all identical and crystallographically equivalent, packed in a two-dimensional hexagonal lattice[15,22]. Concerning the cubic Q phase, six different structures with cubic symmetry have thus far been identified, and their structures have been determined unambiguously[6,24] by using a recently proposed pattern recognition approach[5]. We adopt here the nomenclature introduced by Luzzati's group: a cubic phase is called $Q^n$, where Q stands for cubic and n is the number of the relative space group, according to the International Tables of Crystallography.

The structure of the $Q^{230}$, $Q^{224}$ and $Q^{229}$ phases, reported in Fig. 10, can each be described in terms of two three-dimensional networks of joined rods, mutually intertwined and unconnected the rods are respectively linked coplanarly three by three, tetrahedrally four by four and cubically six by six. As originally pointed out by Luzzati and coworkers[6], these three structures are bicontinuous (i.e. both the water and the hydrocarbon media are continuous throughout the structure) and can be visualised as 3-D topological generalizations of the lipid bilayer. The structure of the $Q^{212}$ phase is related to that of $Q^{230}$ one of the two networks of rods is preserved, while the other is replaced by a lattice of closed micelles. Also, this phase consists of two continuous disjoined media, one apolar, containing the micelles, and the other polar the $Q^{212}$ can be visualised as a 3D generalization of the lipid monolayer.

The structure of the last two cubic phases, the $Q^{223}$ and $Q^{227}$, have been the object of controversial reports and only very recently seem to have been determined unambiguously[10,28]. The two phases consist of two types of closed and disjoined micelles embedded in a continuous matrix (Fig. 11) these structures are non-bicontinuous. It is interesting to notice that examples of the $Q^{230}$ phase have been observed both of type I or II, depending on the chemical composition of the system. Instead, all the known examples of phases $Q^{212}$, $Q^{224}$, $Q^{227}$, and $Q^{229}$ belong to type II and those of phase $Q^{223}$ to type I. Concerning phase behaviour, it must be observed that the most fundamental parameter which controls polymorphic phase behaviour has been identified in the interfacial curvature[9,10,22,28] (Fig. 12); the polymorphism of lyotropic systems appears as a direct consequence of the different types of interactions (crudely schematized in Fig. 13) existing between the polar head groups (through electrostatic forces) and the paraffinic chains (through van der Walls forces). In fact, in amphiphilic systems, head and chains are coupled in the same molecule and neither of them is able to independently minimise its free energy with respect to its cross-sectional area; a compromise is achieved by introduction of symmetry within the structure[22]. Sadoc and Charvolin[28] have discussed how the frustrations in bilayers lead to polymorphism in lyotropic liquid crystals.

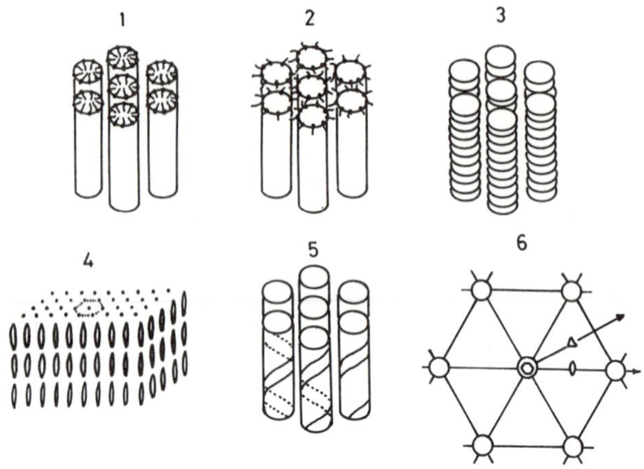


Fig.9. Structure of hexagonal phases formed by different kinds of molecules. 1, 2, type I and type II hexagonal phases formed by lipid-containing systems; 3, discotic hexagonal phase formed by disc-shaped molecules; 4, smectic B phase formed by thermotropic mesogens; 5, hexagonal phase formed by helical biological polymers. The position of the symmetry elements is indicated in 6. (From Ref. 37).

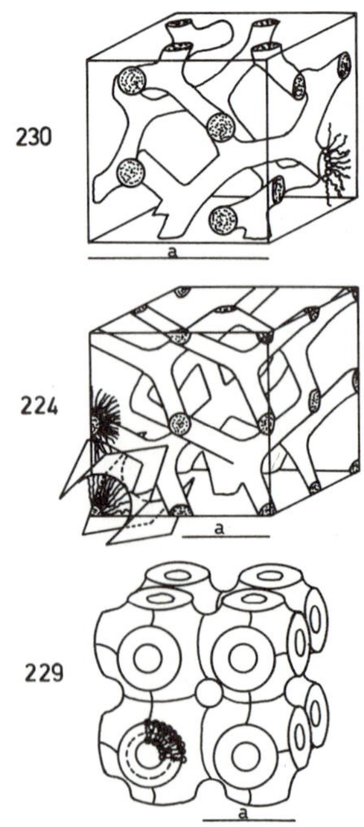


Fig.10. Structures of the inverse bicontinuous cubic $Q^{230}$, $Q^{224}$ and $Q^{229}$ phases. (From Ref. 22).

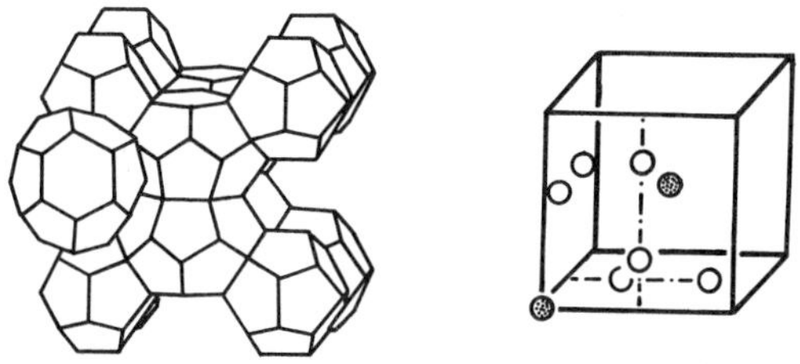

Fig.11. Structure of the $Q^{223}$ phase. The structure can be described as a packing of distorted polyhedra (in this case 2 12-hedra and 6 14-hedra per unit cell) that fill the space (left side). The position of the polyhedra is represented in the right box by filled and open circles, respectively. (From Ref. 28).

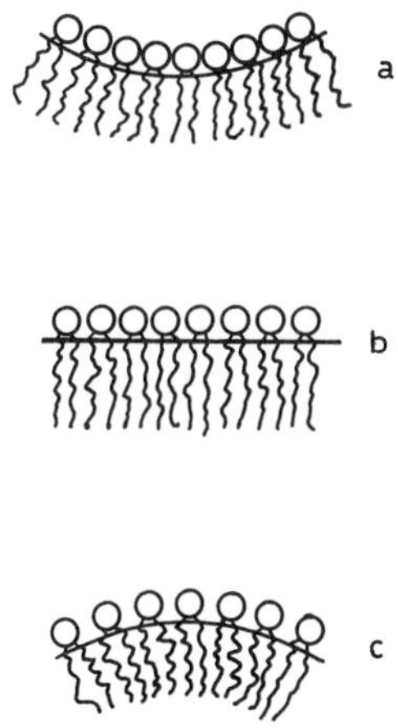


Fig.12. Spontaneous curvature of lipid monolayers. From a to c: negative, zero and positive mean curvature. (From Ref. 22).

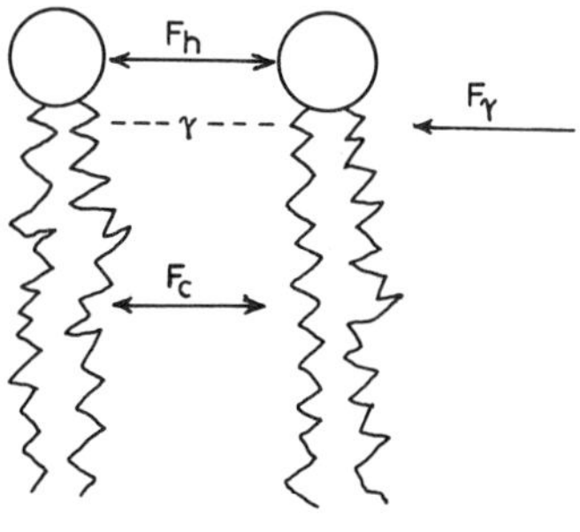


Fig.13. Schematic representation of the lateral forces between lipid molecules which determine the spontaneous curvature of the monolayer. $F_h$, usually repulsive, arises from steric, hydrational and electrostatic effects; $F_\gamma$ which tends to minimize the interfacial area, arises from hydrophobic effects; $F_c$, repulsive, is due to the thermally activated cis/trans rotations of the chains. (From Ref. 22).

### *3.2. Phase Equilibria and Phase Diagrams*

For lipid-containing systems, phase transitions may be induced either by varying the temperature or the water concentration[6,15,22]. It is interesting to note at once that for some systems, such as ionic amphiphiles with only one hydrophobic chain, variation of concentration is the more important driving force for inducing transitions between mesophases, so that phase boundaries are close to vertical (the corresponding generalised phase diagram is reported in Fig. 14[22]). On the contrary, when the amphiphile is non-ionic or zwitterionic, and its molecule presents two aliphatic chains, temperature appears to be the most important variable, and many of the phase boundaries appear closer to horizontal than to vertical. In this case, an increase in the temperature plays a role qualitatively similar to that of a decrease in the water content[22].

Concerning the phase diagram reported in Fig. 14, it must be observed that this phase sequence is in agreement with most binary systems studied to date, although as a general rule a given system will not form all of the phases shown[22,29]. As indicated above, the case where phase transitions are driven principally by changes in water content is found in many charged lipid systems, such as alkali carboxylic soaps. In reality, a certain temperature dependence is always observed, so that the phase boundaries can deviate from vertical. Noteworthy is the case of lysophospholipids, which display a similar behaviour even when they are uncharged[30]: as an example, the phase diagram of lauroyl lysophosphatidyl choline is reported in Fig. 15. In general, phase diagrams of such a kind are observed when the lipid presents only a single chain and is considerably less hydrophobic than the corresponding diacyl compounds[22]. In fact, the case of transitions driven predominantly by temperature is more typical of some diacyl zwitterionic phospholipids.

Some examples extracted from Ref. 22 (where the reader will find the original references) are presented in Fig. 15. The phase diagram for saturated $C_{20}$ diacyl phosphatidyl ethanolamine indicates that the L$\alpha$ phase is only stable over a short range of temperature, while the H phase occurs at high temperatures at all water concentrations out to the excess water region. If, in the lipid polar head, we exchange the ammonium group for a trimethyl ammonium group there is a drastic effect on the phase diagram: as an example, we consider the $C_{14}$ saturated diacyl phosphatydil choline/water system. In this case, as observed for most other chain lengths, only a fluid lamellar L$\alpha$ phase is seen when the chains are melted, except at very low water contents.

Variation of the length of the aliphatic chains can also have profound effects on phase diagrams. For example, in the case of phosphatidyl ethanolamines with short chains, additional phases appear between the lamellar and the inverse hexagonal phase, as Fig. 15 shows for $C_{12}$ dialkyl phosphatidyl ethanolamine. It is very interesting that these phases include a number of bicontinuous cubic phases $Q^{230}$, $Q^{224}$ and $Q^{229}$, but other unidentified nonlamellar phases probably also occur. Referring again to the generalised phase diagram reported in Fig. 14, we recall that a great deal of effort has been spent in recent years in trying to understand even the qualitative features of such phase sequences (reviewed in Ref. 22. See also Ref. 29). In part, this work has been hampered by the difficulty of establishing the structure of the intermediate phases, but perhaps the major problem lies in identifying and quantifying the important molecular interactions which are responsible for mesophase stability. In particular, the many intermediate phases which can be found in locations "a", "b", "c" and "d" of the phase diagrams are only poorly understood. The majority of these phases detected to date are cubic: the ones observed in regions "c" and "d" are of type I, whereas the ones in regions "a" and "b" are the inverse, namely type II. Furthermore, the cubic phases in region "d" are based on an ordered packing of anisotropic micellar aggregates, whereas the phases detected in regions "b" and "c" present structures based upon bicontinuous interfacial headgroup regions.

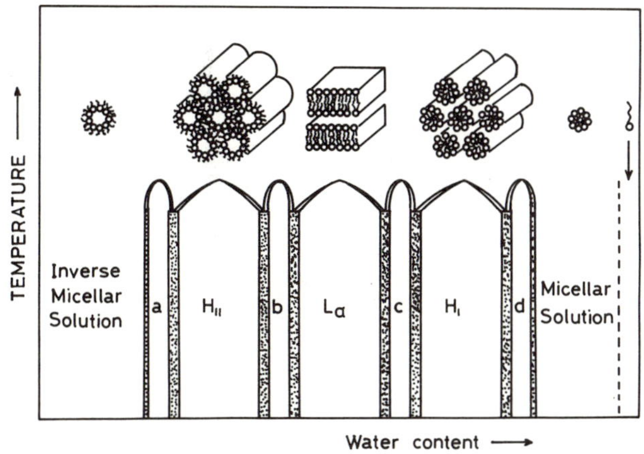


Fig.14. Generalised lyotropic binary phase diagram of lipid-containing systems where phase transitions are mainly determined by changes in water concentration. (From Ref. 22).

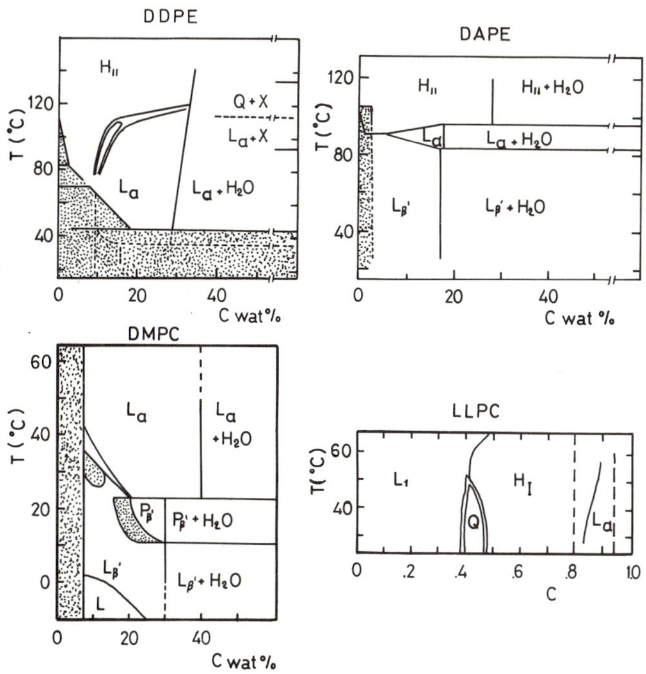


Fig.15. Examples of lyotropic binary phase diagrams. DDPE, didodecyl - phosphatidyl - ethanolamine; DMPC, dimyristoyl - phosphatidyl - choline; DAPE, diarachinoyl - phosphatidyl - ethanolamine; LLPC, lauroyl - lysophosphatidyl - choline. The dotted lines and shaded areas indicate either insufficiently analysed or biphasic regions. C is the lipid weight concentration. (From Ref.s 22 and 30).

Finally, cubic mesophases in region "a" might be expected to consist of ordered lattices of inverse micellar aggregates, although no such structures have as yet been determined[22].

A crucial problem is that of analysing the underlying factors which determine phase behaviour. However, even if the mean and Gaussian curvatures of the polar-apolar interfaces play an important role[9,22], in our opinion the number of systems that have been explored is still too small to draw general conclusions. In any case, a much larger number of papers has been devoted to the structural and thermodynamical analysis of non-lamellar phases, placing particular emphasis on their geometrical and topological characteristic[9,10,22,31].

### 3.3. *Chemical Composition, Phase Behaviour and Cubic Phase Occurrence*

Concerning cubic phases, we will offer a few comments about the relation between composition and structure. In particular, we will refer to several systems studied very recently by Luzzati's group[5,6]: they carefully analysed the phase diagrams of MO/water, Cytochrome C/MO/water, PLPC/water and DTAC/water (see Fig. 16), in order to determine the structure of all the observed cubic phases by applying their recently proposed pattern recognition approach[5]. Concerning the phase diagrams, we note in particular that the presence of Cytochrome C, a soluble protein, induced the appearance of new cubic phases. This protein is not the only one that interacts with monoglycerides and with other lipids to yield well organised phases: Ericsson et al. have also reported cubic phases in a variety of protein/MO/water systems[32], and cubic phases have been observed by one of the authors (P. M.) in systems containing myoglobin, catalase and lysozyme. However, only the phase $Q^{212}$ seems to require protein: all the other cubic phases have been observed in binary lipid water systems[6,22,23].

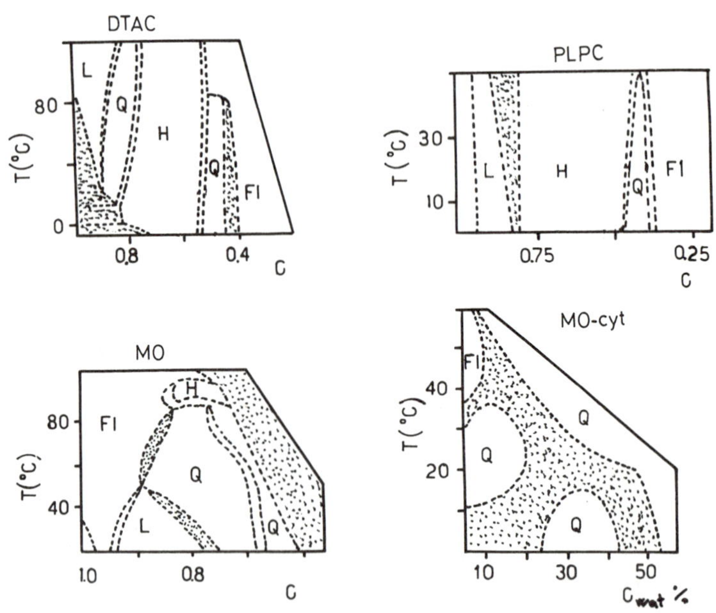


Fig.16. Examples of lyotropic phase diagrams. DTAC, dodecyl - trimethyl - ammonium chloride; MO, monoolein; PLPC, palmitoyl - lysophosphatidyl - choline; MO-cyt, monoolein and Cytochrome C. The dotted lines and shaded areas indicate either insufficiently analysed or biphasic regions. C is the lipid weight concentration. (From Ref. 6).

Simple qualitative arguments can be used to explain certain aspects of the phase behaviour. As reported in Ref. 6, the notion of relative bulkiness of the polar and the apolar moieties of the lipid molecules, introduced originally to explain the phase sequences and to justify the very existence of structures of either type I or II[33], can be used to analyse the four phase diagrams presented in Fig. 16. We consider first the cases of DTAC and PLPC, two lipids whose structure elements are of type I. Reducing the water content has the effect of decreasing the surface-per-chain ($S_{ch}$): therefore, dehydration of the hexagonal phase eventually leads to an area-per-chain becoming so small that phase H gives way to phase L. In the intermediate hydration range, the chains may be too flexible to fold inside the volume compatible with the value of $S_{ch}$ if the structure is hexagonal, and too stiff if the structure is lamellar; breaking up the cylinders into the short rods of a $Q^{230}$ phase may have the effect of reconciling these two conflicting constraints[6]. This process was originally called the cumulative strain effect[33]; later Sadoc and Charvolin introduced the equivalent notion of "frustration"[28].

When, on the contrary, water is added to phase H, the increase of $S_{ch}$ eventually leads to the formation of micelles; in the intermediate region, the conflict between the area-per-chain and the flexibility of the tails may be reconciled by a structure containing very closed and anisotropic micelles. This phase sequence is observed in many detergents; in lysolecithin either the phase $Q^{223}$ or the phase $Q^{230}$ is observed, according to the nature of the hydrocarbon chains[30]. If we consider lipids whose structure elements are of type II, for example MO, we note that the phase sequence in the order of increasing water content is $Q^{230}$ to $Q^{224}$ in keeping with the notion that at constant $S_{ch}$, the polar to apolar volume ratio increases as the number of the joining rods increases (3 in $Q^{230}$ and 4 in $Q^{224}$)[6]. The same argument applies to the transition from $Q^{224}$ to $Q^{229}$ observed at low temperature in the MO/Cytochrome C/water system (4 joining rods in $Q^{224}$ and 6 in $Q^{229}$). In the case of $Q^{212}$, the presence of structures containing both rods and micelles seems to be related to the particular chemical composition. Moreover, the $Q^{212}$ phase is thus far unique among the lipid phases with disordered chains in displaying a non-centrosymmetric structure; apparently the chirality is related to the presence of the protein and to the very low water content[6].

## 4. Phase Behaviour of Lyomesophases Formed by Guanosine Derivatives

### *4.1. Lyotropic Polymorphism of DNA Derivatives : the Hexagonal Phase*

It has been known for some time that DNA and RNA solutions can spontaneously undergo transition to a liquid crystalline state above a critical concentration[11]. Although the critical concentrations are usually high (about 50 mg/ml in the case of short DNA molecules), it is interesting to note that the local concentration of DNA *in vivo* can also be quite large, and liquid crystal formation may possibly play some role in packaging DNA in some biological systems[35]. In recent years, the increasing number of papers devoted to such systems demonstrates the importance of this subject for the physical-chemical and biological sciences (see, for example, Ref. 34).

The first report that high molecular weight DNA might form liquid crystalline phases was made in 1959 by Luzzati and Nicolaieff[35], who recognised two liquid crystalline phases including one of hexagonal type, and by Robinson[36], who suggested that the other phase is cholesteric. To date, it is well known that numerous helical biological polymers, such as polynucelotides, polypeptides and polysaccharides, form lyotropic liquid crystalline phases whose structures depend on the polymer concentration[11,34,37]. In particular, when the polymer content increases, cholesteric spherulites appear within the isotropic phase and, at higher concentrations, the cholesteric regions expand to occupy the entire preparation.

When the concentration increases further, the cholesteric organization disappears and is replaced by a hexagonal phase. The textures and defects of these phases, observed by polarizing optical microscopy, have been described in detail and compared with those exhibited by classical (i.e. amphiphilic) lyotropic systems[11,37]. It is interesting to note that in the case of the hexagonal phase, new textures were observed, which relates to the problem of the coexistence of hexagonal and helical ordering[37].

Concerning the structure of the phase which are observed, we have previously showed that hexagonal liquid crystalline phases are classically formed by amphiphilic compounds in water. Within an appropriate range of temperature and concentration, the molecules form columns which align in parallel and often show a hexagonal order in cross-section[22]. More recently, discoid molecules consisting of a rigid core surrounded by flexible paraffinic chains have been synthesised (see, for example Ref. 38). Notably, these molecules tend to stack in more or less ordered columns which are arranged in a hexagonal array, and which are able to slide longitudinally with respect to one another: such phases are called hexagonal discotic.

Hexagonal ordering has also been found within smectic layers of thermotropic liquid crystals[39]; the hexagonal order often persists coherently through several layers, and in this case the phase can be considered to be a true crystal. In the case of the DNA-water system, the structure strongly resembles those seen in amphiphiles and discotics: the elongated molecules are aligned in parallel and form a hexagonal array. Each molecule is able to translate longitudinally and to rotate about its long axis[37]. A sketch of all these hexagonal structures appears as Fig. 9. The cholesteric phase can also be described as a "supermolecular" organization[11]; in fact, the individual DNA helices are the building structures of this mesophase, which is characterised by a cholesteric superhelix.

### *4.2. Phase Transitions Induced by Variable Concentrations*

To consider the analysis of the phase behaviour of these biological polymers, we refer here to some particular derivatives, a series of polyguanosines, which have been recently investigated in our laboratory[40-42]. In fact, the study of the liquid crystalline properties of such systems appears useful for an understanding of the behaviour of polynucleotides in living systems. Moreover, the unusual features of guanosine derivatives are invested with biological significance in several cellular processes[12].

The guanosine moiety is unusual among nucleic acid bases because of its ability to self-associate in aqueous solutions into stable and regular structures[43]. This self-structuring phenomenon is shown clearly in the formation of temperature-dependent anisotropic gels in acidic 3'- and 5'-guanosine monophosphate solutions, and in the formation of stable regularly ordered complexes of the dianion in neutral and alkaline 5'-GMP solutions. These structures involve the formation of planar hydrogen-bonded tetramers, shown in Fig. 17, that are stacked in a helical fashion.

In 1988, Spada and coworkers analysed using optical polarising microscopy the textures obtained by peripherical evaporation of an isotropic solution. They reported that the sodium salt of the dinucleoside phosphate G2 (whose chemical formula is shown in Fig. 18) when dissolved in water forms two mesophases, the more dilute solutions exhibiting typical cholesteric textures and the more concentrated mixtures showing herringbone patterns similar to those reported for hexagonal amphiphilic mesophases[12]. Further investigations based on X-ray diffraction experiments confirmed these observations[40]. The phase sequence determined at constant temperature (20°C) was the following

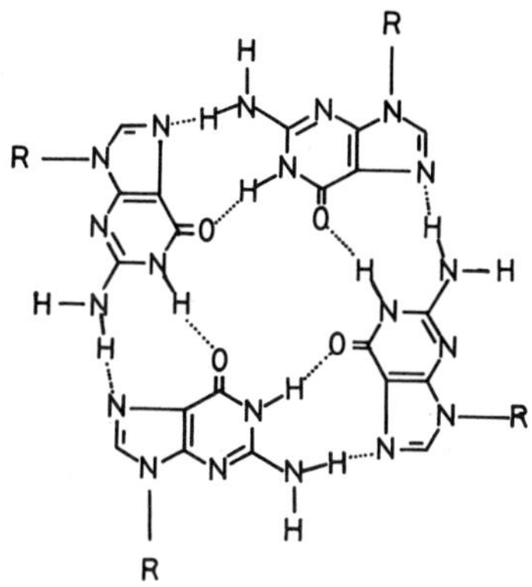

Fig.17. Planar tetrameric arrangement of four guanosine bases bonded in a Hoogsteen mode. (From Ref. 40).

$$I \xrightarrow{2.5\%} N^{*} \xrightarrow{18\%} H \xrightarrow{65\%} K$$

where I, N*, H and K denote isotropic, cholesteric, hexagonal and crystalline phases, and the concentration is expressed as the weight of the guanosine derivatives divided by the total weight.

In addition, structural information was obtained directly, so that it was possible to propose a structural model for the mesophase[40]. It is useful here to summarise the X-ray diffraction results two typical diffraction patterns obtained from samples containing respectively 10% and 50% of G2 are reproduced in Fig. 19[40]. In the upper curve, only a very broad and weak band is present in the small angle region ($2\Theta < 10°$, $2\Theta$ being the scattering angle), as is expected for a cholesteric or a nematic liquid crystalline phase; on the contrary, the high angle region is characterised by the presence of a broad band, apparently due to the excess of water, and by a superimposed sharp peak centred at a spacing of $(3.3Å)^{-1}$.

The small angle profile relative to the more concentrated sample, as given in the lower curve of Fig. 19, consists of a very strong and narrow peak coupled with few weak peaks.

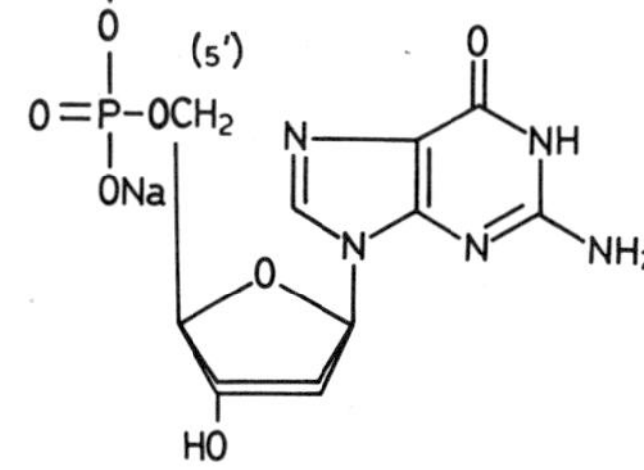

Fig.18. 2′-deoxyguanylyl-(3′-5′)2′-deoxyguanosine (G2) sodium salt.

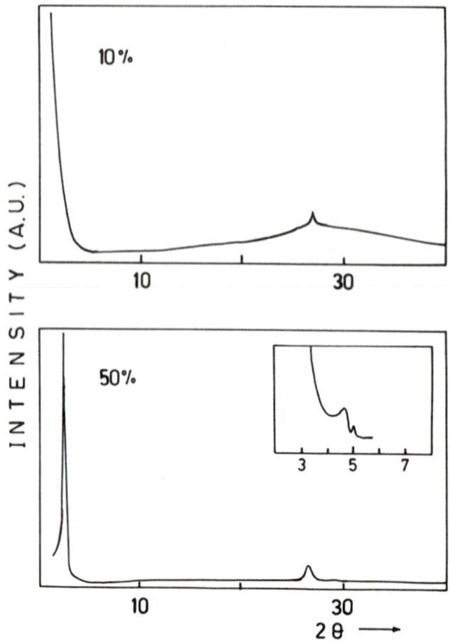


Fig.19. X-ray diffraction patterns of G2/water samples in cholesteric (upper, C=10% w/w G2) and hexagonal (lower, C=50% w/w G2) phases. (From Ref. 40).

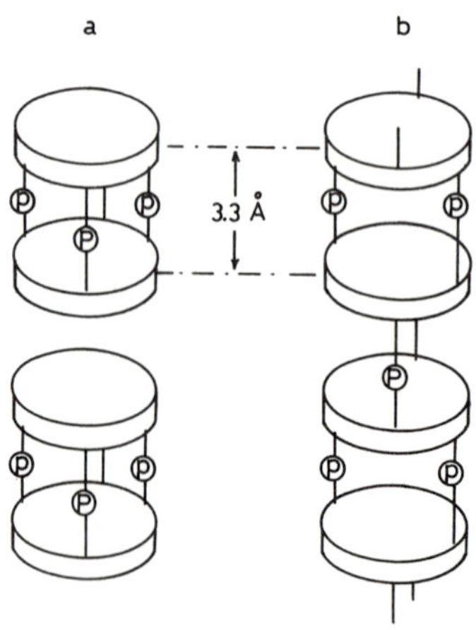


Fig.20. Sketch of two possible models for the rod structure in the case of the G2 derivative. Left: piling of discrete pseudooctamers; right: continuous array of cross-linked tetramers. The circles indicate the phosphate bridge while the disks represent the guanine tetramers. (From Ref. 40).

The ratios of the peak spacings in unit S (S denoting 2 sinΘ/λ, where λ is the X-ray wavelength) are $1:\sqrt{3}:\sqrt{4}:\sqrt{7}$ thereby indicating a two-dimensional hexagonal lattice. Using the convention usually adopted in lipid crystallography[15], the spacings of the symmetry permitted reflections in the case of a 2-D hexagonal symmetry are given by

$$S_{hk} = (1/a)\,(h^2 + k^2 - h\,k)^{1/2}$$

where h and k are the Miller indices of the reflection and a is the unit cell dimension. The high angle region remains characterised by the presence of the sharp peak at 3.3Å. That is, for changes in the composition, the position of the small angle signals changes as a linear function of the water content, whereas the peak at 3.3Å does not show any detectable dependence. Corresponding to the transition from the hexagonal to the crystal phase, the diffraction profile exhibits sharp reflections superimposed to the hexagonal high intensity peak[40]. This pattern indicates the presence of the more ordered crystalline phase. However, since the relative intensity of the peak due to the hexagonal order appears to decrease as the water content further decreases, a large biphasic region apparently precedes the appearance of the pure crystalline phase[40].

As reported in Ref. 40, both the magnetic behaviour and the X-ray diffraction results appear consistent with a structure consisting of chiral rod-shaped aggregates, with negative diamagnetic anisotropy, arranged as a stacked array of planar tetramers formed by Hoogsten-bonded guanosine moieties, as schematically reported in Fig. 20. It is notable that this "chiral cylinder" is a structural unit that can form either a cholesteric or a hexagonal mesophase, depending on the water content. At high dilutions, these aggregates give rise to isotropic solutions; when the concentration increases, these solutions become cholesteric mesophases characterised by weak correlation between the rods. Upon further decreasing the water content, we obtain hexagonal mesophases in which the rods are arranged in a highly ordered array.

The simple relationship which links the hexagonal lattice constant with the concentration of G2 merely reflects the dilution of the aggregates in the array, and this suggests that water molecules are present only outside the rods[40]. Furthermore, the reported constancy of the diffraction profile with temperature[40] confirms the rigidity of the aggregates. In addition, the value of the rod diameter, calculated using simple chemical relations from the concentration and unit cell dimension, and its observed insensitivity to the water concentration[40], clearly indicates that the cross section of the aggregates may be determined by a constant and discrete unit such as the guanosine tetramers[43].

The "chiral cylinder" model also explains the presence of the 3.3Å high angle peak, which appears related to the distance between the planar tetramers of guanosines inside the columns. The occurrence of this peak also in the cholesteric phase indicates that the columns of stacked guanosines are already present in dilute solutions. A very recent investigation using small angle neutron scattering seems to suggest that such aggregates still exist even in the isotropic phase, their length continuing to increase as the water content decreases.

In order to analyse the self-ordering properties of polyguanosines, several other derivatives have been analysed by X-ray diffraction, optical microscopy and circular dichroism[41,42]. In particular, it has been observed that the "monomer", 2′-deoxyguanosine- 5′-monophosphate, G1, the "trimer", G3, the "tetramer", G4, the "hexamer", G6 and also the cyclic "dimer", G2cyc form cholesteric and hexagonal phases. While the overall structure of the columnar aggregates are similar to each other and to that of G2 (see the models reported in Fig. 21), differences were observed in the handedness of the cholesteric phases and in the concentration values at which phase transitions occur[41,42]. The following phase sequences were determined at 20°C

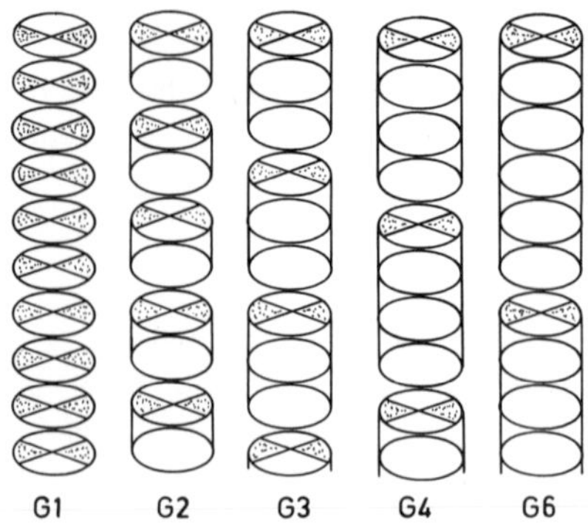


Fig.21. Sketch of the overall structure of the columnar aggregates for the guanosine derivatives investigated. The phosphate bridges are indicated by vertical lines, while the disks represent the Hoogsten-bonded tetrameric planes (From Ref. 41).

$$G1 \qquad I \xrightarrow{30\%} N^* \xrightarrow{40\%} H$$

$$G3 \qquad I \xrightarrow{8\%} N^* \xrightarrow{25\%} H$$

$$G4 \qquad I \xrightarrow{13\%} N^* \xrightarrow{35\%} H$$

$$G6 \qquad I \xrightarrow{18\%} N^* \xrightarrow{35\%} H$$

$$G2cyc \qquad I \xrightarrow{40\%} H$$

The critical concentration at which the cholesteric phase appears increases in the order G2, G3, G4, G6 and G1[41]. In the case of the G2cyc derivative[42], the results indicate that only the H phase is stable at room temperature, since the cholesteric phase is observed only in biphasic systems. For the "homogeneous" series G2, G3, G4 and G6, which do not possess a free phosphate end group, the isotropic to cholesteric phase transition concentration seems to follow a definite pattern related to the ratio of negative charge to guanine units[41].

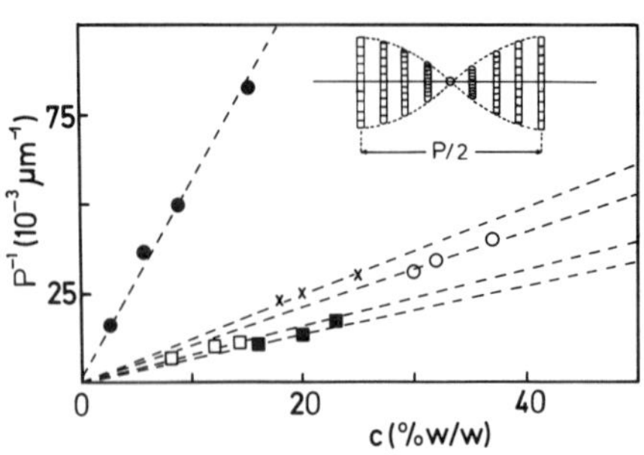


Fig.22. Plot of the inverse pitch, measured at 20°C and as a function of concentration, for the cholesteric solutions of some guanosine derivatives: G1, open circles; G2, filled circles; G3, open squares; G4, filled squares; G6, crosses (from Ref.s 40 and 41). The inset shows the cholesteric superhelix organization.

In fact, this ratio, which is indicative of both the electrostatic interaction and the hydrophilic/hydrophobic balance, increases in the sequence G2, G3, G4 and G6. Obviously, the smaller this ratio, the easier the formation of liquid crystalline phases: G2, where this ratio is the lowest, originates the cholesteric phase at higher dilution, while G6, in which this ratio is the highest, gives rise to the cholesteric phase only at high concentration.

On the contrary, the behaviour of G1 departs from the trend exhibited by the other derivatives[41]: this can be explained noting that for monomeric derivatives, the hydrophilicity/hydrophobic balance is certainly not the factor which determines their ability to form liquid crystalline phases. Instead, a series of experiments on the 2′-deoxyguanosine-3′-monophosphate and on the isobutyl ester of G1, strongly indicates that both the location of the phosphate group and the stereochemical constraints imposed by the presence of a substituent play major roles[41]. Concerning the G2cyc derivative, it is notable that, despite the steric constraint imposed by the presence of a second phosphate bridge, G2cyc gives rise to liquid crystalline phases with a mechanism similar to its linear counterpart[42]: however, these constraints, which have little effect on the ability to form chiral cylindrical aggregates, nevertheless noticeably affect its stereochemistry, as the structure of the rod is left-handed.

The effect of dilution on the lattice parameter of the hexagonal phase can also be analysed comparing the different guanosine derivatives[41]. While the behaviour of G1, G3 and G6 are similar, showing a moderate linear dependence of the intercolumnar distance on the concentration, in the case of G2 this dependence is dramatic. This trend is confirmed in a completely independent way from measurements of the pitch of the cholesteric phases as a function of concentration[41]. As expected, we observe an increase of the pitch with dilution, i.e., as the distance between the rods increases and their chiral interactions decrease, (see Fig. 22). However, the concentration dependence for G2 is much more pronounced than for the other derivatives. This anomalous behaviour is likely to be related to its more pronounced hydrophobic character: its rod-shaped aggregate interacts to a lesser extent with water, which acts merely as a diluting solvent, whereas in the case of the other derivatives specific interactions seem to be operative.

Considering the handness of the cholesteric superhelix, G1 and G2 give rise to a left-handed cholesteric, while G3, G4 and G6 give the opposite handed phases[41]. A tentative explanation has been formulated, assuming that the characteristics of the cholesterics, and in particular their handness, are connected to the stereochemistry of the chiral columnar aggregates. Each individual column is maintained by two general types of forces: phosphodiester bridges and non-covalent interactions. The former are dominant in G6, while the latter are exclusive for G1. The handness of the chiral aggregates imposed by the two forces is not necessarily the same. On the basis of CD measurements in isotropic solutions, for G1 the non-covalent interactions seem to generate a left-handed structure, while for G6 the covalent forces generate a right-handed structure (as in the polyG fibres). Therefore, it appears obvious that the two cholesterics must present opposite handness. For G2, G3 and and G4, both forces exist and in these cases the handness of the cholesteric is determined by the strongest contribution[41].

ACKNOWLEDGEMENTS. We are grateful to P. Laggner for communicating unpublished observations. We also thank S. Franguelli, E. Mosca and S. Polenta for technical assistance in the preparation of this Chapter.

## References

1. J.A. N. Zasadzinski, J. Schneir, J. Gurley, V. Elings and P. K. Hansma, Science **239**: 953 (1988).

2. G.S. Smith, E. B. Sirota, C. R. Safinya and N. A. Clark, Phys. Rev. Lett. **60**: 813 (1988).
3. M.P. Hentschel and F. Rustichelli, Phys. Rev. Lett. **66**: 903 (1991).
4. P. Laggner and M. Kriechbaum, Chemistry and Physics of Lipids,in print (1991).
5. V. Luzzati, P. Mariani and H. Delacroix, Makromol. Chem. Macromol. Symp. **15**: 1 (1988).
6. P. Mariani, V. Luzzati and H. Delacroix, J. Mol. Biol., **204**: 165 (1988).
7. V. Luzzati, A. Gulik, T. Gulik-Krzywicki and A. Tardieu, in "Lipids and Membranes: Past, Present and Future", J. A. F. Op den Kamp, B. Roelofsen & K. W. A. Wirtz, eds., pp. 137-151, Elsevier Science Pub., Amsterdam (1986).
8. G. Lindblom and L. Rilfors, Biochim. Biophys. Acta **988**: 221 (1989).
9. International Workshop on Geometry and Interfaces, J. Physique (Coll.) C7-**51** (1990).
10. J.F. Sadoc and J. Charvolin, Acta Cryst. A **45**: 10 (1989).
11. F. Livolant, J. Physique **47**: 1605 (1986).
12. G.P. Spada, A. Carcuro, F.P. Colonna, A. Garbesi and G. Gottarelli, Liquid Crystals **3**: 651 (1988).
13. H. Delacroix, P. Mariani and T. Gulik-Krzywicki, J. Physique (Coll.) C7-**51**: 119 (1990).
14. J.M. Seddon, J.L. Hogan, N.A. Warrender and E. Pebay-Peyroula, Progr. Colloid Polym. Sci. **81**: 189 (1990).
15. V. Luzzati, in "Biological Membranes", D. Chapman, ed., Academic Press, London, **1**: 71-123 (1968).
16. G. Albertini, M.G. Ponzi-Bossi and F. Rustichelli, Chapter 30 of this volume.
17. M.J. Janiak, D.M. Small and G.G. Shipley, J. Biol. Chemistry **254**: 6068 (1979).
18. R.E. Goldstein and S. Leibler, Phys. Rev. Lett. **61**: 2213 (1988).
19. M.P. Hentschel and R. Hosemann, Mol. Cryst. Liq. Cryst. **94**: 291 (1983).
20. D. Marsh, Biochemistry **19**: 1632 (1980).
21. B.G. Tenchov, H. Yao and I. Hatta, Biophys. J. **56**: 757 ( 1989).
22. J.M. Seddon, Biochim. Biophys. Acta **1031**: 1 (1990).
23. K. Fontell, Colloid Polym. Sci. **268**: 264 (1990).
24. V. Luzzati, P. Mariani and T. Gulik-Krzywicki, in "Physics of Amphiphilic Layers", J. Meunier, D. Langevin and N. Boccara, eds., pp. 131-137, Springer-Verlag, Berlin, (1987).
25. V. Luzzati and A. Tardieu, Ann. Rev. Phys. Chem. **25**: 79 (1974).
26. A. Tardieu, V. Luzzati and F.C. Reman, J. Mol. Biol. **75**: 711 (1973).
27. V. Luzzati and P.A. Spegt, Nature **215**: 701 (1967).
28. J. Charvolin and J.F. Sadoc, Colloid Polym. Sci. **268**: 190 (1990).
29. D.M. Small, "The Physical Chemistry of Lipids", Plenum Press, New York (1986).
30. P.O. Eriksson, G. Lindblom and G. Arvidson, J. Phys. Chem. **91**: 846 (1987).
31. L.E. Scriven, Nature (London) **263**: 123 (1976).
32. B. Ericsson, K. Larsson and K. Fontell, Biochim. Biophys. Acta 729: **23** (1983).
33. V. Luzzati and F. Husson, J. Cell Biol. **12**: 207 (1962).
34. Y.M. Yevdokimov, S.G. Skuridin and V.I. Salyanov, Liq. Cryst. **3**: 1443 (1988).
35. V. Luzzati and A. Nicolaieff, J. Mol. Biol. **1**: 127 (1959).
36. C. Robinson, Tetrahedron **13**: 219 (1961).
37. F. Livolant and Y. Bouligand, J. Physique **47**: 1813 (1986).
38. S. Bonsignore, G. Cometti, E. Dalcanale and A. Du Vosel, Liquid Crystals **8**: 639 (1990).
39. G.W. Gray and J.W. G. Goodby, " Smectic liquid crystals. Textures and Structures", Leonard Hill, ed., Glasgow and London, Heyden & Son, Inc., ed., Philadelphia (1984).
40. P. Mariani, C. Mazabard, A. Garbesi and G.P. Spada, J. Am. Chem. Soc. **111**: 6369 (1989).
41. S. Bonazzi, M. Capobianco, M. Miranda De Morais, A. Garbesi, G. Gottarelli, P. Mariani, M.G. Ponzi-Bossi, G.P. Spada and L. Tondelli, J. Am. Chem. Soc. **113** :5809 (1991).
42. S. Bonazzi, M. Miranda De Morais, A. Garbesi, G. Gottarelli, P. Mariani and G.P. Spada, Liquid Crystals **10**: 495 (1991).
43. S.B. Zimmerman, J. Mol. Biol. **106**: 663 (1976).

Chapter 30

# MODIFICATIONS OF THE PHASE TRANSITIONS OF MODEL MEMBRANES BY EXOGEN MOLECULES

G. ALBERTINI, M.G. PONZI-BOSSI and F. RUSTICHELLI

## 1. Introduction

This Chapter discusses the interaction of model membranes, which are organized as lamellar lyotropic liquid crystal phases, with exogen molecules such as cholesterol, anaesthetics, drugs and aminoacids.

A series of recent experiments, using such techniques as X-ray, neutron diffraction and differential scanning calorimetry, investigated the modifications induced by these molecules on the phase transitions of model membranes. These effects depend on the type of lipids and exogen molecules and on their concentration, and can be divided into three categories

a) suppression of phase transitions
b) induction of phase transitions
c) modification of phase transitions.

The phospholipids, the principal components of biological membranes, are molecules consisting of a polar head group (the hydrophilic part) and one or two hydrocarbon chains (the hydrophobic part); as regards structural characteristics and related chemical-physics properties[1], they belong to the lyotropic liquid crystal family. An example of a phospholipid molecule, DSPC, is given in Fig. 12b. In fact, when mixed with water, quite a large number of different phases are observed as a function of lipid concentration and/or temperature, ranging from spherical micelles to more ordered structures[2] having one, two or three-dimensional positional order (lamellar, hexagonal and cubic phases, respectively).

Considering in particular mixtures having excess water, some natural and synthetic lecithins, such as egg lecithin or synthetic distearoyl-phosphatydil-choline (DSPC) and dipalmitoyl-phosphatydil-choline (DPPC), give rise to the formation of multilamellar liposomes, which can be described as a regular stacking of lipidic lamellae separated by water: the structure organization is shown in Fig. 1. Every lipidic lamella has a very strong structural, functional and biochemical similarity with the well known cellular bilayer membrane[3] and it appears to consist of a hydrocarbon sheet covered by polar groups of the lipidic molecules. Fig. 2 presents an electron microscopic image of a fractured multilamellar liposome: the regular stacking of the

G.Albertini - Dipartimento di Scienze dei Materiali e della Terra, Università di Ancona, Italy.

M.G.Ponzi-Bossi and F.Rustichelli - Istituto di Fisica Medica, Facoltà di Medicina e Chirurgia, Università di Ancona, Italy.

*Phase Transitions in Liquid Crystals*, Edited by S. Martellucci and A.N. Chester, Plenum Press, New York, 1992

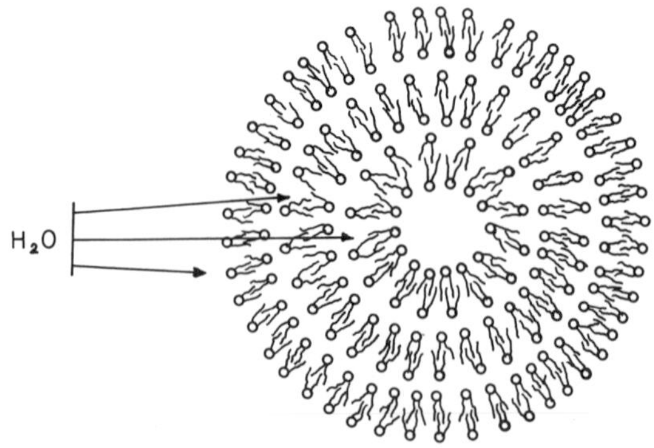


Fig.1. Schematic representation of the structural organization of multilamellar liposomes. The hydrophilic group is represented by an open circle and the paraffinic chains by a wriggle.

layers is quite evident. The liposomes are considered to be very useful models for studying the functions and the properties of biological membranes[4] and are the principal subjects of the investigations reported in this Chapter.

## 2. Basic Properties of Syntetic Lecithin Lyotropic Liquid Crystals

Differential scanning calorimetry (DSC) curves obtained by heating multilamellar liposomes of lecithin with two identical hydrocarbon chains display two endothermic peaks, which refer to the socalled pre- and main-transitions[5].

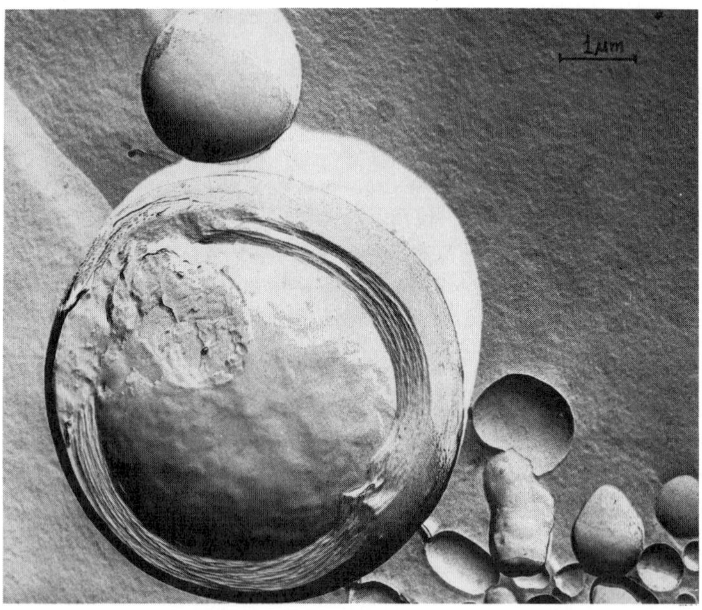


Fig.2. Freeze-etching electron micrographs showing a soya lecitin multilayered vesicle. The specimen was fractured and Pt shadowed at 150°C.

In particular, the first peak is at low enthalpy and corresponds to the phase transition from the gel state Lβ′ phase to the Pβ′ lamellar phase, while the second, which is at higher enthalpy, corresponds to the phase transition from the Pβ′ phase to the Lα liquid crystalline lamellar phase (see Fig. 4 for more structural details). In particular, Fig. 3 presents the thermogram obtained for DPPC liposomes in excess water. It must be noticed that when DSPC liposomes are considered, the phase transition temperatures are about 12 °C higher, however the thermograms are similar in shape[5].

Fig. 4 shows X-ray diffraction profiles obtained in these different phases in the case of DPPC liposomes: from the position of the low angle diffraction peaks, it can be easily deduced that all phases are lamellar.[6] The X-ray diffraction profiles for DSPC liposomes are very similar; however, the peak positions change because the unit cell dimensions (in particular, the interlayer repetition distance, i.e., the distance between the lamellae given by the sum of a lipidic bilayer and of a water layer thickness) are different due to the longer DSPC lipidic molecule.

A schematic representation of the structural lipid/water organization corresponding to the three different lamellar phases is given in the inserts of Fig. 4 and Fig. 5. In particular, in the Lβ′ phase, the lattice is 1-dimensional; the repeat distance at 25 °C in excess water is about 64 Å and 68 Å for the DPPC and DSPC liposomes, respectively. In the Pβ′ phase, the lipidic bilayers are organized in a 2-dimensional lattice, as the lamellae are distorted by a periodic ripple; the corresponding interlamellar distances are higher than in the Lβ′ phase. Finally, in the Lα phase, the lattice is 1-dimensional and in this case as in the Lβ′ phase the structural elements are planar lipidic sheets separated by water layers; the interlamellar distances decrease to values lower than those observed in the Lβ′ phase, this giving a characteristic pattern when the repeat distance is plotted as a function of temperature (see for example Ref. 7).

Information about the chain organization inside the lipidic layer can be obtained from the pattern of high angle diffracted X-ray intensity. For example, in the Lβ′ phase, two peaks are observed: one is narrow and very intense while the other appears to be broad and diffuse. This pattern was interpreted[8] by assuming a pseudohexagonal packing of the chains, which are stiff and tilted with respect to the lamellar plane (Fig. 5). Considering the geometry of packing and the position of the two peaks, it is possible to determine the two characteristic distances between the lipidic chains $l_1$ and $l_2$[9].

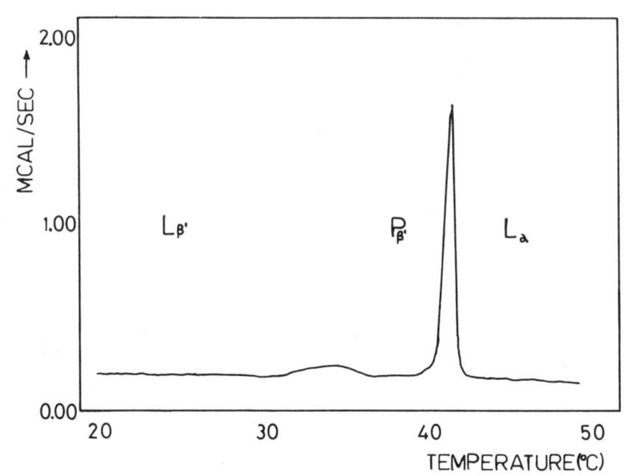


Fig.3. Differential scanning calorimetry (DSC) heating endotherms of fully hydrated samples of dipalmitoyl-phosphatidyl-choline (DPPC) multilamellar liposomes.

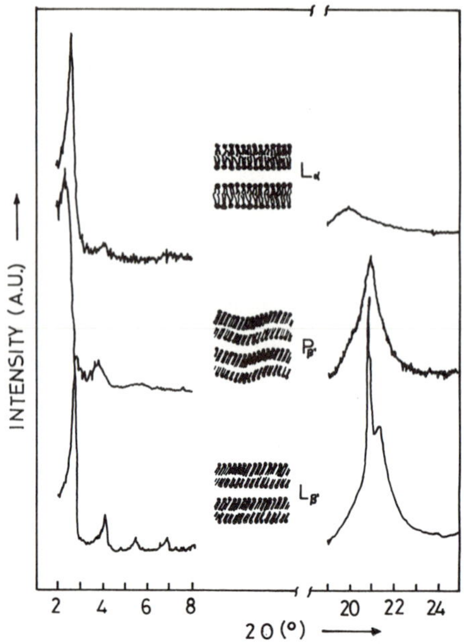


Fig.4. Low-angle (left side) and high-angle (right side) X-ray diffraction patterns from fully-hydrated samples of dipalmitoyl-phosphatidyl-choline (DPPC) multilamellar liposomes below the thermal pre-transition (lowest curves), intermediate between the two transitions (middle curves) and above the chain melting transition (upper curves). The inserts show a schematic representation of the structure of the three different lamellar phases.

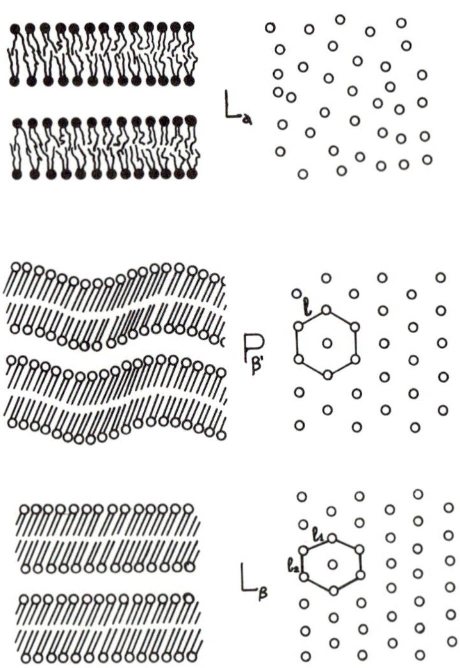


Fig.5. Schematic representation of the structural organization of the three different lyotropic mesophases of aqueous dispersions of lipids.

In the Pβ′ phase, only a single peak is observed in high angle X-ray diffraction: it corresponds to a hexagonal packing[8] of the chains, which are stiff and tilted with respect to the lipidic membrane. Finally, in the higher temperature Lα phase, the chains are melted: only a very broad and diffuse peak is detectable in the high angle region of the X-ray diffraction patterns.

## 3. Modifications Induced by Cholesterol on Model Membrane Phase Transitions

Studies of the interactions of membranes with cholesterol are very interesting (see for example Ref. 10); for example, in animals the permeability and the stability of cell membranes are modified by this molecule. However, in some pathological conditions, cholesterol can be found in lipidic deposits like those evidenced in atherosclerosis and in the xanthomas and xanthelasmas skin diseases. It should also be noted that very high quantities of cholesterol can be incorporated into a phospholipid bilayer: a unit molar ratio, i.e. one molecule of cholesterol every molecule of lecithin, can be attained without functional alteration of the membrane.

To obtain information about the modifications induced by cholesterol on phospholipid bilayer phase transitions, a study was recently performed of the structural and thermodynamical behaviour of the DSPC multilamellar system in excess water at different cholesterol concentrations, varying the temperature from 20 to 70 °C[11].

The phase diagram which was found appears to be complex; in particular, some new phases, induced by the cholesterol, were observed. At room temperature, for example, when the cholesterol concentration is about 5%, an Lβ phase, i.e. a lamellar phase with non-tilted stiff chains[12], exists; while at higher concentrations a Γ phase, i.e. a melted chain lamellar phase very similar to the Lα previously described, is found.

For the particular molar ratio of 1 molecule of cholesterol for every 2 molecules of phospholipids the transition from the Γ to the Lα phase was studied[13].

Fig. 6a shows the repeat thickness D within the liposomes as a function of temperature,

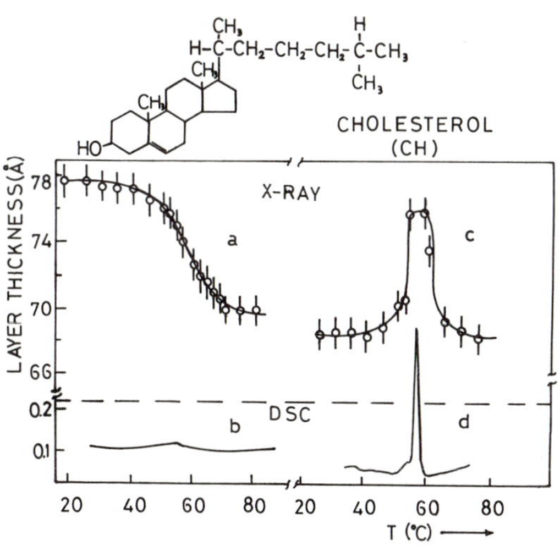


Fig.6. Layer thickness (a, c) and endothermic heat flow (b, d) as a function of temperature for liposomes of pure DSPC (c, d) and liposomes with cholesterol/lecithin molar ratio = 35:65 (a, b). (From Ref. 13). The molecule of cholesterol is shown at the top of the figure.

for a mixture of lecithin and cholesterol in the molar ratio 65/35, through the transition towards the Lα phase, as obtained from the position of the low-angle peak (see Fig. 4) by applying the Bragg law. This thickness is the sum of a lipid thickness (D) and a water thickness ($D_w$) (see Figs. 1 and 4). Constant values are observed for temperatures lower than 40 °C (D = 77.5 ± 0.8 Å) and for temperatures higher than 70 °C (D = 70.0 ± 0.7 Å).

A progressive reduction from 77.5 Å to 70.0 Å is observed in the temperature range from 40 °C to 70 °C. Within his range an endothermic transition peak is observed in the calorimetric curve, as reported in Fig. 6b). The peak enthalpy is 0.4 ± 0.1 cal per gram of sample. For comparison, the corresponding data obtained from pure DSPC in water are reported in Fig. 6c and 6d.

In order to better understand the phenomena occurring at the transition to Lα phase in presence of cholesterol, it is useful to also look at high-angle X-ray diffraction patterns. Measurements were also taken from aligned samples, having the layer normal perpendicular to the impinging beam, by using a flat cell; these measurements showed that both at room temperature and at 60 °C the wide angle X-ray reflections are located on the equatorial region, indicating that the chains are perpendicular to the layers.

Two main contributions can be distinguished in the high-angle diffraction patterns reported in Fig. 7 and obtained at high temperature from the phase having melted chains: a large peak at higher angle from water, and a sharper peak at lower angle which arises from scattering by the melted chains[14].

The shape of the water peak deduced directly from a sample of pure water was fitted to a Gaussian curve:

$$I_w(2\Theta) = I_o \exp[(2\Theta - 2\Theta_o)/\sigma]^2$$

where $2\Theta_o$ = 28.5° and $\sigma$ = 7.6°. The $I_o$ value depends on the quantity of water present in the sample.

The experimental data concerning the chain peak were matched to theoretical curves obtained by considering a hexagonal chain packing with short-range positional order within the layers. These theoretical curves were obtained by using a method derived from that reported in Ref.s 15,18.

Specifically, a Lorentzian shape was attributed to the in-plane structure factor

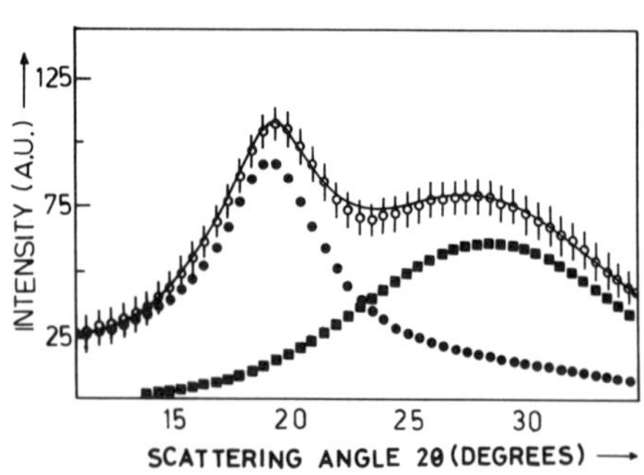


Fig.7. High-angle X-ray diffraction at a temperature T = 64 °C: ϕ = experimental data for DSPC/cholesterol mixture (65/35 molar ratio); ■ = water contribution; • = theoretical curve for chain scattering (correlation length ξ = 8 Å, $Q_o$ = 1,347 Å );— = theoretical curve for the sum of water and lipid scattering. (From Ref. 13).

$$f(Q_p) = 1/[(Q_p - Q_o)^2 + \xi^{-2}]$$

where $\xi$ is the in-plane correlation length, $Q_o$ is the (100) vector in reciprocal space and $Q_p$ is the in-plane component of the scattering vector.

The best-fit curves were obtained by varying the $Q_o$ and $\xi$ values. Fig. 7 shows an example of the good agreement between the theoretical and experimental curves obtained at a temperature T = 64 °C.

At lower temperatures no agreement is found between theoretical and experimental data unless a new peak, with an assumed Lorentzian in-plane structure factor, is added at a lower scattering angle (2Θ = 18.5°) (Fig. 8). The best fit without the new peak, for a temperature T=35°C, is reported in the insert of Fig. 8.

Two structural quantities were deduced for each peak by using the above-reported data analysis: the correlation length and the $Q_o$ value, as a function of temperature. From the $Q_o$ value the interchain distance $l$ was obtained by assuming hexagonal packing $l = 4\pi/(\sqrt{3}Q_o)$.

Fig. 9b reports the correlation length as a function of temperature for the peak at larger scattering vectors (Γ phase) until the transition to the Lα phase occurs. The corresponding peak widths Δd are reported in Fig 9a. For the other peak (Γ′ phase according to our subsequent discussion) the uncertainties are quite large: the correlation length seems to be temperature independent and to have the value $\xi = (6.0 \pm 0.5)$ Å.

Additional information which can be derived from Fig. 9 is the considerable disorder characterizing the in-plane molecular packing of cholesterol and DSPC molecules in the Γ phase, in agreement with what was previously found for this system by X-ray diffraction[19] and for the cholesterol DPPC system by other techniques[20].

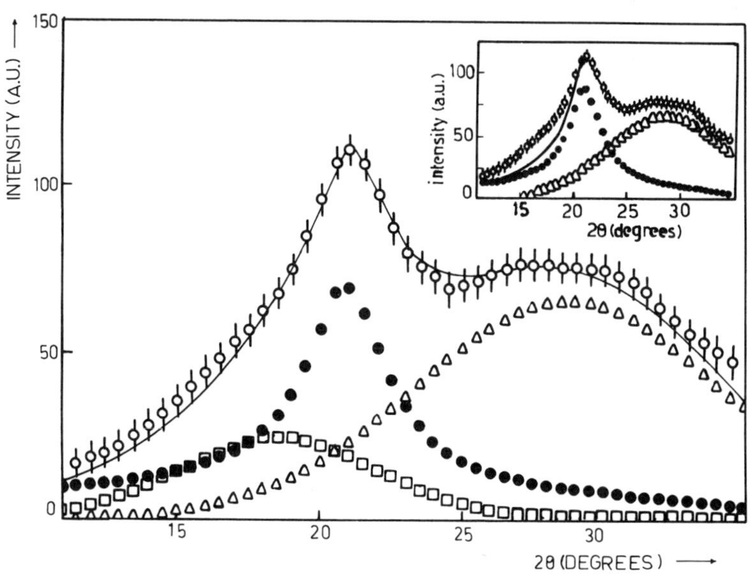


Fig.8. High-angle X-ray diffraction at a temperature T=35°C: ϕ experimental data for DSPC/cholesterol mixture (65/35 molar ratio); △ = water contribution; • = theoretical curve for chain scattering (correlation length $\xi = 14$ Å, $Q_o = 1.459$ Å$^{-1}$) Γ phase ; □ =scattering from the Γ′ phase;— = theoretical curve for the sum of scattered intensities from water, Γ and Γ′ phases. In the insert the continous line is obtained without considering the scattered intensity from the Γ′ phase. (From Ref. 13).

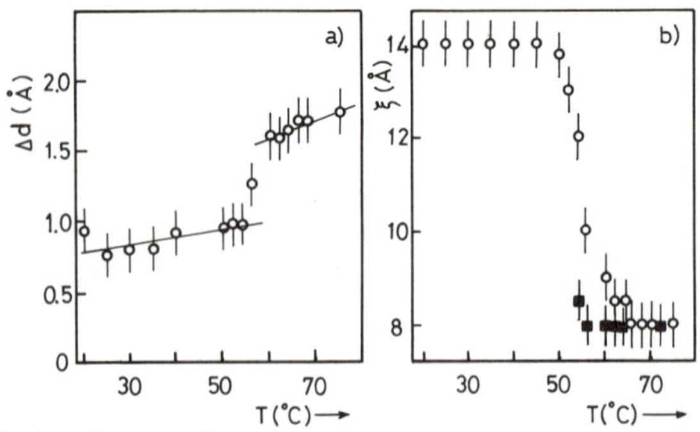


Fig.9. a) Width of high-angle X-ray diffraction peak in $s^{-1}$ units as a function of temperature, for a 35/65 cholesterol/DSPC molar ratio; b) ξ values corresponding to the data of fig.9a) in the temperature range from 20 to 75 °C (ϕ) and to data of pure DSPC in Lα phase (■). (From Ref. 13).

Referring to the corresponding large angle peaks of pure DPPC in the different phases, it can be concluded that the interchain positional order of the Γ phase is considerably lower than in the Lβ′ and Pβ′ phases of pure lecithin, but higher than in the Lα liquid crystalline phase. The highly disordered nature of the Γ phase is also confirmed by the low-enthalpy peak associated with the transition to the Lα phase, as compared with the high peak associated with the transition Pβ′ → Lα in pure DSPC (Fig. 6).

As mentioned above, the data analysis of Fig. 8 suggests the presence of a very broad peak at $2\Theta = 18.5°$ in order to obtain a satisfactory fit of the experimental data. This need does not exist in the Lα phase, as appears from Fig. 7. The existence of this peak and its very large width suggests the presence inside the bilayers of another phase, characterized by a larger disorder as compared to the Γ phase, and by a larger intermolecular spacing. This larger spacing suggests that this phase is richer in cholesterol as compared with the Γ phase, the lateral average cross-section of cholesterol being larger than that of a lipid chain. The authors conclude that it may be the Γ′ phase mentioned in Ref. 20.

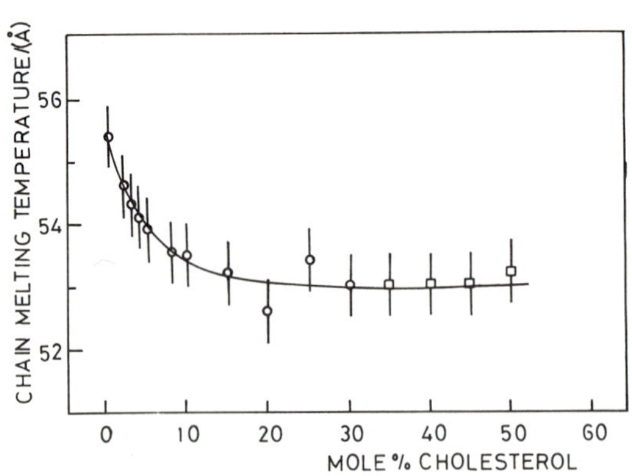


Fig.10. Chain melting temperature vs. cholesterol concentration for DSPC aqueous dispersions (the line is a guide for the eye). (From Ref. 22).

The angular position of the X-ray diffusion peak associated with the Γ′ phase coincides with the position of the Lα phase peak, so that in principle one might doubt whether the Γ′ phase really transforms into the Lα phase at the transition, or remains unchanged. However, since to the limit of experimental sensitivity and resolution the Γ′ phase seems to be characterized by the same overall layer thickness of 77.5 ± 0.8 Å characteristic of the Γ phase, whereas the La phase is characterized by a thickness of 70 ± 0.7 Å, one should expect the presence of two peaks in the low-angle diffraction region if the Γ′ phase remained unchanged.

In fact, both diffractometer results and X-ray diffraction photographs show only one diffraction peak, both in the Γ + Γ′ phase coexistence region and in the Lα phase. Consequently, it can be concluded that at low temperature two phases are present, one cholesterol-poorer (Γ) and the other cholesterol-richer (Γ′); and that both of these transform into a single liquid crystalline Lα phase due to an increase in the cholesterol diffusion coefficient, which makes the persistence of phase segregation at high temperature impossible. In Fig. 10 the chain melting transition temperature is shown as a function of the cholesterol concentration, as obtained from the position of calorimetric peaks[21]. This temperature is the transition temperature to the liquid-crystalline phase Lα which is almost a smectic A phase, the structure of which in the presence of cholesterol is shown in Fig. 11. For cholesterol molar fraction greater than 30% but no greater than 50%, no calorimetric peak was observed for a heating rate of 2.5 °C/min the values reported for these concentrations refer to a 10 °C/min heating rate.

A decrease of transition temperature with increasing cholesterol content can be seen in the range of concentrations between 0% and 50% cholesterol molar fraction. No calorimetric peak was observed between room temperature and 80 °C for higher cholesterol concentrations.

This last experiment confirms the fluidifying effect of cholesterol.

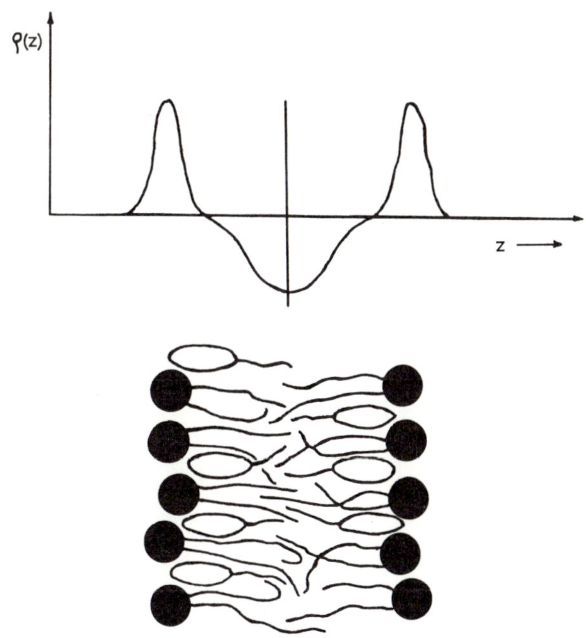


Fig.11. Schematic view of the Lα phase structure of the lipid lamellae containing cholesterol. The black dots represent the polar groups and the wriggles the hydrocarbon chains; the cholesterol molecules are represented as open ellipses. The corresponding electron density profile is also shown.

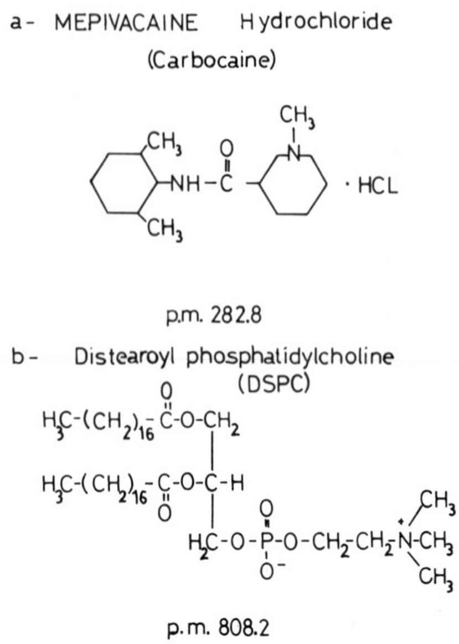


Fig.12. The local anaesthetic mepivacaine (a) and the lipid DSPC (b).

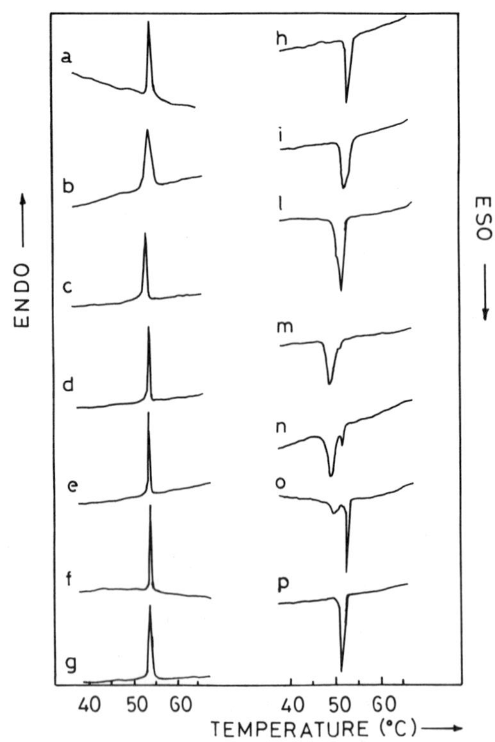


Fig.13. Calorimetric curves obtained on heating (left) and on cooling (right) liposomes containing mepivacaine molar ratio R = O (a, b), 0.1 (h, i), 0.9 (c, l), 3.3 (d, m), 1 (e, n), 5 (f, o), 5.5 (g, p). (From Ref. 23).

## 4. Modifications Induced by Anaesthetics on Model Membrane Phase Transitions

Several studies by different techniques have recently been performed on the modifications induced by anaesthetics on phase transitions in model membranes, i.e. in aqueous dispersions of synthetic phospholipids[23,24].

It is well known that, a local anaesthetic blocks membrane conduction, decreasing or preventing the high increase of membrane permeability to sodium ions[25]: this permeability depends on the sodium channels, the activity of which is related to the action of lipid-protein complexes. Anaesthetics exert a non-specific action on the lipid part of these complexes[26]. This action believed to take place on the lipid portion and be non-specific, since a linear dependence was found between the strength of anaesthetics and their oil solubility[27,28], in particular their solubility inside the hydrophobic chains of the lipids[28].

For these reasons, investigations on the interactions between local anaesthetics and lipids have importance for understanding how anaesthetics act.

In Ref. 23 we investigated the influence of a local anaesthetic (Mepivacaine) on liposomes of 1,2-distearoyl-3-sn-phosphatidylcholine (DSPC). Fig. 12 shows the Mepivacaine and the DSPC molecules.

Fig. 13 reports the calorimetric curves obtained on heating and on cooling liposomes containing different concentrations of anaesthetic. Two peaks appear on heating the control sample (Fig. 13a) one at about 51 °C, corresponding to the so-called pre-transition ($L\beta' \rightarrow P\beta'$ transition), the other at about 55 °C which corresponds to the so-called main transition ($P\beta' \rightarrow L\alpha$ transition) in which the aliphatic chains melt[29,30].

With increasing anaesthetic concentration, both the temperature and the width of these peaks change and, furthermore, the pre-transition peak is no longer detectable at the highest concentrations (Fig. 13b-g).

In more quantitative terms, Fig. 14 reports the temperatures of the two transitions, as obtained from the position of the maximum of the calorimetric peak, as a function of the molar ratio R between mepivacaine and lecithin.

No detectable variation is observed for concentrations lower than $R = .01$, while a decrease is observed for concentrations higher than this value but lower than $R = .5$; for still higher anaesthetic concentrations the pre-transition disappears and the main transition is at a temperature $T = 53.5 \pm .5$ °C.

Fig.s 15 and 16 give the transition enthalpies per gram of lecithin for the two peaks, as obtained by heating liposomes containing different anaesthetic concentrations. In the case of the pre-transition, the enthalpy decreases for increasing concentration until the transition is no longer detectable (Fig. 15). For the main transition (Fig. 16) the enthalpy variation is $12.5 \pm 1$ cal per gram of lecithin for molar ratios less than .1 and $16 \pm 1$ cal/g for higher values.

Upon cooling (Fig. 13 h-p) the behaviour is more complex. The splitting into two peaks indicates the occurrence of two phases with different concentration of anaesthetic. A more detailed discussion can be found in Ref. 23.

In conclusion, we confirmed that anaesthetics interact with the lipid part of membranes, and that introduction of anaesthetic molecules in lecithin - water systems results in decreasing the main transition cooperativity (as can be deduced from the widening of the corresponding calorimetric peak). Moreover, a fluidifying action of the anaesthetic on the membrane can be deduced from the lowering of the main transition temperature, which is a melting transition for the aliphatic chains.

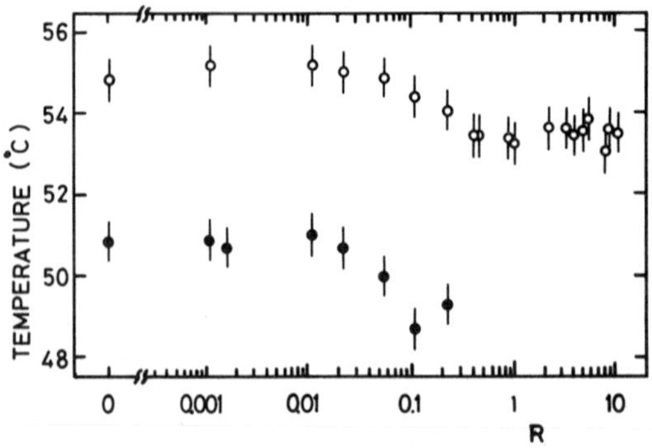


Fig.14. Pre-transition (●) and main transition (ɸ) temperatures obtained on heating, as functions of the molar ratio R of mepivacaine to DSPC. (From Ref.23).

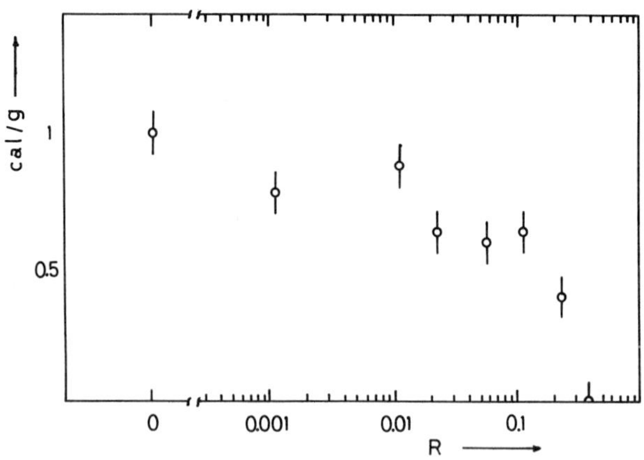


Fig.15. Pre-transition enthalpy per gram of lecithin, as a function of the molar ratio R of mepivacaine to DSPC (From Ref. 23).

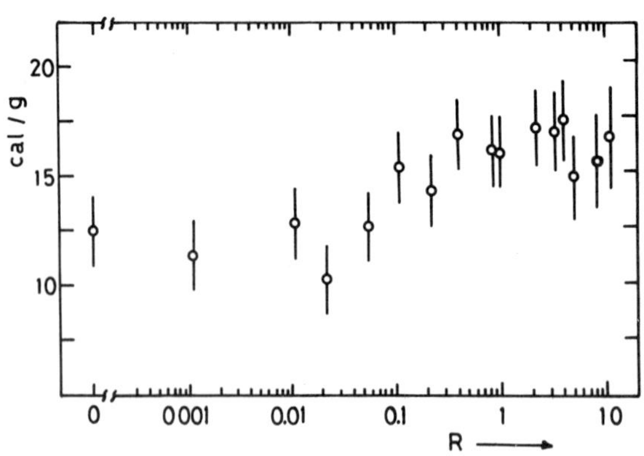


Fig.16. Main transition enthalpy per gram of lecithin, as a function of the molar ratio R of mepivacaine to DSPC. (From Ref. 23).

## 5. Influence of Drugs on Model Membrane Phase Transitions

The first step in the absorption of a drug molecule in a living cell involves its interaction with lipids. The drug initially present in the aqueous solution at the interface with the cell membrane generally dissolves in the lipid matrix, migrates to the other side of the lipid bilayer and enters the aqueous compartment of the cell where it manifests its action [32]. For this reason, investigations of lipid-drug interactions are of paramount importance to biologist as well as to liquid-crystal scientists. Moreover, considerable interest has recently been shown in using this approach to transport therapeutic drugs through carrier liposomes. In this Section we will consider two examples of the modifications induced by drugs in phase transitions of liposomes, specifically the cases of propranolol and azelaic acid. Propranolol is a prototype β-adrenoceptor blocking agent which is widely used in the treatment of arrhythmia, since it exhibits considerable efficacy in the management of cardiac rhythm disorders[32,33]. However, its bioavailability is low due to a high hepatic extraction ratio or an extensive hepatic first-pass metabolism[34]. It is, therefore, desirable to encapsulate this drug in a carrier system which can protect it from immediate biodegradation and deliver it into heart tissue. Lipids are thought to be attractive carrier systems because of their inherent biocompatibility[35]. In particular, multilamellar vesicles of dipalmitoyl phosphatidylcholine (DPPC) with or without cholesterol have been considered as carrier systems[36]. ESR spin labelling, differential scanning calorimetry (DSC), NMR techniques, X-ray diffraction and optical microscopy were used to investigate modifications of the lipid dispersion phase transition induced by propanolol at varying concentrations[37,38]. Here, only the main features of the thermodynamical and structural modifications will be presented.

DSC calorimetric scans were obtained in the temperature range 290-330 K. DPPC liposomes without the drug (R = 0) showed characteristic peaks at 309 and 314.5 K; Fig. 17,

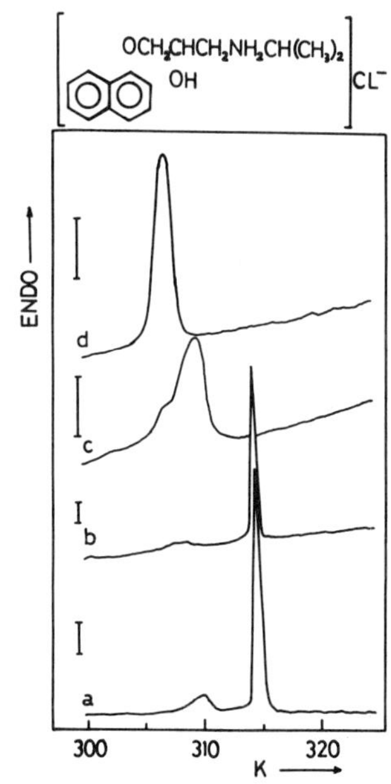


Fig.17. Calorimetric heating scans for DPPC liposomes containing different concentrations of the drug propanolol (R = propanolol/DPPC molar ratio) R=0 (a), R=0.001 (b), R=0.1 (c) and R=0.6 (d). The bar in each scan corresponds to 0.025 cal/(g.s). Scan rate 2.5 K/min. (From Ref. 38). The propranolol-HCl molecule is shown at the top of the figure.

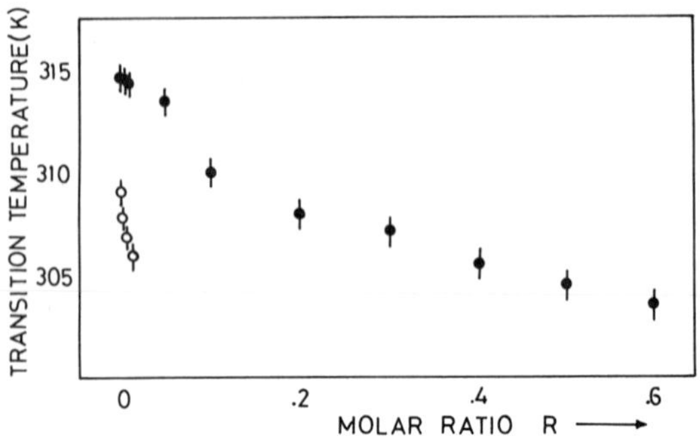


Fig.18. Calorimetric peak maximum temperatures vs. the propanolol to DPPC molar ratio R. DPPC/water = 1/3 w/w . ● = higher temperature peak; ○ = lower temperature peak. (From Ref. 38).

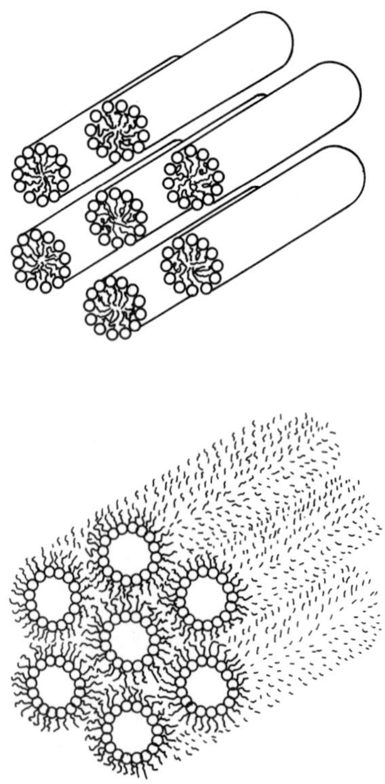

Fig.19. Schematic representation of two possible lyotropic hexagonal phases.

curve a) shows the phase transitions $L\beta' \rightarrow P\beta'$ and $P\beta' \rightarrow L\alpha$ (chain melting transition or main transition). Liposomes containing low concentrations of propranolol ($R < 0.08$) exhibited similar behaviour, but the transitions occurred at lower temperatures (Fig. 17, curve b). Increasing the propranolol content further ($R > 0.08$) resulted in the disappearance of the $L\beta' \rightarrow P\beta'$ transition, whereas the main transition peak appears to be broadened and shifted to lower temperatures (Fig. 17, curves c and d ): X-ray diffraction data confirmed that this peak corresponds to the chain melting transition. The full width at half maximum of the main transition peak increases, reaching a maximum near $R = 0.1$, and then decreases. Fig. 18 presents the temperatures of the two phase transitions, as a function of the drug-to-DPPC molar ratio R.

Concerning the structural modifications induced by propranolol and observed by X-ray diffraction, one should take note of the destruction of liposome organization induced by drug concentrations > 1, and a phase transition to a hexagonal phase of the type shown in Fig. 19, occurring at a concentration $R = 2.2$. In fact, the induced phase transition is deduced from Fig. 20, obtained at room temperature; part a) of Fig. 20 shows the X-ray diffraction pattern obtained for the propranolol concentration $R = 0.6$ where the diffraction peaks angular positions are in the ratios 1:2:3 typical of lamellar organization, and part (b) shows the pattern obtained for R=2.2 with the angular peak positions in the ratio $1:\sqrt{3}:2:\sqrt{7}$, typical of a hexagonal structure.

Fig. 21, showing the X-ray diffraction pattern obtained by using a diffractometer, confirms this interpretation in the low angle region and provides information on the melted state of the lipidic chains, as the high angle peak is very broad. The cell parameter of the hexagonal structure deduced from Fig. 21 is $a = 5.85$ nm. This last result is in agreement with a previous NMR study[39] on the interaction of propranolol with egg yolk lecithin vesicles in NaCl solutions.

For lower concentrations the presence of drug does not disturb the lamellar arrangement; however, it imparts fluidity to lipids, almost doubles the lamellar repeat spacing, increasing the water thickness, and induces a β conformation of the chains, which thus become perpendicular to the lamellae and seem to be interdigitated. In addition, the calorimetric and X-ray results suggest that the propranolol-HCl molecule interacts with the phospholipid head groups and slightly penetrates into the hydrocarbon chain region. Let us now consider the case of azelaic acid. Dicarboxylic acid forms a special class of molecules consisting of a straight chain with carboxyl groups at both ends[40], which exhibits lamellar liquid crystalline phases at high temperature and/or in water when in soap form. Azelaic acid (Fig. 22) is a 9 carbon atom saturated dicarboxylic acid having special biological properties and therapeutic applications: in particular, it has been found that this acid is an inhibitor of tyrosinase[41] as well as thioredoxin reductase (an enzyme involved in the biosynthesis of DNA)[42]. From a therapeutic point of view, it has been shown that a cream containing 20% azelaic acid exhibits benefical effects on hyperpigmentary disorders such as toxic melanoderma, melasma[43], lentigo maligna[44] and malignant melanoma[45], all without hypochromia or other deleterious effects. Moreover, this drug has been found to be very effective in acne treatment, with additional qualification as an antikeratinizing agent[46]. Finally, azelaic acid seems to adversely affect the growth of abnormal cells, but not the proliferation of their normal counterparts[47].

Several studies have been published describing clinical observations, but little is known regarding the exact mechanism of action (see for example Ref.s 41-47 and references therein). Because it has been suggested that surface lipids play a significant role in the mechanism of action of azelaic acid[48], a study was recently performed on the interaction of this azelaic acid with phosphatidylcholine using several different experimental techniques: differential scanning calorimetry (DSC), nuclear magnetic resonance, X-ray diffraction and optical microscopy[49]. Only the principal thermodynamical and structural results will be reported here.

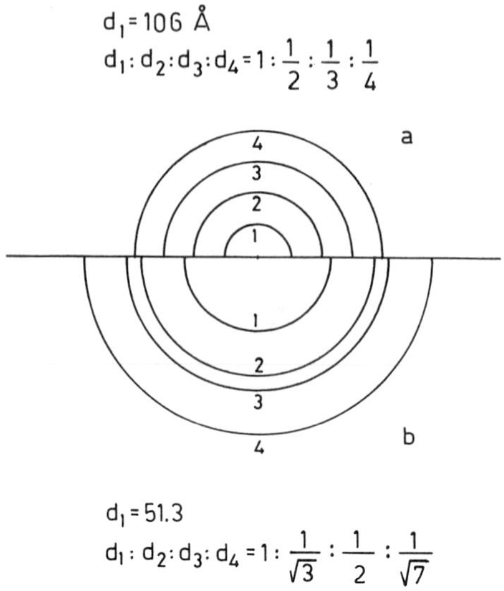


Fig.20. Low-angle X-ray scattering of DPPC propanolol from samples with: (a) R=0.6. Ratios 1 : 1/2 : 1/3 : 1/4 in the d values, obtained for the different X-ray diffraction peaks by the Bragg equation; and, (b) R=2.2. Ratios $1:1/\sqrt{3}:1/2:1/\sqrt{7}$ in the d values.

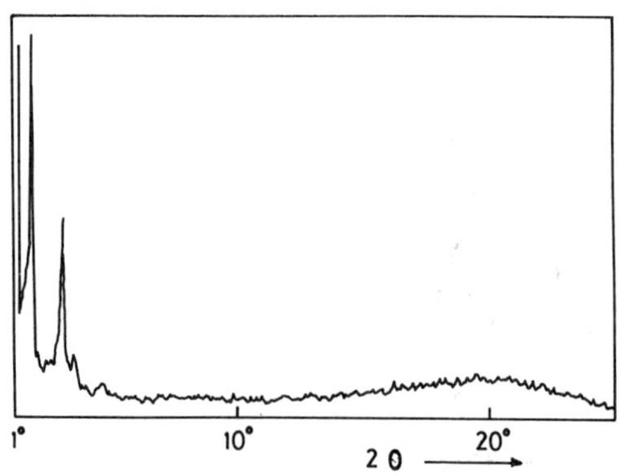


Fig.21. X-ray diffraction pattern obtained for a DPPC-propanolol sample at R = 2.2, T = 303 K. Lower scale takeoff angle 2Θ. (From Ref. 38).

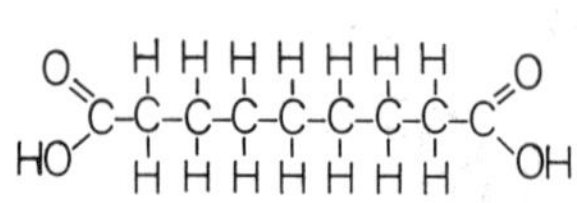

Fig.22. Chemical formula of the drug azelaic acid.

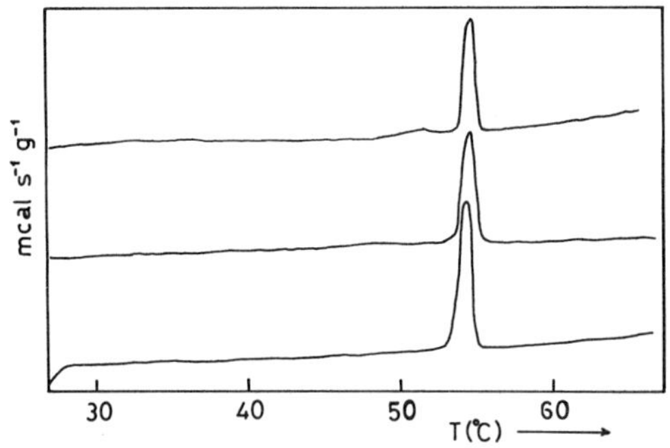


Fig.23. Calorimetric heating scans of DSPC-azelaic acid mixtures in excess water (scan rates 2.5°C/min). (From Ref. 49). Azelaic acid to DSPC molar ratio R = 0.02, middle curve; R = 0.2, lower curve.

Some DSC calorimetric scans obtained by heating samples are depicted in Fig. 23. At low molar ratios of azelaic acid (R < 0.2) the heating thermograms exhibit two well-separated transition peaks which, in analogy to pure DSPC liposomes, can be associated with the $L\beta' \rightarrow P\beta'$ and $P\beta' \rightarrow L\alpha$ phase transitions; the former is the so-called pre-transition while the latter is the so-called main transition, in which chain melting takes place. At higher drug concentrations the pre-transition peak completely disappears, while chain melting occurs at slightly lower temperature.

All samples, at all temperatures investigated, show the characteristic X-ray diffraction patterns of lamellar phases. Some typical low and high angle diffraction patterns observed at 25 °C for DSPC samples in excess-water condition and in the absence or in the presence of azelaic acid (R = 0.1 and R = 0.2) are shown in Fig. 24 and 25.

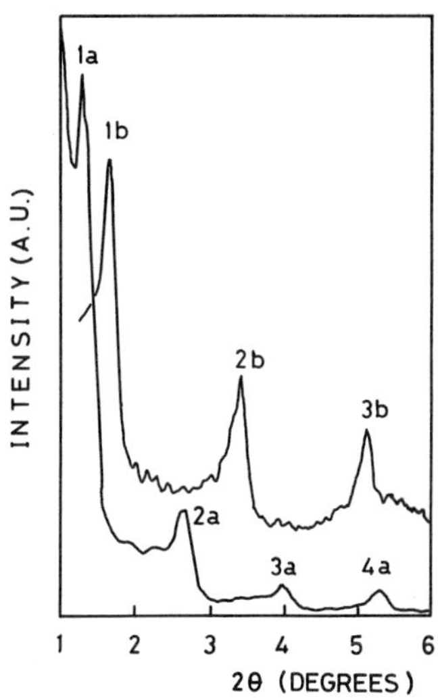


Fig.24. Low-angle X-ray patterns from DSPC-azelaic acid mixtures in excess water at 25°C. Lower curve R = 0, upper curve R = 0.2. (From Ref. 49).

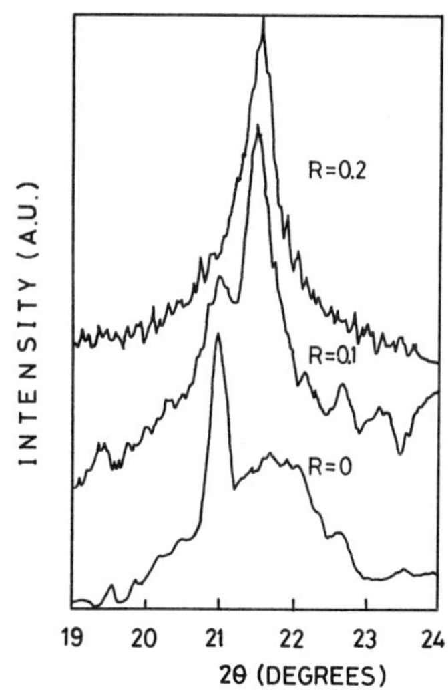


Fig.25. High-angle X-ray diffraction patterns from DSPC-azelaic acid mixtures in excess water at 25°C. Lower curve R = 0.0; middle curve R = 0.1; upper curve R = 0.2. (From Ref. 49).

The very appearance of these profiles, without going into detailed analysis, is suggestive of significant differences: in the low-angle region, the peaks corresponding to the different lamellar diffraction orders are shifted towards higher 2Θ values in the presence of high concentrations of azelaic acid. The corresponding lamellar thickness determined by applying the Bragg equation, plotted in Fig. 26 as a function of azelaic-acid concentration, changes at 25 °C from 68 Å (R = 0) to 52 Å (R > 0.1). The high-angle pattern, which for the pure DSPC Lβ′ phase consists of closely spaced high- and low-intensity peaks (Fig. 25, lower profile) changes to a simple, sharp, shifted peak as shown in the upper part of Fig. 25 (R = 0.2). According to Luzzati and coworkers[50], this simple sharp peak corresponds to the vanishing of the head group tilt.

The abrupt change in lamellar repeat spacing occurring at high molar ratio of azelaic acid (Fig. 26) was interpreted, with the help of the other experimental measurements, as a phase transition produced by azelaic acid from the conventional Lβ′ phase (see Fig. 4) to an "interdigitated phase", a model for which is shown in Fig. 27: each acid molecule is represented by a reclining <<U>>. In fact, dicarboxylic acids are known to orient at the polar/apolar interface with both carbonyl groups in the water, thus producing a hairpin-like configuration[40].

A check of the validity of the proposed model is provided by Fig. 28, where the electron density distribution along z (z being the direction perpendicular to the lipid layers) as deduced from the low angle X-ray diffracted intensities by the traditional methods of crystallography (upper curve), is compared with the electron density deduced from the models of Fig. 27 (lower curves): the agreement is quite satisfactory. We recall that the electron density profile (z) is obtained from the intensity I(h) of the h reflection trough the equation

$$(z) = \sum_{n=-\infty}^{+\infty} F(h) \cos 2\pi h z \, .$$

## 6. Interaction Between Aminoacids and Model Membranes

Among the aminoacids there is particular interest in tryptophan, since the fluorescence of its indole residue has been widely used to study protein structure and membrane protein environments[51,52]; moreover, the amino acid has been tentatively used, trapped in lecithin liposomes, as an antisickling agent[53,54].

Two very recent investigations[55,56] addressed the structural and thermodynamic modifications induced by the amino acid tryptophan on the lamellar phases of the dipalmitoyl phosphatidylcholine-water system at pH 7.4.

By using electron spin resonance, X-ray diffraction, differential scanning calorimetry and fluorescence spectroscopy, experimental evidence was obtained that tryptophan was incorporated into the model membrane. The results showed that in spite of the high value of the partition coefficients between water and organic solvents[57], the amino acid interacts mainly with the membrane polar region at concentrations lower than about 20 mM; at higher concentration, the amino acid also interacts with the paraffinic moiety, suggesting a concentration-dependent localization. The presence of tryptophan (TRP) induces a decrease in local rigidity, the order parameter and polarity, as determined by ESR measurements. Moreover, the widening of the calorimetric peak indicates a decrease in the main transition cooperativity, which correlates well with the previous observations. The increase in layer thickness at low TRP concentration is attributed mainly to the increase in the aqueous component within the liposomes.

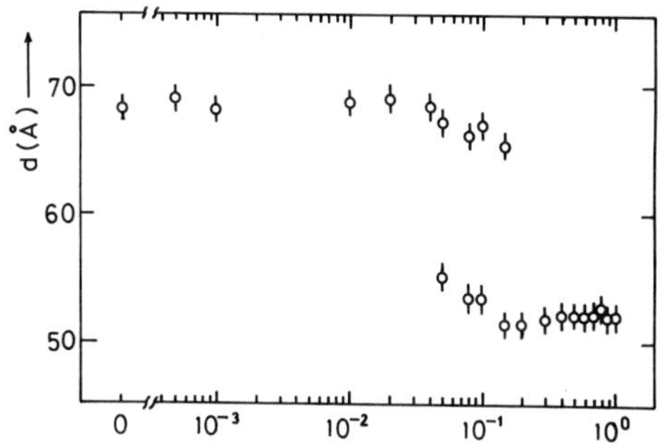


Fig.26. Lamellar repeat spacing d, vs. azelaic-acid molar ratio R (T = 25°C). (From Ref. 49).

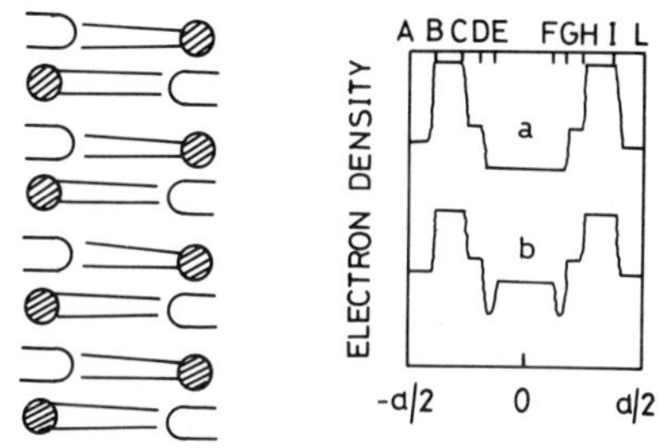


Fig.27. Azelaic acid and DSPC molecular interaction model at high drug molar ratio: the lipid is represented by a dashed circle with two straight lines and the acid by a reclining <<U>>. The two step model curves correspond to the molecular arrangement when the electrons of the terminal methyl groups are delocalized along the chains (upper function - curve a) and fixed in the terminal position (lower function - curve b). (From Ref. 49).

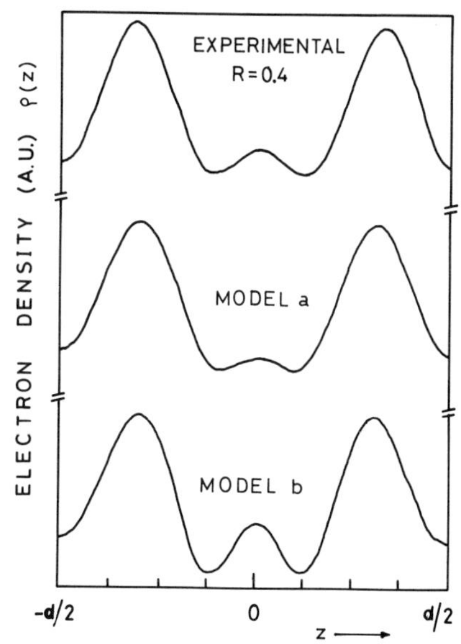


Fig.28. Electron density distribution of the <<gel phase>> of azelaic acid and DSPC in excess water condition at 25°C. Upper curve R = 0.4 (experimental); middle curve step model a; lower curve step model b (see text for additional information). (From Ref. 49).

Again, this increase is related to the presence of TRP molecules near the polar heads. Although these changes in thermodynamic and structural properties were observed, both the main transition and pre-transition still exist at the highest TRP concentration tested.

Finally, we wish to point out that, since the TRP fluorescence is sensitive to the phase changes of lecithin, the indole residue of the amino acid could be a good probe for monitoring lipid-protein interaction in natural and model membranes.

## 7. Conclusions

Due to the relatively complex nature of phase changes in aqueous dispersions of single molecule lecithins, the investigation of exogenous molecules such as cholesterol, anaesthetics, drugs and aminoacids incorporated into these model membrane systems, provides considerable useful information through the analysis of the modifications induced in the phase transitions.

This information refers in particular to the localization of exogenous molecules in the model membrane and to the mechanism of interaction, with useful implications both in biology and medicine. The liquid-crystalline nature of cell membranes plays an essential biological role, in the sense that these supramolecular organizations should be sufficiently fluid ("liquid") to allow diffusion and penetration of the molecules which are necessary for the metabolic processes, and sufficiently compact ("crystalline") in order to act as an efficient boundary of the cell.

ACKNOWLEDGEMENTS. We are grateful to S. Franguelli, E. Mosca and S. Polenta for technical assistance in the preparation of this Chapter.

### References

1. G.H. Brown and J.J. Wolken, "Liquid Crystals and Biological Structures," Academic Press, NY (1979).
2. V. Luzzati and A. Tardieu, Ann. Rev. Phis. Chem. **25**: 79 (1974).
3. S.J. Singer and G.L. Nicolson, Science **175**: 720 (1972).
4. D. Chapman, in "Biological Membranes", D. Chapman and D. H. F. Wallach, eds., Academic Press, N.Y., **2**: 91 (1973).
5. J. Stumpel, H. Eibel, and A. Nicksch, Biochim. Biophys. Acta **727**: 246 (1983).
6. V. Luzzati, in "Biological Membranes", D. Chapman, ed., Academic Press, NY, p. 71 (1968).
7. M.J. Janiak, D.M. Small and G.G. Shipley, J. Biol. Chem. **254**: 6068 (1979).
8. M.J. Janiak, D.M. Small and G.G. Shipley, Biochemistry **15**: 4575 (1976).
9. A. Watts, K. Harlos and D. Marsh, Biochim. Biophys. Acta **645**: 91 (1981).
10. N. P. Franks, J. Mol. Biol. **100**: 345 (1976).
11. G. Albertini, A. Ambrosini, B. Dubini M.G. Ponzi-Bossi and G. Bossi, Il Nuovo Cimento 12D **9**: 1309 (1990).
12. A. Tardieu, V. Luzzati and F.C. Reman, J. Mol. Biol. **75**: 711 (1973).
13. G. Albertini, G. Bossi, B. Dubini, M.G. Ponzi-Bossi and F. Rustichelli, Il Nuovo Cimento 1OD: 1385 (1988).
14. J. Stumpel, H. Eibl and A. Nicksch, Biochim. et Biophys. Acta **727**: 246 (1983).
15. J. J. Benattar, F. Moussa and M. Lambert, J. Phys. (Paris) **44**: 1371 (1980).
16. J. J. Benattar, F. Moussa, M. Lambert and C. Germain, J. Phys. (Paris). Lett. **12**: 67 (1981).
17. G. Albertini, E. Fanelli, S. Melone, F. Rustichelli and G. Torquati, Solid State Commun. **49**: 1143 (1984).
18. G. Albertini, S. Melone, G. Poeti, F. Rustichelli and G. Torquati, Mol. Cryst. Liq. Cryst. **101**: 121 (1984).
19. T. J. McIntosh, Biochim. et Biophys. Acta **513**: 43 (1978).
20. B.R. Lentz, D.A. Barrow and M. Hoechli, Biochemistry, **19**: 1943 (1980).
21. W. Knoll, G. Schmidt, K. Ibel and E. Sackmann, Biochemistry **24**: 5240 (1985).
22. G. Albertini, G. Bossi, B. Dubini, M.G. Ponzi-Bossi and F. Rustichelli, Mol. Cryst. Liq. Cryst. **179**: 383 (1990) .
23. G. Albertini, A. Ambrosini, V. Baldassarre, F. Rustichelli, P.L. Sapelli and R. Sconocchia, Stomat. Lomb., anno I, **2**: 59 (1987).

24. C. Gobbi, "Effetti strutturali e termodinamici dell'anestetico locale xilocaina su membrane modello di lecitina". Thesis, University of Ancona, 1988.
25. M.B. Feinstein, Journal of General Physiology, **48**: 357 (1964).
26. P.R. Cullis, A.P. Hornby and M.J. Hope, in "Progress in Anesthesiology," B. Raymond Fink, ed., Raven Press, New York (1980).
27. K. H. Meyer, Trans. Faraday Soc., **33**: 1062 (1937).
28. P. Seeman, Pharmacol. Rev., **24**: 583 (1972).
29. J. Stumphel, H. Eibl and A. Nicksch, Biochimica et Biophysica Acta, **727**: 246 (1983).
30. M.J. Janiak, D.M. Small and G.G. Shipley, Biochemistry, 21, **15**: 4575 (1976).
31. "Drug Treatment Principles and Practice of Clinical Pharmacology and Therapeutic," G. S. Avery, ed., Adis Press, Churchill Livingston, 2nd edition (1980).
32. D.T. Mason, A.N. De Maria, R. Amsterdam, R. Zelis and R.A. Massumi, Drugs, **5**: 261 (1973).
33. B.N. Singh, Drugs, **15**: 218 (1978).
34. G. Johnson and C.G. Regardh, Clinical Pharmacokinetics, **1**: 233 (1976).
35. A. Bienvenue and J. Philippot, in "Les liposomes applications therapeutiques," Ed. F. Puisieux and J. Delatbre, Paris, p.149 (1985).
36. N. Dousset and L. Douste-Blazy, in "Les liposomes applications therapeutiques", Ed. F. Puisieux and J. Delatbre, Paris, p.41 (1985).
37. G. Albertini, C. Donati, F. Rustichelli and R.S. Phadke, Il Nuovo Cimento, 1OD **11**: 1399 (1988).
38. G. Albertini, C. Donati, R.S. Phadke, M.G. Ponzi-Bossi and F. Rustichelli, Chemistry of Physics and Lipids, **55**: 331 (1990).
39. L.E.G. Eriksson, Biophys. Chem. **26**: 9 (1987).
40. D. Small, in "The Physical Chemistry of Lipids," D.J. Hanahan, ed., Plenum Press, New York, N.Y. Chapt.8 (1986).
41. S. Passi, M. Picardo, G. Mingrone, A. Brethnach and M. Nazzaro-Porro, Acta Derm. Venereol., Suppl. **143**: 8 (1987).
42. K.U. Schallreuter, M.R. Pittelkow and J.M. Wood, J. Invest. Dermatol., **88**: 516 (1987).
43. M. Nazzaro-Porro and S. Passi, G. Ital. Dermatol., **113**: 401 (1978).
44. M. Nazzaro-Porro, S. Passi, L. Balus, A.S. Breathnach, B. Martin and G. Morpurgo, J. Invest. Dermatol. **72**: 296 (1979).
45. M. Nazzaro-Porro, S. Passi, G. Zina, A. Bernengo, A.S. Breathnach, S. Gallagher and G. Morpurgo, Lancet, **109** (1980).
46. A. Mayer Da Silva, H. Gollnick, M. Detmar, J. Gassmuller, A. Parry, R. Muller and C.E. Orfanos, Acta Derm. Venereol., Suppl. **143**: 20 (1987).
47. A.S. Breathnach, E.J. Robins, M. NazzaroPorro, S. Passi and M. Picardo, Acta Derm. Venereol., Suppl. **143**: 62 (1987).
48. M. Nazzaro-Porro, S. Passi, M. Picardo, R. Mercantini and A.S. Breathnach, J. Invest. Dermatol **87**: 108 (1986).
49. G. Bossi, B. Dubini, P. Mariani, M.G. Ponzi-Bossi, F. Rustichelli and R.S. Phadke, Il Nuovo Cimento 12D **9**: 1293 (1990).
50. R.C. Chaatelier, P.S. Rogers, K.P. Ghiggino and W.H. Saawyer, Biochem. Biophys. Acta **776**: 75 (1984).
51. S.S. Lehrer in: R.F. Chen and H. Edelhoc (Eds), Biochemical Fluorescence Concepts, Vol.2 Marcel Dekker, New York **515**, (1976).
52. C.T.A. Aquaye, M. Wilcheck and M. Goreki, Biochem. Int. **4**: 221 (1982)
53. J. Kumpati, Biophys. Res. Commun. **105**: 482 (1982).
54. G. Albertini, E. Bertoli, G. Curatola, P. Mariani, F. Rustichelli and G. Zolese, Chemistry and Physics of Lipids, **50**: 143 (1989).
55. P. Mariani, A. Colotto and G.Albertini, Chemistry and Physics of Lipids, **55**: 283 (1990).
56. M.C. Block, L.L. Van Deenen and J. De Gier, Biochim. Biophys. Acta, **433**: 420 (1976).

# INDEX

# Index